Lecture Notes in Civil Engineering

Volume 260

Lecture Notes in Civil Engineering (LNCE) publishes the latest developments in Civil Engineering—quickly, informally and in top quality. Though original research reported in proceedings and post-proceedings represents the core of LNCE, edited volumes of exceptionally high quality and interest may also be considered for publication. Volumes published in LNCE embrace all aspects and subfields of, as well as new challenges in, Civil Engineering. Topics in the series include:

- Construction and Structural Mechanics
- Building Materials
- Concrete, Steel and Timber Structures
- Geotechnical Engineering
- Earthquake Engineering
- Coastal Engineering
- Ocean and Offshore Engineering; Ships and Floating Structures
- Hydraulics, Hydrology and Water Resources Engineering
- Environmental Engineering and Sustainability
- Structural Health and Monitoring
- Surveying and Geographical Information Systems
- Indoor Environments
- Transportation and Traffic
- Risk Analysis
- Safety and Security

To submit a proposal or request further information, please contact the appropriate Springer Editor:

– Pierpaolo Riva at pierpaolo.riva@springer.com (Europe and Americas);
– Swati Meherishi at swati.meherishi@springer.com (Asia—except China, and Australia, New Zealand);
– Wayne Hu at wayne.hu@springer.com (China).

All books in the series now indexed by Scopus and EI Compendex database!

M. S. Ranadive · Bibhuti Bhusan Das ·
Yusuf A. Mehta · Rishi Gupta
Editors

Recent Trends in Construction Technology and Management

Select Proceedings of ACTM 2021

Volume 2

Springer

Editors
M. S. Ranadive
Department of Civil Engineering
College of Engineering Pune
Pune, Maharashtra, India

Bibhuti Bhusan Das
Department of Civil Engineering
National Institute of Technology Karnataka
Mangalore, Karnataka, India

Yusuf A. Mehta
Department of Civil and Environmental Engineering
Rowan University
Glassboro, NJ, USA

Rishi Gupta
Department of Civil Engineering
University of Victoria
Victoria, BC, Canada

ISSN 2366-2557 ISSN 2366-2565 (electronic)
Lecture Notes in Civil Engineering
ISBN 978-981-19-2147-6 ISBN 978-981-19-2145-2 (eBook)
https://doi.org/10.1007/978-981-19-2145-2

This Springer imprint is published by the registered company Springer Nature Singapore Pte Ltd.
The registered company address is: 152 Beach Road, #21-01/04 Gateway East, Singapore 189721, Singapore

Contents

Recent Trends in Concrete Technology

Self-healing Behavior of Microcapsule-Based Concrete 3
B. S. Shashank and P. S. Nagaraj

Durability Properties of Fibre-Reinforced Reactive Powder Concrete 15
Abbas Ali Dhundasi, R. B. Khadiranaikar, and Kashinath Motagi

Performance of Geopolymer Concrete Developed Using Waste Tire Rubber and Other Industrial Wastes: A Critical Review 29
Dhiraj Agrawal, U. P. Waghe, M. D. Goel, S. P. Raut, and Ruchika Patil

Thermal Behaviour of Mortar Specimens Embedded with Steel and Glass Fibres Using *KD2* Pro Thermal Analyser 43
P. Harsha Praneeth

Modulus of Elasticity of High-Performance Concrete Beams Under Flexure-Experimental Approach 57
Asif Iqbal A. Momin, R. B. Khadiranaikar, and Aijaz Ahmad Zende

Impact of Phase Change Materials on the Durability Properties of Cementitious Composites—A Review 71
K. Vismaya, K. Snehal, and Bibhuti Bhusan Das

Comparative Study of Performance of Curing Methods for High-Performance Concrete (HPC) 83
Sandesh S. Barbole, Bantiraj D. Madane, and Sariput M. Nawghare

Development of a Mathematical Relationship Between Compressive Strength of Different Grades of PPC Concrete with Stone Dust as Fine Aggregate by Accelerated Curing and Normal Curing ... 101
N A G K Manikanta Kopuri, S. Anitha Priyadharshani, and P. Ravi Prakash

Using Recycled Aggregate from Demolished Concrete to Produce Lightweight Concrete ... 111
Abd Alrahman Ghali, Bahaa Eddin Ghrewati, and Moteb Marei

Recent Trends in Construction Materials

A Study in Design, Analysis and Prediction of Behaviour of a Footbridge Manufactured Using Laminate Composites—Static Load Testing and Analysis of a Glass Fibre Laminate Composite Truss Footbridge ... 125
Col Amit R. Goray and C. H. Vinaykumar

State-of-the-Art of Grouting in Semi-flexible Pavement: Materials and Design ... 135
Hemanth Kumar Doma and A. U. Ravi Shankar

Utilization of Agro-Industrial Waste in Production of Sustainable Building Blocks ... 149
S. S. Meshram, S. P. Raut, and M. V. Madurwar

Effect of Exposure Condition, Free Water–cement Ratio on Quantities, Rheological and Mechanical Properties of Concrete ... 161
Mahesh Navnath Patil and Shailendra Kumar Damodar Dubey

Development of Sustainable Brick Using Textile Effluent Treatment Plant Sludge ... 185
Uday Singh Patil, S. P. Raut, and Mangesh V. Madurwar

Utilization of Pozzolanic Material and Waste Glass Powder in Concrete ... 201
Lomesh S. Mahajan and Sariputt R. Bhagat

Recent Trends in Construction Technology and Management

Integrating BIM with ERP Systems Towards an Integrated Multi-user Interactive Database: Reverse-BIM Approach ... 209
M. Arsalan Khan

Application of Game Theory to Manage Project Risks Resulting from Weather Conditions ... 221
Abd Alrahman Ghali and Vaishali M. Patankar

Environmental Impact Analysis of Building Material Using Building Information Modelling and Life Cycle Assessment Tool 233
Kunal S. Bonde and Gayatri S. Vyas

Enhancing the Building's Energy Performance through Building Information Modelling—A Review 247
Dhruvi Shah, Helly Kathiriya, Hima Suthar, Prakhar Pandya, and Jaykumar Soni

Analysis of Clashes and Their Impact on Construction Project Using Building Information Modelling 255
Samkit V. Gandhi and Namdeo A. Hedaoo

Predicting the Performance of Highway Project Using Gray Numbers 271
Supriya Jha, Manas Bhoi, and Uma Chaduvula

COVID-19—Assessment of Economic and Schedule Delay Impact in Indian Construction Industry Using Regression Method 283
Soniya D. Mahind and Dipali Patil

Comparison of Afghanistan's Construction and Engineering Contract with International Contracts of FIDIC RED BOOK (2017) and NEC4—ECC 299
Mohammad Ajmal and C. Rajasekaran

Comparative Analysis of Various Walling Materials for Finding Sustainable Solutions Using Building Information Modeling 315
Amey A. Bagul and Vasudha D. Katare

Studies on Energy Efficient Design of Buildings for Warm and Humid Climate Zones in India 327
Santhosh M. Malkapur, Sudarshan D. Shetty, Kishor S. Kulkarni, and Arun Gaji

Recent Trends in Environmental and Water Resources Engineering

Spatial Variability of Organic Carbon and Soil pH by Geostatistical Approach in Deccan Plateau of India 351
N. T. Vinod, Amba Shetty, and S. Shrihari

Hydro-geo Chemical Analysis of Groundwater and Surface Water Near Bhima River Basin Jewargi Taluka Kalburgi, Karnataka 361
Prema and Shivasharanappa Patil

Conjunctive Use Modeling Using SWAT and GMS for Sustainable Irrigation in Khatav, India 373
Ranjeet Sabale and Mathew K. Jose

Modelling and Simulation of Pollutant Transport in Porous Media—A Simulation and Validation Study 387
M. R. Dhanraj and A. Ganesha

Adsorptive Removal of Malachite Green Using Water Hyacinth from Aqueous Solution 401
Sayali S. Udakwar, Moni U. Khobragade, and Chirag Y. Chaware

Sediment Yield Assessment of a Watershed Area Using SWAT 415
Prachi A. Bagul and Nitin M. Mohite

"NDVI: Vegetation Performance Evaluation Using RS and GIS" 425
A. Khillare and K. A. Patil

Comparison of Suspended Growth and IFAS Process for Textile Wastewater Treatment 437
Sharon Sudhakar, Nandini Moondra, and R. A. Christian

Innovative Arch Type Bridge Cum Bandhara for Economical and Quick Implementation of Jal Jeevan Mission 449
P. L. Bongirwar and Sanjay Dahasahasra

Green Synthesis of Zinc Oxide Nanoparticles and Study of Its Adsorptive Property in Azo Dye Removal 467
C. Anupama and S. Shrihari

Anthropogenic Impacts on Forest Ecosystems Using Remotely Sensed Data 481
Gaurav G. Gandhi and Kailas A. Patil

Seasonal and Lockdown Effects on Air Quality in Metro Cities in India 497
K. Krishna Raj and S. Shrihari

Inter-Basin Pipeline Water Grid for Maharashtra 511
Raibbhann Sarnobbat, Pritam Bhadane, Vaibhav Markad, and R. K. Suryawanshi

Removal of Heavy Metals from Water Using Low-cost Bioadsorbent: A Review 527
Praveda Paranjape and Parag Sadgir

Adsorptive Removal of Acridine Orange Dye from Industrial Wastewater Using the Hybrid Material 547
Vibha Agrawal and M. U. Khobragade

Basin Delineation and Land Use Classification for a Storm Water Drainage Network Model Using GIS 561
Kunal Chandale and K. A. Patil

Prediction of BOD from Wastewater Characteristics and Their Interactions Using Regression Neural Network: A Case Study of Naidu Wastewater Treatment Plant, Pune, India 571
Sanket Gunjal, Moni Khobragade, and Chirag Chaware

Floodplain Mapping of Pawana River Using HEC-RAS 579
Tejas R. Bhagwat and Aruna D. Thube

Performance Evaluation of Varying OLR and HRT on Two Stage Anaerobic Digestion Process of Hybrid Reactor (HUASB) for Blended Industrial Wastewater as Substrate 597
Rajani Saranadagoudar, Shashikant R. Mise, and B. B. Kori

Analysis of Morphometric Parameters of Watershed Using GIS 603
Bhairavi Pawar and K. A. Patil

Mapping Ground Water Potential Zone of Fractured Layers by Integrating Electric Resistivity Method and GIS Techniques 613
R. Chandramohan and B. Kesava Rao

Sustainable Development in Circular Economy: A Review 629
Mohnish Waikar and Parag Sadgir

Biogas Generation Through Anaerobic Digestion of Organic Waste: A Review 641
Vaishali D. Jaysingpure and Moni U. Khobragade

Recent Trends in Geotechnical Engineering

A Cost Comparison Study of Use of Cased and Uncased Stone Column in Marshy Soil 653
Starina J. Dias and Wilma Fernandes

Review on Field Direct Shear Test Methodologies 665
Kakasaheb D. Waghmare and K. K. Tripathi

Numerical Prediction of Tunneling Induced Surface Settlement of a Pile Group 673
B. Swetha, S. Sangeetha, and P. Hari Krishna

Numerical Study of Pile Supported Embankment Resting on Layered Soft Soil 685
Uzma Azim and Siddhartha Sengupta

A Review on Application of NATM to Design of Underground Stations of Indian Metro Rail 715
Sandesh S. Barbole, M. S. Ranadive, and Apurva R. Kharat

Slope Stability Analysis of Artificial Embankment of Fly Ash and Plastic Recycled Polymer Using Midas GTS-NX 729
Prajakta S. Chavan, Kalyani G. Patil, and Sariput M. Nawghare

Lateral Capacity of Step-Tapered Piles in Sand Deposits 739
K. V. S. B. Raju, K. S. Rajesh, L. Dhanraj, and H. C. Muddaraju

Strength and Dilatancy Behaviour of Granular Slag Sand 753
K. V. S. B. Raju and Chidanand G. Naik

Parametric Study of the Slope Stability by Limit Equilibrium Finite Element Analysis 767
Prashant Sudani, K. A. Patil, and Y. A. Kolekar

Model Testing on PET Bottle Mattress with Aggregate Infill as Reinforcement Overlaying on Fly Ash Under Circular Loading 779
Shahbaz Dandin, Mrudula Kulkarni, and Maheboobsab Nadaf

Recent Trends in Structural Engineering

Sustainable Project Planning of Road Infrastructure in India: A Review 799
Appa M. Kale and Sunil S. Pimplikar

Behaviour of Space Frame with Innovative Connector 805
Pravin S. Patil and I. P. Sonar

Seismic Analysis of Base Isolated Building Frames with Experimentation Using Shake Table 819
Mahesh Kalyanshetti, Ramankumar Bolli, and Shashikant Halkude

Critical Study of Wind Effect on RC Structure with Different Permeability 839
Ankush Asati and Uday Singh Patil

Performance Assessment of SMA-LRB Isolated Building Structure Due to Underground Blast-Induced Ground Motion 861
Sonali Upadhyaya, Narendranath Gogineni, and Sourav Gur

Research Progress on the Torsion Behavior of Externally Bonded RC Beams: Review 875
Rajesh S. Rajguru and Manish Patkar

Study on Static Analysis and Design of Reinforced Concrete Exterior Beam-Column Joint 887
Yogesh Narayan Sonawane and Shailendrakumar Damodar Dubey

Refined Methodology in Design of Reinforced Concrete Shore Pile: A Design Aid 897
Mahesh Navnath Patil and Shailendra Kumar Damodar Dubey

Investigating the Efficacy of the Hybrid Damping System for Two-Dimensional Multistory Building Frame Using Time History Analysis ... 919
A. P. Kote and R. R. Joshi

Thermal Buckling Analysis of Stiffened Composite Cutout Panels ... 935
K. S. Subash Chandra, T. Rajanna, and K. Venkata Rao

Effect of Isolated Wind Incidence on Local Peak Pressure ... 949
Supriya Pal, Ritu Raj, and S. Anbukumar

Investigation of Performance of Perforated Core Steel Buckling Restrained Brace ... 961
Prajakta Shete, Suhasini Madhekar, and Ahmad Fayeq Ghowsi

A Method for Evaluating Maximum Response in Multi-storied Buildings Due to Bi-directional Ground Motion ... 973
P. B. Kote, S. N. Madhekar, and I. D. Gupta

Finite Element Analysis of Piled Raft Foundation Using Plaxis 3D ... 991
Anupam Verma and Sunil K. Ahirwar

FRP Strengthened Reinforced Concrete Beams Under Impact Loading: A State of Art ... 1001
Swapnil B. Gorade, Deepa A. Joshi, and Radhika Menon

Effect of Lateral Stiffness on Structural Framing Systems of Tall Buildings with Different Heights ... 1015
A. U. Rao, Sradha Remakanth, and Aditya Karve

Free Vibration Response of Functionally Graded Cylindrical Shells Using a Four-Node Flat Shell Element ... 1031
R. B. Dahale, S. D. Kulkarni, and V. A. Dagade

The Behaviour of Transmission Towers Subjected to Different Combinations of Loads Due to Natural Phenomenon ... 1047
Devashri N. Varhade and R. R. Joshi

Fragility Assessment of Mid-Rise Flat Slab Structures ... 1061
B. P. Dhumal and V. B. Dawari

Seismic Response of RC Elevated Liquid Storage Tanks Using Semi-active Magneto-rheological Dampers ... 1073
Manisha V. Waghmare, Suhasini N. Madhekar, and Vasant A. Matsagar

Virtual Testing of Prototypes Using Test Frame Designed for Lateral Load ... 1089
Suyog Nikam and I. P. Sonar

Crack Simulation and Monitoring of Beam-Column Joint by EMI Technique Using ANSYS 1101
Tejas Shelgaonkar and Suraj Khante

The Impact of Perforation Orientation on Buckling Behaviour of Storage Rack Uprights 1115
Kadeeja Sensy, Ashish Gupta, K. Swaminathan, and J. Vijaya Vengadesh Kumar

Modelling Interfacial Behaviour of Cement Stabilized Rammed Earth Using Cohesive Contact Approach 1127
T. Pavan Kumar Reddy and G. S. Pavan

Four Node Flat Shell Quadrilateral Finite Element for Analysis of Composite Cylindrical Shells 1135
V. A. Dagade and S. D. Kulkarni

Exact Elasticity Analysis of Sandwich Beam with Orthotropic Core 1149
Ganesh B. Irkar and Y. T. LomtePatil

Flexural Fatigue Analysis of Cross Ply and Angle Ply Laminates 1171
Sammed Patil and Y. T. LomtePatil

Recent Trends in Transportation and Traffic Engineering

Review on Mechanisms of Bitumen Modification: Process and Variables 1185
N. T. Bhagat and M. S. Ranadive

Alkali Activated Black Cotton Soil with Partial Replacement of Class F Fly Ash and Areca Nut Fiber Reinforcement 1193
B. A. Chethan, A. U. Ravi Shankar, Raghuram K. Chinnabhandar, and Doma Hemanth Kumar

Development of Road Safety Models by Using Linear and Logistic Regression Modeling Techniques 1205
Krantikumar V. Mhetre and Aruna D. Thube

Finite Element Analysis for Parametric Study of Mega Tunnels 1227
Shilpa Kulkarni and M. S. Ranadive

Development of Financial Model for Hybrid Annuity Model Road Project 1245
Pratiksha B. Gilbile and Gayatri S. Vyas

Analysis of Perpetual Pavement Design Considering Subgrade CBR, Life-Cycle Cost, and CO_2 Emissions 1257
Saurabh Kulkarni and M. S. Ranadive

Laboratory Investigation of Lateritic Soil Stabilized with Arecanut Coir Along with Cement and Its Suitability as a Modified Subgrade 1273
B. A. Chethan, B. M. Lekha, and A. U. Ravi Shankar

Pavement Analysis and Measurement of Distress on Concrete and Bituminous Roads Using Mobile LiDAR Technology 1287
Prashant S. Alatgi and Sunil S. Pimplikar

Laboratory Study on New Type of Self-consolidating Concrete Using Fly Ash as a Pavement Material 1295
Bhupati Kannur and Hemant Chore

FTIR Analysis for Ageing of HDPE Pyro-oil Modified Bitumen 1311
H. P. Hadole and M. S. Ranadive

Utilization of Aluminium Refinery Residue (ARR), GGBS and Alkali Solution Mixes in Road Construction 1329
Nityanand S. Kudachimath, Raviraj H. Mulangi, and Bhibuti Bhusan Das

About the Editors

Prof. M. S. Ranadive is currently working as Professor and Head, Department of Civil Engineering, College of Engineering Pune (COEP), Pune, Maharashtra. He obtained his B.E. (Civil) from the University of Pune, Maharashtra; M.E. (Civil) From Shivaji University, Kolhapur and Ph.D. from the University of Pune. His major areas of research interests include quality monitoring of pavements, use of anti-stripping agents in bituminous concrete, effective use of bio-oil obtained by pyrolysis of municipal solid waste in flexible pavement, continuous pavement monitoring through dynamic responses by instrumentation. He has published 22 papers in leading international journals, 20 papers in national journals and in all there are more than 100 papers on his account. Professor M. S. Ranadive was a Guest Editor for the Journal of Performance of Constructed Facilities, ASCE. He is a member of American Society of Civil Engineers, Member of Indian Roads Congress, the Life Member of the Indian Geotechnical Society, Member of Indian Concrete Institute, and many more. He is a reviewer for various journals like *Journal of Materials in Civil Engineering*, ASCE, *International Journal of Construction Management and Economics*, Germany; *Journal of Transportation Engineering*; *International Journal of Pavement Engineering*, Taylor and Francis; *Journal of Building Engineering*, Elsevier Publication; *International Journal of Innovative Infrastructure Solutions*, Springer and many more.

Dr. Bibhuti Bhusan Das is currently working as an Associate Professor at the Department of Civil Engineering, National Institute of Technology Karnataka (NITK), Surathkal, Mangalore, India. He obtained his B.Tech. (Civil) from Orissa University of Agriculture and Technology, Orissa; M.Tech. from Indian Institute of Technology (IIT) Delhi in Construction Engineering and Management; Ph.D. from Indian Institute of Technology (IIT) Bombay and Post-doctoral from Lawrence Technological University, Southfield, Michigan, USA. His major areas of research interests include concrete technology, sustainable construction, building materials. He has published more than 40 research papers in and SCI Scopus indexed journals. Dr. Bibhuti Bhusan Das has edited four books namely *Sustainable Construction and*

Building Materials, *Recent Developments in Sustainable Infrastructure*, *Smart Techniques for Sustainable Development*, and *Recent Trends in Civil Engineering* which were published by Springer Nature Singapore. Dr. Bibhuti Bhusan Das is a member of the Indian Concrete Institute, Chennai, India; American Society for Testing and Materials; Prestressed Concrete Institute, Chicago; American Society of Civil Engineers and various other reputed Societies. He is also a reviewer for reputed journals such as *Journal of Materials in Civil Engineering*, ASCE, *Canadian Journal of Civil Engineering, Construction and Building Materials*, Elsevier Publications, and many others.

Prof. Yusuf A. Mehta is currently a Professor at the Department of Civil and Environmental Engineering, Rowan University, USA; and Director, Centre of Research and Education in Advanced Transportation Engineering Systems. He obtained his B.S. from the University of Bombay, India; M.S. from the University of Oklahoma, Norman; and Ph.D. from Pennsylvania State University. Since coming to Rowan, Dr. Mehta, has received approximately $25 million dollars of external funding in pavements and materials. He has extensive experience working on several research projects with New Jersey, Florida, Wisconsin and Rhode Island departments of transportations, Federal Highway Administration and Federal Aviation Administration, and Department of Defense. He has also led the effort to acquire the Heavy Vehicle Simulator (HVS) that can simulate 10-20 years of traffic in a few years. Dr. Mehta has received several teaching awards, such as ASCE-NJ Educator of the Year Award, May 2014 and the 2012 Louis J. Pignataro Memorial Transportation Engineering Education Award. The award was for outstanding record of achievement in transportation engineering research, and undergraduate and graduate engineering education. He has received the Mid-Atlantic American Society of Engineering Education Section Distinguished Teaching Award, West Point, 2008. He has also received the faculty research achievement award in 2014. Under the direction of Dr. Mehta, CREATEs has expanded its capabilities to integrate research in Intelligent Transportation Systems, transportation safety, geotechnical engineering, cementitious materials, and Structural engineering by collaborating with faculty members at Rowan University. These collaborations will allow CREATEs to seek research funding in the above mentioned areas. The CREATEs award is approximately $37 M since its inception. CREATEs provides hands-on experience to sixty undergraduate and graduate students in various fields of transportation. This expansion has allowed CREATEs to conduct research, education, and outreach in all the above mentioned fields.

Dr. Rishi Gupta is a professor at the Department of Civil Engineering, University of Victoria, Canada where he also leads the Facility for Innovative Materials and Infrastructure Monitoring (FIMIM). He obtained his Diploma from Bombay Technical Board (India); B.E. (Civil) from Pune University (India); M.A.Sc. in Civil Engineering from University of British Columbia and Ph.D. in Civil Engineering (Materials) from University of British Columbia. His major areas of research interests include shrinkage and self-sealing characteristics of concrete and development of 'crack-free' cement composites, structural health monitoring (SHM) of infrastructure, durability and corrosion studies of reinforced concrete, sustainable construction technologies, and advanced materials for structures. Dr. Rishi Gupta holds several patents to his name. He has published more than 70 papers in refereed international journal publications and more than 50 refereed conference publications. Dr. Rishi Gupta has been awarded UVic's REACH award for excellence in Undergraduate Research Enriched teaching, June 2020; Drishti award for Innovation in Science and Technology, 2020; Nominated for Medal for distinction in Engineering Education, Engineers Canada award, 2020; Recipient of EGBC's President's Awards for Teaching Excellence in Engineering and Geoscience Education, 2019; Awarded fellowship with Engineers Canada or Geoscientists Canada, February 2017. Dr. Rishi Gupta is a Member of the Board of Examiners at Engineers and Geoscientists of British Columbia (since July 2018); Editorial board member (2010–2017), EGBC Burnaby/NW Chair (2011-2012); Vice-Chair (2009–2011), Volunteer (2007–to date), EGBC-Burnaby/New Westminster Branch; UVic faculty liaison for EGBC's Victoria Branch (Since May 2015); Member of The Centre for Advanced Materials and Related Technology (CAMTEC); University of Victoria, (Since October 2013) and Board member of Habitat for Humanity Victoria and many other respected boards. He is a reviewer for *Construction and Building Materials*, Elsevier Publications, ASCE *Journal of Materials in Civil Engineering*, *Journal of American Society of Testing and Materials*, *Canadian Journal of Civil Engineering*, *Construction Materials*, Institute of Civil Engineers, UK, and many more.

Recent Trends in Geotechnical Engineering

A Cost Comparison Study of Use of Cased and Uncased Stone Column in Marshy Soil

Starina J. Dias and Wilma Fernandes

Abstract Over the past decade, there is a lot of growth in infrastructure and now it is becoming very difficult to find suitable soils for construction purpose, hence this has forced geotechnical engineers to take up the challenge of constructing structures even on weak soft soils. The characteristics of soft clay are highly compressible low shear strength and low permeability. General construction problems in this soil are insufficient bearing capacity, instability on excavation and excessive post-construction settlement. Among various technologies implemented by engineers, stone columns are the most beneficial method for the modification of weak soil. For the present study, at a bridge site in Goa, we encountered extremely soft soil where a sufficient number of stone columns were used, thus in this study, an attempt was made to reduce the number of stone columns by wrapping the granular pile with geotextile which will enhance the performance of stone column by increasing the bearing capacity which will help in reducing the number of stone columns. This proposed improvement by encasement with a geotextile was done by using a finite element software called Plaxis 2D. It can be seen that for a given area, the stone column numbers were reduced from twelve to nine columns.

Keywords Geotextile · Stone columns · Soft soil · Encased stone column

S. J. Dias (✉)
Civil Engineering Department, Don Bosco College of Engineering, Fatorda, Goa, India
e-mail: starina23@gmail.com

W. Fernandes
Civil Engineering Department, Goa College of Engineering, Farmagudi, Goa, India
e-mail: wilma@gec.ac.in

M. S. Ranadive et al. (eds.), *Recent Trends in Construction Technology and Management*, Lecture Notes in Civil Engineering 260,
https://doi.org/10.1007/978-981-19-2145-2_50

1 Introduction

Ground improvement is one of the most applicable methods for improving soil condition and humankind till date has pursued ground improvement techniques to improve soil condition. This method is the most useful and economic than other kinds. Ground improvement has numerous methods and a suitable method is selected based on the soil type, bearing condition and structure type. Several ground improvement techniques have been developed to convert weak soil into a stratum of desired strength and compressibility as per the design requirements with adequate safety. The soil can be improved by using various ground improvement techniques like Dry soil mixing, Dynamic compaction, Injection systems, Vibro compaction, Vacuum Consolidation, Heating, Ground freezing, Soil nailing, Stone columns, etc.

Nowadays, in Goa, land is slowly getting exhausted and there is a decrease in good construction sites. There are a number of soil types in Goa and each one has unique characteristics like colour, texture, structure, and mineral content. There are certain types of soils that are problematic as they are collapsible, liquefiable and some soils are even expansive in nature. One of these types of soil is weak compressible soil like soft clay and marshy soil which is a classic example of soil found along the coastal areas in India and along the perimeters of river bays. In places like Goa, soft soils are often found near the mouths of rivers and can be especially troublesome. The problems of soft soil are low bearing capacity, low shear strength, high settlements and low permeability. Hence, these soils have to be improved in order to construct building, storage tanks, embankments, etc. Hence, ground improvement techniques in recent times have now become a necessity and thus this particular field is branded as a fast-improving one. In the present study, by using a finite element software called PLAXIS we are studying the effect of encasing a Stone column with geosynthetic.

2 Concept of Stone Column

2.1 Introduction of Stone Columns

Stone column is basically a technique where granular material is compacted in long cylindrical holes that are mostly used for improving the soil characteristic of soft clay. When a stone column is subjected to vertical loading, it undergoes a tremendous vertical compression caused by the lateral bulging of the aggregates mostly seen in the top portion of the column. Figure 1 shows the behaviour of a stone column where we can see that the column bulges almost four times the diameter of the column and the granular material present in the stone column penetrates into the surrounding soft soil and transfers the stress to the surrounding soil. The bulge, in turn, increases the lateral stress within the clay which provides additional confinement for the stones.

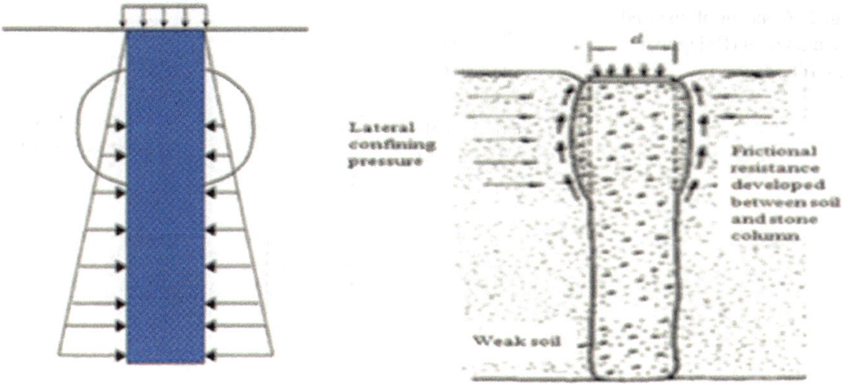

Fig. 1 Behaviour of a stone column (Rajagopal K.—IIT Madras)

The passive pressure from the neighbouring clay allows the column to resist the vertical load on the stone column. It can be said that the strength of the surrounding clay is the main factor that contributes to the load-carrying capacity of the stone column (Malarvizhi and Ilamparuthi).

3 Concept of Encased Stone Column (ESC)

3.1 Introduction of Encased Stone Columns

Mostly stone columns have found to be effective for soils having undrained strength that is Cu $\geq$ 15 kN/m^2, but for very soft soil having undrained shear strength that is Cu $\leq$ 15 kN/m^2, the ordinary stone column technique may not be feasible either due to excessive bulging of the stone column in the clay or due to clay clogging in the pores of the aggregates of the stone column. Hence, when such a situation arises, the performance of a stone column can be enhanced by using a high tensile strength geosynthetic that is placed around the stone column and such stone columns are termed as called as Geosynthetic Encased Stone Column [ESC]. Basically, this method is an extension of the well-known stone or sand column foundation improvement techniques. Encased stone columns consist of inserting continuous, seamless, high strength geotextile tubes into soft soil with a mandrel. The tube is then filled with either sand or fine gravel to form a column with a high bearing capacity. Their load carrying capacity varies from 20 to 40 tonnes (Murugesan S. and Rajagopal K).

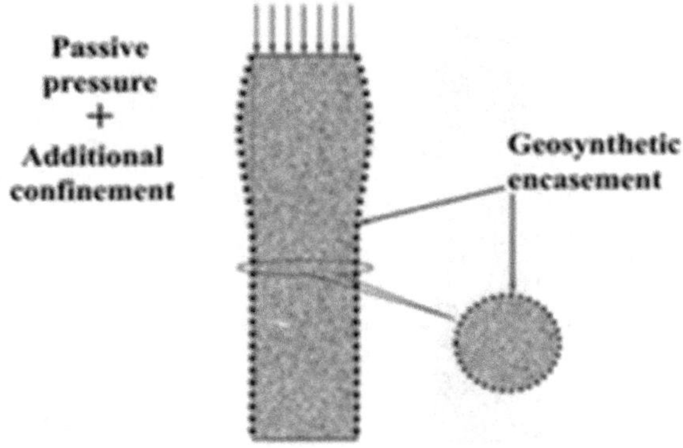

Fig. 2 Concept of encased stone column (Rajagopal K.—IIT Madras)

3.2 Schematics of the Encased Stone Column

Generally, it is seen in a stone column that the bearing capacity of the treated ground is mostly derived from the passive resistance offered by the surrounding soil due to lateral bulging of stone column material under applied loads, but the behaviour of geosynthetic encased stone column is thought out to be entirely different from that of the conventional stone column. The vertical deformations and the load distribution between the encased stone column and the soft soil are defined by the tensile strength and the stiffness of the encasing material used for encasement. Hence, the encasement provides passive pressure and additional confinement to the stone column. Figure 2 explains the concept of encased stone columns.

3.3 Site Details

The present study is chosen in the Tiswadi taluka of Goa at a site named Manduras seen in Fig. 3. At this site, in the past, there existed a small bridge that was used by the locals to commute to the other side of the tributary of the Zuari river, which had collapsed in due course of time. Hence, a single-lane bridge construction was proposed. This bridge, if constructed, could be of use to connect this small village of Mandur to Kundaim which lies close to it, and would reduce the travelling distance mainly from Vasco-Agacaim–Mandur to Kundaim, which is a prominent industrial estate of Goa. Table 1 shows the Travel distance which the proposed single lane bridge would reduce.

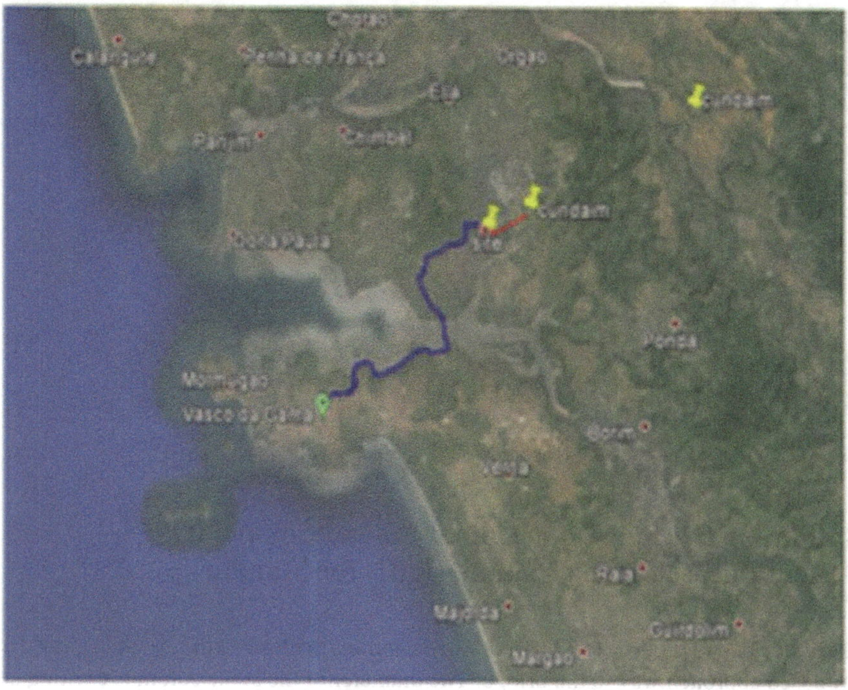

Fig. 3 Distance from Vasco to Kundaim with proposed bridge

Table 1 Travel distance which the proposed single lane bridge would reduce

Destination	Travel time (minutes)	Distance (Kilometer)	Route
Margao to kundaim	49	30.1	Via Borim, Priol
Vasco to Kundaim	53	34.8	Via Agacaim, Neura, Ella, Corlim
Vasco to Kundaim	36	20.84	Via Agaciam, Neura, Proposed Dongrim-Madur bridge

The present bridge site in Goa falls along the floodplains of the tidal estuary and the soil encountered here is mostly soft marine clay. The report of the soil investigation carried out at the site suggested very weak soft saturated soil having poor strength and having SPT potential (N) betweens $N = 1$–3. Also, the water table was encountered at ground level. Presently, the soil is improved by stone columns in the western approach way to the bridge site, resulting in a hundred and six stone columns, thus an attempt is made in reducing the number of stone columns by wrapping the granular pile with geotextile. It may be noted that geosynthetics are being used nowadays to a great extent in the environment of increasing land scarcity (Fig. 4).

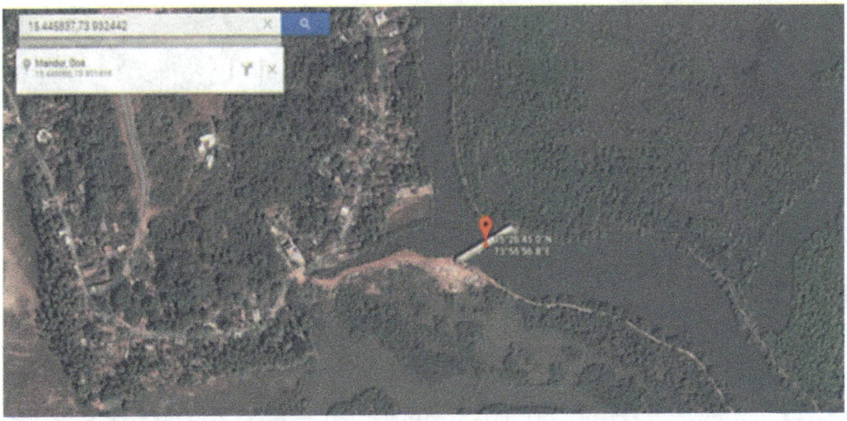

Fig. 4 Coordinates showing the proposed bridge at Mandur site

4 Parametric Study on Ordinary Stone Column (OSC) and Encased Stone Column (ESC) for Mandur

It was decided to propose to construct stone columns to improve the ground below the proposed approach way to the bridge. The diameter of the stone columns existing at the site is 900 mm, length of the stone column is 1500 mm, centre-to-centre spacing that was used was 2700 mm and load of 165 kN was used on the column. The geosynthetic used in the Plaxis software is used to show the effect of encasement and is modelled as a geogrid element having tensile stiffness only. The input parameters required by the software for clay and sand are given in Table 2.

Table 2 Material properties used in PLAXIS software

Properties	Clay	Stone
γ_{unsat} (unsaturated unit weight)	15 kN/m^3	17 kN/m^3
γ_{sat} (saturated unit weight)	18 kN/m^3	18 kN/m^3
K_X (Permeability in x direction)	2×10^{-4} m/day	12 m/day
K_Y (Permeability in y direction)	1×10^{-4} m/day	6 m/day
E (Elastic Modulus)	2000 kN/m^2	30,000 kN/m^2
ν (Poissons ration)	0.45	0.3
C (cohesion)	9.81 kN/m^2	0 kN/m^2
ϱ (friction angel)	0°	40°
Ψ (Dilation angel)	0°	10°
LL (%)	39	
PL (%)	21.52	
PI (%)	17.48	

In the present case, at Mandur, taking an area of 57.6 m^2, there exist twelve ordinary stone columns at centre-to-centre spacing of 2700 mm. An attempt is made to reduce the number of stone columns in the said area with appropriate encasement stiffness and also check its cost feasibility (Fig. 5).

The centre-to-centre spacing of the stone column was varied from the existing spacing of 2700–3500 mm along with the stiffness to achieve the optimum spacing and appropriate stiffness. The geogrid material found satisfactory for Mandur is a biaxial geogrid called TechGrid Biaxial (TGB)-40, having a stiffness of 333.33 kN/m. Below is the property of the geogrid (Table 3).

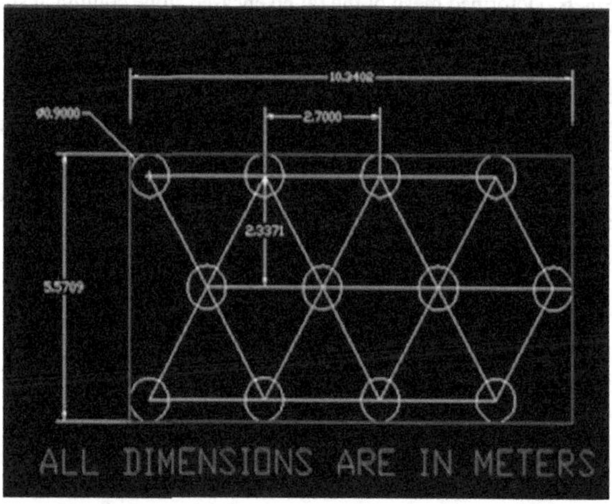

Fig. 5 Present case at Mandur-Twelve ordinary stone columns at center to center spacing of 2700 mm for an area of 57.6 m^2

Table 3 Specification for tensile geogrid biaxial (TGB-40)

Property		TGB-40	Test method
Ultimate tensil strength	MD	40 kN/m	ASTM
	CD	40 kN/m	D 6637
Elongation at maximum load	MD	12%	EN
	CD	12%	SO 10,319
Phyical property			
Apperture size (+/− 3 mm)	25 × 25		
Roll dimensions			
Roll length	50 m/100 m		
Roll width	2.5 m/3.9 m/5.0 m		
MD—Machine direction	CD-cross direction		

Source Techfab–India

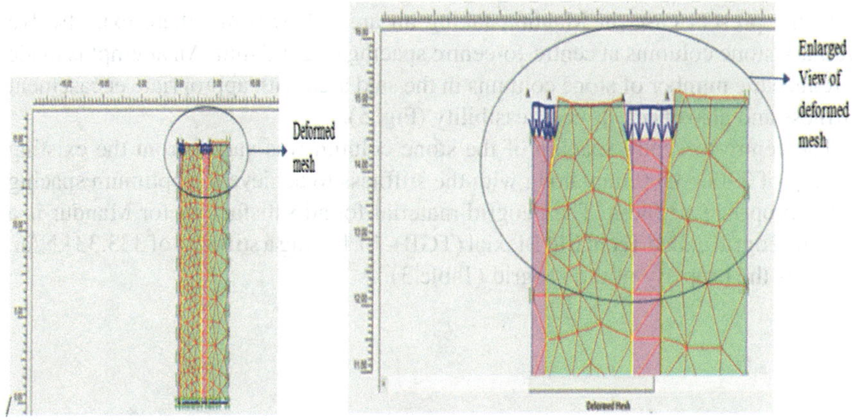

Fig. 6 Deformed mesh of the failed encased stone column at spacing of 3500 mm center to center. *Source* Plaxis-output

The following is a figure of a deformed mesh of a failed ESC at a spacing of 3500 mm (Fig. 6).

The following is a figure of a deformed mesh of a failed ESC at a spacing of 3200 mm (Figs. 7 and 8).

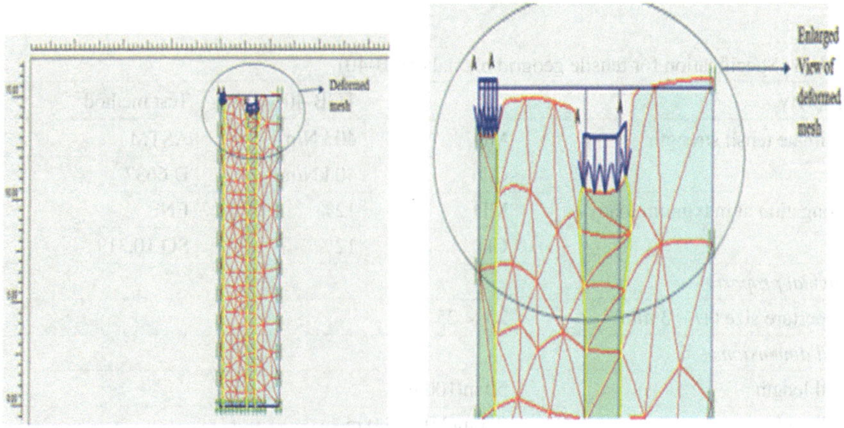

Fig. 7 Deformd mesh of the successfully passed encased stone column using a stiffness of 333.33 kN/m at a spacing of 3200 mm center to center. *Source* Plaxis-output

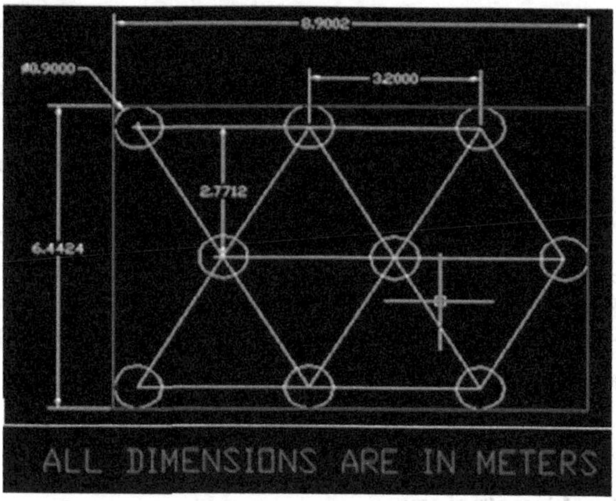

Fig. 8 After analysis-nine stone columns at center-to-center spacing of 3200 mm with a geogrid stiffness of 333.33 kn/m for an area of 57.35 m^2

After doing the analysis in the software with the said geogrid, it is found that for an area of 57.35 m^2, nine stone columns at centre-to-centre spacing of 3200 mm with a geogrid of stiffness of 333.33 kN/m were found satisfactory for a given load of 165 kN. It can be noted that the stiffness of encasement plays a role in transferring the stresses to the surrounding soil. It was seen that, for the given stiffness of 333.33 kN/m, the spacing of the encased stone column could be increased to 3200 mm centre-to-centre, but for the same stiffness, the spacing could not be increased more than 3500 mm as the encasement is not able to sustain the given load.

5 Cost Feasibility

The costing was done to determine the cost factor between OSC and ESC. The Table 4 has the breakup of the cost involved in setting up a single OSC.

Table 4 Nature of work and cost

Sr. No.	Nature of work	Units of measurement	Cost (rupees)
1	Cost of drilling-which includes bailing out of soil, operator charges	per meter	3000
2	Quantity of stone 6–40 mm	cum	1400
3	Labour cost per person	per day	500
4	Rental of drilling machine	per day	4000
5	Geogrid-TGB-40	per meter	200

Assuming it takes three days to construct a single stone column of diameter 900 mm and length of 1500 mm and taking into consideration eight labourers per day, we arrive at a cost of approximately Rs. 75,000/- for a single OSC. Using geogrid-TGB-40 for a single encased stone column for an outer circumferential area of 42.5 m^2 and using the same criteria mentioned above, we arrived at a cost of approximately Rs. 83,500/-.

In an area of 57.6 square metres, an OSC having twelve columns has a cost of Rs. 9,00,000/-, while an ESC for an equivalent area has nine columns, having a cost of Rs. 7,51,500/-. Hence, by comparing the cost of construction of an ordinary stone column and Encased stone column, it was found that the cost reduces by 16.5% for an Encased stone column.

6 Conclusions

It can be concluded for the present site where stone columns are used in the embankment for the approach way for the single-lane bridge that:

By encasing the stone column by a geogrid of stiffness of 333.33 kN/m and maintaining a centre-to-centre spacing of 3200 mm for the existing diameter at the site, the stone columns reduce from twelve to nine columns for a given area. Generally, the spacing between stone columns is generally 2–3 times the diameter, but after analysis, it may be seen that for a particular stiffness of encasement, the spacing between the stone columns could be increased to 3.5 times the diameter, hence reducing the number of constructed stone columns in an equilateral pattern for the said area.

Also, upon comparing the cost of construction of an ordinary stone column and encased stone column for the area of 57.6 m^2, it was found that the cost reduced by 16.5%. Hence, considering the type of soil at the site, it would be advantageous to use this type of encasing method of stone column to reduce the overall cost of the project.

Bibliography

1. Murugesan S, Rajagopal K (2010) Studies on the behavior of single and group of geosynthetic encased stone columns. J Geotech Geoenviron Eng 136(1):130
2. Nayak S, Vibhoosha MP, Bhasi A (2019) Effect of column configuration on the performance of encased stone columns with basal geogrid installed in lithomargic clay. Int J Geosynth Ground Eng

3. Murugesan S, Rajagopal K (2006) Geosynthetic-encased stone columns: numerical evaluation. Geotext Geomembr 24:349–358

Review on Field Direct Shear Test Methodologies

Kakasaheb D. Waghmare and K. K. Tripathi

Abstract This paper reviews the techniques used for in situ testing of soil shear strength. Conventional laboratory direct shear test has problems like the simulation of field conditions, particle size restriction, etc. Therefore, many researchers have tried to perform in situ testing for the shear strength of soil to get reliable results. But because of the soil's heterogeneous nature and heavy designs of the set-up, these tests are not popular. The study will help to identify the advantages and limitations of existing in situ direct shear testing set-ups currently being used. With the help of this review, it has been concluded that there is a need for simpler and strain-controlled in situ direct shear testing on the soil.

Keywords Shear strength · In situ direct shear test · Cohesion · Angle of internal friction

1 Introduction

Assessment of the bearing capacity of soil is mainly influenced by its shear strength (c–ϕ) parameters. Various tests, such as triaxial test, vane shear test, direct shear test, etc. are available for the determination of bearing capacity from shear strength parameters. Whereas tests such as standard penetration test and plate load test are available for the determination of bearing capacity from settlement criteria. The present study focuses on the study of work done on In situ direct shear test standards and some other modified methods. Laboratory direct shear test is one of the simplest and fastest tests to know the shear strength parameters of soil. But the field direct shear test is not very popular due to the requirements of the heavy set-up, the problems of applying shear and normal load, the test controls, etc. Laboratory tests are carried

K. D. Waghmare (✉)
College of Engineering Pune, Pune, India
e-mail: kdwaghmare3@gmail.com

K. K. Tripathi
Civil Engineering Department, College of Engineering Pune, Pune, India

M. S. Ranadive et al. (eds.), *Recent Trends in Construction Technology and Management*, Lecture Notes in Civil Engineering 260,
https://doi.org/10.1007/978-981-19-2145-2_51

out under controlled conditions, which may differ from actual on-site conditions. Due to technical constraints such as scale effect and quality of sampling, there are considerable potential limitations to the on-field testing techniques in the form of sampling and testing anomalies. Therefore, the study aims to compare the various methods and techniques used for in situ shear testing to better understand their accuracy, consistency, and adoptability. In this study, standard procedures like the ASTM code tests and I.S. code tests, modifications in the ASTM and I.S. code field direct shear tests have been explained and compared.

2 Objective

To study various methods of field direct shear test of soil and do a comparative assessment of their suitability for wide-scale adoption.

3 Literature Survey

The various techniques as recommended by the ASTM code and the I.S. code and their respective modifications have been studied and published. The following section includes brief notes on such previous investigations carried out for in situ direct shear tests to measure shear strength and shear strength parameters of soil.

3.1 Field Direct Shear Strength Test to Measure Rock Discontinuities (ASTM D 4554-02)

In this technique of field shear strength testing, ASTM D 4554-02 recommends that during the performance of the in situ shear test on rectangular-shaped blocks of rock, the blocks to be isolated on all surfaces, except for the shear plane surface. The size of blocks for which this technique is deemed suitable is of size 700 mm × 700 mm × 350 mm. The detailed set-up of the equipment is shown in Fig. 1. Care should be taken that the blocks are not to be disturbed during preparation operations and that the base of the block shall coincide with the plane to be sheared. A normal load is applied perpendicular to the shear plane, and then a lateral load is applied to induce shear along the plane of discontinuity. This technique requires the placement of the test specimen in concrete, which considerably increases the test's conduct duration. While the technique is useful for checking the discontinuities of rock, it does not help in checking the soil's shear strength.

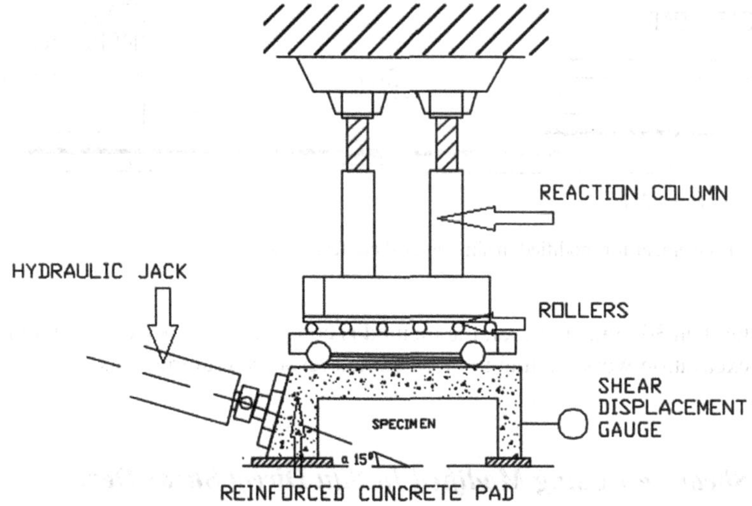

Fig. 1 Typical arrangement of equipment for in situ direct shear test

3.2 Direct Shear Test for Soils Containing Gravel IS 2720-1979 Part 39/2

Using this technique, as recommended in IS 2720-1979 part 39/2, the shear strength of soils containing gravel and cobblestone can be determined using the direct shear test. In this technique, a 1500 mm^2 box is lowered into the ground with the help of a hydraulic jack. The soil around the box is excavated step-by-step till the box reaches the desired position. After the digging is complete, an assembly is placed on the top of the blocks for the application of normal load. This assembly is made up of rolled steel joists, wooden sleepers and sandbags. The shear load is applied with a controlled hydraulic jack and proving ring arrangement, taking reaction from the adjacent box (Fig. 2). This method can be used even for soils where the particle size

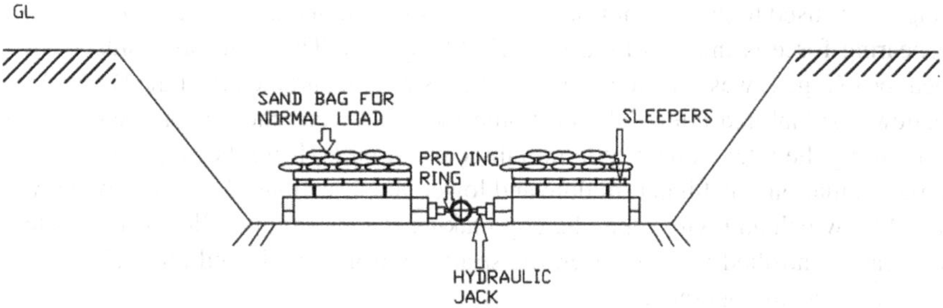

Fig. 2 Typical arrangement I.S. code method for in situ direct shear test

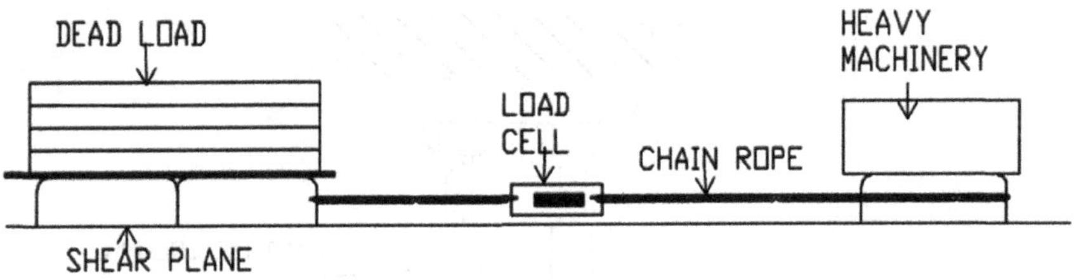

Fig. 3 Arrangement for modified in situ direct shear test device

is greater than 30 mm. However, the method is complicated to perform as it requires heavy excavation work leading to more human resource requirements.

3.3 Shear Test Using Modified In Situ Direct Shear Device

Liu [3] proposed a modified version of the direct shear on-field test device similar to the I.S. code version. In this test, the upper part of the shear box, as presented in the IS Code test, is replaced by the lattice shear box. The lattice shearing box frame is placed directly on the ground. The box is then pulled with a flexible chain while a dead load is applied onto the shearing frame. Validation is done by carrying out tests on trials with similar gradations under normal stress (up to 880 kPa) (Fig. 3). As can be understood, this test has no control of strain rate, as the shear load is applied by pulling the chain with heavy machinery.

3.4 Field Testing of Shear Strength of Soil with the Help of an Open-Sided Shear Box

A paper by M. Cross, July 2010, presents an open-sided shear box with a 300 mm^2 basal shearing zone and two 300 mm × 150 mm horizontal shearing areas. Lead weights are used to provide normal loads that simulate natural geostatic forces, and a shearing force is induced by a winch/cable system. This field open-sided direct shear box (Fig. 4) was used to assess the shear strength parameters of the Derbyshire shallow residual and colluvial soil. Some locations were chosen for shear testing by dividing the total slope profile into three equal parts, from the slope crest to the slope foundation, the higher, middle and lower slope sections. The soil sample was pulled by winch and wire rope, thereby making it a strain-controlled test. This test is a strain-controlled test, however, the strain control is crude and unprecise as the winch is manually operated.

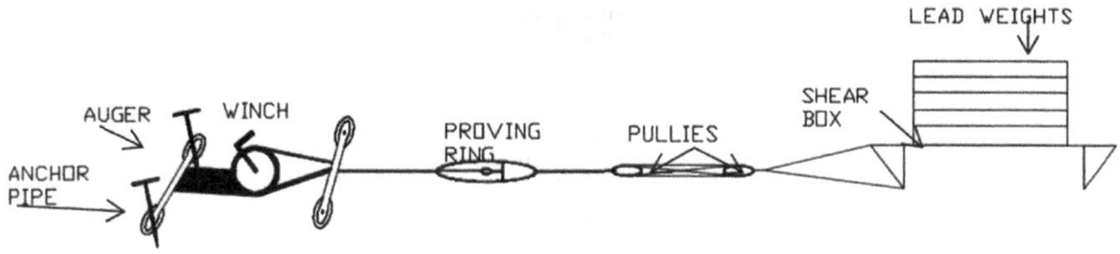

Fig. 4 Schematic view of open-sided shear box test

3.5 *Shear Testing Using Modified Standard Operation of Rock Piles*

Another modification of the direct shear test technique was used to measure the rock pile materials' cohesion and friction angle. Two test apparatuses had been designed, 300 and 600 mm^2 Shear boxes. The test apparatus also included a manufactured roller plate, a metal top plate, shear dial gauges with support services and two hydraulic pistons and cylinders with a maximum hydraulic cylinder of 70 MPa (10,000 psi) for normal loading (Fig. 5).

The Questa Rock Pile Stability Study SOP 82v4 made use of this method. This Questa Rock Pile Stability Study SOP 82v4 was developed to provide technical guidance and practices to test rock pile materials during mines' environmental investigations. The test aimed to determine how much cohesion exists in the rock pile material and to check the weathering impact on the material's shear strength properties, such as cohesion and friction angle. A trench was excavated with the help of a shovel up to a specified layer of rock. With the help of a nail board and nail cage, the required shape was given to the rock pile to fit into the shear box. A roller plate was then placed over the rock pile and a dead load was applied with the help of machinery

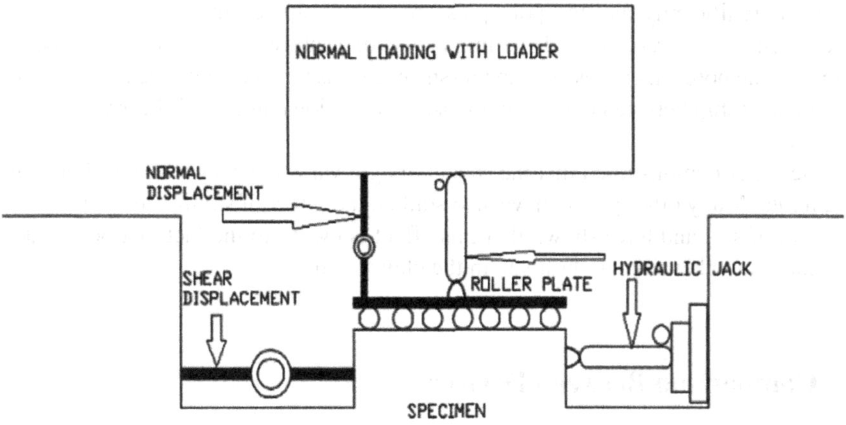

Fig. 5 Modified in situ direct shear test technique of rock piles

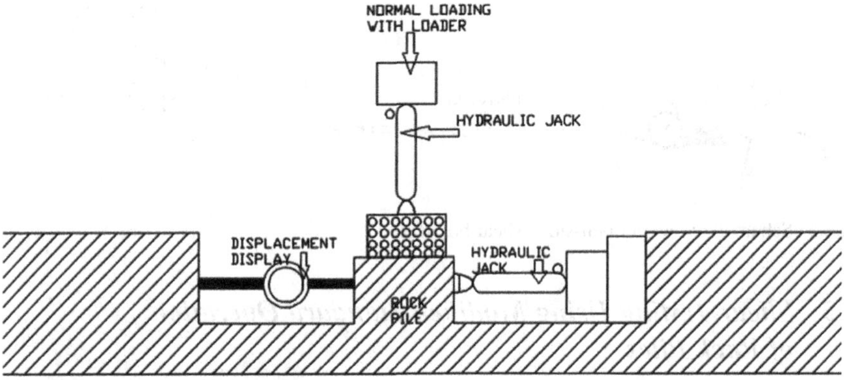

Fig. 6 Arrangement of SOP for testing rock pile

or dead plates. The normal load was applied with the help of a hydraulic jack and a horizontal shear load was applied with the help of another hydraulic jack. As can be seen, this method of testing soil shear strength is a stress-controlled test and requires heavy support (Fig. 6).

3.6 Effect of Strain rate on Shear strength of soil.

This paper [7] shows the relationship between shear strength and shear rate of the soil. The test was carried out on the Ring Shear Test apparatus. The test consists of eight stages with four different shear rates: 0.01, 0.1, 1 and 10 mm/s. The specimens were saturated and consolidated at 100 kPa and the total normal stress was maintained at 100 kPa during the tests. After the residual energy was formed, the shear rate was reduced significantly and the pore pressure was dissipated under wet conditions. The specimen was re-extracted at 1 mm per second (phase 2). This shearing process removed the pore water pressure, and re-shearing was repeated until the eighth stage. The relationship between the remaining active collision angles and shear values was recorded.

The paper demonstrated that the shear strength value changes with the change in strain rate. Many other papers have also studied the effect of strain rate on the shear strength of soil and have shown that this effect is owing to the fact that pore water pressure rate takes time to dissipate in the clayey soil.

4 Comparison Between Devices

Table 1 shows the tabular comparison of the above literature and shows the limitations of the same.

Table 1 Comparison of various field direct shear test

Method	Type of soil test	Mould size	Test type	Limitations
Field direct shear strength test to measure rock discontinuities (ASTM D 4554-02)	Rock	700 × 700 mm	Stress controlled test	(1) The test requires the emplacement of the test specimen in concrete
Direct shear test for soils containing gravel I.S. 2720-1979-part 39/2	Gravel	1500 × 1500 mm	Stress controlled test	(1) Heavy set-up (2) Require a lot of excavation
Shear test using modified in situ direct shear device by Liu [3]	Coarse-grained	122.5 × 122.5 mm	Stress controlled test	(1) Heavy construction machinery (2) Disturbance in soil (3) Calculations complicated
Field testing of shear strength of soil with help of an open-sided shear box	Coarse-grained	300 × 300 mm	Partially strain-controlled test	(1) Sample disturbance is not eliminated (2) Shallow depth testing
Shear testing using modified standard operation of rock piles (Questa Rock Pile Stability Study SOP 82v4)	Rock pile	300 × 300 mm and 600 × 600 mm	Stress controlled test	(1) Heavy set-up (2) Require more than five labourers (3) Normal load with a front end loader

5 Discussion

From the results shown in Table 1, it can be seen that the most available techniques and methods that have been studied and recommended were stress-controlled, and even in the strain-controlled tests, the strain control is crude and partial. It is also seen that the strain rate influences the shear strength of soil and hence poor control of strain rate while conducting the tests can lead to inaccurate results. It is worth noting here that the most popular test, i.e. the laboratory direct shear test, is a strain-controlled test. The reason for wide-scale acceptability and adoption of laboratory direct shear test is its simplicity, speed of conduction, consistency and reliability of results. Most of the trials of the above literature have been done on the rock piles and very few tests or modifications have been done on soil samples. It is also observed that no tests have introduced a testing depth of more than one metre. All the present in situ tests require heavy set up and human resources to perform, thereby making them expensive.

6 Conclusion

In light of the above literature review and comparative analysis of different types of tests for shear strength, it can be concluded that further research and experimentation are required in order to make field shear testing of soil more acceptable in routine practice. Considering that the current practices are heavy set-ups, it is necessary to devise lightweight set-up tests to perform field direct shear tests on the soil. As the strain rate is found to influence the results of shear strength testing, it must be ensured that the newly devised tests have better, preferably complete strain control. Most existing field tests entail a high requirement of manpower, especially labourers. This dependence on manual labour needs to be reduced in order to make the tests efficient, cost-effective and easily replicable. The above factors will ensure that the field direct shear testing technique has all the virtues of a laboratory direct shear test device, thus making it widely acceptable and adoptable.

References

1. *ASTM D 4554-02* (2006) Standard test method for in situ determination of direct shear strength of rock discontinuities. ASTM, Philadelphia
2. *I.S. 2720-1979 part XXXIX/2*, Direct shear test for soils containing gravel. Bureau of Indian Standards, New Delhi
3. Liu S (2009) Application of in situ direct shear device to shear strength measurement of rockfill materials. Water Sci Eng 2(3):48–57
4. Cross M (2010) The use of a field open-sided direct shear box for the determination of the shear strength of shallow residual and colluvial soils on hillslopes in the south Pennines, Derbyshire. North West Geogr 10(2):8–18
5. Fakhimi A, Boakye K, Sperling DJ, McLemore VT (2007) Development of a modified in situ direct shear test technique to determine shear strength parameters of mine rock piles. Geotech Testing J 31(3):1–5
6. Standard Operating Procedure (2008) In-situ direct shear tests. Questa Rock Pile Stability Study, SOP 82v4
7. Saito R, Fukuoka H, Sassa K (2006) Experimental study on the rate effect on the shear strength. Disaster mitigation of debris flows, slope failures, and landslides, pp 421–427

Numerical Prediction of Tunneling Induced Surface Settlement of a Pile Group

B. Swetha, S. Sangeetha, and P. Hari Krishna

Abstract Tunnel construction below or adjacent to the piles of Existing Building Structure (EBS) will affect the performance and eventually the stability of piles due to ground deformation which may result in the settlement of piles. A two-dimensional finite element analysis using PLAXIS 2D was performed to study the behavior of a pile group due to tunnel construction beneath the ground. A 15-noded plane strain model was used to simulate the soil structure. The Mohr–Coulomb (MC) model was used to simulate the soil–structure interaction at the tunnel–soil interface. An embedded beam row element was used to simulate the pile embedded in mixed soil conditions. An isotropic elastic plate element was used for modeling the existing building. A twin tunnel circular in shape was considered to be constructed underneath the building. Different tunnel locations are considered vertically and horizontally away from the building. A contraction volume loss of 0.5% during tunnel construction was adopted for analysis. The settlement of the building was found to decrease with an increase in tunnel depth from the ground surface. The settlement was also observed to decrease on moving the tunnels horizontally away from the building at a given depth. Compared to tunnel depth increment, an increase in horizontal distance between tunnels and the building has a greater influence on decreasing settlement. A 45° pressure distribution line was drawn to understand the settlement decrement pattern. Tunnel cross-sectional area overlapping with the influence region of building load was determined for each twin tunnel location. The settlement of the building was found to increase or decrease similar to the changes in the overlapping area. Differential settlements of pile group were observed in all the tunnel locations. But the difference in settlement of individual piles was found to decrease with an increase in the horizontal distance and vertical depth from the pile group.

B. Swetha (✉) · S. Sangeetha
VNR Vignana Jyothi Institute of Engineering and Technology, Hyderabad, Telangana 500090, India
e-mail: b.swetha1828@gmail.com

P. Hari Krishna
Department of Civil Engineering, NIT, Warangal 506004, India

M. S. Ranadive et al. (eds.), *Recent Trends in Construction Technology and Management*, Lecture Notes in Civil Engineering 260,
https://doi.org/10.1007/978-981-19-2145-2_52

Keywords Pile group settlement · Existing building settlement · Tunneling · Numerical analysis · PLAXIS 2D

1 Introduction

Tunneling is being considered the best option to avoid increasing traffic congestion in urban areas. Tunnels can be used to take heavy traffic from one point in city to another so that local roads can be freed up. In practical, tunnels can rebuild the city, generate returns in long term by letting networks of roads to be born-again, and collectively improve the livability of whole urban areas. For high-rise buildings supported by deep foundations, the construction of tunnel induces ground movements, which, in turn, affect the bearing capacity as well as the settlement of the existing piles [1]. The surface settlements during the tunneling and optimum tunnel depth could be predicted based on the numerical methods that are accurate [2], economical, and time-saving. The present work was aimed to evaluate the effect of tunneling on an existing building with piled foundation [3, 4] by adopting the numerical finite element method, PLAXIS 2D [5]. For its to avail the saving time, men and machine power and accuracy for observing the deformation of soil mass while tunneling. There has been considerable research examining the behavior of the soil–tunnel–pile interaction and the possible damage to an existing piled foundation caused by tunneling. The objective of this study is to investigate the surface settlement behavior of piled foundations due to twin tunnel construction.

2 Methodology

2.1 Data Collection

A study by Raghavendra et al. [6] has given the following salient features of the Bangalore underground metro tunnel:

- The total length of the Bangalore metro line is 42.3 km, which is divided into two corridors. The first one runs along the east–west corridor extending up to 18.1 km and the other running along the north–south corridor is 24.2 km long.
- The focus of this study is confined to the underground section with a twin tunnel for a length of 4.8 km along the east–west section (See Fig. 1).
- Tunnels are bored using an inner diameter of 5.6 m which is reinforced with a concrete lining of thickness 280 mm.

The properties such as Young's modulus, Poisson's ratio, and density, for the different layers of soil in Bangalore, along with their depths, and some of the elastic properties are taken from a study conducted by Sitharam and Anbazhagan [8] based on Multichannel analysis of surface waves (MASW) survey and tabulated as shown

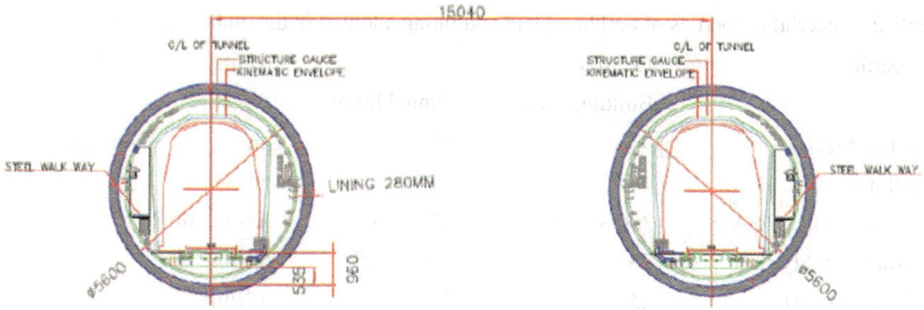

Fig. 1 Cross section of Bangalore Metro Underground Tunnel [6, 9, 10]

Table 1 Input properties of soil layers adopted in the numerical analysis

Properties	Identification				Units	Model
	Layer 1	Layer 2	Layer 3	Layer 4		
Drainage type	Drained	Drained	Drained	Drained	–	Mohr–coulomb
Unsaturated unit weight (γ_{unsat})	15	15	16	26	kN/m^3	
Saturated unit weight (γ_{sat})	16	16	17	26	kN/m^3	
Young's modulus (E')	325×10^3	111×10^3	408×10^3	523×10^3	kN/m^2	
Poisson's ratio (v')	0.3	0.3	0.3	0.2	–	
Cohesion (C_{ref})	5	20	15	4000	kN/m^2	
Friction angle (φ)	35	30	35	42		
R_{inter}	1.0	1.0	1.0	1.0	–	

in Table 1. The properties of the foundation, tunnel lining, and pile foundation are given in Tables 2 and 3.

2.2 Finite Element Modeling and Boundary Conditions

Finite element plane strain model with 15 nodded triangular elements was used for modeling the building which is supported by a group of 6 piles of length 10 m with a spacing of 3 m, carrying a line load of 80 kN/m^2, and twin tunnels with 5.6 m diameter and 15 m c/c are constructed under the preexisting building which was modeled to study the settlement behavior of the pile foundation. Preexisting elements like Elastic plate element, embedded beam row [7], and interface structures were adopted to simulate the Existing Building Structure (EBS). The model dimensions were 80

Table 2 Material properties of building plate and lining adopted in the numerical analysis

Properties	Identification		Units	Model
	Building plate	Tunnel lining		
Diameter (D)	–	5.8	m	Elastic
Thickness (t)	1	0.28	m	
Normal stiffness (EA_1)	300×10^6	351×10^6	kN m²/m	
Flexural rigidity (EI)	25×10^6	966×10^6	kN/m	
Unit weight (γ)	25	25	kN/m³	
Poisson's ratio (ν)	0.15	0.15	–	

Table 3 Material properties of the pile

Properties	Identification model pile (embedded beam row)	Units	Model
Diameter (D)	0.3	m	Elastic
Predefined beam type	Massive circular beam	–	
Young's modulus (E)	20×10^6	kN/m²	
Unit weight (γ)	25	kN/m³	
Poisson's ratio (ν)	0.15	–	
$L_{spacing}$	3	m	

× 40 m. Figure 2 shows the two-dimensional finite element model. The boundary conditions of the geometry model are enabled in staged construction mode. The boundary conditions are as follows, Boundary X_{min}: horizontally fixed, Boundary X_{max}: horizontally fixed, Boundary Y_{min}: fully fixed, Boundary Y_{max}: free.

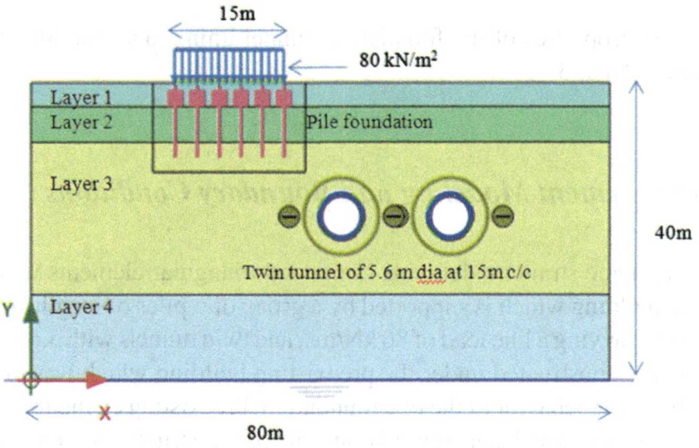

Fig. 2 The geometric model of pile foundation and twin tunnels

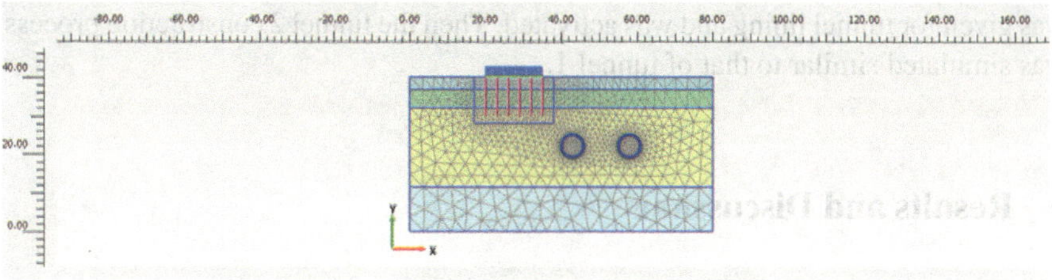

Fig. 3 The finite element mesh of the numerical model

An automated mesh was generated in PLAXIS 2D, with fineness ranging from fine to very fine mesh. In the model, very fine mesh was enabled near structures and tunnels to obtain more accurate results, and the fine mesh for the remaining model (Fig. 3) gives the automated mesh generated in numerical analysis.

2.3 *Procedure of Simulation and Analysis in Staged Construction*

Staged construction is a powerful PLAXIS feature that enables a realistic simulation of construction and excavation processes by activating and deactivating soil clusters and elements, application of loads, etc. This procedure allows a realistic assessment of stresses and displacements as caused (PLAXIS 2D manual), for example, soil excavation during an underground construction project.

The flowchart given in Fig. 4 shows the procedure adopted for the simulation of the model having twin tunnels construction under EBS with pile foundation.

Initially, only soil layers were in an activated state and the remaining structures were deactivated. Then the piles and building plate were activated and loads were applied to the plate. Construction of tunnel 1 was simulated by activating the tunnel lining and deactivating the soil cluster inside tunnel 1. A contraction value of 0.5%

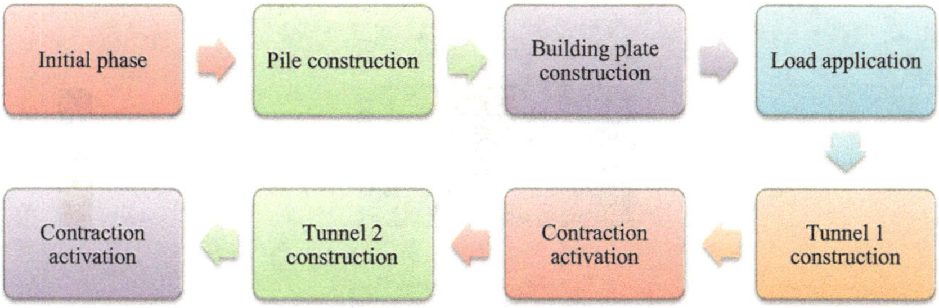

Fig. 4 Flow chart showing procedure used for simulation and analysis of project

was given for tunnel lining and was activated. Then the tunnel 2 construction process was simulated similar to that of tunnel 1.

3 Results and Discussions

The results obtained through the output window after the calculation in staged construction mode are discussed in the section. Figure 5 shows the deformation of mesh due to tunneling activity and Fig. 6 gives the intensity of settlements near tunnels and building plates.

To find the effect of twin tunnel location on pile group, the location of the tunnel was shifted in the transverse direction and vertical direction from the first pile of foundation. A contraction value of 0.5% was given to account for volume losses as per FHWA-NHI-09-010. The settlement values of the building plate were found to vary with different tunnel locations. Table 4 represents the settlement values for the tunnel location from exactly below the building and just beside the pile to 6 m away from the edge pile, and vertically 15 m from GL up to 21 m from GL. Figure 7 represents the variations of surface settlements along with tunnel depth and were explained based on the tunnel area under the influence of building load.

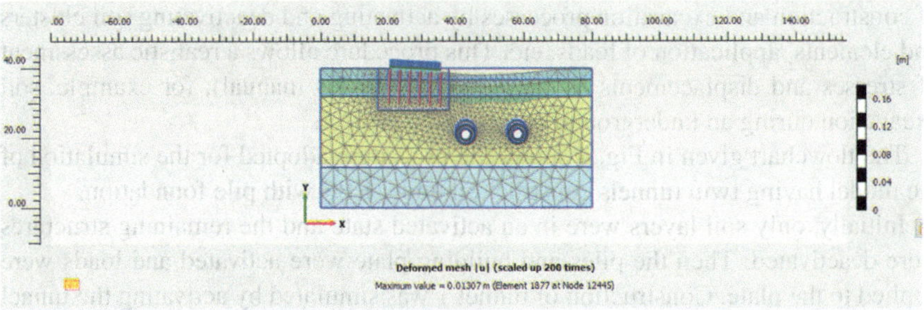

Fig. 5 Deformed mesh after performing calculation

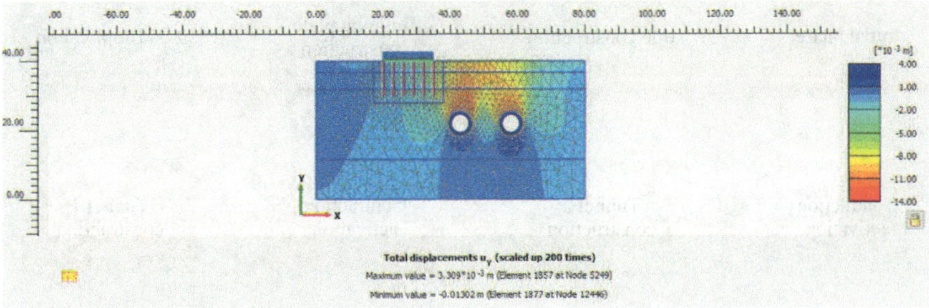

Fig. 6 Maximum settlement after calculations

Table 4 Maximum surface settlements (mm) values of building plate for different tunnels locations

Depth below ground surface (m)	Distance of tunnels from the edge of pile foundation							
	Exactly below the EBS	0 m	1 m	2 m	3 m	4 m	5 m	6 m
15	12.8	9.8	8.7	7.5	6.6	5.4	4.4	3.7
16	12.5	9.7	8.7	7.8	6.5	5.8	4.5	3.9
17	12.2	9.5	8.6	7.5	6.5	6	4.9	4.1
18	11.8	9.3	8.4	7.6	6.9	5.9	4.9	4.1
19	11.7	9.3	8.4	7.5	6.7	5.8	5.1	4.4
20	11.4	8.9	8.3	7.5	6.7	6.1	5.1	4.5
21	11	8.9	8.2	7.3	6.5	5.8	5.2	4.6

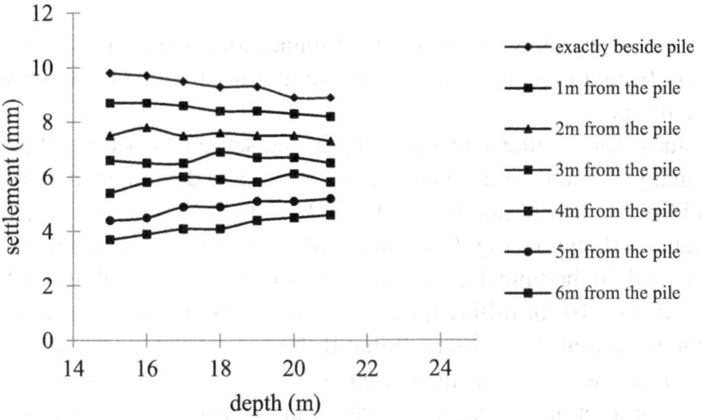

Fig. 7 Maximum surface settlement (mm) values of building plate for different tunnels locations

Figure 8 shows one of the graphs obtained for varying tunnel depths and depicts the differential settlement of piles due to varying tunnels location at 18 m depth below ground level. Similar graphs were drawn for various other depths. Table 5 gives the

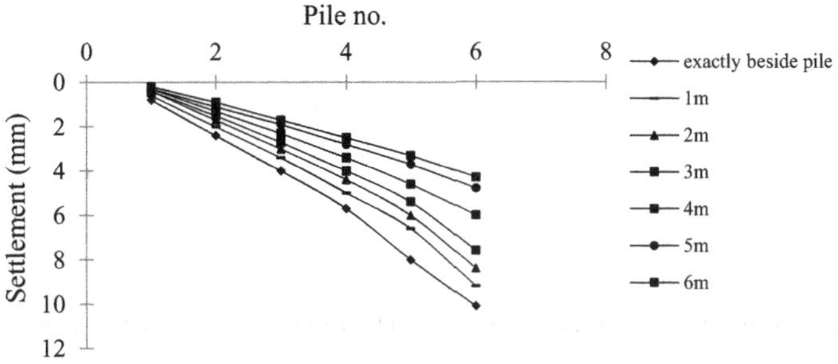

Fig. 8 Differential settlement of piles for different tunnel locations at 18 m below GL

Table 5 Distortion values of piles for 0.5% contraction value

Depth below GL (m)	Distortion values of pile for different tunnel locations						
	Beside the edge pile (0 m)	1 m	2 m	3 m	4 m	5 m	6 m
15	1/1415	1/1562	1/1829	1/2173	1/2727	1/3488	1/4054
16	1/1470	1/1612	1/1764	1/2272	1/2542	1/3409	1/3947
17	1/1546	1/1648	1/1875	1/2205	1/2459	1/3191	1/3846
18	1/1612	1/1744	1/1875	1/2083	1/2631	1/3333	1/3658
19	1/1648	1/1764	1/1973	1/2205	1/2678	1/3191	1/3658
20	1/1785	1/1875	1/2027	1/2238	1/2678	1/3125	1/3658
21	1/1851	1/1948	1/2142	1/2380	1/2727	1/3125	1/3571

calculated distortion values of piles for all tunnel locations. The distortion values were varying from a maximum of 1/1415 to a minimum 1/4054 and are within the permissible limits.

In this study, for a better understanding of the settlement values obtained from numerical analysis and to understand the increment and decrement pattern of settlements, a 45° stress distribution line is drawn from the last pile tip of the pile group to hard strata as shown in Fig. 9. For example, tunnels at 5 m horizontal distance are taken to explain the tunnel area coming under the influence of the building load area shaded in Fig. 10 for different tunnel depths. Similar drawings were made for other locations as well. It is observed that if the area of the tunnel cross section is increasing within the stress line, the settlements were observed to be increasing. This may be the reason for increasing settlement value with depth, beyond 3 m horizontal distance from the edge pile. If the area of a tunnel under the influence zone remains

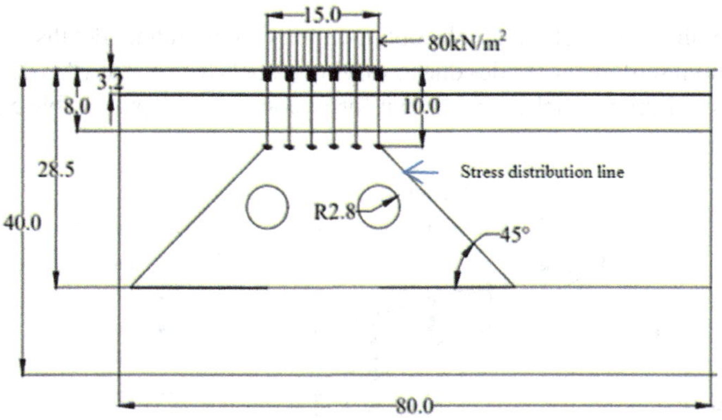

Fig. 9 A typical model showing the stress distribution line and dimensions of the model (All dimensions are in m)

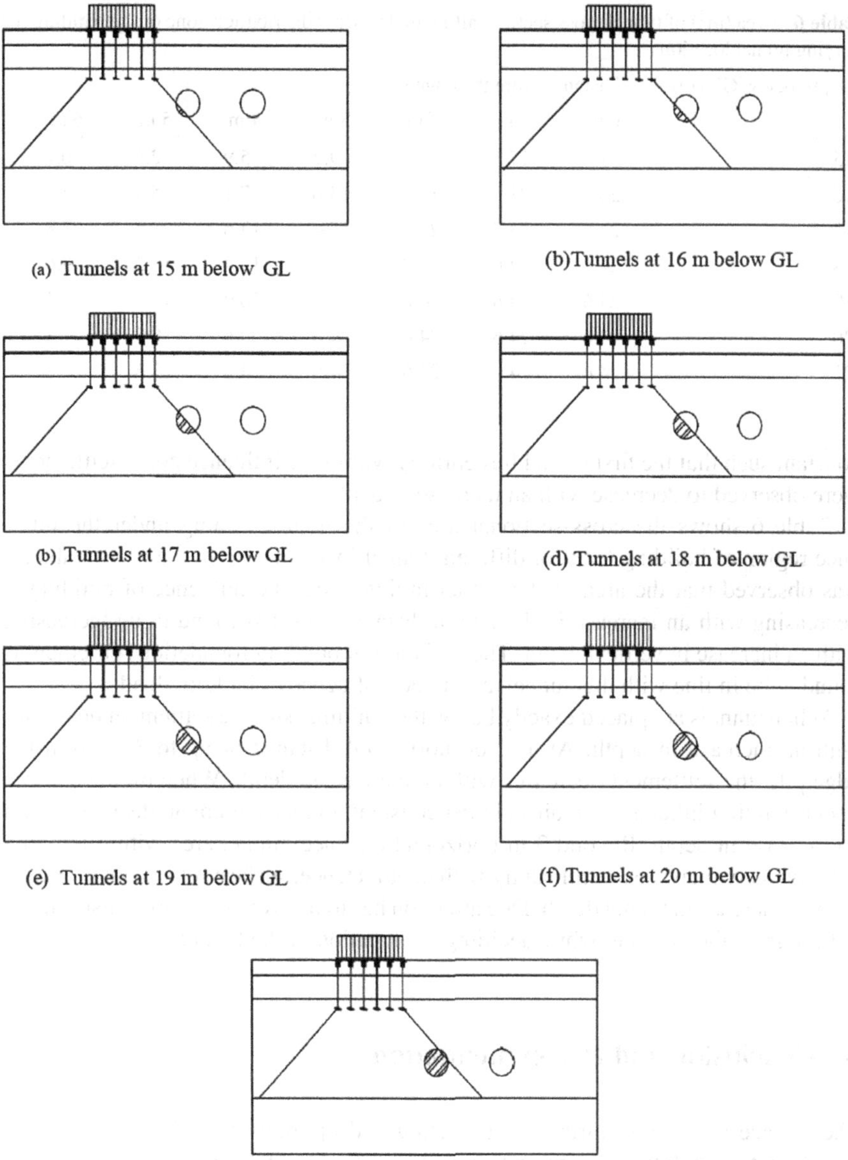

(a) Tunnels at 15 m below GL

(b)Tunnels at 16 m below GL

(b) Tunnels at 17 m below GL

(d) Tunnels at 18 m below GL

(e) Tunnels at 19 m below GL

(f) Tunnels at 20 m below GL

(g) Tunnels at 21 m below GL

Tunnels at 5m from the edge pile

Fig. 10 Tunnel area within stress distribution region for 5 m horizontal distance from the edge pile

Table 6 Area (m^2) of tunnel cross section falling under stress distribution zone of the building for varying tunnel locations

Depth below GL (m)	Distance from the edge pile						
	0 m	1 m	2 m	3 m	4 m	5 m	6 m
15	21	17.5	13.7	9.7	5.9	2.6	0.2
16	23.7	21	17.5	13.6	7.1	5.9	2.5
17	24.6	23.7	21	17.4	13.6	9.6	5.8
18	24.6	24.6	23.7	20.9	17.4	13.5	9.6
19	24.6	24.6	24.6	23.7	20.9	17.4	13.5
20	24.6	24.6	24.6	24.6	23.6	20.9	17.3
21	24.6	24.6	24.6	24.6	24.6	23.6	20.8

constant such that the first tunnel lies entirely within the influence zone, settlements were observed to decrease with an increase in depth.

Table 6 shows the cross-sectional area of the tunnel coming under the influence region of building load for different tunnel locations. From the above table, it was observed that the area of the tunnel falling under the influence of building is decreasing with an increase in the lateral distance, but it is found to be increasing with an increase in vertical depth. The variation of building foundation settlement is found to be in line with this tunnel area under influence of building load.

When tunnels are placed exactly below the building, surface settlement decreases with an increase in depth. At a given horizontal distance of up to 3 m from the edge pile, the settlement decreases with an increase in depth. When tunnel c/s area under building influence remains almost constant, tunnel settlement decreases with an increase in depth. Beyond 3 m horizontal distance, tunnel area within building load intensity increases significantly with depth. Hence, surface settlement increases with an increase in tunnel depth. Due attention has to be given to the load distribution pattern from the building while deciding the location of the tunnel.

4 Conclusion and Recommendations

The surface settlement during the tunneling and optimum tunnel depth could be based on the analytical, empirical or numerical methods. The present work was aimed to evaluate the effect of tunneling on surrounding piled foundation structures by observing the deformation of soil mass while tunneling.

From the above study, the conclusions arrived are as follows:

- Tunnel construction causes significant settlement on the surface structure supported by a pile foundation. Settlement decreases with an increase in horizontal distance from the EBS and decreases with an increase in depth of the tunnel.

- Tunnel-induced settlement pattern of existing building structure follows the tunnel area under influence of existing building load.
- Beyond 3 m horizontal distance, the computed settlements are found to increase in depth as much of the tunnel cross-sectional area comes under the building load distribution zone.
- Differential settlements were observed on piles and are within allowable limits.
- Tunnel-induced settlement pattern of existing building structure follows the tunnel area under influence of existing building load.
- Beyond 3 m horizontal distance, the computed settlements are found to increase in depth as much of the tunnel cross-sectional area comes under the building load distribution zone.
- Care must be taken while constructing the tunnel underneath the immediate vicinity EBS foundation. The zone of influence to create surface settlement must be decided considering vertical load distribution from the EBS.
- In order to maintain the deformation of EBS within the permissible limit, the tunnel has to be located outside the building load distribution region. In the absence of an alternate route, preventive measures are to be followed to secure EBS from undesirable structural damage.

References

1. Pastore A, Bhadauria SS (2017) Numerical modelling of building response to underground tunnelling—a case study of Chennai Metro. Int J Civil Eng 4(10):7–17
2. Raja K, Premalatha K, Hariswaran S (2015) Influence of tunneling on adjacent existing pile foundation. Int J Eng Res And4(08):477–483
3. Khabbaz H, Gibson R, Fatahi B (2019) Effect of constructing twin tunnels under a building supported by pile foundations in the Sydney central business district. Undergr Space (China) 4(4):261–276
4. Basile F (2014) Effects of tunnelling on pile foundations. Soils Found 54(3):280–295
5. El-kasaby EA (2017) Numerical analysis of tunnel boring machine in soft ground. Jokull J
6. Raghavendra V, Jose S, Arjun Shounak GH, Sitharam TG (2015) Finite element analysis of underground metro tunnels. IJCIET 6(2):06–15. ISSN 0976-6316
7. Sluis J, Bos W (2013) Validation and application of the embedded pile row. Plaxis, (December), 10–13
8. Sitharam TG, Anbazhagan P (2006) Measurements of dynamic properties and soil profiling using multichannel analysis of surface waves. In: Invited Keynote in 4th Karl Terzaghi Memorial Workshops (November 2016), p 6
9. U.S. Department of Transportation Federal Highway Administration 'Technical manual for design and construction of road tunnels—civil elements' Publication No. FHWA-NHI-09-010 March 2009
10. PLAXIS 2D Manual for General Information, Reference and Scientific Manual

Numerical Study of Pile Supported Embankment Resting on Layered Soft Soil

Uzma Azim and Siddhartha Sengupta

Abstract Adaption of construction on soft soil is coming into the picture with the increased demand for developing infrastructure. Since soft grounds are characterized by low permeability and high compressibility as well as low shear strength, the construction needs to be accompanied by ground strengthening methods. This paper studies settlement study of a 4 m high pile supported embankment resting on layered soft soil with respect to change in depth of pile penetration, the influence of altering slope and geometry of the embankment, effect of varied pile spacing and surcharge, variation of mesh size involved in simulation and effect of simulating model. Commercially available finite element software PLAXIS-2D was employed in the investigation. Increasing the depth of penetration of piles from 6 to 9 m resulted in a deduction in maximum displacement in the underneath soft soil layers from 0.3052 to 2.982 m due to transmission of load to a comparatively deep stiff soil layer. Decreasing the spacing between the piles from 4 m c/c spacing to 2 m c/c spacing, lowered the deformation in the soft soil layer and the time taken for consolidation due to homogenization of soil particles caused due to surcharge load and efficient load transfer platform in combination with friction acting on the surface of the piles. Also, widening the embankment and increasing the number of piles resulted in a decrease in the total displacement because of the larger area endured under an efficient load transfer platform. With regard to mesh size, fine mesh types with r_e of 0.50 gave accurate output but resulted in increased computational time due to the involvement of higher densification of elements. Therefore, a medium mesh size composed of r_e equivalent to 1.00 was employed in order to obtain desirable results in nominal time. Modified Cam Clay model (MCC) and Soft Soil model (SS) were used to perform a comparative study to observe the characteristics of soft soil with respect to the total displacement and dissipation of excess pore pressure. The total displacement when Modified Cam Clay (MCC) model was used to simulate the problem was 0.3052 m; modeling with the Soft Soil (SS) model with decreased stiffness resulted in an over prediction of the settlement to almost 3.154 m. This was attributed to the difference in the way Modified Cam Clay (MCC) and Soft Soil (SS)

U. Azim · S. Sengupta (✉)
Civil & Environmental Engineering, Birla Institute of Technology, Mesra, Ranchi 835215, India
e-mail: siddhartha@bitmesra.ac.in

M. S. Ranadive et al. (eds.), *Recent Trends in Construction Technology and Management*, Lecture Notes in Civil Engineering 260,
https://doi.org/10.1007/978-981-19-2145-2_53

model represented the failure surface on the left side of the critical state line and also on the modification in the compression indices involved in the calculation of simulating parameters. Further, the MCC model incorporates the capability of modeling the elasto-plastic behaviour of the soil medium which assists in efficiently analyzing the simulated model under hardening, softening and critical state behaviour.

Keywords Pile supported embankment · Finite element · Modified Cam Clay model · Soft Soil model · Excess pore water pressure

1 Introduction

With the increasing need for developing infrastructure in the era of paucity in land availability and increasing land costs results in the unavailability of stiff grounds for creating a firm foundation for huge roadway and railway networks. Thereby it's high time to look for sustainable construction on unpropitious ground conditions such as soft soil. But the risks involved with soft ground comprising of low bearing capacity, inadequate shear strength, high permeability and compressibility raises obstacles during and post construction activities. In order to rectify the instability of embankments during the construction phase and to safeguard long service life, various researches have been done to make soft soils compatible for construction by accompanying with other conventional construction methods or by creating a medium for transferring embankment and surcharge loads to stronger soil stratum at a deeper depth.

Al-Neami et al. [1] observed that relative density of sand was found to have a proportional effect on bending moments of pile group whereas pile spacing to diameter ratio had less effect on the same. Bagavasingam [2] analyzed the behaviour of soft clays undergoing consolidation. Sudden high settlement occurred with a sudden increase in stress whereas the application of constant stress for a considerable time leads to a decrease in the rate of instant settlement. Borthakur and Dey [3] analyzed the group capacity of micropile in weak clayey soil and concluded that the settlement to micropile cap size ratio to be the most appropriate parameter to predict the load carrying capacity and micropile cap position with respect to the ground position to be the least effective parameter. Dang et al. [4] concluded that Fibre Reinforced Load Transfer Platform (FRLTP) thickness enhanced Stress Concentration Ratio (SCR) between columns and surrounding soil thereby decreasing lateral deformation and effectively improving embankment system stability. Esmaeili et al. [5] compared two experimental models of a high embankment with and without micropiles reinforcement with respect to the bearing capacity of the embankment fill material, axial strains in the micro piles and resulting displacements. Pham and Dias [19] numerically inspected pile supported embankment by employing FLAC-3D software and concluded that in this case soil arching is the main contributor towards reducing settlements which acts on the pile cap thereby causing a deduction in the load reaching the soft subsoil layer. The arching was computed by the parameter named efficacy given by $E_a = \frac{P_c}{(\Upsilon H+q)A}$, where P_c is the load acting upon the pile cap due to arching, H denotes the embankment height, A is the area influenced due to piles, q represents surcharge and Υ is the unit weight of soils composing the embankment.

Fattah et al. [6] carried out experiments to record the response of pile models under the presence of cavities. Juran and Guermazi [9] studied the effect of compacted sand column reinforcement in soft soils on the settlement by performing laboratory experiments and observed that the group effect, consolidation and replacement factor caused a substantial reduction in settlement and vertical stress concentration on columns. Karim et al. [10] carried out creep analysis of soft foundation soil underlying geogrid reinforced embankment associated with Pre-fabricated Vertical Drains (PVD's) using the Modified Cam Clay (MCC) model. They distinguished between two creep analyzes:- Creep A analysis employed an elliptical yield surface and Creep B analysis was based on upgraded and modified yield surface criteria. Creep B analysis gave more accurate results for lateral displacement profile, displacement profiles and critical settlement profiles with about 3% over prediction as compared to 7% in creep A analysis. Kumar et al. [13] investigated stability, the effect of height and slope angle, stresses, strains and displacement occurring in embankments constructed of laterite soil in the Karnataka region. The results showed that strains decreased and stresses remained constant as embankment height was increased. For a constant height, steep slopes gave higher stresses and strains. Ghosh et al. [8] evaluated embankment resting on soft soil supported by columns injected with concrete. Consolidation analysis was performed to take into account the variation in soil permeability with respect to void ratio along with settlements and lateral movements in the subsoil layer. It was seen that the settlements in the columns decreased with depth and the columns resting on stiff clay layer experienced maximum shear forces and bending moments and arching caused the column heads to carry higher amount of load. Pham et al. [21] performed an analytic and numerical 3D discrete element analysis to study the contrivances involved in pile supported embankments. They concluded discrepancies between both the analysis in envisaging stress reduction ratio, efficacy and differential settlement due to the complex load transfer mechanism involved.

Moni and Sazzad [17] analyzed homogeneous and layered slopes with the shear strength reduction (SSR) technique. Rise to a certain level was seen in factor of safety (FOS) with an increase in point of application of surcharge from the crest of soil slope and beyond that the effect of surcharge remained constant. Sadaoui and Bahar [23] concluded that Young's modulus of stone columns relied upon the compaction and lateral confinement of improved soil. Said et al. [24] performed settlement analysis for soft soil reinforced by a floating soil–cement column and it was seen that the reinforcement had led to an increase in the bearing capacity and stiffness of the composite ground thereby reducing settlements. Sazzad and Haque [25] studied the effect of surcharge acting on a homogeneous slope and observed an inverse proportional relation between the factor of safety (FOS) and slope angle whereas high FOS values were obtained for low surcharge loads. Tai et al. [28] performed a parametric analysis which showed that by increasing foundation stiffness, time taken for completion of 90% degree of consolidation decreases, differential settlement reduces and the steady stress concentration ratio increases. Zhang et al. [35] prepared 3D numerical models to carry out a comparison analysis between the deformation caused due to uniform and localized load acting on the pile supported embankment and it was observed that the soil arching effect acting on the pile cap enhanced during uniform loading conditions resulting in reduced settlements whereas it tarnished under localized loading conditions causing failure. Pham and Dias [20] prepared 25 full scale models and performed 6 series of experiments in order to analytically analyze the design of pile supported embankments. They concluded that the analytical design methodologies correctly interpret the deformation in the case of cohesionless fill soils as compared to cohesive fill materials and it is less appropriate to be employed in the case of low rise embankments. These deformities are due to inconsistent assumptions involved with respect to the subsoil support and arch shape.

Wen et al. [29] found post grouting to be effective in the improvement of skin friction of soft clay by 48.7% and 30.8% under compression and tension respectively. Wu et al. [30] concluded that maximum horizontal displacement occurred amidst the toe and center line of embankment resulting in bending failure of piles. Yan-hui et al. [32] performed experiments on soft ground enhanced by the vacuum-surcharge preloading method and concluded the rate of surcharge application to have a great impact on ground deformation. Soil parameters such as void ratio, Poisson's ratio and compression modulus also had a profound influence on the deformation characteristics. Zhang and Wang [34] prepared centrifuge model tests to observe slope stability strengthened by piles and concluded that reduced pile spacing enhanced the factor of safety of the reinforced slope. Zhang et al. [33] investigated the effect of spacing amid piles on surrounding soil press, press on top of piles and pile strain by preparing 3 different centrifugal models of cement-fly ash-gravel (CFG) pile composite foundation. The foundation was found to undergo shear failure with large pile spacing and was affected by pile quality, embankment height and cushion strength. Zhou et al. [36] summarized factors affecting the stability of uplift resistance of foundations in soft clay which included the soil's anisotropy with respect to strength, depth of soil under foundation, the type, size and shape of foundation pit. Pham [18] analytically studied geosynthetic reinforced pile supported embankment in order to understand the interaction amidst the linked elements such as foundation, granular platform, fill soil and geological media around the embankment foundation. They suggested that the overall efficacy of the system can be improved by augmenting the degree of consolidation in the subsoil, increasing the height of the embankment, rising the angle of friction of the embankment soils and reducing the subsoil modulus.

Based on the study done in the literature review with principal findings summarized in Table 1, it can be observed that with the rise in the demand to improve infrastructure and increase in land costs due to scarcity of land availability, it is an urgent need to check for alternatives for construction on soft soil preferably accompanied with other conventional construction methods. It can be seen that studies have been done to observe the stability of structures on different types of soils and to check the variance of the factor of safety under different circumstances. Research needs to be done in the area of deformation analysis underlying an embankment constructed on layered soft soil strata. A study on pile supported embankments over soft soils

Table 1 Summary of literature review

S. No.	Name of author	Description	Principal findings
1	Esmaeili et al. [5]	Prepared models with and without micropiles to observe the difference in bearing capacity of the embankment and displacements of the foundation and the embankment	Embankment settlement reduced by 40–45%, FOS increased by 30% and bearing capacity of high embankment increased by 65%
2	Fattah et al. [6]	Carried out experiments to record the response of pile models under the presence of cavities	Presence of cavities in close proximity to piles caused a reduction in soil density and shaft friction along the pile thereby decreasing the pile failure load.
3	Fırat et al. [7]	Investigated the effect of soil type, height and water table level on embankment stability	Concluded that the fill properties for embankment and sub base highly affected stability. Medium semi stiff clay gave high factor of safety than soft clay fill and increasing the height of embankment lowered the factor of safety
4	Ghosh et al. [8]	Evaluated embankment resting on soft soil supported by columns injected with concrete and performed deformation and consolidation analysis	Settlements in the columns decreased with depth and the columns resting on a stiff clay layer experienced maximum shear forces and bending moments and arching caused the column heads to carry higher amount of load
5	Laskar and Pal [14]	Researched on consolidation effects due to increase in surcharge	Increasing surcharge pressures reduced the values of compression index and coefficient of consolidation for silty clayey soil
6	Leeworthy and Asr [15]	Performed a comparison study between LEM and FEM slope analysis	Each loading condition was found to affect slope stability contrarily due to varied load distribution on soil and instances of displacement decrease with an increase in load indicated a local failure in both the analysis

(continued)

Table 1 (continued)

S. No.	Name of author	Description	Principal findings
7	Manna et al. [16]	Numerically studied model slopes prepared from Yamuna sand	Observed a similar slip failure with bulging in all the models and concluded slope stability to be dependent on slope angle, soil cohesion and surcharge loads
8	Moni and Sazzad [17]	Analyzed homogeneous and layered slopes with the shear strength reduction (SSR) technique	Rise to a certain level was seen in FOS with an increase in point of application of surcharge from the crest of soil slope and beyond that the effect of surcharge remained constant
9	Pham [18]	Analytically studied geosynthetic reinforced pile supported embankment in order to understand the interaction amidst the linked elements such as foundation, granular platform, fill soil and geological media around the embankment foundation	The overall efficacy of the system can be improved by augmenting the degree of consolidation in the subsoil, increasing the height of the embankment, rising the angle of friction of the embankment soils and reducing the subsoil modulus
10	Pham and Dias [19]	Numerically inspected pile supported embankment composed of both cohesive and non-cohesive soils in order to incorporate the study of fill cohesion effect, the height of the embankment and ground elastic modulus	Soil arching effect was seen to be influenced by the fill soil cohesion by increasing the loading efficacy. The ratio of the embankment height to pile spacing, subsoil stiffness and embankment fill soil properties are the crucial parameters required to be considered while designing the depth and spacing of the piles supporting the embankment
11	Pham and Dias [20]	Prepared 25 full scale models and performed 6 series of experiments in order to analytically analyze the design of pile supported embankments	The analytical design methodologies correctly interpret the deformation in the case of cohesionless fill soils as compared to cohesive fill materials and it is less appropriate to be employed in the case of low rise embankments. These deformities are due to inconsistent assumptions involved with respect to the subsoil support and arch shape

(continued)

Table 1 (continued)

S. No.	Name of author	Description	Principal findings
12	Rajput et al. [22]	Analyzed the deformation under load in soft soil strengthened with sand piles	Bearing capacity increased to 3.25 times in comparison to that of unreinforced soil at the optimum spacing of 2.5 times the piles diameter
13	Souria et al. [27]	Experimented with various pile group configurations and analyze their impact on the lateral resistance of pile groups using numerical simulations	Battered piles had the highest group efficiency of 70–80%, followed by 60–85% for the mixed pile group and 40–80% for the vertical pile group
14	Xie et al. [31]	Performed research on combined retaining wall structure (CRS) to support an embankment constructed on a steep slope of soft soil	CRS platform effectively reduced the bending moment acting on piles and thereby prevented potential sliding

should be performed considering their impact on hastening construction and diminishing deformations. This paper addresses to the study of deformation characteristics in a pile supported embankment resting on layered soft soil strata has been studied by varying simulating parameters and doing a comparison analysis. The comparisons have been done with respect to the penetration depth of piles, the consequence of variation in geometry and slope of the embankment, spacing between piles, the effect of surcharge on the settlement characteristics, effect of employed mesh size and the model used for simulation of the soft soil.

2 Methodology

In the paper, finite element based software, PLAXIS-2D has been employed to analyze the deformation characteristics in order to study pile supported embankments resting on layered soft soil. The plot for simulation was set by fixing, the horizontal extent of 40.0 m, soft soil depth of 9.0 m with each soil layer of 3 m each and embankment height of 4.0 m. The contours for the simulation were set as $x_{min} = 0.0$ m, $x_{max} = 40.0$ m, $y_{min} = -9.0$ m, $y_{max} = 4.0$ m. The road embankment having a crest width of 20 m and side slope of 1:3 was supported by 9 pile elements with a spacing of 2 m amidst the piles (Fig. 1). The top most layer, very soft clay (shown in blue), had the weakest geotechnical properties underlain by comparatively stiffer layers of Soft clay (1) and Soft clay (2) (shown in green and yellow respectively). The properties used for the simulation were referred from [12] and are listed in Table 2. The water table was located at 1 m below the ground level in order to take into

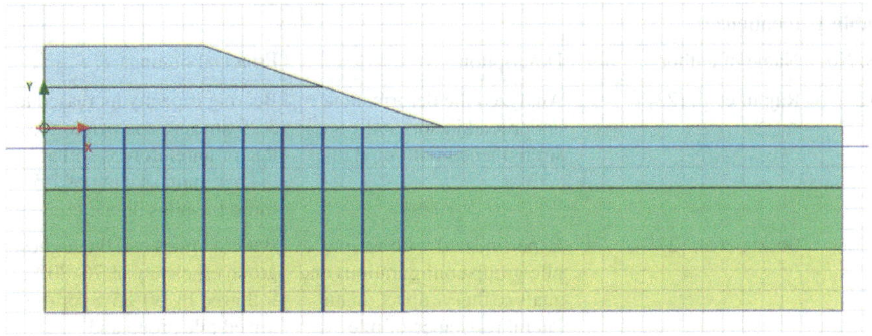

Fig. 1 Geometrical modeling of an embankment resting on pile reinforced soft soil up to base depth

Table 2 Parameters considered for embankment resting on pile supported soft soil strata

Description	Embankment fill	Very soft clay	Soft clay (1)	Soft clay (2)	Pile
Model	Mohr–Coulomb (MC)	Modified Cam Clay	Modified Cam Clay	Modified Cam Clay	Linear-elastic
Drainage	Drained	Undrained	Undrained	Undrained	–
γ_{unsat} (kN/m^3)	15	8.19	12.55	10.76	–
γ_{sat} (kN/m^3)	18.90	12.82	14.72	14.27	–
k_x (m/day)	4.32E–3	3.825E–4	1.3E–3	2.125E–4	–
k_y (m/day)	2.16E–3	1.912E–4	6.5E–4	1.062E–4	–
Λ	–	0.075	0.057	0.068	–
K	–	0.025	0.032	0.03	–
M	1.2	1.2	1.2	1.2	–
c' (kN/m^2)	10	7	10	5	–
Φ'(°)	30	27	30	30	–
υ'	0.3	–	–	–	0.3
R_{inter}	1	1	1	1	1
EA (kN/m)	–	–	–	–	7.5E6
EI (kN-m)	–	–	–	–	1E6
E (kN/m^2)	8500	–	–	–	–

Source Kasim et al. [12]. "Simulation of Safe Height Embankment on Soft Ground using Plaxis." *Asia–Pacific Chemical, Biological & Environmental Engineering Society*, Vol 5, 152–156. And Sengupta and Azim [26]. "Numerical modelling of a pile-supported embankment". Constrofacilitator consideration the dissipation of excess pore pressure caused during the installation of pile elements during the staged construction method.

3 Load Transfer Mechanism in Pile Supported Embankment

The load transfer mechanism in an embankment resting on pile supported soft soil comprises of two steps of shearing resistance and soil arching. The load coming from the superstructure is partly repelled by the shearing resistance of the soft soil. Due to the stiffness difference between the materials used in pile caps and that of adjacent soil, there is a difference in the load taking capacity causing a larger concentration of load at the pile caps in comparison to the soil present in the foundation amid the pile caps which undertakes reduced stress in contrast to the overburden stress, ultimately resulting in the differential settlement at the foundation level. The stresses in granular embankment are redistributed and re-oriented to form arches. This mechanism transfers embankment load above arches and external load to the piles. This mechanism is referred to as 'soil arching' which decreases the load reaching the soft soil layers thereby reducing the settlement in the soft soil layers.

4 Numerical Simulation

After fixing the contours and incorporating the properties of the soils and pile elements, plane strain analysis was used to model the pile supported embankment in order to study the displacement and stress characteristics in the pile along with lateral movement exhibited by soil due to load application. In order to perform precise finite element calculations, the model was medium meshed and was discretized into 150 elements by keeping the relative element size (r_e) as 1.

To investigate the impact of depth of pile penetration, the embankment was made to rest on 9 m layered soft soil with the depth of each layer being 3 m. Two pile depths of 6 m and 9 m were modeled and studied (refer Fig. 2a, b) so as to study the deformation characteristics at the mid soil layer depth and the stiffest bottom most soft soil layer respectively. Consolidation analysis was adopted for the calculation of stresses in the staged construction mode of the software by taking into account the water table situated 1 m below the ground level.

To analyze the influence of variance in slope and geometry of the embankment, two embankment models were prepared to rest on the soft soil layers having the same properties. The first model incorporated an embankment with a slope of 1:3 and crest width of 40 m supported by 9 pile supports with amid pile spacing of 2 m and depth of 9 m (refer Fig. 2b). The second model was prepared with an embankment slope of 3:5 and a crest width of 60 m. Accordingly, the number of pile supports got increased

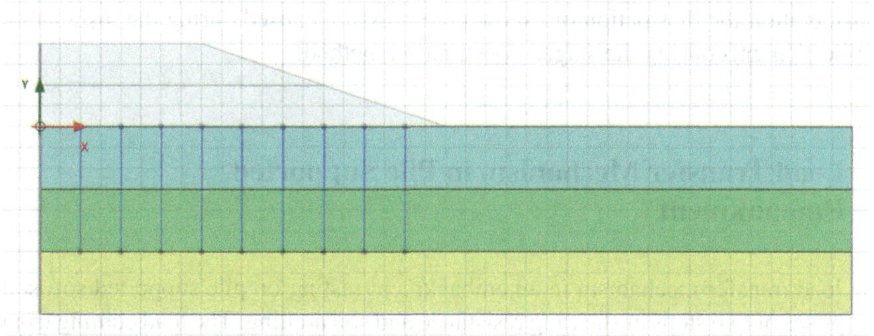

(a)

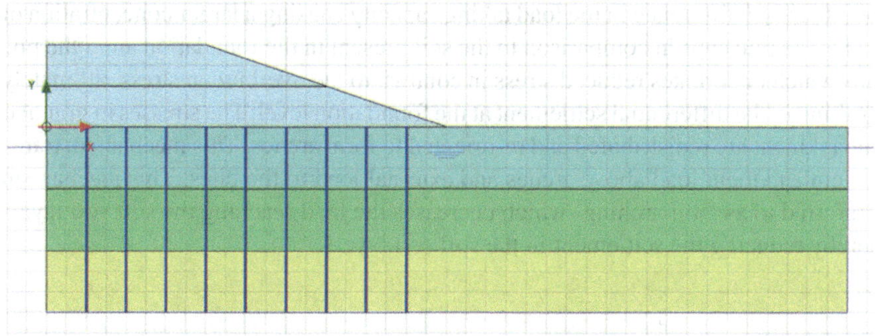

(b)

Fig. 2 Simulating geometry to study effect of depth of pile penetration **a** pile depth of 6 m **b** pile depth 9 m

to 17 with 2 m centre to centre spacing (refer Fig. 3). Water level was considered to be positioned at 1 m below the surface level so as to carry out consolidation analysis and incorporate the dissipation of excess pore pressure.

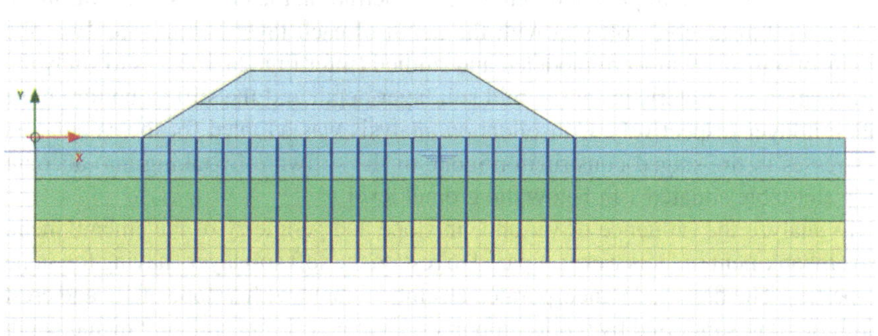

Fig. 3 Modeling of embankment with slope 3:5

Apart from the self-weight of the embankment with 1:3 slope and crest width of 40 m acting on the layered soft soil, the effect of surcharge loaded upon the embankment was taken into account along with change in pile spacing. In the case of 2 m pile spacing, 9 pile elements were considered with pile depth of 9 m and surcharge in the form of static line load was increased gradually from 10 to 150 kN/m (refer Fig. 4a) and the structure was found to fail at load of 150 kN/m. 5 pile elements of depth 9 m were considered when pile spacing was increased to 4 m and static line load was varied from 10 to 120 kN/m (refer Fig. 4b) and settlement failure was observed to occur at 100 kN/m in this case.

After drawing the geometry and assigning properties to the soil and pile elements in the soil stratigraphy mode, the finite element software PLAXIS-2D involves meshing of the structure in order to perform a precise calculation. In contemplation to study the influence of varying mesh sizes on the deformation characteristics, excess pore pressure variation and time taken for consolidation, 5 mesh sizes, namely, very coarse mesh, coarse mesh, medium mesh, fine mesh and very fine mesh, were analyzed and compared based on the meshing parameter target element size or average element size (I_e) which is globally required to generate a mesh. In the used

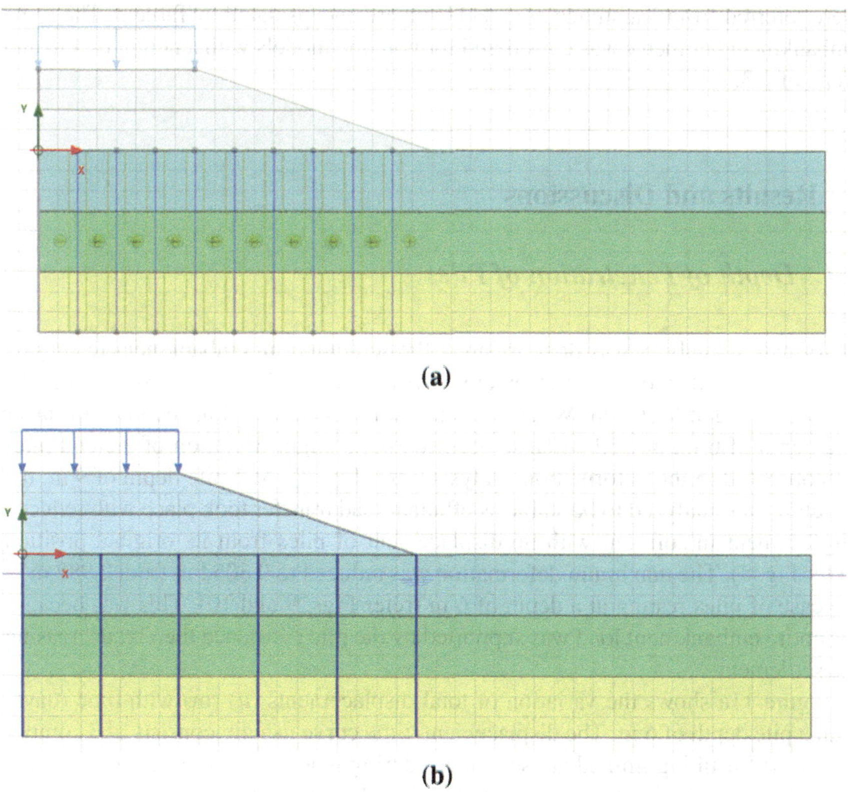

Fig. 4 Simulating model of embankment with surcharge **a** 2 m pile spacing **b** 4 m pile spacing

software, it was calculated using the outer boundary dimensions, $x_{\min}$, $x_{\max}$, $y_{\min}$ and $y_{\max}$, as:

$$I_e = r_e \times 0.06 \times \sqrt{(x_{\max} - x_{\min})^2 + (y_{\max} - y_{\min})^2} \tag{1}$$

I_e is based on a parameter called 'relative element size factor (r_e)' referring to the ratio of the sides of the triangular mesh adopted in the software according to which the number of elements into which the model will be discretized is decided. Discretization of the model in very coarse meshing was keeping r_e as 2.00 (Fig. 5a), r_e as 1.33 for the coarse mesh (Fig. 5b), r_e was kept at 1.00 for the medium mesh (Fig. 5c), fine mesh adopted r_e as 0.67 (Fig. 5d) and r_e was kept 0.5 for very fine mesh (Fig. 5e).

To study the effect of models adopted to simulate the behaviour of soft soil, Modified Cam Clay (MCC) and Soft Soil (SS) model were employed in the analysis to model the layered soft soil. In the case of the MCC model, 3 soil layers were adopted (refer Fig. 1) and the properties involved in the analysis are summarized in Table 2, whereas for the SS model, 2 soil layers were used, Soft clay-3 and Soft clay-4 (shown in green and yellow respectively) (refer Fig. 6) and the properties were referred from Karstunen et al. [11] and are encapsulated in Table 3. The same embankment geometry was considered for both the models with a height of 4 m and slope of 1:3.

5 Results and Discussions

5.1 Depth of Penetration of Piles

Deformation analysis was done to study the settlement in embankment resting on layered soft soil with pile reinforcement resting up to 6 m of the soil strata and up to a base depth of 9 m. When the piles were made to rest at the mid soil layer at a depth of 6 m, no soil collapse occurred but a slight deviation of the end piles was observed in the deformation analysis (refer Fig. 7). At a pile depth of 9 m, the structure was analyzed to be stable as efficient load transfer took place with reduced displacement in soft soil with no displacement of piles from its original position (refer Fig. 8). The maximum deformation was reduced to 0.3052 m from 2.982 m in the case of piles resting at a depth of 6 m (refer Figs. 9 and 10). This was because the entire embankment load was supported by the pile resistance thereby decreasing the settlement.

Figure 11a shows the variation of total displacements (u) (m) with time (days) with a pile depth of 6 m. The displacement–time curve usually consists of 3 components: (a) Resulting immediate settlement during load application (b) deformation caused due to consolidation as the transpiration of excess pore pressures occurs under loads (c) Secondary consolidation settlement is restrained by the structure and soil

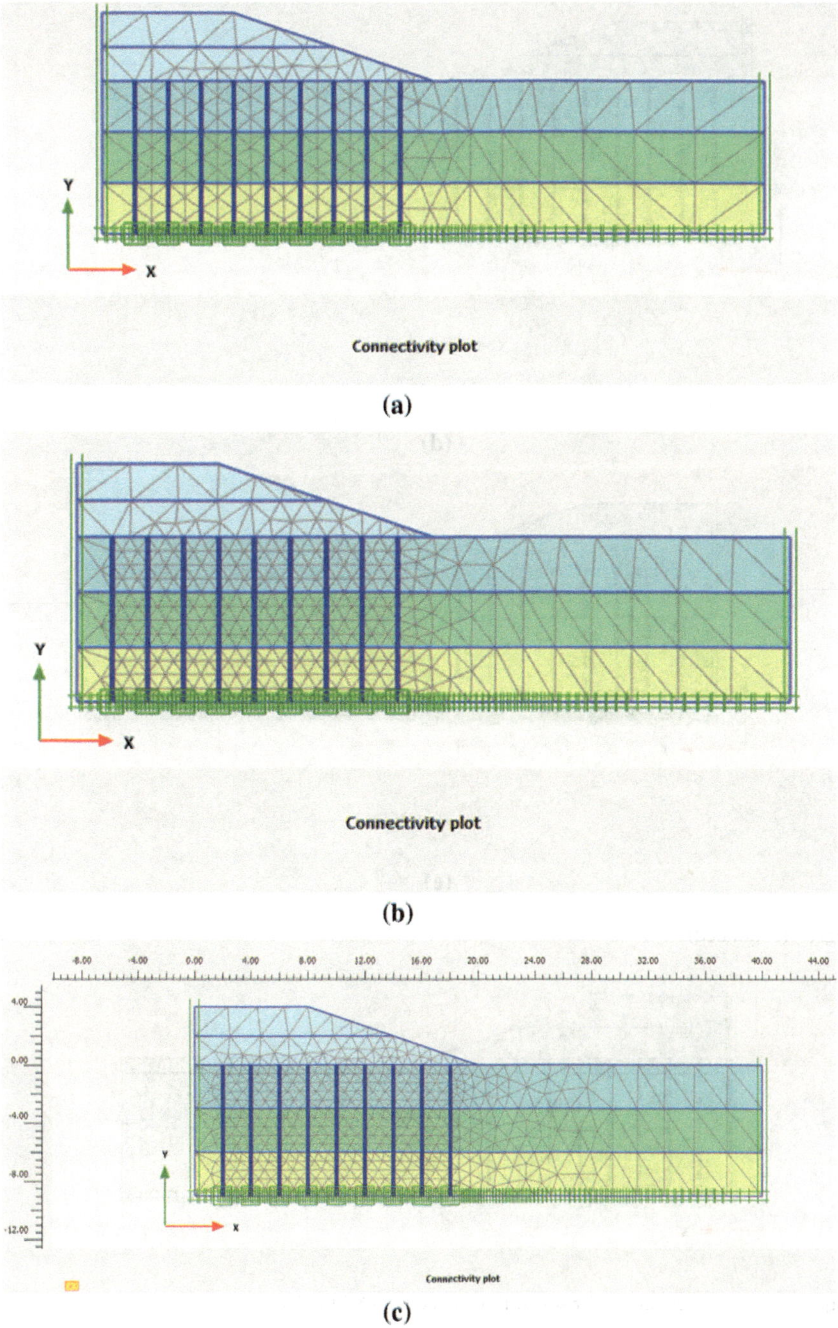

Fig. 5 Meshing in PLAXIS 2-D **a** very coarse mesh **b** coarse mesh **c** medium mesh **d** fine mesh **e** very fine mesh

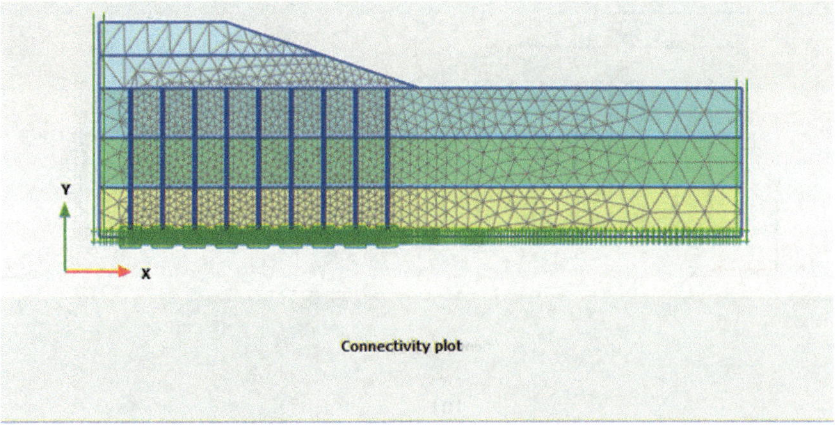

(d)

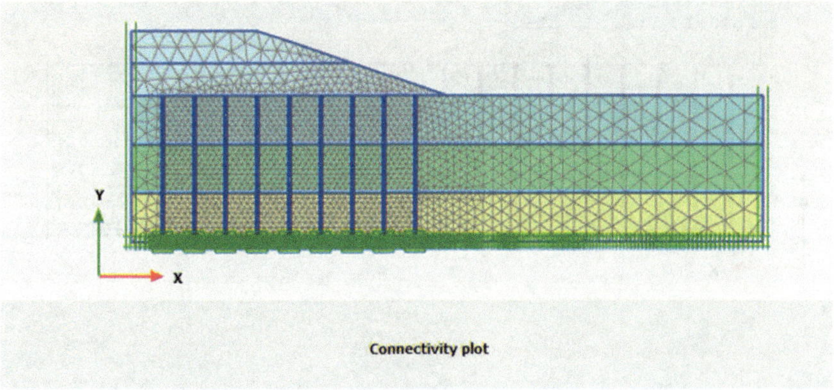

(e)

Fig. 5 (continued)

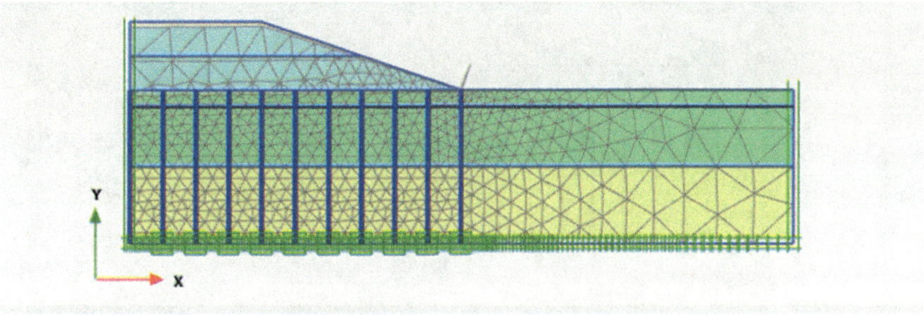

Fig. 6 Simulating geometry for soft soil modeled with SS model

Table 3 Parameters considered for embankment resting on pile supported soft soil strata simulated using Soft Soil model

Description	Embankment fill	Soft clay-3	Soft clay-4	Pile
Model	Mohr–Coulomb	Soft Soil	Soft Soil	Linear-elastic
Drainage	Drained	Undrained	Undrained	–
γ_{unsat} (kN/m^3)	15	12.55	10.76	–
γ_{sat} (kN/m^3)	18.90	14.72	14.27	–
k_x (m/day)	4.32E–3	From data set	From data set	–
k_y (m/day)	2.16E–3	From data set	From data set	–
c' (kN/m^2)	10	0	0	–
Φ' (°)	30	30	30	–
υ'	0.3	0.2	0.2	0.3
R_{inter}	1	–	–	1
E (kN/m^2)	8500	–	–	–
λ*	–	0.24	0.24	–
κ*	–	0.01	0.01	–
K_o^{NC}	–	0.67	0.5	–
EA (kN/m)	–	–	–	7.5E6
EI (kN-m)	–	–	–	1E6

Source Karstunen et al. [11] "Modeling the behaviour of an embankment on soft clay with different constitutive models." *International Journal for Numerical and Analytical Methods in Geomechanics*, Vol 30, 953–982

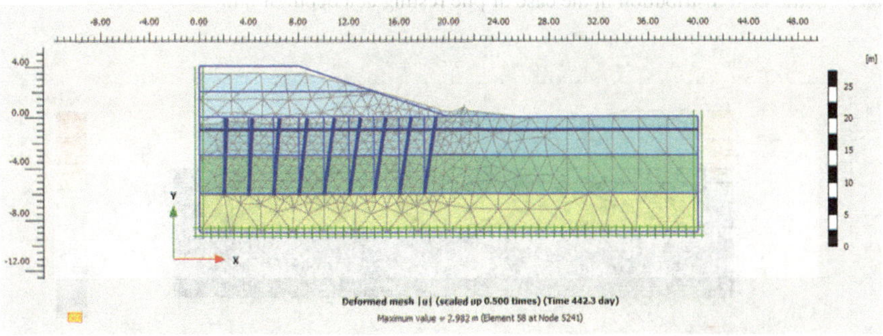

Fig. 7 Deformed mesh at pile depth of 6 m

composition. In this case, a small deviation from the original trend is observed due to the displacement of the end piles while taking the load. The graph for 9 m pile depth depicted a perfect smooth curve for the immediate settlement indicating that the load acting on the embankment was efficiently engrossed by the piles followed by the curve reaching a constant state to depict the secondary settlement state depicting complete dissipation of the excess pore water pressures (refer Fig. 11b).

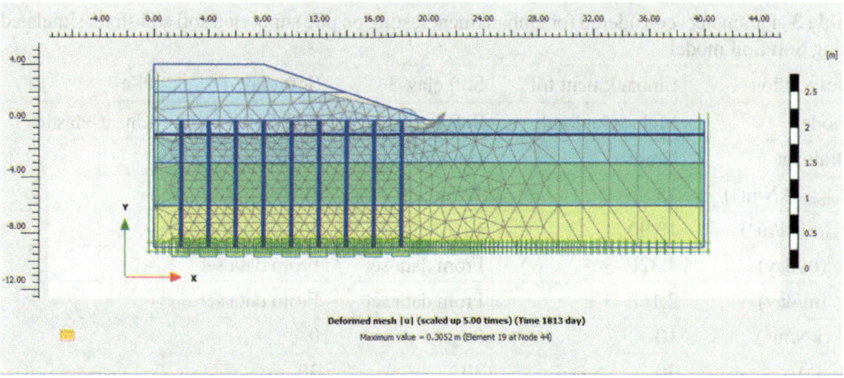

Fig. 8 Deformed mesh at pile depth of 9 m

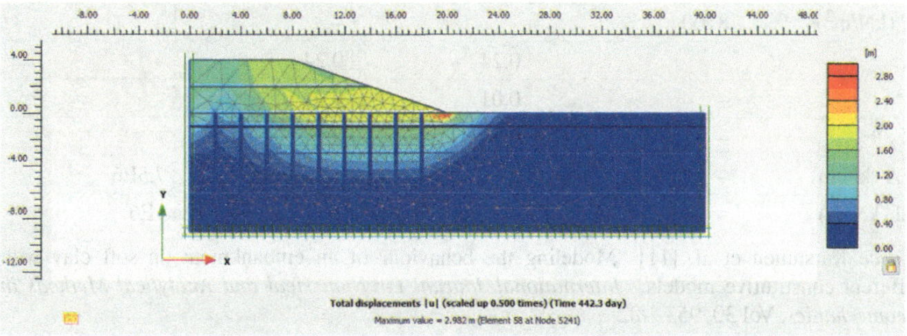

Fig. 9 Settlement distribution in the case of pile resting at a depth of 6 m

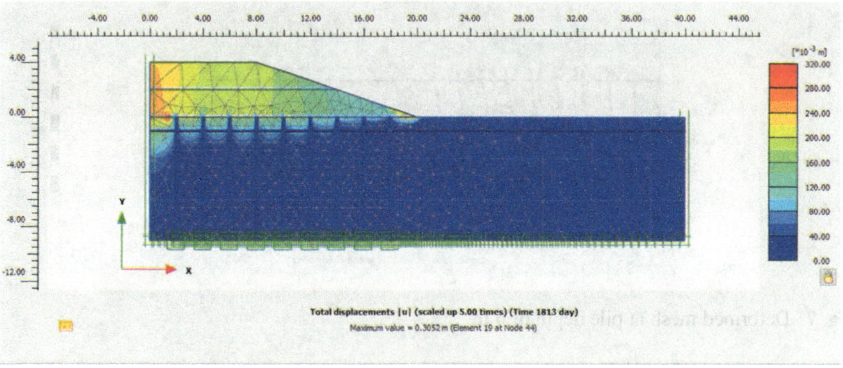

Fig. 10 Settlement distribution in the case of pile resting at a depth of 9 m

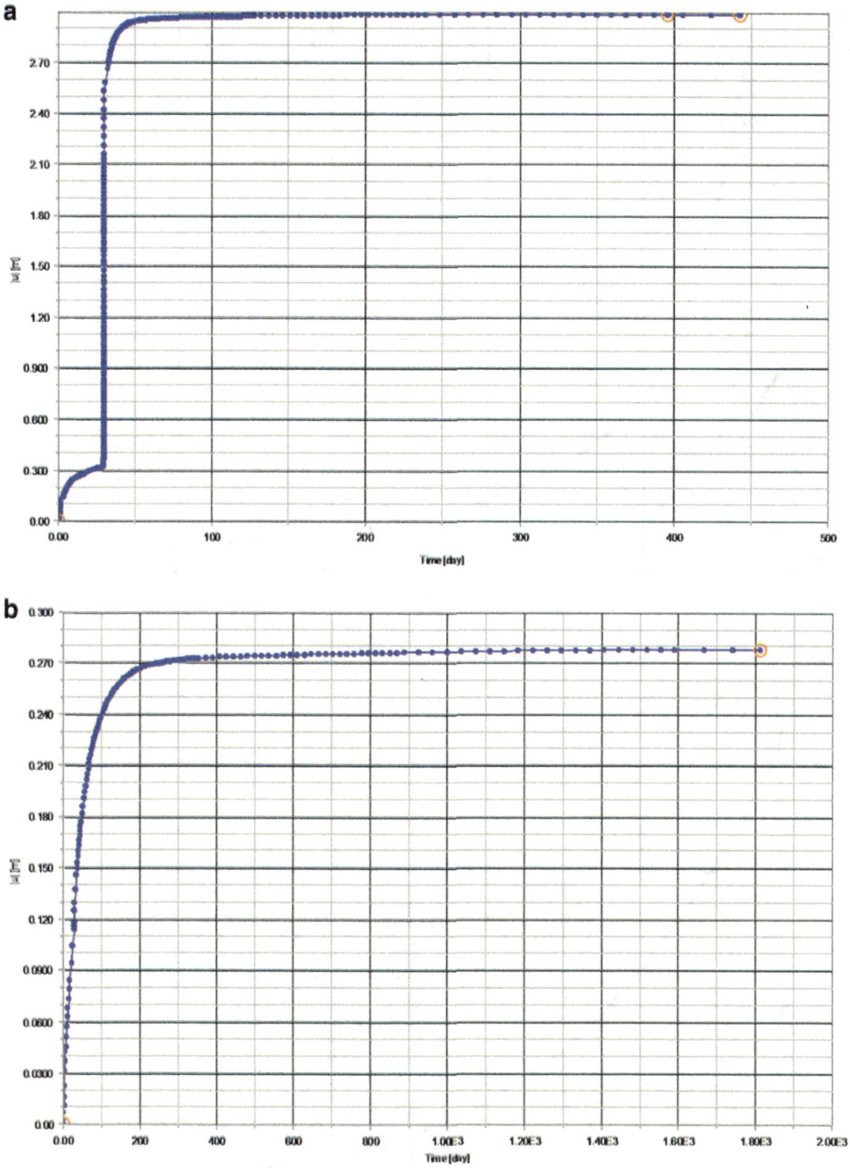

Fig. 11 Variation of total displacements (u) (m) with time (days): **a** 6 m pile depth **b** 9 m pile depth

Figure 12a, b illustrates alteration of excess pore pressure with time for pile depth of 6 m and 9 m respectively. The two peaks in Fig. 7a, symbolizes the peaks of the construction phase and consolidation phase respectively. In the case of 6 m pile depth, the maximum excess pore pressure reached was 20 kN/m^2 during the initial days of construction and 50 kN/m^2 in the consolidation phase. It took almost

a

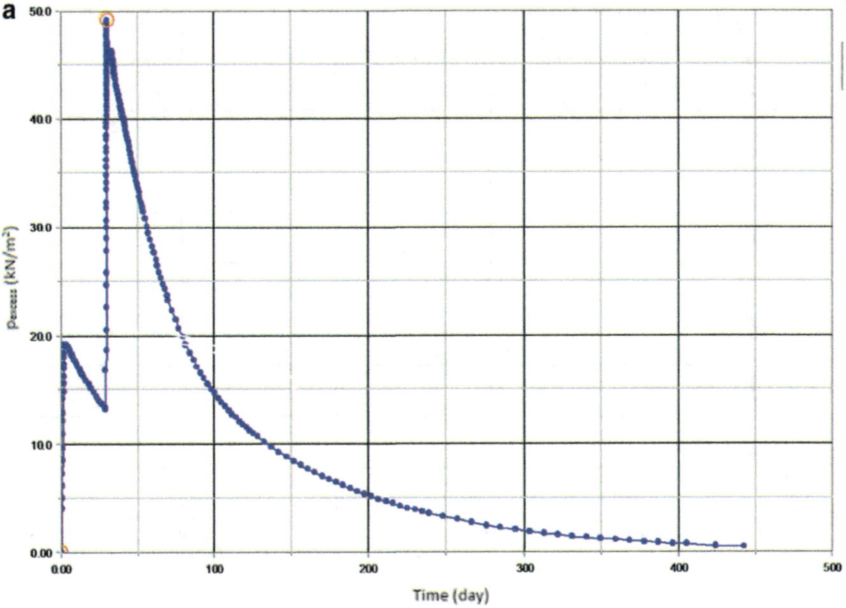

b

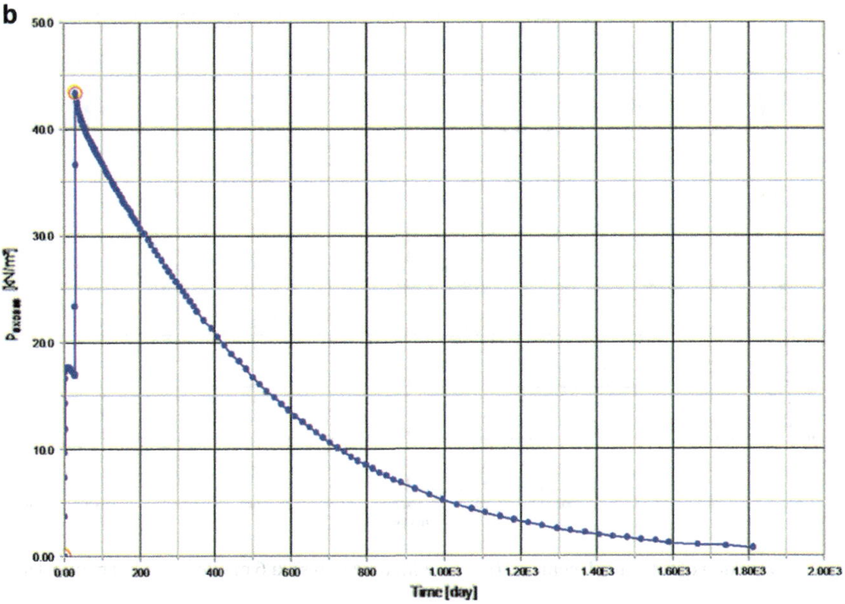

Fig. 12 Variation of excess pore pressure (p_{excess}) (kN/m^2) with time (days): **a** 6 m pile depth **b** 9 m pile depth

442.3 days to complete the consolidation phase in this simulation. The difference in excess pore pressure peaks between the construction phase and the consolidation phase was due to increased weight from the superstructure which leads to a rise in the rate of consolidation. For 9 m pile depth, the peak excess pore pressure reached during the construction phase was around 19 and 43 kN/m^2 for the consolidation phase. 1813 days were taken according to the simulation for complete transpiration of excess pore pressures thereby reducing the deformation in the soft soil strata.

Owing to the results, it can be deduced that the depth and type of soil below the level at which the pile reinforcement rests affect the settlement characteristics. This is because of the geotechnical properties of the soft soil and the strength anisotropy affecting the stability of uplift resistance offered by the soil adjoining at the pile. When the piles were at 6 m, the piles did not rest at the stiffest layer and acted as floating piles and as the load from the embankment was increased in the staged construction mode, displacement of the end piles occurred to take up the load and complete the simulation. When the piles were made to rest at 9 m, there was a reduction in the maximum displacement, as increasing the length of pile resulted in homogeneous dissemination of the settlement in the soft soil attributable to the efficient load transfer platform which caused the entire embankment load to be taken by the pile resistance.

5.2 Effect of Geometry and Slope of the Embankment

In pursuance of analyzing the effect of variation in geometry and slope of the embankment, the results of the above model of the embankment with a side slope of 1:3 (refer Fig. 1) was compared with the model of the embankment with a side slope of 3:5 supported by 17 piles and resting on layered soft soil strata bearing the same geotechnical properties as summarized in Table 2. The maximum deformation in the case of embankment with a 3:5 slope was 0.1420 m (refer Fig. 13) and 0.3052 m for embankment with a 1:3 slope (refer Fig. 10). Figure 14 depicts the distribution of plastic points in the simulation which illustrate the concentration of stresses with respect to

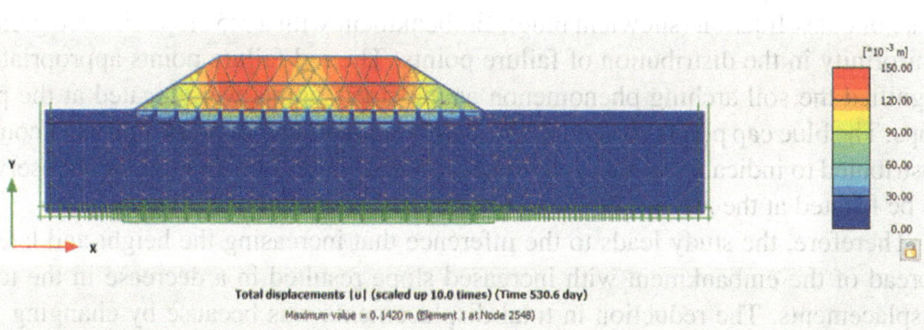

Fig. 13 Deformation analysis of embankment with 3:5 slope

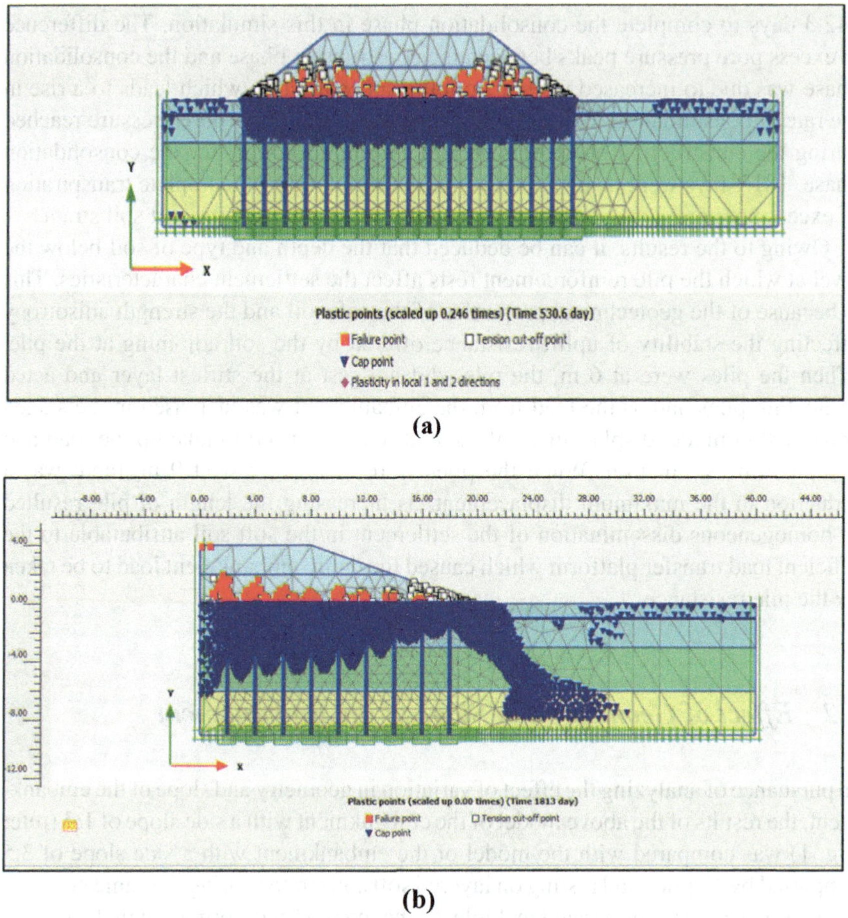

Fig. 14 Location of failure points **a** 3:5 slope **b** 1:3 slope

critical failure points (shown in red), tension cut-offs (shown in white blocks) and state of consolidation (shown in blue). Embankment with a 3:5 slope showed greater uniformity in the distribution of failure points. The red failure points appropriately signified the soil arching phenomenon as the stress points were located at the pile caps. The blue cap points indicating the state of consolidation in also homogeneously distributed to indicate the p_{excess} dissipation. The tension cut off points was observed to be located at the zones of increased slope in the figure.

Therefore, the study leads to the inference that increasing the height and lateral spread of the embankment with increased slope resulted in a decrease in the total displacements. The reduction in total displacements was because by changing the geometry of the embankment, the width increased thereby increasing the number of pile supports passing through the soft soil. This increased the area available for soil arching with more efficient load transfer mechanism which resulted in evenly distribution of the load. Researchers have also reported that the soil arching effect is responsible for the decrease in the deformation of the soft soil layers enhances with the increase in the embankment dimensions up to a certain limit.

5.3 *Effect of Spacing Between Piles and Surcharge*

Owing to the study of effect of variation of pile spacing on the deformation and time taken for consolidation, a 4 m high embankment was made to rest on pile supported soft soil with 2 and 4 m centre to centre spacing. Static line load acting on the embankment was increased in increments. Table 3 summarizes the results obtained by varying the pile spacing and surcharge for maximum deformation and time taken for consolidation.

It was observed that the settlement and settlement rate increases with the increase in the spacing between the piles. Increasing the number of piles by decreasing the centre to centre spacing from 4 to 2 m gave the advantage of about application of 50 kN/m static line load. This can be related to the overlapping of soil pressures resulting from lateral friction acting on the piles due to decreased spacing causing densification of soil adjoining the piles thereby magnifying the pile load taking capacity. Further, reducing the spacing between piles caused an increase in the number of piles for the same embankment width which resulted in an efficient load transfer platform by increasing the interaction mechanism between the embankment material, the piles extending from the load transfer platform to the stiff subsoil and the underlying soil medium. The homogenization of soil particles caused due to surcharge load and efficient load transfer platform in combination with friction acting on the surface of the piles accounts for the deduction in the maximum deformation in the case of pile spacing of 2 m.

From Table 4, it was also observed that the time taken for consolidation is reduced when 2 m spacing was adopted. This is because with the increase in surcharge pressure and decreased pile spacing increases the speed of pore water extraction from soil.

Table 4 Comparison of pile spacing and surcharge variation

Load (kN/m)	Maximum deformation		Time taken for consolidation	
	2 m c/c pile spacing	4 m c/c pile spacing	2 m c/c pile spacing	4 m c/c pile spacing
10	0.219 m	0.3725 m	605.5 days	1198 days
50	0.3253 m	1.319 m	670.2 days	1255 days
100	0.4353 m	25.38 m (failed)	726.9 days	1334 days (failed)
120	16.67 m	29.64 m (failed)	827.1 days	49.82 days (failed
150	37.22 m (failed)	–	days (failed)	–

This in turn causes re-arrangement of soil particles instigating a depletion in the void ratio of the adjoining soil. This triggers a decline in the lateral movements of the consolidating soil particles along with deduction in the lateral pore water movement which reduces the time taken for consolidation. Therefore, the decreased compression and rate of consolidation in the case of increased surcharge in 2 m spacing increases the load taking capacity with less settlement in the soft soil layer.

5.4 Effect of Mesh Size

Regarding the element distribution, there are 5 distinct mesh sizes, namely, very coarse, coarse, medium, fine and very fine, the analysis of which are discussed below (refer Table 5).

In finite element numerical analysis, different mesh sizes accounts to precision in the estimation of physical and mechanical properties of the simulated model. Figure 15a–e illustrates the deformation analysis in the different mesh sizes. It can be seen that the fine meshing with decreased element size shows a better concentration of stresses and accurate deformation (refer Table 5). It can be clearly observed that mesh size relates to the accuracy of the results. Simulating a model in fine meshing resulted in critically investigating each node for the geotechnical properties of the soils along with the interface, the pile behaviour and the configuration of each meshing node. Therefore, fine meshing gave highly accurate results due to decreased element size and increased number of elements for the simulation analysis. The only drawback associated with fine meshing is that it intensifies the complexity thereby increasing the time required for computation and therefore fine meshing is utilized when highly precise results are required. It can also lead to over-estimating the results. Coarse meshing resulted in less accurate estimation for deformation and p_{excess} analysis due to increased element size and therefore required less computational time. Therefore,

Table 5 Mesh comparison

Mesh type	Relative element size	No. of elements	Maximum deformation (m)	Maximum p_{excess} (kN/m^2)	Minimum p_{excess} (kN/m^2)	Time taken for consolidation (days)
Very coarse	2.00	30–70	0.2459	6.954E−3	−0.8987	564.8
Coarse	1.33	50–200	0.2430	3.781E−3	−0.9565	644.6
Medium	1.00	90–350	0.3052	5.754E−3	−0.9229	1813
Fine	0.67	250–700	0.3019	0.9858E−3	−0.9780	774
Very fine	0.50	500–1250	0.2245	0.7456E−3	−0.8990	477.1

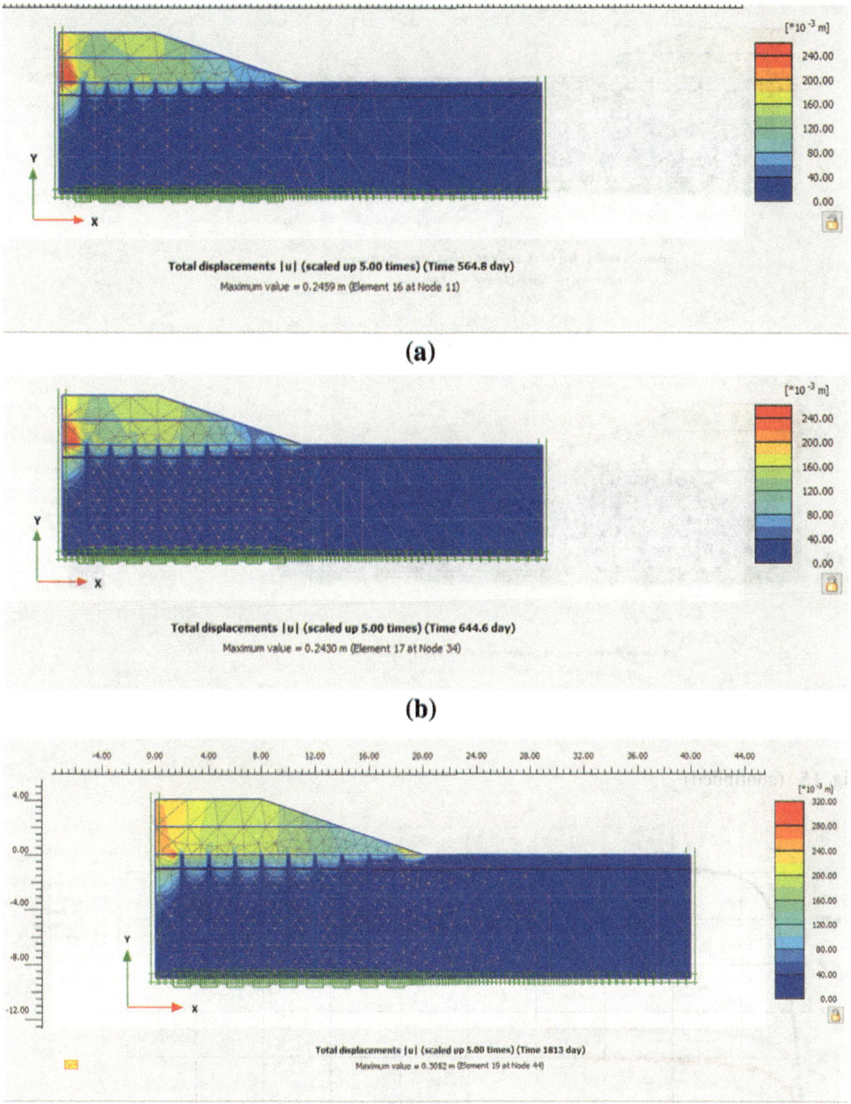

(c)

Fig. 15 Deformation analysis in various mesh sizes: **a** very coarse meshing **b** coarse meshing **c** medium meshing **d** fine meshing **e** very fine meshing

coarse meshing employed a large element size thereby reducing the finite element model's size and can be utilized for rough and quick estimation of results.

Figures 16 and 17 depict the variation of displacement (u) (m) with time (days) and excess pore pressure (kN/m^2) with time (days) respectively. Increasing mesh

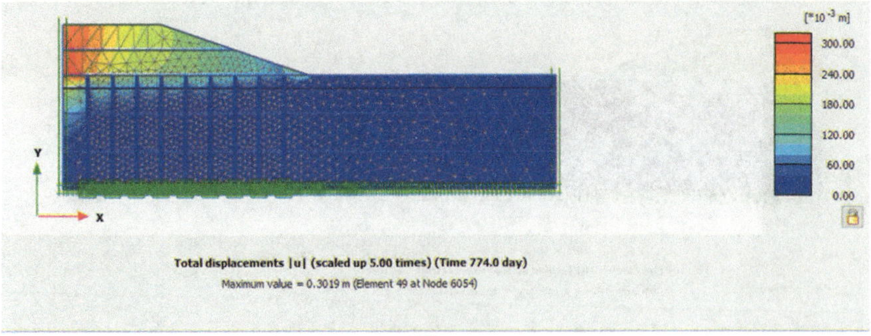

(d)

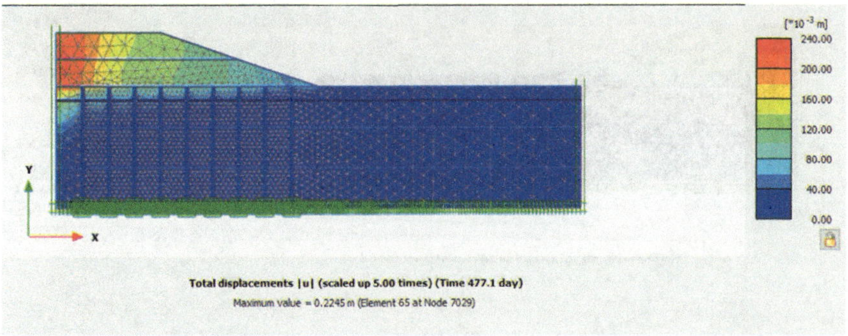

(e)

Fig. 15 (continued)

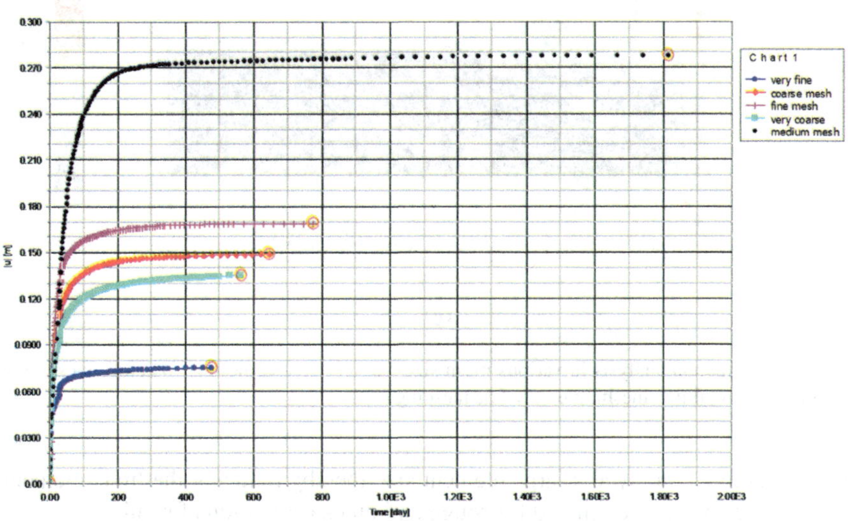

Fig. 16 Variation of displacement (u) (m) with time (days)

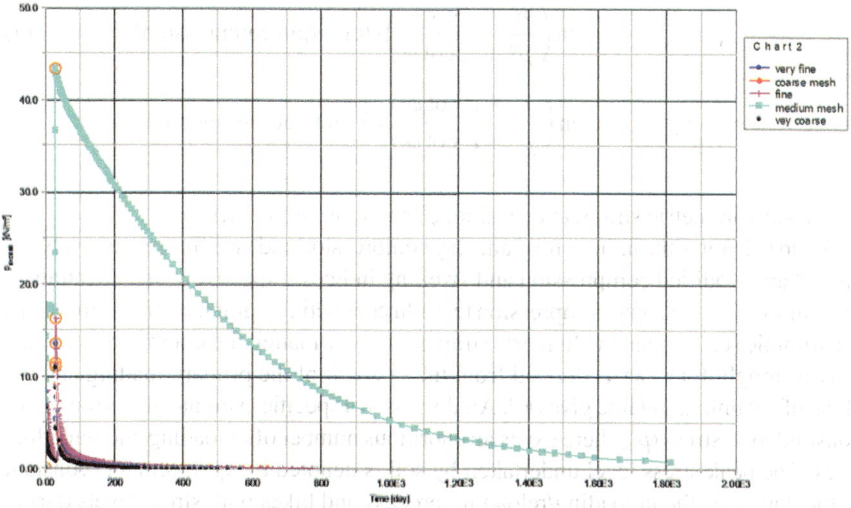

Fig. 17 Variation of excess pore pressure (p_{excess}) (kN/m^2) with time (days)

sizes showed a decrease in maximum deformation in the pile supported embankment resting on layered soft soil strata. It can be concluded that increasing mesh density leads to over-estimation of analysis whereas large element size leads to under-estimation of results in less computational time. Therefore, it is advisable to employ medium mesh to achieve the average results in normal computational time.

5.5 *Effect of Simulating Models*

MCC and SS models were adopted to study the effect of simulating models as both the models are designed to be used for soft consolidated clays. The difference lies in which the used parameters are computed. There exists a logarithmic relationship between void ratio (e) and mean effective stress (p') in MCC model as:

$$e - e^{o} = -\lambda \ln\left(p'/p_{o}\right)(\text{for isotropic compression under virgin condition}) \quad (2)$$

$$e - e^{o} = -\kappa \ln\left(p'/p_{o}\right)(\text{for isotropic unloading/reloading}) \quad (3)$$

where e: final void ratio, e^{o}: initial void ratio, p': mean effective stress, p_{o}: stress corresponding to initial void ratio.

The SS model considers volumetric strain (ε_{v}) to vary logarithmically with changes in and mean effective stress (p') as:

$$\varepsilon_v - \varepsilon_v^o = -\lambda^* \ln\left(\frac{p' + c \cot\varphi}{p^o + c \cot\varphi}\right) \text{(for virgin compression)} \quad (4)$$

$$\varepsilon_v - \varepsilon_v^{eo} = -\kappa^* \ln\left(\frac{p' + c \cot\phi}{p^o + c \cot\phi}\right) \text{(for unloading/reloading)} \quad (5)$$

where ε_v: volumetric strain, c: cohesion, φ: angle of internal friction.

λ and κ denotes the indices for Cam clay compression and swelling respectively. λ^* and κ^* are Modified compression and swelling indices respectively and are attained from unloading isotropic compression test. On concocting a graph of mean stress on a logarithmic scale against volumetric strain for clayey matter, the graph is proximated to two straight lines. λ^* is derived from the gradient of the primary loading and the slope of the unloading line gives κ^*. Analogous to a specific estimate of isotropic pre-consolidation stress (p_p) there exists an enormous number of unloading and reloading lines. The peak stress level undertaken by soil is denoted by p_p. It remains constant in the course of the unloading/reloading process and hikes with stress levels during primary loading which causes irreparable plastic volumetric strains.

Table 6 enlists the comparative results obtained after modeling the soft soil layers with the two models. Simulation with the MCC model showed the settlement of the embankment material with maximum deformation of 0.3052 m (refer Fig. 10) whereas modeling using the SS model resulted in a maximum deformation of 3.154 m (refer Fig. 18). This can be due to the parameter of modified compression index (λ^*) and modified swelling index (κ^*) (refer Eqs. 2–5) been incorporated for the calculations in the SS model which employs volumetric strain for computation whereas the void ratio is used for calculating λ and κ in MCC model. Therefore, the inference drawn indicates that the SS model over-predicted the deformation and did not predict the settlement pattern in the simulation appropriately. The difference arises in the assumptions involved in the computation of the deformation behaviour on the left fringe of the critical state line. The MCC is based on elasto- plastic deformation characteristics in which the nonlinear elasticity and critical state conditions are administered by the volumetric strains thereby taking the incremental strains and deformations into account. Moreover, MCC predicts the deformation by applying

Table 6 Simulation model comparisons

Analysis	Modified Cam Clay (MCC) model	Soft Soil (SS) model
Total displacement (m)	0.3052	3.154
Maximum excess pore pressure (kN/m^2)	5.754E−3	−0.9229
Minimum excess pore pressure (kN/m^2)	0.06719	−80.68
Time taken for complete consolidation (days)	1813	957.3

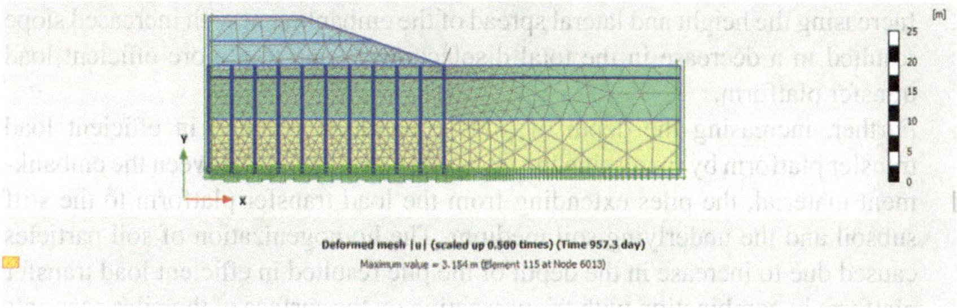

(a)

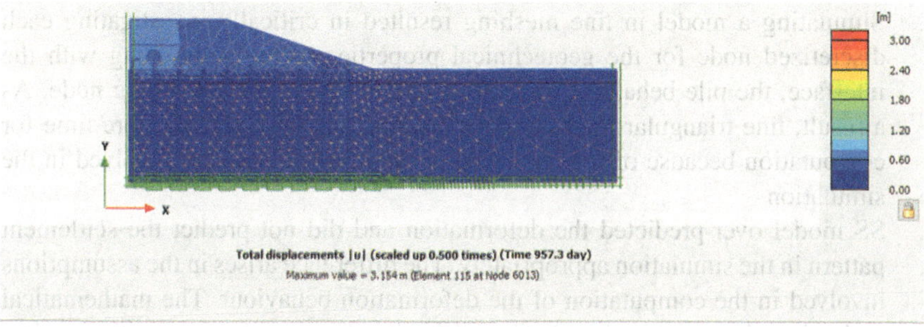

(b)

Fig. 18 **a, b** Deformation analysis in the case of SS model

notable strain softening whereas in the SS model the behaviour is controlled by Mohr–Coulomb failure slope and is perfectly plastic.

6 Conclusions

In this paper, pile supported embankment resting on layered soft soil was numerically studied by incorporating results of varying parameters such as pile depth, pile spacing, surcharge, slope and geometry of embankment and simulating models. The main results obtained from the analysis are summarized herewith:

1. Settlement and settlement rate increase with the increase in the spacing between the piles. This can be related to the overlapping of soil pressures resulting from lateral friction acting on the piles due to decreased spacing causing densification of soil adjoining the piles thereby enhancing the piles load taking capacity. By reducing the spacing between the piles, the rate of load penetration increases but pile-soil stress shearing ratio and negative skin friction acting on the piles decreases which result in a decrease in the load acting on the pile thereby increasing the load taking capacity by the piles.

2. Increasing the height and lateral spread of the embankment with increased slope resulted in a decrease in the total displacements due to a more efficient load transfer platform.
3. Further, increasing the depth of piles penetration resulted in efficient load transfer platform by increasing the interaction mechanism between the embankment material, the piles extending from the load transfer platform to the stiff subsoil and the underlying soil medium. The homogenization of soil particles caused due to increase in the depth of the pile resulted in efficient load transfer platform in combination with friction acting on the surface of the piles accounts for the deduction in the maximum deformation when the pile rested at 9 m.
4. Simulating a model in fine meshing resulted in critically investigating each discretized node for the geotechnical properties of the soils along with the interface, the pile behaviour and the configuration of each meshing node. As a result, fine triangular meshing gave accurate results but took more time for computation because of the increased density and complexity involved in the simulation.
5. SS model over-predicted the deformation and did not predict the settlement pattern in the simulation appropriately. The difference arises in the assumptions involved in the computation of the deformation behaviour. The mathematical reckonings involved in the MCC model critically consider the elasticity, plasticity and critical state conditions which assist in accurately interpreting the settlement in the soft soil medium.

References

1. Al-Neami MA, Samueel ZW, Al-Noori M (2019) The influence of pile groups configuration on its stability in dry sand under lateral loads. IOP Conf Ser Mater Sci Eng 579(1). IOP Publishing
2. Bagavasingam T (2015) Secondary consolidation and the effect of surcharge load. Int J Eng Res Technol 4(11):695–699
3. Borthakur N, Dey AK (2020) Evaluation of group capacity of micropile in soft clayey soil from experimental analysis using SVM-based prediction model. Int J Geomech 20(3)
4. Dang LC, Dang C, Khabbaz H (2017) Numerical modeling of embankment supported by fiber reinforced load transfer platform and cement mixed columns reinforced soft soil. International Congress and Exhibition "Sustainable Civil Infrastructures: Innovative Infrastructure Geotechnology". Springer, Cham
5. Esmaeili M, Nik MG, Khayyer F (2013) Efficiency of micro piles in reinforcing embankments. Institution of Civil Engineers-Ground Improvement, vol 167(2), pp 122–134
6. Fattah MY, Al Helo KH, Abed H (2017) Load distribution in pile group embedded in sandy soil containing cavity. Korean Soc Civ Eng J Civ Eng 22(2):509–519
7. Fırat S, Işık NS, Yamak S (2018) Effect of height and water table level on stability analysis of embankments. In: 3rd International Sustainable Buildings Symposium (ISBS), Geotechnique Journal (2018). https://doi.org/10.1007/978-3-319-63709-9_43
8. Ghosh B, Fatahi B, Khabbaz H, Nguyen HH, Kelly R (2021) Field study and numerical modelling for a road embankment built on soft soil improved with concrete injected columns and geosynthetics reinforced platform. Geotext Geomembr 49:804–824. https://doi.org/10.1016/j.geotexmem.2020.12.010

9. Juran B, Guermazi A (2013) Settlement response of soft soil reinforced by compacted sand columns. J Geotech Eng 114(8):930–943
10. Karim MR, Manivannan G, Gnanendran CT, Lo SR (2011) Predicting the long-term performance of a geogrid-reinforced embankment on soft soil using two-dimensional finite element analysis. Can Geotech J 48(5):741–753
11. Karstunen M, Wiltafsky C, Krenn H, Scharinger F, Schweiger HF (2006) Modeling the behaviour of an embankment on soft clay with different constitutive models. Int J Numer Anal Meth Geomech 30:953–982
12. Kasim F, Marto A, Othman BA, Bakar I, Othman MF (2013) Simulation of safe height embankment on soft ground using Plaxis. Asia-Pacific chemical, biological & environmental engineering society, vol 5, pp 152–156. https://doi.org/10.1016/j.apcbee.2013.05.027
13. Kumar A, George V, Marathe S (2016) Stability analysis of lateritic soil embankment sub-grade using plaxis-2D. Int J Res Civ Eng 2(1). https://www.researchgate.net/publication/313442092
14. Laskar A, Pal SK (2017) The effects of different surcharge pressures on 3-D consolidation of soil. Int J Appl Eng Res 12(8):1610–1615
15. Leeworthy T, Asr AA (2020) Effects of various surcharge loading conditions on the stability of soil slopes. Loughborough's Research Repository. https://repository.lboro.ac.uk/articles/conference_contribution/
16. Manna B, Rawat S, Zodinpuii R, Sharma KG (2014) Effect of surcharge load on stability of slopes—testing and analysis. Electron J Geotech Eng 19:3397–3409. http://www.ejge.com/Index_ejge.htm
17. Moni MM, Sazzad MM (2015) Stability analysis of slopes with surcharge by LEM and FEM. Int J Adv Struct Geotech Eng 4(4):216–225
18. Pham TA (2020) Analysis of geosynthetic-reinforced pile-supported embankment with soil-structure interaction models. Comput Geotech 121:103438
19. Pham TA, Dias D (2021a) Comparison and evaluation of analytical models for the design of geosynthetic-reinforced and pile-supported embankments. Geotext Geomembr 49(3):528–549
20. Pham TA, Dias D (2021b) 3D numerical study of the performance of geosynthetic-reinforced and pile-supported embankments. Soils Found. https://doi.org/10.1016/j.sandf.2021.07.002
21. Pham TA, Tran Q, Villard P, Dias D (2021) Geosynthetic-reinforced pile-supported embankments—3D discrete numerical analyses of the interaction and mobilization mechanisms. Eng Struct 242:112337
22. Rajput D, Kumar R, Jain PK, Chandrawanshi S (2016) Load-settlement behaviour of soft soil reinforced with sand piles. Int Res J Eng Technol (IRJET) 3(11). https://www.irjet.net/volume3-issue11
23. Sadaoui O, Bahar R (2017) Field measurements and back calculations of settlements of structures founded on improved soft soils by stone columns. Eur J Environ Civ Eng, 1–27. https://doi.org/10.1080/19648189.2016.1271358
24. Said KN, Rashid AS, Osouli A, Latifi N, Yunus NZ, Ganiyu AA (2018) Settlement evaluation of soft soil improved by floating soil cement column. Int J Geomech 19(1) (2018). https://doi.org/10.1061/(asce)gm.1943-5622.0001323
25. Sazzad MM, Haque MF (2014) Effect of surcharge on the stability of slope in a homogeneous soil by FEM. In: 2nd International conference on advances in civil engineering. https://www.cuet.ac.bd/icace2/
26. Sengupta S, Azim U (2020) Numerical modelling of a pile-supported embankment. Constrofacilitator. https://www.constrofacilitator.com/numerical-modelling-of-a-pile-supported-embankment/
27. Souria A, Abu-Farsakha MY, Voyiadjis GZ (2020) Evaluating the effect of pile spacing and configuration on the lateral resistance of pile groups. Mar Georesour Geotechnol. https://doi.org/10.1080/1064119X.2019.1680780
28. Tai P, Indraratna B, Rujikiatkamjorn C (2020) Consolidation analysis of soft ground improved by stone columns incorporating foundation stiffness. Int J Geomech 20(6). https://doi.org/10.1061/(ASCE)GM.1943-5622.0001668

29. Wen L, Kong G, Li Q, Zhang Z (2020) Field tests on axial behaviour of grouted steel pipe micropiles in marine soft clay. Int J Geomech 20(6). https://doi.org/10.1061/(ASCE)GM.1943-5622.0001656
30. Wu JT, Ye X, Li J, Li GW (2019) Field and numerical studies on the performance of high embankment built on soft soil reinforced with PHC piles. Comput Geotech 107:1–13. https://doi.org/10.1016/j.compgeo.2018.11.019
31. Xie M, Zheng J, Zhang R, Cui L, Miao C (2020) Performance of a combined retaining wall structure supporting a high embankment on a steep slope: case study. Int J Geomech 20(6). https://doi.org/10.1061/(ASCE)GM.1943-5622.0001644
32. Yan-hui GE, Shu-cai LI, Chuan-fu L, Guo-fu S, Wang MB (2009) The settlement analysis of soft ground improved by vacuum-surcharge preloading method. In: International conference on engineering computation. https://doi.org/10.1109/ICEC.2009.67
33. Zhang J, Guo C, Xiao S (2012) Analysis of effect of CFG pile composite foundation pile spacing on embankment stability based on centrifugal model tests. Appl Mech Mater 178–181:1641–1648
34. Zhang G, Wang L (2017) Simplified evaluation on the stability level of pile-reinforced slopes. Soils Found 57:575–586. https://doi.org/10.1016/j.sandf.2017.03.009
35. Zhang Z, Tao F, Han J, Ye G, Liu L (2020) Numerical analysis of geosynthetic-reinforced pile-supported embankments subjected to different surface loads. In:Geo-congress 2020: engineering, monitoring, and management of geotechnical infrastructure. American Society of Civil Engineers, Reston, VA
36. Zhou A, Shen H, Sun J (2017) Effects of inserted depth of wall penetration on basal stability of foundation pits. AIP Conf Proc 1839:020012-1–020012-9

A Review on Application of NATM to Design of Underground Stations of Indian Metro Rail

Sandesh S. Barbole, M. S. Ranadive, and Apurva R. Kharat

Abstract This paper describes the advanced concepts used for the underground construction of tunnels i.e., the New Austrian Tunneling Method (NATM) in Indian Metro Rail. Many areas of Indian cities are highly congested and populated. Hence, elevated metro rail construction is not always feasible. Therefore, the Metro Rail system in India is elevated as well as underground also. Underground tunnel construction below highly congested urban area is another challenge in itself. It was constructed by using various tunneling methods. Some most popular methods used for tunnel construction in urban areas are Tunnel Boring machines (TBM), Cut & Cover Method, NATM, etc. The construction of a complete tunnel by using a single method does not only consume more time but also results in uneconomical construction. NATM is commonly adopted on both sides of underground stations for providing a safe opening for TBM launching and outbreaks. It is also used for providing cross-passage between up-line and down-line tunnels. In this paper, the methodology of applications of NATM in Indian Metro Rail is explained. The NATM method has been used to construct all underground metro tunnels in India. A comparative aspect of the construction of an underground tunnel by using NATM in Pune Metro and Mumbai Metro is also discussed in this paper. In addition, challenges like heavy groundwater seepage, application of NATM in soft ground, etc. are presented here.

Keywords New Austrian Tunneling Method (NATM) · Cut & Cover Method · Tunnel Boring Machine (TBM) · Metro stations · Cross passage

S. S. Barbole (✉) · A. R. Kharat
Rail & Metro Technology, College of Engineering Pune (COEP), Pune, India
e-mail: barboless19.pgdrmt@coep.ac.in

A. R. Kharat
e-mail: kharata19.pgdrmt@coep.ac.in

M. S. Ranadive
Civil Engineering Department, College of Engineering Pune (COEP), Pune, India
e-mail: msr.civil@coep.ac.in

M. S. Ranadive et al. (eds.), *Recent Trends in Construction Technology and Management*, Lecture Notes in Civil Engineering 260,
https://doi.org/10.1007/978-981-19-2145-2_54

1 Introduction

The first metro rail system in India was Kolkata Metro, which started in 1984. As technology was not much developed in the nineteenth century, 16.45 km of underground construction was constructed in 14 years using the 'Cut & Cover Method' and 'Shield Tunneling'. 'Elattuvalapil Sreedharan', is a Metro Man of India who took great efforts and constructed this masterpiece. His great efforts and hard work resulted in the development of the metro rail grid in many Indian cities. Delhi Metro was the first advanced and modern metro that started its operation in 2002. It was constructed using modern technological methods like TBM and NATM. Most preferably, TBM is used for metro tunnel construction. But, for Station Construction, cross-passage construction, exit, and launching of TBM at both ends of stations, NATM is used. Furthermore, Chennai Metro, Bangalore Metro, Mumbai Metro, and Pune Metro were also constructed with these technologies and advanced concepts were used for the underground metro network [1, 2]. Currently, Kolkata Metro Rail Corporation (KMRC) is constructing underwater metro. The metro will run under the foothills of the river [3]. Now, the yellow line of the Delhi Metro is the longest underground tunnel in India with a total length of around 33–35 km with 34 metro stations [4].

Pune Metro Rail Project consists of two routes, currently, which are under construction. The first route is Pimpri Chinchwad (PCMC)—Swargate and another is Vanaz (Kothrud)—Ramvadi, the first route consists of an 8.2 km length of underground metro tunnel construction. It is constructed by using NATM, TBM, and the Cut & Cover Method. There is a total of five underground stations namely Swargate, Mandai, Budhwar Peth, Civil Court, and Shivajinagar, which are constructed by using the Cut & Cover Method. In Mumbai Metro Rail Project, the aqua line (line three) is a complete underground metro tunnel passing from Cuffe Parade Station (South Terminal Station) to SEEPZ station (North Terminal Station), which is of length 33.5 km. It is also constructed by using NATM, TBM, and the Cut & Cover Method. A description of all underground metro is given in Table 1.

Table 1 Underground Metro Rail in India

S. No.	Underground Metro Rail	Present length (KM)	Under-construction (KM)
1	Delhi Metro	48.06	41.044
2	Chennai Metro	31.3	45.81
3	Kolkata Metro	10.8	–
4	Banglore Metro	8.82	21.386
5	Mumbai Metro	–	33.5
6	Pune Metro	–	8.2

1.1 NATM Concept

The New Austrian Tunneling Method (NATM) appeared between the years 1957–1961. NATM was invented by Von Rabcewice, Pacer, and Muller-Salzburg. Their main idea was to use conventional tunneling methods by following the principles of the observation method to apply the support system. The NATM requires minimum loss of strength to avoid the softening of rock mass and it is achieved by minimum distortion of ground. Parallelly, a sufficient amount of deformation of the ground is necessary to achieve a sufficient amount of strength of rock mass. In addition, there is no use of thick and stiff lining in NATM [5].

NATM is a misnomer as it is not a method but it is a concept with uniformity and sequence. NATM is used by continuous conduction of ground parameters while driving the tunnel. In NATM, the response of the already excavated portion of the tunnel is continuously monitored, and according to results, ahead tunnel construction is carried out [6]. NATM is often referred to as the "design as you go" approach of tunneling. It is not a bunch of excavation methods and support systems. More precisely it can be described as a "design as you monitor" approach but it depends on observation of surrounding rock mass, produced stresses inlining, etc. In this method, a tunnel is constructed using the open face excavation method and support system as wire mesh, rock bolt, lattice girder, etc. with shotcrete lining to provide ground support. The main aim of this method is to the utilization of strength of the surrounding ground. NATM is very useful in adverse and difficult geological ground conditions. It proves an economical as well as a less time-consuming method of tunneling in unfavorable conditions. In congested urban areas, it is the safest method for underground tunnel construction.

1.2 Elements of NATM

Following are the main elements of NATM:

1. Mobilization of the strength of rock mass.
2. Protection of tunnel by using shotcrete lining.
3. Measurements of deformations in the lining.
4. Providing flexible supports by a combination of rock bolts, wire mesh, and lattice girder.
5. Closing of inverts and creating load-bearing rings.
6. Contractual measurements for changes if necessary.
7. Rock mass classification for determination of support parameters.

1.3 Principles of NATM

Following are some of the major principles on which the NATM works:

1. First and most important principle is to take maximum advantage of the surrounding rock mass.
2. Most of the original rock strength should be retained, so the cutting of rock should not be less or more.
3. Loosening of surrounding rock mass must be avoided as maximum as possible.
4. Create triaxial condition i.e., the load distribution should continue in all directions.
5. Minimum deformation of surrounding rock mass to form supporting envelopes.
6. Preliminary support must be provided at the right moment i.e., not so early or not too late.
7. Calculation of time factor for rock mass and rock mass plus preliminary support.
8. To avoid large deformations or loosening of rock, shotcrete is provided to get active support.
9. For the least bending moment provide thin and flexible preliminary support.
10. Avoid sharp corners in profile as they develop stresses and damages surrounding rock.
11. Release the seepage pressure by drainage and filters.

2 Literature Review

Xen Ren et al. (2019), have described a case study on Beijing Subway's extensive expansion plan, the refined model is applied to Beijing Subway Line 6 project and a series of risk control measures were taken. The risk control system was shown to be effective and ensure the success of tunnel excavation. This model is believed to serve as a valuable reference for future subway construction in Beijing using NATM in the congested urban area [7].

Aejaz Ahad et al. (2019), have presented a paper on a case study on Udhampur-Srinagar-Baramulla Rail Link Project (USBRL Project) and this project hence was monitored by PMO (Prime Minister's Office). The steps in the construction of this tunnel by using the NATM method were explained according to the geology of the Himalayan Region in the described manner. Each step is described with construction details and the latest technologies used in this project. The focus is on construction steps, geological mapping, 3D Mapping, etc.

Young-Chul Han et al. (2013), have described a new tunnel load evaluation method (Ground Lining Interaction model) that was presented and discussed while considering the rock grade, tunnel depth, and in-situ stress ratio. Moreover, a new equation that evaluates the tunnel load acting on the concrete lining is proposed using the results obtained from the GLI model.

Palmstrom et al. [8], states that NATM involves a combination of several tunneling aspects from ground characterization via rock mechanics and tunnel design to construction principles, rock support design, and monitoring during tunnel excavation. The main principle of the method is, however, utilization of the bearing capacity of the ground surrounding the tunnel. This is practically done by letting the weak ground around the tunnel deform in a controlled way by applying a flexible rock support.

Wayne Clough et al. (1992), have described that the construction process of NATM tunnels is analyzed using the finite element method. The softening technique, currently used for the analysis of NATM tunnels, is improved and applied to a parametric study of ovoid transportation tunnels constructed in undrained clays. Based on this study, a simplified method is provided for the preliminary design of the NATM tunnel support [9].

3 Geological and Geotechnical Properties

Geology plays a very significant role in the cost of a tunnel. It is important to frequently gather necessary data of tunneling after each small step. It is collected by pilot tunneling or probe drilling method. During the excavation process, the stability of the surrounding ground depends on strength of rock mass, faults zones, discontinuities in the ground, available groundwater, etc. Mostly, the worst conditions occur in soft ground, which may travel, flow, squeeze or swell.

Pune City is a part of Deccan volcanic providence consisting of layers of basalt rock. Around, 7–45 m thickness of layers of basalt is occupying more than 45% of the region of Pune. The top layer is vesicular basalt rock and the bottom layer consists of massive basalt rock in this area [10]. It is a very good quality rock for construction purposes. Mumbai's geology is somewhat different from that of Pune. Mumbai is also part of the Deccan plateau. Parent rocks of Mumbai are Basalt, Breccia and Tuff. In Mumbai, the strength of rock varies in a similar grade of rock in various locations and groundwater is close to ground level [11].

Generally, geological and geotechnical parameters are studied by drilling boreholes along the alignment of the tunnel including laboratory as well as field tests. The number of boreholes and test pits is excavated, soil samples are taken out and the groundwater level is checked. Field tests conducted on-site are the Standard Penetration Test (SPT) and Plate Load Test (PLT). Laboratory tests conducted are uniaxial & triaxial tests, direct shear test, consolidation test, sieve analysis, Atterberg limit test, and chemicals tests on a soil sample. After studying these tests, the properties of geological and geotechnical parameters are determined.

4 Construction Sequence of NATM

Below diagram shows the steps involved in NATM tunnel construction (Fig. 1).

Sequence of NATM tunneling is as follows:

1. Profile Making

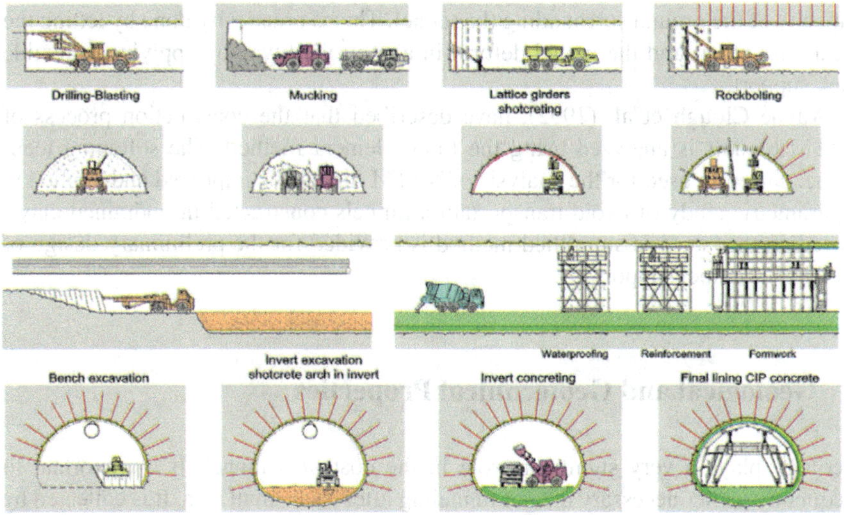

Fig. 1 NATM sequence [12]

2. Face Drilling
3. Charging and Blasting
4. De-fuming
5. Mucking
6. Scaling
7. Geological Face Mapping
8. Shotcrete Protection
9. Lattice Girder
10. Fore Poling
11. 3D Monitoring
12. Initial Lining (Shotcrete)
13. Rockbolt
14. Grouting.

5 Applications of NATM in Metro Rail

NATM is used in the underground construction of every metro rail project in India. Two metro rail systems are selected which have different geology namely Pune Metro and Mumbai Metro. In the below table (Table 2), the comparison of the application of NATM in tunneling is given briefly.

In underground metro construction, the beginning of the NATM tunnel mostly starts with the earlier construction of a vertical shaft or with a station box. The shaft is used for access of people, material, and equipment, and for the removal of muck. A typical cross-section of NATM or face of tunnel is divided into several small portions,

Table 2 Comparison of NATM in Pune and Mumbai Metro Rail Project (8, 9, 11, 12, 13, 14)

Description	Pune Metro	Mumbai Metro
Geology	Basalt Rock (Strong to moderately strong rock)	Basalt, Breccia, Tuff (Moderately strong to weak rock)
Shape of tunnel	Horizontal egg shape, 22.35 m × 7.50 m	Circular shape
Excavation method	Heading and Benching method (3 heading + 1 Benching) with Drilling and Blasting method	Heading and benching method with mechanical excavation
Explosive type	Emulsion explosive (83 and 23 mm cartridge with electric detonator)	Not used
Fore poling pipes	Dia. 101.6 and 5.4 mm thick (only at face)	Dia. 101.6 and 5.4 mm thick
Shotcrete	Wet mix, M30 grade and cylindrical strength—125 MPa	Wet mix, M30 grade and cylindrical strength—125 MPa
Rockbolt	Sn-bolts and Self-drilling (SD) Bolts, 25 mm dia.	Sn-bolts, 25 mm dia.
Wire Mesh	150 × 150 mm × 6 mm dia. of wire	150 × 150 mm × 6 mm dia. of wire
Lattice Girder	Not used	Used
Grouting	w/c ration = 0.35–0.7	w/c ratio = 0.35–0.45
Major problems	High groundwater seepage	Weak geology, risk of collapse of rockmass, water seepage with high pressure

which are excavated one by one i.e., heading and benching method as shown in Fig. 2. In Pune Metro Project, excavation is conducted in a sequence of three headings and one benching. It is proceeding in small sequential portions because; the entire face

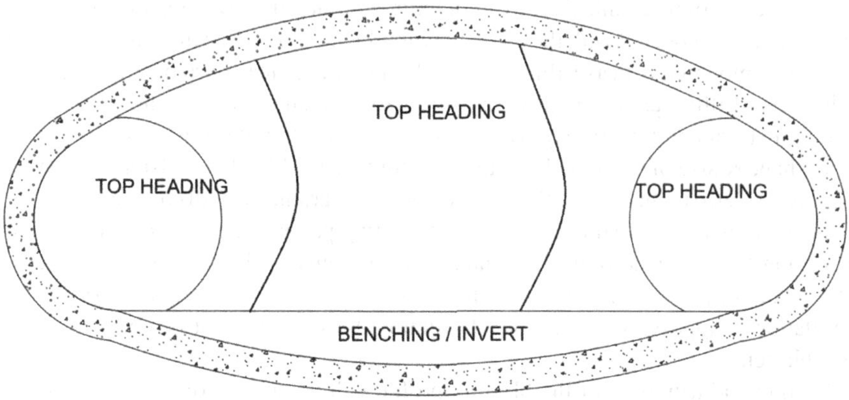

Fig. 2 Typical NATM profile for two-way metro tunnel

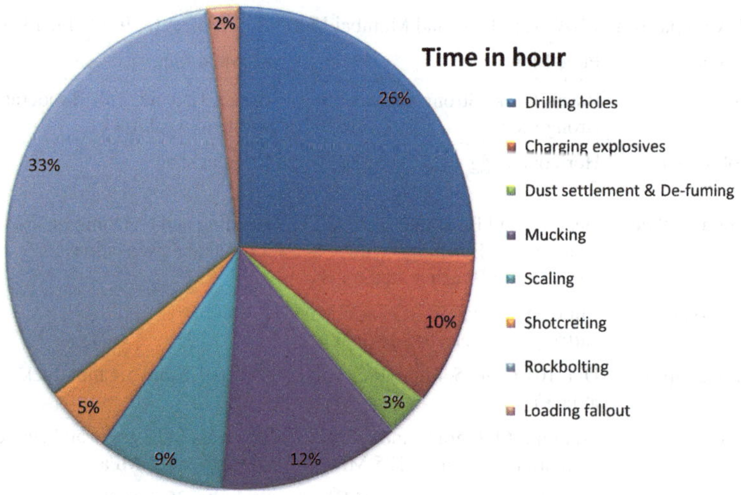

Graph 1 Activity time analysis

area is too large for simultaneous blasting. This simultaneous blasting may lead to the generation of a huge amount of vibrations, which will damage structures in the surrounding locality. The umbrella fore-poling method is used only when rock strength is weak. Either it is commonly used at the face of a tunnel to avoid the risk face falling at the beginning of tunnel operation. In Pune Metro NATM work, fore-poling is used only at the face of the tunnel.

In hard rock, 'Drilling and Blasting Methods' are typically used and on the weak ground, mechanical excavators like 'Roaheaders' are used along with the heading and benching method. Basalt is typically found in Pune region, hence in Pune Metro underground construction drilling and blasting method is used for tunnel excavation. However, in Mumbai rock is basalt, breccia & tuff, which are weak rocks leading to the application of mechanical excavation for metro tunneling. Drilling of holes as per blast patterns conducted with the use of 'Hydraulic jumbos'. This machine has one or more arms mounted with a drilling rig, which can drill vertical, horizontal, and angled holes. The drilling pattern is different for every sequential face of the tunnel. The drilling pattern and type of explosive is depended on strength of the rock, stand up time, shape & size of the tunnel, depth of the tunnel, etc. Mostly emulsion explosives are used in metro tunneling in urban areas. The blast parameters will change according to the geology at face. All pre-determined blasting patterns are subjected to change according to rock conditions encountered in the tunnel. Blasting shall take place, only when all people will discharge from the tunnel. The time required for tunnel ventilation is approx. 45 min. After de-fuming, blast holes must be checked alone by a blaster.

Scaling and trimming of the blasting profile will be done to remove loose rocks resulting from the blast and ensures the excavated profiles conform to the design

Table 3 Activity sequence and time

S. No.	Activities	Time in hour
1	Drilling holes	6.61
2	Charging explosives	2.66
3	Dust settlement and de-fuming	0.75
4	Mucking	3.23
5	Scaling	2.28
6	Shotcreting	1.21
7	Rockbolting	8.60
8	Loading fallout	0.58
	Total Time =	**25.92**

profile of the tunnel. Then all excavated material or muck is collected and transported out of the tunnel. Usually, geological mapping is executed before shotcreting. Depending on the data collected by geological mapping, the parameters of the support system are decided. After every excavation step, shotcreting is performed with high pressure on the tunnel surface. Shotcrete supports wire mesh, rock bolt, lattice girder, etc. It serves as arching support around the face of the tunnel. Shotcrete is usually applied in three layers, the first layer and last three layers are reinforcing with wire mesh and without wire mesh respectively. Combining all three layers makes the primary lining wall of the tunnel. After sufficient time rock bolting, lattice girder, etc. are installed at specified locations. Lattice girders are typically used of a crown shape or circumferential and are mostly used when rock quality is poor. It provides assistance in tunnel profiling and in achieving correct shotcrete thickness. In danger zones, a support system must be placed before geological mapping. The support system changes according to the strength of the rock and surrounding conditions. About 40% cost of NATM tunnel construction is required for support systems. The activities in NATM construction and the time required for it is shown below (Table 3) and analyzed graphically.

Graphical representation on the previous page shows that rock bolting and drilling require maximum time in all NATM activities. More than 50% of the time is required only for these two activities. The time required for rock bolting or drilling is totally depending on strength of the rock. More the strength of the rock more will be the time required for both activities. For the blasting purpose, a number of holes are required to drill as per the drilling pattern. One drilling pattern includes 100–1000 holes, which consumes huge time. Similarly, rock bolts are inserted in sidewalls as well as in the roof of the tunnel. So many numbers of holes are drilled into the wall and roof of the tunnel. Grouting in all holes also needs more time.

6 Comparison of NATM with Other Methods of Tunnel Excavation

There is a variety of tunneling methods used in this world. There is nothing like any single method that is the best method of tunneling. The selection of tunneling method is mostly depending on the geology of the site, choice of technology, available funds, etc. The most advanced and commonly used methods of tunneling are TBM, NATM, and the Cut & Cover Method. In the given tables (Tables 4 and 5), a comparison of NATM with TBM and Cut & Cover is described.

6.1 NATM Versus TBM

In this table (Table 4), a comparative analysis of the New Austrian Tunneling Method (NATM) with Tunnel Boring Machine (TBM) is described.

Table 4 NATM versus TBM

Description		NATM	TBM
Cross-section		Any shape can be achieved	Limited to circular shape only
Longitudinal curves		Any tunneling radius can be achieved	Tunneling radius is limited
Requirement of geological investigation		Good amount of investigation required	Way more detailed investigation is required. Lack of investigation may lead to a breakdown
Requirement of ventilation		High because of blasting activity	Less
Mechanism		Not completely mechanized	Completely mechanized
Electricity requirement		Moderately high	Very high
Automatization		Lacking as many equipments are handled manually	Completely automatized
Time	Preparation time	3–4 months	1–1.5 years
	Familiarization time	Almost negligible	3–4 months
	Performance time	2–3 to 12–15 m/day	10–50 m/day, largely depend on geology and surrounding conditions
	Breakdown time	Can be solved within a limited time	Lead to a major delay in project
Cost		Not so high cost	Very high cost

Table 5 NATM versus TBM

Description	NATM	Cut & Cover method
Diversion of traffic	Not much required	Compulsory required
Impact on environment	Less	More as it is open to ground surface
Freedom of alignment	Provides more options for an alignment	Limited options for alignment due to traffic, surface infrastructure, etc.
Dust and noise production	Less as it is surface	More as it is open to surface
Production of construction and demolition (C & D) material	Lesser quantity as compared to Cut & Cover Method	Larger quantity is needed to excavate up to the ground surface
Depth suitability	Can use as any depth	Not suitable for very deep excavations
Skilled manpower	Highly required	Not much as it is frequently in use from a long time
Cost	Less costly	More costly
Time	Consumes less time as compared to the Cut & Cover Method	Consumes more time as the volume of excavation is more

Even though TBM is costly, and required high technology and skilled people, it is preferred for metro tunneling. Because it has almost two to five times more speed of tunnel construction than other methods. However, NATM is used where highly unfavorable conditions are present.

6.2 NATM Versus Cut and Cover Method

In this table (Table 4), a comparative analysis of the New Austrian Tunneling Method (NATM) with the Cut and Cover Method is described.

NATM has many advantages over the Cut and Cover Method. NATM is a safe, less time-consuming, and economical method in an urban locality. It does not disturb surface traffic as well as there is a small possibility of disturbance or damage to the surrounding structures due to settlement.

7 Groundwater Seepage in Tunnel

It is very necessary to control groundwater seepage as continuous groundwater seepage can trigger fractures in the rock and will result in ground settlement. It will damage buildings and infrastructure in the surrounding locality. Groundwater seepage is a problem, which cannot be avoided in tunneling. Under the presence of high groundwater table, it is very difficult to control the seepage of water. The small seepage of water can trigger the closure of pores and fractures. It will cause heavy damage to the tunnel and surrounding structures.

7.1 Countermeasures for Groundwater Seepage

Water leakage is depending on various physical attributes of the rock joints. In high water ingress areas, pre-injection or grouting is the only reasonable solution, as this process seals open joints in the rock before the water starts to flow. The objective of grouting is to fill the open voids using a liquid matter that will be harder later. Grouting creates a dry tunneling environment, which reduces the tunneling cycle and total construction time. Ultimately it leads to saving the cost of the project.

There are two main functions for the pre-grouting of tunnels:

(a) To prevent the ingress of water into the tunnel,
(b) To make the working of tunneling more precise. So that there will be an improvement in the rock mass quality.

There are many conditional situations in which severe water leakage problems are faced in shotcrete-lined or unlined water tunnels, which not only cause a reduction in the stability of rock mass but also create a water loss in tunneling systems. Therefore, the leakage leads to high safety risks and thus leads to an incremental economical loss of the project. Whereas, the application of the full concrete lining for the support causes a rise in the time and cost of the construction. So to reduce the leakage and prevent the rise in construction cost and time pre-grouting injection in the tunnels is the most suitable solution.

It can be categorized into either cement or chemical grout. There are various methods of grouting like open-end tube grouting, Tube a manchette grouting, double packer grouting, etc. Nowadays the use of the new technique has been increased i.e., Pre Excavation Grouting (PEG). PEG provides effective prevention of ingress of groundwater in advance of the excavation. It not only provides a dry underground environment but also improves ground stability. Modern PEG includes high-pressure injection of non-bleeding stable grout with low viscosity and fixes the water-cement ratio. In addition, the use of suitable Microfine Cement and Colloidal Silica in grouting is also quite useful and effective to avoid problems associated with groundwater.

7.2 *Interruptions to Groundwater Flow and Countermeasures*

The underground transportation system solves the problems of traffic above the surface. However, it may cause other problems in the surrounding environment. The big size of the tunnel may interrupt the flow in the aquifers at the sites. Sometimes, in NATM tunnels, walls are of higher length, or in the Cut and Cover method, deep walls secant piles are constructed. These deep and long walls of the tunnel interrupt the flow of groundwater from upstream to downstream at the site. It causes the rise and level down of groundwater at upstream and downstream sides, respectively. This change in the nature of the flow of groundwater affects the surrounding environment. It may result in excess seepage of groundwater in a tunnel. The change also leads to dehydration or the death of plants on the downstream side.

Groundwater flow prevention is very important to prevent these problems associated with flow interruption. It is necessary to collect interrupted water upstream and transport it through siphon or pipes laid in or beneath the structure and is recharged downstream. Another method is to collect groundwater in a tunnel inside the sump and discharge it at the downstream side with pumps.

8 Conclusions

Observation plays a very significant role in NATM tunnel construction as the outcomes of observations decide the NATM support system. Observations depend on data obtained from geological mapping and 3D monitoring. Analysis of tunnel profile at every face at regular intervals is a key to safe and productive NATM tunneling. NATM allows a higher amount of flexibility for fixing the alignment in a congested urban region. NATM is a safe, economic, and time-consuming method of tunnel construction. It plays a vital role in providing a safe opening for TBM launching and outbreaks. It is also used for the construction of the cross passage. The problem of high ingress of underground water is observed in every Indian Metro underground tunnel construction. The problem of water ingress can be solved by the use of applications of effective grouting.

References

1. Chaudhary P, Saha G (1995) Calcutta Metro—construction by "Cut & Cover" and "Shield Tunnelling" methods, ICI Bulletin No. 53
2. METRO RAIL: FACTS & HISTORY, Indiatimes, December 2014. https://www.indiatimes.com/culture/travel/metro-rail-facts-history-313969
3. Kolkata set to start India's first Underwater Metro, Metro Rail News, January 2020. https://www.metrorailnews.in/kolkata-set-to-start-indias-first-underwater-metro/

4. India's Longest Underground Tunnels of Metro Lines, Walkthrough India. http://www.walkthroughindia.com/walkthroughs/trains/indias-longest-underground-tunnels-of-metro-lines/
5. Springer, Chapter—7, New Austrian Tunneling Method. https://doi.org/10.1007/3-540-28500-8_7
6. Boldini D, Lackner R, Mang H (2005) Shotcrete interaction of NATM tunnels with high overburden. J Geotech Geoenviron Eng. American Society of Civil Engineering (ASCE)
7. Ren X, Han Y, Fu C, Liu B, Du W (2019) Management for Beijing subway tunnel construction using the New Austrian tunneling method: a case study. In: International conference on transportation and development. American Society of Civil Engineering (ASCE)
8. Palmstrom A, Stpndiat (1993) New Austrian Tunneling Method (NATM), FJELL-SPREAIKK/GEOTEKNIKK
9. Leca E, Wayne Clough G (1992) Design for NATM tunnel support in soil. J Geotech Eng. ASCE
10. Groundwater of Pune: an over-exploited and ungoverned lifeline, South Asia Network on Dams, Rivers, and People (SANDRP), July 2016
11. What Lies Beneath The Earth, METRO CUBE—Mumbai Metro Rail Corporation Newsletter, issues February 2018–December 2018
12. Ahmad A, Ashirwad N, Sinha M (2019) New Austrian Tunneling Method (NATM) in Himalayan Geology: emphasis on execution cycle methodology. Int J Eng Res Technol (IJERT) 8(06). ISSN: 2278-0181
13. Detailed Project Report (DPR) of Pune Metro Rail Project
14. Detailed Project Report (DPR) of Mumbai Metro Rail Corporation Ltd (MMRCL)
15. Pune Metro:Official site of Pune Metro Rail Project. www.punemetrorail.org
16. Official site of Mumbai Metro Rail Corporation Ltd (MMRCL). www.mmrcl.com
17. Metro dig tells the story of the ground beneath our feet, December 2016. http://timesofindia.indiatimes.com/articleshow/55881989.cms?utm_source=contentofinterest&utm_medium=text&utm_campaign=cppst

Slope Stability Analysis of Artificial Embankment of Fly Ash and Plastic Recycled Polymer Using Midas GTS-NX

Prajakta S. Chavan, Kalyani G. Patil, and Sariput M. Nawghare

Abstract Artificial embankments are constructed to make provisions for infrastructure such as roads, railways, canals and to protect low lying ground from flooding. The slope stability analysis of the embankments should be carried out to know the nature of the underlying material. Careful study and consideration of topographic features along with the site investigation, testing, sampling, modelling monitoring and slope stability analysis is required for the construction of any artificial embankment. In this study, an artificial embankment of fly ash and recycled plastic polymer (RPP) is analysed. The plastic recycled polymers were mixed with fly ash at different proportions to inspect its influence on the slope stability of the artificial embankment. The fly ash was mixed with various proportions of plastic recycled polymers, i.e., with 0, 25, 50 and 75%. In this regard, the laboratory study included Atterberg limits, standard proctor test, unconfined compressive strength, and direct shear test. The various parameters (modulus of elasticity, cohesive strength, internal angle of friction, unit weight, saturated unit weight) of embankment mix were obtained from the experimental testing. The slope stability of the artificial embankment of fly ash and recycled plastic polymers analysed using MIDAS GTS-NX software. MIDAS GTS-NX is a FEM based modelling software which carries out slope stability analysis by the strength reduction method. Embankment models of only fly ash and fly ash with recycled plastic polymer mix have been accounted and simulated using MIDAS GTS-NX software to understand the failure mechanism and the changes in factor of safety. The maximum factor of safety is obtained at a mix percentage of 50% of recycled plastic polymer with fly ash. Further increase in the polymer content decreases the overall factor of safety of the artificial embankment.

P. S. Chavan (✉) · K. G. Patil · S. M. Nawghare
Civil Engineering Department, College of Engineering Pune, Pune, India
e-mail: chavanprajakta21@gmail.com

K. G. Patil
e-mail: kalyanigpatil12@gmail.com

S. M. Nawghare
e-mail: smn.civil@coep.ac.in

M. S. Ranadive et al. (eds.), *Recent Trends in Construction Technology and Management*, Lecture Notes in Civil Engineering 260,
https://doi.org/10.1007/978-981-19-2145-2_55

Keywords Slope stability · Finite element method · Fly ash · Plastic recycled polymer · MIDAS GTS-NX

1 Introduction

Slope stability problems are extensively important in engineering applications and notably within the construction and operation of surface transportation facilities. The stability of the fill slopes, natural slopes, embankments, and excavations are the crucial issues not only for geotechnical designing but also important in assessing the effect of these phenomenon on safety of humans. Slopes fail consequently due to the soil strength and faulty slope geometry. In some cases, the slope cannot sustain disturbing forces resulting from soil weight and surcharge loads, high groundwater conditions or seismic loading. In order to quantify stability of any slope, the general factor of safety is appropriated. Factor of safety (F.O.S.) is ratio of the sum of maintaining forces (friction, cohesion of the material) to the sum of disturbing forces (gravitation and filtration forces). Based on the value of this ratio, we can approximate the stability of the slope or embankment into consideration as follows:

$$\text{F.O.S} = \frac{\text{Shear strength of soil}}{\text{Shear stress required for sliding}} \tag{1}$$

- F.O.S. < 1—unstable slope/unstable embankment
- F.O.S. = 1—slope/embankment in temporal stability
- F.O.S. > 1—stable slope/stable embankment.

Conversely, in India, the disposal of fly ash has become an issue of concern [1]. Our country largely depends on thermal power plants for production of energy. Such thermal power plants are largely coal based. Hence, fly ash disposal issue will be the topic of apprehension for the years to come. The residual fly ash is disposed of in places, which creates a menace of land management and danger to human health. Plastic waste is also becoming extremely threatening to the environment as they are generated in high quantities and jeopardize the environment and its inhabitants [2, 3]. Therefore, this study is intended considering both these issues. The stability of the artificial embankment of plastic polymer and fly ash thus proposed is analysed by using MIDAS GTS-NX software and the failure reports are studied.

Table 1 Fly ash: physical properties

S. No.	Property	Value
1	Specific gravity	2.18
2	Maximum dry unit weight	12.20 kN/m^3
3	Optimum Moisture Content (OMC)	18.6%
4	Young's Modulus of elasticity (E)	4200 kN/m^2
5	Cohesion (c)	23.14 kN/m^2
6	Angle of internal friction (ϕ)	20.4^0
7	Coefficient of uniformity (cu)	15
8	Coefficient of curvature (cc)	1.45
9	Poisson's ratio (μ)	0.27
10	Unit weight (γ)	14.47 kN/m^2
11	Saturated unit weight (γ_{sat})	16.41 kN/m^2
12	D_{60}	0.045 mm
13	D_{30}	0.014 mm
14	D_{10}	0.003 mm
15	Liquid limit	23.6%
16	Plastic limit	NP

2 Materials

2.1 Fly Ash

Fly ash is a waste by-product extracted by mechanical or electrostatic precipitator, obtained from coal-based power plants. Fly ash consists of alumina, silica, and iron. Fly Ash of class F was obtained from Tata thermal power plant, Trombay, Mumbai for the experimental program. The fly ash sample was screened through a 12 mm sieve, to get rid of the foreign materials, vegetation, and other deleterious material. The samples were dried in the oven for about 24 h before any further usage [4]. Table 1 represents the physical properties of the fly ash used.

2.2 Plastic Recycled Polymer

Plastic recycling is the process of recovering waste plastic material and converting them into useful products. Basically, all types of plastic can be recycled except those made from recycled plastics are usually unrecyclable. Plastic Recycled Polymers (PC) employed in the research lab work were obtained from R R Petroplast Pvt. Ltd., Pune. These are produced by reprocessing plastic waste from organic

Table 2 Specification of plastic recycled polymers

S. No.	Parameter	Specification
1	Type	Recycled plastic polymer
2	Colour	White and Yellow
3	Diameter (mm)	3.0
4	Specific gravity (G)	2.154
5	Density (kg/m^3)	1300

compounds producers, molders, extruders, fabricators, and utilization corporations. The properties of recycled plastic polymer (RRP) are mentioned in Table 2.

3 Methodology

3.1 Laboratory Testing

In order to obtain parameters required for slope stability analysis, series of tests were conducted on fly ash with various proportions of recycled plastic polymer, i.e., 25, 50, and 75%. During the conduction of experimental laboratory tests, following Indian Standard codes [5] were followed:

(i) Standard Proctor Test—IS: 2720 (Part 7)—1980
(ii) Unconfined Compressive Strength (UCS) Test—IS: 2720 (Part 10)—1991
(iii) Direct Shear Test—IS: 2720 (Part 13)—1983
(iv) Unconsolidated Undrained Test—IS: 2720 (Part 11)—1993
(v) Permeability Test—IS: 2720 (Part 17)—1986
(vi) Liquid limit and Plastic limit test—IS: 2720 (Part 5)—1985.

3.2 Midas GTS-NX

The experimental results are typically accurate and reliable since they replicate the precise behaviour of the tested structure. It is an incontrovertible fact that the experimental testing of structures is time overwhelming and expensive. Therefore, it is desirable to rely on numerically simulated models which might replace the experimental investigation in a fairly accurate manner. Finite Element Method (FEM) could be a sensible choice for the numerical simulation of complex structures. Midas GTS-NX software is based on the finite elements method (FEM). It is accustomed to simulate actual phenomena occurring within the soils. Midas GTS-NX software uses the Strength Reduction methodology to deduce the kind and reason behind slope failure [6]. In the strength reduction methodology, the shearing strength of the soil material is decreased gradually, till the point where convergence does not occur [7,

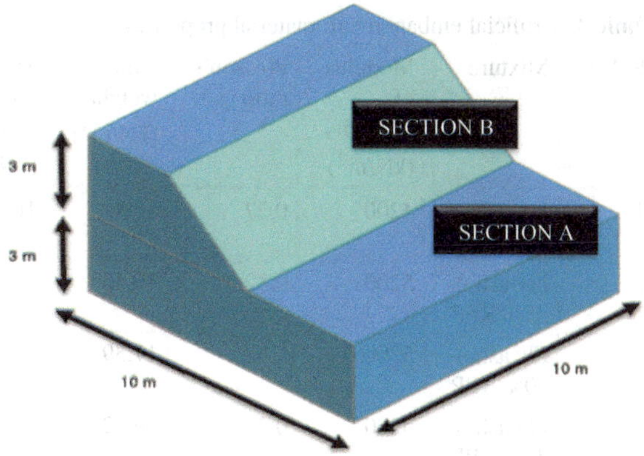

Fig. 1 Embankment model

8]. This point is taken into consideration and the maximum decrease in the strength of that soil material is considered as the least factor of safety.

4 Embankment Model Description

4.1 Geometry of Embankment

Geometric details of the embankment are shown in Fig. 1. The overall length of section A and B is 1000 mm. The width of section A is 1000 mm, and its depth is 300 mm. The top width of section B is 300 mm, and the bottom width is 600 mm. The angle of slope is 45°.

4.2 Material Properties

The summary of properties of the materials comprising the artificial embankment used for the stability analysis are mentioned in Table 3.

Table 3 Artificial embankment: material properties

S. No.	Mixture	Modulus of elasticity (kN/m^2)	Poisson's ratio (μ)	Unit weight (kN/m^3)	Saturated unit weight (kN/m^3)	Cohesion 'c' (kN/m^2)	Friction angle 'Ø'
1	Fly ash only	4200	0.27	14.47	16.41	23.14	20
2	Fly ash + 25% RPP	5200	0.32	16.45	16.98	17.65	27
3	Fly ash + 50% RPP	5200	0.32	17.39	16.97	12.94	39
4	Fly ash + 75% RPP	5200	0.32	16.32	16.98	16.47	29

Table 4 Midas GTS-NX slope stability analysis results

S. No.	Material	Factor of safety (FOS)	Maximum displacement (mm)
1	Fly ash only	0.9672	451.343
2	Fly ash + 25% RPP	1.2680	693.701
3	Fly ash + 50% RPP	2.1148	122.212
4	Fly ash + 75% RPP	1.4910	427.823

5 Results and Discussion

5.1 Software Results

The factor of safety obtained from the software are given in Table 4. The displacement of the artificial embankment for the various mix proportions of fly ash and recycled plastic polymer are shown in Figs. 2, 3, 4 and 5.

5.2 Graphical Results

Graph 1 shows the variation of the embankment material mix with respect to factor of safety predicted by the software.

From the results, it is evident that the amount of recycled plastic polymer influences the factor of safety of the fly ash embankment. The maximum factor of safety is obtained at a mix percentage of 50% of plastic recycle polymer with fly ash. Further

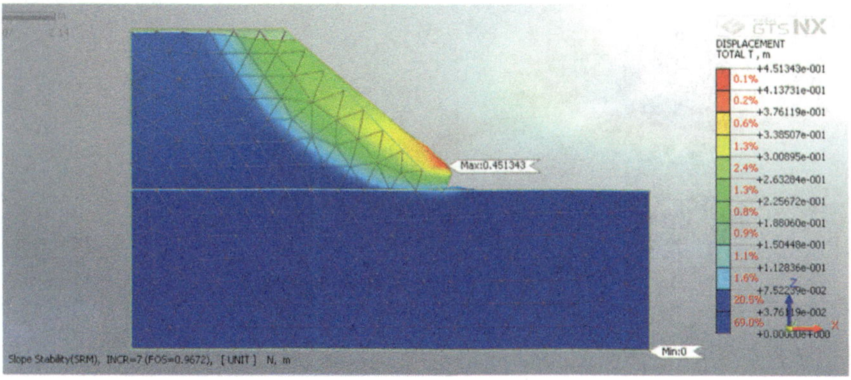

Fig. 2 Total displacement (in m) for fly ash only model

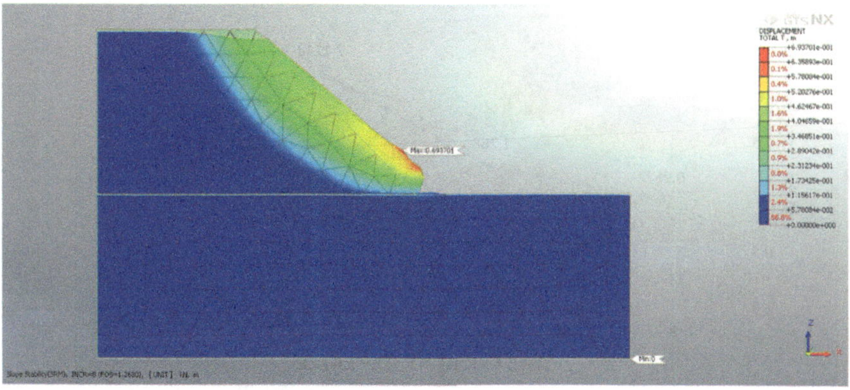

Fig. 3 Total displacement (in m) for FA and 25% RPP model

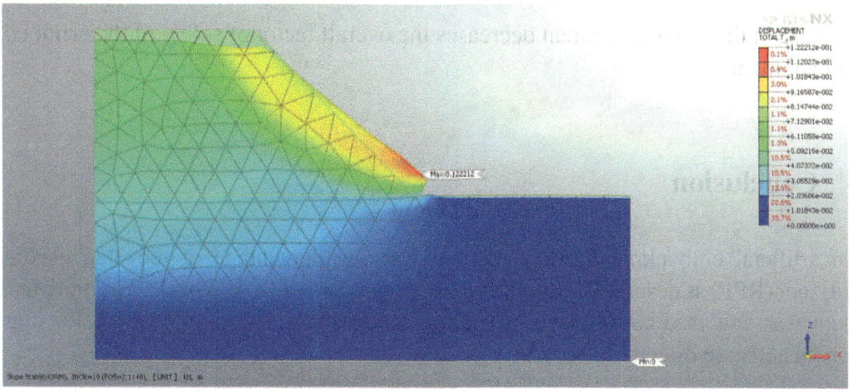

Fig. 4 Total displacement (in m) for FA and 50% RPP model

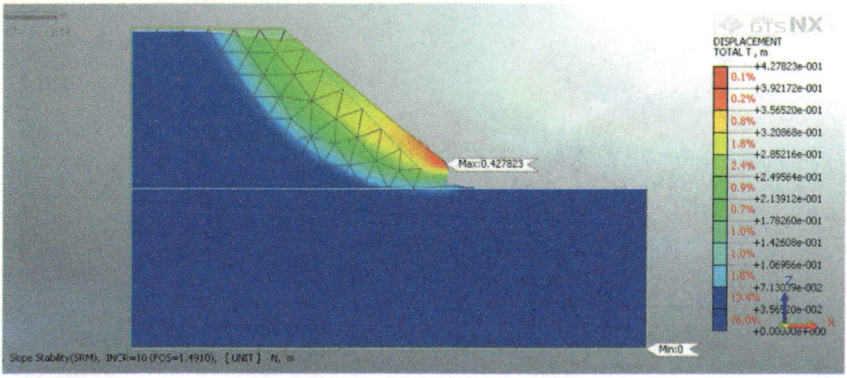

Fig. 5 Total displacement (in m) for FA and 75% RPP model

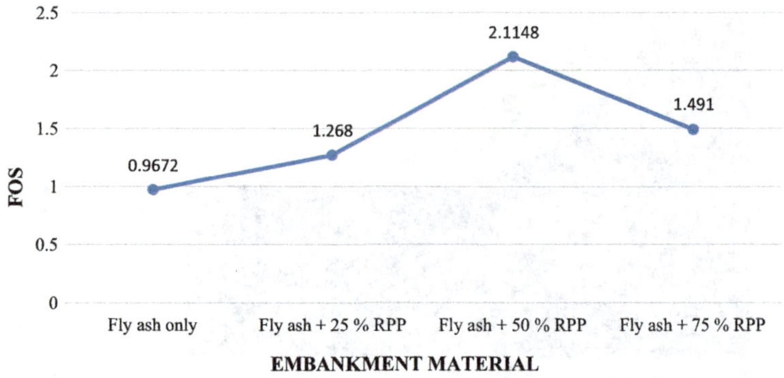

Graph 1 Embankment mix material versus factor of safety

increase in the polymer content decreases the overall factor of safety of the artificial embankment.

6 Conclusion

An artificial embankment of fly ash and varying percentage of recycled plastic polymer (RPP) as described in Sect. 3 was modelled in Midas GTS-NX. The embankment was analysed based on its properties as described in Sect. 2. The following conclusions are drawn:

1. Embankment made of fly ash only is not stable and requires admixtures/additives to improve the stability.
2. Fly ash along with 50% of recycled plastic polymer gives the maximum factor of safety in comparison with other mix percentages.

3. Thus, 50% recycled plastic polymer by weight of fly ash, acts as ideal mix percentage for achieving optimum factor of safety for the proposed embankment.
4. Finally, we can stabilize slopes, make fly ash slopes stable and steeper with the use of recycled plastic polymer. Such slopes take large loads and result in lower deformations, due to its improved confinement and inclusion property.

References

1. Salunkhe T, Nawghare S, Padade A, Mandal JN (2014) Stabilization of fly ash slope using plastic recycled polymer. Int J Sci Eng Res 5(7):248–252. ISSN 2229-5518
2. Salunkhe TV, Mandal JN (2014) Behavior of fly ash at different mix ratios with plastic recycled polymers. Int J Res Appl Sci Eng Technol (IJRASET) 2(III):190–198. ISSN: 2321-9653
3. Awoyera PO, Adesina A (2020) Case study: plastic wastes to construction products: status, limitations, and future perspective. Case Stud Constr Mater 12:30–41
4. IRC: SP: 58-1999, Guidelines for use of fly ash in road embankments. The Indian Roads Congress, New Delhi, India
5. IS 2720 (1983) Methods of test for soils, Part 1-14, 17, Bureau of Indian Standards, New Delhi, India
6. Goyal T, Sonthwal VK (2017) A review on slope stability analysis by strength reduction method using Midas GTS. Int J Innovative Res Sci Eng Technol (IJIRSET) 6(8):17050–17053
7. Saikia R, Kanti Dey A (2016) Slope stability analysis of slides at Sonapur using strength reduction method. In: Proceedings of Indian geotechnical conference IGC2016, pp 15–17
8. Ankur N, Sonthwal VK (2018) Assessment of a failed slope on Mandi Pathankot highway, Himachal Pradesh using Midas GTS NX. Int J Tech Innovation Mod Eng Sci (IJTIMES) 4(7):561–564. e-ISSN: 2455-2585

Lateral Capacity of Step-Tapered Piles in Sand Deposits

K. V. S. B. Raju, K. S. Rajesh, L. Dhanraj, and H. C. Muddaraju

Abstract Many structures on pile foundations, especially in the marine environments are subjected to lateral loads from the impact of berthing ships, sea waves, wind, etc. Piles are mostly used with straight-sided walls or circular ones. However, in recent days enlarged, top cross-section, step-tapered piles are found to be of interest and advantageous over traditional deep foundation systems. Step-tapered pile foundation is a relatively new structure and is such a kind of pile foundation having its cross-section decreased in steps. In this study, the characteristics of step-tapered piles were established from the experimental investigation, under lateral loadings. The tests were carried out on straight and step-tapered piles of different segments, 650 mm long embedded in homogenous and layered sand deposits. The piles were instrumented along their length to know their behaviour along with the depth. In order to examine the ultimate lateral capacity and the deflection, the tests were carried out to failure. The test results indicated a non-linear response and a substantial increase in lateral capacity of the pile with decreased deflections for step-tapered piles. The provision of providing steps or segments for piles with five segments resulted in an increase in the load capacity ranging between 44 and 73% in a layered condition and 42–63% in a homogenous sand condition. Providing segments shows a significant effect on load-bearing capacity. Step-tapered piles experienced lesser deflection at the pile head and also along the shaft when compared to the uniform diameter pile and they have better material distribution leading to overall efficiency.

Keywords Step-tapered piles · Homogenous sand · Layered sand · Ultimate lateral capacity · Deflections

---

K. V. S. B. Raju · K. S. Rajesh (✉) · L. Dhanraj · H. C. Muddaraju
Department of Civil Engineering, UVCE, Bangalore University, Bengaluru 560056, India
e-mail: rajeshksra@gmail.com

M. S. Ranadive et al. (eds.), *Recent Trends in Construction Technology and Management*, Lecture Notes in Civil Engineering 260,
https://doi.org/10.1007/978-981-19-2145-2_56

1 Introduction

Piles are slender structural members in a foundation that transfers the load from the superstructure to a stiffer medium. Structures on pile foundation, especially in an offshore environment, are subjected to both axial and lateral loads, and the ultimate capacity under lateral loads is important and maybe the controlling factor in foundation design. Nowadays, the shape of the pile can be varied from pile to pile. One such is a step-tapered pile foundation, a relatively new structure and in its kind cross-section decreased in steps.

Step-tapered piles feature favourable formation applicability and lateral load-carrying capability, indicating that they are flexible in layout and may have significant economic benefits. They were broadly utilized in latest years, especially in large bridge pile foundations in deep water, such as the pile foundations of the main piers of the Sutong Yangtze River Bridge. These piles have various advantages in terms of driving efficiency, flexibility, installation, load-carrying capacity, inspection, etc. Also, a wide range of configurations is available to best satisfy subsoil conditions and loading requirements.

2 Literature Review

The literature on step-tapered piles is still not systematic or comprehensive. Ismael [1] conducted a load test on straight and step-tapered bored piles in the sand under compression. The analysis included both the base capacity and the shaft friction at ultimate loads. The results showed that step-tapered piles have large lateral resistance with an increase in the axial bearing capacity compared to straight piles with the same base area. Ghazavi and Lavasan [2] studied the bearing capacity under axial loads of cylindrical, step and tapered piles. The results indicated an increase in the capacity of both step and tapered piles. Also, the bearing capacity was less than 10% between stepped and fully tapered piles.

The behaviour of step-tapered and straight bored piles in calcareous sand by lateral loads was examined in field tests by Ismael [3]. The piles were 5.3 m long embedded 5 m below the ground level. The straight piles were 0.2 and 0.3 m in diameter and step-tapered piles are 0.3 m along with the upper 10, 20 and 40% segment changing to 0.2 m for the lower section. Test results indicated a non-linear response and a substantial increase in lateral capacity, and at working loads, the deflections for the step-tapered piles were less compared with the straight shaft piles.

Load transfer mechanism of step tapered hollow pile analyzed by Ming et. al [4], Fang and Ming [5] studied the deformation and load-bearing characteristics of step tapered piles under lateral load using physical model tests and theoretical analysis. The results indicated that a step-tapered pile are more effective against lateral loads. The behaviour of step-tapered bored piles under static lateral loading in the sand was tested at a particular site in Kuwait [6]. Results indicated an increase in load-carrying capacity and deflections decreased at specified applied loads to the step-tapered piles due to the strengthening of the top sections of the piles. Tavaroli and Ghazavi [7] studied the same length and volume of stepped and tapered piles as offshore piles driven by hammer blows and concluded that an increase in shear strength parameters of soil increases the bearing capacity in step-tapered piles and this is suitable for stiff soil strata.

Since there is limited knowledge of the performance of these kinds of piles under lateral loads, more research and investigation need to be performed to know the behaviour of these piles in homogenous and layered soils under horizontal loads.

3 Test Preparation

3.1 Description of Equipment

The test tank was made of mild steel and the walls of the tank were 10 mm thick with dimensions 1.2 m × 0.75 m × 1.5 m. In order to apply the lateral load, the tank is provided with a hand-winch arrangement. The load applied to the model piles was measured using a proving ring. The pile head deflection along the length was recorded using sensitive dial gauges with the least count of 0.01 mm.

3.2 Soil Medium

In the present investigation commercially available dry sand was chosen as a soil medium. Laboratory tests were conducted as per IS code and the properties are presented in Table 1.

The tests were conducted in two different soil systems, i.e., homogenous and layered systems. In the first case, the compactness of sand is loose and dense (Fig. 1). In the second case, the compactness of sand is medium dense between loose and dense states, respectively. Another case is loose between dense and medium dense states, respectively (Fig. 2). The relative densities for the loose, medium and dense states were 30%, 50% and 70%, respectively, for the tests on model piles in a sand medium.

3.3 Model Piles

The dimension of piles was selected as per the similitude law proposed by Wood [8]. The model piles selected were made of aluminium, and a scale factor of 1/18 is adopted. This simulates a prototype pile made from steel of Young's modulus $E =$ 210 GPa, diameter = 600 mm and length = 12 m.

Table 1 Properties of soil sample

Properties		Values
Specific gravity (G)		2.64
Particle size	D_{10}	0.5
	D_{30}	0.7
	D_{60}	1.1
Uniform coefficient, C_u		3.2
Curvature coefficient, C_c		0.89
Classification		Poorly graded sand (SP)
Maximum unit weight ($\gamma_{d,max}$), e_{min}		16.19 kN/m^3, 0.62
Maximum unit weight ($\gamma_{d,min}$), e_{max}		13.63 kN/m^3, 0.9
Peak friction, ϕ_{peak}		28° ($D_r = 30\%$) 34° ($D_r = 50\%$) 39° ($D_r = 70\%$)

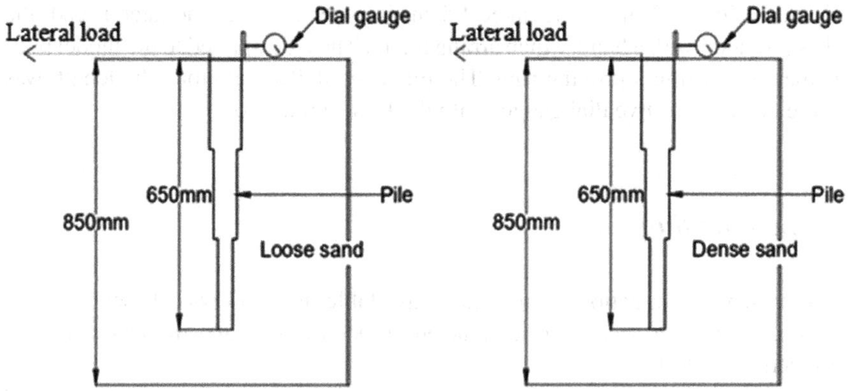

Fig. 1 Homogenous sand system

In order to produce the step-tapered piles, the pile is divided into appropriate increments of length of constant diameter having the same total perimeter area as that of a cylindrical pile. The model pile dimensions and schematic view are shown in Table 2 and Fig. 3.

4 Test Procedure

The placement of sand plays an important role to achieve reproducible density. In this study, sand raining technique is adopted. Initially, sand was allowed to fall freely

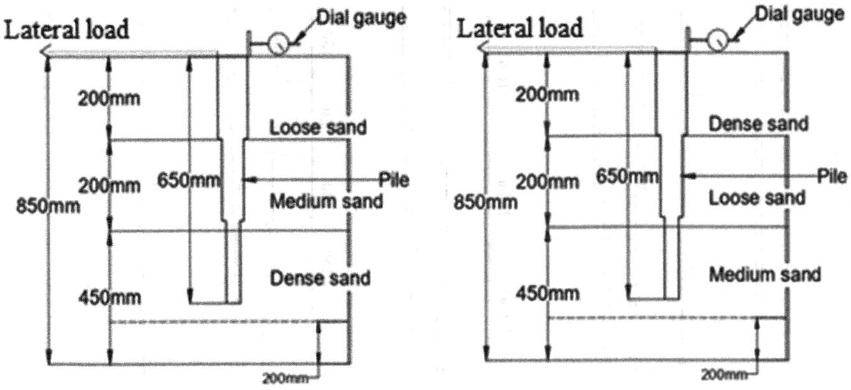

Fig. 2 Layered sand system

Table 2 Model pile dimensions

Pile type	Length of pile (mm)	Number of segments	Length of each segment (mm)	Diameter of each segment (mm)	Wall thickness of the pile (mm)	% increase in diameter of each segment
Straight (SP)	650	0	650	27	5	0
Step-tapered	650	2	325	40	5	48
				27	5	0
Step-tapered	650	3	216.67	40	5	48
				33.5	5	24
				27	5	0
Step-tapered	650	5	130	40	5	48
				36.75	5	36
				33.5	5	24
				30.25	5	12
				27	5	0
Straight (SP)	650	0	650	40	5	48

in the tank continuously through the raining mesh from a particular height and the density with which the sand is deposited is determined. For a particular height of fall, the density that can be achieved is predetermined. The schematic view of the experimental setup is shown in Fig. 4.

The raining platform initially was positioned directly over the tank to the required height to achieve the desired density of sand. The sand was allowed to fall freely into the tank. At the pile base, the surface of the sand is levelled with the help of a straight edge. The pile was placed with the help of a clamp; the pile is held vertical. Again the sand was placed at the required height. The raining platform position was changed

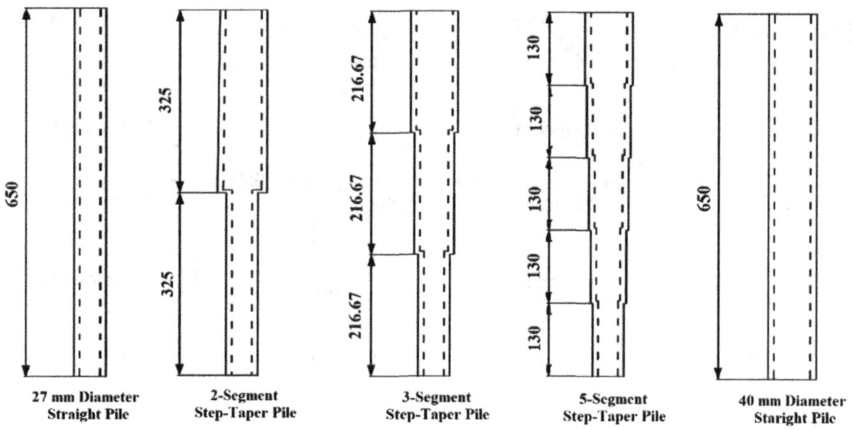

Fig. 3 Schematic view of model piles

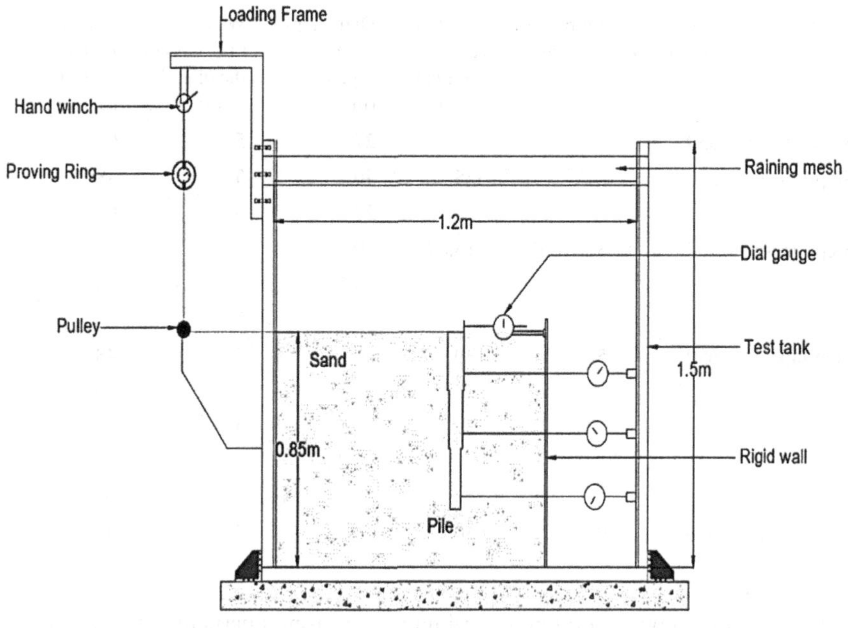

Fig. 4 Schematic representation of test setup

regularly to the needed height to make sure that the intended density is achieved. Dial gauges were attached to the pile throughout the length to know the deflection along with the depth and corrected to zero lateral displacements. The corresponding loads and lateral displacements were measured and recorded during the load test. The test has been continued till the proving ring reading becomes constant or starts decreasing or up to a minimum displacement of 25 mm. The test is repeated with different pile combinations and the corresponding load–displacement graphs were plotted and the capacities of piles were determined. For each model test, the soil was removed from the tank and was replaced with the required depth and density.

5 Results and Discussions

Soil layers and the number of segments of step-tapered piles on lateral resistance of straight and step-tapered piles were tested. In the present study, at a lateral displacement of 25 mm, failure is considered. The corresponding load–deflection response graphs were plotted for all straight and step-tapered piles. The following section discusses the obtained results in detail.

5.1 Load–Deflection Response

Load–deflection curves for all the model piles in loose sand are shown in Fig. 5. It shows that enlarging the upper part of the pile section for the step-tapered piles has significantly decreased the pile head deflection. Enlarging pile diameter from 27 to 40 mm for upper 50% of the length of pile has reduced the head displacement with increased lateral capacity from 20.36 to 28.9 N, approximately a 45% increase. Also, corresponding to an enlargement of 100%, the diameter of 40 mm, there was no significant improvement in deflection and the load capacity was around 29 N, indicating no increase in capacity over the 50% enlargement.

The pile-head deflection curve for straight and step-tapered piles is shown in Fig. 6. It can be seen that step-tapered piles showed lesser pile head deflection than a

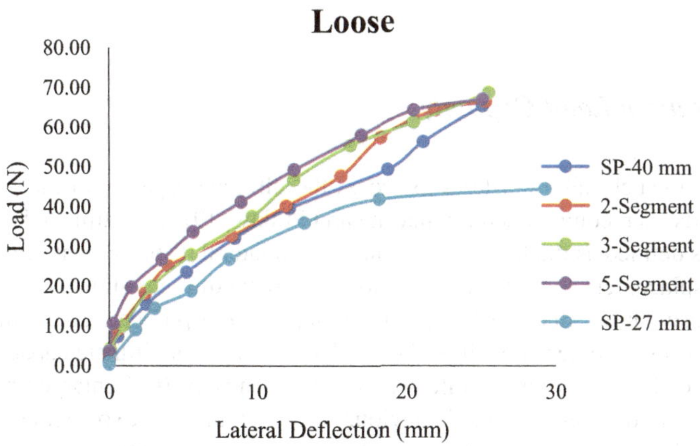

Fig. 5 Load–deflection curve in loose sand

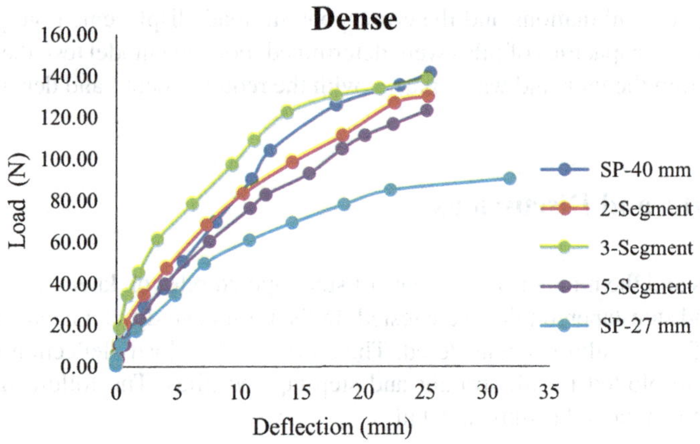

Fig. 6 Load–deflection curve in dense sand

straight or cylindrical pile. An increase in pile diameter to the upper 50% of the pile length decreased head deflection with an increased load-carrying capacity of around 63%. But the case is not so when the enlargement was 100% since the load capacity is reduced with the deflection.

The five-segmented pile showed better performance compared to other piles (Fig. 7a). The pile-head deflection is reduced for step-tapered piles than the straight piles. In the latter case, loose between dense and medium-dense state, Fig. 7b, comparatively the three-segmented pile showed better performance. Also with enlarging the pile cross-section, the head deflection is increased with decreased pile's load-carrying capacity. However, it can also be noted from the figure that as the number of segments increases, the deflection is reduced. This increment is limited to piles up to three segments, beyond which increment had no beneficial effect.

5.2 *Ultimate Load Capacity*

There are no such failure criteria to access the ultimate capacity of the pile under lateral loads, as recommended by Sharma and Prakash [9]. The ultimate lateral load capacity is defined as the load corresponding to a lateral displacement of 6.25 mm.

The displacement of 6.25 mm in the loose sand condition is shown in Fig. 8. The figure clearly shows that the five-segment step-tapered pile carries the maximum ultimate load than other piles. It is about a 10% increase in ultimate capacity of the straight pile of 40 mm diameter pile, which corresponds to 100% enlargement. There is a significant difference in the behaviour of SP-40 mm and two-segment piles. It shows that by increasing the steps in pile models, the load-carrying capacity was found to be increased.

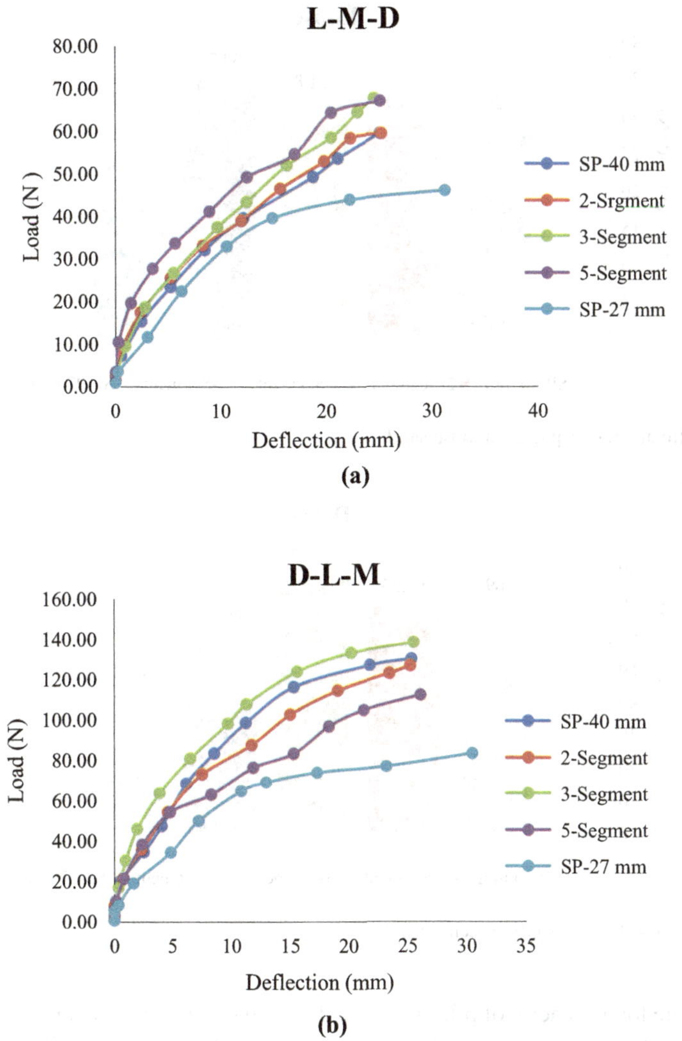

Fig. 7 Load–deflection curve in layered sand

The ultimate load capacity of straight and step-tapered piles for the dense sand conditions is shown in Fig. 9. It can be noted that a three-segment step-tapered pile carries a maximum ultimate lateral capacity than other piles and an increase of 12.5% in the lateral capacity than a pile of 40 mm diameter, which corresponds to 100% enlargement. Also beyond three segments, there will be no significant improvement in pile's lateral capacity.

In this case of medium-dense between loose and dense state, Fig. 10a, a five-segment step-tapered pile carries the maximum ultimate load than other piles. It is about a 25% increase in 40 mm diameter for a straight pile. Figure 10b shows

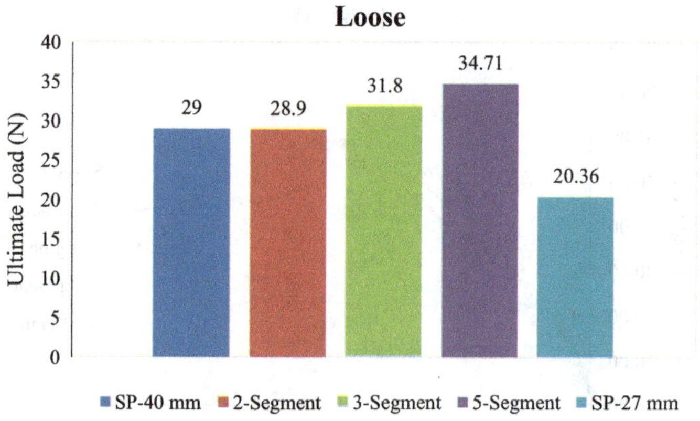

Fig. 8 Ultimate load capacity in loose sand

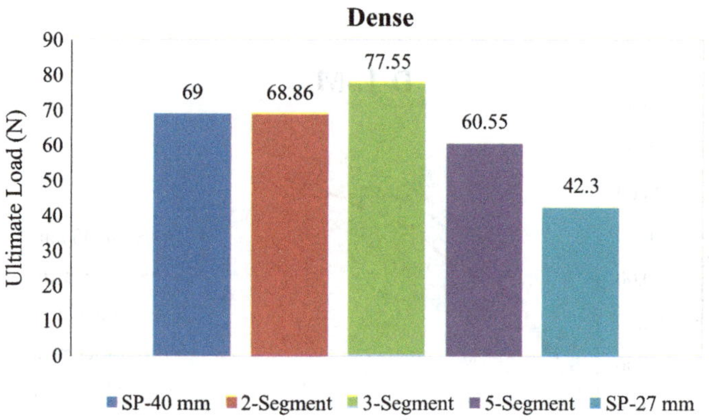

Fig. 9 Ultimate load capacity in dense sand

the ultimate load capacity of piles in dense–loose–medium dense sand, respectively. It can be noted from the figure that the three-segment step-tapered pile carries the maximum ultimate load than other piles. The increment in the load capacity is around 17% compared to that of a straight pile of 40 mm diameter.

The deflection for all the piles is reduced in dense sand conditions compared to loose sand cases. Also, the same can be observed for the top dense conditions than top loose conditions (layered sand case), respectively. Compared to the straight pile, step-tapered piles experienced lesser pile head deflection. This is due to the increased stiffness of step-tapered piles due to changes in cross-sectional area. In order to know the effectiveness under lateral load conditions of step-tapered piles, the ultimate load capacity was expressed in terms of pile load capacity ratio, given in Table 3. The ratio of capacity is defined as the load-carrying capacity of step-tapered piles to the load-carrying capacity of a straight pile of 27 mm diameter.

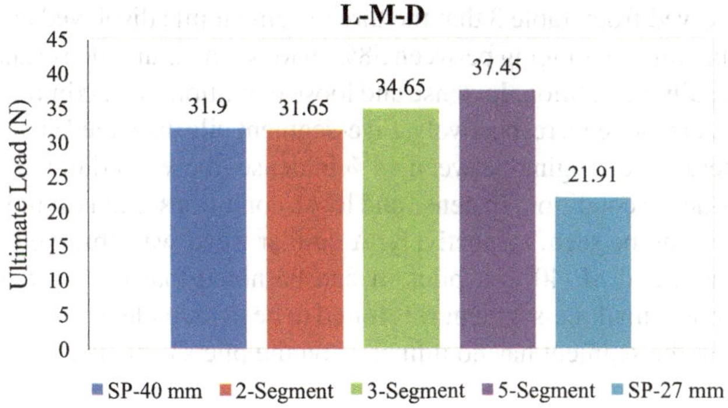

(a)

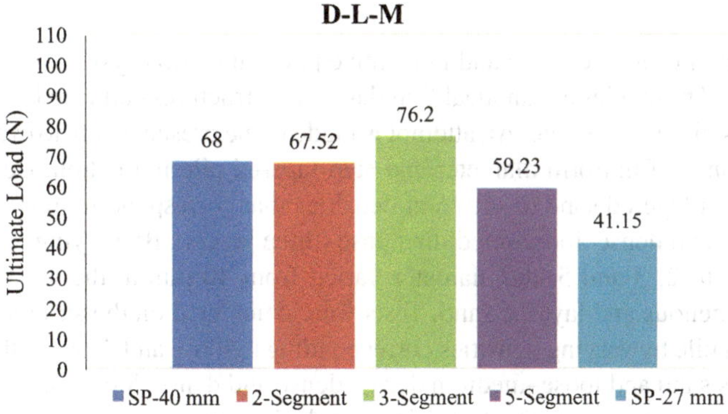

(b)

Fig. 10 Ultimate load capacity in layered sand

Table 3 Ultimate lateral load capacity and ratio of capacity

Pile	Condition	SP-27 mm	2-Seg	3-Seg	5-Seg	SP-40 mm
Lateral load capacity (N)	Loose	20.36	28.9	31.8	34.71	29
	Dense	42.3	68.86	77.55	60.55	69
	L-M-D	21.91	31.65	34.65	37.45	31.9
	D-L-M	41.15	67.82	76.2	59.23	68
Ratio of capacity	Loose		1.42	1.56	1.7	1.42
	Dense		1.63	1.83	1.43	1.63
	L-M-D		1.44	1.58	1.73	1.46
	D-L-M		1.65	1.85	1.44	1.65

It can be observed from Table 3 that the three-segment pile displayed an increase in the ultimate lateral load ranging between 58% in loose–medium–dense and 85% in dense–loose–medium condition. In dense and loose conditions, a maximum increase of 83% and 56% can be seen, respectively. Five-segment pile displayed an increase in the ultimate lateral load ranging between 44% in dense–loose–medium and 73% in loose–medium–dense condition. In dense and loose conditions, a maximum increase of 43% and 70% can be seen, respectively. A similar trend was observed for other piles, two-segment and SP-40 mm piles. It can be noted that for top dense sand conditions, the pile with three-segment was found to be favourable. However, beyond this, increment in the segment has no influence on the pile's capacity.

6 Conclusions

Step-tapered piles are versatile and innovative pile foundation system for accessing lateral loads. These piles are an ideal foundation for structures carrying larger lateral loads, jetties, lighthouses, etc. An attempt is made in the present study to evaluate the lateral response of uniform diameter and step-tapered pile in the homogenous bed of sand and in layered sand of different densities that correspond to loose–medium dense–dense and dense–loose–medium dense (three layers). By varying the number of segments to 2, 3 and 5, the diameter varied from 40 mm at the top to 27 mm in the homogenous and layered sand. Tests were conducted on the straight pile and step-tapered pile by varying densities corresponding to loose and dense in the case of homogenous sand and loose–medium dense–dense and dense–loose–medium dense in the case of layered sand. The following conclusions were drawn from the study:

1. The response of the pile under the lateral loads was affected by soil fabric, soil density, and shape of pile characteristics.
2. The provision of providing steps or segments for piles with five segments resulted in an increase of the load capacity ranging between 44 and 73% in layered condition and 42 and 63% in homogenous sand condition. Providing segments shows a significant effect on load-bearing capacity.
3. It was observed that when compared to a uniform diameter pile, step-tapered pile carries more lateral load and lesser pile-head deflection. Also in the case of uniform diameter piles in homogenous sand beds, the dense state of sand carries the more lateral load when compared to the loose state of sand. In a loose state of sand as the number of segments increases lateral load was found to increase. In the case of a dense state of sand, the lateral load increases significantly till three segments, and thereafter for five segments, it was found to decrease because as the number of segments increases the overall surface area was found to decrease and the corresponding mobilization of passive resistance. Also in the case of the loose bed of sand, there will be slippage of sand particles surrounding the pile, thereby decreasing the adhesion between the pile–soil interface.

4. It was observed that in layered sand when compared to uniform diameter pile, step-tapered pile carries more lateral load and lesser pile-head deflection. In layered sand when compared to loose–medium dense–dense state, the strata consisting of dense–loose–medium dense state was observed to offer more lateral resistance and lesser pile-head deflection due to mobilization of more passive resistance in dense strata (top one-third of pile length), and it was also observed that three segments carry more lateral load than five segments due to overall decrease in surface area in the case of five segments and decrease of pile–soil adhesion.
5. Compared to the straight pile, step-tapered piles experienced lesser deflection at the pile head and also along the shaft. Also, step-tapered piles have better material distribution leading to overall efficiency.
6. The present study is helpful in accessing foundations for the structures where larger lateral loads are anticipated, such as tall buildings, wharves, and structures in the marine environment.

References

1. Ismael NF (2003) Load tests on straight and step tapered bored piles in weakly cemented sand. Field measurements in geomechanics. ISBN 9058096025
2. Ghazavi M, Lavasan AA (2006) Bearing capacity of tapered and step-tapered piles subjected to axial compressive loading. In: 7th International conference on coastal, ports & marine structures (ICOPMAS)
3. Ismael NF (2006) Analysis of lateral load tests on step tapered bored piles in calcareous sands. ASCE, GeoCongress
4. Ming H, Song J, Dexiang X, Tao D, Xing S, Tao F (2018) Load transfer mechanism and theoretical model of step tapered hollow pile with huge diameter. Chin J Rock Mech Eng2018(10):2370–2383
5. Fang T, Huang M (2019) Deformation and load-bearing characteristics of step-tapered piles in clay under lateral load. Int J Geomech 19(6):04019053
6. Ismael NF (2010) Behavior of step tapered bored piles in sand under static lateral loading. J Geotech Geoenviron Eng 136:669–676
7. Tavasoli O, Ghazavi M, Driving behavior of stepped and tapered offshore piles due to hammer blows. Mar Geores Geotechnol. ISSN:1064-119X (P) 1521-0618
8. Wood DM (2004) Geotechnical modelling. CRC Press
9. Prakash S, Sharma HD (1990) Pile foundations in engineering practice. Wiley, New York, NY

Strength and Dilatancy Behaviour of Granular Slag Sand

K. V. S. B. Raju and Chidanand G. Naik

Abstract This paper focuses on the shear strength and dilatancy properties of granular slag sand. Direct shear tests were conducted on sample slag sand collected from the JSW steel plant, Toranagallu of Bellary district. The different relative densities at which tests were conducted are 20%, 50%, and 80% respectively, with corresponding unit weights are 14.58 kN/m^3, 15.27 kN/m^3, and 16.03 kN/m^3, respectively. All the direct shear tests were continued till the shear strain equals 40% in order to achieve the critical state. The dilatancy and shear strength characteristics of the slag sand sample were observed at different relative densities under different normal stresses ranging from 50 to 400 kPa. It was found that both the friction angle and dilatancy angle reduced with the decrease in the applied normal stress. Also based on a series of test results critical state frictional angle was evaluated, which is independent of normal stress and relative density. For the chosen slag sand sample, a correlation is established between the peak friction angle and the critical state friction angle in the present work. The study can be useful in assessing the stability of geostructures, wherein granular slag sand is used to enhance engineering performance. The established correlation was finally compared with those established by various researchers.

Keywords Dilatancy · Peak friction angle · Density index · Correlations · Critical state · Slag sand

K. V. S. B. Raju · C. G. Naik (✉)
UVCE, Bangalore University, Bangalore 560056, India
e-mail: naik.aec@gmail.com

M. S. Ranadive et al. (eds.), *Recent Trends in Construction Technology and Management*, Lecture Notes in Civil Engineering 260,
https://doi.org/10.1007/978-981-19-2145-2_57

1 Introduction

Almost all granular materials undergo volume change when they are subjected to shearing. This behaviour is termed dilatancy. While the factors causing this dilation behaviour of soil are very well known, there is no practical solution based on in situ conditions to measure the dilation angle. The strength of the granular materials is generally denoted by two important factors, namely, critical state friction angle and peak friction angle. The present work focuses on the shear strength and dilation characteristics of slag sand. The experimental programme consists of numerous direct shear experiments on slag sand maintained at different relative densities. In terms of the stress–dilatancy relationship and shear strength parameters, test results were interpreted. The main aim of the present study is to find a correlation between the peak friction angle (φ_p), dilation angle (ψ_p), and critical state friction angle (φ_{cv}) for the chosen slag sand sample.

2 Literature Review

Several researchers have worked on the strength and dilation characteristics of various granular materials; a few of them are highlighted here.

In order to establish a correlation between stress and dilatancy of coarse aggregate used for base course construction in Poland, Katarzyna [1] employed a large direct shear apparatus. The standard pressures used for the experimental work were 55.5–59.6, 130.9–133.9, and 254.6–258.2 kPa [1]. At a horizontal displacement of 60 mm, all experiments were interrupted (i.e. about 12% of the sample diameter). For tested coarse granular soil, the critical state frictional angle was estimated to be $\varphi_p = 41.2°$ and the measured critical state frictional angle was found to be independent of compaction and soil moisture condition. Wang et al. [2] have investigated the shear strength of sandstone–mudstone mixture by conducting numerous direct shear tests in a laboratory [2]. For the study, lightly weathered blocks of rock collected from a mountain situated near the Yangtze River in Chongqing of China were used. Six different combinations of mixtures were used for the study, namely, 0, 20, 40, 60, 80, and 100% mudstone. Effective vertical normal stresses were applied in the range between 50 and 400 kPa. Variations of shear stress against relative horizontal displacement were plotted for the study material. Four major influence factors were recognised, namely, the distribution of particle size, particle composition of mudstone in the mixture, initial dry bulk unit weight, and water content. Dry sand maintained at various density indices (i.e. 20, 50, and 80%) was subjected to a direct shear test by Raju and Khan [3]. Vertical normal stress ranging from 50 to 400 kPa was applied to the prepared sample. It was observed that with an increase in effective normal stress, the peak frictional angles and dilation angles were found to decrease [3]. Also, it was found that an increase in density leads to an increase in peak friction angle and dilation angle [3]. Also, it is found that as the grain size of sand decreases from coarse to fine, there is a substantial reduction in peak frictional angle, critical state friction angle, and dilation angle. In the study, critical state friction angle was found for the three different types of graded sand. It was found that φ_{cv} for coarse sand was 35.34°, φ_{cv} for medium sand was 27.07°, and φ_{cv} for fine sand was 24.08°. Bolton [4] explored shear strength and dilatancy characteristics of almost 17 different types of sands maintained at various densities and normal stresses. By performing axisymmetric and plane-strain tests, a variation in shear stress and volumetric strains with respect to axial strains was found. The new term "relative density index" has been established in terms of density index and effective stress level. Also, he has indicated a much simpler correlation among φ_{p}, φ_{cv}, and ψ_{p}. The established correlations are mentioned below:

$$\varphi_{\mathrm{p}} = \varphi_{\mathrm{cv}} + 0.8\psi_{p} \quad (1)$$

$$\varphi_{\mathrm{p}} = \varphi_{\mathrm{cv}} + 5\,\mathrm{IR}\ \text{for plain strain condition} \quad (2)$$

$$\varphi_{\mathrm{p}} = \varphi_{\mathrm{cv}} + 3\,\mathrm{IR}\ \text{for triaxial condition} \quad (3)$$

The correlation between dilatancy index (IR), relative density (DR), and effective stress (σ_{v}) is given by the relationship:

$$\mathrm{IR} = \mathrm{DR}(Q - \ln(\sigma_{\mathrm{v}})) - R \quad (4)$$

In the above expression, Q and R are constants; vertical normal stress (σ_{v}) is expressed in kPa and relative density (DR) in decimal number. According to observations made by [4], the values of Q and R were taken as 10 and 1, respectively. Later another researcher, Salgado et al. (2000) took values of Q and R as 9 and 0.49, respectively [5]. Jayant [6] studied the characteristics of Bangalore sand and provided empirical relations mentioned as follows [6]:

$$\varphi_{\mathrm{p}} = \varphi_{\mathrm{cv}} + 0.932\psi_{\mathrm{p}} \quad (5)$$

$$\varphi_{\mathrm{p}} = \varphi_{\mathrm{cv}} + 3.5\,\mathrm{IR}\ \text{for plane-strain condition} \quad (6)$$

Strength–dilatancy relationships of almost 17 types of sands were extensively studied by Bolton and other researchers.

For the present experimental study, slag sand is taken as study material. The scarcity of natural building materials has necessitated the use of alternative materials. In most applications requiring fine aggregate, slag sand is used and is especially suitable for structural fills and embankments, road base layers, asphalt hot mix, and flowable lightweight fill.

The established strength–dilatancy relation plays an important role in:

1. modelling of stress–strain behaviour of slag sand
2. the design of shallow foundation placed on slag sand fill
3. slope stability calculation of embankment made up of slag sand.

3 Materials and Methods

3.1 Materials Description

Iron ore, fluxes (dolomite and/or limestone), and iron scrap along with coke or fuel are fed into a blast furnace during the production of iron. Carbon monoxide is released as the coke undergoes combustion, which reduces the ore to molten iron. Molten slag is floating on the upper surface of molten iron. Molten iron is channelised to a vessel after separating it from molten slag using a weir. Cooling and solidification of molten slag are done by rapid water quenching. This process leads to the formation of sand size fragments with friable material very similar to clinker called slag sand. The slag sand used in this analysis was collected from the JSW Steel Plant, Toranagallu, Bellary district.

3.1.1 Properties of Slag Sand

The slag sand sample selected for the present study comprised sub-angular grains and was found to be poorly graded as per the Indian Standard [7]. The properties of chosen slag sand sample are mentioned in Table 1. The slag sand sample used in the present study is shown in Fig. 1. The scanning electron microscope photographs of slag sand taken from NITK, Surathkal are shown in Figs. 2 and 3.

Table 1 Properties of slag sand

Parameter		Slag sand
D_{10} (mm)		0.25
D_{30} (mm)		0.41
D_{60} (mm)		0.57
Coefficient of uniformity, C_u		2.28
Coefficient of curvature, C_c		1.18
Specific gravity, G		2.654
% Uncompacted volume of voids		41.48% (Sub-angular)
Relative density, I_d %	20%	14.58 kN/m^3
	50%	15.27 kN/m^3
	80%	16.03 kN/m^3

Fig. 1 Slag sand sample

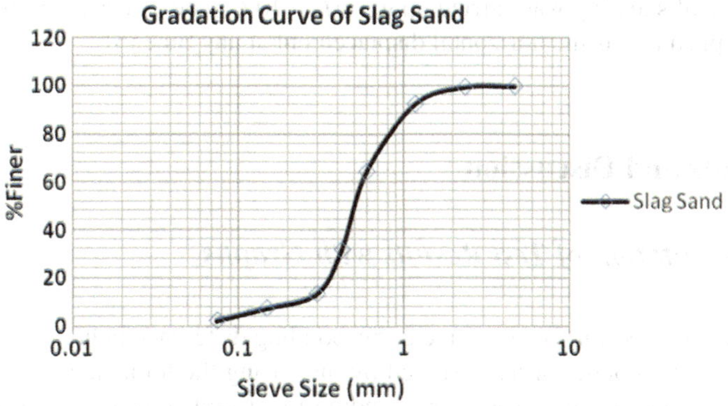

Fig. 2 Gradation curve

Fig. 3 A microscopic view of a portion of the slag sand sample by SEM

3.2 Direct Shear Tests

The slag sand sample selected for the study was subjected to a series of direct shear tests under five different vertical effective normal stresses between 50 and 400 kPa. Three distinct density indices were maintained, namely, 20% for loose state, 50% for medium dense state, and 80% for the very dense state. The required density was achieved by either raining the material from constant height (to achieve a loose to medium dense state) or by a tamping technique using a fixed number of blows (to achieve dense–very dense state) [3]. A shear box of size 60 mm × 60 mm × 25 mm was used for the present study. A uniform strain rate of 0.05 mm/min was maintained. The process of shearing was carried up to $u/H = 40\%$, where H is the initial height of the sample and u is the horizontal displacement at any time [3].

4 Results and Discussion

4.1 Presentation of Test Results with Graphs

The magnitude of shear force and the corresponding difference in height of the slag sand sample was continuously observed by increasing the horizontal displacement at regular interval of time. The corresponding test results conducted on slag sand sample are indicated in Figs. 5, 6 and 7 in terms of (i) variation in stress ratio (Ph/Pv) with shear strain (u/H), and (ii) variation in volumetric strain (v/H) with u/H, where Ph and Pv are the magnitude of horizontal and vertical force, respectively [3]. From these plots the values of friction angles (φ) and dilatancy angles (ψ) were determined using the following expressions [3]:

$$\varphi = \tan^{-1}(\text{Ph}/\text{Pv}) \tag{7}$$

$$\psi = \tan^{-1}(\delta v/\delta u) \tag{8}$$

In the above expressions, φ_p and ψ_p are the peak friction and peak dilation angles, respectively. Figure 9 shows φ_p and ψ_p variations with normal stress (σ_v) for the chosen slag sand. Critical observations made by researchers [3, 7] from a similar study were referred for Figs. 5, 6, 7 and 10.

4.2 *Correlation Between φ_p and ψ_p*

A linear relationship is established between peak friction angle and peak dilation angle as shown in Fig. 8. As a critical state friction angle (φ_cv), the intercept produced corresponding to the zero-dilation state is taken. It is a unique parameter, which remains independent of density, stress level, and type of test conducted. It depends only on the grain size and minerals comprising the granular material. The values φ_p were plotted against corresponding values of ψ_p, for different values of density index (DR) and σ_v values.

The relationship between ψ_p and φ_p is better describable by the following equation. All data points are shown in Fig. 8.

$$\varphi_\text{p} = \varphi_\text{cv} + 1.393\psi_\text{p} \tag{9}$$

Figure 8 shows that the value of φ_cv for the selected sample of slag sand equals 27.40°.

From Fig. 8 and regression analysis it is evident that Eq. (9) is suitable to express the correlation between shear strength and dilatancy of slag sand with $R^2 = 0.88$, where R^2 is the coefficient of correlation. Hence, from the regression of φ_p on ψ_p, it is observed that both the φ_p and ψ_p values are positively correlated. From Fig. 8, it is found that $\varphi_\text{cv} = 27.40^\text{o}$ (i.e. stress ratio Ph/Pv = 0.52). Also, an achievement of the critical state is observed close to Ph/Pv = 0.52 in all tests at a shear strain range of 35–40%, which is seen in Figs. 5a, 6a, and 7a.

4.3 Correlation Between φ_p and σ_v

The plot of φ_p versus σ_v is shown in Fig. 9. It is evident that σ_v increases as the φ_p value reduces. The following Eqs. (10) and (11) are given by Bolton [4] and Kumar [6], respectively. Kumar [6] took the values of Q and R as 10 and 1, respectively.

$$\varphi_p = \varphi_{cv} + 5\,\mathrm{IR} \tag{10}$$

$$\varphi_p = \varphi_{cv} + 3.5\,\mathrm{IR} \tag{11}$$

In the above expressions, IR is termed as dilatancy index, whose magnitude is dependent on density index (DR) and effective vertical normal stress (σ_v). The expression for determining IR is mentioned below:

$$\mathrm{IR} = \mathrm{DR}(Q - \ln(\sigma_v)) - R \tag{12}$$

σ_v expressed in kPa and DR in decimal.

The following relationship is suitable for the chosen slag sand sample:

$$\varphi_p = \varphi_{cv} + 3.204\,\mathrm{IR} \tag{13}$$

where $\mathrm{IR} = \mathrm{DR}(Q - \ln(\sigma_v)) - R$.

σ_v expressed in kPa and DR in decimal.

Figure 10 shows the prediction of ($\varphi_p - \varphi_{cv}$) by different formulae given by various researchers against measured values of ($\varphi_p - \varphi_{cv}$) for the granular materials. It is clearly indicated from the graph that the recommendations proposed by Salgado [5] and Bolton [4] falls on the higher end. On the other hand, the estimation from Eq. (13) seems to be better. Additionally, from the regression analysis, it is found that the value of coefficient of correlation, i.e., R^2 is equal to 0.963. From this, it can be interpreted that there exists a strong positive correlation between the estimated values ($\varphi_p - \varphi_{cv}$) and measured values of ($\varphi_p - \varphi_{cv}$) for the present data.

5 Conclusions

On the basis of shear tests conducted on slag sand at various density and stress conditions, correlations have been suggested close to those established by Kumar et al.

Fig. 4 SEM image of the surface texture of slag sand

[6], Salgado et al. [5] and Bolton [4]. The φ_p value can be estimated corresponding to any known values of φ_{cv} and DR at any magnitude of applied vertical stress. Furthermore, to estimate the value of φ_p an expression correlating ψ_p with φ_{cv} and φ_p is provided from the present experimental work. The expressions suggested fit the test findings well. On the basis of test data, it can be concluded that (i) reduction in σ_v values results in an improvement in both ψ_p and φ_p values. (ii) The value of φ_{cv} is independent of density and stress levels (Fig. 4).

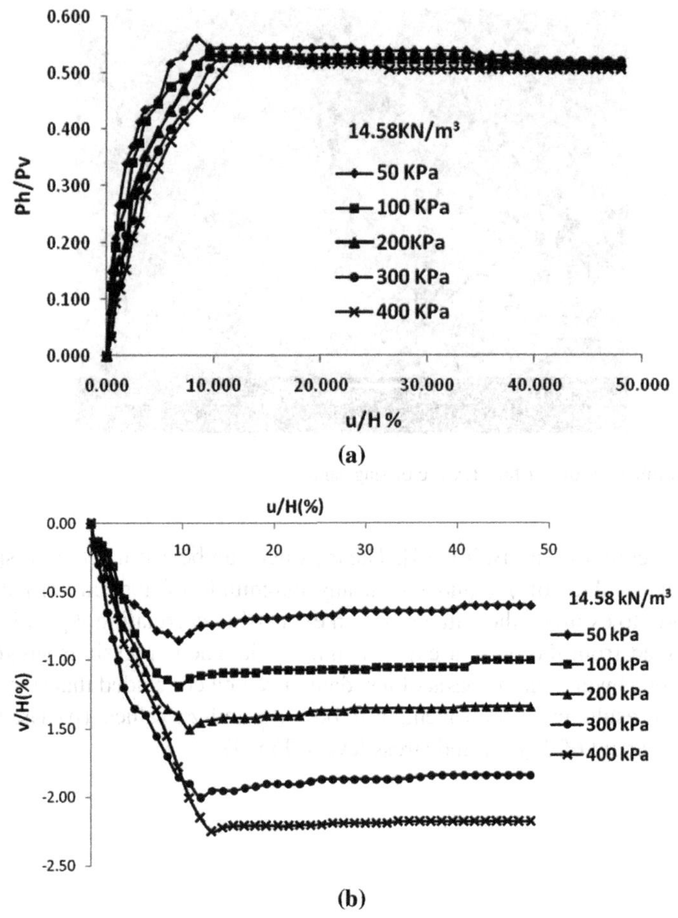

Fig. 5 For $\Upsilon = 14.58$ kN/m^3 slag sand, the observed variation of **a** stress ratio (Ph/Pv) with shear strain (u/H), and **b** volumetric strain (v/H) with shear strain (u/H)

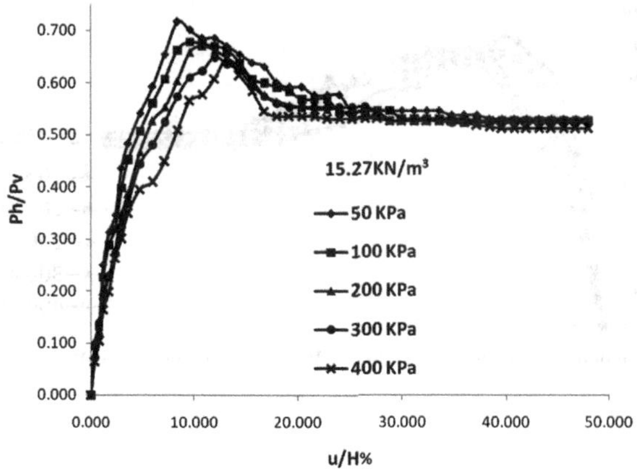

(a)

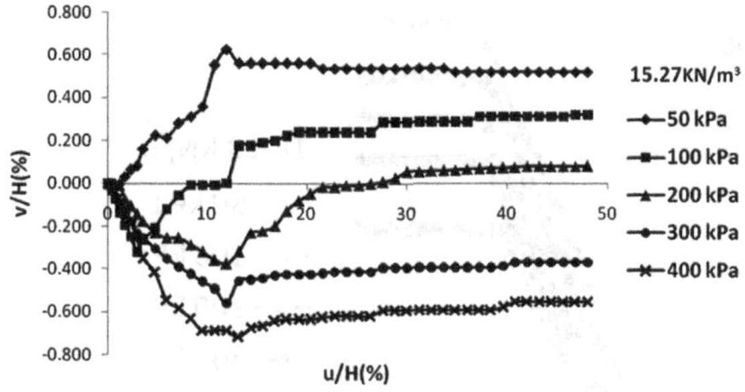

(b)

Fig. 6 For $\Upsilon = 15.27$ kN/m^3 slag sand, the observed variation of **a** stress ratio (Ph/Pv) with shear strain (*u/H*), and **b** volumetric strain (*v/H*) with shear strain (*u/H*)

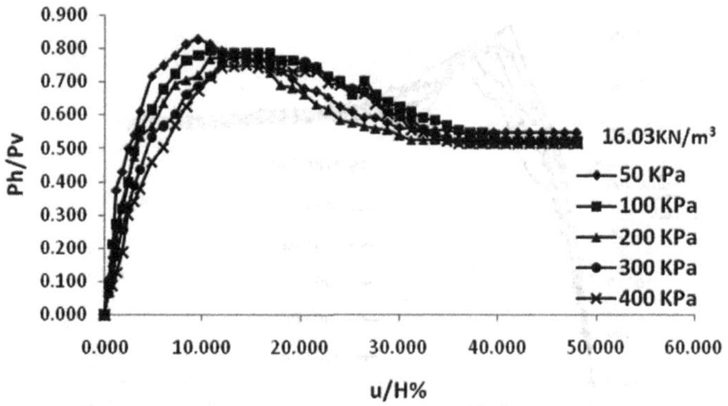

(a)

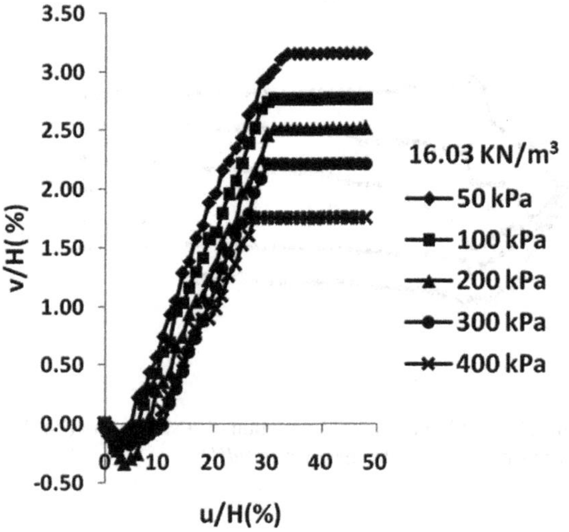

(b)

Fig. 7 For $\Upsilon = 16.03$ kN/m^3 slag sand, the observed variation of **a** stress ratio (Ph/Pv) with shear strain (u/H), and **b** volumetric strain (v/H) with shear strain (u/H)

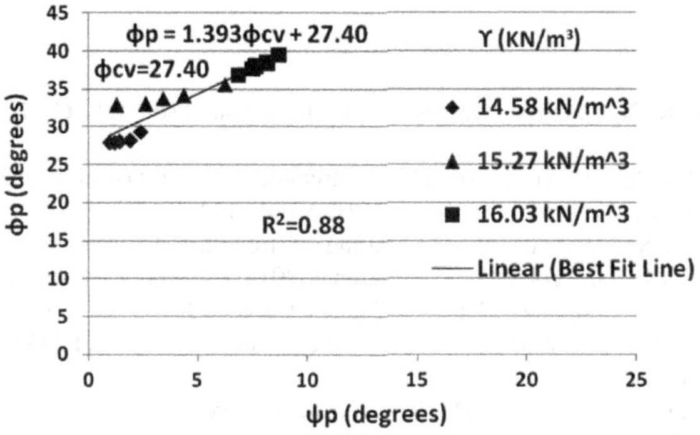

Fig. 8 Established relationship between peak friction angle (φ_p) and maximum dilation angle (ψ_p)

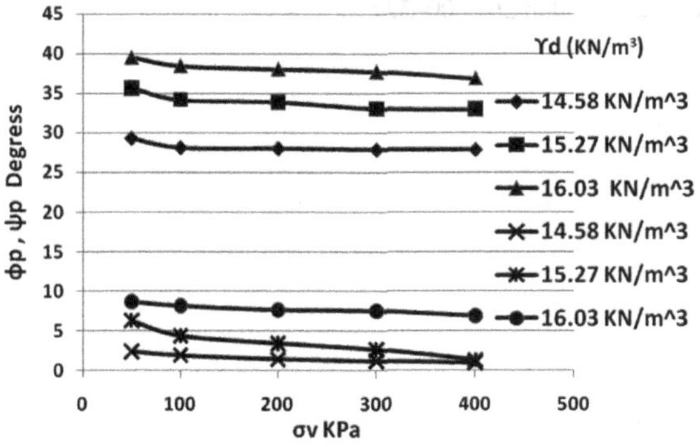

Fig. 9 The variation of φ_p and ψ_p with σ_v for slag sand

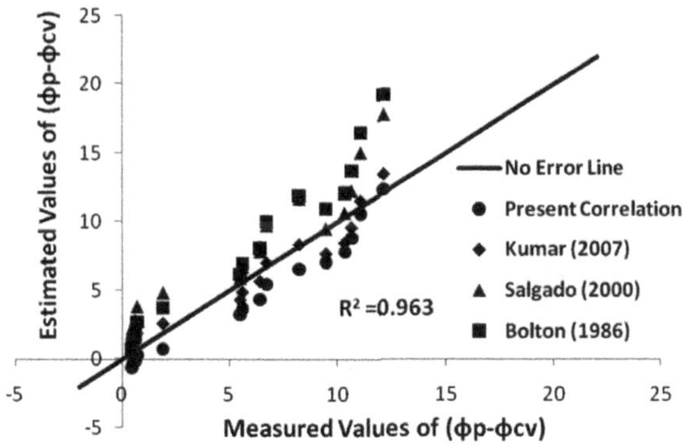

Fig. 10 The prediction of (φ_p–φ_{cv}) by different formulae against measured values of (φ_p–φ_{cv}) for all the tests for slag sand

References

1. Dolzyk-Szypcio K (2019) Direct shear test for coarse granular soil. Int J Civ Engi 17(12):1871–1878
2. Wang J-J, Guo J-J, Bai J-P, Wu X (2018) Shear strength of sandstone–mudstone particle mixture from direct shear test. Environ Earth Sci 77(12), Article 442
3. Raju KVSB, Khan S (2014) The effect of grading on strength and dilatancy parameters of sands. In: Proceedings of Indian geotechnical conference-2014, December 18–20, 2014, Kakinada
4. Bolton (1986) The strength and dilatancy of sands. Geotech J 36(1):65–78
5. Salgado, Bandini, Karim (2000) Shear strength and stiffness of silty sand. J Geotech Geoenviron Eng, ASCE, 451–462
6. Kumar J, Raju KVSB, Kumar A (2007) Relationships between rate of dilation, peak and critical state friction angles. Indian Geotech J 37(1):53–63
7. IS:1498-1970, Classification and identification of soils for general engineering purposes
8. Rowe PW (1962) The stress-dilatancy relation for static equilibrium of an assembly of particles in contact. Proc R Soc London 269:500–527

Parametric Study of the Slope Stability by Limit Equilibrium Finite Element Analysis

Prashant Sudani, K. A. Patil, and Y. A. Kolekar

Abstract Slope stability analyses of both natural and artificial slopes are an issue and need to be taken care of as they can affect both economic and infrastructure development. Water is the main factor that often leads to slope instability. Numerous studies are available in this direction, and it shows that the presence of the water harms the shear parameter, and it should increase the groundwater table, which further leads to loss of shear strength. The slope's stability depends on the shear parameter, cohesion, and angle of internal friction. If we can develop the relationship of this parameter with the instability, it can monitor the slope. This research's prime objective is to analyze the effect of the different controlling parameters on slope instability. The stability analysis has been done through the limit equilibrium finite element slope stability model. The slope was modeled in the commercial package GeoStudio SEEP/W, slope/w, and analyzes the stability. The stability factor was evaluated utilizing fully coupled flow deformation analysis compared with the various controlling parameters such as cohesion and angle of internal friction. For slope stability analysis, slopes were modeled with only one type of soil called the homogeneous fill. The influence of each parameter on slope instability was analyzed, and the results are discussed in this paper. It was also studied whether the relationship between the slope's instability and the controlling parameter existed. The research output shows that all chosen parameters' value negatively affects slope stability if their variations were favorable.

Keywords Slope stability · Landslide warning · Finite element analysis

P. Sudani (✉)
College of Engineering Pune, SPPU, Pune, India
e-mail: sudaniprashant93@gmail.com

K. A. Patil · Y. A. Kolekar
Civil Engineering Department, COEP, Pune, India

M. S. Ranadive et al. (eds.), *Recent Trends in Construction Technology and Management*, Lecture Notes in Civil Engineering 260,
https://doi.org/10.1007/978-981-19-2145-2_58

1 Introduction

Rainfall-induced slope instability is widespread worldwide, and it depends upon the geology, topography, and geotechnical properties of the slope strata; instability leads to several hazardous effects on human beings and infrastructure development. There are numerous cases of massive destructive landslides present in the literature worldwide, triggered by intense rainfall [9–11, 13].

In a tropical country like India, heavy rainfall after the hot and humid climate condition leads to the formation of the residual soils in a state of unsaturated previously due to lower groundwater table in the summer season [6]. In unsaturated soil, the suction pressure is present, which increases the slope's shear strength by suction effect in the slope's soil strata. During the rainfall, precipitation water is infiltrating into the slope geometry through the voids of the soil, and it depends on the properties of the material forming the slope. Infiltrating water leads to saturating the ground, and hence the suction will be reduced soon; moisture has a significant effect on the shear parameter, which further results in lowering the shear strength and increase in the stress; the factor of safety is reduced. When saturation reached 100%, it would reduce the suction pressure to zero and increase positive pore water pressure [4, 5]. Fredlund found that matric suction is the main parameter responsible for the slope's instability [3]. The development of the instability through the formation of the slip surface depends on more than one factor, including slope geometry (height, width, and inclination of the slope), topography and morphology of the area, stratification of the slope, material properties of the soil consisting of the slope, effect of the moisture, geotechnical analysis, and mechanical behavior. Since the slope's failure resulting from the heavy rainfall is in massive debris fall, it can make colossal destruction to both life and properties. Thus the precise physical analysis for the slope stability is essential.

In the present study, by keeping various hazardous landslides (such as the Malin landslide of Maharashtra, which caused an entire village burial and reported 161 lives losses [1, 2, 8, 12], into the center, the controlling parameter was analyzed. According to Rahardjo et al. [7], the slope geometry and the initial water table determine the initial safety factor, and the actual failure conditions are much affected by rainfall characteristics and properties of the soils on the slope. Ering et al. also presented the rainfall slope instability phenomenon in his study by back analysis assuming the slope as a fully saturated and on the verge of the failure factor of safety. They concluded that the rainfall effect on the slope has significant losses in the shear strength which can be evaluated through shear parameter cohesion and angle of the internal friction. Hence here in this study, the extensive analytical analyses were made to find out the effect of this crucial parameter on the slope stability.

To analyze the factor of safety with varying the controlling factors, the soil's geometry and profile were very simple to complex in this research. Slope stability was analyzed using the limit equilibrium finite element analysis using the GeoStudio Seep/W, Slope/W module. For the material model, the Mohr–Coulomb model was used in the analysis, and for stability analysis, phi-ci reduction analysis was employed for better results. The slope was modeled and analyzed by adopting two different cases, often presented in the field. In the first case, the entire slope was made up of one material, a so-called homogeneous slope, which can help stimulate the slope's result. While on the second case, the slope was modeled in the two-layer with different soil properties and analyzed. For analysis purposes, the slope stability controlling parameters, i.e., cohesion and angle of internal friction, were varied and the slope instability effect was observed.

2 Slope Geometry and Material Properties

In this study, slope geometry was used, as shown in Fig. 1. The inclination of the slope 26.56° (2:1) is considered. The height of the slope is 60 m. The groundwater table was encountered at the base, which was then varied to the top of the slope to analyze the groundwater table's effect on the slope instability. The Morgenstern-price method is used to analyze the slope stability as this method can simulate the effect of both vertical and horizontal interaction force between the slice.

The surface soil's behavior was modeled as linear and elastic–plastic soil material using the Mohr–Coulomb material model.

The surface soil behavior was modeled as linear and elastic–plastic soil material using the Mohr–Coulomb material model and analyzed using soil parameters derived from undrained tests. The dilatancy effect on the slope was not considered in the physical model simulation and soil dilation angle is negligible as per the empirical relationship ($\Psi = \phi - 30°$).

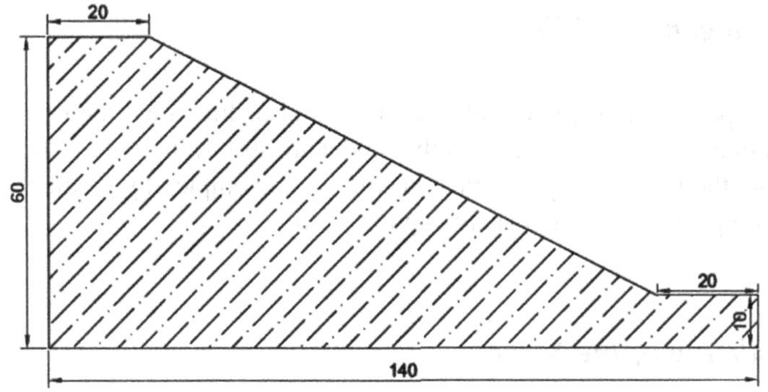

Fig. 1 Geometry of the slope model and initial groundwater condition

Table 1 Properties of the soil were used in the study

Soil type	γ_{unsat}	γ_{sat}	K_x	K_y	E	v	C	ϕ
Clayey loam	15.8 kN/m^3	17.8 kN/m^3	0.4 kN/m^3	0.4 m/day	2000 m/day	0.33 kN/m^2	25 kN/m^2	40°

2.1 Finite Element Method (FEM)

Finite element analysis is a technique used to solve differential equation problems in engineering science (Vinod et al. 2020). Finite element analysis is best suited for mechanical and physical analysis as it can solve the problem by discretizing the large elements into smaller ones and then analyze them. FEA is a computational method used for engineering design and analysis purposes. The method used the mesh generation, which was then calculated based on the program coded with FEM. The basic theory in the analysis of finite elements is (1) discretization of the area being scrutinized in FE. These discrete constituents are thought to be interconnected exclusively at the joints, which are called nodes. (2) The use of familiarizing polynomials defines a field variable's difference within an element (Vinod et al. 2020). For slope stability analysis through FEM, there are two approaches to deal with the breakdown of the instability. The first is by building a gravity load, and the other is to diminish the quality characteristics of the soil mass, also known as ci-phi reduction techniques. The "ci-phi reduction approach" for slope stability analysis is received in a fantastic commercial package known by GeoStudio.

2.2 Homogeneous Fill

Here the slope was used to be with the same kind of fill called the homogeneous fill. The properties of the soil used in the model are shown in Table 1. The analysis started from the initial soil properties, making them complex by gradually varying the controlling parameters and analyzing the yield.

2.3 Modeling of the Slope

A sample study has been carried out on assumed slope geometry. The slope was modeled with a homogeneous fill of the clayey loam. The slope's current condition was determined using limit equilibrium finite element slope stability modeling through the SLOPE/W program. The influence of the stability with the cohesion, angle of internal friction, and groundwater table soil was also examined.

The modeling sequence includes the path like geometry input—material definition—mesh generation—boundary condition—loading condition—construction sequence definition—analyze case—check results.

3 Result and Discussion

The entire study here divides into four cases, including (i) cohesive soil would analyze where the angle of internal friction would be zero and would observe the effect of the cohesion on the slope instability, (ii) frictional soil where cohesion would be made zero and would observe the effect of the friction angle on the slope stability, (iii-a) cohesive frictional soil with constant cohesion and varying the frictional angle, (iii-b) cohesive frictional soil with constant frictional angle and varying the cohesion, and (iv) effect of the rainfall simulation on the slope.

3.1 Case 1: Cohesive Soil, Where $\Phi = 0$ and Different Values of Cohesion

In this case, cohesive soil is modeled in the SLOPE/W to know the effect of the cohesion on the slope stability. Here, we used to vary the cohesion from 45 to 15 kN/m^2. The results of the simulation are shown in Table 2 and Fig. 2.

Observation of case one where the cohesion effect was analyzed, and the result is shown in Table 2. It shows that as the value of the cohesion reduced from its base value and the slope's safety factor got disturbed significantly, which often seems during the rainfall infiltration.

Table 2 Effect of the cohesion on the slope stability

S. No.	ϕ	C	FOS
1	0	45	1.384
2	0	40	1.231
3	0	35	1.077
4	0	30	0.922
5	0	25	0.772
6	0	20	0.615
7	0	15	0.462

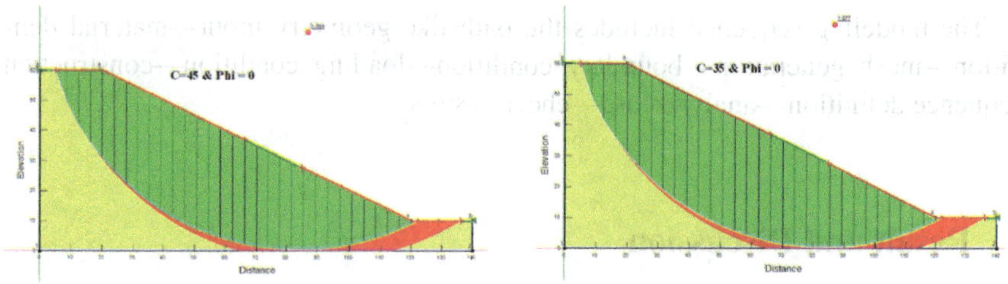

Fig. 2 SLOPE/W output for case, cohesion = 45 kN/m^2 and $\phi = 0$ (Left), and cohesion = 35 kN/m^2 and $\phi = 0$ (Right)

3.2 Case 2: Frictional Soil, Where C = 0 and Varying Frictional Angle

In this case, cohesive soil is modeled in the SLOPE/W to know the frictional angle's effect on the slope stability. Here, we used to vary the value of the frictional angle from 45° to 15°. The results of the simulation are shown in Table 3 and Fig. 3.

The output of the analysis shows that the internal friction angle has a remarkable effect on slope stability. Results show that the variation of the slope's safety factor with the variation of the friction angle is linear.

Table 3 Effect of the friction angle on the slope stability

S. No.	*C*	ϕ	FOS
1	0	45	2.002
2	0	40	1.680
3	0	35	1.402
4	0	30	1.156
5	0	25	0.772
6	0	20	0.729
7	0	15	0.536

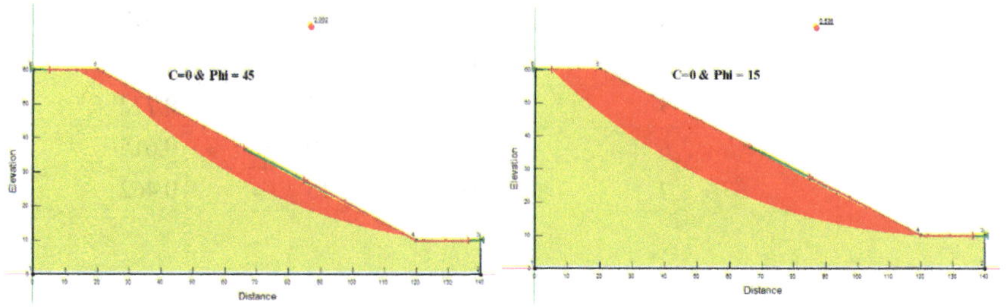

Fig. 3 SLOPE/W output for case, cohesion = 0 and $\phi = 45°$ (Left), and cohesion = 0 and $\phi = 15°$ (Right)

3.3 Case 3 (a): Cohesive Frictional Soil with Constant Cohesion and Varying Frictional Angle

In this case, frictional cohesive soil is modeled in the SLOPE/W to know the frictional angle's effect on slope stability. Here, in this case, the cohesion's value would make constant as 25 kN/m^2. Here, in this case, we used to vary the value of the frictional angle from 45° to 0°. The results of the simulation are shown in Table 4 and Fig. 4.

The result of variation of the friction angle in cohesive frictional soil has a severe effect on the slope instability compared to the cohesion.

Table 4 Effect of the friction angle on the slope stability in cohesive frictional soil

S. No.	*C*	ϕ	FOS
1	25	45	3.474
2	25	40	3.077
3	25	35	2.723
4	25	30	2.401
5	25	25	2.110
6	25	20	1.842
7	25	15	1.591
8	25	10	1.316
9	25	00	0.772

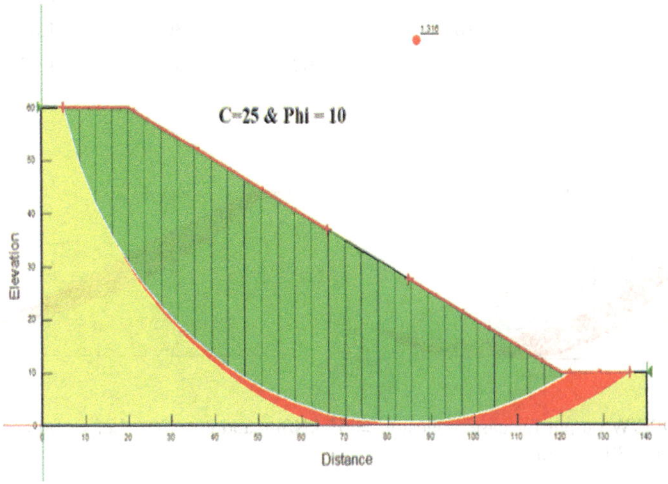

Fig. 4 SLOPE/W output for case, cohesion = 25 kN/m^2 and $\phi = 10°$

3.4 Case 3 (B): Cohesive Frictional Soil with Constant Frictional Angle and Varying Cohesion

In this case, frictional cohesive soil is modeled in the SLOPE/W to know the effect of the cohesion on the slope stability with C–ϕ soil. Here, in this case, the value of the friction angle would make constant as 30°. We used to vary the cohesion value from 55 to 0 kN/m^2. Results of the simulation are as shown in Table 5 and Fig. 5.

It has been seen that the cohesion can change negatively after coming into contact with the water at a faster rate than the friction angle. Hence, in this case, the cohesive frictional soil is considered for the study with varying cohesion, and it is observed that the safety factor makes their value from nearly five to less than one.

Table 5 Effect of the cohesion on the slope stability in cohesive frictional soil

S. No.	C	ϕ	FOS
1	55	30	3.460
2	50	30	3.283
3	45	30	3.106
4	40	30	2.929
5	35	30	2.751
6	30	30	2.574
7	25	30	2.401
8	20	30	2.221
9	15	30	2.026
10	10	30	1.825
11	00	30	1.156

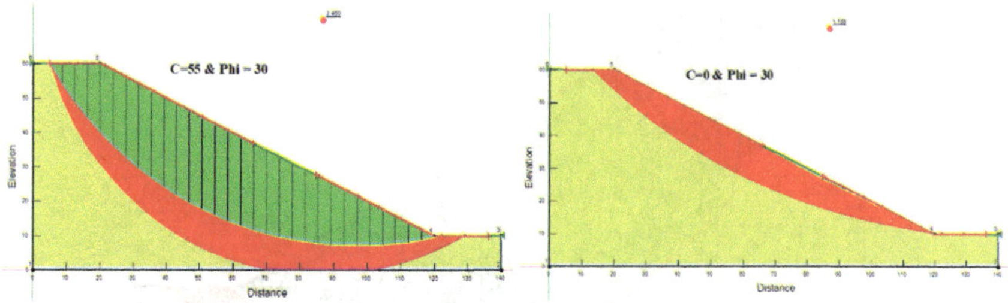

Fig. 5 SLOPE/W output for case, cohesion = 55 kN/m^2 and ϕ = 30° (Left), and cohesion = 0 kN/m^2 and ϕ = 30° (Right)

3.5 Case 4: Rainfall Simulation in the Slope

Here in Case 4, we simulate the rainfall in the slope by applying the flux on the slope's surface at a 9 and 90 mm/h rate. The slope is modeled with the composite fill with geometry, as shown in Fig. 6. A two-layered slope profile is used to simulate the rainfall infiltration effect on the slope stability. For the hydrological simulation of the slope, the SEEP/W module of GeoStudio is used, and then SLOPE/W is used to keep SEEP/W as the parent simulation so that hydrological modeling can be incorporated into the slope stability analysis. The soil properties of the modeled two-layered slope are shown in Table 6.

As shown in Fig. 6 of the modeled slope, different boundary conditions are used to analyze slope stability. Zero boundary flux is provided on the 9-2-1-6 node of the slope on that boundary, and the rain will not infiltrate. 5-4-3-8-9 is open to get the rainfall, so boundary flux with 9 mm/h is provided here. This can vary according to the expected rainfall and infiltration properties of the modeled slope. For the sleep surface entry and exit method, the GeoStudio is used to simulate.

Further, for the SLOPE/W module Morgenstern-price method is used to analyze the problem. SEEP/W, SLOPE/W combination yields the FoS for the composite two-layered soil as 1.710 and 0.846 for 9 mm/h and 90 mm/h rainfall simulation, respectively, which indicates the transformation of instability of the safe slope. The output of the same is shown in Fig. 7.

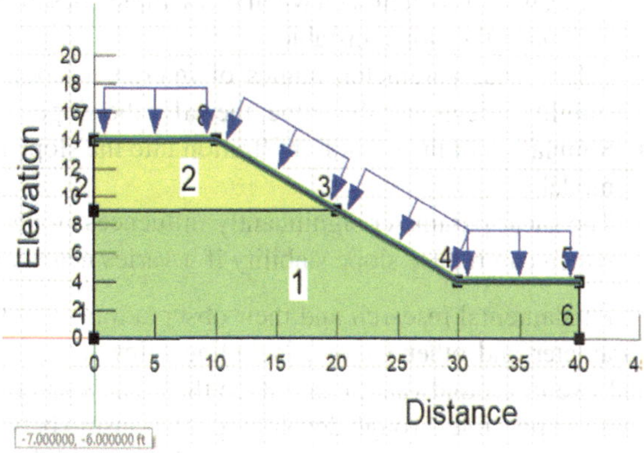

Fig. 6 Modeled slope geometry with rainfall simulation on the slope

Table 6 Material properties used for Case 4 rainfall simulation

Soil type	γ_{unsat}	γ_{sat}	K_x	K_y	E	v	C	ϕ
Upper layer	17	17.8	0.4	0.4	1500	0.25	4	20
Lower layer	18 kN/m^3	19 kN/m^3	0.4 kN/m^3	0.4 m/day	2500 m/day	0.23 kN/m^2	8 kN/m^2	25 Degree

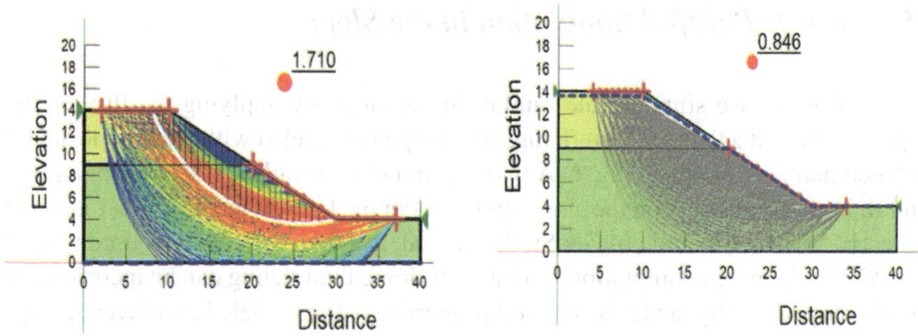

Fig. 7 Rainfall simulation SEEP/W, SLOPE/W output with 9 mm/h (Left) and 90 mm/h (Right) rainfall simulation

4 Conclusions

The parametric study for instability status of the land slope found helpful to understand and assess the effect of the stability governing parameters i.e., cohesion, angle of internal friction and rainfall. Results of the study represents the how the stability could vary once the shear strength parameters decay their values, which is seems often during saturation of the soil. Inference of the research are as presented below:

- The output of the analysis shows that as cohesion decreased from their undrained value, which is often seen when soil sample got saturated, the safety factor reduced and made the slope unstable.
- Other shear parameter, angles of internal friction also greatly influence slope stability; the higher the value, the safer the slope.
- Simulation of the rainfall infiltration into the slope model yields the hydrological model.
- The shear parameter significantly influences slope stability, especially cohesion, which can reduce slope stability if it varies negatively in the soil.

Fundamental research and their observation show that the variation of the shear parameter and other factors, including rainwater infiltration, saturation evolution, and rise in groundwater, can reduce the slope's safety. Monitoring of the landslide could be possible through developing relationship between the shear parameters, and saturation of the soil, which further could be possible to link with the stability of the landmass.

Acknowledgements We are thankful to AICTE-NDF Scheme (Application number 60400) for providing a research fellowship. We are also thankful to the College of Engineering Pune for providing a research facility.

References

1. Ering P, Sivakuma Babu GL (2016) Probabilistic back analysis of rainfall induced landslide—a case study of Malin landslide, India. Eng Geol 208:154–164
2. Ering P, Kulkarni R, Kolekar Y, Murty Dasaka S, Sivakuma Babu GL (2015) Forensic analysis of Malin landslide in India. IOP Conf Ser Earth Environ Sci 26(1)
3. Fredlund DG, Xing A (1994) Equations for the soil-water characteristic curve. Can Geotech J 31(4):521–532
4. Keefer DK, Wilson RC, Mari RK, Brabb EE, Brown Iii WM, Ellen SD, Harp EL, Wieczorek GF, Alger CS, Zatkint RS (1987) During heavy rainfall (NovEMBER)
5. Krahn J (2004) Stability modeling with SLOPE/W an engineering methodology
6. Rahardjo H, Li XW, Toll DG, Leong EC (2001) The effect of antecedent rainfall on slope stability. Geotech Geol Eng 19(3–4):371–399
7. Rahardjo H, Ong TH, Rezaur RB, Leong EC (2007) Factors controlling instability of homogeneous soil slopes under rainfall. 133(December):1532–1543
8. Sarvade SM, Sarvade MM, Khadatare PS, Kolekar MR (2017) 30/7 Malin landslide: a case study (October 2014)
9. Schuster RL, Fleming RW (1986) Economic losses and fatalities due to landslides. Bull Assoc Eng Geol 23(1):11–28
10. Schuster RL, Highland LM (2007) The third Hans Cloos lecture. Urban landslides: socioeconomic impacts and overview of mitigative strategies. Bull Eng Geol Env 66(1):1–27
11. Shah MV, Sudani P (2020) Strength and deformation characteristics of laterite rock with different rock matrix's using triaxial system, 85–102
12. Singh TN, Singh R, Singh B, Sharma LK, Singh R, Ansari MK (2016) Investigations and stability analyses of Malin village landslide of Pune district, Maharashtra, India. Nat Hazards 81(3):2019–2030
13. Thiebes B, Bell R, Glade T, Jäger S, Mayer J, Anderson M, Holcombe L (2014) Integration of a limit-equilibrium model into a landslide early warning system. Landslides 11(5):859–875

Model Testing on PET Bottle Mattress with Aggregate Infill as Reinforcement Overlaying on Fly Ash Under Circular Loading

Shahbaz Dandin, Mrudula Kulkarni, and Maheboobsab Nadaf

Abstract The consumed PET bottles which are readily cellular restrict lateral movement. Comprehensive studies surface every day with alarming statistics on the most pressing environmental issues. By recycling, an attempt is made to lay down the effective procedure for reinforcement in fly ash fills, for which the subsurface and mountains are excavated for the purpose of backfill in the infrastructure construction. The objective of this work is to know the practical behavior of fly ash (a coal combustion residue) backfill overlaying reinforcement of consumed water bottle mattresses which are customized with reasoning and bottle pockets filled with various aggregate grading. By placing 1000 and 500 ml PET bottle mattress on fly ash bed, a sequence of plate load tests would be performed. The varying parameters considered for these series are four varying cellular infill aggregate grading (A_d, A_3, A_6, and A_{12}); two different PET bottle cell aperture sizes (P_d 50 mm and P_d 73 mm); and three varying sizes of circular footing plate for loading ($B' = 80$, 160, and 240 in mm). A total of 36 trial procedures are noted to analyze the retort of the system. As a result, the reinforced fly ash system indicated a higher bearing capacity than the non-reinforced condition by 432%; the optimum ratio of the diameter of the cell (P_d) to the aggregate grading (A) is $P_d/A = 16$, and also it is known that the effective ratio of the diameter of the cell (P_d) to the width of footing (B') values should be lower than 0.5 times the footing width (B') for achieving the highest possible stability and reliability under vertical loading.

Keywords Plate load test · PET bottle mattress · Fly ash · Cellular reinforcement · Aggregate infill

---

S. Dandin (✉) · M. Kulkarni
School of Civil Engineering, MIT World Peace University, Pune, India
e-mail: shahbaz.dandin@mitwpu.edu.in

M. Nadaf
Department of Civil, ADCET, Astha, Sangli, India

M. S. Ranadive et al. (eds.), *Recent Trends in Construction Technology and Management*, Lecture Notes in Civil Engineering 260,
https://doi.org/10.1007/978-981-19-2145-2_59

Abbreviations

CD	Consolidated drain
PET	Polyethylene terephthalate
P_d	Diameter of the cell pocket
ϕ	Angle of internal friction
c	Apparent cohesion
A_d	Quarry dust as aggregate
A_3, A_6, A_{12} (3/6/12)	Size of aggregate in mm
PLT	Plate load test
F	Footing plate size
B	Width of the PET bottle mattress reinforcement
D_b	Buried depth from a top surface of cell mattress
Q_{ult}	Ultimate bearing capacity
δ_{ult}	Ultimate deformation
B'	Width of the footing plate
A	Aggregate

1 Introduction

The reinforcement of backfill material in various civil engineering projects like retaining wall, low-lying areas, plinth filling, and reinforced earth wall are experienced based on the performance, stability, and usage in practice, as well as in prototype model research. The consumption of natural resources for earth fills, majorly in infrastructure construction, is leaving a serious impact on the environment [28], Nevertheless, the utilization of fly ash as a concern is to be effectively replaced with a conventional system of backfilling proposal. The performance is based on the following factors, which include the material used, type of installation, mode of practice, drainage pattern, and type of loading. Previous research works worked on the effects, behavior, and bearing capacity of fly ash backfill, with and without providing reinforcement [2, 12, 15, 16, 27, 30, 31]. The alarming statistical data of water bottle generation is booming with greater demand [13]. The overflowing landfill capping system with millions of tons of discarded water bottles has been an environmental hazard. The usage and recycling have been effectively settled up with design procedures as of the conventional procedure of earth filling for stability. Among all the reinforcement concerning function, the geocell reinforcement is more advantageous and beneficial; it has reduced settlement and has increased the bearing capacity [28]. Such honeycomb structured reinforcement is to be replaced by a consumed water bottle mattress as the functional role is similar. Some researchers have worked on the laboratory experimentations on large and small-scale model studies on the parameters like settlement, bearing capacity resistance, and stability with various procedures laying down, like the studies by Ghosh and Subbarao [1], Kim and Prezzi [22], Xu and Shi [17] experimentally remarked that the fly ash as a mass can be worked in geotechnical engineering application.

Choudhary et al. [12] provided geogrid reinforcement in fly ash slope under strip loading. The different series of patterns of reinforcement by geogrid were placed with varying layer thickness within fly ash with respect to burial depth for varying footing widths. The load versus settlement characters were observed experimentally and concluded that the increase in geogrid layers within the fly ash fill enhances the bearing capacity and sustains the fly ash slope. Moreover, the alternative reinforcement was experimented with by the author Nadaf et al. [27] with plastic bottle arrangement in fly ash slope by considering 60° to the horizontal. The parameters considered were coverage ratio, varying edge distance with a series of strip footing widths and cell heights of 10, 20, and 30 mm, and wall thickness of the cell was 0.15 mm of the plastic bottles. The fly ash slope was placed with different burial depths and layers of bottle strips were placed by considering the coverage ratio for achieving the desired density of fly ash bed with steel strips mixture. The vertical strip loading was experimented; results and observations were made; and bearing capacity incremental behavior was observed with an increase in coverage ratio, reducing the spacing of burial depth. A 30–60% settlement reduction was observed compared with the unreinforced fly ash slope. Tavakolo Mehrjardi et al. [33] experimentally concluded that the subgrade with weak strength over overlayed reinforcement ends up with higher settlement exerting more tension through the reinforcement plane. Dutta and Mandal [15] examined the efficiency of unreinforced and reinforced fly ash backfill that underlays soft clay, with particular emphasis on the behavior of the fill material. Different types of techniques were adopted in the tank study; fly ash column was embedded into marine clay with fly ash column perforated with plastic encasement of 105 mm diameter with inner wrapping with jute geotextile (end bearing column). A layer of woven jute geotextile is placed as a function of separators between marine clay and fly ash fill in a geocell mattress made up of plastic bottles and experimented with impinged to vertical loading of a circular plate of varying sizes. The study determines maximum efficiency for the group column situation and efficiency of triangular and square patterns. They observed that the four columns had good footing capacity than the single and three columns. The literature reviewed indicates that encasing the stone columns with suitable geosynthetic reinforcement enhances its capacity to a greater extent. Ram Rathan et al. [30] have noted that minimizing the coverage ratio of reinforcement reduces load-bearing capacity. Fly ash as cell infill material is less efficient as it causes larger backfill material sliding down due to its smaller grain size alongside the cell walls during settlements, which in turn restricts the effectiveness of bottle mattress on increasing bearing capacity [24]. The consumed water bottles should also be considered for laboratory testing [15].

Based on the above-stated views on the reuse of fly ash and PET bottle in mass quantity, the following areas have been identified and considered in the present research. The response of plastic bottle mattress reinforced fly ash bed is directly affected by coverage area, so laying down plastic bottle mattress (100% coverage) is considered in the present research. The primary study aims to prove that the behavior of compacted fly ash bed ($H = 750$ mm) subjected to vertical loading with varying sizes of footing plate ($B' = 80$, 160, and 240 in mm); to examine the compression behavior of fly ash bed overlying customized plastic bottle mattress with varying ratios of cell diameter (P_d 50 mm and P_d 73 mm) to the aggregate size (A_d, A_3, A_6, and A_{12}) and also to decide the ratio of the diameter of a plastic bottle (P_d) to the width of the footing (B').

In view of the fact that, as claimed by Maghadas et al. [24], the ratio of burial depth of geocell to the loading plates diameter was considered 0.1 and Tavakoli et al. [34] claimed that the loading plate's diameter was selected as 0.25 times the width of geocell (in our case it is B) which is considered in this present research.

2 Test Article

2.1 *Fly Ash*

Class "F" Grade II fly ash is considered for this experimental study. Fly ash was collected from Nashik thermal power station, Eklahare, Maharashtra, India. The physical characteristics and chemical properties of the fly ash are summarized in Tables 1 and 2.

Table 1 Physical and mechanical characteristics of fly ash

Properties		Value
Color		Tan-brownish
Specific gravity (G_S)		2.1
Fine sand size particles		20%
Silt size particles		70%
Clay size particles		10%
Liquid limit; W_L		26%
Plasticity; W_P		–
Dry unit weight; γ_d		13.5 kN/m^3
Optimum moisture content; OMC		18.5%
Classification	IS 3812–1981	Grade II
	ASTM C618	Class F
The angle of shearing resistance; Triaxial-CD test (φ)		12.04°
Apparent cohesion; Triaxial-CD test (c)		7 kN/m^2

Table 2 Major chemical characteristics of fly ash (X-ray fluorescence)

Components	Measure (%)
SiO_2	56
Al_2O_3	25
Fe_2O_3	8
CaO	2
LOI	4

2.2 Cellular Reinforcement-Consumed PET Bottles

The polyethylene terephthalate bottles considered for this study have the cells of diameter and height of 50/100 and 73/150 in mm, mentioned aspect ratio between 2 and 2.2 (500 and 1000 ml mineral water bottle), respectively, and are collected from our university garbage yard. There are various diameters of water bottles which are available for packed mineral water from manufacturers. Among all the sizes most widely consumed are of 500 and 1000 ml.

Figure 1a, b illustrates a view of a customized PET bottle, individually and a group of bottles forming a mattress with a pocket diameter of 50 mm (P_d 50); P_d is the "diameter of the pocket". The aspect ratio of the cell pocket's diameter to the medium grain size varies from 3.4 to 36.7. To make a cellular mattress, plastic water bottles of 500 and 1000 ml capacity with 50 and 73 mm diameter are used in the tank study. These bottles were cut from the cap side with a height of 100 mm and

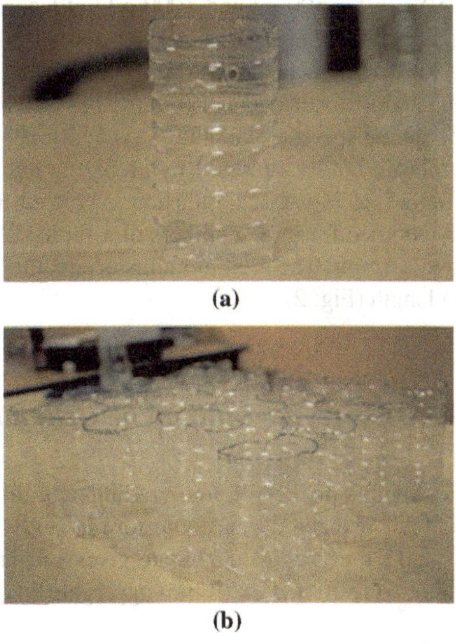

Fig. 1 **a** A view of the customized single plastic bottle with aspect ratio = 2. **b** A view of the customized PET bottle mattress

Table 3 Characteristics of a plastic water bottle

Reinforcement	Property	Average value
PET bottle (strip of 15 × 2 cm) ASTM D638-2014	Tensile strength (*ISO 527-1; Dutron testing machine*)	150.57 MPa
	Tensile modulus	2349.63 MPa
	Wall thickness	0.25–0.30 mm

Fig. 2 Testing setup of PET bottle strip of 15 cm × 2 cm in *Dutron testing machine*

150 mm for 500 ml and 1000 ml bottles, respectively. Holes of diameter 5 mm at equal spacing intervals from the side and bottom face were made with the help of a soldering machine for drainage purposes. These bottles were joined together by a manual 26/6 stick stapler to form a mattress. The engineering properties of the PET bottle are specified in Table 3. The ultimate tensile strength and tensile modulus of a plastic bottle are tested by preparing the strip of 15 × 2 cm, firmly fixed in the frame (see Fig. 2), and tension is created in Dutron testing machine. The reading is recorded for three numbers of samples till failure, and the mean value was taken into account. The total modulus is equal to (tensile strength × total length)/change in length (Fig. 2).

2.3 Aggregate Infill in the Cellular PET Bottle Pockets

For cellular pocket fill, four grading of uniform aggregate with the size of A_d, A_3, A_6, and A_{12} is taken into account for the study, underlaying fly ash bed. Further, more than 12 mm size of aggregate gives self-lateral restrain because of a higher void ratio, so restricted to 12 mm aggregate (A_{12}) as cellular fill material. The physical characteristics of these pocket infill materials are classified as per the Indian Standard Soil Classification System (ISSCS) and are described in Table 4.

A_d is stone quarry dust as aggregate; A_3 is $\leq$3 mm aggregate grading; A_6 is $\leq$6 mm aggregate grading; A_{12} is $\leq$12 mm aggregate grading (Fig. 3).

Table 4 Physical properties of aggregate used as infill in PET bottle mattress

Description	Aggregate quarry dust	Aggregate 3 mm	Aggregate 6 mm	Aggregate 12 mm
Specific gravity	2.6	2.52	2.43	2.33
Coefficient of curvature, C_c	0.62	1.24	1.03	0.93
Coefficient of uniformity, C_u	13.8	2.7	2.13	1.3
Medium grain size, D_{50} (mm)	–	4	5.9	12.4
Water content	–	–	–	–
Fractured particles (%)	–	90	88	85
Crushing value (%)	–	26	29	34
Soundness (%) $NaSO_4$	–	7	9	9

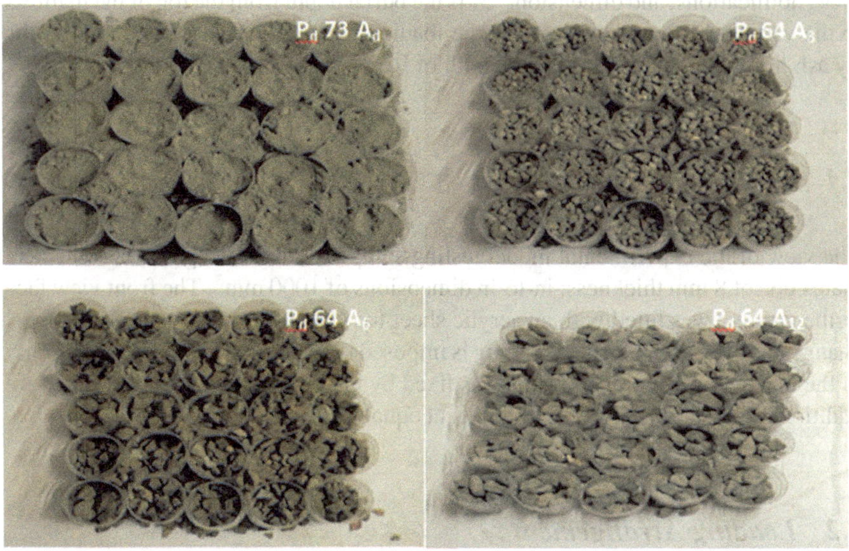

Fig. 3 A view of the plastic bottle mattress "P_d 73" with cellular pocket infill with 3, 6, 12 in mm and quarry dust aggregate grading

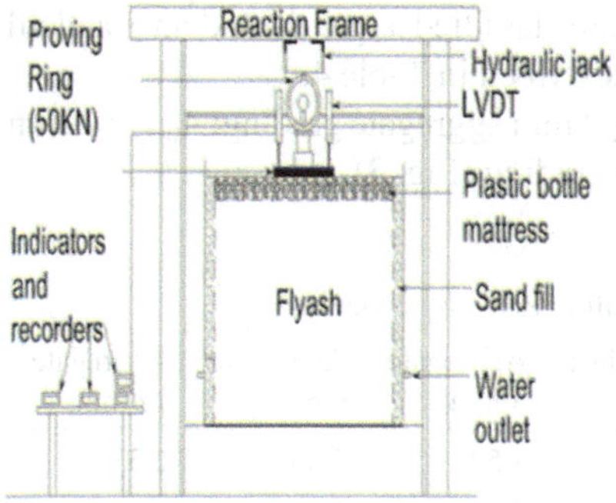

(a)

(b)

Fig. 4 **a** Schematic representation of the test setup with reinforcement (not to scale). **b** Model test tank experimental setup in the laboratory

3 Experimental Setup, Instrumentation, and Test Approach

Tankage was used to execute the experimental test series. It exemplifies a schematic view, notifications, and dimension; "z" is the burial depth from the top of the mattress to the vertical loading surface; B' is the loading plate width; H is the height of the fly ash fill in the tank; and B is the width of PET bottle mattress.

3.1 Geometry of the Tankage

The view of the experimental model testing setup is shown in Fig. 4a, b. Mild steel plate cube of 8 mm thickness; in-to-in dimensions of 1000 mm^3. The front view face of the tankage was fitted with an acrylic sheet for visual observation and behavioral change has been seen. Vertical loading is imposed till the failure to ensure the rigidity of the tankage. The complete cube was fixed by the bolting system. Sidewalls were additionally supported by steel columns at equal spacing.

3.2 Loading Arrangement

The reaction loading frame was fabricated, as shown in Fig. 4b. A hydraulic jack was supported by a reaction beam and a detachable setup was made on the jack for proving ring placement; in turn, the proving ring had been given a detachable connection to the square loading plate of 20 mm thickness. The three different sizes of circular loading plate were considered, i.e., 80, 160, and 240 in mm such that the vertical loading is imposed on the center of the tankage; the size of the circular loading plate was chosen such that the stress should be within the tankage, i.e., semi-infinite condition. Selig and Mckee [32] and Chummer [13] stated with their experimental analysis that in the foundation bed the failure wedge is extended over a distance of about 2–2.5 times the width of the footing from the center. The uniform vertical load is imposed via a hydraulic jack and the proving ring measures the applied pressure having the capacity of 50 kN; uniform incremental loading is maintained throughout the series of the test as per the laid procedure for plate load test (PLT) IS:1888-1982. In the entire reinforcement experiment series or pattern on fly ash fill, the incremental loading has increased to workable stress up to 179 kN/m^2, but it was not the same in all the testing program patterns. If the failure pattern was not observed, the deformation was restricted to 40 mm settlement or 0.25 times the diameter of the loading plate. The LVDTs were placed diagonally on the loading plate to avoid tilting of footing with an accuracy of 0.01%; due to the diagonal placing of LVDTs, tilt was easily known so that appropriate balance can be maintained and an average of deformation can be noted on each incremental loading.

3.3 *Fly Ash in the Tankage*

The tankage is filled with fly ash at the height of 750 mm, which is compacted in three layers of equal height with respect to an optimum moisture content of 18.5%. Each layer is compacted with the compaction effect of around 600 kJ-m^3 to achieve the maximum dry density of 13.5 kN/m^3 as per IS-2720 part7. In order to ensure maximum, a dry density proctor needle is used as obtained from the calibration curve (water content in % vs. penetration resistance in kN/m^2). All the sides of the tankage are filled with coarse material for drainage purpose and compaction is also achieved effortlessly. Once the desired height of 750 mm of fly ash fill is achieved, then the implementation of the plastic bottle mattresses is introduced.

3.4 *Bottle Pocket Infill*

Once the customized bottle mattresses are placed on a compact fly ash fill surface, covering the area of 0.75 m^2, the aggregates are poured into the bottle pocket and after the desired height of the aggregate is achieved, i.e., the height of the bottle plus 20 mm (z, the burial depth), the compaction is carried out by vibratory plate to achieve relative density (D_r) of 84%.

4 Test Program

The series of reinforcement patterns overlaying fly ash bed were performed on cellular pocket diameter, loading plate area, and aggregate grading to check the failure behavior on loading. Plate load tests of 36 trials are carried out as mentioned in Table 4. Examination parameters were a cellular bottle of diameters 50 and 73 mm; the aggregate of 3, 6, and 12 in mm and quarry dust, respectively, and loading plate size (B') is 80, 160, and 240 mm. Each trial is experimented with to investigate the influence of one parameter, whereas the diverse parameter continued to be persistent. Results were drawn on load–settlement. While conducting the test procedures few test trials were reiterated to ensure the reliability of the result. The present research considers the conclusions of Tafreshi and Dawson [24] that the buried depth of cellular reinforcement (geocell) with varying burial trials was analyzed and concluded that the buried depth should be 0.1 times of loading plate width and also considered the remark of the research by Tavakoli et al. [34] that the geocell mattress should be laid such that it must be four times the vertical loading plate width; the length of bottle cell size is considered as constant with length by diameter (L/D) ratio of 2:2.2 (Table 5).

5 Results and Discussions

The following results and analysis are related to the bearing capacity of the foundation in the fly ash bed without reinforcement, plastic bottle fly ash bed reinforced by P_d 50 and P_d 73. Figure 5 outlines the bearing capacity for each reinforced and unreinforced fly ash foundation which is on load–settlements. The feedback could be contracting due to poorly graded aggregate size and backfill surface having different diameters of loading areas.

Series 1: Fly ash fill as foundation bed without reinforcement. The basic findings of this experimental tank study are to know the behavior of fly ash subjected to vertical loading with varying sizes of footing plate, the fly ash as foundation bed, load versus settlement data were recorded. As predicted initially the behavioral observation report was similar. The increase in footing plate size and incremental values of Q_{ult} were observed, as shown in Fig. 6. The settlement was drawn down rapidly within the small loading of a maximum of 41.6 kN/m^2 for three cases. As the vertical deformation was observed the particles were showing lateral movement toward the phase of the tank which ensures and teaches us that confinement is needed to sustain higher loads. The above observation was seen as similar in all the three footing plate trials. The graph seen in Fig. 6 named load–settlement behavior for fly ash fill is to be represented as Q_{ult} by the use of Table 6 so that values of Table 6 are achieved. Let us recall Terzaghi's bearing capacity equation that as we increase the breadth of the footing, the bearing capacity increases. The results were not so shocking as they were predicted (Figs. 7 and 8).

Series 2: Fly ash fill as foundation bed overlaying P_d 50 PET bottle mattress with various aggregate grading within the cell. The coverage area of 100% laid as plastic bottle mattress was more efficient than reducing the coverage area. The bottles were used with two intentions in mind; lateral confinement and end bearing

Table 5 Testing program

	Test trials				Reinforcing condition	Footing plate diameter (mm) B'	Burrial depth (D_b) $= 0.1 \times B'$ (mm)	Width of reinforcement $(B) = 4 \times B'$ (mm)	Fly ash fill height (H) (mm)
	Aggregate grading								
	A_d	A_3	A_6	A_{12}					
Series 1	1	10	19	28	Unreinforced	80	–	–	750
	2	11	20	29		160	–	–	750
	3	12	21	30		240	–	–	750
Series 2	4	13	22	31	Reinforced by P_d 50	80	8	320	750
	5	14	23	32		160	12	480	750
	6	15	24	33		240	15	600	750
Series 3	7	16	25	34	Reinforced by P_d 73	80	8	320	750
	8	17	26	35		160	12	480	750
	9	18	27	36		240	15	600	750

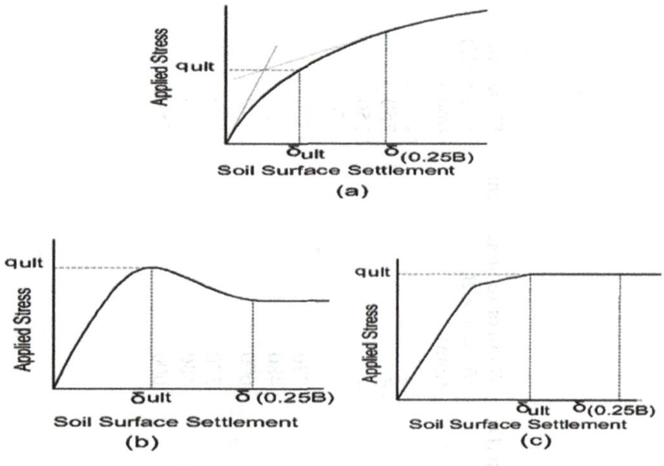

Fig. 5 Behavior of load versus settlement curve and representation of q_{ult}. **a** Toughening action, **b** soften action, **c** flexible ideal plane action

Fig. 6 Load–settlement behaviors for fly ash fill bed (*F is the footing plate*)

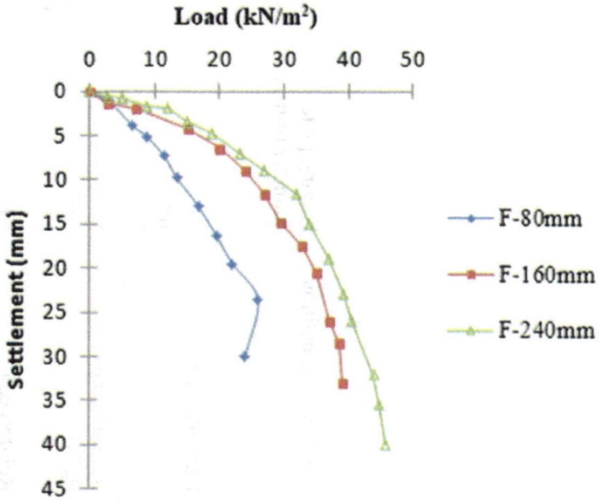

Table 6 Fly ash fill as foundation bed values of load and settlement

Loading plate diameter (mm)	Q_{ult} kN/m^2	δ_{ult} (mm)
80	21.3	28.5
160	38.3	32.6
240	41.6	35.2

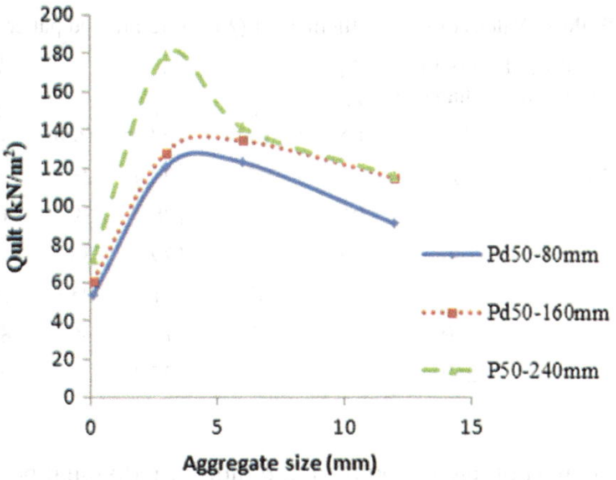

Fig. 7 Response of Q_{ult}—aggregate size (reinforced condition P_d 50)

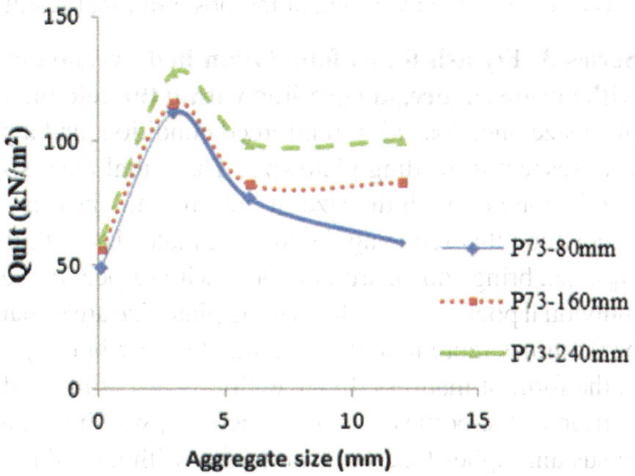

Fig. 8 Response of Q_{ult}—aggregate size (reinforced condition P_d 73)

to the aggregate as an infill into the pocket so that aggregate won't penetrate into the fly ash and, in turn, sustain less load and rapid settlement. The learnings of vertical loading behavior of plastic bottle mattresses by varying ratio of the diameter of cell aperture (P_d) to the aggregate size (P_d/A) were obtained 16, i.e., 16 times of aggregate size to that of an aperture of the bottle. Table 6 shows the values of load versus settlement with the reinforced condition of two aperture sizes of bottle with footing plate diameters 80, 160, and 240 in mm and fly ash in micron as cell infill (Table 7).

A_3 showed greater values compared to other sizes of aggregate infill. A_d, quarry dust infill, also showed excellent bearing resistance against loading but slight heaving was observed on the surface that leads to large settlement, as the loading plate was circular with a maximum of 240 mm, so heaving was predicted in quarry dust infill because of the lower weight of surcharge pocket infill, but it was not the same in other cases, due to larger deadweight of heavy aggregate heave was negligible. Once

Table 7 Values of load–settlement at Q_{ult} for reinforced patterns of P_d 50 and P_d 73

Reinforced condition	Footing diameter (mm) B'	A_d		A_3		A_6		A_{12}	
		Q_{ult} kN/m^2	δ_{ult} (mm)	Q_{ult} kN/m^2	δ_{ult} (mm)	Q_{ult} kN/m^2	δ_{ult} (mm)	Q_{ult} kN/m^2	δ_{ult} (mm)
P_d 50	80	83	35	120.4	28	122.8	26.2	90.8	28
	160	91	33	128	39	134	30	114.8	38
	240	109	40	179	33	140/8	44	115.6	33
P_d 73	80	89	32	111.2	25.6	76.4	29	58.8	25
	160	97	30	115.2	36	82.4	29.2	83.2	36
	240	101	37	127.2	32	98.8	32.8	100	32

the most efficient is known, the ratio of individual pocket cell size to the aggregate size (P_d/A) is observed as 16. In short, small size aggregate grading ≤3 mm screen was effective when we look at the observational results.

Series 3: Fly ash fill as foundation bed overlaying P_d 73 PET bottle mattress with various aggregate grading within the cell. Figure 9 describes the response of plate size once loaded in reinforced condition, and a gradual increment is observed with respect to loading plate size. The initial plate size is 80 mm, and as the plate size is doubled with the size of 160 mm, the bearing capacity increases. Now it is understood that reducing the loading plate size with the phase as aggregate reduces Q_{ult}, i.e., brings down around 200% with respect to the maximum Q_{ult}. The ratios of individual pocket size to the loading plate size are remarked. The loading plate should be 0.5 times more than the individual plastic bottle pocket when taken into account in the form of mattress. In the entire fly ash reinforced condition pattern the bottom portion of the bottles was not punctured, which means the bottle has the capability to sustain higher loading. Due to this without failure the entire load was taken by aggregates and lateral support by plastic bottles; this was a good combination needed to sustain higher stress (Fig. 10).

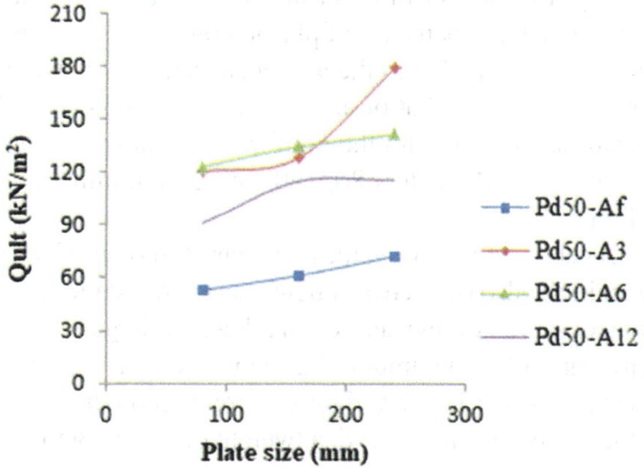

Fig. 9 Response of plate size once loaded (reinforced condition P_d 50)

Fig. 10 Response of plate size once loaded (reinforced condition P_d 73)

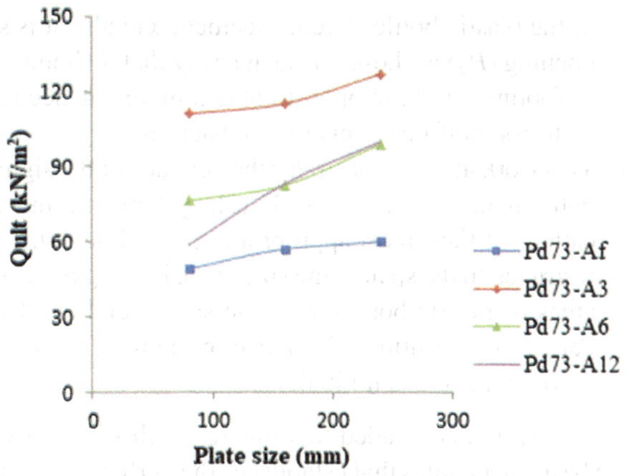

Limitations The model tankage is 1 m^3, where the surface area was 1 m^2 (100 cm × 100 cm). The loading plate diameter (B') was restricted to 240 mm due to the plain strain footing concept, in the view of the semi-infinite boundary condition; the stress shall be within the fly ash foundation. The aggregate grading considered in this research as infill material in PET bottle pockets is limited to A_{12} because >12 mm aggregate can self withstand the lateral movement once loaded vertically; in short, no cellular support is required. The saturated condition of fly ash fill with and without reinforcement is not practiced due to the lack of prerequisites held on drainage arrangement while installation of model tankage in the laboratory.

6 Conclusions

In plate load tests, a series of trials are carried out to evaluate the consumed plastic bottles as reinforcement underlying fly ash bed as these stated materials (fly ash and PET bottle mattress) are the major contributors to lower the environmental hazards. An attempt to utilize these materials has a positive impact on replacing the conventional earth filling. This view of transportation of fly ash from the nearby thermal power stations will be economical in utilization for infrastructure constructions.

The detailed results obtained from this experimentation with respect to systems are presented below:

- It is an upright method to intensify the bearing capacity and reduce settlement. In some situations, proper selection of effectual parameters for cellular reinforcements can boost the reinforced fly ash backfill up to 5.2 times when compared with non-reinforcement; the mass utilization of fly ash compaction bed is effective and resistant to shear failure.
- By maintaining the footing width (B') in the range of 13–27 (20 on average) times, the aggregate size (A) of $\leq$3 mm pocket fill material has benefitted in the usage

of the plastic bottle as reinforcement and also it is suggested that the size of cell opening (P_d) is chosen in such a way that it should be less in size than 0.67 times of footing width, in order to obtain more balanced and dependable plastic bottle mattress reinforcements fly ash backfill.

- This work also demonstrates the delicacy of the significant variables on the plastic bottle mattress-reinforced fly ash system and thus allows stimulating the applications of their most appropriate span. Though the experimentation results were restricted to the span of the size of the loading plate, just two types of cross-section areas of plastic bottles with the same height and four types of gravel grading. Particular execution of practice in the field must only be made after taking into account the above limitations.

Lastly, it is concluded that this research work provides vision and awareness to the basic mechanics that establishes the performance of waste plastic bottle mattress as reinforcement underlying partially saturated fly ash fills under vertical loading.

Funding and Acknowledgements This experimental research was funded by the Directorate of Minorities, Bengaluru, India. Grant ID: PhD/CR-35/2018-19. We are also thankful to Abhishek Mandlik, Onkar Madane, Rushikesh Lokhande, Shadab Shakil, and Raunak Walekar—UG students, School of Civil Engineering, MITWPU, Pune, India for the assistance in the overall journey of the experimentation.

References

1. Anand A, Sarkar R (2020) A probabilistic investigation on bearing capacity of unsaturated fly ash. J Hazard Toxic Radioact Waste (ASCE) 24(4):06020004. https://doi.org/10.1061/(ASCE)HZ.2153-5515.0000547
2. Kumara A, Mandal JN (2017) Effect of reinforcement on multi-tiered fly ash wall. Transportation Geotechnics and Geoecology, TGG 2017, 17–19 May 2017, Saint Petersburg, Russia. https://doi.org/10.1016/j.proeng.2017.05.072
3. ASTM (1995) Standard test method for measuring pH of soil for use in corrosion testing. ASTM G51-95, West Conshohocken, PA
4. ASTM (2010a) Standard test method for measuring mass per unit area of geotextiles. ASTM D5261, West Conshohocken, PA
5. ASTM (2011a) Standard test method for breaking force and elongation of textile fabrics (strip method). ASTM D5035-11, West Conshohocken, PA
6. ASTM (2010b) Standard test methods for specific gravity of soil solids by water pycnometer. ASTM D854-10, West Conshohocken, PA
7. ASTM (2011b) Standard test method for consolidated drained triaxial compression test for soils. ASTM D7181-11, West Conshohocken, PA
8. ASTM (2011c) Standard test method for tensile properties of geotextiles by the wide-width strip method. ASTM D4595-11, West Conshohocken, PA
9. ASTM (2012a) Standard specification for coal fly ash and raw or calcined natural pozzolan for use in concrete. ASTM C618-12, West Conshohocken, PA
10. ASTM (2012c) Standard test method for measuring nominal thickness of geosynthetics. ASTM D5199-12, West Conshohocken, PA
11. Biabani MM, Indraratna B, Ngo NT (2016) Modelling of geocell-reinforced subballast subjected to cyclic loading. Geotext Geomembr 44(4):489–503

12. Choudhary AK, Jha JN, Gill KS (2010) Laboratory investigation of bearing capacity behavior of strip footing on reinforced fly ash slope. Geo text Geomembr 28(4):393–402
13. Chummer AV (1972) Bearing capacity theory from experimental results. J Soil Mech Found Div ASCE 98(12):1311–1324
14. Capital D (2011) Plastics commodity to custom products: *Redefining perception*. Mumbai, India
15. Dutta S, Mandal JN (2016) Model studies on geocell reinforced fly ash bed overlying soft clay. J Mater Civil Eng. https://doi.org/10.1061/(ASCE)MT.1943-5533.0001356,04015091
16. Dutta S, Mandal JN (2017) Model studies on encased fly ash column-geocell composite systems in soft clay. J Hazard Toxic Radioact Waste Manage. https://doi.org/10.1061/(ASCE)HZ.2153-5515.0000353,04017001
17. Xu G, Shi X (2018) Characteristics and applications of fly ash as a sustainable construction material: a state-of-the-art review. Resour Conserv Recycl 95–109
18. Ghosh KP, Mukherjee K, Saha S (2015) Fly ash of thermal power plants: review of the problems and management options with special reference to the Bakreshwar thermal power plant, Eastern India. Int J Geol Earth Environ Sci 5(2):74–91
19. Gill KS, Choudhary AK, Jha JN, Shukla SK (2013) Experimental and numerical studies of loaded strip footing resting on reinforced fly ash slope. Geosynth Int 20(1):13–25
20. Hegde A, Sitharam TG (2015) Joint strength and wall deformation characteristics of a single-cell geocell subjected to uniaxial compression. Int J Geomech. https://doi.org/10.1061/(ASCE)GM.1943-5622.0000433,4014080
21. Hegde AM, Sitharam TG (2014) Effect of infill materials on the performance of geocell reinforced soft clay beds. Geomech Geoeng Int J 10(3):163–173
22. Kim B, Prezzi M (2008) Evaluation of the mechanical properties of class-F fly ash. Waste Manag 28(3):649–659
23. Kim B, Prezzi M, Salgado R (2005) Geotechnical properties of fly and bottom ash mixtures for use in highway embankments. J Geotech Geoenviron Eng. https://doi.org/10.1061/(ASCE)1090-0241(2005)131:7(914),914-924
24. Moghaddas Tafreshi SN, Dawson AR (2010b) Comparison of bearing capacity of a strip footing on sand with geocell and with planar forms of geotextile reinforcement. Geotext Geomembr 28(1):72–84
25. Nadaf MB, Dutta S, Mandal JN (2016) Fly ash as backfill material in slopes using waste pet bottles as reinforcement. In: Proceedings of 6th IconSWM, International Society of Waste Management, Air and Water, Kolkata, India, pp 1209–1215
26. Nadaf MB, Mandal JN (2014) Triaxial behavior of steel grid reinforced fly ash. In: Proceedings of Indian Geotechnical Conference, Indian Geotechnical Society, Kakinada, India, pp 50–54
27. Nadaf MB, Mandal JN (2017) Model studies on fly ash slopes reinforced with planar steel grids. Int J Geotech Eng 11(1):20–31
28. NPCS (NIIR Project Consultancy Services) (2020) Pet bottle recycling. http://www.niir.org/profiles/profile/2045/pet-bottle-recycling.html (Dec. 9, 2020)
29. Ram Rathan Lal B, Mandal JN (2012) Feasibility study on fly ash as backfill material in cellular reinforced walls. Electron J Geotech Eng 17(J):1437–1458
30. Ram Rathan Lal B, Mandal JN (2014a) Behavior of cellular reinforced fly-ash walls under strip loading. J Hazard Toxic Radioact Waste Manage, 45–55. https://doi.org/10.1061/(ASCE)HZ.2153-5515.0000201
31. Ram Rathan Lal B, Mandal JN (2014b) Model tests on geocell walls under strip loading. Geotech Test J 37(3):477
32. Dash SK, Majee A (2021) Geogrid reinforcement for stiffness improvement of railway track formation over clay subgrade. ASCE 21(9). https://doi.org/10.1061/(ASCE)GM.1943-5622.0002128
33. Tavakoli Mehrjardi G, Behrad R, Moghaddas Tafreshi SN (2019) Scale effect on the behavior of geocell-reinforced soil. Geotext Geomembr 47(2):154–163
34. Yoo C (2001) Laboratory investigation of bearing capacity behavior of strip footing on geogrid-reinforced sand slope. Geotext Geomembr 19(5):279–293

Recent Trends in Structural Engineering

Sustainable Project Planning of Road Infrastructure in India: A Review

Appa M. Kale and Sunil S. Pimplikar

Abstract Road infrastructure in India is growing speedily. Vision (2030) anticipates a quality, dependable, sustainable, resilient infrastructure, and sound technology. Road infrastructure can be built by using the sustainability concept. A conventional road causes environmental and social issues as well as affects the economy of road infrastructure. To overcome this problem sustainable planning is a very prominent solution. By reviewing research papers of the previous 30 years, life cycle cost analysis seems to be a key method for achieving required growth. Further, along with life cycle cost analysis, various other techniques such as Project Definition Rating Index (PDRI) and Project Rework Reduction Tool (PRRT) suggested by the researchers are also inferred.

Keywords Road infrastructure · Sustainable development · Life cycle cost analysis · Sustainable planning · Sustainable development goals

1 Introduction

After the Paris Convection of the United Nations Climate Change Conference (2015), the usage of 17 development goals for a developing nation like India has opened the doors just as difficulties. The Government of India, by and large, and the Ministry of Road Transport and the Highway (MoRT&H) have specifically focused on an absolute fulfillment of 200,000 km of national highways by 2022, including on venture of 7 trillion for the commitment of other roads and highways throughout the following 5 years.

India is practically seeing a proceeded with advancement in sustainable development. The improvement business has been an onlooker to a strong advancement wave

A. M. Kale (✉) · S. S. Pimplikar
Dr. Vishwanath Karad, MIT World Peace University, Pune 411038, India
e-mail: appakale297@gmail.com

S. S. Pimplikar
e-mail: sunil.pimplikar@mitpune.edu.in

M. S. Ranadive et al. (eds.), *Recent Trends in Construction Technology and Management*, Lecture Notes in Civil Engineering 260,
https://doi.org/10.1007/978-981-19-2145-2_60

fuelled by huge spending on housing, streets, ports, water supply, and air terminal progression. The advancement division has enlisted twofold digit improvement in the midst of the several years and its proposal as a degree of GDP has extended stunningly when diverged from the latest decade. The Planning Commission of India has proposed a theory of around US$ 1 trillion in the Twelfth 5-year plan (2012–2017), which is twofold that of the Eleventh 5-year plan.

The report on Indian urban infrastructure and services MoUD (2011) investigated the three primary elements affecting income contrarily, for example, deficit in granting projects according to design, setback in financing, and wasteful venture execution which adds to a US$ 200 million misfortune to GDP by 2017 (100 + 20 + 80).

India's logistic infrastructure is lacking, unfit, and ill-designed to help the expected development pace of 7–8% throughout the following decade. Expected 2.5-overlap development in dismay traffic will additionally press India's infrastructure.

Gaining from the past and embracing worldwide accepted procedures seek after a calculated foundation system that limits speculation, expands cost-effectiveness for clients, and is energy productive.

2 Literature Review

Vision (2030) expects a quality, reliable, sustainable, and resilient infrastructure which is affordable as well as even more having increased resource use efficiency, and use of clean, environmentally sound technology, and industrial processes, further demanding a focus on greater research and development efforts as well as restoring to the innovative approaches (Achieving Sustainable Development in India: The study of Financial Requirement and Gaps 2015) [21].

NHAI with 40% equity and the remaining 60% annuity model investment from Canada, Malaysia, Japan, and Asian Development Bank is accelerating the infrastructure development in India. Sapatnekar et al. [31] have emphasized the need of regulating infrastructure development in India based on the modeling and anticipating various risks arising due to the gap existing between the planning and execution, adopting world standards as against their effective implementation, improper coordination and communication among various stakeholders, uncontrolled wrong professional practices, and the like.

The report on Indian Urban Infrastructure and Services MoUD (2011) has analyzed the three main factors impacting revenue negatively, such as shortfall in awarding projects as per plan, shortfall in funding, and inefficient project execution which contribute to a US$ 200 million loss to GDP by 2017 (100 + 20 + 80).

India's logistic infrastructure is lacking, unfit, and ill-designed to help the expected development pace of 7–8% throughout the following decade. Expected 2.5-overlap development in dismay traffic will additionally press India's infrastructure.

Gaining from the past and embracing worldwide accepted procedures seek after a calculated foundation system that limits speculation, expands cost-effectiveness for clients, and is energy productive [23].

It examines how India's present logistic infrastructure is lacking to meet its development goal and gauge the current and future grouping of freight traffic streams in the country to characterize strategic prerequisites and monetary ramifications (McKinsey & Company 2015).

Life cycle analysis appears to be a very promising solution which will enable to reduce the aforementioned gap [1, 4, 5, 7, 8, 10–12, 14, 15, 17, 24–27]. Nevertheless, for determining the research gap the following points need to be considered.

These include:

1. Clarifying enhancing and achieving optimum combination of life cycle cost analysis and value engineering and an objective of developing comprehensive guidelines or standards, thus enabling the inclusion of hidden and social costs [5].
2. Raghuram et al. [10] have recommended measures to reduce time and cost overrun related to environmental clearance and land acquisition being identified as a major reason for delays in the road sector. Researchers further suggest scope for improvement in the use of technology, transparent assessment, appointment of an independent trustworthy land evaluator, decentralization in decision making, and special capacity building so as to crack global business opportunities.
3. Extension of the sustainable engineering infrastructure model suggested by Okon et al. [13] for applicability under Indian conditions and for further facilitating decision making based on risk [22, 28 (Chi, Bunker, Teo 2016) and return evaluation, stakeholder's perspective, and sustainability issues in the highway project.
4. Ikpe and Hammon [15] have recommended a potential area for further research in construction industries with respect to establishing effective health and safety on construction sites.
5. Alam et al. [17] have presented a qualitative model which identifies the impact levels of various life cycle analysis components associated with road projects, thus establishing a comprehensive system boundary for life cycle analysis of the road. Researchers have suggested for the work in regards to the quantitative assessment of parameters so as to facilitate objectivity in decision making.
6. Buneo et al. [18] recognize significant difficulties in improving life cycle assessment of transport projects, like the requirement for normalization and using between the fleeting accumulation of substantial and immaterial environmental, social and economic impacts, fixing the markdown rate for its maintainability thing distinguished as the zone of additional exploration. Specialists further accentuate that particular attributes and worries of society have not been remembered for and utilized by practical examination instruments.

7. Mishra et al. [20] suggest further research to identify critical success factors converting current issues of PPP project as essential for the sustainability of PPP of road projects.
8. Godinho and Dias, Alam [14, 24] has incorporated the impact of road user's greenhouse gases (RUG factor), has recommended the determination of global warming potential (GWP) based on life cycle analysis, has used a research framework in determining the road sustainability index, suggested a universal concept of standard carbon vehicle (SCV), further advocated that road agencies develop albedo matrix different types of pavement surfacing based on the age of payment, for assessing GWP on real-time field measurements, and has finally observed that infrastructure rating schemes show inadequate consideration to life cycle aspect of road infrastructure in terms of credit items and weights.
9. Babashamsi et al. [25] have suggested including preventive maintenance treatment as a part of strategic management for bettering the performance of existing pavement life cycle cost analysis models, further quantifying them, and if possible, standardizing the value associated with user cost.
10. Castro [26] accentuates consideration of risk and sensitivity analysis in the existence cycle cost examination to build the degree of dependability and certainty assessed model.
11. Yu et al. [29] suggest the use of sustainable project planning (SPP) which has been used for predicting project success of Chinese construction projects for other sectors as well (applicability to road sector performance in the Indian context may be explored).

3 Conclusion

1. Based on the literature review conducted, it appears that sustainable highway infrastructure can be achieved with help of Sustainable Project Planning Model (SPPM).
2. This in turn is expected to contribute to social, environmental, and economic aspects of the society.

References

1. Fwa, Sinha (1991) Pavement performance and life cycle cost analysis. Am Soc Civ Eng, 1–5
2. Cottell (1993) Facilitating sustainable development: Is our approach correct? J Prof Issues Eng Educ Pract ASCE, 21–24
3. Feis (1994) Sustainable development issues: industry, environment, regulation and competition. J Infrastruct Syst ASCE, 51–59
4. Ravirala, Grivas (1995) State increment method of life cycle cost analysis for highway management. J Infrast Syst ASCE, 151–159

5. Arditi, Messiha (1999) Life cycle cost analysis (LCCA) in municipal organizations. J Infrastruct Syst ASCE, 1–10
6. Horvath, Matthews (2004) Advancing sustainable development of infrastructure systems. J Infrastruct Syst ASCE, 55–63
7. Feng, Wang (2008) Integrated cost–benefit analysis with environmental factors for a transportation project: case of Pinglin interchange in Taiwan. J Urban Plann Dev, 161–167
8. Gho, Yang (2009) Extending life-cycle costing analysis for sustainability considerations in australian road infrastructure projects, Rethinking sustainable development: planning, infrastructure engineering, design and managing urban infrastructure. Faculty of Built Environment and Engineering, Queensland University of Technology, Queensland University of Technology, Brisbane, Queensland, pp 324–331
9. Raghuram, Bastian, Sundaram (2009) Mega projects in India: environmental and land acquisition issues in the road sector. Indian Institute of Management Ahmedabad, pp 2–15
10. Madanu et al (2010) Life-cycle cost analysis of highway intersection safety hardware improvements. J Transp Eng ASCE, 25–33
11. Okon, Ekpo, Elhag (2010) Cost-benefit analysis for accident prevention in construction projects. J Transp Eng ASCE, 103–111
12. Gho, Yang (2010) Responding to sustainability challenge and cost implications in highway construction projects. J Transp Eng ASCE, 236–245
13. Okon, Ekpo, Elhag (2010) A sustainable engineering infrastructure model for the 21st century. In: Proceedings of the world congress on engineering, pp 35–46
14. Godinho P, Dias J (2012) Cost-benefit analysis and the optimal timing of road infrastructure. J Transp Eng ASCE, 99–107
15. Ikpe, Hammon (2012) Cost-benefit analysis for accident prevention in construction project. J Constr Eng Manag ASCE, 64–69
16. Hwang, Tan (2012) Sustainable project management for green construction: challenges, impact and solutions. In: Ciob construction conference, WCC, pp 39–47
17. Alam, Kumar, Dawes (2012) Life cycle analysis for sustainability assessment of road projects. CIB World Building Congress, pp 76–88
18. Buneo, Vassello, Cheung (2013) Sustainability assessment of transport infrastructure projects: a review of existing tools and methods, 1–28
19. Meyer, Kassa (2013) Sustainable streets and highways: an analysis of green roads rating systems. Georgia Transportation Institute Report, pp 1–23
20. Mishra, Narendra, Kar (2013) Growth and infrastructure, investment in India: achievements, challenges, and opportunities. Econ Anal, 51–70
21. Envision 2030 (2015) 17 Goals to transform the world for persons with disabilities, Department of Economic and Social Affairs, United Nations-Disability, pp 57–81
22. Mohapatra (2015) An economic analysis of improvement of road infrastructure: a case study. Eur Acad Res, 14637–14650
23. Achieving Sustainable Development in India: The Study of Financial Requirement and Gaps, Build Resilient Infrastructure, Promote Inclusive and Sustainable Industrialization and *Foster Innovation Report*, 5–34
24. Alam (2016) Developing life cycle environmental indicators for road infrastructure. Ph.D. thesis, Queensland University of Technology, Australia, pp 10–28
25. Babashamsi et al (2016) Sustainable development factors in pavement life-cycle: highway/airport review. Sustainability, 3–8
26. Castro (2016) Life cycle cost analysis of road infrastructures using LCC Am/Qm software. Tecnico Lisboa, 1–9
27. Chi, Bunker, Teo (2017) Measuring impacts and risks to the public of privately operated toll road projects by considering perspectives in cost-benefit analysis. J Transp Eng ASCE, 77–76
28. Agarchand (2017) Sustainable infrastructure development challenges through Ppp procurement process: Indian perspective. Int J Proj Manage, 53–59
29. Yu, Zhu, Yang, Wang, Sun (2018) Integrating sustainability into construction engineering projects: perspective of sustainable project planning. Sustainability, 1–17

30. Demand for Grants 2018–19 Analysis; Road Transport and Highways, PRS Legislative Report, pp 1–6
31. Sapatnekar, Patnaik, Kishore (2018) Regulating infrastructure development in India. NIPFP Working Paper Series, pp 23–33

Behaviour of Space Frame with Innovative Connector

Pravin S. Patil and I. P. Sonar

Abstract The hollow circular tube is the most efficient section of space structure because of its lightweight and load resistance. Flatted end tube with a bolted connection is being the most economical connection with fast assembly, easy to transport, easy to maintain and easy to repair for space truss. This flatted end tube is connected with welded plates by bolting jointly, provided that welded plates are in the direction of force to avoid eccentricity in the space truss. A connector is an extremely important part of a prefabricated system and the final commercial of structural connections of all types has been the subject of intensive research during the last decade. Investigation into the behaviour of flatted, flattened and tapered flattened end tube is the main objective of the present work. Flatted, flattened and tapered end tube of 3-D space structure with an innovative connector is modelled in Catia. The performance is checked by Ansys analysis and results are discussed. The research on flattened and tapered end tube connector for steel tube space structure is the main contribution to this research work. The flattened and tapered flattened end tube connector shows better performance as compared to flatted end tube connector.

Keywords Flatted end tube · Flattened end tube · Tapered flattened end tube · Deflection of space frame

---

P. S. Patil (✉)
Assistant Professor @ Department of civil Engineering, DY Patil University Ambi Pune, Pune, Maharashtra, India
e-mail: pravinspatil154@gmail.com

I. P. Sonar
Associate Professor @ Department of Civil Engineering, College of Engineering Pune, Pune, India

M. S. Ranadive et al. (eds.), *Recent Trends in Construction Technology and Management*, Lecture Notes in Civil Engineering 260,
https://doi.org/10.1007/978-981-19-2145-2_61

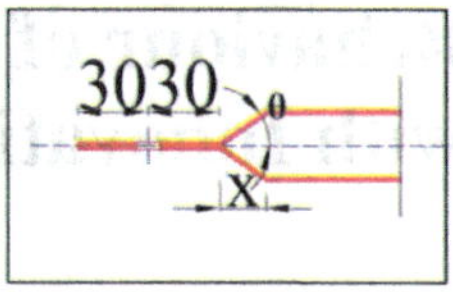

Fig. 1 Flattened end tube

1 Introduction

End flattened connectors are a type of bolted connection. The hollow sections used for space frame elements are pressed and flattened at their ends. The flattened ends are connected with a single bolt. This type of connector is economic and gives a pleasing appearance in the space frame. This type of connector is more common in Brazil, where it is used in many large textile industries, churches, sports gymnasiums, and soccer stadiums and in infrastructures such as warehouses for grain storage, bus and railway stations and airports. The structural components are end flattened structural elements of a bolt. Since the structural elements are less, it is easy to assemble. For space structure flatted, flattened and tapered flattened members have been used. The flatted end tube is the hollow circular tube that has been pressed at the ends ($\theta = 90$ as shown in Fig. 1). In flattened end tube give a particular angle for smooth load path analysis ($\theta = 30$ as shown in Fig. 1). Tapered flattened end thickness has to be increased at the end to reduce the stresses, and after this, it should be called a tapered flattened end tube.

From the literature review, this paper contains a newly developed space truss system with reduced cost but the same behaviour and ease of construction [1]. Two new approximately analysing methods favourable for dynamic analysis of space trusses have been presented in this paper [2]. The paper introduces a space truss system newly developed as an attempt to reduce the cost of space trusses without compromising their behaviour or ease of construction [3]. In this paper, efficient neural networks are analyzed for the design of a double-layer grid. Square diagonal on diagonal with a span varying 26.5–75 m was considered [4]. The results conclude that 68% raise in local collapse is associated with the serviceability limit state and a 17% increase in global collapse is associated with the ultimate limit state in the truss load-carrying capacity [5, 6]. The structural behaviour up to the collapse of reinforced eccentric and simple eccentric nodes was discussed on the basis of the experimental. To improve the structural load-carrying capacity and minimize deflection, a structural reinforcement is created [7].

2 Innovative Connector

The innovative connector has been employed consisting of the use of tubular members with flatted and flattened ends, but connected to a device composed of welded steel

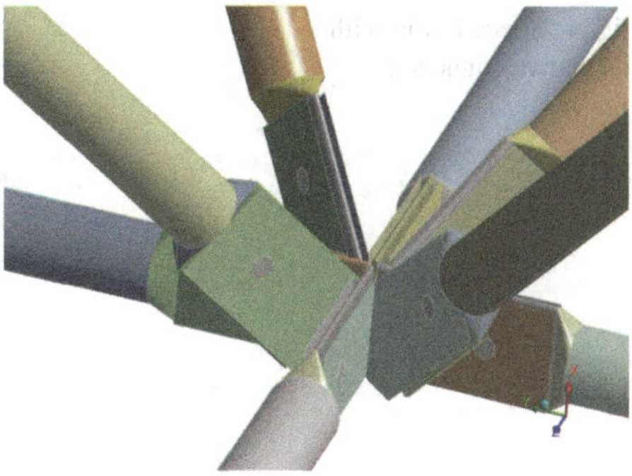

Fig. 2 Innovative connector

plates in the direction of force referred to in Fig. 2. This innovative connector allows for the elimination of node eccentrics.

3 Assembly of Space Frame Connector

After creating all the necessary component parts, each model is assembled in Catia Assembly. Each member is assembled to form a node. In each assembly, the datum plane is given three references in *x*, *y* and *z* to match its exact position. After assembling the horizontal and inclined (bracing) parts, the node is bolted. To check if there is an interface between the members in that particular node, then that node is analysed for 'global interface'. When global interface is there, then meshing will not be made in ANSYS. Therefore, if any parts are overlapping, it is to be corrected at each node. Level complete model is assembled as in Fig. 3, checked for interface and exported for analysis in ANSYS, as shown in Fig. 4.

Meshing has done in such a way that there is no connection with each other. That gives unstructured mesh generation for all parts (Fig. 5).

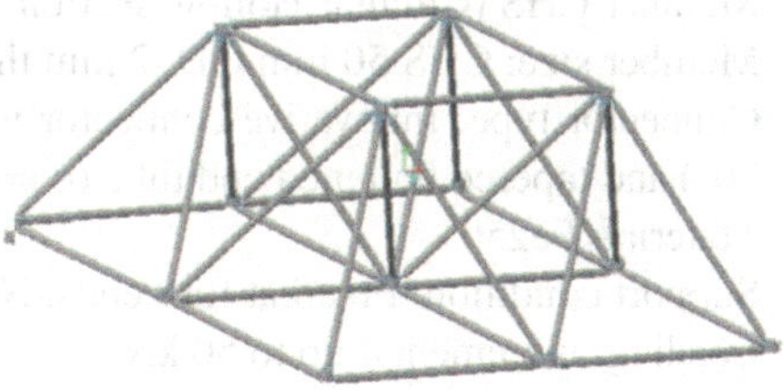

Fig. 3 Space frame

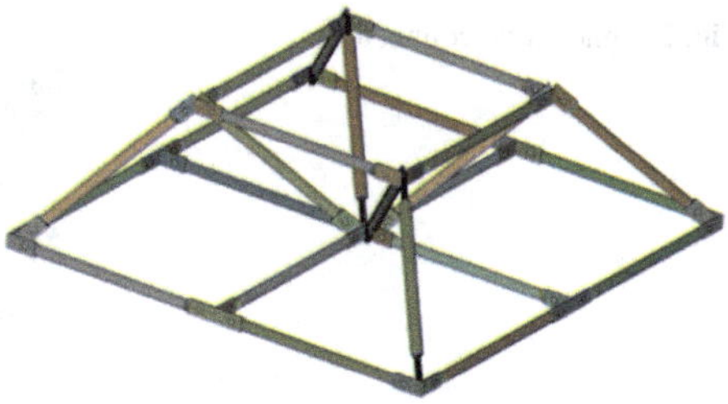

Fig. 4 Space frame with innovative connector

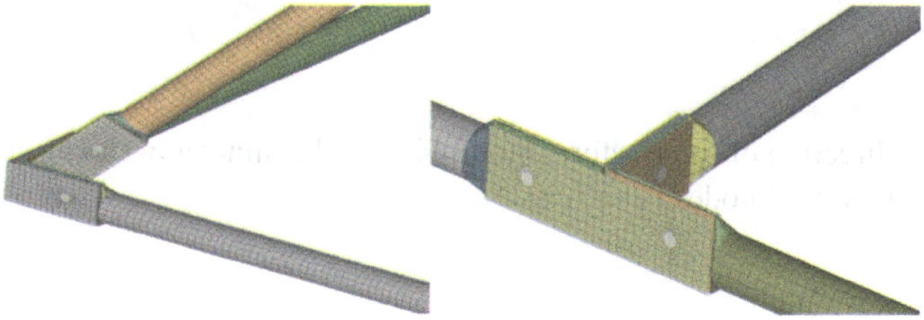

Fig. 5 Meshing of space frame

4 Analysis Setup

To perform a better finite element analysis, our work was done by ANSYS 14.5. However, SAP analysis is made for a space frame without flattening the tube. This is performed to check the member forces and deflection theoretically.

Modelled space frame and analysis setup details are furnished below:

Long span: 2 m
Short span: 2 m
Grid size: 1 m
Grid type: Square on square
Bracing angle: 45°
Member CHS (Circular Hollow section)
Member size: CHS 50 mm OD, 2 mm thick
Connector type: Innovative connector with flatted, 30° flattened end tube (ϴ = 30°) and tapered flattened end tube (ϴ = 30°)
Material: Fe250
Support condition: Fixed at four ends. Node 1, 3, 7 and 9 (Ref Fig. 7)
Loading: incremental up to 50 kN
Bolt diameter: 16 mm (Fig. 6).

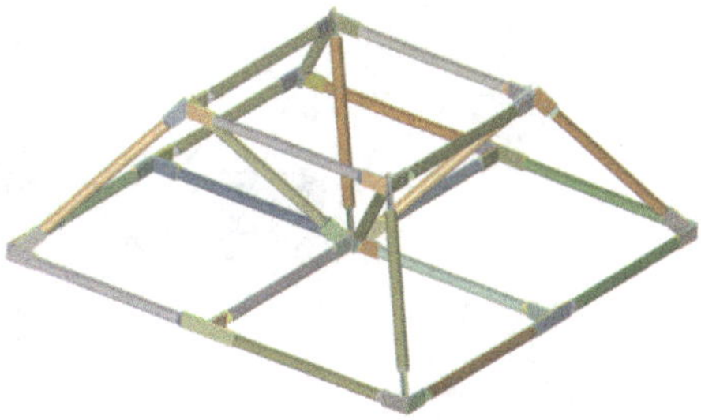

Fig. 6 Space frame analysis setup

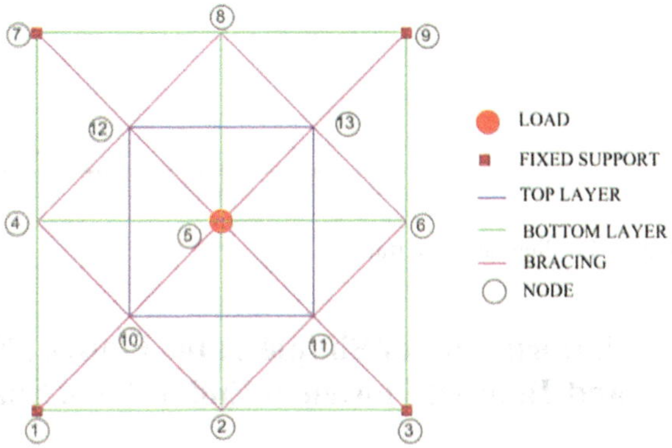

Fig. 7 Space frame analysis setup SAP2000

5 Analysis of Space Frame in SAP2000

Analysis setup is made in SAP2000. AutoCAD line diagram prepared is exported in SAP. For all the lines, the geometry of the hollow section with a size of 50 mm outer diameters and 2 mm thickness is assigned. Material steel with a yield strength of 250 MPa (Fe250) is assigned to all the tubes. End nodes in the bottom layer are fixed, while the centre node is loaded from 5 to 50 kN. The analysis model in SAP is shown in Fig. 7.

6 Loading on Space Frame

Load is applied in node 5 (Fig. 8). Properly the load is applied in the bolt fixed at node 5 in 'y' direction. In ANSYS, the y direction acts as the vertical direction. Load is applied from 5 to 50 kN in an incremental order of 5 kN. Totally, in 10 s, 10 loads were applied. This is performed to study the behaviour of the space frame for an incremental load.

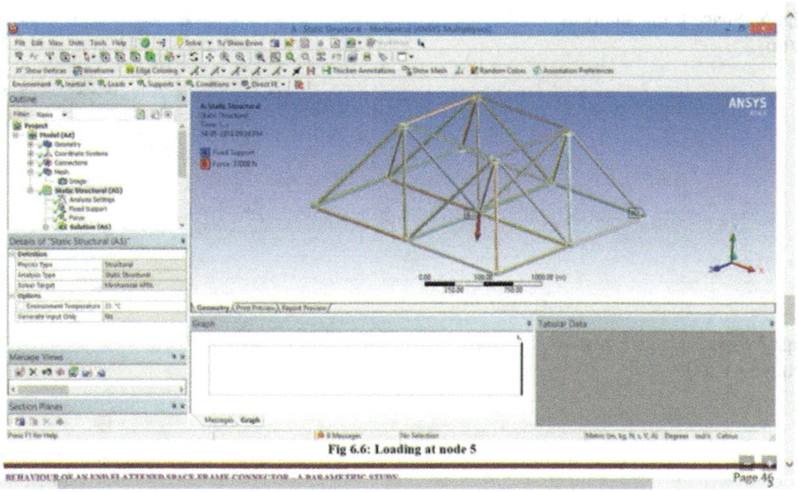

Fig. 8 Loading on space frame

7 Performance of Simple Tube, Flatted, Flattened and Tapered Flattened End Tube of Space Frame

In any structure, the displacement of nodes must be less. Due to local effects, a structure shall experience positive or negative deflections. Globally, a structure with minimum deflection is considered as a safe structure.

7.1 Deflection for Flatted, Flattened and Tapered Flattened End Tube

Load vs displacement shows the elastic performance of a structure. It depends on the material property of the structure. Our space frame is of steel Fe250, so yielding is expected. In Fig. 9, load versus deflection of the 2 mm thick model is seen. It is observed that flatted end tube model experiences more deflection than other models. Minimum deflection is very essential to select a good model since deflection is a criterion of limit state of serviceability. When deflection is less, then the structure is stiff enough to resist loads. It is observed that flatted end tube experiences more deflection than flattened and tapered flattened end tubes. If we see the slope of load–deflection curve for flattened and tapered flattened model, it is linear, with an inclination nearly equal to 85°. Minimum deflection is very essential to select a good tube since deflection is a criterion of limit state of serviceability. When deflection is less, then the tube is stiff enough to resist loads. The flattened end tube model experiences less deflection (load application point of flatted end tube).

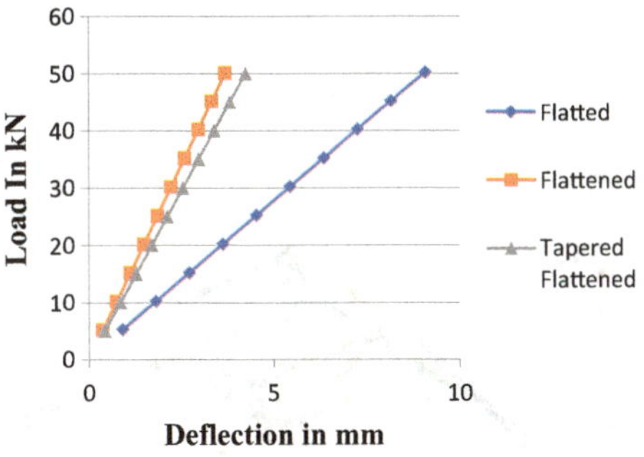

Fig. 9 Load versus deflection of space frame

From load–displacement figures, minimum deflection is seen in flattened end tube. It shall be justified that when load path is smooth in a member then secondary effects like torsion and moment in that member will be less which gives negative effects and reduces deflection. At flattened end tube, the load path is smooth; therefore, these models experience less deflection. Other models have a sudden change in load path due to flattening angle, which increases the secondary effects. So, these models experience much displacement.

The deformed shape of the space frame of flatted end tube with an innovative connector for incremental loading is shown in Fig. 10. The red colour shows maximum deflection in the space frame.

The deformed shape of flattened end tube with an innovative connector for incremental loading is shown in Fig. 11. In this tube, the proper angle for smooth load path analysis is given at the end of the tube.

The deformed shape of tapered flattened end tube with an innovative connector for incremental loading is shown in Fig. 12. End thickness has been increased to reduce the stresses (Table 1).

8 Comparisons of Flatted, Flattened and Tapered Flattened End Tube

A comparison between flatted, flattened and tapered flattened models for the 5–50 kN is plotted in Fig. 13. In this comparison, at all loadings from 5 kN to 50 kN, the flattened end tube model experiences less deflection, while flatted end tube undergoes maximum deflection, so the flattened end tube gives good performance compared to flatted and tapered flattened end tube.

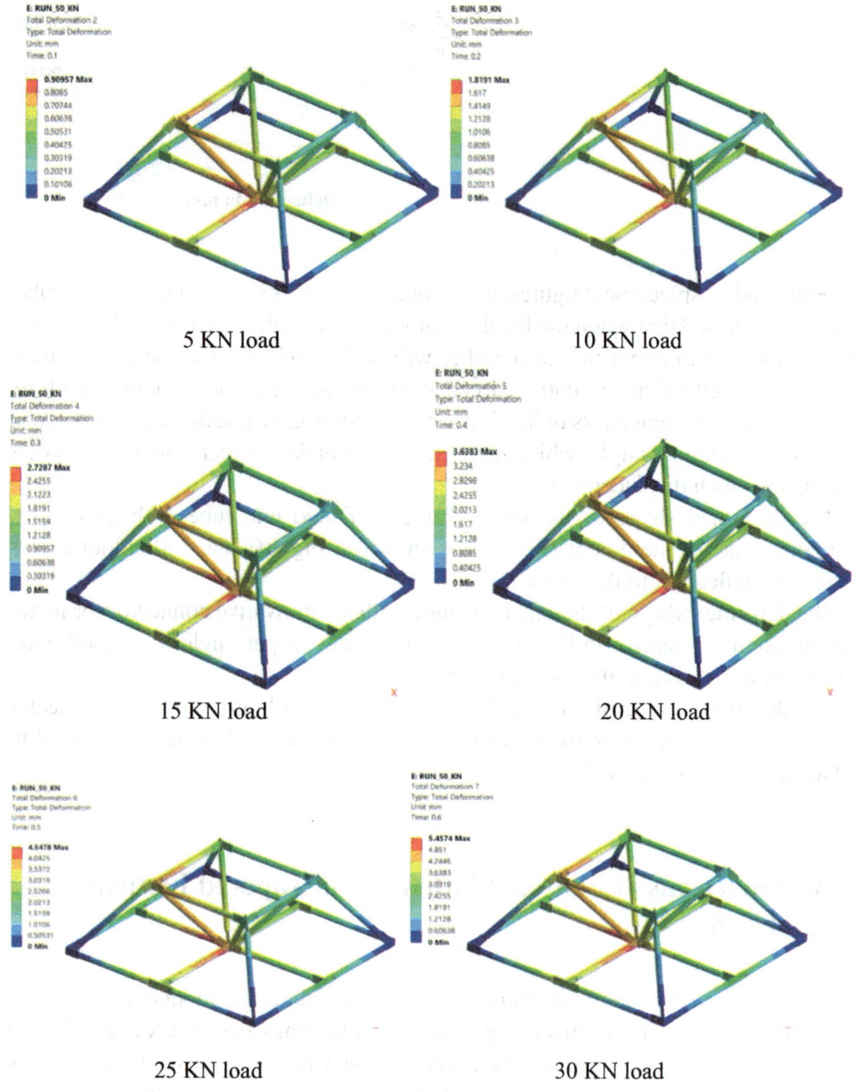

Fig. 10 Deformed shape of flatted end tube

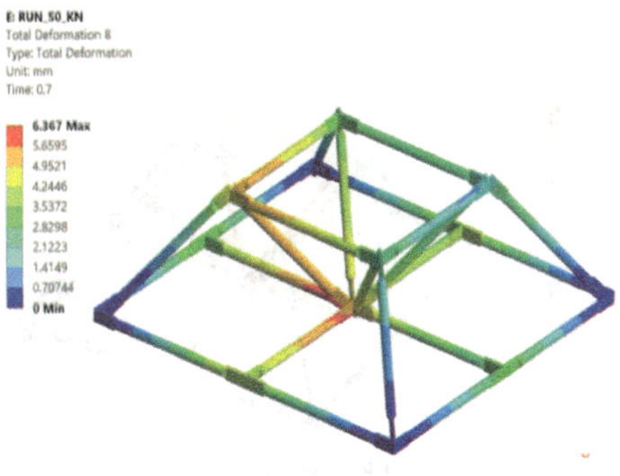

35 KN load

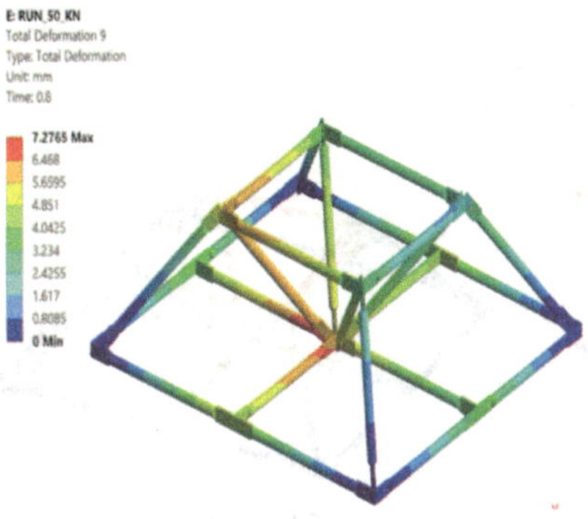

40KN load

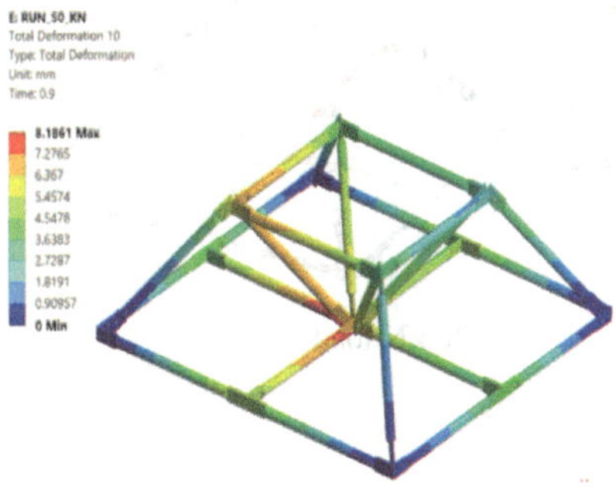

45 KN load

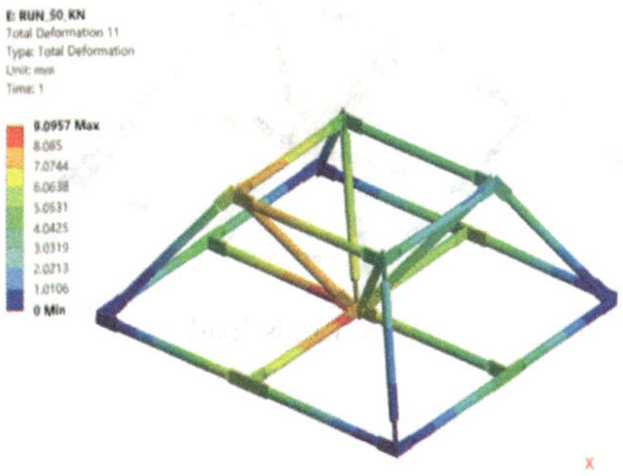

50 KN load

Fig. 10 (continued)

9 Conclusion

Space frame with end-flattened hollow sections connected with one bolt is easy to manufacture and economic. But, it does have some disadvantages like nodal eccentricities which produce secondary effects. In the case of space structure to avoid eccentricity and for easy connections, it is suggested to use an innovative connector. In the space frame model with an innovative connector, flattened end tube gives good performance compared to tapered flattened and flatted end tube.

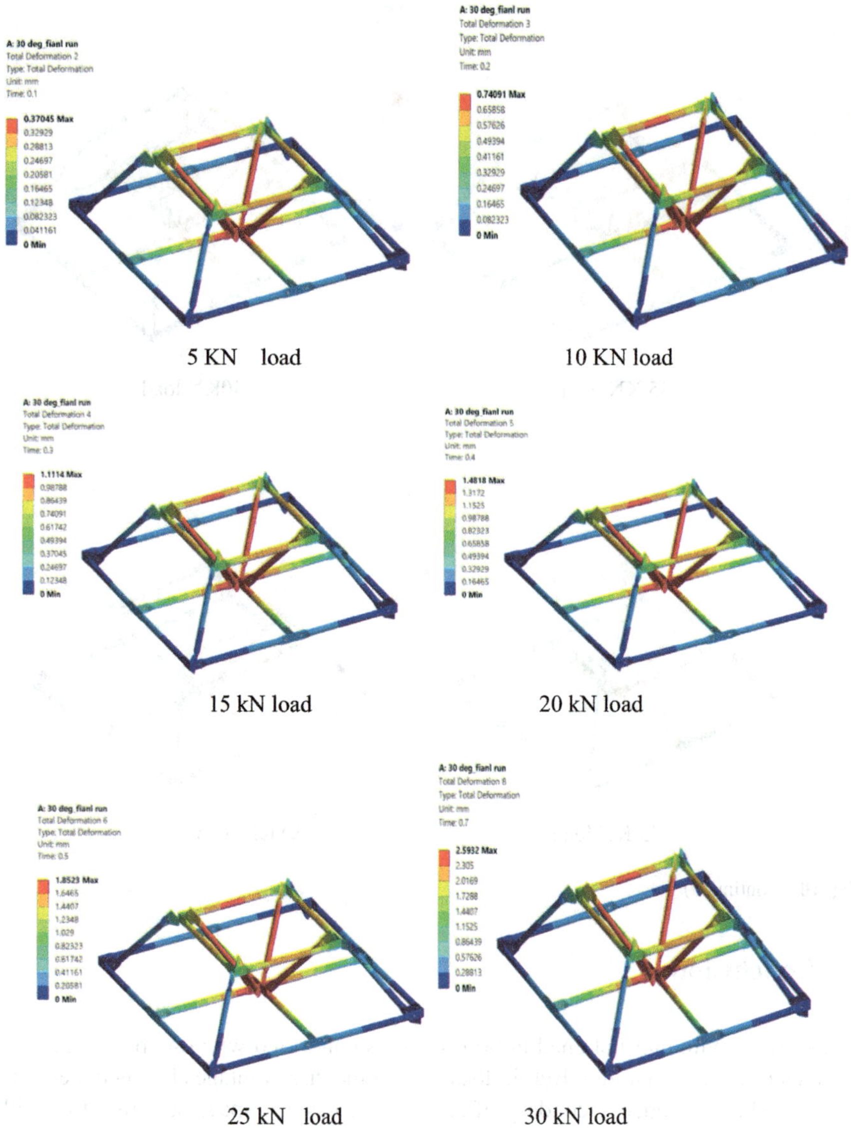

Fig. 11 Deformed shape of flattened end tube

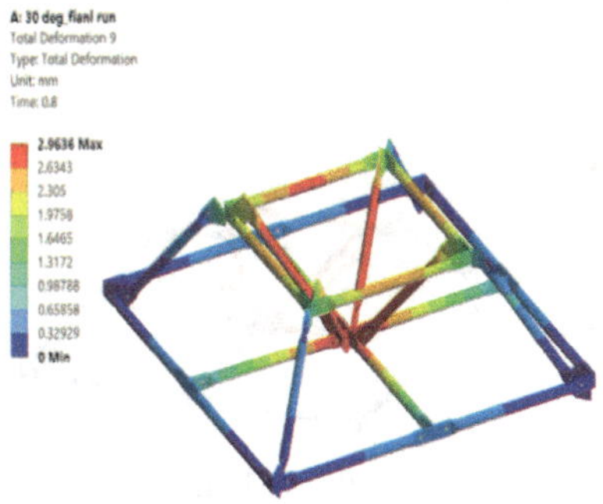

35 kN load

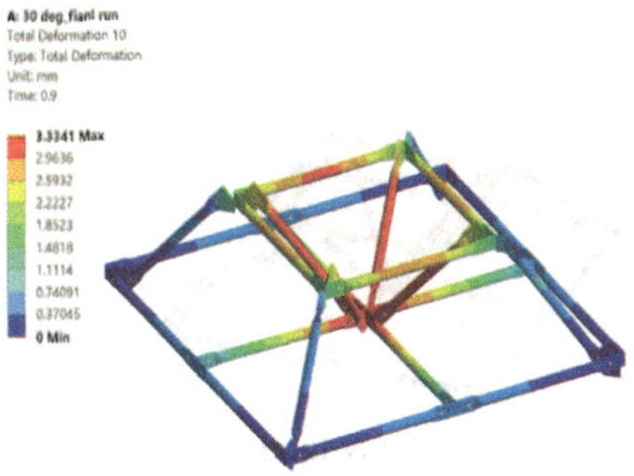

40 kN load

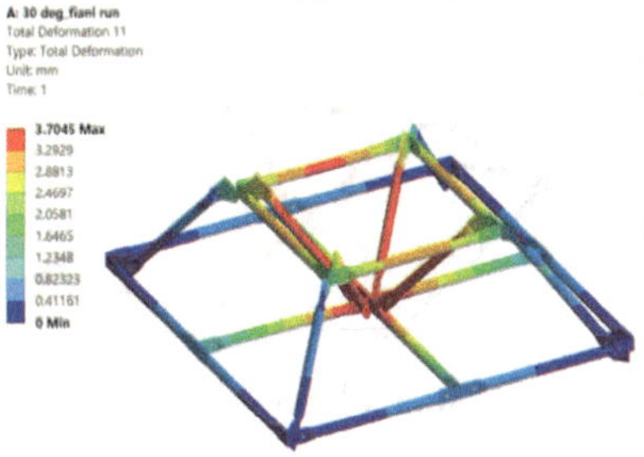

45 kN load

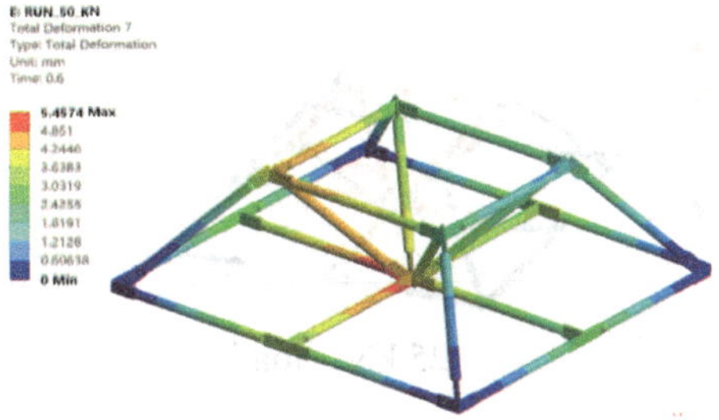

50 kN load

Fig. 11 (continued)

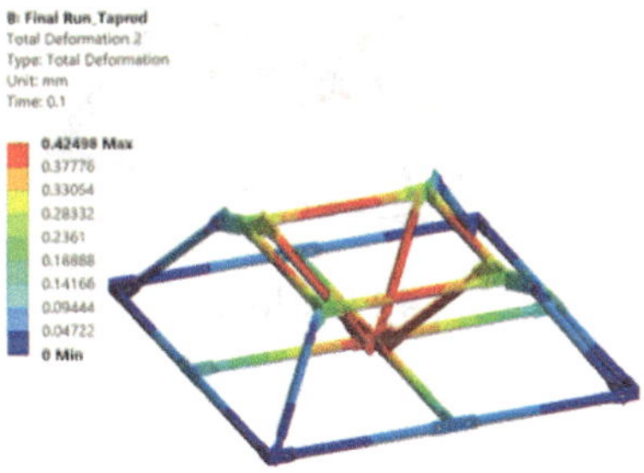

5 kN load

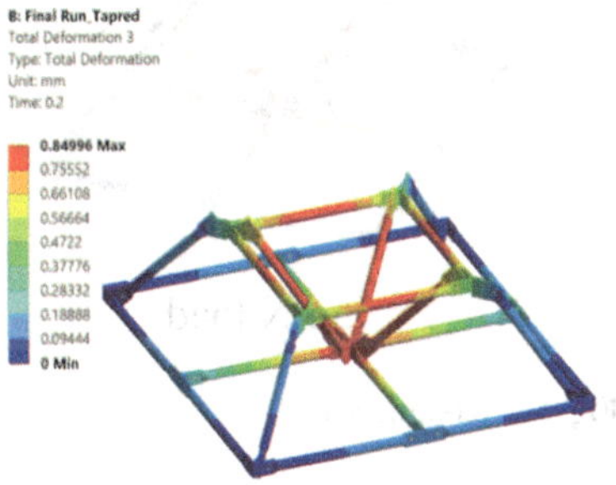

10 kN load

Fig. 12 Deformed shape of tapered flattened end tube

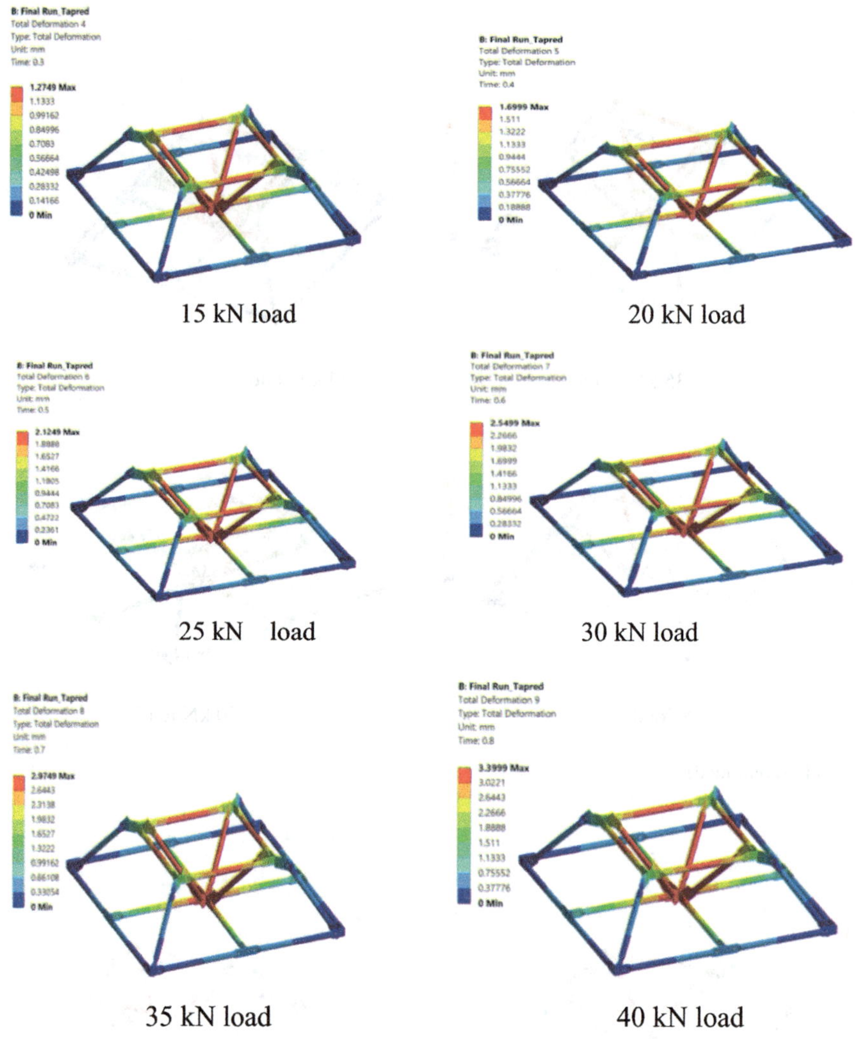

Fig. 12 (continued)

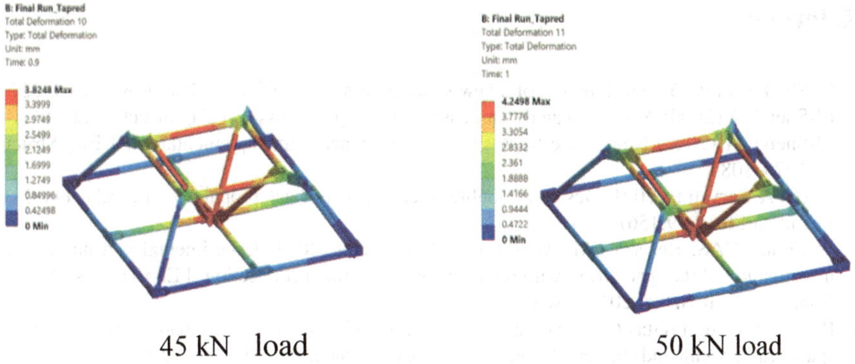

Fig. 12 (continued)

Table 1 Deflection of space frame for flatted, flattened and tapered flattened end tube

End tube	Deflection in mm									
	5 kN	10 kN	15 kN	20 kN	25 kN	30 kN	35 kN	40 kN	45 kN	50 kN
Simple tube	0.013	0.027	0.04	0.054	0.067	0.081	0.094	1.08	1.21	1.35
Flatted end tube	0.909	1.819	2.728	3.638	4.547	5.457	6.367	7.276	8.186	9.095
Flattened end tube	0.3704	0.740	1.111	1.481	1.852	2.593	2.963	3.334	3.704	3.704
Tapered flattened end tube	0.424	0.849	1.274	1.699	2.124	2.549	2.974	3.399	3.824	4.2498

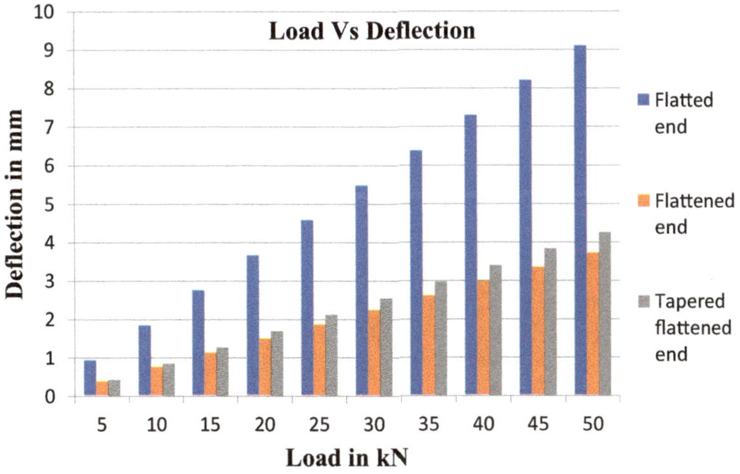

Fig. 13 Load versus deflection of flatted, flattened and tapered flattened end tube

References

1. El-Sheikh A (1996) Development of a new space truss system. J Constr Steel Res 37:205–227
2. El-Sheikh A (2000) Approximate dynamic analysis of space trusses. J Eng Struct 22:26–38
3. El-Sheikh A (2000) New space truss system from concept to implementation. J Eng Struct 22:1070–1085
4. Kaveh A, Servati H (2001) Design of double layer grid using back propagation neural networks. J Comput Struct 79:1561–1568
5. de Freitas CAS, Bezerra LM, Araujo RM, Araújo GM (2013) Experimental and numerical investigation of the space-truss with reinforce of the stamped connection. J Exp Trusses, Roofs, Connections; ICSDEC 2012 ASCE 2013
6. Bezerra LM, de Freitas CAS, Matias WT, Nagato Y (2009) Increasing load capacity of steel space trusses with end- flattened connections. J Constr Steel Res 65:2197–2206
7. de Andrade SAL, Vellasco PCGDS (2005) Tubular space trusses with simple and reinforced end-flattened nodes-an overview and experiments. J Constr Steel Res 61:1025–1050

Seismic Analysis of Base Isolated Building Frames with Experimentation Using Shake Table

Mahesh Kalyanshetti, Ramankumar Bolli, and Shashikant Halkude

Abstract Base isolation is the most effective and widely used technique for protecting a structure against seismic forces. The base isolation substantially decouples a structure from its base resting on the shaking ground, thus safeguarding the integrity of the structure by making them sustain the seismic forces without damage and also protect the lives of occupants. In the present study, an investigation is done to assess the seismic response of the structure with conventional fixed base and isolated base conditions. The isolation is configured with an elastomeric rubber isolator analyzed and designed in accordance with the International Building Code, IBC:2000. The analytical study is performed using ETAB application software and the experimentation is done using a uniaxial shake table on 1:12.5 scaled-down steel model developed using similitude laws corresponding to G + 3 and G + 5 prototype building frame. The study is carried out by applying three time histories, namely El-Centro, Uttarkashi and Indo-Burma. The fixed base condition is simulated in the laboratory by directly bolting the model on the shake table. The isolated base condition is simulated by mounting the model on an elastomeric rubber isolator. The study reveals that the acceleration produced in the base-isolated structure is reduced in the range of 40–60% in comparison with the fixed base condition. Pseudo spectral acceleration corresponding to fixed and isolated base conditions is obtained, which reveals that the time period of base-isolated structure increased to almost 2.5 s from 0.5 s. The shift in time period resulted in depletion in various parameters such as energy demand, pseudo acceleration and reinforcement requirement. The cumulative energy demand throughout the event of an earthquake is also studied. The reduction in energy demand for the isolated base condition is observed to be 50–90% lesser than in the fixed base condition. The reinforcement demand is reduced to 50% in isolated base condition. Thus, the study reveals that base isolation reduces the seismic demand of structure and enhanced safety, thereby reducing the cost of the structure.

M. Kalyanshetti (✉) · R. Bolli
Civil Engineering Department, Walchand Institute of Technology, Solapur, Maharashtra, India
e-mail: mgkalyanshetti@gmail.com

S. Halkude
Walchand Institute of Technology, Solapur, Maharashtra, India

M. S. Ranadive et al. (eds.), *Recent Trends in Construction Technology and Management*, Lecture Notes in Civil Engineering 260,
https://doi.org/10.1007/978-981-19-2145-2_62

Keywords Shake table · Base isolation · Scaled-down model · Elastomeric rubber · Time history · Time period · Pseudo acceleration

1 Introduction

The base isolation reduces earthquake forces by prolonging the time period of building. Different types of isolation systems used are friction pendulum bearing (FPB), elastomeric bearing, high damping bearing and lead core bearing. Elastomeric isolation system is used widely. Design of structure conventionally in seismic hazard regions would result in heavy costs. Provision of base isolation will prolong the time period, hence seismic demand will be reduced. The peak lateral acceleration was reduced by over 40% during dynamic tests carried out by Falborski and Jankowski [1]. Base isolation does not make a building earthquake-proof, but it can make the structure more flexible to control its frequency from dangerous resonance range [2]. The reduction in acceleration reduces the forces acting upon each story and the structure will move relatively less. The whole building frame during the earthquake tries to remain rigid. Base isolation reduced bending moment almost by four times as compared to fixed base structure [3]. This reduces the demand for reinforcement in the structure, hence making it cost-effective in heavy seismic zones. The base isolation is not effective in high-rise buildings as it produced large overturning moments and subsequently more tensile forces at the base [4]. Considering the effect of overturning and tilting, the tension resisting device arrangement is made to restrict the model from overturning and tilting. Low stiffness rubber isolators provide satisfactory structural behavior than moderate or high stiffness rubber isolator [5]. The elastomeric rubber bearing with hardness IRHD 60 is provided under each column which provides better stability and damping during the event of an earthquake, and also it can sustain the huge axial loads acting on it [6].

2 Objectives

The study is carried out to investigate the response of structure corresponding to fixed and isolated base conditions by carrying out analytical and experimental studies subjected to various time histories. Following are the specific objectives:

i. To study seismic performance of structure subjected to various time history motions;
ii. To study the effect of base isolation systems on seismic performance of building frame analytically using application software and experimentally using shake table;
iii. To know the effect of isolation for two different configurations of building frames;

Table 1 Details of prototype R.C. building frame

S. No.	Particulars	Details	
1	Configuration	G + 3	G + 5
2	Height of story (m)	3.5	3.5
3	Concrete grade	M 25	M 25
4	Steel grade	Fe415	Fe415
5	Width of bay (both directions) (m)	4	4
6	Thickness of slab (m)	0.15	0.15
7	Column size	0.45 m × 0.3 m	0.40 m × 0.25 m (above third story) 0.45 m × 0.45 m (up to third story)
8	Beam size	0.4 m × 0.23 m	0.4 m × 0.23 m

iv. To study the energy demand for the structures considering fixed and isolated base conditions;
v. To study the change in time period of the structure due to base isolation;
vi. To compare analytical results with experimental results to validate the experimental setup.

3 RC Building Frame Used for the Study

The study is carried out on G + 3 and G + 5 building frame with a single bay in both directions. In view of the feasibility of the experimental study, an appropriate geometry of the building frame is considered which is shown in Table 1.

4 Scaled-Down Steel Model

The important aspect of an experimental study is to configure an experimental model, which will produce identical behavior to that of the prototype structure. It is also observed that every aspect of the structure is not possible to be scaled due to payload capacity limitation of the shake table. But it should not be considered as a matter of big concern, since the objective of the test is not to qualify the prototype structure by testing but to study the behavior of structures, in general, subjected to dynamic loading. The most important task in scaling down is to achieve "dynamic similarity" so that the prototype and model experience homologous forces. For this purpose, the method suggested by Meymand [7] is adopted. Many of the scaling factors can be achieved by three principal test conditions. The first condition is a 1-g environment for testing, which ensures prototype and model accelerations to be the same. Secondly, the density of prototype and model should be similar by maintaining

Table 2 Scale factors

Parameters	Mass density	Acceleration	Frequency	Time	Length	EI	Strain
Scale factor	1	1	$S^{-1/2}$	$S^{1/2}$	S	S^5	1

other components of the scaling relations. Thirdly, appropriate scaling relation for natural frequency of prototype and model should be used. The scaling conditions are expressed in terms of the geometric scaling factor (S) for length, mass, acceleration, density and time.

4.1 Model Scale Factor

Selection of geometric scale factor is the critical step in the modeling process. In view of shake table specifications and the arrangements of holes on the table to hold the specimen, the spacing between two columns is kept as 0.32 m, which will lead to a linear scale factor of 'S' = 4.0/0.32 = 12.5 (corresponding to 4 m spacing in prototype structure). Therefore, a geometric scaling factor of 1:12.5 is adopted, and accordingly height, length, and width of the model are determined as 1.12 m, 0.32 m and 0.32 m, respectively. Table 2 shows various scale factors used for various parameters.

4.2 Scaling of Building Frames

The scaled-down models of both the building frames are prepared based on similitude laws. As a representative case, the scaled-down procedure of G + 3 building frame is discussed here.

The relationship between the natural frequency of model and prototype corresponding to $S = 12.5$ as per Table 2 is given in Eq. (1).

$$\begin{aligned} f_{\mathrm{m}}/f_{\mathrm{p}} &= S^{-1/2} \\ &= 3.54 \end{aligned} \tag{1}$$

Using the application software (modal analysis) natural frequency of the G + 3 prototype building is found, $f_{\mathrm{p}} = 1.4792$ Hz. Therefore, according to Eq. (1), the required frequency of the model is 5.24 Hz.

The density of the prototype structure (ρ_{m}) is 264 kg/m^3. Therefore, as per the second condition, the mass of the model (M_{m}) is calculated as

$$\begin{aligned} M_{\mathrm{m}} &= \rho_{\mathrm{m}} \times V_{\mathrm{m}} \\ &= 30.269\,\mathrm{kg} \end{aligned} \tag{2}$$

Table 3 Details of scaled-down model

S. No.	Particulars	Details	
1	Configuration	G + 3	G + 5
2	Height of story (mm)	280	280
3	Steel grade	Fe250	Fe250
4	Width of bay (mm)	320	320
5	Thickness of slab (mm)	4	3
6	Column size	10 mm × 10 mm	12 mm × 12 mm

The column and slab dimensions of scaled-down model are determined so that the weight of the model nearly equals to 30.269 kg and also it satisfies stiffness and flexural rigidity as required by simulated laws and produces a natural frequency of 3.54 Hz. The details of both the scaled-down models are presented in Table 3.

4.2.1 Scaling of Time History Input Motion for Scaled-Down Model

The study is carried out on time histories relevant to the Indian and global contexts. Two Indian time histories which are generally used by many researchers, i.e., Indo-Burma and Uttarkashi are used as these have produced high-intensity shaking, resulting in the destruction of thousands of homes. In the global context, El-Centro time history is very significant as it is characterized as a typical moderate-sized destructive event with a complex energy release. Thus, in view of all the above, the study is carried out on three time histories, namely Indo-Burma, Uttarkashi and El-Centro. It is essential to scale the input time history motion for the experimental study. The input provided to shake table is displacement time history. Hence in the case of scaling for the input motion of the shake table, the displacement and time has to be scaled with a linear scale factor. The representative calculations are shown below for El-Centro time history. Figure 1 shows the original time history motion of El-Centro 1940 earthquake with maximum displacement input of 131 mm. The scale factor for displacement and time is calculated in accordance with Table 2. The scale factor for displacement is 'S', i.e., 12.5 and the scale factor for time is '$S^{1/2}$', i.e., 3.535. Accordingly, the time history is scaled which is shown in Fig. 2. Similarly, Indo-Burma and Uttarkashi time histories are scaled.

5 Experimental Setup on Shake Table

The servo-hydraulic uniaxial shake table size is 2 m × 2 m with a maximum payload capacity of 30 kN with a frequency range of 0.01–50 Hz. The experimental setup

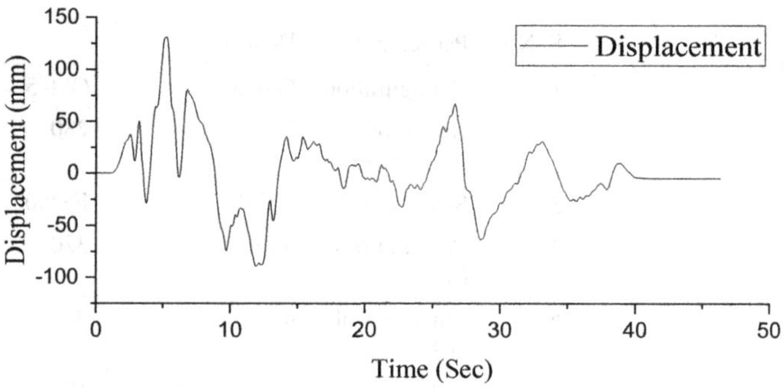

Fig. 1 Original El-Centro time history

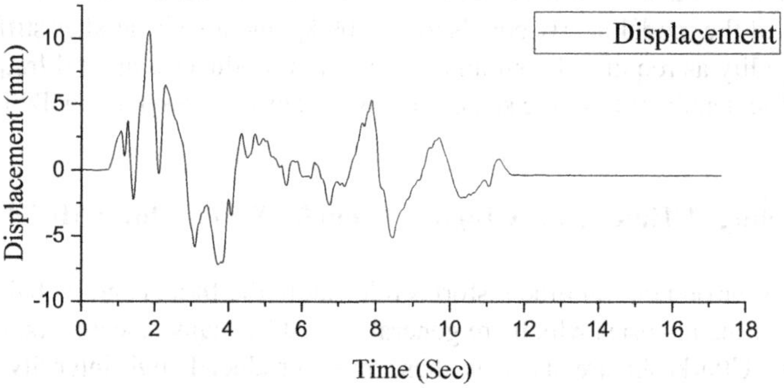

Fig. 2 Scaled El-Centro time history

corresponding to fixed and isolated base conditions is developed in the laboratory described in the following sections.

5.1 Fixed Base Condition

The fixed base condition is established by directly bolting the scaled-down model to the platform of the shaking table. Four accelerometers are used to record the acceleration. Accelerometer no. 1 is placed at the actuator of the shake table. Accelerometer nos. 2, 3 and 4 are used to record accelerations of various floor levels. The fixed base setup for G + 3 model is shown in Fig. 3. A similar setup is done for G + 5 model.

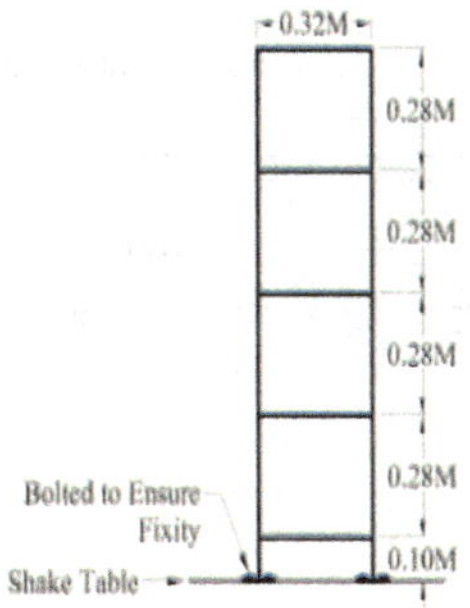

(a) Typical details of G+3 model

(b) Experimental setups of G+3

Fig. 3 Fixed base condition setup

5.2 Isolated Base Condition

Elastomeric rubber is used as an isolator. These are placed below each column footing. The size of the elastomer is determined based on International Building Code:2000 [8]. The details of the isolator are presented in the following subsections.

5.2.1 Design of Elastomeric Rubber Isolator

Seismic isolation systems are difficult to design and require complex analysis to produce the ideal design for the structure which is meant to achieve the desired target time period [9]. International Building Code (IBC) made provisions for the guidelines and design criteria for the isolation systems. The IBC:2000 is referred for design. The horizontal effective stiffness of the rubber required to achieve the target time period of 2.5 s is obtained by making various trials in the computer application. The effective stiffness for G + 3 model is worked out to be 105 kN/m. Similar calculations are done for G + 5 model.

The thickness of rubber is governed by the design displacement and it is calculated referring the Sect. 1623.2.2.1 of IBC:2000. Usually we need to achieve this displacement at about 150% shear strain. The dimensions of rubber for model worked out to be 35 mm × 35 mm in plan and 15 mm in thickness. The design of isolator for G + 5 models is also done similarly.

5.2.2 Laboratory Setup of Isolated Base Model

The experimental setup for isolated base condition is done by decoupling the structure from the shaking table. The rubber pad of the required size is prepared. The wooden

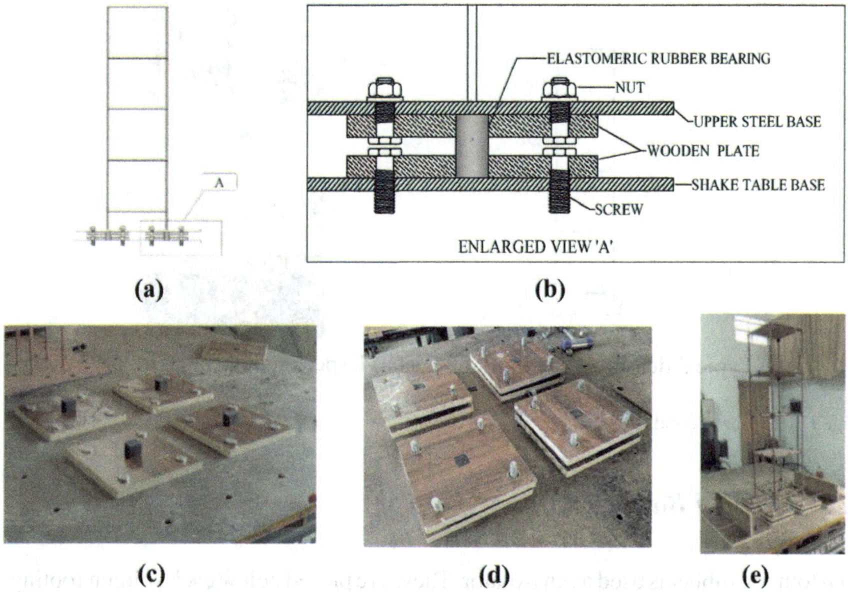

Fig. 4 **a** Typical sectional elevation of isolated base setup. **b** Enlarged detailed view 'A' of setup of isolator at base. **c** Rubber isolator without top plate at base. **d** Rubber isolator with top plate at base. **e** G + 3 isolated base model on uniaxial shake table

planks under each column have been used to hold the rubber isolator properly. The detailed setup is shown in Fig. 4a–e.

6 Analytical Study

The analytical study is carried out using the software ETABS which provides integrated features to do time history analysis on building frames and extract the results quickly and precisely. The original time history motions are given as input for the prototype structure shown in Fig. 5c. The fixed and isolated base G + 3 model and G + 5 model developed using application software are shown in Fig. 5a, b, respectively.

7 Results

The scaled-down models of G + 3 and G + 5 buildings with the fixed base condition and isolated base condition are subjected to El-Centro, Indo-Burma and Uttarkashi time histories. The models are fitted with an accelerometer at all the stories to

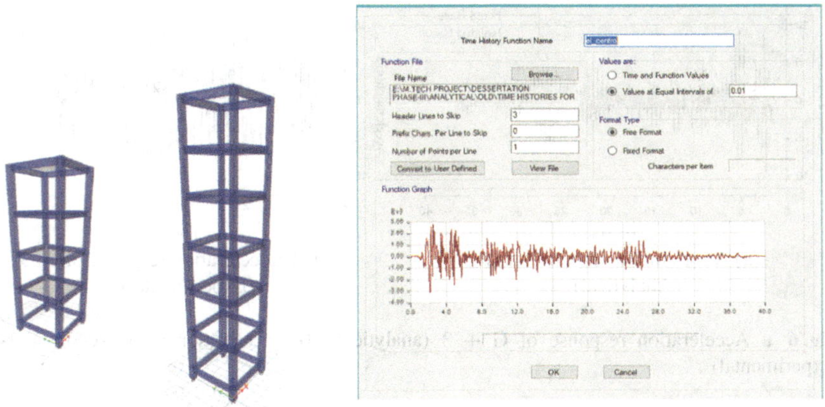

(a) (b) (c)

Fig. 5 **a** 3D model of G + 3 building. **b** 3D model of G + 5 building. **c** El-Centro time history function definition in ETABS

record the acceleration. The acceleration responses are presented below for fixed and isolated base conditions.

7.1 Response of G + 3 Building

Time history responses of roof story are shown in Sect. 7.1.1 and the response of all stories are illustrated in Sect. 7.1.2 for G + 3 building.

7.1.1 Comparison of Roof Acceleration of G + 3 Building Frame

The analytical and experimental results of roof acceleration for El-Centro time history for G + 3 model are presented in Fig. 6a, b, respectively. The response of the experimental study is presented on scaled time. The acceleration time history of the analytical study is presented in Fig. 6a, which shows that the maximum acceleration is @9.64 m/s^2, whereas for isolated base condition maximum acceleration is @3.41 m/s^2. Thus, the acceleration of the isolated base is reduced by almost 64% that of acceleration corresponding to the fixed base. The acceleration time history of the experimental study is presented in Fig. 6(b), which shows that the maximum acceleration is @10.70 m/s^2, whereas for isolated base condition maximum acceleration is @4 m/s^2. Thus, the acceleration of isolated base is reduced by almost 62% that of acceleration corresponding to the fixed base. It is also revealed that the acceleration obtained by analytical and experimental studies is within close agreement. The acceleration produced in the experimental study is almost 9–11% higher than the

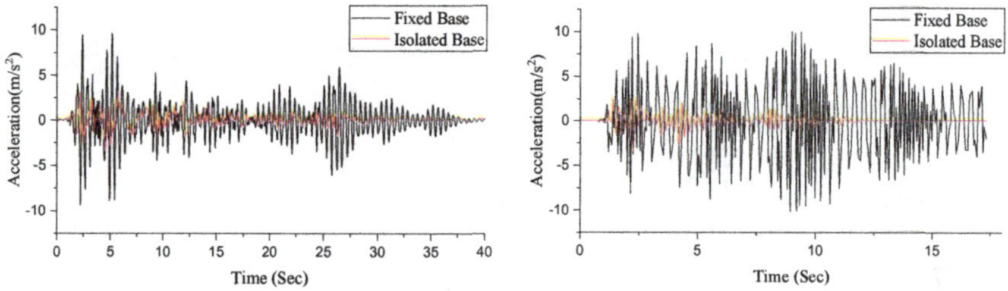

(a) Acceleration response of G+3 (Analytical)

(b) Acceleration response of G+3 (Experimental)

Fig. 6 **a** Acceleration response of G + 3 (analytical). **b** Acceleration response of G + 3 (experimental)

analytical study, which may be considered reasonably marginal. Thus, it is inferred that the experimental setup is simulated precisely in the laboratory.

7.1.2 Variation of Acceleration of All the Stories of G + 3 Building

The acceleration response of all stories subjected to all three time histories corresponding to fixed and isolated base are presented in Table 4 and subsequently shown in Fig. 7. The reduction in roof acceleration for isolated base condition in analytical study is almost 65% for El-Centro (from 9.64 to 3.42 m/s^2), 73% for Indo-Burma (from 4.40 to 1.27 m/s^2) and 64% (from 7.19 to 2.60 m/s^2) for Uttarkashi time history, shown in Fig. 7a, b, c, respectively. Similarly, in experimental study, reduction in acceleration for isolated base condition is almost 66% for El-Centro (from 10.70 to 3.59 m/s^2), 73% for Indo-Burma (from 4.88 to 1.34 m/s^2) and 66% (from 7.98 to 2.73 m/s^2) for Uttarkashi time history. So, it is revealed that the base isolation is causing almost 64–73% reductions in acceleration.

It is observed from Fig. 7a–c that the accelerations increase with a higher rate (almost nonlinear) in the case of fixed base condition, whereas in the case of isolated base condition the acceleration is observed to be increasing with a lower rate (almost

Table 4 Variation of acceleration of G + 3 building for all time histories

Acceleration for G + 3 building (m/s^2)												
Time history	El-Centro				Indo-Burma				Uttarkashi			
Story	IV	III	II	I	IV	III	II	I	IV	III	II	I
FA	9.64	7.62	6.09	4.00	4.40	3.47	2.78	2.50	7.19	5.68	4.54	4.20
FE	10.70	8.46	6.76	4.30	4.88	3.85	3.08	2.66	7.98	6.30	5.04	4.33
IA	3.42	2.70	2.16	1.80	1.27	1.01	0.80	0.76	2.60	2.05	1.64	1.45
IE	3.59	2.84	2.27	2.00	1.34	1.06	0.84	0.78	2.73	2.15	1.72	1.52

FA Fixed analytical; *FE* Fixed experimental; *IA* Isolated analytical; *IE* Isolated experimental

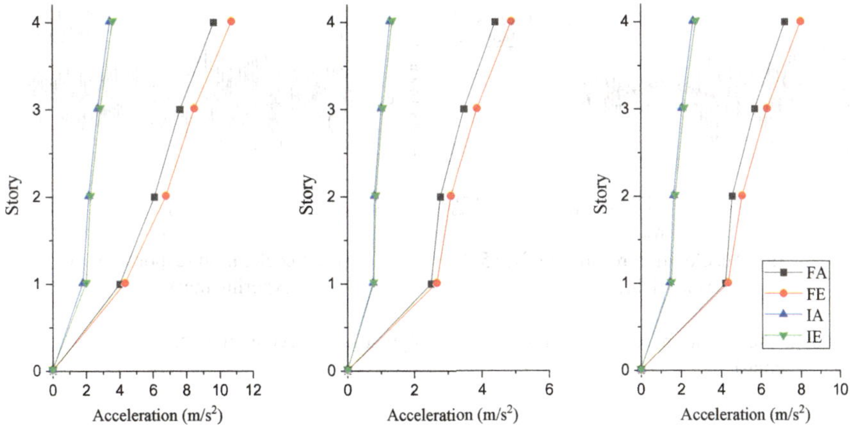

(a). El-Centro Response **(b).** Indoburma Response **(c).** Uttarkashi Response

Fig. 7 Variation of acceleration for G + 3 building

linear). The same trend is observed in analytical and experimental studies. In the fixed base condition, the difference in acceleration for the experimental and analytical studies is observed to be in the range of 10–12%, whereas in the isolated base condition it is in the range of 4–6%.

7.2 Response of G + 5 Building

Time history responses of roof story are shown in Sect. 7.2.1 and the response of all stories are illustrated in Sect. 7.2.2 for G + 5 building.

7.2.1 Comparison of Roof Acceleration of G + 5 Building Frame

The analytical and experimental results of roof story acceleration for El-Centro time history for G + 5 model are presented in Fig. 8a, b, respectively. The response of the experimental study is presented on scaled time. The acceleration time history of the analytical study is presented in Fig. 8a, which shows that the maximum acceleration of the fixed base is @9.17 m/s^2, whereas for the isolated base condition the maximum acceleration is @3.64 m/s^2. Thus, the acceleration of the isolated base is reduced by 61% that of acceleration corresponding to the fixed base. The acceleration time history of the experimental study is presented in Fig. 8b, which shows that the maximum acceleration of the fixed base is @10.59 m/s^2, whereas for the isolated base condition maximum acceleration is @4.35 m/s^2. Thus, the acceleration of the isolated base is reduced by almost 60% that of the acceleration corresponding to the

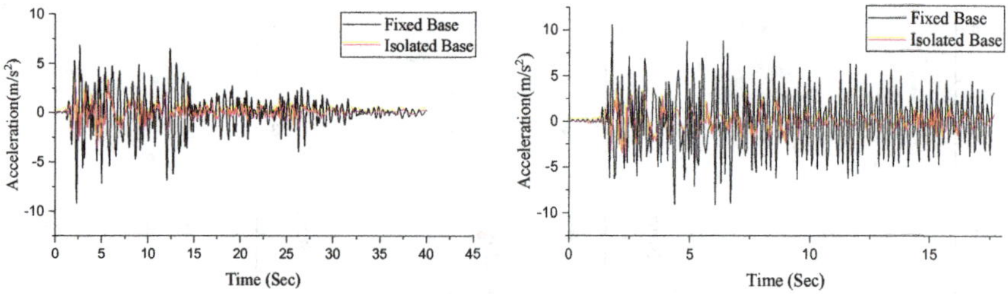

(a). Acceleration response of G+5 (Analytical)

(b). Acceleration response of G+5 (Experimental)

Fig. 8 **a** Acceleration response of G + 5 (analytical). **b** Acceleration response of G + 5 (experimental)

fixed base. The acceleration obtained from the analytical and experimental studies is in close agreement. The acceleration produced in the experimental study is almost 15–20% higher than the analytical study.

7.2.2 Variation in Responses of All the Stories of G + 5 Building

The acceleration responses of all stories subjected to all three time histories corresponding to the fixed and isolated base conditions are presented in Table 5 and subsequently shown in Fig. 9. The reduction in roof acceleration for isolated base condition in the experimental study is almost 59% for El-Centro (from 10.59 to 4.35 m/s^2), 65% for Indo-Burma (from 4.43 to 1.57 m/s^2) and 48% (from 5.67 to 2.97 m/s^2) for Uttarkashi time history, shown in Fig. 9a–c, respectively. Similarly, in analytical study, reduction of acceleration for isolated base is almost 60% for El-Centro (from 9.17 to 3.64 m/s^2), 65% for Indo-Burma (from 3.80 to 1.32 m/s^2) and 55.71% (from 5.60 to 2.48 m/s^2) for Uttarkashi time history, shown in Fig. 9a–c, respectively. So, it is revealed that the base isolation is causing almost 48–65% reductions in acceleration. In the fixed base condition, the difference in acceleration

Table 5 Acceleration response of G + 5 building for all time histories

Acceleration for G + 5 building (m/s^2)																		
Time history	El-Centro						Indo-Burma						Uttarkashi					
Story	VI	V	IV	III	II	I	VI	V	IV	III	II	I	VI	V	IV	III	II	I
FA	9.2	7.0	5.9	5.3	3.3	3.0	3.8	2.9	2.7	2.4	1.7	1.5	5.6	5.4	5.1	4.4	3.3	3.0
FE	10.6	8.2	6.9	6.0	3.9	3.3	4.4	3.3	2.5	2.0	2.5	1.7	5.7	5.5	5.2	5.0	3.7	3.3
IA	3.6	3.6	3.5	3.3	3.2	2.5	1.3	1.3	1.3	1.3	1.2	1.2	2.5	2.5	2.5	2.5	2.5	2.2
IE	4.4	4.1	4.1	3.5	3.3	2.7	1.6	1.4	1.4	1.4	1.3	1.3	3.0	2.9	2.8	2.7	2.7	2.4

FA Fixed analytical; *FE* Fixed experimental; *IA* Isolated analytical; *IE* Isolated experimental

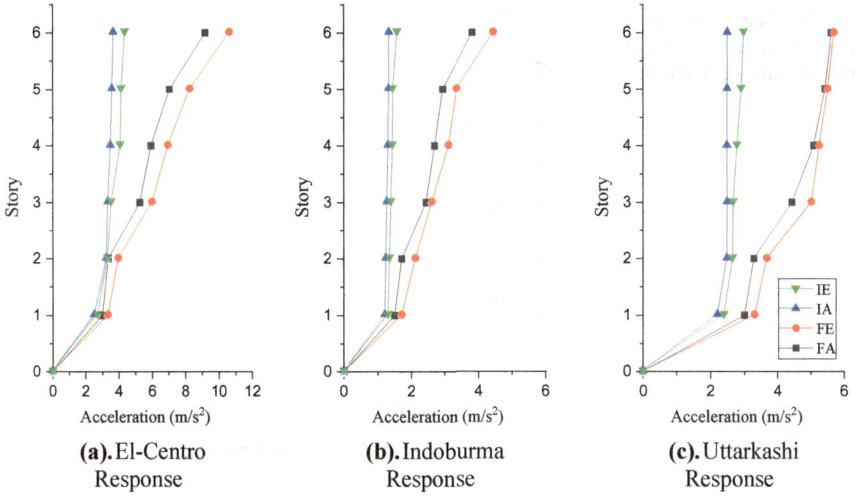

Fig. 9 Variation in acceleration for G + 5 building

for the experimental and analytical studies is observed to be in the range of 10–12%, whereas in the isolated base condition it is 4–6%.

7.3 *Pseudo Spectral Acceleration*

Response spectrum curves for the three time histories have been obtained for both the buildings at different damping ratios from 0 to 10%. The time period for isolated model has to be shifted to the desired target time period to which the structure is designed. The isolator is designed considering the IBC:2000 code which recommends achieving a target time period of 2.5 s for having little or no ductility demand.

The typical plot showing pseudo spectral acceleration on Y-axis vs the time period on X-axis for G + 3 building subjected to El-Centro is shown in Figs. 10 and 11 for fixed and isolated base conditions, respectively. It is observed that the time period is 0.5 s and peak pseudo acceleration is 190 m/s^2 for G + 3 fixed base building shown in Fig. 10, which is after isolation observed to be approximately 2.5 s with peak velocity being 13 m/s^2 as shown in Fig. 11. Hence the isolation of the base increased the time period of the structure to have little or no ductility demand.

The PSA for all time histories for G + 3 building is shown in Figs. 12, 13 and 14 for fixed and isolated base conditions, respectively. The PSA is plotted at different damping levels. The maximum pseudo acceleration is observed at 0% and 2% damping for El-Centro time history shown in Fig. 12. A reduction of 88–97% in peak pseudo acceleration is observed. The time period is consistent and it is close to 2.5 s for isolated base model.

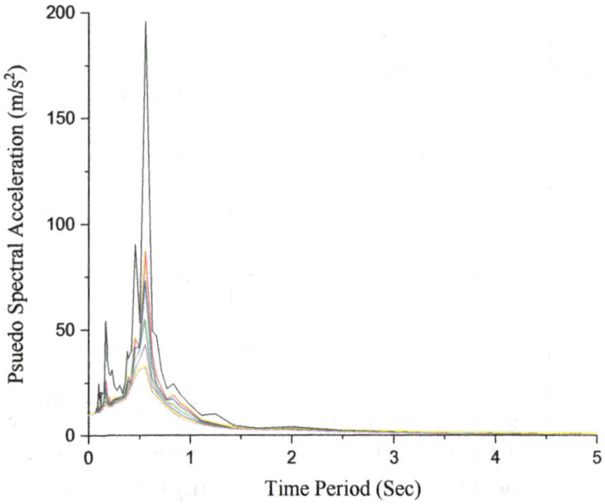

Fig. 10 Pseudo spectral acceleration for G + 3 fixed base building (El-Centro)

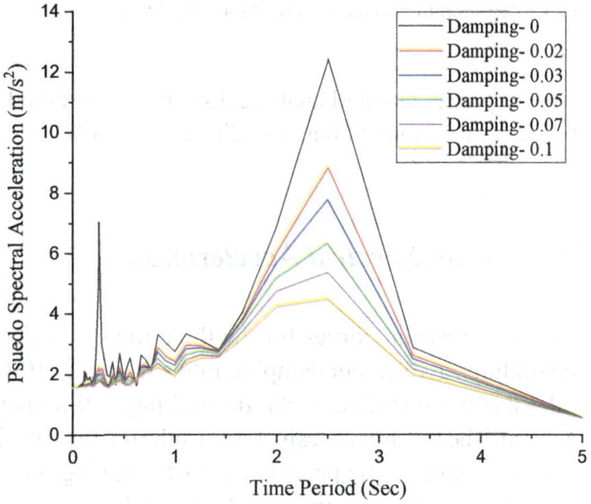

Fig. 11 Pseudo spectral acceleration for G + 3 isolated base building (El-Centro)

The PSA for G + 5 building for all time histories is shown in Figs. 15, 16 and 17. The maximum pseudo acceleration is observed at 0 and 2% damping for El-Centro time history. A reduction of 82–97% in peak pseudo acceleration is observed and shown in Fig. 15. The pseudo acceleration is observed to be 13% more in G + 5 building than in G + 3 building. However, the isolated base model was producing almost similar results in peak accelerations. The time period is shifted approximately to 2.5 s for isolated base model, resulting in reduced accelerations and increased time period.

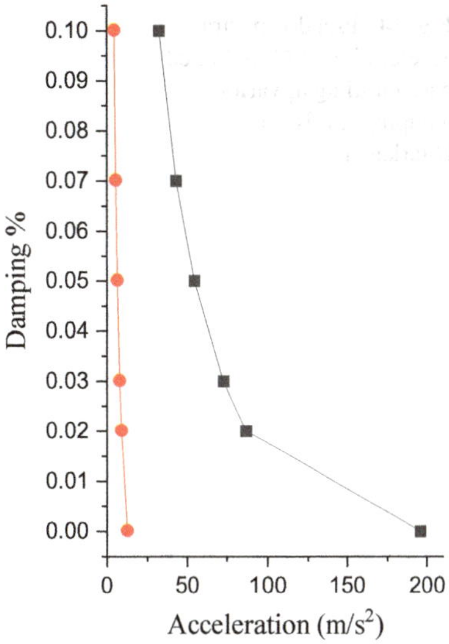

Fig. 12 Pseudo spectral acceleration of G + 3 isolated base building at various damping levels for El-Centro

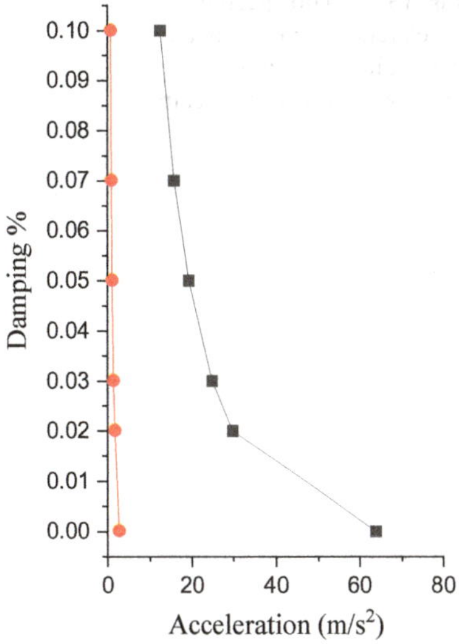

Fig. 13 Pseudo spectral acceleration of G + 3 fixed base building at various damping levels for Indo-Burma

7.4 Energy Demand

The conventional structures have evolved a lot to resist the seismic forces and are more capable of distributing the load evenly while maintaining a sustained energy demand. In the regions where an earthquake has more dominance than any other load, the conventional structure even designed very efficiently will require higher energy, hence the demand will be higher. The most efficient solution to sustain an earthquake while maintaining low energy demand is the introduction of an energy dissipation

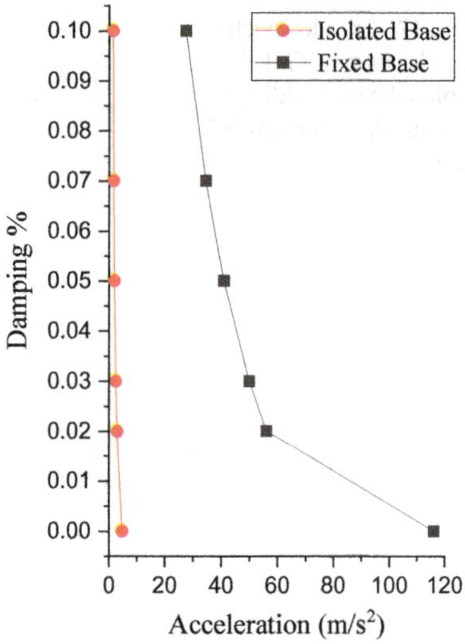

Fig. 14 Pseudo spectral acceleration of G + 3 fixed base building at various damping levels for Uttarkashi

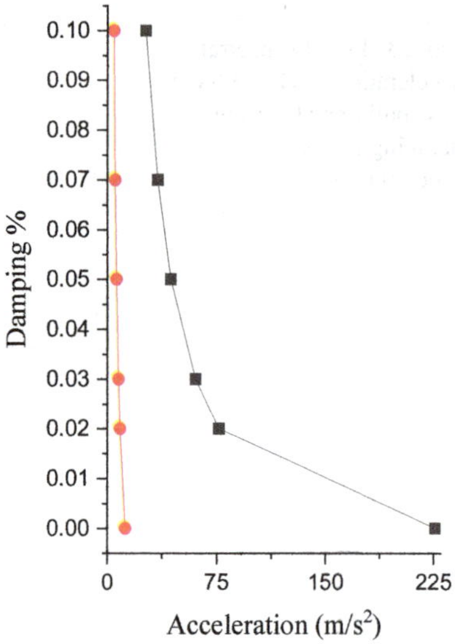

Fig. 15 Pseudo spectral acceleration of G + 5 fixed base building at various damping levels for El-Centro

device (i.e., base isolation). The energy demand for both the conventional, i.e., fixed base and isolated base conditions have been found in the analytical study in ETABs and compared.

The typical plot showing the energy demand of G + 3 building for El-Centro time history is shown in Figs. 18 and 19 for fixed and isolated base conditions, respectively. It is observed that the fixed base building is demanding almost 78% more energy than the isolated base condition.

Fig. 16 Pseudo spectral acceleration of G + 5 isolated base building at various damping levels for Indo-Burma

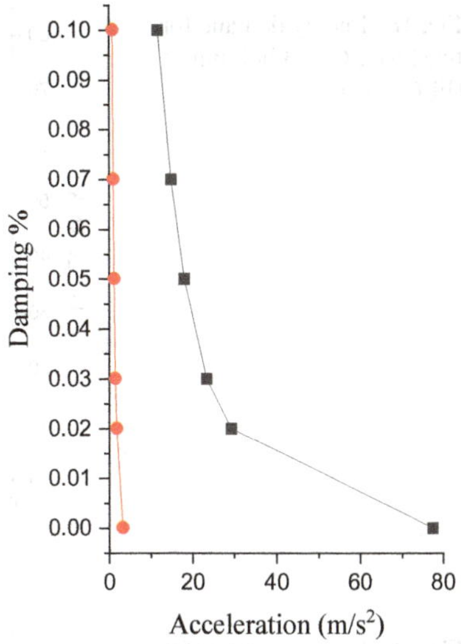

Fig. 17 Pseudo spectral acceleration of G + 5 fixed base building at various damping levels for Uttarkashi

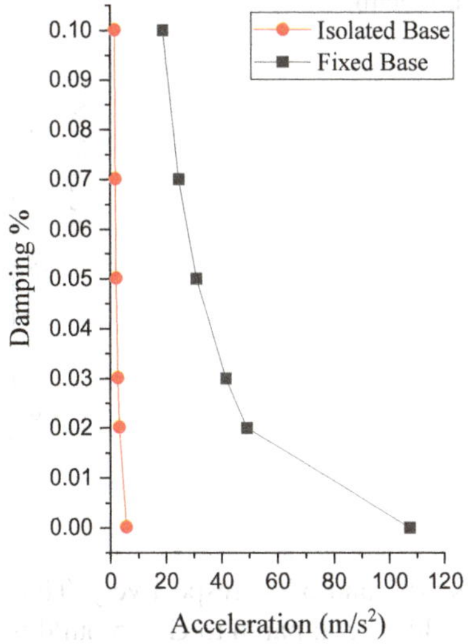

The cumulative energy required to sustain an earthquake through the entire event for all time histories is shown in Fig. 20 for G + 3 building and in Fig. 21 for G + 5 building. It is observed that the energy demand in the fixed base condition is 78.57% more for El-Centro time history for G + 3 building shown in Fig. 20. Similarly, for Indo-Burma and Uttarkashi approximately 90% more energy is required as compared to the isolated base condition.

For G + 5 building the energy demand is reduced by 44.23% compared to the fixed base condition shown in Fig. 21. The reduction in Indo-Burma and Uttarkashi

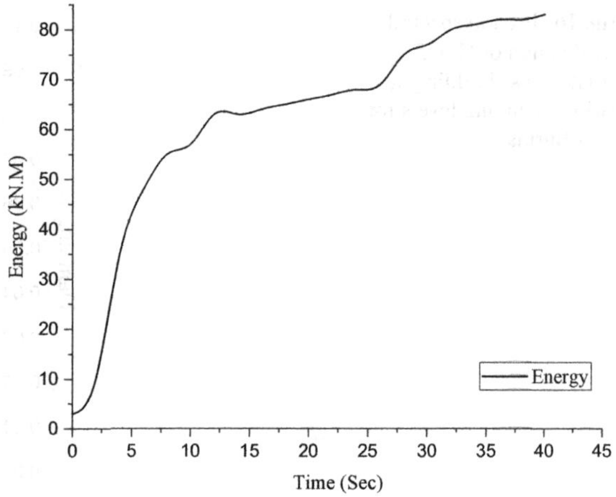

Fig. 18 Energy demand for fixed base G + 3 building (El-Centro)

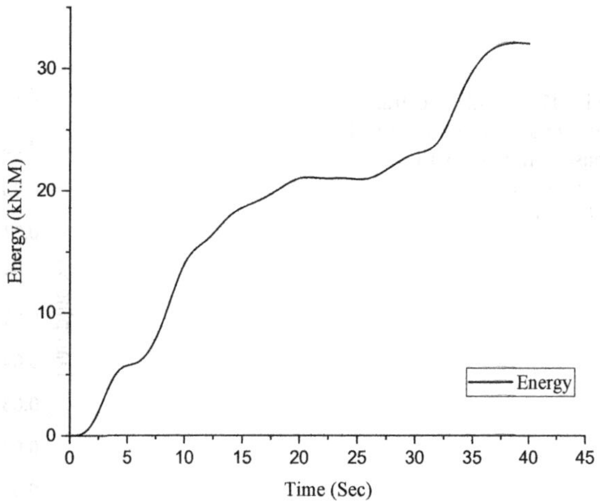

Fig. 19 Energy demand for isolated base G + 3 building (El-Centro)

is 90% and 68%, respectively. The reduction in energy demand is more for G + 3 building compared to G + 5 building.

8 Conclusion

In this study, the effect of base isolation is investigated experimentally and analytically for two scaled-down steel building frames of G + 3 and G + 5 which are designated as low-rise and high-rise buildings, respectively. The results of each building are compared and the conclusions are drawn as follows:

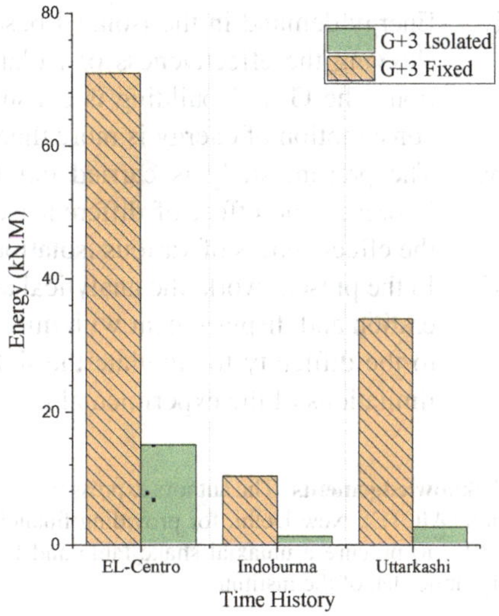

Fig. 20 Energy demand for G + 3 building

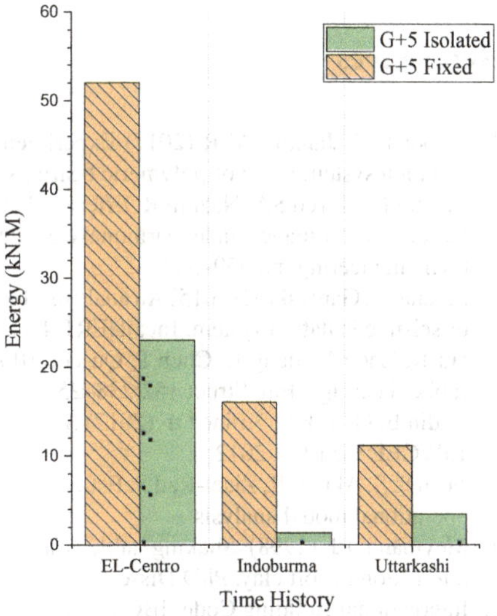

Fig. 21 Energy demand for G + 5 building

i. Base isolation is observed to be distinctly effective as acceleration has been reduced by almost 60% and displacement is reduced by @40%. This trend is observed to be almost the same for all time histories.
ii. Base isolation effect is more significant in low-rise buildings as the percentage reduction in the acceleration and displacement is higher than high-rise building. So, it is inferred that base isolation is beneficial in low-rise buildings.
iii. The base isolation increases the time period so as to have no or little ductility demand which is achieved at all damping levels.

iv. Energy demand in the isolated base buildings is reduced by almost 40–90% showing the effectiveness of isolation to keep the structure from deterioration. The G + 3 building is consuming less energy and also a reduction in consumption of energy is more than that of G + 5 building.
v. The present study is carried out by using an elastomeric rubber isolator. However, the effect of different isolation systems may be studied to access the effectiveness of various isolation methods.
vi. In the present work, the analytical and experimental study produces the acceleration and displacement with minimal difference which is quite obvious due to the difficulty to simulate the field conditions in the laboratory due to the limitations of the experimental facility.

Acknowledgements The authors express their gratitude to All India Council for Technical Education (AICTE), New Delhi, for providing financial assistance under Research Promotion Scheme (RPS) to procure a uniaxial shake table and to create an experimental facility in the structural dynamics lab of the institute.

References

1. Falborski T, Jankowski R (2017) Experimental study on effectiveness of a prototype seismic isolation system made of polymeric bearings. MDPI Appl Sci 7(808):7–16
2. Zahura FT, Javed SA, Naznin R. Effect of base isolation & different bracing system to improve building performance under earthquake excitations. In: International conference on advances in civil engineering, pp 359–363
3. Seranaj A, Garevski M (2015) An analysis of reinforced concrete buildings with different location of seismic isolation system. Int J IJERT 4(1). ISSN: 2278-0181
4. Hu K, Zhou Y, Jiang L, Chen P, Qu G (2017) A mechanical tension-resistant device for lead rubber bearings. Eng Struct 152:238–250
5. Aydin E, Ozturk B, Kilinc OF (2012) Seismic response of low-rise base isolated structures. In: 15WCEE LISBOA 2012
6. Maalek S, Akbari R, Ziaei-Rad S. Estimation of elastomeric bridge bearing shear modulus using operational modal analysis
7. Meymand PJ (1998) Shaking table scale model tests of nonlinear soil-pile-superstructure interaction in soft clay. PhD Dissertation, University of California, Berkeley
8. International Building Code, IBC:2000
9. Naeim F, Kelly JM. Design of seismic isolated structures: from theory to practice

Critical Study of Wind Effect on RC Structure with Different Permeability

Ankush Asati and Uday Singh Patil

Abstract Wind load is one of the significant loads that affect the design of a structure. Therefore, to analyze and design a building considering the effect of wind load, it is necessary to calculate the wind speed and pressure for different heights of buildings with different geometry. In the current building codes and design standards, internal stresses are insufficiently defined, primarily due to the complexities involved in their assessment, but play an important role in the design of buildings and cladding. The category of the building with a single dominant opening is usually considered critical for wind load design. However, the correct assessment of internal pressure behavior for buildings with cross-openings of realistic wall porosities is essential. In this paper, the reinforced concrete framed structure is analyzed for three classes of structure, i.e., Class A, B, and C with the constant base dimension; with varying heights of the structure; having four bays in each direction, and ceiling height of 3 m. All the models are analyzed for three different basic wind speeds, i.e., 33, 44, and 55 m/s with three different permeabilities, i.e., <5, 5–20, and >20% for different terrain categories specified in respective standards. The influence of the wind load response in terms of axial force, bending moment, and peak deflections are calculated and compared. It is observed that the effect of terrain category and permeability is not significant in centrally located columns. However, in corner columns, the axial forces are not constant. It increases with the increase in wind speed, permeability, and reduces from terrain category 1–4. Also, significant co-relation between forces and displacement was observed for less than 5% permeability, wind speed of 33 m/s, and terrain category 1.

Keywords Permeability · RC structure · Terrain category · Wind load

A. Asati · U. S. Patil (✉)
Department of Civil Engineering, Yeshwantrao Chavan College of Engineering, Nagpur, India
e-mail: patil.udaysingh4@gmail.com

M. S. Ranadive et al. (eds.), *Recent Trends in Construction Technology and Management*, Lecture Notes in Civil Engineering 260,
https://doi.org/10.1007/978-981-19-2145-2_63

1 Introduction

Structural designers refer to relevant standards on wind loads while designing tall buildings for wind loads to arrive at correct values of wind forces that will be acting at different floor levels of the structure. One of the significant loads that affect the design of the structure is wind load. To analyze the structure considering the effect of wind load, it is necessary to calculate the wind speed and pressure for different heights of buildings and different geometry. Various major standards and codes give us the procedures and coefficients required for the calculation of wind load for particular parameters, like the importance of building, surrounding terrain, the topography of building size, pressure, and force coefficients for different geometric shapes and sizes of buildings. (IS: 875-Part-3,1987) provides the values of coefficient of external pressure for rectangular and square plan shape clad buildings with rectangular corners for wind incidence angle normal to the surfaces. As per the historical wind speed data, India is divided into several zones, and design wind speed is considered according to the wind map of India. In the current building codes and design standards, internal stresses are insufficiently defined, primarily due to the complexities involved in their assessment, but play an important role in the design of buildings and cladding. While designing the structure subjected to wind load, generally wall with a foremost opening is considered important but the correct assessment of internal pressure behavior for structure with cross-openings of realistic wall porosities is also essential. Many researchers have used various methods to evaluate the equivalent static wind load. Ranjitha et al. analyzed buildings with different shapes for zone I and IV of India from linear static and dynamic analysis. Naved Ahmed and Rajeev Banerjee studied the effect of wind load on buildings with different H/B ratios. Sidhh et al. compared IS 875 Part 3: 1987 and IS 875 Part 3: 2015 on an existing structure. Nakul et al. carried out an analytical study to calculate the R factor for RCC and industrial structures to check its feasibility as given in Indian standards. In the present study, the critical study of the effect of wind loading on various sizes and shapes of the building with different permeability as per IS 875-part 3 is considered. The wind loads were applied to three different structures of class A, class B, and class C. The influence of the wind load response in terms of axial force, bending moment, and peak deflections is calculated and compared. As per the Indian standards, the static load is calculated for various structures and the acting forces depend largely on the basic wind speed, terrain category, class of structure, the height of the structure, and permeability. There is also a need to develop a mathematical model and investigate if there is a close relationship between the coefficient specified in various tables in the code. Accordingly, the problem statement of the present study is to critically analyze structure subjected to wind load with variable input parameters such as wind velocity, class of structure, the height of the structure, permeability and to investigate the possibility of development of co-relation/mathematical model among the forces and displacement. The results of a parametric study will show the variation in the forces of the structural member of the RC buildings with the change in the permeabilities and terrain categories. Thus, subsequently will help in the efficient design of structures that are both cost-effective and safe.

2 Literature Review

Raj and Ahuja studied the variations of base shear, base moments, and twisting moments with wind incidence angle on 3 models, i.e., A, B, and C [1–18]. On a building structure, wind load is maximum when it is exposed to a maximum area. It is found that with the change in the wind incidence angle, wind loads get affected and modified accordingly. It is also observed that for the same floor area of the building the shape of the building affects the wind load to a great extent. Thus compared to the square shape of the building, the percentage increase in the values of these forces depends on the building's cross-sectional shapes. Verma et al. carried out an experimental investigation on a wind tunnel with a plan of square shape and tall structure with boundary layer flow.

The values of coefficients of wind pressure were obtained from the various pressure point on the surfaces of the wall of the model. The goal of the research was to produce and create data for the designers to be able to safely design structural frames and wall claddings subjected to wind loads [19]. The internal airflow field has been found to have a major effect on the values of the internal pressure. It is seen that this field is not uniform for a wall that has porosity configurations greater than 10% and openings that are located on the neighboring walls. For the building of category 3, the values of peak internal pressure coefficients were found to surpass design values. Stathopoulos et al. analyzed models of low-rise structures with various geometries and experimentally evaluated the induced internal pressures in it for different volumes. It is found that magnitudes of internal pressures obtained are lesser than the values of local external pressure. Internal pressure coefficients for windward are found to be positive and found to be negative or zero for high porosity with small openings. A good relation between components of external and internal pressures has been established. Seavhai and Narayanan compared the value of roof pressures of the low-rise rectangular plan building with a dominant, non-dominant opening, and canopies with the help of 4 standards, namely, MS 1553-2002, EC 1991-1-4:2005, IS 875-3-1987, and BS 6399-2:1997 on various roof types. The maximum wind speed of 33.5 m/s as stated in MS 1553-2002 was used for the analysis. Reference height of buildings, averaging time for wind speed, and pressure coefficient values are some of the factors that are found to vary with different codes. Chauhan et al. estimated the wind forces acting on tall buildings for the design of structural elements. Using the Force coefficient method and Gust Factor Method, the influence of static and dynamic velocity fluctuations on the along wind loads for the structures was determined. A 60 and 120 m buildings are analyzed by ETABS Software with 4 different Terrain Categories and 6 basic wind speeds. Holmes et al. compared 3 buildings using 15 different codes and standards from the Asia–Pacific Region. The comparisons showed varying degrees of agreement.

3 Mathematical Model and Analytical Parameter

3.1 Computational Model

In this study, an analysis of the multi-storied building subjected to wind forces for all-terrain categories is carried out. 3-D model is prepared for G + 5, G + 14, and G + 24 multi-storied buildings in STAAD-Pro. The reinforced concrete frame structure is modeled for three classes of structure, i.e., Class A, B, and C having constant base dimensions with varying heights of structure; having four bays in each direction, and a ceiling height of 3 m. All the models are analyzed for basically three different basic wind speeds, i.e., 33, 44, and 55 m/s with three different permeability, i.e., <5, 5–20, and >20%, for four different terrain categories. The structural details used in the present study are cross-sectional dimensions of columns (450 × 450) mm, dimensions of beams (300 × 500) mm, 150 mm slab thickness, infill walls in the buildings are assumed to be 230 mm thick, storey height is 3 m, live load considered as 3 KN/m^2, plan size is 16 m × 16 m. Wind parameters considered are terrain category: 1, 2, 3, and 4; risk coefficient: 1; wind in *X*-direction and *Z*-direction. For analysis, the corner column and center column of the building are considered.

4 Results and Discussions

The member forces, bending moment, and displacement are calculated for a load combination of 1.2(DL + LL + WL), for Class A, Class B, and Class C structures.

4.1 Class A Structure

From Figs. 1, 2, 3, 4, 5, 6, 7, 8, 9 and 10, it can be seen that the axial force in the corner column for class A structure is increasing with the increase in wind speed and permeability. Also, it reduces from terrain category 1 to terrain category 4. The shear force FY and FZ in the corner column for class A structure are found to increase with an increase in wind speed with an increase in permeability. Shear force value is found to reduce from terrain category 1 to terrain category 4. The displacement in the *X*, *Y*, and *Z* directions in the corner column for class A structure is increasing with an increase in wind speed, permeability and reduced from terrain category 1 to terrain category 4. Since the corner columns are more exposed to wind thus the maximum response in *X*, *Y*, and the *Z* directions is observed that explains the increase in the displacement in the corner columns.

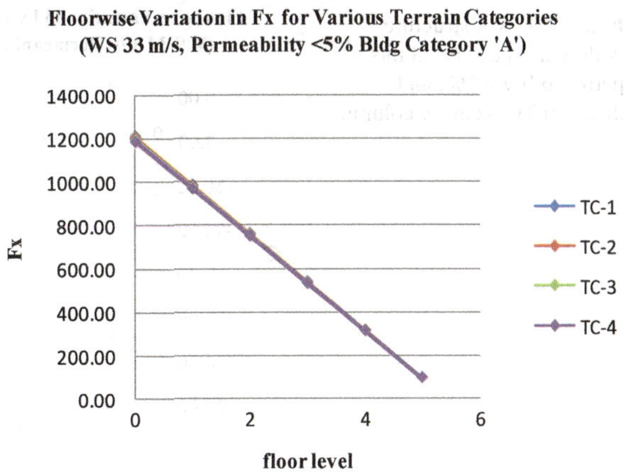

Fig. 1 Class A structure with wind speed = 33 m/s, permeability < 5%, and different TC—corner column

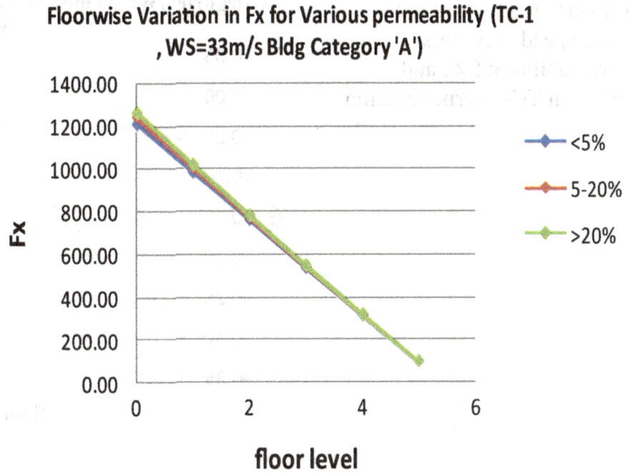

Fig. 2 Class A structure with TC-1, wind speed = 33 m/s, and different permeability—corner column

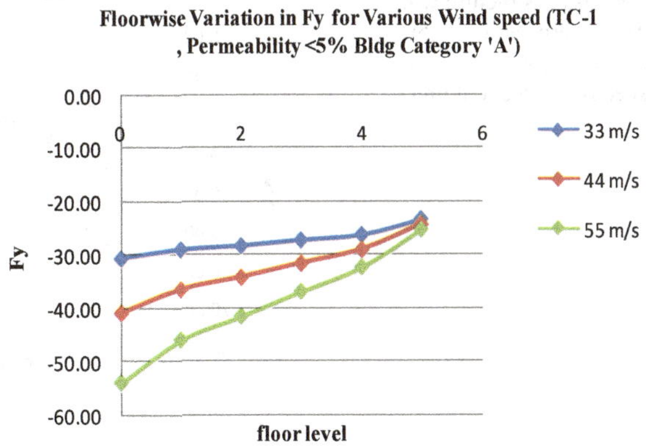

Fig. 3 Class A structure with TC-1, permeability < 5%, and different wind speeds—corner column

Fig. 4 Class A structure with wind speed = 33 m/s, permeability < 5%, and different TC—corner column

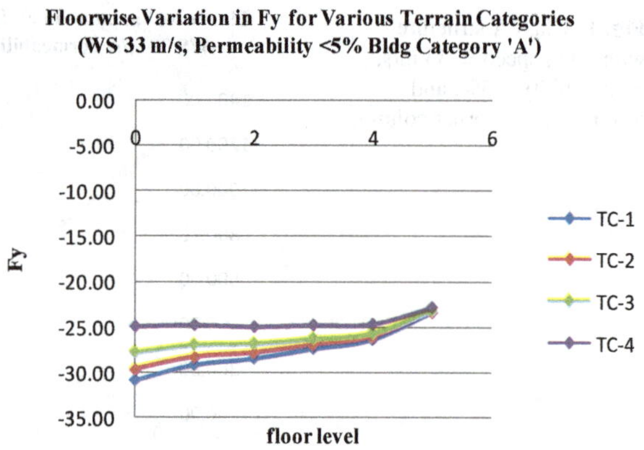

Fig. 5 Floor level and MX for class A structure with wind speed = 33 m/s, permeability < 5%, and different TC—corner column

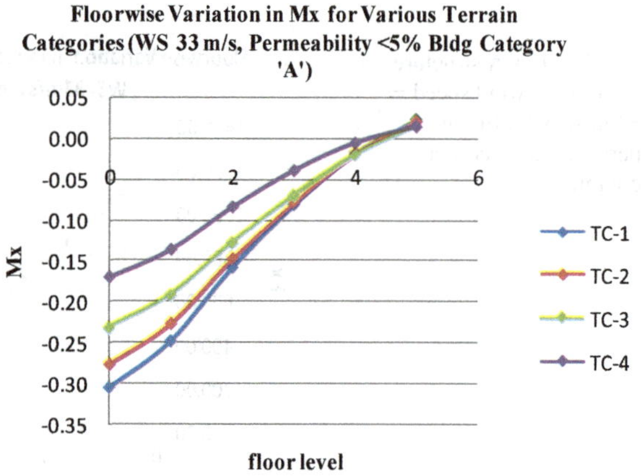

Fig. 6 Floor level and MY for class A structure with TC-1, permeability < 5%, and different wind speeds—corner column

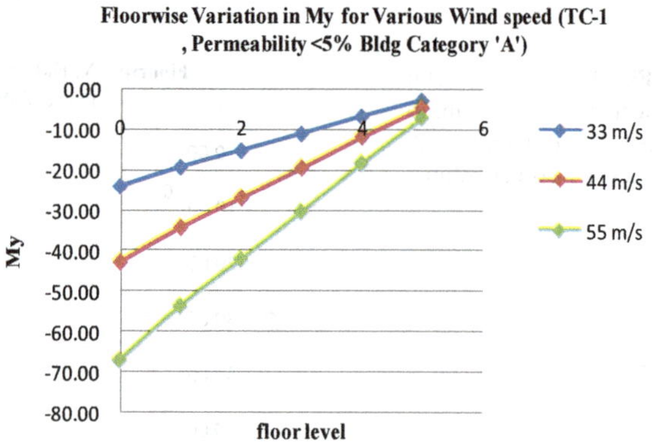

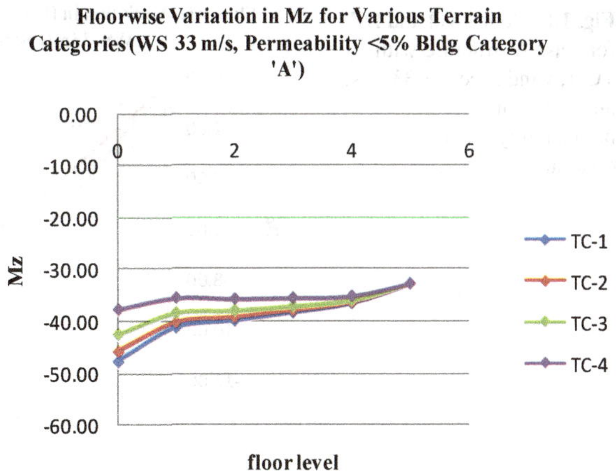

Fig. 7 Floor level and MZ for class A structure with wind speed = 33 m/s, permeability < 5%, and different TC—corner column

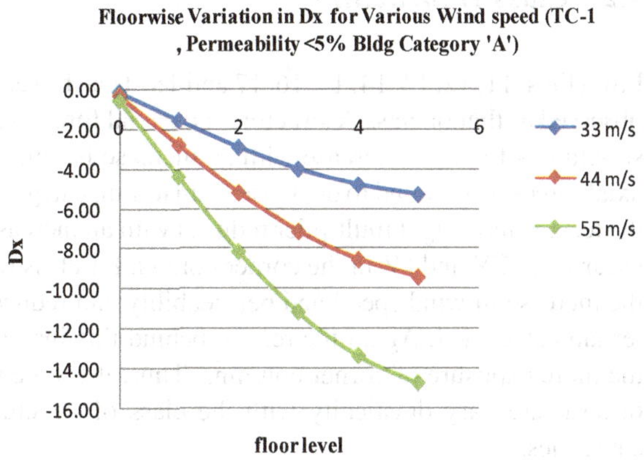

Fig. 8 Floor level and DX for class A structure with TC-1, permeability < 5%, and different wind speeds—corner column

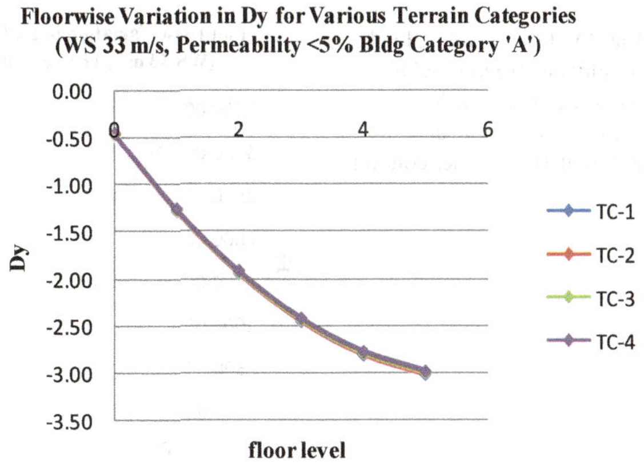

Fig. 9 Floor level and DY for class A with wind speed = 33 m/s, permeability < 5%, and different TC—corner column

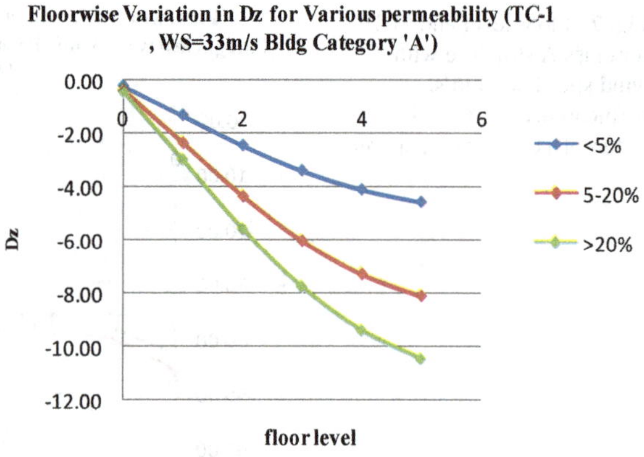

Fig. 10 Floor level and DZ for class A structure with TC-1, wind speed = 33 m/s, and different permeability—corner column

4.2 *Class B Structure*

From Figs. 11, 12, 13, 14, 15, 16, 17 and 18, it can be seen that a similar trend has been observed as that of class A structure, i.e., axial force in the corner column for class B structure is found to increase with an increase in wind speed and permeability. The axial forces were found to decrease from terrain category 1 to terrain category 4 since the terrain and height multiplier reduces with an increase in the terrain category. The shear force FY and FZ in the corner column for class B structure has increased with the increase in wind speed and permeability and reduced from terrain category 1 to terrain category 4. Again the reason behind this increased response is attributed to the more exposure of corner columns. Thus, it can be inferred that dynamic effects of structure vary drastically with the class of structure, permeability, and terrain categories.

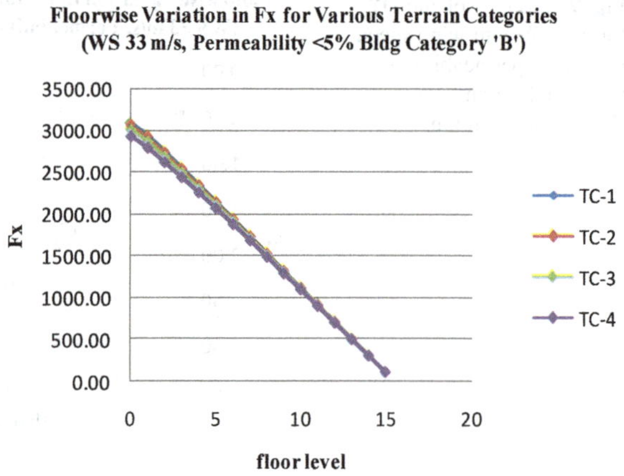

Fig. 11 Floor level and FX for class B structure with wind speed = 33 m/s, permeability < 5%, and different TC—corner column

Fig. 12 Floor level and FY for class B structure with wind speed = 33 m/s, permeability < 5%, and different TC—corner column

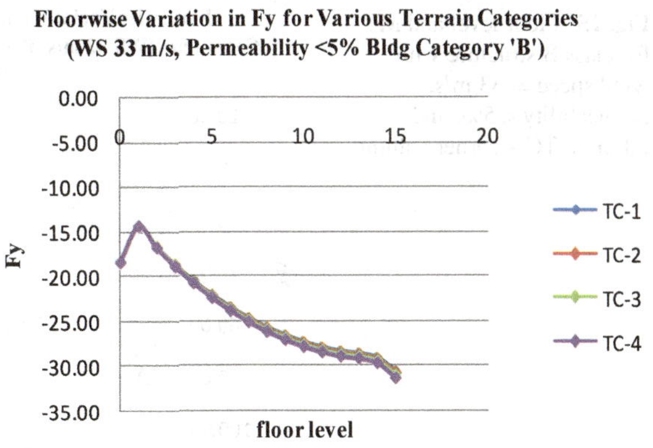

Fig. 13 Floor level and FZ for class B structure with wind speed = 33 m/s, permeability < 5%, and different TC—corner column

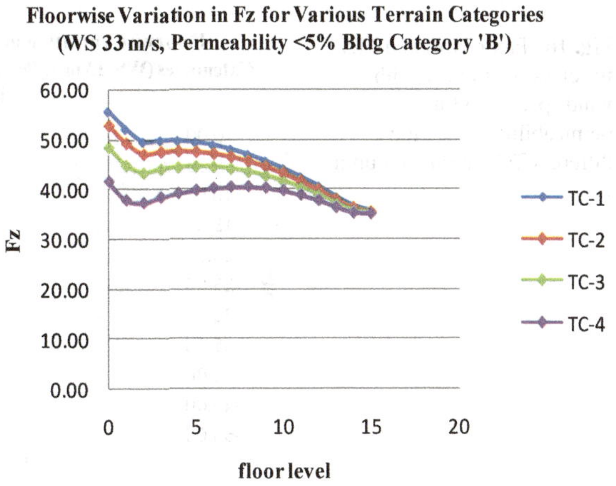

Fig. 14 Floor level and MX for class B structure with wind speed = 33 m/s, permeability < 5%, and different TC—corner column

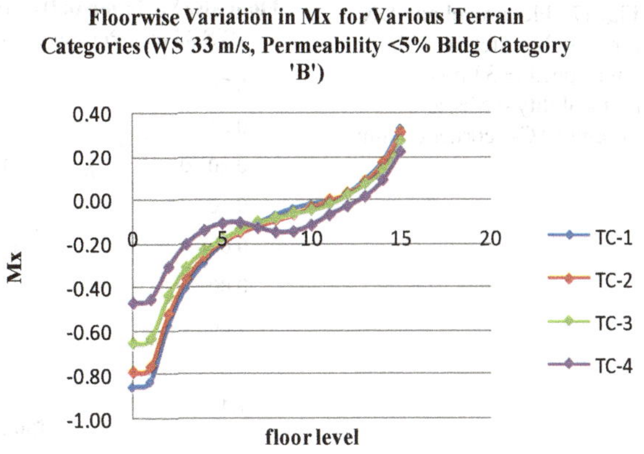

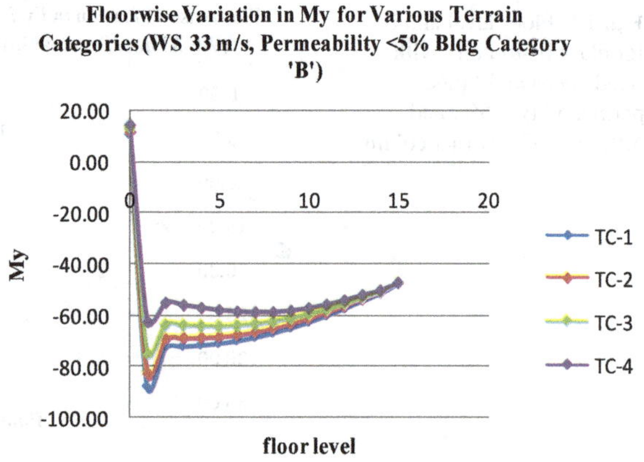

Fig. 15 Floor level and MY for class B structure with wind speed = 33 m/s, permeability < 5%, and different TC—corner column

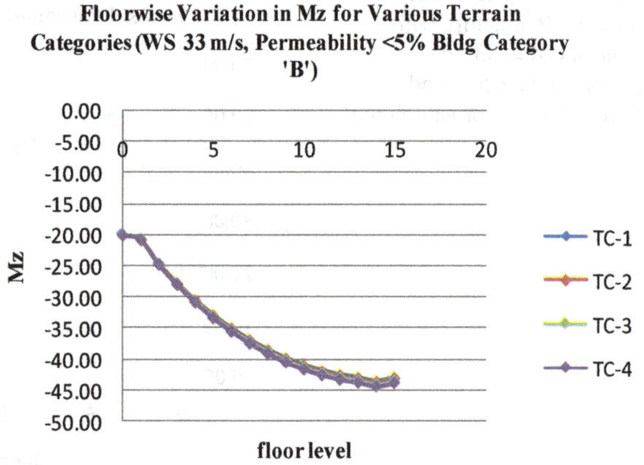

Fig. 16 Floor level and MZ for class B structure with wind speed = 33 m/s, permeability < 5%, and different TC—corner column

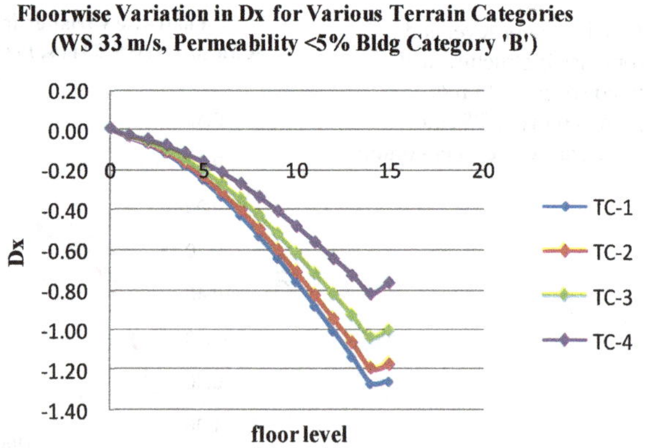

Fig. 17 Floor level and DX for class B structure with wind speed = 33 m/s, permeability < 5%, and different TC—corner column

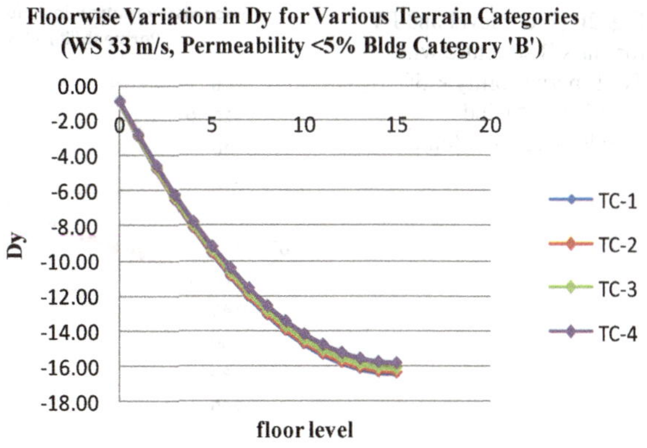

Fig. 18 Floor level and DY for class B structure with wind speed = 33 m/s, permeability < 5%, and different TC—corner column

4.3 Class C Structure

Figures 19, 20, 21, 22, 23, 24, 25, 26, 27, 28, 29 and 30, axial force in the corner column for class C structure is found to increase with the increase in wind speed, permeability and started reducing from terrain category 1 to terrain category 4. The shear force FY and FZ in the corner column for class C structure are increasing with the increase in wind speed, permeability, and reducing from terrain category 1 to terrain category 4. Thus, it can be inferred that the variation in wind speed with height is also influenced by the roughness of the ground, and so differs for each terrain category (IS: 875(Part3): Wind Loads on Buildings and Structures—(Proposed Draft and Commentary)).

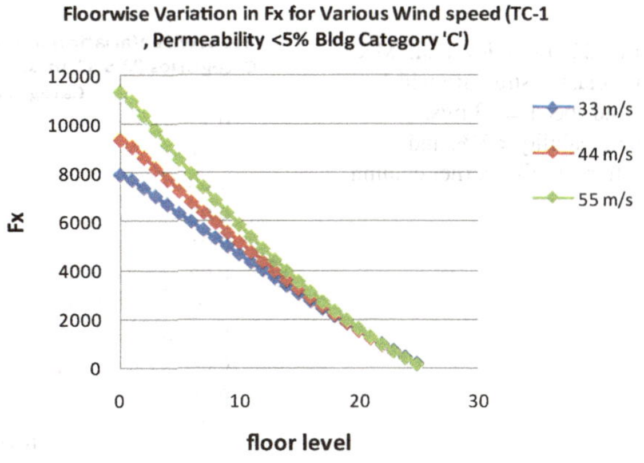

Fig. 19 Floor level and FX for class C structure with TC-1, permeability < 5%, and different wind speeds—corner column

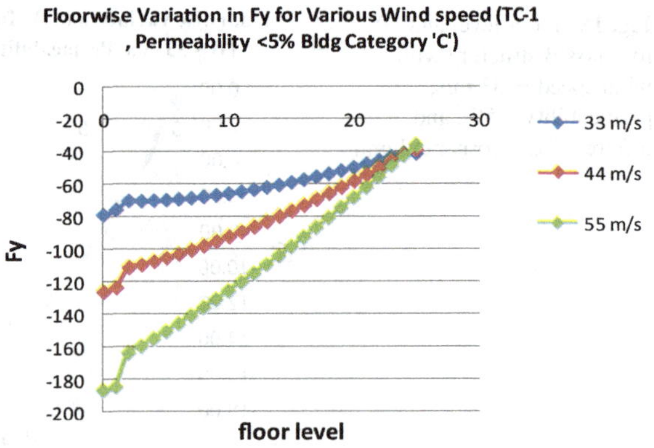

Fig. 20 Floor level and FY for class B structure with TC-1, permeability < 5%, and different wind speeds—corner column

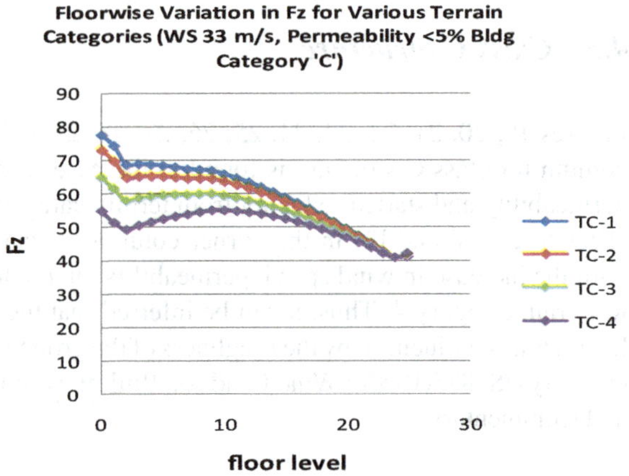

Fig. 21 Floor level and FZ for class C structure with wind speed = 33 m/s, permeability < 5%, and different TC—corner column

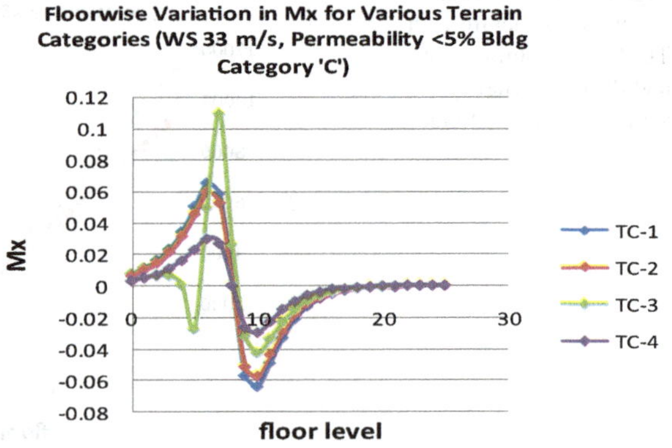

Fig. 22 Floor level and MX for class C structure with wind speed = 33 m/s, permeability < 5%, and different TC—corner column

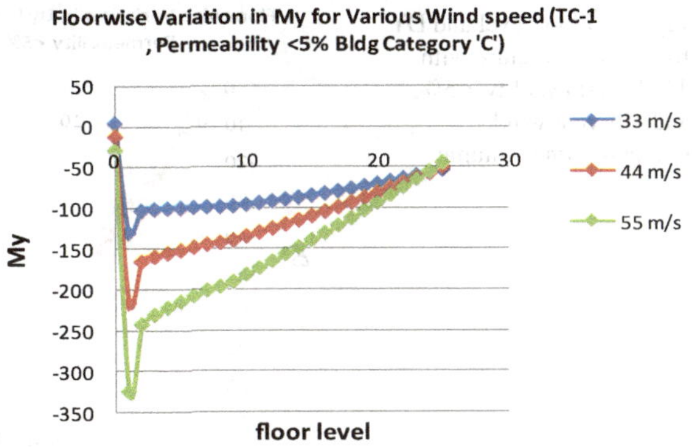

Fig. 23 Floor level and MY for class C structure with TC-1, permeability < 5%, and different wind speeds—corner column

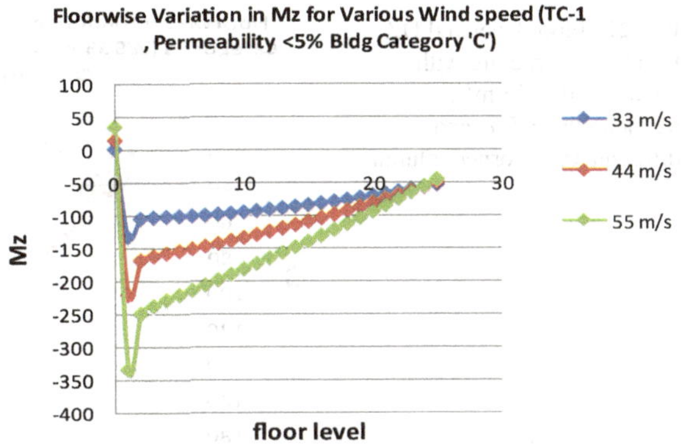

Fig. 24 Floor level and MZ for class C structure with TC-1, permeability < 5%, and different wind speeds—corner column

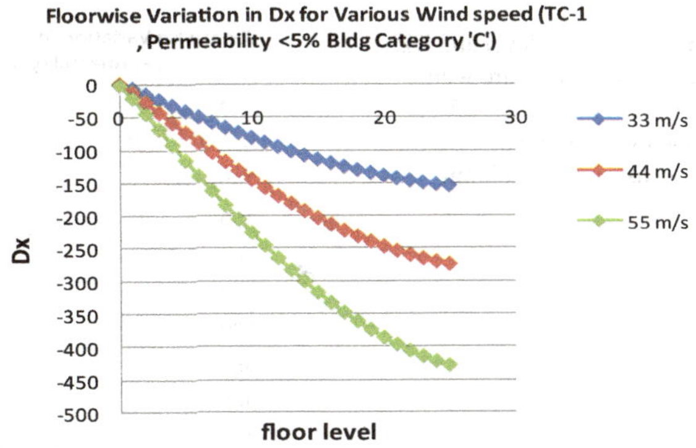

Fig. 25 Floor level and DX for class C structure with TC-1, permeability < 5%, and different wind speeds—corner column

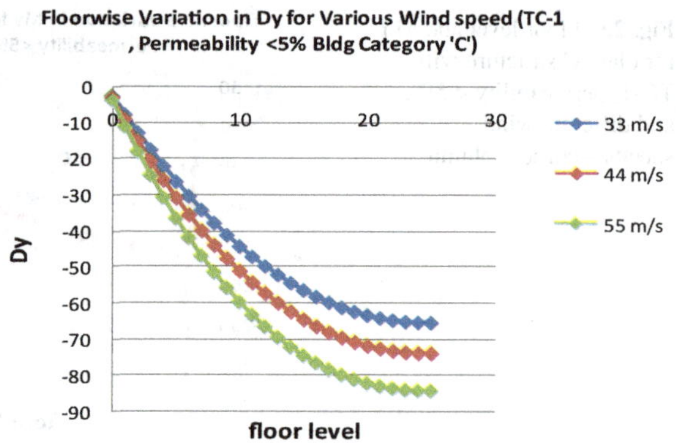

Fig. 26 Floor level and DY for class C structure with TC-1, permeability < 5%, and different wind speeds—corner column

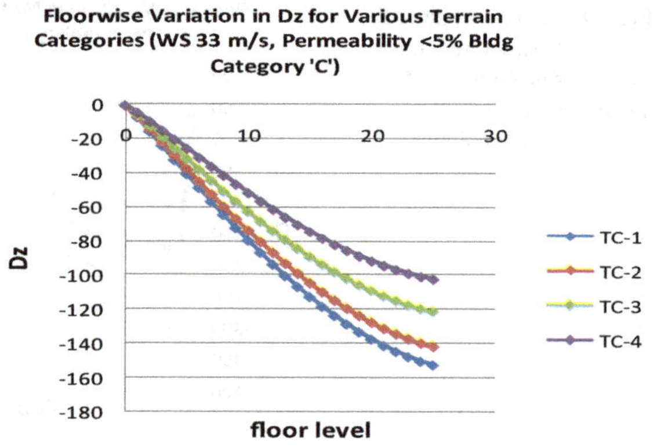

Fig. 27 Floor level and DZ for class C structure with wind speed = 33 m/s, permeability < 5%, and different TC—corner column

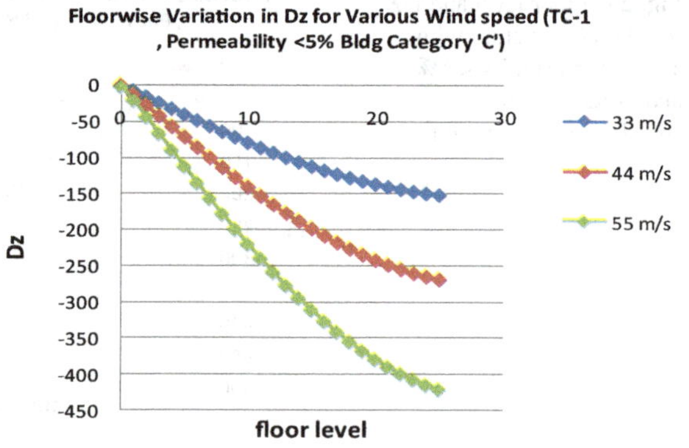

Fig. 28 Floor level and DZ for class C structure with TC-1, permeability < 5%, and different wind speeds—corner column

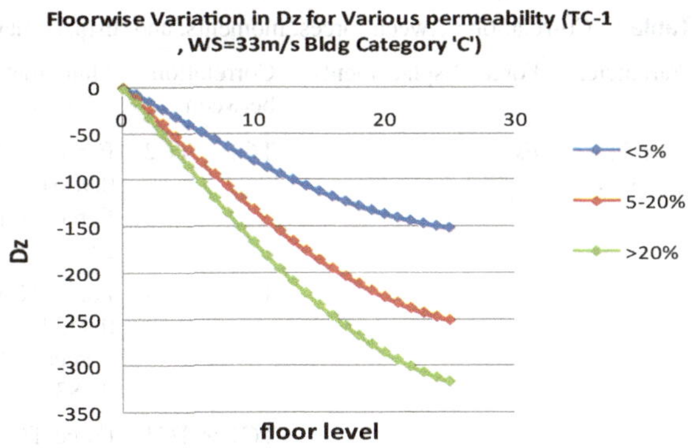

Fig. 29 Floor level and DZ for class C structure with TC-1, wind speed = 33 m/s, and different permeability—corner column

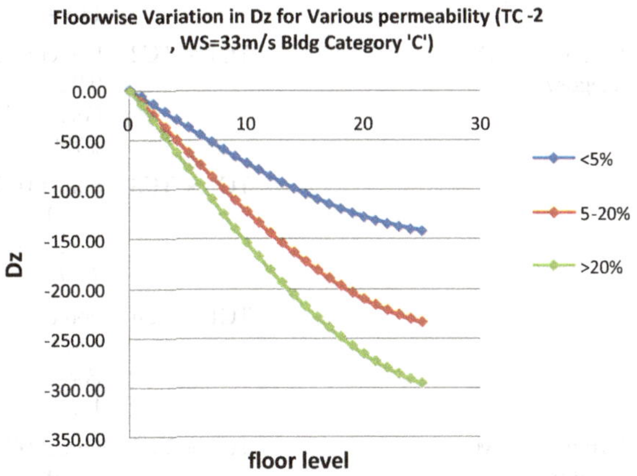

Fig. 30 Floor level and DZ for class C structure with TC-2, wind speed = 33 m/s, and different permeability—corner column

4.4 Co-relation Between Forces, Moments, Displacement with Terrain Categories, Permeabilities, and Wind Speed

To understand the correlation between various parameters like forces, moments, and displacement among the condition specified in IS 875 (Part-3), the data for all buildings are clubbed together. Graphs are plotted, and the trained line equation along with R^2 values was noted. Corresponding values are shown in Tables 1, 2 and 3 and observed that there exists an excellent relation between forces, moments, and displacements with terrain categories, permeabilities, and wind speed.

Table 1 Co-relation between forces, moments, and displacement for terrain categories

Parameter	Force/Displacements	Correlation between	Mathematical relation	R^2 value	Observation
Terrain category	Fx	TC1 × TC2	Force(TC2) = 0.9804 × Force(TC1) + 25.27	0.9999	Excellent correlation
		TC1 × TC3	Force(TC3) = 0.9427 × Force(TC1) + 72.83	0.9991	
		TC1 × TC4	Force(TC4) = 0.906 × Force(TC1) + 116.4	0.997	
Terrain category	Fy	TC1 × TC2	Force(TC2) = 0.925 × Force(TC1) − 2.799	0.999	
		TC1 × TC3	Force(TC3) = 0.790 × Force(TC1) − 7.791	0.996	
		TC1 × TC4	Force(TC4) = 0.639 × Force(TC1) − 13.70	0.978	
Terrain category	Mx	TC1 × TC2	Force(TC2) = 0.900 × Force(TC1) − 0.008	0.999	
		TC1 × TC3	Force(TC3) = 0.741 × Force(TC1) − 0.017	0.993	
		TC1 × TC4	Force(TC4) = 0.492 × Force(TC1) − 0.042	0.909	
Terrain category	My	TC1 × TC2	Force(TC2) = 0.926 × Force(TC1) − 1.027	0.999	
		TC1 × TC3	Force(TC3) = 0.795 × Force(TC1) − 2.032	0.999	

(continued)

Table 1 (continued)

Parameter	Force/Displacements	Correlation between	Mathematical relation	R^2 value	Observation
		TC1 × TC4	Force(TC4) = 0.651 × Force(TC1) − 2.730	0.995	
Terrain category	Dx	TC1 × TC2	Force(TC2) = 0.930 × Force(TC1) + 0.27	1	
		TC1 × TC3	Force(TC3) = 0.799 × Force(TC1) + 0.413	1	
		TC1 × TC4	Force(TC4) = 0.665 × Force(TC1) + 1.308	0.999	
Terrain category	Dy	TC1 × TC2	Force(TC2) = 0.984 × Force(TC1) − 0.078	0.999	
		TC1 × TC3	Force(TC3) = 0.952 × Force(TC1) − 0.249	0.999	
		TC1 × TC4	Force(TC4) = 0.924 × Force(TC1) − 0.313	0.997	

5 Conclusion

Following are the conclusions from the present study:

1. It is observed that the forces bear a constant proportionality ratio with various terrain categories, wind speeds, and different permeability conditions for class A, class B, and class C structures.
2. The moment and displacement were found to vary with a constant ratio for all-terrain categories, wind speed, and different permeability in all classes of structures, i.e., class A, class B, and class C.
3. No significant effect is observed in the axial force in the center column of the structure due to wind loads.
4. Displacements and Moments are significant to wind loads for all-terrain categories, especially for corner columns.

Table 2 Co-relation between forces, moments, and displacement for permeability

Parameter	Force/Displacement	Correlation between	Mathematical relation	R^2 value	Observation
Permeability	Fx	<5% × 5–20%	Force(5–20%) = 1.111 × Force(<5%) − 132.7	0.997	Excellent correlation
		<5% × >20%	Force(>20%) = 1.185 × Force(<5%) − 220.4	0.993	
Permeability	Fy	<5% × 5–20%	Force(5–20%) = 1.615 × Force(<5%) + 20.60	0.985	
		<5% × >20%	Force(>20%) = 2.022 × Force(<5%) + 34.17	0.975	
Permeability	Mx	<5% × 5–20%	Force(5–20%) = 1.673 × Force(<5%) + 2E−05	0.998	
		<5% × >20%	Force(>20%) = 2.122 × Force(<5%) + 0	0.997	
Permeability	My	<5% × 5–20%	Force(5–20%) = 1.649 × Force(<5%) − 0.413	0.999	
		<5% × >20%	Force(>20%) = 2.082 × Force(<5%) − 0.712	0.999	
Permeability	Dx	<5% × 5–20%	Force(5–20%) = 1.649 × Force(<5%) + 0.169	1	
		<5% × >20%	Force(>20%) = 2.077 × Force(<5%) + 0.147	1	
Permeability	Dy	<5% × 5–20%	Force(5–20%) = 1.094 × Force(<5%) + 0.454	0.997	

(continued)

Table 2 (continued)

Parameter	Force/Displacement	Correlation between	Mathematical relation	R^2 value	Observation
		<5% × >20%	Force(>20%) = 1.156 × Force(<5%) + 0.752	0.994	

Table 3 Co-relation between forces, moments, and displacement for wind speed

Parameter	Force/Displacement	Correlation between	Mathematical relation	R^2 value	Observation
Wind speed	Fx	33 × 44	Force(44) = 1.113 × Force(33) − 119.3	0.997	Excellent correlation
		33 × 55	Force(55) = 1.259 × Force(33) − 272	0.991	
Wind speed	Fy	33 × 44	Force(44) = 1.708 × Force(33) + 24.32	0.974	
		33 × 55	Force(55) = 2.615 × Force(33) + 55.39	0.945	
Wind speed	Mx	33 × 44	Force(44) = 1.777 × Force(33) − 4E−05	1	
		33 × 55	Force(55) = 2.774 × Force(33) − 0.00	0.999	
Wind speed	My	33 × 44	Force(44) = 1.778 × Force(33) + 0.022	1	
		33 × 55	Force(55) = 2.776 × Force(33) + 0.027	1	

(continued)

Table 3 (continued)

Parameter	Force/Displacement	Correlation between	Mathematical relation	R^2 value	Observation
Wind speed	Dx	33 × 44	Force(44) = 1.778 × Force(33) − 0.004	1	
		33 × 55	Force(55) = 2.763 × Force(33) − 0.359	0.996	
Wind speed	Dy	33 × 44	Force(44) = 1.094 × Force(33) + 0.284	0.998	
		33 × 55	Force(55) = 1.213 × Force(33) + 0.627	0.993	

5. Corner columns are more significant to all the parameters, viz., axial force, moments, and displacements than center columns for all conditions and terrain categories.
6. Effect of terrain category and permeability is not significant in centrally located columns of the structures.
7. Significant correlation exists between forces and displacement in less than 5% permeability with forces and displacement in other permeability.
8. Significant correlation exists between forces and displacement developed with basic wind speed of 33 m/s with other wind speeds chosen.
9. Significant co-relation exists between forces and displacement developed considering terrain category 1 with that in other terrain categories, however, the R^2 value is more than 0.9 for Mx between terrain categories 1 and 4.

References

1. Raj R, Ahuja AK (2013) Wind loads on cross shape tall buildings. J Acad Indus Res 2(4):111–113
2. Al-Tamimi M, Ahmed N et al (2011) The effects of orientation, ventilation, and varied WWR on the thermal performance of residential rooms in the tropics. J Sustain Dev 4(2):142–149. https://doi.org/10.5539/jsd.v4n2p142
3. Verma SK, Ahuja AK, Pandey AD (2013) Effects of wind incidence angle on wind pressure distribution on square plan tall buildings. J Acad Indus Res 1(12):747–752
4. Siddh SP Dr, Pagaria P, Hiwase PD Dr, Patil U. Comparative study of old and new Indian seismic and wind codes for analysis of existing high rise structure. Int J Adv Sci Technol 29(7):12546–12551

5. Patil U, Kabra N, Hiwase PD Dr, Siddh SP Dr (2019) Analytical study of response reduction factor of RCC structures. Helix Scien Explorer 9(6):5768–5770
6. Hiwase PD Dr, Patil U, Kabra N, Asati A (2019) Response reduction factor for industrial structures. Helix Scien Explorer 9(6):5775–5778
7. StathopouloS T, Surry D, Davenport AG (1980) Internal pressure characteristics of low-rise buildings due to wind action. Zement-Kalk-Gips 1:451–463
8. Kumar BD, Swami BLP. Critical gust pressures on tall building frames-review of codal provisions. Int J Adv Technol Civil Eng 1(2):141–147
9. Wang K, Stathopoulos T (2006) The impact of exposure on wind loading of low buildings. In: Structures congress 2006, American Society of Civil Engineers, pp 1–10. https://doi.org/10.1061/40889(201)9
10. Chauhan HM, Pomal MM, Bhuta GN (2013) A comparative study of wind forces on high-rise buildings as per is 875-iii (1987) and proposed draft code (2011). Glob J Res Anal 2(5)
11. Mehta KC (1984) Wind load provisions ANSI a58.l-1982. ASCE J Struct Eng 110(4), Paper No. 18758
12. Faysal RM (2014) Comparison of wind load among BNBC and other codes in different type of areas. Int J Adv Struct Geotech Eng 03(03):265–271
13. Holmes JD, Tamura Y, Krishna P (2008) Wind loads on low, medium and high-rise buildings by Asia-Pacific codes. In: The 4th international conference on advances in wind and structures, 29–31 May, 2008
14. Das NK. A comparative study of wind load standards. B.C.E Graduate Faculty of Texas Tech University
15. Ambadkar SD, Bawner VS (2012) Behaviour of multistoried building under the effect of wind load. Int J Appl Sci Eng Res 1(4):656–662. https://doi.org/10.6088/ijaser.0020101066
16. Kumar KS (2011) Commentary on the Indian standard for wind loads. In: 13th International conference on wind engineering, Amsterdam, The Netherlands, July 10–15, 2011
17. Ellingwood BR, Tekie PB (1999) Wind load statistics for probability-based structural design. J Struct Eng 125(4):453–463. https://doi.org/10.1061/(ASCE)0733-9445(1999)125:4(453)
18. Ranjitha KP, Khan KN, Kumar NS Dr, Raza SA (2014) Effect of wind pressure on R.C tall buildings using Gust factor method. Int J Eng Res Technol (IJERT) 3(7)
19. Karava P, Stathopoulos T (2012) Wind-induced internal pressures in buildings with large façade openings. J Eng Mech 138(4):358–370. https://doi.org/10.1061/(ASCE)EM.1943-7889.0000296

Performance Assessment of SMA-LRB Isolated Building Structure Due to Underground Blast-Induced Ground Motion

Sonali Upadhyaya, Narendranath Gogineni, and Sourav Gur

Abstract Blast-induced ground motions (BIGM) are highly damaging to structural and infrastructural systems, as it imparts a large amount of energy to the system over a small duration of time. Thus, it is very important to use high-energy dissipating materials in the vibration control device, which will show quick response during blast loading. In this context, shape memory alloy (SMA) material shows its capability of high-energy dissipation and auto-recentering through the phase transformation process. Therefore, the current research work considers base-isolated building structure, and aims to study the vibration control efficiency of SMA supplemented lead rubber bearing (SMA-LRB) under BIGM. Nonlinear time history analyses are executed to estimate the response quantities of interest, and the results are compared with the responses of LRB isolated building structures. To represent the LRB and SMA Bouc-wen and Graesser–Cozzarelli models are used, respectively, and the BIGM input is modeled as an exponentially decaying function with acceleration time history. Finally, to demonstrate superior control efficiency of SMA-LRB over LRB, parametric study results are provided for a wide range of isolator and structure properties, and different loading cases. Study results show that, for a wide range of parameters, the SMA-LRB isolation system can substantially reduce maximum floor acceleration and maximum isolator displacement than the LRB isolation system. Also, it has been found that the presence of SMA significantly reduces the residual deformation of the isolator at the end of BIGM. Thereby, SMA-LRB provides a much higher recentering capability of isolator than convention LRB isolation system.

Keywords BIGM · Building structure · Vibration control · SMA-LRB isolator · Residual displacement

S. Upadhyaya (✉) · N. Gogineni · S. Gur (✉)
Department of Civil and Environmental Engineering, Indian Institute of Technology, Patna, Bihar 801106, India
e-mail: sonali_2021ce21@iitp.ac.in

S. Gur
e-mail: sgur@iitp.ac.in

M. S. Ranadive et al. (eds.), *Recent Trends in Construction Technology and Management*, Lecture Notes in Civil Engineering 260,
https://doi.org/10.1007/978-981-19-2145-2_64

1 Introduction

Over time, blast-induced ground motion (BIGM) gained a lot of importance due to the increased activities of the blast. Though earthquake and the underground blast both produce ground vibration, blast pulses are widely different from the earthquake motion, due to their characteristics of high frequency and large amplitude wave over an extremely small duration. Also, due to the very less distance of epicenter for BIGM, than the seismic motion, the attenuation, as well as the spatial variation of BIGM, are more important, for the same propagation distance. Further, in the case of seismic loading, the structural response is mainly dominated due to the lower modes, without any higher modes' contribution; whereas, during BIGM, the higher frequency modes are primarily excited. In this regard, Su et al. [1] put forward that lead rubber bearing (LRB) when used as a base isolator (BI) is more effective for a higher frequency that matches with the blast pattern of loading. As well as, it has been observed that the LRB has a larger tolerance to peak acceleration as compared to sliding type BI. Other studies Kangda and Bakre [2, 3] use LRB without or with fluid viscous damper for vibration control of regular or vertically irregular buildings under BIGM with seismic loading.

In the case of near-fault earthquake motion (similar to the pulse type of BIGM), which leads to large isolator displacement, traditional elastomeric bearings are not very much suitable as they have no recentring capacities. Therefore, a recent study Gur et al. [4, 5] presented a view that smart material shape memory alloy (SMA) outperforms LRB under the ground motion of near-fault and pulse type. Other studies on SMA supplemented conventional isolation system for buildings and bridges [4–8] show superior seismic vibration control efficiency, under deterministic and stochastic ground motion. A similar comparison study of the performance of various base isolators due to underground BIGM is studied by Mondal et al. [9, 10]. The study result demonstrates that the New Zealand (NZ) type isolation system provides good response reduction with a small amount of peak bearing displacement. However, Mondal et al. [11] found there is a substantial amount of permanent deformation with the NZ bearing system, and this is one of the critical isolator design parameters. Thus, to reduce this permanent deformation or residual deformation, among various materials, SMA is more promising due to their super-elastic effect and its flag shape hysteric curve, which provides great energy absorption capacity. In some other research work [9, 10], a comparison study of SMA mounted RB (SMARB) with NZ bearing is carried out. Interestingly the results of this study showed that the SMARB has negligible residual deformation, whereas it is quite higher and unpredictable for NZ bearing for different blast loading. Various underground blast energy produced during a nuclear explosion, mining activities, military activities, etc., threaten human life and essential instruments. For high-risk facilities like public and commercial building design consideration against extreme events such as blasts is very important [12]. They cause catastrophic failure of structures depending upon their intensity. Thus, the design of structure due to underground blast loading is gaining importance in the engineering field.

In a past study [13], SMA supplemented LRB (i.e., SMA-LRB) employed as BI under the stochastic ground motion and optimal performance is assessed. Other studies [14] focus on understanding the properties of SMA-LRB, investigating peak and residual isolator displacements, ultimate shear resistance, recentering capabilities, and enhancement of energy dissipation capacity. In another study [15], using a new constitutive model called SC-BKH (Self-cantering bilinear kinematic hardening) which gives greater accuracy in predicting self-centering behavior, but the response behavior of any structure isolated via SMA-LRB due to underground BIGM is still unfamiliar. The present study focuses on the vibration suppression efficiency of the SMA-LRB isolator, during underground BIGM, and compares the results with the LRB isolation system.

2 Material Modeling of Isolator

This section will provide the hysteresis behavior of LRB and super-elastic SMA isolation system through various force–deformation (f–δ) behavior.

2.1 Modeling of Lead Rubber Bearing (LRB)

Under the cyclic loading, LRB dissipates energy through yielding and shows a stable bilinear f-δ hysteresis loop. The parametric Bouc-Wen model [4, 5] is used to describe the f-δ hysteresis loop. According to the model, non-linear constitutive relation is stated as

$$F_z(x, Z) = \alpha_z k_z x + (1 - \alpha_z) F_{yz} Z \tag{1}$$

$$\dot{Z} = \left(\frac{\delta}{q_z}\right)\dot{x} - \left(\frac{\gamma}{q_z}\right)|\dot{x}|Z|Z|^{\eta-1} - \left(\frac{\beta}{q_z}\right)\dot{x}|Z|^{\eta} \tag{2}$$

where k_z is the initial elastic stiffness, α_z is the ratio of post-yield to pre-yield stiffness (i.e., rigidity ratio), F_{yz} and q_z are respectively the yield strength and yield displacement, the parameters δ, γ, and β in Eq. (2) are the controlling parameters for the shape of the hysteresis loop. Here, x and $\dot{x}$ are, respectively, the relative displacement and velocity of the yield damper, variable Z is a non-dimensional quantity, which represents the hysteretic behavior of metallic material in the yield damper. δ, γ, β, and η are the parameters controlling the shape of the hysteresis curve. The parameter η controls the transition from elastic to plastic state and parameter β control the nature of the loop.

2.2 Modeling of Shape Memory Alloy (SMA)

Shape Memory Alloys (SMAs) are such materials, which can regain large percentages (almost 8%) of strain upon heating above a critical temperature. It happens due to stress-free and thermally driven phase transition in the microstructure. However, another property of the SMA material, known as super-elasticity (SE), is the primary focus of interest and is mainly related to the phase transformation of microstructure due to applied mechanical loading. During the loading cycle, this austenite (A) phase transforms into the martensite (M) phase through stress-free microstructure orientation, causes a stress plateau. During unloading, martensite SMA recovers its deformed shape through a backward transformation process from martensite (M) to austenite (A) phase and leaves no residual deformation. This process of loading-unloading cycle produces flag shape hysteresis loop, dissipates a large amount of energy, and leaves no residual deformation. Out of several alternatives, for super-elastic SMA, the Graesser–Cozzarelli material model [4, 5] has been predominantly used for the analysis of the dynamic response of the SMA-based control system. The one-dimensional material model is given as

$$\dot{F}_{\mathrm{SMA}}(\dot{x},\beta) = k_{\mathrm{SMA}}\left[\dot{x} - |\dot{x}|\left|\frac{F_{\mathrm{SMA}}-\beta}{F_{y\mathrm{SMA}}}\right|^{(\eta-1)}\left(\frac{F_{\mathrm{SMA}}-\beta}{F_{y\mathrm{SMA}}}\right)\right] \tag{3}$$

$$\beta = \alpha_{\mathrm{SMA}}k_{\mathrm{SMA}}\left[x - \left(\frac{F_{\mathrm{SMA}}}{k_{\mathrm{SMA}}}\right) + f_T|x|^{c'}\mathrm{erf}\big(a'x\big)\right] \tag{4}$$

$$\mathrm{erf}(z) = \frac{2}{\sqrt{\pi}}\int_0^z \left(e^{-t^2}\right)\mathrm{d}t \tag{5}$$

where x is the displacement, F_{SMA} is the restoring force in SMA, k_s is the stiffness of SMA in the austenite phase, $F_{y\mathrm{SMA}}$ is the force causing the forward curve of transformation from the austenite to the martensite phase. Here, α_s is a constant definition of pre- to post-transformation stiffness ratio, a' is a constant controlling the amount of shape recovery during unloading. Parameter η also controls the sharpness of the transformation curve, c' decides the slope of unloading path, β is one-dimensional back stress given by Eq. (4), f_T is a controlling parameter of type and size of hysteresis. Here, $\mathrm{erf}(z)$ is the error function with argument z, as defined by Eq. (5).

2.3 Force Deformation Hysteresis of LRB and SMA

A comparison of typical f–δ characteristics of the lead core of LRB and the SMA wire is shown in Fig. 1. As can be observed, hysteresis energy dissipation of capacity (i.e., area under the hysteresis curve) of lead core and SMA wire are almost comparable.

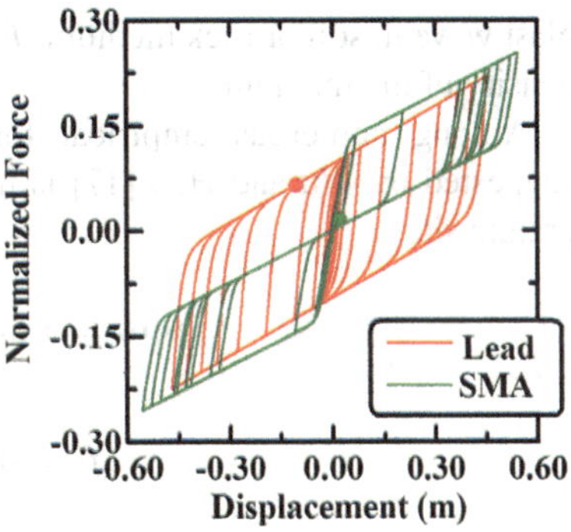

Fig. 1 Comparison of the hysteresis loop of lead core and SMA material

However, compared to the lead core, SMA leaves almost no residual deformation upon unloading. In the case of the lead core, after yielding plastic strain accumulates and thus provides large residual deformation. Whereas for SMA, above austenite finish temperature (As), upon unloading, complete microstructure transforms take place and SMA regains its original shape. This phenomenon has been depicted in the flag shape hysteresis loop and recentering capability of SMA material, which ultimately helps to provide better control efficiency. Thus, such a remarkable behavior of super-elastic SMA material makes it very much pertinent for blast application as the recentring device.

3 Numerical Modeling and Analysis

This section provides the important details of the simulation process in terms of the equation of motion and modeling of the BIGM, as discussed below.

3.1 *Blast Loading*

The present section deals with the modeling of underground blast loading as an exponentially decaying function, suggested by the previous study [16]. Blast load $F(t) = m\ddot{x}_g(t)$, is used and time history records are developed for the ground acceleration, $\ddot{x}_g(t)$ based on charge wets and distance of blast.

$$\ddot{x}_g(t) = -(1/t_d)\bar{\dot{x}}_g \exp(t/t_d) \tag{6}$$

where $\bar{\dot{x}}_g$ is the peak particle velocity (PPV), $t_d = R/C_p$ is the time of arrival. Here, R is the radius of the charge center and $C_p = \sqrt{E/\rho}$ denote the velocity of the blast wave in soil or rock medium, E represents Young's modulus and ρ is the mass density of the medium.

Among numerous empirical formulae of prediction of PPV, here relation suggested by Wu and Hao [17] is used in following Eqs. (7) and (8) considering granite site.

$$\mathrm{PPV} = 2.981 f_1 \left(R/Q^{(1/3)}\right)^{-1.3375} \tag{7}$$

$$f_1 = 0.121(Q/V)^{0.2872} \tag{8}$$

Here, f_1 is the decoupling factor for PPV, Q is Trinitrotoleum (TNT) weight in kilogram, V is the volume of the TNT in cubic meter, and R is the distance in meters. In this study, the distance of charge center R is 50 m and the TNT charge weight Q is 20 tons. The volume of charge chamber V and propagation velocity C_p are considered as 1000 m^3 and 5280 m/s, respectively [17].

3.2 Dynamic Analysis

In the present study, a two-dimensional shear-type building (non-isolated) and isolated via LRB and SMA-LRB has been considered for the analysis, as shown in Fig. 2a, b, respectively. The mechanical model of the LRB and SMA-LRB systems are shown in Fig. 2c, d, respectively.

Since the input blast energy dissipation occurs through the hysteresis curve of LRB or SMA-LRB isolation, thus they will behave non-linearly, and the building will stay linear-elastic domain. With linear superstructure and non-linear isolator behavior, the dynamics building and isolator are written as

$$[M]\{\ddot{x}\} + [C]\{\dot{x}\} + [K]\{x\} = -[M]\{r\}\left(\ddot{x}_g + \ddot{x}_b\right) \tag{9}$$

$$m_b\ddot{x}_b + c_b\dot{x}_b + F_b - (c_1\dot{x}_1 + k_1x_1) = -m_b\ddot{x}_g \tag{10}$$

$$F_b = k_r x_b + F_H(\dot{x}_b, x_b) \tag{11}$$

Here, $[M]$, $[C]$, and $[K]$ are the mass matrix, damping matrix, and elastic stiffness matrices of the building structure; $\{r\}$ is the vector containing stories deformation due to unit deformation of the ground, i.e. $\{r\} = \{1\ 1\ 1\ \ldots\ 1\}^T$; $\{\ddot{x}\}$, $\{\dot{x}\}$ and $\{x\}$ are acceleration, velocity, and displacement vector of the building; $\ddot{x}_g$ and $\ddot{x}_b$ are acceleration due to BIGM and isolator, m_b and $c_b = 4\pi\xi_b m_b/T_b$ are the isolator mass and damping of the isolator, ξ_b and T_b are the damping ratio and time period of the isolator, and F_b is the restoring force of the isolator which consists of two parts liner force due to rubber and non-linear hysteresis force lead core or SMA wires.

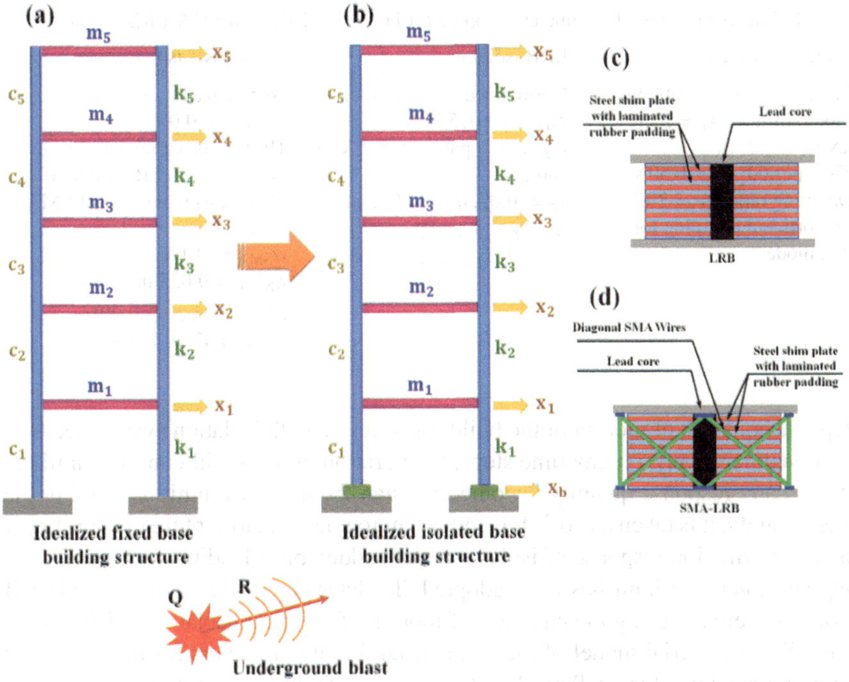

Fig. 2 Model of **a** fixed base and **b** isolated building. Mechanical model of isolator **c** Lead-Rubber-Bearing (LRB) and **d** Shape-Memory-Alloy-Lead-Rubber-Bearing (SMA-LRB)

The time period of the isolator is calculated as

$$T_b = \begin{cases} 2\pi\sqrt{M_t/(k_r + \alpha_z k_z)} & \text{for LRB} \\ 2\pi\sqrt{M_t/(k_r + \alpha_z k_z + \alpha_{\text{SMA}} k_{\text{SMA}})} & \text{for SMA-LRB} \end{cases} \tag{12}$$

where k_r is the rubber stiffness, $\alpha_z k_z$ and $\alpha_{\text{SMA}} k_{\text{SMA}}$ are the post-yield stiffness of lead core and SMA wire, $M_t = m_b + \sum_{i=1}^{n} m_i$ is the total system mass, and m_i is the each floor mass. Here the mass of the isolator is set the same as floor mass, i.e., $m_b = m_i$, stiffness of the lead core and SMA wires are calculated as $k_z = F_{yz}/q_z$ and $k_{\text{SMA}} = F_{y\text{SMA}}/q_{\text{SMA}}$, respectively, and finally using Eq. (12), stiffness of the rubber (k_r) is calculated. The time period and damping ratio of the isolator are adopted as $T_s = 2.00$ s and $\xi_s = 2\%$, respectively. This provides the default value of the total stiffness of the isolator as 177.65 kN/m.

The response of the non-isolated/ isolated building structure under the BIGM excitation is obtained via the step-by-step numerical integration method (Newmark-beta average acceleration technique) with the time step Δt of 0.0001 s. It is well known that the isolation system will dissipate input energy from the blast loading, through a non-linear force deformation hysteresis loop. Thus, the building structure will stay in the linear-elastic domain. Thus, to obtain the correct solution at each time

Table 1 Values of different parameters adopted for buildings, LRB and SMA-LRB isolator

Building structure	LRB isolator	SMA-LRB isolator
Floor mass: $m_i = 3000$ kg Floor stiffness: $k_i = 5830$ kN/m Time period: $T_s = 0.50$ s Damping ratio: $\xi_s = 2\%$ proportional damping for the first mode	Normalized yield strength: $F_{0yz} = 0.075$ Hysteresis parameter of lead core: $\alpha_z = 0.05$, $q_z = 0.025$ m $\beta = 0.5$, $\gamma = 0.5$ $\delta = 1.0$, $\eta = 1.0$	Normalized transition strength: $F_{0SMA} = 0.075$ Hysteresis parameter of lead core: same as LRB isolator Hysteresis parameter of SMA wire: $\alpha_{SMA} = 0.10$, $q_{SMA} = 0.035$ m $a' = 2500$, $c' = 0.001$ $f_T = 0.07$, $\nu = 3.0$

step, the equation of motion of the building structure with isolation system needs to be solved iteratively. At any time steps, the iteration process will continue until the error in any response quantity become less than the tolerance limit δ_{tol}, and in the present study, it is taken as 10^{-5}. It is important to note that, to capture the non-linear force–deformation response of isolator under sudden blast loading such small-time step and tolerance limit has been adopted. To determine the responses MATLAB code is developed using the equation of motion of building, isolator, and BIGM as a non-linear material model of the isolator. Different parameters for the numerical solution are adopted from Table 1 and are also available in other studies [4–8].

4 Results and Discussion

Performance of the isolated and non-isolated building structures is studied through the numerical simulation, during underground BIGM, as mentioned in the above section. The isolation efficiency of LRB and SMA-LRB isolators is demonstrated by comparing the top floor acceleration and isolator displacement. In the parametrical study, maximum top floor acceleration, ultimate and residual isolator displacement are adopted for comparing the vibration control efficiency and performance of LRB and SMA-LRB isolator building structures.

4.1 Numerical Illustration

A typical time history analysis result is shown in Fig. 3, for non-isolated building structures and LRB or SMA-LRB isolated building structures. Figure 3a, b shows responses of top floor acceleration and displacement for the fixed base and isolated (via LRB and SMA-LRB) building structures. Compared to the non-isolated building, the isolated building substantially reduces the top floor acceleration and displacement, but the response control efficiency of SMA-LRB is more than the LRB isolator.

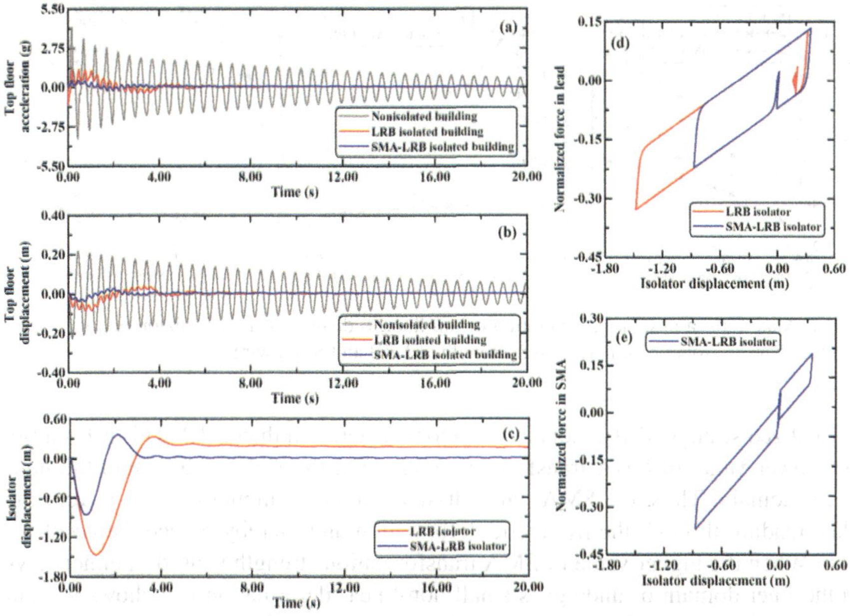

Fig. 3 Response of top floor **a** acceleration and **b** displacement of fixed base, LRB isolated, and SMA-LRB isolated building, **c** isolator displacement. Hysteresis loop of isolator **d** Lead and **e** SMA material

It is because of the added hysteresis loop of the SMA wire, as shown through the hysteresis loop area in Fig. 3e. Displacement time history of isolator response is shown in Fig. 3c, where due to the proximity of the blast site both the LRB and SMA-LRB isolator undergoes a large level of non-linear deformation, and thereby dissipate a significant amount of blast loading energy. Further, this can be observed from the hysteresis loop of lead core and SMA material as shown in Fig. 3d, e, respectively. Another important observation from Fig. 3c is that, in contrast to the LRB, due to the presence of SMA wires in SMA-LRB, it substantially reduces the peak displacement of the isolator, as well as residual displacement of the isolator. Thus, the super-elastic SMA wires in the SMA-LRB isolation system act as the auto-recentering device and thereby reduces the damage possibility of the isolator.

4.2 Parametrical Study

At first, the effect of the isolator strength on the different response parameters is provided in Fig. 4. The maximum value of top floor absolute acceleration (Fig. 4a) increases with increasing normalized strength of LRB and SMA. Isolator peak displacement and residual displacement for LRB and SMA-LRB are provided in Figs. 4b, c. As observed, incorporation of SMA material up to a critical value (i.e.,

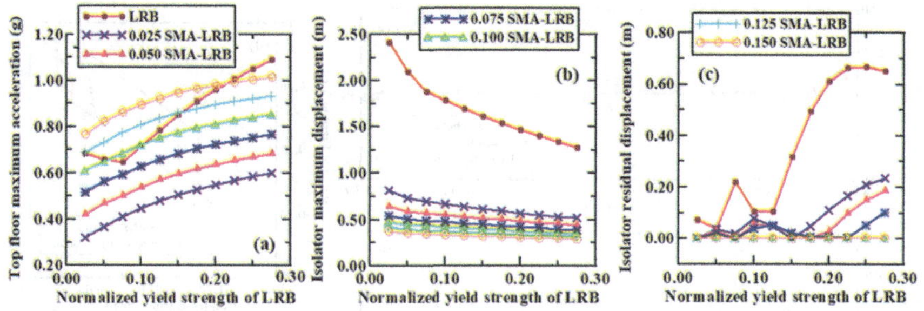

Fig. 4 Maximum **a** absolute acceleration of top floor, **b** displacement of isolator, and **c** residual displacement of isolator for normalized strength of LRB and SMA wire

normalized strength 0.10) reduces top floor acceleration than LRB. This is because, at a lower value of SMA transformation strength, the isolator goes into the non-linear domain. Thus, the SMA wires dissipate a large amount of energy from the blast loading through the hysteretic deformation and thereby reduce the acceleration. When the higher value of SMA transformation strength is used, it either stays in the liner domain or undergoes small non-linear deformation i.e., shows a small hysteresis loop. Therefore, SMA wires do not dissipate a large level of blast loading energy and make the isolator stiff which transfers more blast loading energy to the building and therefore it increases top floor acceleration. Figure 4b shows that the maximum value of isolator displacement decreases with increasing normalized strength of isolators (LRB or SMA-LRB). Figure 4c shows that residual displacement of the isolator reduces with increasing normalized transformation strength of SMA, whereas the LRB isolator shows a high level of residual displacement. The addition of SMA wires makes the SMA-LRB isolator stiff and thus it reduced the peak displacement and residual displacement (due to the shape recovery property of SMA), compared to the LRB isolator.

Figure 5 shows the control efficiency of both the isolation system under different isolator time periods. Figure 5a shows that the maximum value of top floor absolute acceleration decreases with increasing time period of isolator in the case of

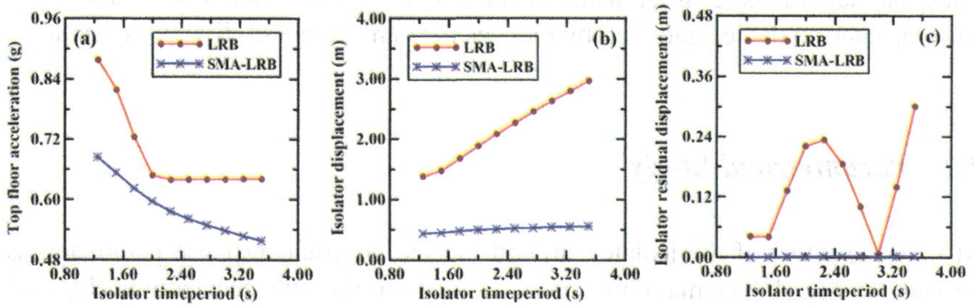

Fig. 5 Maximum **a** absolute acceleration of top floor, **b** displacement of isolator, and **c** residual displacement of isolator for the time period of LRB and SMA-LRB isolator

SMA-LRB, but for LRB it shows initially decreasing and then staying constant. The decrease in acceleration is because of the increased isolator flexibility which transfers less force to the structure. However, in the case of LRB, this effect is not significant after 2 s of the isolator time period. According to Fig. 5b, with an increase in isolator time period, the maximum value of isolator displacement increases at a much higher rate in the case of LRB than the SMA-LRB. Here as the base of the structure becomes flexible, thus this leads to an increase in the building displacement. However, it can be observed that the SMA-LRB isolator is showing better control of displacement than LRB. Figure 5c shows that residual displacement of the isolator is much high for LRB than SMA-LBR for all the values of the isolator time period.

Effect of structure time period on isolation efficiency of the LRB and SMA-LRB isolator is shown in Fig. 6. As it can be observed from Fig. 6a, with the increasing time period of the structure, the maximum value of top floor absolute acceleration drastically increases for LRB isolated buildings than SMA-LRB isolated buildings. Figure 6b shows that the maximum value of isolator displacement stays almost constant with varying structure time periods, but overall the isolator displacement is more for LRB than SMA-LRB. Thus, when the structure time period is increasing and it became more flexible, this can be said from Fig. 6a, b that the SMA-LRB isolator has better control efficiency of reducing both structural and isolator response than only LRB isolated structure. Figure 6c shows that the residual displacement of the isolator is much high for LRB than SMA-LRB for all the values of the structure time period. This is due to the recentering capacity of SMA which can be shown from the hysteresis loop of SMA material in Fig. 3e.

Next, the effect of BIGM characteristics on the vibration control efficiency of LRB and SMA-LRB is studied. The first influence of blast loading intensity on the structure and isolator response has been shown in Fig. 7, where the distance from the blast site is kept as 50 m. Figure 7a show that, with increasing blast loading intensity, the maximum value of top floor absolute acceleration increases for both LRB and SMA-LRB isolated building. As the BIGM act as pulse type loading to structure, thus it creates sudden impact loading, consequently, it excites the higher modes of the structure. Thus, an increase in blast loading intensity leads to a drastic increase in the response of the structure. However, this increase in top floor acceleration is

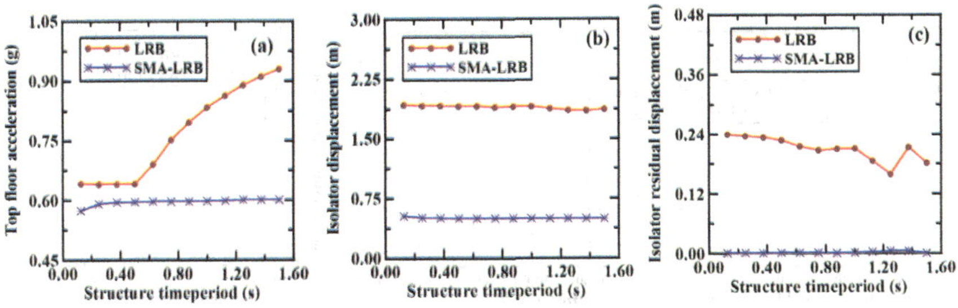

Fig. 6 Maximum **a** absolute acceleration of top floor, **b** displacement of isolator, and **c** residual displacement of isolator for the building structure time period

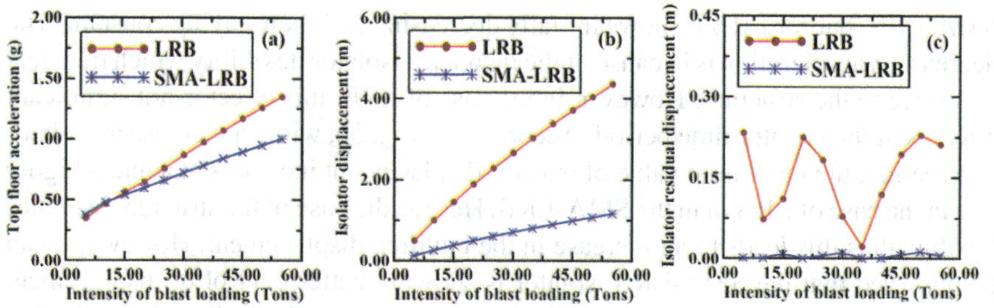

Fig. 7 Maximum **a** absolute acceleration of top floor, **b** displacement of isolator, and **c** residual displacement of isolator for the intensity of blast loading

more for LRB isolated buildings than SMA-LRB isolated buildings. Figure 7b, c shows the variation of isolator peak displacement and residual displacement (for both LRB and SMA-LRB) with blast loading intensity, respectively. It is shown in Fig. 7b that the maximum value of LRB isolator displacement increases largely with increasing blast loading intensity and always much higher SMA-LRB isolator displacement. Figure 7c shows that the residual displacement of the isolator is much high for LRB than SMA-LRB, for all the values of blast loading intensity. It can be observed that even at very high blast intensity, the SMA-LRB isolator leaves no residual displacement. This proves the robustness of SMA material.

Figure 8 shows the effect of the distance from the blast site to the structure on the peak floor acceleration isolated structure and isolator peak and residual displacement for the LRB and SMA-LRB isolation system. Here the intensity of blast loading is kept as 20 tons. Figure 8a shows that, with the increasing blast site distance from the structure, the maximum value of top floor absolute acceleration drastically decreases for both LRB and SMA-LRB isolated buildings. But more reduction can be observed for SMA-LRB isolated building cases than LRB isolated buildings. Figure 8b, c shows the variation isolator peak and residual displacement with varying distance from the blast site, respectively. Figure 8b shows that the maximum value of isolator displacement decreases largely with increasing blast site distance from structure, but more reduction is observed for SMA-LRB than LRB. As provided in the formulae

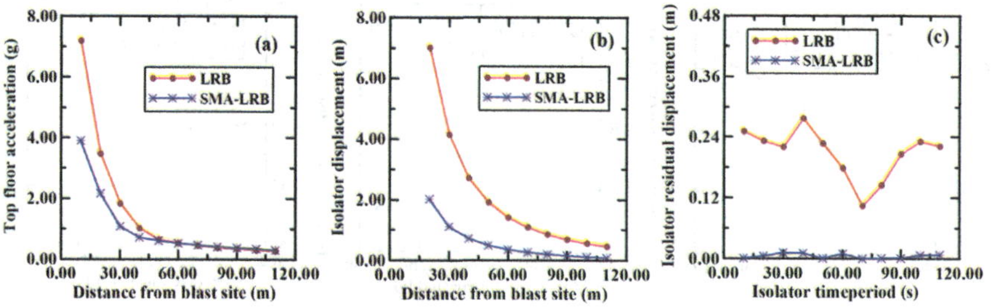

Fig. 8 Maximum **a** absolute acceleration of top floor, **b** displacement of isolator, and **c** residual displacement of isolator for the distance from the blast site

of Eqs. (6) and (7), it can be found that the blast radius R is inversely related to the ground acceleration $\ddot{x}_g$. Thus, with an increase in distance of the blast site, both structural and isolator responses are reducing at a decreasing rate. Figure 8c shows that residual displacement of the isolator is much high for LRB than SMA-LRB for all the values of blast site distance.

5 Conclusion

This paper presents the vibration control efficiency of the SMA-LRB isolator under blast loading and compares the results with the LRB isolation system. To this end, a parametrical study has been performed for varying isolator property, structure–property, and various loading cases. The time history analysis results reflect that the acceleration and displacement response of isolated structures is much lower than non-isolated ones. In addition, it has been observed that the reason for superior control efficiency of SMA-LRB is due to its better energy dissipation capacity, via a large hysteresis loop.

Parametrical study results show that, at the lower value of SMA normalized transformation strength (i.e., 0.1), the SMA-LRB isolator provides higher floor acceleration control efficiency than the LRB isolator. With the increase in the intensity of blast loading the reduction in top floor acceleration and isolator, displacement is much more in the case of SMA-LRB than LRB and this reduction is very significant for smaller loading (Q < 30 tones). Furthermore, the maximum isolator displacement increases with the increased blast loading intensity, but the rate of increase is negative. In contrast to the LRB isolator, the residual displacement of the SMA-LRB isolator is very negligible, irrespective of blast intensity. This is due to the recentring (due to shape recovery property) ability of SMA that makes it very attractive to use in case of such BIGM. For a wide range of isolator time periods and structure time period, the top floor acceleration, isolator displacement, and residual displacement of SMA-LRB are much lower than LRB. Hence, the SMA material behaves very promisingly under blast loading application.

Acknowledgements The authors acknowledge the SERB under the DST Government of India for providing the research funding under the SRG scheme (SRG/2020/000892).

References

1. Su L et al (1989) Comparative study of base isolation systems. J Eng Mech 115(9):1976–1992
2. Kangda MZ, Bakre S (2018) The effect of LRB parameters on structural responses for blast and seismic loads. Arab J Sci Eng 43(4):1761–1776
3. Kangda MZ, Bakre S (2020) Performance evaluation of moment-resisting steel frame buildings under seismic and blast-induced vibrations. J Vib Eng Technol 8(1):1–26

4. Gur S et al (2013) Performance assessment of buildings isolated by shape-memory-alloy rubber bearing: comparison with elastomeric bearing under near-fault earthquakes. Struct Control Health Monit 21(4):449–465
5. Gur S et al (2013) Multi-objective stochastic-structural-optimization of shape-memory-alloy assisted pure-friction bearing for isolating building against random earthquakes. Soil Dyn Earthq Eng 54:1–16
6. Gur S et al (2014) Stochastic optimization of shape-memory-alloy rubber bearing (SMARB) for isolating buildings against random earthquake. Struct Control Health Monit 21(9):1222–1239
7. Gur S et al (2017) Thermally modulated shape memory alloy friction pendulum (tmSMA-FP) for substantial near-fault earthquake structure protection. Struct Control Health Monit 24(11):e2021
8. Mishra S et al (2016) Response of bridges isolated by shape memory–alloy rubber bearing. J Bridge Eng 21(3):04015071
9. Mondal PD et al (2017) Control of underground blast induced building vibration by shape-memory-alloy rubber bearing (SMARB). Struct Control Health Monit 24(10):e1983
10. Mondal PD et al (2017) Performances of various base isolation systems in mitigation of structural vibration due to underground blast induced ground motion. Int J Struct Stab Dyn 17(4):1750043
11. Mondal PD et al (2013) Performance of N-Z systems in the mitigation of underground blast induced vibration of structures. J Vib Control 20(13:2019–2031
12. Ngo T et al (2007) Blast loading and blast effects on structures–an overview. Electron J Struct Eng 7:76–91
13. Shinozuka M et al (2015) Shape-memory-alloy supplemented lead rubber bearing (SMA-LRB) for seismic isolation. In: Probabilistic engineering mechanics, vol 41, pp 34–45
14. Hu JW (2016) Seismic analysis and parametric study of SDOF lead-rubber bearing (LRB) isolation systems with recentering shape memory alloy (SMA) bending bars. J Mech Sci Technol 30(7):2987–2999
15. Dezfuli FH et al (2017) Effect of constitutive models on the seismic response of an SMA-LRB isolated highway bridge. Eng Struct 148:113–125
16. Carvalho EML, Battista RC (2003) Blast-induced vibrations in urban residential buildings. Struct Build 156(3):243–253
17. Wu C, Hao H (2005) Numerical study of characteristics of underground blast induced surface ground motion and their effect on above-ground structures. part i. ground motion characteristics. Soil Dyn Earthq Eng 25(1):27–38

Research Progress on the Torsion Behavior of Externally Bonded RC Beams: Review

Rajesh S. Rajguru and Manish Patkar

Abstract Externally strengthened reinforced concrete beams (RCB) are subjected to torsional loading having different properties of the externally bonded (EB) material and bond conditions. These conditions make it more complex compared to the normal beams. In the previous decade, a substantial number of investigation is been recounted on the utilization of various types of material for strengthening of RC beams under flexural and shear loading using EB method. A very few works have been reported on torsional behavior. This paper represents a state of review of the various materials utilized for strengthening RC beams using the EB practice and their assessment measures. The behavior of the strengthened beams (SB) is contested in relation with load carrying ability under different geometrical, power-driven characteristics of the RC beam, composite wrapping configuration, and different modes of failure. In the present paper, analytical methods described to calculate the torsional response of RC beams strengthened with EB technique are also discussed. This critical review will help to develop a database of strengthening systems of RC beams and will act as footprint for further research.

Keywords Reinforced concrete · Torsions · Strengthening · Mode of failure · External bonding

R. S. Rajguru (✉)
Assistant Professor, Civil Engineering Department, SRES's Sanjivani COE, Kopargaon, Maharashtra, India
e-mail: rajgururs25@gmail.com

Assistant Professor, Civil Engineering Department, Savitribai Phule Pune University, Pune, Maharashtra, India

Research Scholar, Civil Engineering Department, Oriental University, Indore, Madhya Pradesh, India

M. Patkar
Associate Professor, Civil Engineering Department, Oriental University, Indore, Madhya Pradesh, India

M. S. Ranadive et al. (eds.), *Recent Trends in Construction Technology and Management*, Lecture Notes in Civil Engineering 260,
https://doi.org/10.1007/978-981-19-2145-2_65

1 Introduction

Nowadays globalization is at its peak. Many infrastructural projects are going on in the world. It includes high-rise buildings, bridges, and water tanks. Structural members are generally subjected to flexural, compression shear loading and barely subjected to torsion loading. However, RC beam located along the periphery of structures, ring beam of the circular slab, subjected to cantilever loading may have a torsional effect. The behavior of such member and its strengthening is more complicated and need to study. Many researchers have been working on strengthening of RC beam by using externally bonded fiber reinforced polymer (FRP) [1–16]. FRP's good strength to mass proportion, good opposition to chemical and corrosion attack, flexible to use make it more popular for strengthening work [17]. FRP is continuous fiber strip that can be warped on beam surface by using different polymer matrices to transfer stresses. To eradicate the limitations of the well-established FRP strengthening system, a new type of amalgamated method, denoted as fiber reinforced cementitious matrix (FRCM) composite, has been inspected lately. It has been investigated in view of providing a substitute strengthening practice [18]. Numerous forms of fibers have been utilized in FRCM composite practices which include basalt, steel, carbon, glass, and polyparaphenylene benzobisoxazole (PBO). Very few researches are available in the literature on its usage for torsional firming [19].

The last two decades literature has been split into four groups or zones as shown in Fig. 1. From the year 2000 to 2005, only 0.04% of investigation has been done on strengthening of RC beams under torsion. From 2006 to 2010, 2011 to 2015, and 2016 to 2020 progress on the same area was 22.44, 16.33, and 57.14%. Ghobarah et al. in 2002 carried out experimental investigation on glass fibers as well as carbon fibers utilized in the torsional opposition. The outcomes or results revels that completely wrapped beams from all sides executed better than that by utilizing the strips. The investigation suggested that the material is efficiently utilized under 45° orientation of the fibers. From the year 2007, various authors propose various analytical models. Beams reinforced with carbon-Fiber reinforced polymer bars and transverse reinforcement under torsional loading studied in 2014 show good agreement with normal beam [20]. Epoxy adhesive has poor performance as it is allied with fumes with high toxicity and in elevated thermal atmospheres. To overcome this issue, a novel adhesive with cement base has been established and utilized as a substitute to

Fig. 1 Research progress

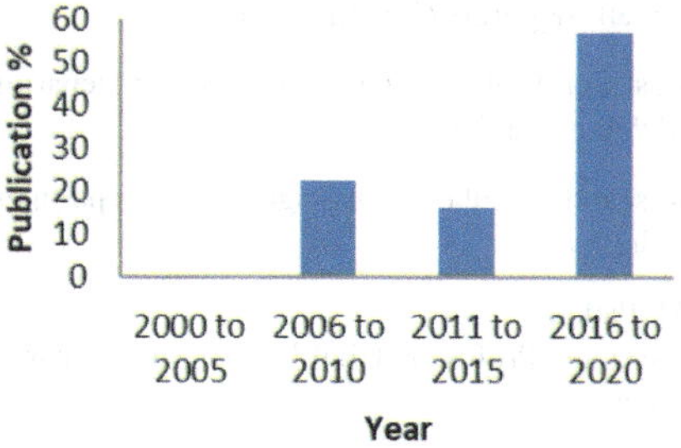

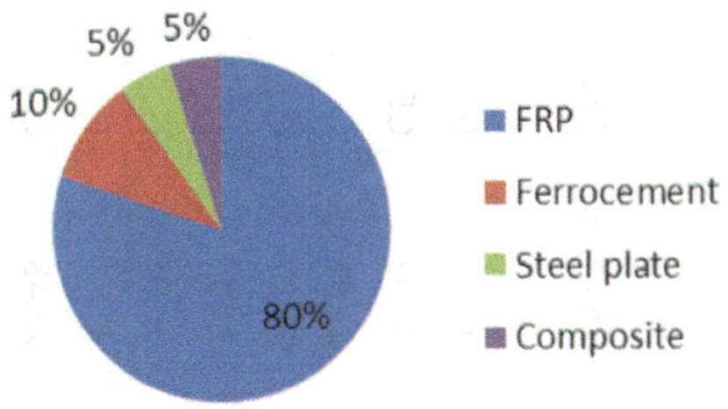

Fig. 2 Strengthening material

epoxy in numerous readings [9]. Figure 2 indicates that many researchers used FRP as strengthening material, although its high cost makes its strengthening process completely an expensive affair. Apart from this Ferrocement is more economical alternative. Still, there is an unavailability of codes for ferrocement compounds [21].

Though many researchers have focused on this issue, still many constraints need to be explored. Ferrocement is known to have good resistance to impact, good ductility, and superior in crack resisting ability [22–31]. High rich mortar matrices in ferrocement laminates seem to improve the considerable positive impact in the load carrying potential and ductility [22, 32] [22]. Revathy et al. [31] carried out experimental investigation on a rich and flowable nanosilica-based cementitious matrices of ferrocement which increase the compressive strength by 7.5 to 50% compared to control cement mortar [32]. CFRP strengthened beam increases cracking and ultimate strength in torsion up to 40% and 75%, respectively, compared to normal beams [33]. To increase the torsional enactment of RC beam members, utilization of high performance fiber reinforced cementitious composite mortar was found to be a beneficial technique [34].

The current paper presents the progress of investigation done on torsional strengthening of RC beams with EB composites. From a comprehensive literature study, a databank of investigational trials is established and debated. The usefulness of the strengthening scheme is inspected in regard to mechanical and geometrical properties, amalgamated wrapping configuration, and its categories. Various approaches to failure of the SB are also conferred in the present study. Later, various analytical systems established to calculate the torsion strength reaction of strengthened RC beams by using EB methods are presented.

2 Experimental Investigation

Table 1 displays the experimental data of beam strengthening under torsion loading. Table 1 summarizes the data related to cross section of specimen along with percentage of reinforcement used, strengthening material used and experimental results obtained. Geometrical properties and reinforcement used in RC strengthened beams may affect the strength of RC beam as compared to control specimens. Strength also depends on percentage of main reinforcement used and center-to-center spacing of stirrups. As compressive strength of concrete, shape, and type of

Table 1 Experimental data base

References	External bounded fiber												Results		
	BN	Sh	*W*	*D*	Pt%	Co	Si	NL	*W*	*S*	*T*	An	Tc kN.m	Tu kN.m	FM
[39]	C1	SR	150	350	0.33	Ca	4	1	–	–	0.165	N	12.01	32.1	F
	C2	SR	150	350	0.33	Ca	2	1	100	100	0.165	N	9.88	24.93	CF
	C3	SR	150	350	0.33	Ca	1	1	100	–	0.165	N	16.75	21.95	C
	C4	SR	150	350	0.33	Ca	2	1	200	100	0.165	N	11.74	28.27	C
	C5	SR	150	350	0.33	Ca	2	1	100	150	0.165	N	10.49	23.96	C
	C6	SR	150	350	0.33	Ca	4	1	100	–	0.165	N	–	30.05	F
	G1	SR	150	350	0.33	Gl	4	1	–	–	0.353	N	12.81	33.81	F
	G2	SR	150	350	0.33	Gl	2	1	100	100	0.353	N	11.23	23.48	C
[7, 40]	CS1	SR	500	350	0.40	–	–	–	–	–	–	–	68.4	62.9	C
	FS050D2	SR	500	350	0.40	Ca	4	2	50	175	0.176	N	73.7	93.8	CF
	FH050D2	BR	500	350	0.94	Ca	4	2	50	175	0.176	N	22.2	87.7	D
[41]	Ra-FC(1)	SR	100	200	0.5	Ca	4	1	–	–	–	N	2.8	4.87	F
	RaS-FS150(2)	SR	100	200	0.5	Ca	4	2	150	150	0.11	N	2.35	4.33	C
	Rb-FC(1)	SR	150	300	0.22	Ca	4	1	–	–	–	N	8.79	10.05	F
	Rb-FS200(1)	SR	150	300	0.22	Ca	4	1	200	200	0.11	N	6.73	9.32	C
	Rb-FS300(1)	SR	150	300	0.22	Ca	4	1	300	300	0.11	N	6.96	7.52	C
	RbS-FS200(1)	SR	150	300	0.22	Ca	4	1	200	200	0.11	N	6.93	9.8	C
[42]	Ra-c	SR	100	200	0.5	–	–	–	–	–	–	–	2.389	–	C
	Ra-F(1)	SR	100	200	0.5	Ca	4	1	–	–	0.11	N	2.8	4.868	F
	Ra-F(2)	SR	100	200	0.5	Ca	4	2	–	–	0.11	N	2.83	6.65	F

(continued)

Table 1 (continued)

References	External bounded fiber												Results		
	BN	Sh	*W*	*D*	Pt%	Co	Si	NL	*W*	*S*	*T*	An	Ꞇc kN.m	Tu kN.m	FM
	Ra-Fs150(2)	SR	100	200	0.5	Ca	4	2	150	300	0.11	N	2.219	3.018	F
	Rb-F(1)	SR	150	300	0.22	Ca	4	1	–	–	0.11	N	8.794	10.05	F
	T-FU(2)	T	150	300	0.22	Ca	3	2	–	–	0.11	N	8.775	9.45	D
[5]	Control	SR	200	300	1.29	–	–	–	–	–	–	N	10.4	16.8	C
	NP3S1	SR	200	300	1.29	CF	3	1	101.6	101.6	0.046	N	11.6	18.1	C
	NP4S1	SR	200	300	1.29	CF	4	1	101.6	101.6	0.046	N	14.3	21.8	C
	NP4C1	SR	200	300	1.29	CF	3	1	–	–	0.046	N	13.7	27.2	C
	NP4C2	SR	200	300	1.29	CF	4	2	–	–	0.046	N	14.5	35.1	C
[25, 26]	U4H	SR	125	250	0.54	FE	3	4	–	–	25	N	6.424	7.68	CFE
	L04H	SR	125	250	2.17	FE	3	4	–	–	25	N	6.675	7.87	CEF
	T04H	SR	125	250	0.54	FE	3	4	–	–	25	N	6.618	8.66	CFE
	CO4H	SR	125	250	2.17	FE	3	4	–	–	25	N	6.71	12.91	CFE
[43]	F1	SR	150	300	0.22	Ar	4	1	–	–	0.25	N	4.235	4.675	C
	F2	SR	150	300	0.22	Ar	4	1	–	–	0.25	N	4.070	4.620	C
	F3	SR	150	300	0.22	Ar	4	1	–	–	0.25	N	4.097	4.207	C

BN = Beam name, *Sh* = Shape, *W* = Width, *D* = Depth, *Pt%* = Percentage main steel, *Co* = Composite, *Si* = Side, *NL* = No. of layers, *S* = Spacing, *T* = Thickness, *An* = Anchorage, *Ꞇc* = Cracking strength, *Tu* = Ultimate strength, *FM* = Failure mode, *Ca* = Carbon fiber, *Gl* = Glass fiber, *CF* = Composite fiber, *C* = Concrete, *F* = fiber, *CF* = Concrete and fiber, *D* = Debonding, *SR* = Solid rectangular, *BR* = Box rectangular, *T* = Tee, *FE* = Ferrocement, *Ar* = Aramid fiber composite fiber changes, the mode of failures also changes accordingly. Data in Table 1 shows that approximately 50% increment in the strength was achieved when beam was strengthened by FRP. Kim et al. [35] conducted eleven experimental tests were conducted to study the effect of reinforcement and cross section on torsional strength. His results revealed that solid cross section and hallow cross section has similar level of strength also smaller amount of torsion reinforcement increased the ductile behavior regardless of cross section [35]. Maalej and Leong [36] stated that the size of beam doesn't have any significant effect on the strengthening proportion and deflection. According to Lee et al. [37] carbon fiber reinforced polymer when used as a strengthening material for deep beam, its length parameter has a noteworthy effect on the performance of deep beams. Sole form of reinforcement is ineffectual in increasing the torsion strength [22]. It is perceived that all the beams enveloped with aramid fiber strips upsurge torsion carrying ability. The FRP strip spacing and torsional moment carrying ability is inversely proportional to each other [38].

Experimental database included total of thirty-five beam specimen details. Out of that 82% of the test were performed on RC beam having 0.22–0.5% main steel as represented in Fig. 3a. The transverse reinforcement in the RC beam samples is found to be less significant in enhancing the torsion strength [34]. Therefore, CFRPs are found to be more practicable strengthening practice.

Copiously enfolded RCB aramid fiber grosses 140% extra ultimate torsional moment, compared with controlled beam [44]. The ultimate torsional strength increased by up to 150% with the utilization of 90° orientated entirely wrapped glass fibers. Composite fibers are also effective in order to increase the torsional capacity of RC members as that of Glass-FRP and Carbon-FRP combinations [5]. The maximum torsional moment is highly sensitive to the area of steel and yield strength parameters used in RC beams [45]. Brittle failure of concrete in RC beam can be reduced by designing RC beam as a under reinforced section. As percentage of transverse reinforcement increases torsional strength and stiffness enhances, regardless of beams size parameters [35]. A steel-below balance section with three side ferrocement covering showed better torsional strength when matched to over reinforced beams [30]. Ferrocement wrapped RC beam without steel in longitudinal direction is not

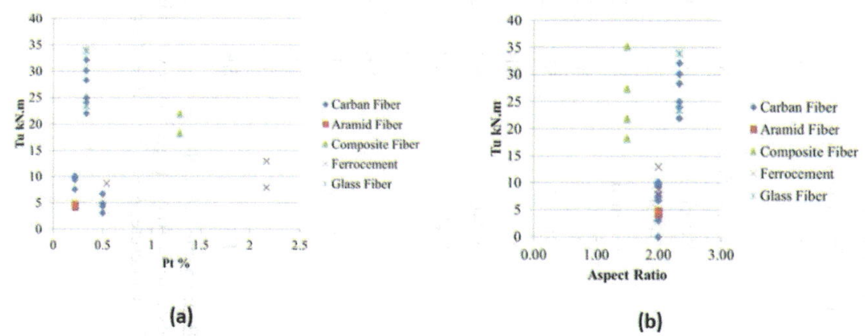

Fig. 3 Graphs showing **a** percentage steel (Pt %) versus Tu and **b** aspect ratio versus Tu

actual mode of improving the torsional strength, while growth in toughness is found to be negligible [46]. Figure 3b shows that strength can be enhanced under any aspect ratio depending upon the type of composite used and warping configuration. Carbon fiber is mostly preferred fiber for various aspect ratios. It might be due to its high tensile strength of up to 3900 MPa and better in mechanical properties over the other fibers. Maximum research work has been carried out on aspect ratio between 1.5 and 2.5 and more research need to be carried out for different aspect ratios.

3 Warping Configuration

Figure 4 shows typical warping configuration used by many researchers. Practically four side warping around the cross section is not feasible although full side warpped strengthening system is the only configuration that shows fiber rupture failure [42]. Therefore, 3-side, 2-side, and 1-side warping configurations have been explored. The prolonged three side jacket bands were very much operative strengthening ways compared to the un-anchored three side jacket bands due to excessive fiber slippage in 2-side and 1- side warping [4, 47, 48]. For 0-degree alignment, the critical strength was found to be not much more than the strength at first crack. FRP rupture of beams is controlled when fibers are oriented in the 90° direction [49]. Though 90-degree alignment of fibers were established to be quite operative in rising the strength (torsion) than 45-degree alignment, still de-bonding of fiber occurs at the end. The SBs in rectangular form and enveloped uninterrupted FRP sheets achieved improved torsional ability compared to SBs with FRP strips [42]. PBO-FRCM composite-strengthened beams show that the 90-degree fiber alignment is far too operative in rising the strength (torsion) than the 0-degree alignment [19]. Database shows that majority of test were conducted on single layer and two layers of fibers. When a more amount of films are provided, the efficacy of strengthening system declines. Properly anchored number of fiber layers forced failure mode in fiber rupture instead of debonding of fibers [19]. The complete wrap of a RCB is quite efficient in strengthening the torsional resistance compared to the beams in which strips of different patterns are used for the purpose of strengthening [39].

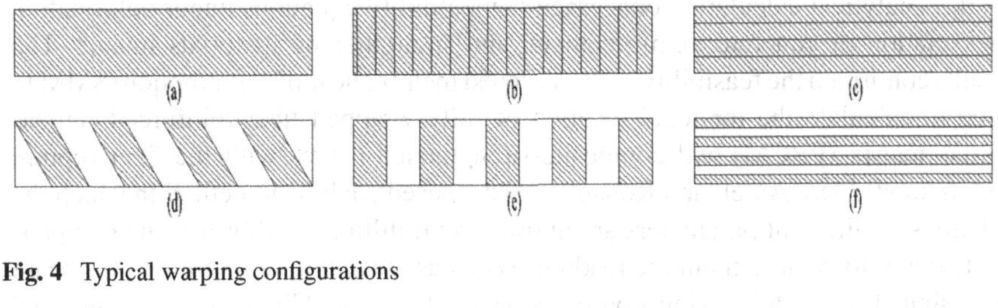

Fig. 4 Typical warping configurations

4 Mode of Failure

Deifalla et al. [4] divided mode of failure into three modes. The first mode includes concrete failure due to crushing caused by compressive loads or slanting cracking caused due to oblique tensile loads. The second mode includes the steel stirrup yielding tailed by end debonding of fiber reinforced polymer and lastly oblique failure. Third mode is characterized by steel yielding tailed by flaking off of fiber reinforced polymer and lastly slanting failure [4]. Though the samples had main fissures that obliquely flew on three verges of the sample, the main depreciations were observed on the bottom part of the back face [50]. The initial fissures were introduced at the intermediate of the lengthier faces [20]. Two layers of FRCM composite with 4-sided continuous wrapping configuration improved the post cracking stiffness and torsional strength quite efficiently compared to single film composites [5]. The fiber reinforced polymer strips break at the final strain in succession to the yielding of strengthening steel rebar under tensile forces [17] PBO-FRCM complex was unsuccessful because of the fiber estrangement at the beam junctions tailed by crushing of concrete [5]. A detachment of the CFRP-sheet was detected at the final value of force after the process of concrete deafening [51]. In the case of plain beams, the crack is originated on the lengthier side, but in this case a mono-latent crack established in the center was responsible for the failure of unwrapped concrete side [26, 30, 46]. This may be due to the slight detention provided by the fiber. Untimely de-bonding of the fiber films happened due to a substantial cracking of concrete [39]. Sometimes it may be because of epoxy bonding complications linked with toxicity of fumes and reduction in performance in elevated thermal atmosphere [9]. The beams in absence of bars and stirrups displayed an unexpected failure and parted in two numbers [52]. Crack width and pattern of this failed specimen sample specifies that the width was 13.6 mm for regulator beam. Beam strengthened with carbon fiber has smaller fissure width of 3.1 mm as matched to beam strengthen with glass fiber was of 4.1 mm [53].

5 Analytical Models

Extraction of analytical outcomes for RC beam subjected to torsion is greater tough work. Analytical calculation concerned compatibility equations under deformation and relation of concrete, reinforcement, and strengthening materials impact. The results confirmed the feasibility of the softened membrane model for torsion (SMMT) used to calculate the torsional reaction of fully wrapped fiber reinforced cementitious matrix (FRCM) and composite-strengthened beams with the fiber rupture failure mode [18]. Ameli and Ronagh [54] proposed analytical method in which the relations of different parameters are allowed by fulfilling equilibrium and compatibility conditions throughout the loading. Whereas the maximum torque of the beam is designed accurately using compression field theory (CFT) [54]. Deifalla and Ghobarah [47] established an analytical prototype constructed on the basics of the modified compression field theory. Model accounted for the effect of various parameters including various strengthening techniques, FRP influence, and various failing types. This model displayed decent relationships with the investigated outcomes [55]. The Generalized Softened Variable Angle Truss-Model is reconstructed in a integrated way and efficiently utilized to reinforced concrete and plain concrete beam. (Constantin E. Chalioris) has proposed distinct truss model, which uses Fiber reinforced polymer restrained concrete interaction of the EB Fiber reinforced polymers to the torsional ability. Accountable compression and tension interaction with Fiber reinforced polymer strengthened concrete are showed in the projected model [56].

6 Conclusions and Future Scope

The key findings from this comprehensive review are mentioned as follows.

1. The experimental data shows that maximum performance of RC beam under torsion was achieved by providing four side wrapping configuration and is explored by many researchers.
2. Fibers with 0-degree, 45-degree, and 90-degree alignments with respect to the longitudinal axis of beam are examined. Fibers wrapped with 90° orientations have good agreement with the result. 45°-orientated fibers are also good in diagonal crack arresting. Fiber with 0° orientation has less contribution in improving the torsional strength.
3. Fibers wrapped in strips are not much effective as compared to the fully wrapped continuous fibers; however, they prove effective in arresting the cracks.
4. Failure of strengthening beam occurred either due to failure of concrete or failure of fibers. Failure of concrete lead to the de-bonding of fibers.
5. Literature survey revealed that Analytical models such as softened membrane model for torsion (SMMT), Softened Variable Angle Truss-Model (GSVATM), and special truss model were proposed to strengthened RC beam under torsion.

Present work can be extended in the following areas to know in-depth behavior of RC beam strengthened with externally bonded (EB) composites:

1. Very few studies were noticed on proper anchorage system, which can effectively resist the slippage between fiber and concrete joint with three side wrapping configuration.
2. Additional work is needed on torsional strengthening with FRCM composite to know its response under FRCM strengthening beams with different fiber types.
3. Experimental work can be carried out on large size beam in large scale to evaluate the effect of size on externally strengthened RC beam under torsion.
4. Future investigation can be carried out on different strengthening configurations along with different types of composite FRP'.
5. Analytical model on RC beam strengthened with FRCM composite needs to be explored.

References

1. Sundarraja MC, Rajamohan S (2009) Strengthening of RC beams in shear using GFRP inclined strips—an experimental study. Constr Build Mater 23(2):856–864. https://doi.org/10.1016/j.conbuildmat.2008.04.008
2. Almusallam TH, Al-Salloum YA (2001) Ultimate strength prediction for RC beams externally strengthened by composite materials. Compos Part B Eng 32(7):609–619. https://doi.org/10.1016/S1359-8368(01)00008-7
3. Anania L, Badalà A, Failla G (2005) Increasing the flexural performance of RC beams strengthened with CFRP materials. Constr Build Mater 19(1):55–61. https://doi.org/10.1016/j.conbuildmat.2004.04.011
4. Deifalla A, Awad A, Elgarhy M (2013) Effectiveness of externally bonded CFRP strips for strengthening flanged beams under torsion: an experimental study. Eng Struct 56:2065–2075. https://doi.org/10.1016/j.engstruct.2013.08.027
5. Alabdulhady MY, Sneed LH, Carloni C (2017) Torsional behavior of RC beams strengthened with PBO-FRCM composite—an experimental study. Eng Struct 136:393–405. https://doi.org/10.1016/j.engstruct.2017.01.044
6. Rizzo A, De Lorenzis L (2009) Behavior and capacity of RC beams strengthened in shear with NSM FRP reinforcement. Constr Build Mater 23(4):1555–1567. https://doi.org/10.1016/j.conbuildmat.2007.08.014
7. Hii AKY, Al-Mahaidi R (2006) Experimental investigation on torsional behavior of solid and box-section RC beams strengthened with CFRP using photogrammetry. J Compos Constr 10(4):321–329. https://doi.org/10.1061/(asce)1090-0268(2006)10:4(321)
8. Gao B, Kim JK, Leung CKY (2004) Experimental study on RC beams with FRP strips bonded with rubber modified resins. Compos Sci Technol 64(16):2557–2564. https://doi.org/10.1016/j.compscitech.2004.05.016
9. Al-Bayati G, Al-Mahaidi R, Kalfat R (2016) Experimental investigation into the use of NSM FRP to increase the torsional resistance of RC beams using epoxy resins and cement-based adhesives. Constr Build Mater 124:1153–1164. https://doi.org/10.1016/j.conbuildmat.2016.08.095
10. Rizzo A, De Lorenzis L (2009) Modeling of debonding failure for RC beams strengthened in shear with NSM FRP reinforcement. Constr Build Mater 23(4):1568–1577. https://doi.org/10.1016/j.conbuildmat.2008.03.009
11. Maalej M, Bian Y (2001) Interfacial shear stress concentration in FRP-strengthened beams. Compos Struct 54(4):417–426. https://doi.org/10.1016/S0263-8223(01)00078-2
12. Guenaneche B, Krour B, Tounsi A, Fekrar A, Benyoucef S, Adda Bedia EA (2010) Elastic analysis of interfacial stresses for the design of a strengthened FRP plate bonded to an RC beam. Int J Adhes Adhes 30(7):636–642. https://doi.org/10.1016/j.ijadhadh.2010.06.003
13. Goudar SK, Shivaprasad KN, Das BB (2019) Mechanical properties of fiber-reinforced concrete using coal-bottom ash as replacement of fine aggregate. In: Sustainable construction and building materials. Springer, Singapore, pp 863–872
14. George RM, Das BB, Goudar SK (2019) Durability studies on glass fiber reinforced concrete. In: Sustainable construction and building materials. Springer, Singapore, pp 747–756
15. Srikumar R, Das BB, Goudar SK (2019) Durability studies of polypropylene fibre reinforced concrete. In: Sustainable construction and building materials. Springer, Singapore, pp 727–736
16. Yadav S, Das BB, Goudar SK (2019) Durability studies of steel fibre reinforced concrete. In: Sustainable construction and building materials. Springer, Singapore, pp 737–745
17. Gao B, Leung CKY, Kim JK (2007) Failure diagrams of FRP strengthened RC beams. Compos Struct 77(4):493–508. https://doi.org/10.1016/j.compstruct.2005.08.003
18. Alabdulhady MY, Aljabery K, Sneed LH (2019) Analytical study on the torsional behavior of reinforced concrete beams strengthened with FRCM composite. J Compos Constr 23(2):04019006. https://doi.org/10.1061/(asce)cc.1943-5614.0000927

19. Alabdulhady MY, Sneed LH (2018) A study of the effect of fiber orientation on the torsional behavior of RC beams strengthened with PBO-FRCM composite. Constr Build Mater 166:839–854. https://doi.org/10.1016/j.conbuildmat.2018.02.004
20. Mohamed HM, Chaallal O, Benmokrane B (2015) Torsional moment capacity and failure mode mechanisms of concrete beams reinforced with carbon FRP bars and stirrups. J Compos Constr 19(2):04014049. https://doi.org/10.1061/(asce)cc.1943-5614.0000515
21. Kaish ABMA, Jamil M, Raman SN, Zain MFM, Nahar L (2018) Ferrocement composites for strengthening of concrete columns: a review. Constr Build Mater 160:326–340. https://doi.org/10.1016/j.conbuildmat.2017.11.054
22. Adhikari A, Bagal DK, Student BT (2019) Torsional strength of ferrocement ' U ' wrapped RC beams: a comparative study. Int J Appl Eng Res 14(13):146–155
23. Mandal A, Nigam MK (2018) A review report on mechanical properties of ferrocement with cementitious materials. Int J Eng Res Technol 1(9):46–54
24. Deepak MS, Surendar M, Aishwarya B, Beulah Gnana Ananthi G (2020) Bending behaviour of ferrocement slab including basalt fibre in high strength cement matrix. Mater Today Proc:8–11. https://doi.org/10.1016/j.matpr.2020.08.074
25. Behera GC, Rao TDG, Kameswara CB (2014) Study of post-cracking torsional behaviour of high-strength reinforced concrete beams with a ferrocement wrap. Slovak J Civ Eng 22(3):1–12. https://doi.org/10.2478/sjce-2014-0012
26. Behera G, Dhal M (2018) Torsional behaviour of normal strength RCC beams with ferrocement 'U' wraps. Facta Univ—Ser Archit Civ Eng 16(1):1–16. https://doi.org/10.2298/fuace160514001b
27. Li B, Lam ESS, Wu B, Wang YY (2013) Experimental investigation on reinforced concrete interior beam-column joints rehabilitated by ferrocement jackets. Eng Struct 56:897–909. https://doi.org/10.1016/j.engstruct.2013.05.038
28. Ganesan N, Indira PV, Irshad P (2017) Effect of ferrocement infill on the strength and behavior of RCC frames under reverse cyclic loading. Eng Struct 151:273–281. https://doi.org/10.1016/j.engstruct.2017.08.031
29. Shaaban IG, Shaheen YB, Elsayed EL, Kamal OA, Adesina PA (2018) Flexural characteristics of lightweight ferrocement beams with various types of core materials and mesh reinforcement. Constr Build Mater 171:802–816. https://doi.org/10.1016/j.conbuildmat.2018.03.167
30. Behera GC, Rao TDG, Rao CBK (2016) Torsional behaviour of reinforced concrete beams with ferrocement U-jacketing—experimental study. Case Stud Constr Mater 4:15–31. https://doi.org/10.1016/j.cscm.2015.10.003
31. Revathy J, Gajalakshmi P, Aseem Ahmed M (2020) Flowable nano SiO_2 based cementitious mortar for ferrocement jacketed column. Mater Today Proc 22:836–842. https://doi.org/10.1016/j.matpr.2019.11.020
32. Sankar K, Shoba Rajkumar D (2020) Experimental investigation on different high rich cement mortar for ferrocement application. Mater Today Proc 22:858–864. https://doi.org/10.1016/j.matpr.2019.11.033
33. Al-Mahaidi R, Hii AKY (2007) Bond behaviour of CFRP reinforcement for torsional strengthening of solid and box-section RC beams. Compos Part B Eng 38(5–6):720–731. https://doi.org/10.1016/j.compositesb.2006.06.018
34. Kim J, Kwon M, Seo H, Lim J (2015) Experimental study of torsional strength of RC beams constructed with HPFRC composite mortar. Constr Build Mater 91:9–16. https://doi.org/10.1016/j.conbuildmat.2015.05.018
35. Kim MJ, Kim HG, Lee YJ, Kim DH, Lee JY, Kim KH (2020) Pure torsional behavior of RC beams in relation to the amount of torsional reinforcement and cross-sectional properties. Constr Build Mater 260:119801. https://doi.org/10.1016/j.conbuildmat.2020.119801
36. Maalej M, Leong KS (2005) Effect of beam size and FRP thickness on interfacial shear stress concentration and failure mode of FRP-strengthened beams. Compos Sci Technol 65(7–8):1148–1158. https://doi.org/10.1016/j.compscitech.2004.11.010
37. Lee HK, Cheong SH, Ha SK, Lee CG (2011) Behavior and performance of RC T-section deep beams externally strengthened in shear with CFRP sheets. Compos Struct 93(2):911–922. https://doi.org/10.1016/j.compstruct.2010.07.002

38. Kandekar SB, Talikoti RS (2018) Study of torsional behavior of reinforced concrete beams strengthened with aramid fiber strips. Int J Adv Struct Eng 10(4):465–474. https://doi.org/10.1007/s40091-018-0208-y
39. Ghobarah A, Ghorbel MN, Chidiac SE (2002) Upgrading torsional resistance of reinforced concrete beams using fiber-reinforced polymer. J Compos Constr 6(4):257–263. https://doi.org/10.1061/(asce)1090-0268(2002)6:4(257)
40. Hii AKY, Al-Mahaidi R (2006) An experimental and numerical investigation on torsional strengthening of solid and box-section RC beams using CFRP laminates. Compos Struct 75(1–4):213–221. https://doi.org/10.1016/j.compstruct.2006.04.050
41. Chalioris CE (2007) Behavioural model of FRP strengthened reinforced concrete beams under torsion. In: Proceedings of the 1st Asia-Pacific conference on FRP in structures, APFIS 2007, vol 1, no December 2007, pp 111–116
42. Chalioris CE (2008) Torsional strengthening of rectangular and flanged beams using carbon fibre-reinforced-polymers—experimental study. Constr Build Mater 22(1):21–29. https://doi.org/10.1016/j.conbuildmat.2006.09.003
43. Kandekar SB, Talikoti RS (2020) Torsional behaviour of reinforced concrete beams retrofitted with aramid fiber. Adv Concr Constr 9(1):1–7. https://doi.org/10.12989/acc.2020.9.1.001
44. Kandekar SB, Talikoti RS (2019) Torsional behaviour of reinforced concrete beam wrapped with aramid fiber. J King Saud Univ—Eng Sci 31(4):340–344. https://doi.org/10.1016/j.jksues.2018.02.001
45. Ilkhani MH, Naderpour H, Kheyroddin A (2019) A proposed novel approach for torsional strength prediction of RC beams. J Build Eng 25(January):100810. https://doi.org/10.1016/j.jobe.2019.100810
46. Behera GC, Gunneswar RTD, Rao CBK (2013) Torsional strength of ferrocement 'U' wrapped normal strength beams with only transverse reinforcement. Procedia Eng 54:752–763. https://doi.org/10.1016/j.proeng.2013.03.069
47. Deifalla A, Ghobarah A (2010) Full torsional behavior of RC beams wrapped with FRP: analytical model. J Compos Constr 14(3):289–300. https://doi.org/10.1061/(asce)cc.1943-5614.0000085
48. Alabdulhady MY, Sneed LH, Abdelkarim OI, ElGawady MA (2017) Finite element study on the behavior of RC beams strengthened with PBO-FRCM composite under torsion, vol 179. Elsevier Ltd
49. Panchacharam S, Belarbi A (2002) Torsional behavior of reinforced concrete beams strengthened with FRP composites. First FIB Congr., no. September, pp 1–11, 2002, [Online]. Available: http://rb2c.mst.edu/media/research/rb2c/documents/torsional.pdf
50. Jeng CH, Chao M, Chuang HC (2019) Torsion experiment and cracking-torque formulae of hollow prestressed concrete beams. Eng Struct 196(July):109325. https://doi.org/10.1016/j.engstruct.2019.109325
51. Capozucca R, Nilde Cerri M (2002) Static and dynamic behaviour of RC beam model strengthened by CFRP-sheets. Constr Build Mater 16(2):91–99. https://doi.org/10.1016/S0950-0618(01)00036-8
52. Chalioris CE, Karayannis CG (2009) Effectiveness of the use of steel fibres on the torsional behaviour of flanged concrete beams. Cem Concr Compos 31(5):331–341. https://doi.org/10.1016/j.cemconcomp.2009.02.007
53. Tibhe SB, Rathi VR (2016) Comparative experimental study on torsional behavior of RC beam using CFRP and GFRP fabric wrapping. Procedia Technol 24:140–147. https://doi.org/10.1016/j.protcy.2016.05.020
54. Ameli M, Ronagh HR (2007) Analytical method for evaluating ultimate torque of FRP strengthened reinforced concrete beams. J Compos Constr 11(4):384–390. https://doi.org/10.1061/(asce)1090-0268(2007)11:4(384)
55. Bernardo LFA, Andrade JMA (2020) A unified softened truss model for RC and PC beams under torsion. J Build Eng 32(April):101467. https://doi.org/10.1016/j.jobe.2020.101467
56. Shen K, Wan S, Mo YL, Jiang Z (2018) Theoretical analysis on full torsional behavior of RC beams strengthened with FRP materials. Compos Struct 183(1):347–357. https://doi.org/10.1016/j.compstruct.2017.03.084

Study on Static Analysis and Design of Reinforced Concrete Exterior Beam-Column Joint

Yogesh Narayan Sonawane and Shailendrakumar Damodar Dubey

Abstract In reinforced cement concrete, the beam-column joint is always considered as critical structural elements. Therefore, the analysis, design and construction of beam-column joint are indeed a very important task. Due to the geometrical conditions and the basic complexity of the load transfer mechanism, the design code on seismic behavior of moment-resisting frame received many conflicting reviews on the beam-column joint design to date. In fact, the overall structural performance of the entire building components in an earthquake majorly depends on the load-bearing capacity of the beam-column joint. During such an earthquake action, the beam-column becomes more vulnerable and unable to achieve its desired efficiency. The earthquake so far has shown that there are two major failures that occur at the beam-column joint such as (i) Joint shear failure (ii) End anchorage failure due to inadequate reinforcement detailing. This paper focuses on the experimental study of exterior beam-column joint casted as per IS 456:2000 and tested on loading frame.

Keywords Beam-column joint · Anchorage mechanism · Joint shear failure · End anchorage failure

1 Introduction

1.1 Beam-Column Joint

In reinforced cement concrete frame structure, beam-column joint is a really complicated and insecure area, where the beam and column elements come together in all three directions [1], such a joint always confirms the continuation of the entire

Y. N. Sonawane (✉)
KBCNMU, Jalgaon, Maharashtra, India
e-mail: yogeshsonawane789@gmail.com

S. D. Dubey
Department of Civil Engineering, SSVPS, Dhule, Maharashtra, India

M. S. Ranadive et al. (eds.), *Recent Trends in Construction Technology and Management*, Lecture Notes in Civil Engineering 260,
https://doi.org/10.1007/978-981-19-2145-2_66

structure and helps to transfer various forces that are induced at the end of all structural members. Therefore, considering the safety of the building, in places where there is an earthquake-affected area, the buildings are to be designed in such a way that to minimize their damage after the earthquake. The structural behavior of reinforced cement concrete moment-resisting frame structure seen in the latest earthquake around the world has emphasized the performance of the beam-column joints [2]. Therefore, over the few last decades, the researches on the reinforced concrete beam-column joints have been focused on its seismic behavior, because such joints are a major source of energy dissipation. In order to maintain proper connection of all the members after the earthquake, it is necessary to create the proper design of the beam-column joint. If not designed properly, then due to load reversal can lead to a drastic reduction in both strength and stiffness. Still in India, the design guidelines for structural steel and reinforced concrete construction practices used in earthquake-resistant construction have not been effectively revised. Studies so far have shown that the deficiency of these beam-column joints is mainly due to insufficient transverse reinforcement and anchorage capacity in the beam-column joints. Due to the lack of reinforcement details, the constituent material is partly strong, resulting in inadequate structural performance with beam-column joint ability to carry limited strength even in moderate earthquakes.

1.2 Classification of Reinforced Concrete Beam-Column Joint

In reinforced cement concrete moment-resisting frame structure majorly three types of the beam-column joint can be recognized are as follows as shown in Figs. 1 and 2.

(a) Exterior beam-column joint
(b) Interior beam-column joint
(c) Corner beam-column joint.

According to ACI 352 R-02, classifications of the beam-column joint based on various loading conditions are Type-1 and Type-2.

Type-1 beam-column joint is made of a member designed to satisfy ACI 318-02 strength requirement for a member without significant inelastic deformation, i.e., without considering special ductility requirement. Type-1 is to design for resist gravity load and normal wind load.

Type-2 connection, frame members are designed to maintain a constant strength under structural deformation reversal into the inelastic range. This joint is specially designed to resist lateral load due to earthquake, blast, and cyclonic windfall [4].

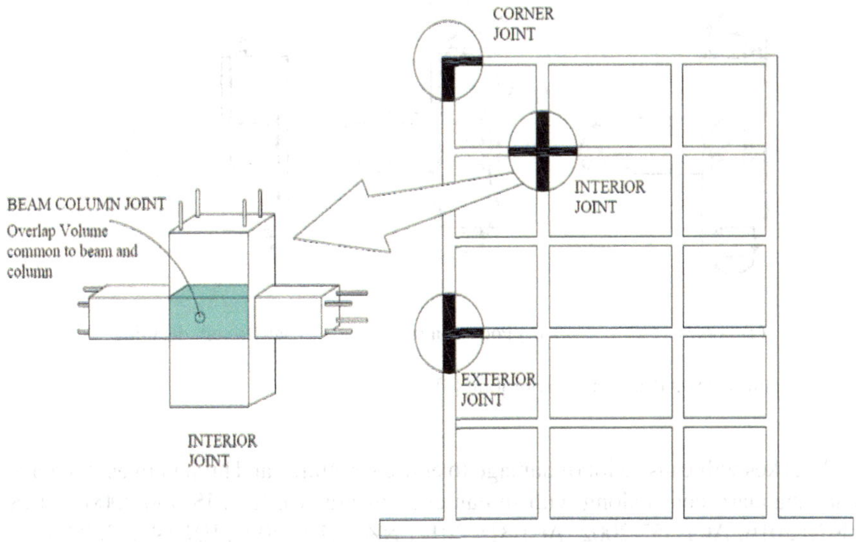

Fig. 1 Beam-column joints are critical part of reinforced framed structure

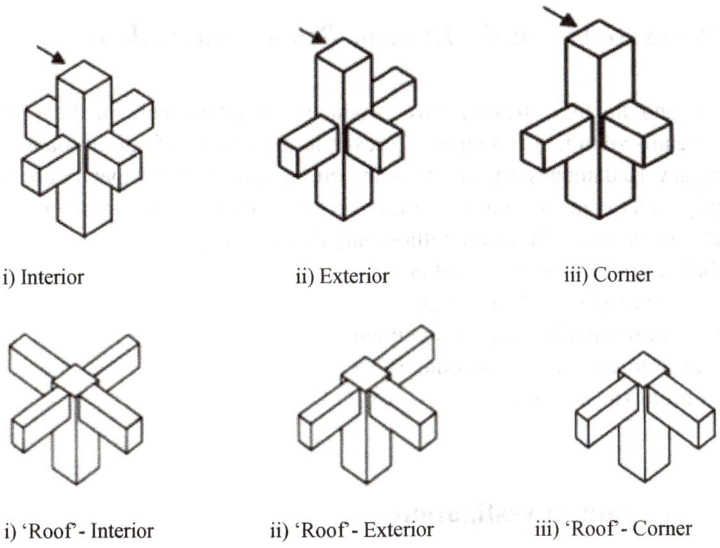

Fig. 2 Classification of Joints in moment-resisting frame structure. *Source* ACI-352R-02 [3]

2 Review of Codes

In India, the Indian Standards Code does not include the necessary recommendations for a reinforced concrete beam-column joint. If such joint does not design properly,

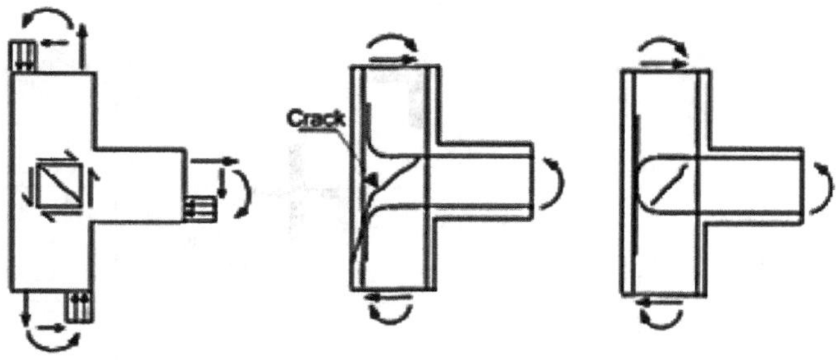

(a) Forces (b) Poor detailing (c) Satisfactory detailing

Fig. 3 Reinforcement details at exterior joint [8]

earthquakes will cause a lot of damage to entire structures and human lives. We have many standard codes along with Indian code of practice (i.e., IS 456:2000 and IS 13920:2016, ACI 352-2002, ACI 318.2011, NZS 3101:2006, EN 1998:2003, etc., association with the beam-column joints [5, 6, 7]

2.1 Forces Acting on the Exterior Beam-Column Joint

The shear and flexural stresses have produced simultaneously in the reinforced concrete beam-column core region. To resist large amount of lateral loads without occurring severe damages, the strength and energy dissipation capacity are needed. It is highly uneconomical to design a reinforced concrete framed structure for most potential seismic ground motion without any damage (Fig. 3).

Symbol refers to stress resultant are
Cc = Compression in the concrete
Cs = Compression in the reinforcement
T = Tension force in reinforcement
V = Sum of shear stress.

3 Experimental Investigation

3.1 Design and Detailing of Exterior Beam-Column Joint Specimen as Per IS—456:2000 for M20 Concrete

The test specimen of exterior beam-column joint has column cross section 225 mm × 150 mm and height 800 mm connected with cantilever beam portion having the

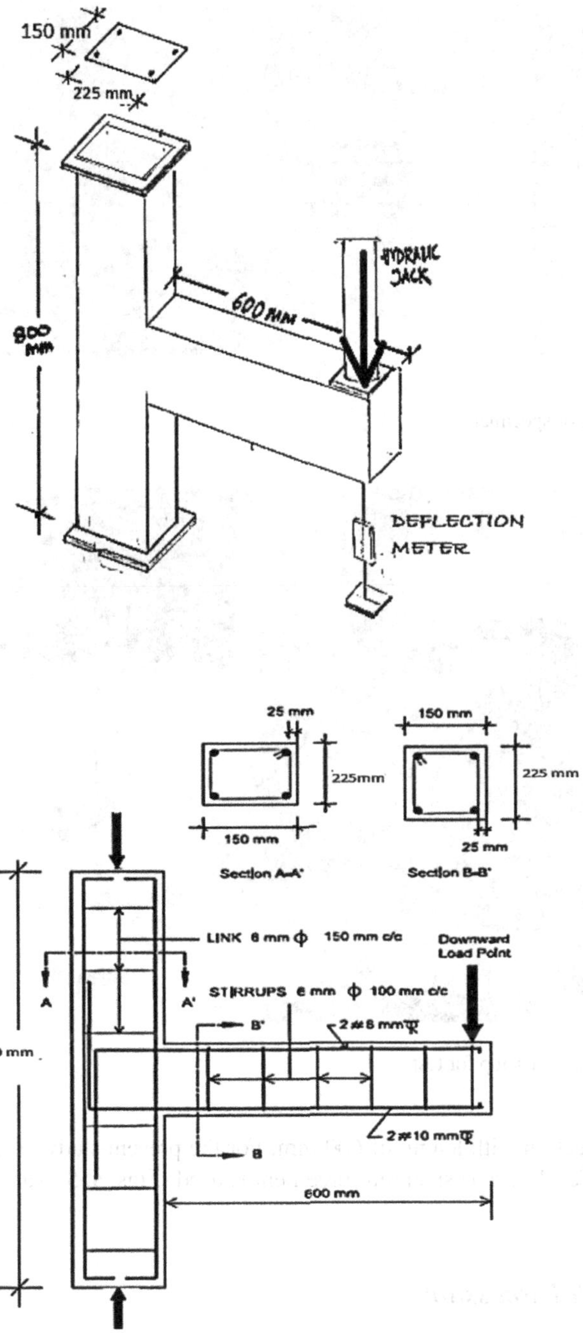

Fig. 4 Specimen details for test

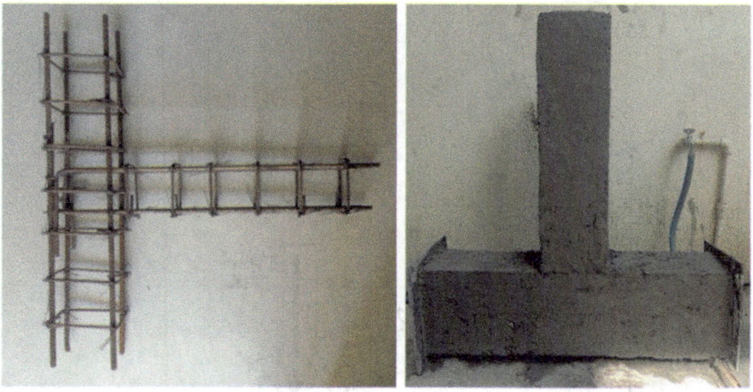

Fig. 5 Casting of specimen

Fig. 6 Experimental setup for test

same cross section with length of 600 mm. For the present study grade of concrete is M20 and Fe 415 grade steel reinforcement is used (Figs. 4, 5, 6 and 7).

3.2 *Result Discussion*

See Table 1 and Fig. 8.

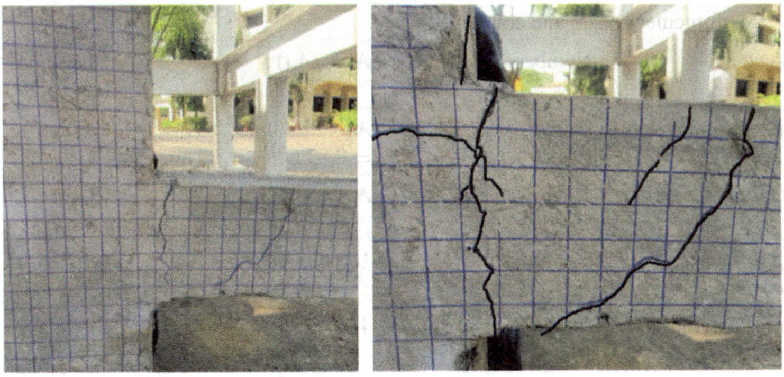

Fig. 7 Crack pattern

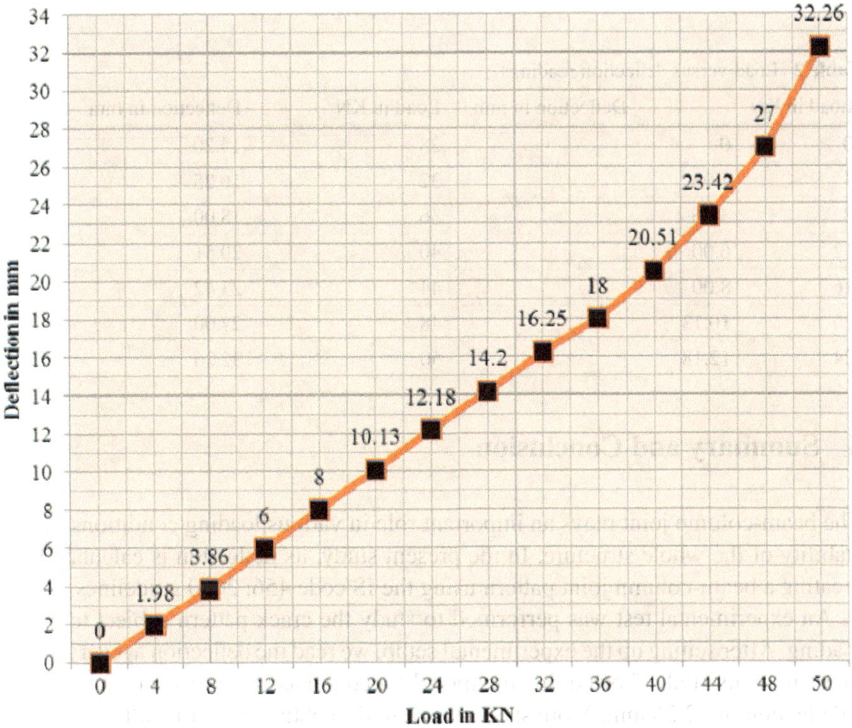

Fig. 8 Load versus deflection graph

Table 1 Reinforcement-beam and column [9]

Reinforcement	Numbers	Area (A_{st})	Clause no. (IS456:2000)
Tension reinforcement in beam	2 bars of 10 mm Φ	$A_{st} = 157\ mm^2 > A_{st\ min} = 0.85\ bd/fy = 69\ mm^2$	Cl. 26.5.1.1 (a)
Shear reinforcement in beam	2-legged 6 mm Φ	$A_s = 56.52\ mm^2$	Cl. 26.5.1.6
Spacing of shear reinforcement	6 mm Φ stirrups	100 mm c/c	Cl. 26.5.1.5
Anchorage length (L_d)	for the tension rod 10 mm	470 mm	Cl. 26.2.1
Reinforcement in column (longitudinal)	5 bars of 12 mm Φ	$A_{sc} = 452\ mm^2 > 0.8\ bd = 270\ mm^2$	Cl. 26.5.3.1 (a)
Lateral ties in column	6 mm Φ	150 mm c/c	Cl. 26.5.3.2 (c)

Table 2 Load versus deflection readings

Load in KN	Deflection in mm	Load in KN	Deflection in mm
0	0	28	14.20
4	1.98	32	16.25
8	3.86	36	18.00
12	6.00	40	20.51
16	8.00	44	23.42
20	10.13	48	27.00
24	12.18	50	32.26

4 Summary and Conclusion

The beam-column joint plays an important role in various loading conditions as the stability of the whole structure. In the present study, its deflection is calculated by creating a beam-column joint pattern using the IS code 456: 2000 guidelines.

An experimental test was performed to study the crack pattern subject to static loading. After setting up the experimental setup, we read the deflection at every 4 KN intervals. This study showed that the model failed a maximum load of 50 KN with a deflection of 32.26 mm. More study is required to obtain proper results.

References

1. Kaliluthin AK, Kothandaraman S, Suhail Ahamed TS (2014) A review on behavior of reinforced concrete beam-column joint. Int J Innov Res Sci Eng Technol 3(4):11299–11312
2. Rajaram P, Murugesan A (2010) Experimental study on behaviour of interior RC beam column joints subjected to cyclic loading. Int J Appl Eng Res Didigul 1(1):49–59
3. ACI-ASCE Committee 352R-02, Recommendations for design of beam-column connections in monolithic reinforced concrete structure
4. American Concrete Institute (ACI) (1985) Building code requirements for structural concrete and commentary. ACI 318-95, Farmington Hills, Mich
5. IS 13290:2016 (2016) Ductile design and detailing of reinforced concrete structures subjected to seismic forces-code of practice (First Revision), BIS
6. IS 13290:1993, Ductile design and detailing of reinforced concrete structures subjected to seismic forces-code of practice
7. IS 1893 (Part 1):2016 (2016) Criteria for earthquake resistant design of structures (Sixth Revision), BIS
8. Uma SR, Prasad AM, Seismic behavior of beam column joints in reinforced concrete moment resisting frames. Final report—a earthquake codes
9. Indian standard code of Plain and Reinforced Concrete—Code of Practice IS 456 (2000) Bureau of Indian Standards, New Delhi
10. Bindhu KR, Jaya KP (2010) Strength and behaviour of exterior beam column joints with diagonal cross bracing bars. Asian J Civ Eng (Build Hous) 11(3):397–410
11. Padmanabham K, Rambabu K (2019) Design improvements of non-seismically detailed R/C beam-column joints during seismic transformation. Int J Tech Innov Mod Eng Sci 5(3):701–709
12. Dhake PD, Patil HS, Patil YD (2015) Role of hoops on seismic performance of reinforced concrete joints. In: Proceedings of the Instituion of Civil Engineers-Structures and Buildings, vol 168, no 10, pp 708–717 (Magazine of Concrete Research)
13. Jain SK, Murty CVR (2005) Proposed draft provisions and commentary on ductile detailing of RC structures subjected to seismic forces, Final report—a earthquake codes

Refined Methodology in Design of Reinforced Concrete Shore Pile: A Design Aid

Mahesh Navnath Patil and Shailendra Kumar Damodar Dubey

Abstract Shore piles with rock anchors are considered as distinct types of retaining walls. The idea behind the erection of shore piles is to protect structures surrounding the construction site from collapse due to the formation of a slip circle. A very limited guidelines and solutions are available in the existing literature for the design of shore piles. The earth pressure distribution differs along the depth of excavation from ground level depending upon the density of soil and occurrence of the water table below the ground surface. This paper represents a finite element analysis of shore piles with Staad *Pro.* V8i. Experimental investigation of earth pressure and surcharge load on shore piles from opposite sides of excavation with two platforms in the case of breaking loads with case study are discussed. A parametric study was conducted to enable a discussion of the effects of the distribution of active lateral earth pressure and surcharge and fixity generated due to rock anchors on maximum flexural moment generated on shore pile, which is analyzed as a rectangular beam. Updated practice is suggested to enrich the manual calculation with software aid in the bending moment of the wall.

Keywords Shore piles · Rock anchors · Finite element analysis · Staad pro. V8i · Earth pressure

1 Introduction

The deep excavation in the middle of the civic requires an alternate arrangement to handle the lateral pressure of surrounding structures. The lateral pressure is normally distributed in the form of a pressure bulb underneath soil [1]. An artificial structure such as a retaining wall is constructed to bear the lateral load of backfill and surcharge.

M. N. Patil (✉)
Research Scholar, Kavyitri Bahinabai Chaudhari North Maharashtra University, Jalgaon, India
e-mail: m.patil123@gmail.com

S. K. D. Dubey
Professor and Head of the Civil Engineering Department SSVPS BSD COE, Dhule, India
e-mail: dubey.dhule@gmail.com

M. S. Ranadive et al. (eds.), *Recent Trends in Construction Technology and Management*, Lecture Notes in Civil Engineering 260,
https://doi.org/10.1007/978-981-19-2145-2_67

In the case of large plan dimensions and deep excavations, a retaining wall is not a wise choice and we have to adopt heavy structures such as shore piles [2]. Shore piles are the segment of interlocked piles designed to carry lateral and/or hydrostatic pressure arising due to the formation of slip circles, formed due to differences in elevations because of excavation, dredging, backfilling, or a combination of these [3]. A sealed shore pile provides a safe and economical solution in the places with the intrusion of groundwater to inhibit the seepage into the construction site and minimize the risk of settlement. The shore piles provide the anchorage with surrounding soil through rock anchors.

When excavation exceeds 6–10 m below ground level, a fencing of interlocking shore piles is needed to resist heavy lateral loads [2]. It includes active earth pressures, passive earth pressures, surcharge loads, and hydrostatic pressure [4]. In such conditions, reinforced shore piles are the best suited as the boundary of excavation. Moreover, the cross-sectional properties and spacing of shore piles depend on present site conditions such as geotechnical properties of soil, depth of water table below ground level, and magnitude of surcharged lateral pressure [2]. Specification of shore piles should include general instruction issued by the project manager, architect, structural engineer, and owners, setting out of shore pile position on site, boring and piling, protection of boreholes against collapse, specification on concrete and reinforcement to be used, and specification of the casing for shore piles. Onsite shore piling and its design should be done in compliance with IS code guidelines.

This paper presents a methodology to design shore piles. The theoretical analysis is a complex process; thus, for precise and speedy analysis, the case study is considered and modeled in STAAD. Pro V8i. The shore pile is analyzed for two different conditions. The first condition is non-yielding support at the top (in place of rock anchor) with spring supports at the base end and the second condition is yielding support at the top (in place of rock anchor) with fixed support at the bottom (midpoint of the support socket length). The analysis includes the lateral loading conditions from the opposite side of the excavation. This study provides guidelines for the design of shore piles as the direct procedure is not available anywhere.

2 Specifications for Shore Piles

Onsite construction of shore piles should be done in compliance with the following guidelines.

2.1 General

The drawings shall be read in conjunction with relevant architectural drawings, soil investigation reports, survey drawings, services drawings, tender conditions, specifications and item descriptions in the bill of quantities, and instructions issued by the project manager, architect, structural engineer, and owners. Any discrepancy among any of the above shall be immediately brought to the notice of the project manager before proceeding to execute the work. No claim shall be entertained if the work is carried out without obtaining a guideline decision in case of discrepancy. Piling design and detailing have been carried out based on the information given in the soil investigation report. In case the contractor discovers site conditions differing from those mentioned in the soil report, they shall be instantly brought to the notice of the structural engineer through the project manager before proceeding to execute the work. Failure to obtain clarification/approval to proceed for execution in such cases shall impose the responsibility of any consequences arising out of taking unilateral decision solely on the contractor.

2.2 Setting Out

The contractor shall follow the architectural centerline drawings to set out the column centers from the given baseline. The contractor shall obtain written approval for the setting out of column centers from the project manager as per the architectural drawings. Before commencing with boring for any pile, removable casings shall be driven up to ground level and their center lines checked jointly by the contractors and Project Manager.

2.3 Boring and Piling

The construction of the shore pile should be carried out as per guideline of IS: 2911 similar to the pile foundation. The casting of shore piles must be cast in situ. The sequence of construction of shore piles should be such that the previously constructed piles should not be affected by the construction of subsequent ones. The sequence of construction should be decided by contractors and proceed with the work only after obtaining approval.

2.4 Allowances and Tolerances

The shift or eccentricity of piles from their theoretical position shall not be more than 75 mm for piles up to and including 1200 mm. Out of plumb of the piles shall not be more than 2%. All consequences of noncompliance with these requirements shall be to the contractor's account.

2.5 Protection of Bore Whole Sides

The contractor should be responsible to protect the sides of the borehole and prevent them from collapsing. Bentonite slurry should be used to stabilize the walls of the boreholes by circulating bentonite slurry if required. Bentonite shall be solution based and conform to relevant IS codes. Bentonite shall be used after mixing with water, and a thick slurry shall be generated with a specific gravity of 1.08–1.10 and a pH value of 6–9. Boring shall be carried out as per the recommendations of the geotechnical investigation report and as per the results obtained during the first pile test results. The depth of the boring shall depend upon desired capacity. The contractor shall be responsible to keep the site clean by carting away mud while boring is in progress. In the case of marine clay, liquid sand, and any other soil in which bentonite clay cannot be used for stabilizing the sides of boreholes, mild steel liners shall be used for the entire length of the borehole. M.S. liner shall be sufficiently strong such that it withstands the soil and water pressure for the full length of the bore even when concrete is not poured. The liner shall not form any gaps and shall be placed properly to withstand all forces while installation, concreting, and retrieval operations.

2.6 Reinforcement

All reinforcement used in the shore piles shall be high yield strength deformed bars with a yield strength of 500 N/mm^2 conforming to IS 1786. All steel shall be tested as per standard norms for physical and chemical tests. Reinforcement should be as per the drawing issued by the structural engineer. The binding wire shall be galvanized. Reinforcement shall be protected against corrosion if exposed to the atmosphere by any standard method approved by the project manager. Clear cover to main reinforcement shall be 50 mm made up of PVC/Plastic material. Provisions of IS 1786:2008 [5], IS 228:1992 [6], IS 1387:1993 [7], and IS 1599:1985 [8] should be followed.

2.7 Concrete

Concrete shall conform to the specifications as per IS 456:2000 [9]. Admixtures such as superplasticizer shall be used for workability requirements. Concrete shall be deposited in the pile using a tremie pipe.

2.8 Dewatering

Dewatering shall be carried out continuously or as directed by the project manager in order to keep the borehole completely dry to retain the consistency and composition of concrete poured in the bores.

2.9 Casing

Temporary steel casing of adequate thickness shall be used if necessary to avoid the sides from collapsing and to maintain the alignment and position of the pile. Use of permanent liners may be permitted if the soil condition demands their use. All such liners shall be prescribed by the contractor in advance and permission obtained from the project manager before installing them.

2.10 Material

2.10.1 Cement

53 grade ordinary Portland cement is recommended. Some basic parameters such as strength, setting, and time should be tested after receiving a batch of cement. Onsite field test should be performed on cement if it is stored over for a longer period as per instructions of the project manager.

2.10.2 Coarse Aggregate

Coarse aggregate shall be obtained from authorized and approved quarries. Aggregate shall be clean, free from vulnerable impurities which affect the strength as well as durability characteristics of the concrete and reinforcement. The coarse aggregates shall be uniformly graded with the size of all the particles should be greater than 4.75 mm. The limits of the impurities shall be within the permissible limit given by IS 2386:1963 [10] and IS 383:2016. The soundness of aggregate, impact value, abrasion value, and crushing value shall be as per norms and limits laid down by IS 383:2016. The source of supply of coarse aggregates is never allowed to change without permission and approval of the competent authority [11].

2.10.3 Fine Aggregate

Grading of fine aggregates used shall be restricted to the size of particles less than 4.75 mm and should lie within zone 1st and 2nd as per IS 383:2016. It should be free from the impurities which affect the durability of concrete. It should not contain impurities which can attack and corrode steel reinforcement. Depending on the situation, crushed sand may be added to natural sand within the permissible limit [12].

2.10.4 Water

Water used while concreting shall be potable water free from impurities such as alkalis and salts which adversely affect the useful life of the concrete and may cause deterioration of reinforcement within the concrete. Water should not affect the hydration reaction of cement and consequently hardening properties of concrete. The level of harmful impurities within water should be limited to the restriction prescribed by IS 456:2000 [9]. Testing of water should be done in accordance with IS 3025 [13]. The average compressive strength of a 15 cm mortar cube should vary more than 10% as compared with a mortar cube prepared using distilled water in accordance with IS516:1959. The pH value of water shall be within the range of 6–9.

2.10.5 Admixture

Admixture confirming to IS 9103:1999 [14] (Clause no. 5.5 of IS 456) shall be permitted to be used without affecting the strength and durability characteristics of concrete. Admixtures shall be free from impurities (e.g. chlorides, sulphates, etc.) which deteriorate concrete. Reinforcement should not be affected by admixtures used within concrete. The admixtures may be accelerators or retarders depending upon situations. If more than one admixtures are suggested at a time, then their compatibility must be assured.

2.10.6 Cover Block

Cover blocks shall be made up of non-corrosive material such as mortar, plastic, and stones. The strength of the cover block shall be nearly equal to the strength of concrete used for shore piles. If cover blocks are made up of concrete or mortar, they should be fully cured before concreting.

2.10.7 Binding Wire

Galvanized iron wires of 18 gauge shall be used as winding wire. It shall not be corroded. It shall not contain a composition that may obstruct the bonding of reinforcement with concrete.

2.10.8 Ready Mix Concrete (R.M.C.)

In the construction of shore piles, concrete from ready mix concrete plants is preferred as mass concreting is required to be done and quality should be assured. Proportions and quality of ingredients of concrete should be as per IS 10262:2019 [15].

The continuous flow of concrete shall be required to be maintained as mass concreting is required at the time of shore piling. A stationary or mobile concrete pump shall be deployed for maintaining the continuous flow of concrete. All the accessories of the pump such as bends and chutes shall be organized in such a manner that concreting shall not be interrupted.

2.11 Bentonite

Boring is to be done before concreting of shore piles. The sides of the boreholes are required to be protected from collapse. Bentonite slurry plays a vital role in stabilizing the walls of the boreholes. Water-based slurry of bentonite clay shall be used for stabilizing walls of boreholes.

2.12 Delivery and Storage of Reinforcement

Care should be taken while loading and unloading steel on site, and it should not bend. Rebar shall be covered with a coating of some suitable material such as turpentine to protect it against corrosion. Reinforcement shall be not stored directly on the ground, and it can be isolated from the ground with the use of a timber rack or by plain cement concrete surface. The minimum yield strength of reinforcement shall not be less than 500 N/mm^2. Rebar of ISI mark shall be preferred.

2.13 Design of Shore Piles

This section consists of the design of shore piles for the basement of the construction site. The design comprises 750 mm diameter shore piles at 800 mm c/c spacing. A 20 KN/m^2 of the live load has been assumed to represent the vehicular load (construction activity). Water pressure is not considered, as there is a 50 mm gap between the shore piles, which will allow the water to seep through it. The angle of internal friction within soil particles is 300 [16].

3 Calculation for Earth Pressures

Based on the thickness of different layers of soil, the calculation of lateral earth pressure is shown below [17].

3.1 Layer-I Sand or Backfill Material, Ground Level to 1.5 m, I.E. 0–1.5 m

With reference to Fig. 1

$$\text{Coefficient of active pressure}(K_a) = (1 - \sin\emptyset)/(1 + \sin\emptyset)$$

where $\emptyset$ = Angle of internal friction $= 30^0$

$$\text{Dry unit weight of soil}(\gamma d) = 18\ \text{kN/m}^2$$

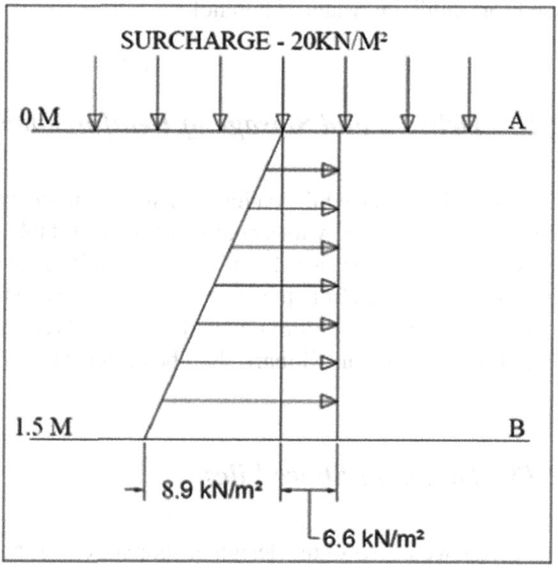

Fig. 1 Soil pressure in zone-I (0–1.5 m)

$$\text{Surcharge(q)} = 20\,\text{kN/m}^2$$

$$\text{Pressure at top(A)} = K_a \times q = 0.33 \times 20 = 6.6\,\text{kN/m}^2$$

$$\begin{aligned}\text{Pressure at bottom(B)} &= (K_a \times q) + (K_a \times \gamma \times H) \\ &= (0.33 \times 20) + (0.33 \times 18 \times 1.5) \\ &= 15.50\,\text{kN/m}^2\end{aligned}$$

3.2 Layer-II Stiff Clay (1.5–5.5 m)

With reference to Fig. 2

$$\text{Coefficient of active pressure}(K_a) = 0.33$$

$$\text{Dry unit weight of soil}(\gamma_d) = 18\,\text{kN/m}^2$$

$$\text{Surcharge(q)} = 20\,\text{kN/m}^2$$

$$\gamma_{\text{submerged}} = 8\,\text{kN/m}^2$$

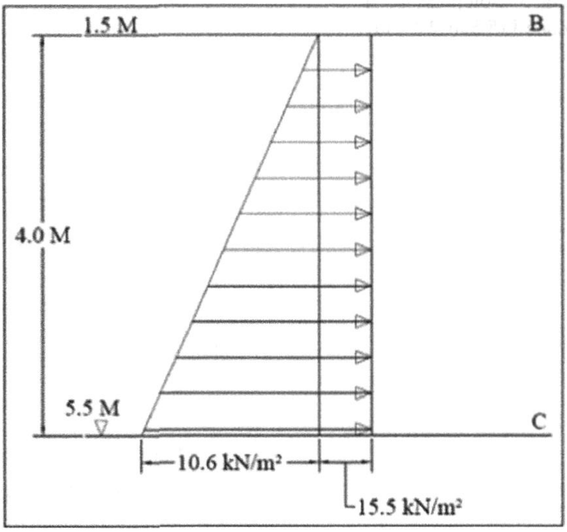

Fig. 2 Soil pressure in Zone-I (1.5–5.5 m)

$$\text{Pressure at top} = 15.50\,\text{kN/m}^2$$

$$\begin{aligned}\text{Pressure at bottom(C)} &= (K_a \times \gamma_{sub} \times \text{H}) + 15.50\\ &= (0.33 \times 8 \times 4) + 15.50\\ &= 26.10\,\text{kN/m}^2\end{aligned}$$

As water table depth is at 1.5 m from ground level, water Pressure (hydrostatic pressure) is calculated as follows:

$$\gamma_w = 10\,\text{kN/m}, \ \text{Water pressure} = \gamma_w \times \text{H} = 10 \times 4 = 40\,\text{kN/m}^2$$

3.3 *Layer III-Highly Weathered Volcanic Breccia with Shale (5.5–12.60 m)*

With reference to Figs. 3 and 4

Angle of internal friction is given 45° as per soil testing report

$$\text{Coefficient of active pressure}(K_a) = (1 - \sin\text{Ø})/(1 + \sin\text{Ø}),$$

where Ø = 45°

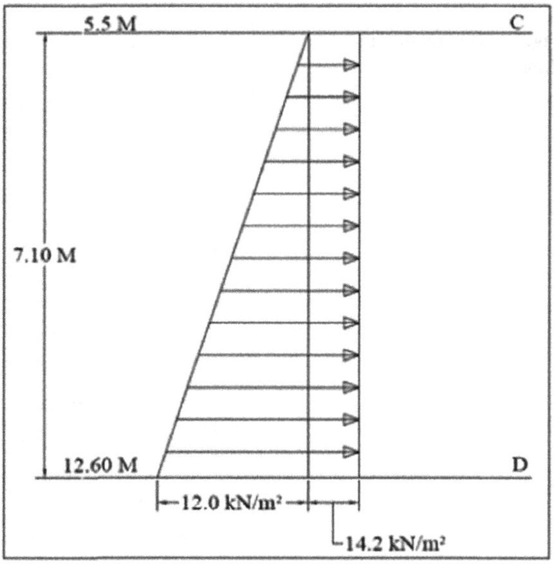

Fig. 3 Soil pressure in Zone-I (5.5 m–12.6 m)

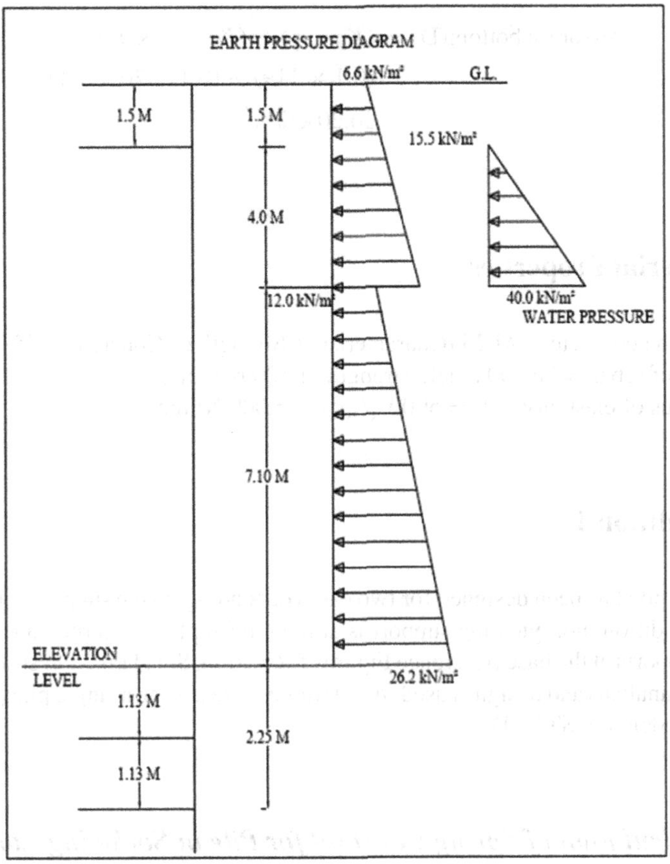

Fig. 4 Soil pressure distribution from ground level to socket length

$$\text{Coefficient of active pressure}(K_a) = 0.1$$

$$\text{Dry unit weight of soil}(\gamma_d) = 20\,\text{kN/m}^2\,(\text{For Rock})$$

$$\text{Surcharge(q)} = 20\,\text{kN/m}^2$$

$$\begin{aligned}\text{Surcharge(q)} &= 20 + (18 \times 1.5) + (8 \times 4) + (10 \times 4)\\ &= 119\,\text{kN/m}^2\end{aligned}$$

$$\begin{aligned}\text{Pressure at top(C)} &= K_a \times q\\ &= 0.1 \times 119 = 12\,\text{kN/m}^2\end{aligned}$$

$$\begin{aligned}\text{Pressure at bottom(D)} &= (K_a \times q) + (K_a \times \gamma \times H)\\ &= (0.1 \times 119) + (0.1 \times 20 \times 7.1)\\ &= 26.20\,\text{kN/m}^2\end{aligned}$$

4 Material Properties

Grade of the concrete = M 25 (Characteristics Strength of Concrete = 25 N/mm^2)

Grade of rebar = Fe 500 (Yield strength of 500 N/mm^2)

Modulus of elasticity = E = 5000 $\sqrt{f_{ck}}$ = 25,000 N/mm^2

5 Condition 1

The shore pile has been designed for two cases depending on the support conditions. In this condition, non-yielding support is at the top (in place of a rock anchor) and spring supports at the base (reference Figures 5, 6 and 7). Staad Pro. V8i software for structural analysis and design is used. A 2D frame representing a single pile (750 mm dia.) is modeled in STAAD.

5.1 Calculation of Spring Constant for Pile in Socketing Zone

Subgrade Modulus (as per soil testing report) = 37,500 KN/m^3

Three spring supports (joints) are assigned in the region of socket length

Spring constant for central spring support (joint)

$$\begin{aligned}&= \text{exposed area of shore pile} \times \text{subgrade modulus}\\ &= \text{socket length} \times \text{diameter of pile} \times \text{subgrade modulus}\\ &= 1.125 \times \text{Pile Diameter} \times 37{,}500\\ &= 1.125 \times 0.75 \times 37{,}500 = 31640.625\,\text{kN/m}\end{aligned}$$

Assigned to end support in the region of socket length in STAAD. Pro Modeled as spring constant

Spring constant for external joint

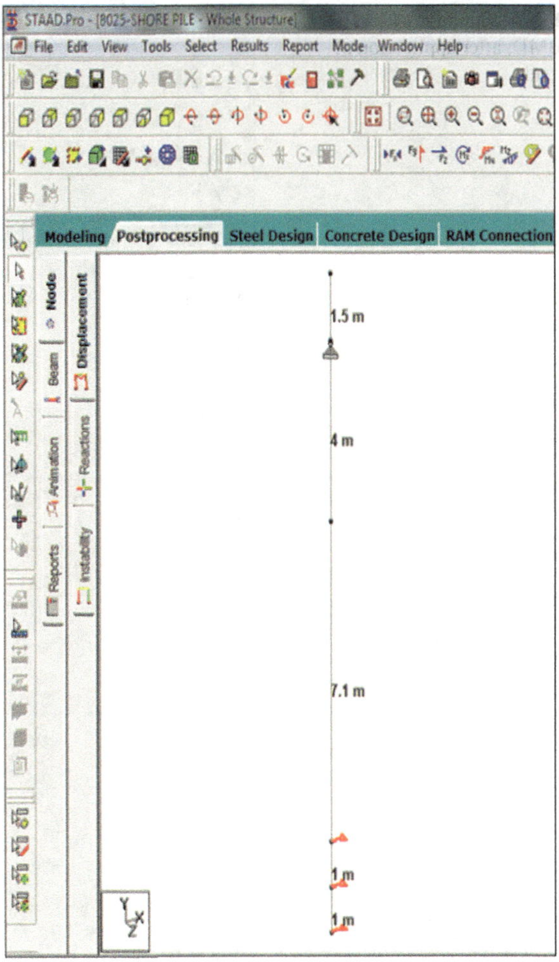

Fig. 5 Condition 1-Staad. Pro. model

= exposed area of shore pile × subgrade modulus

= socket length × diameter of pile × subgrade modulus

$= 1.125/2 \times \text{Pile Diameter} \times 37500$

$= (1.125/2) \times 0.75 \times 37{,}500 = 15820.31\ \text{kN/m}$

Assigned to central support in the region of socket length in Staad Model as spring constant

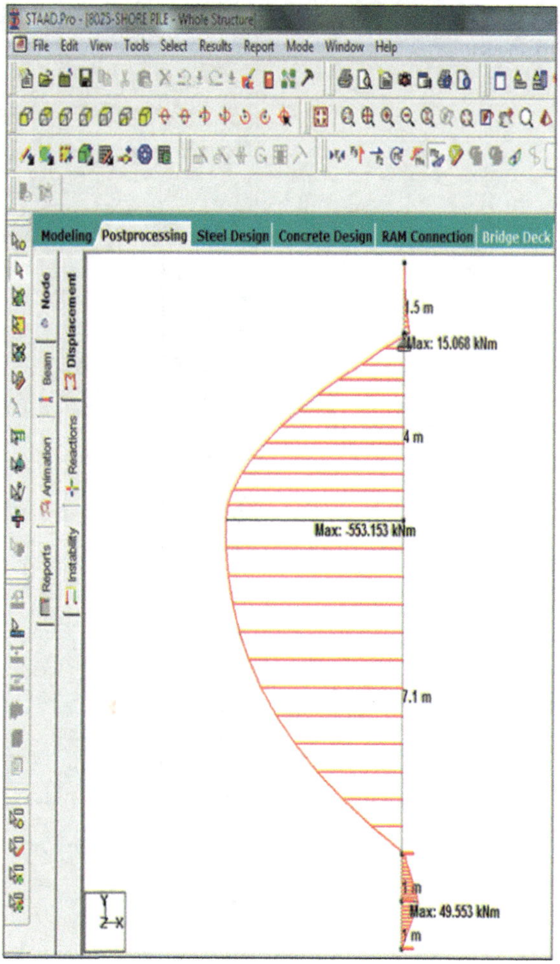

Fig. 6 Condition 1 free BMD, after application of load

5.2 *Design Calculations*

Detail calculations for reinforcement of shore piles on both the faces of piles, i.e. face exposed to excavation and its opposite face are elaborated in the following sections.

5.2.1 Design for Flexure

As in the STAAD model, loads are calculated for 1 m strip, thus to convert to the actual load on the pile multiplying by 0.8 m

$$\text{Maximum B. M. (Positive)} = 0.8 \times 554 = 444\ \text{kN-m}$$

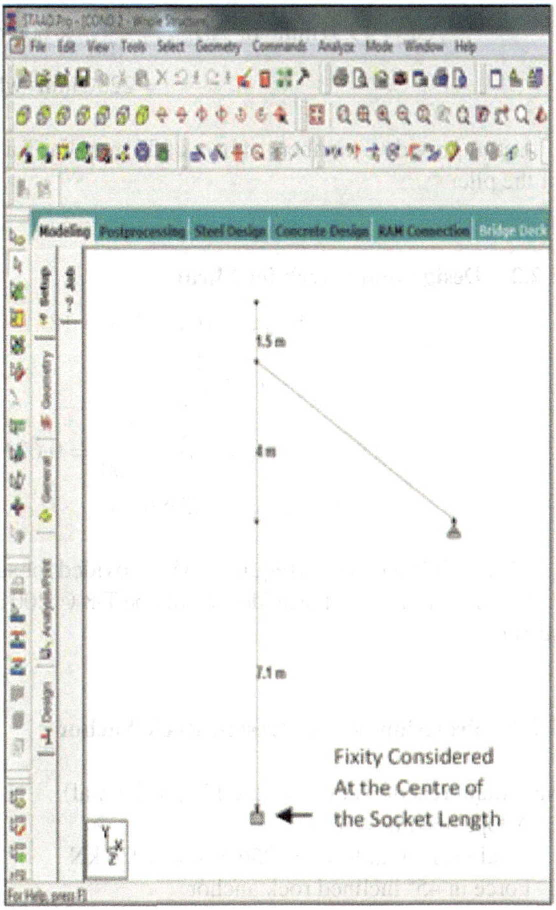

Fig. 7 Condition 2 Staad. Pro. model

$$\text{Area of 750 diameter pile} = \pi/4 \times 7502 = 441786\,\text{mm}^2$$

For design purpose, a rectangular beam with width 500 mm and depth 600 mm is considered in place of the 750 Ø piles

$$\frac{\text{Mu}}{\text{b.d}^2} = \frac{444 \times 10^6}{500 \times 600^2} = 2.47$$

From SP 16 [18], Table no.3, Page No. 49, the Percentage steel P_t can be calculated as $P_t = 0.65$

$$P_t = \frac{100.\text{A}_{\text{st}}}{\text{b.d}}$$

$$\text{Ast} = 1950\,\text{mm}^2 \text{ or } \text{Ast}_{\text{req}} = \frac{0.85}{100} \times 600 \times 500$$
$$= 2250\,\text{mm}^2 \text{ whichever is greater}$$

Let us provide 8 numbers of bars of 20 mm dia. Tor steel is on the excavated side of the pile.

5.2.2 Design and Check for Shear

$$V_{\text{u max}} = 0.8 \times 233 = 187\,\text{kN}$$
$$\tau_v = \frac{V_u}{\text{b.d}}$$
$$\tau_v = \frac{187 \times 10^3}{500 \times 600} = 0.623\,\text{N/mm}^2$$
$$\tau_{\text{permissible}} = 0.5\,\text{N/mm}^2$$

Shear reinforcement is required to be provided for surplus stress of 0.123 N/mm^2.

Shear reinforcement provided should be T-8 @ 200 mm c/c in the form of circular stirrups.

5.2.3 Provision of Pre-stressed Rock Anchor

Maximum reaction at top = 256 kN (horizontal)

Adopt spacing = 4 m c/c

Total horizontal force = 256 × 4 = 1024 kN

Force in 45° inclined rock anchor

COS 45° = 1024/r,

r = 1024/COS 45°

r = 1448 kN

≈ 145 T

According to calculations, one can provide 150 T rock anchor at (45°) with 4 m c/c.

6 Condition II

As discussed in the previous section, the whole analysis is divided into two conditions. Let us discuss the conditions in detail. Yielding support at the top (in place of rock anchor) and fixed support at the bottom (midpoint of the support socket length). In this case, instead of considering a rigid pinned support at the top, we have modeled a tie rod (which represents the rock anchor) with the fixity at the anchor length [19].

With reference to Figs. 8, 10, 11 and 12
Maximum Negative BM = 0.8 × 866 kN-m = 693 kN-m

$$\frac{Mu}{b.d^2} = \frac{693 \times 10^6}{500 \times 600^2} = 3.85$$

$$Ast_{req} = \frac{1.43}{100} \times 500 \times 600 = 4290\,\text{mm}^2$$

Let us provide T 25 − 8 Bars + T16 − 2 Bars
Let us provide 8 numbers of bars of 25 mm dia. bars and 2 numbers of bars of 16 mm dia. bars tor steel on opposite sides of excavation side in shore piles.

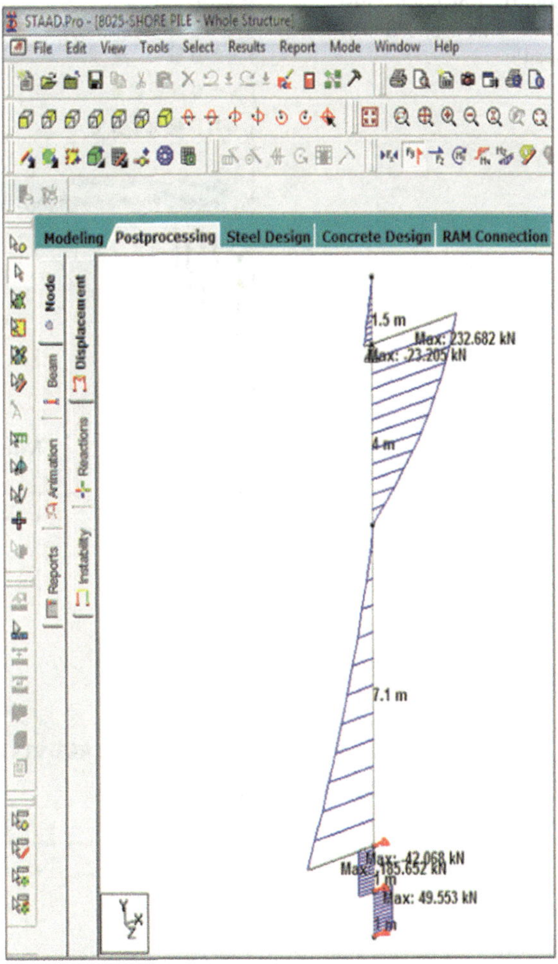

Fig. 8 Condition 1 SFD after application of load

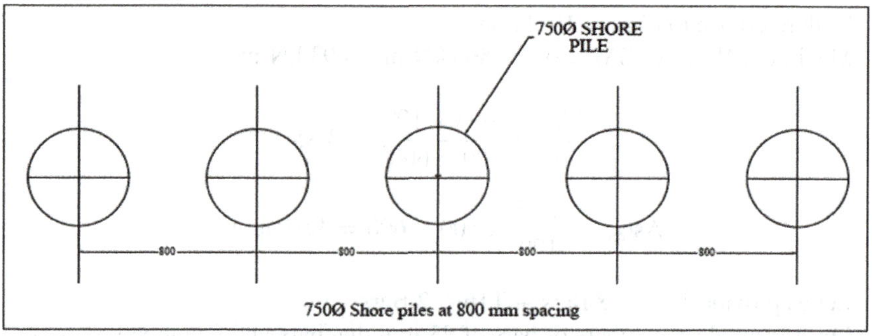

Fig. 9 750 mm diameter shore piles at 800 mm c/c (site condition)

Fig. 10 Condition 2 free BMD after application of load

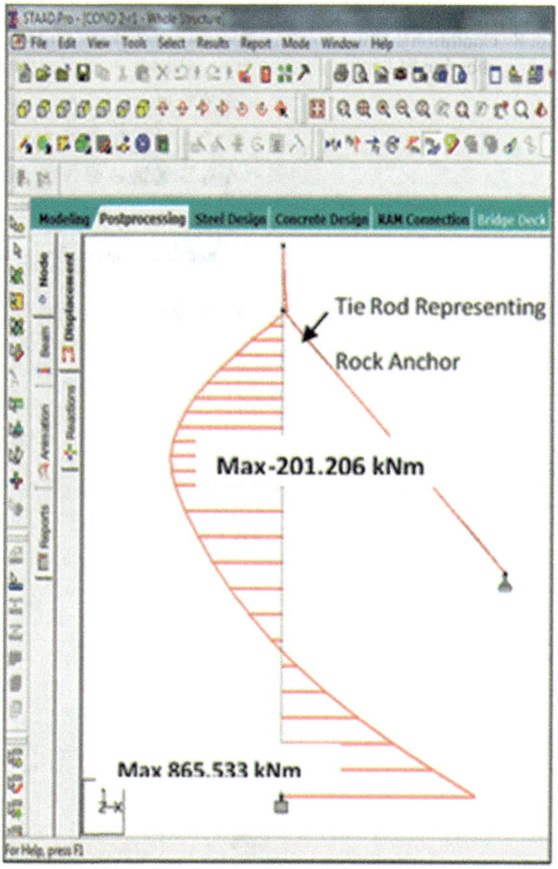

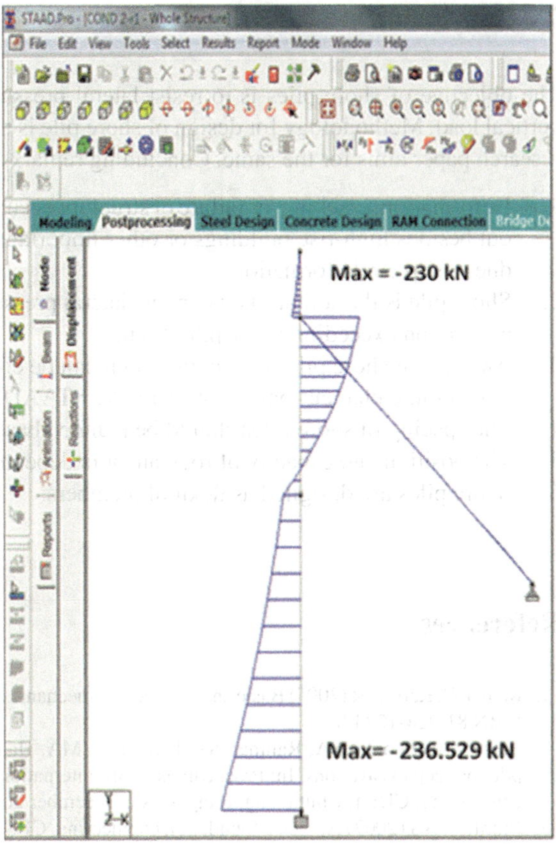

Fig. 11 Condition 2 SFD after application of load

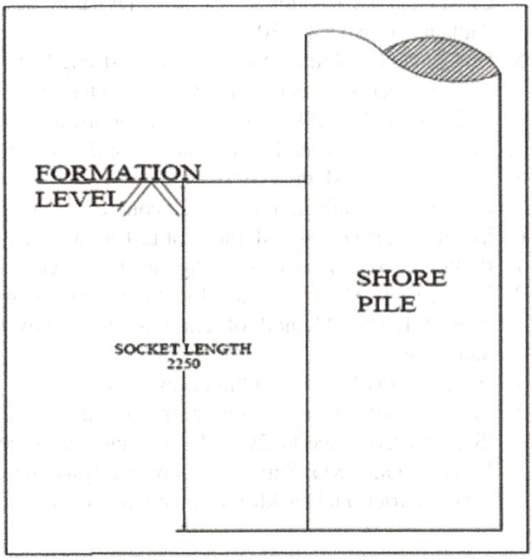

Fig. 12 Representation of socket length

7 Conclusion

The objective of shore piles is to resist lateral pressure in addition to support the vertical load. Methodology for design of shore pile is not elaborated anywhere. This research paper aims for the same. Concluding remarks are as follows.

1. Deformation, cracking, or failure of structure occurs if large excavation is carried out besides high-rise buildings or other important structures such as highways due to slip circle formation.
2. Shore pile is the solitary key to resist lateral pressure due to soil adjacent to the excavation exceeding the depth of 6 m.
3. Analysis of shore piles is tedious, and it can be effectively and accurately done with a finite element-based software like STAAD. Pro V8i.
4. The spacing of shore piles should be planned based on design calculations.
5. The position and capacity of rock anchors depend on the designed lateral load.
6. Shore piles are designed as flexural members.

References

1. Ranjan G, Rao ASR (2005) Basic and applied soil mechanics. New age international publishers, ISBN 81-224-1223-8
2. Basit MA., Shouki, SA, Rahman SMR, Rahman MA, Hanafi BM (2016) Analysis of shore pile for deep excavations. In: Proceedings of 3rd international conference on advances in civil engineering, CUET, Chittagong, Bangladesh, December 2016, pp. 21–23
3. Murthy VNS (2007) Advance foundation engineering. CBS publishers and distributors, ISBN: 9788123915074
4. Choudhury D, Chatterjee S (2006) Dynamic active earth pressure on retaining structures. Sadhana, 31(6):721–730
5. IS 1786:2008, Indian Standard high strength deformed steel bars and wires for concrete reinforcement— specification (fourth revision)
6. IS 228 (Part 15): 1992, Methods for chemical analysis of steels.
7. IS 1387:1993, General requirements for the supply of metallurgical materials
8. IS 1599–1985, Method for bend test
9. IS 456:2000, Plain and reinforced concrete-code of practice.
10. IS 2386 (Part I)-1963, Methods of test for aggregates for concrete
11. IS 383: 2016, Coarse and fine aggregates for concrete-specification
12. IS 516:1959, Methods of test for strength of concrete
13. IS 3025-1:1987: Methods of sampling and test physical and chemical) for water and wastewater Sampling
14. IS 9103:1999 Concrete admixtures—specification
15. IS 10262:2019, Concrete mix proportioning—guidelines (Second Revision)
16. IS 2911 (Part 1/Sec 1): 2010, Design and construction of pile foundations-code of practice
17. Mittal S, Garg KG, Saran S (2006) Analysis and design of retaining wall having reinforced cohesive frictional backfill. Geotechnical Geological Eng 24(3):499–522

18. SP 16:1980, Design aids for reinforced concrete to IS 456-1978
19. Koopialipoor M, Murlidhar BR, Hedayat A, Armaghani DJ, Gordan B, Mohamad ET (2020) The use of new intelligent techniques in designing retaining walls. Eng Computers 36(1):283–294

Investigating the Efficacy of the Hybrid Damping System for Two-Dimensional Multistory Building Frame Using Time History Analysis

A. P. Kote and R. R. Joshi

Abstract Nonlinear analysis is the response of nonlinear structure by the base simulation method. The simulation consists of numerical techniques and mathematical modeling. In the dynamic analysis, we need the consideration of elastic as well internal forces along with energy dissipating devices such as damping. The most powerful technique is the time history analysis for the determination of the nonlinear dynamic response of structures; in this paper, nonlinear dynamic analysis of multistorey steel moment-resisting frame is found out by using different cases such as V.F.D. only, B.R.B. only, and Hybrid damping models (both V.F.D. and B.R.B.). It is found that by using the above-mentioned models, the base shear is reduced when it is compared with a steel moment-resisting frame without any P.E.D. The joint displacement was decreased by 80% and inter-story drift was reduced to 40%.

Keywords Time history analysis · Viscous fluid damper · Buckling restrained braces · Steel moment resisting frame · Hybrid damping · SAP2000

1 Introduction

The safety of life is an important criterion for a structural engineer while designing any structure. Gupta and Krawinkler (1999) critically mentioned that to provide consistency and predictability, there is a necessity for seismic analysis and design of the structure. The seismic codes provide very simplified criteria for simple seismic design but the practical design of any structure involves complications. Seismic design of structure requires various parameters to be taken into consideration before actual design consideration. The earthquake has peculiar characteristics such as amplitude, duration, and frequency. The performance-based seismic design (PSBD) is emerging to be efficient to capture the inelastic behavior of materials and to reduce the effects of earthquakes installation of dampers which are energy dissipating devices used in the lateral force-resisting system of the structure. The passive

A. P. Kote (✉) · R. R. Joshi
Department of Civil Engineering, College of Engineering Pune, Pune, India
e-mail: apk21.civil@coep.ac.in

M. S. Ranadive et al. (eds.), *Recent Trends in Construction Technology and Management*, Lecture Notes in Civil Engineering 260,
https://doi.org/10.1007/978-981-19-2145-2_68

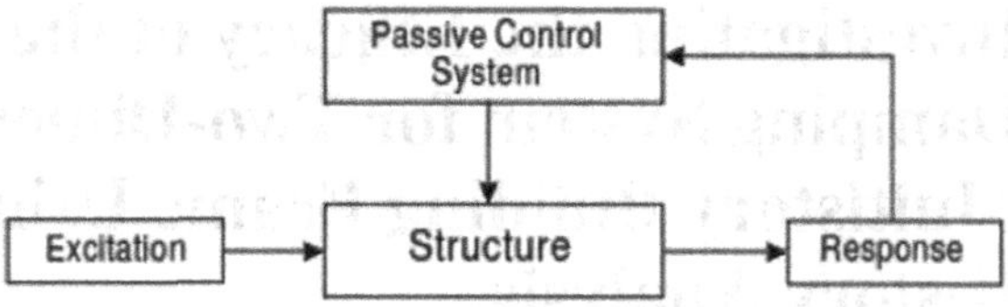

Fig. 1 Workflow of passive hybrid damping system. *Source* Earthquake resistant and design of steel structures

hybrid dampers as shown in Fig. 1 absorb a large amount of energy due to earthquake so it does not undergo inelastic deformation. It also reduces the inter-story drift. The various dampers such as viscous fluid dampers, buckling restrained braces, viscoelastic dampers, and friction dampers are very useful in reducing energy dissipation during an earthquake. Mostly in a highly prone area where the earthquake is a risk factor, the use of a damper is very useful. When there is a heavy earthquake, structure undergoes inelastic deformations; therefore, it depends on ductility and hysteretic energy dissipation to avoid such damage to structures as mentioned by Jinbiao [6].

2 Passive Damping in the Buildings

The purpose of using a passive energy system is to prevent damage to structural components. Damping systems reduce the amplitude of vibration. The passive control system is designed to dissipate a huge portion of earthquake input energy. This depends on the inherent properties of the basic nature. This can also be used for a wide variety of structural components. Buckling restrained braces, viscous fluid dampers (H. K. Miyamoto, A. S. J. Gilani, and A. Wada), and hysteretic dampers (V.F.D. and B.R.B.) are some of P.E.D. types devices. Reference [13] has done a lot of research on buckling restrained braces and has devised B.R.B. for use. The vibratory system has some energy that is dissipated during the motion. The rate of decreasing the amplitude depends upon the amount of damping. V.F.D. works on the principle of dissipation of energy since the fluid is flowing through the orifice as shown in Fig. 2 as given by Taylor and devices (2007) below; in this mechanism, the fluid flows from the chamber. There is another type of energy dissipation system known as buckling restrained braces (B.R.B.) which is having a steel core, bond preventing layer, and a casing. Buckling restrained braces as shown in Fig. 3 are designed to withstand cyclic loadings, and this prevents buckling under axial compression. B.R.B. can absorb a large amount of energy during cyclic loading as shown in Fig. 4 in case of an earthquake. B.R.B. provides a large amount of ductility.

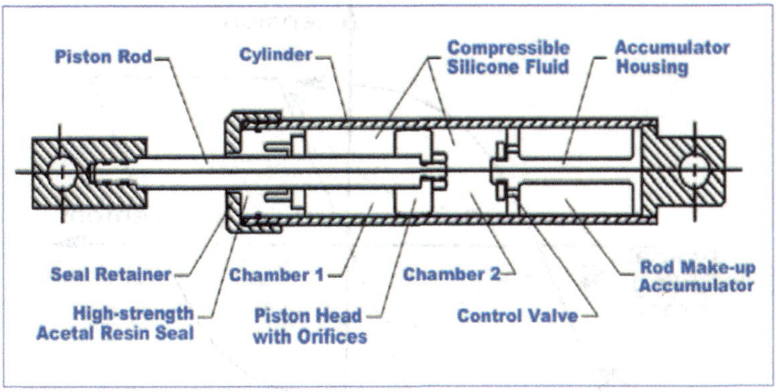

Fig. 2 Viscous fluid damper. *Source* Taylor Devices Inc.

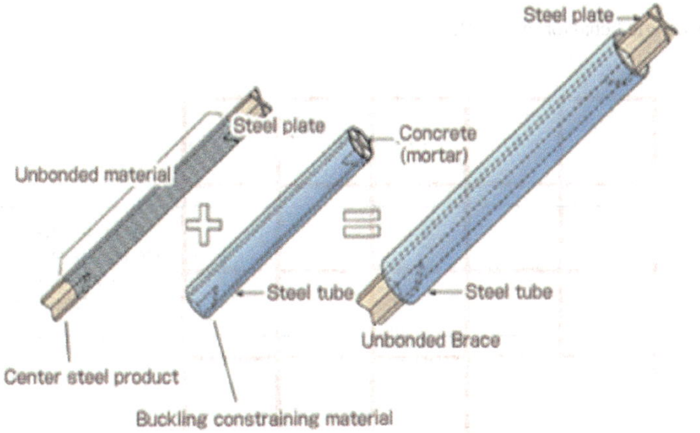

Fig. 3 Schematic sketch of B.R.B.

3 Example Building for Seismic Analysis

A three-dimensional steel moment-resisting frame is considered with 5 bays in the x- and y-directions with 5 m spacing each. The seismic analysis is performed for an intermediate frame as shown in Fig. 5 due to symmetry. The considered building is analyzed and designed for a dead load of 3.0 kN/m^2 and a live load of 5 kN/m^2 along with a floor finish load of 1.5 kN/m^2 on the intermediate story and 3 kN/m^2 on the roof floor. The time history analysis of the building is performed using SAP2000 with bilinear material properties. The material properties for the steel frame considered are given below.

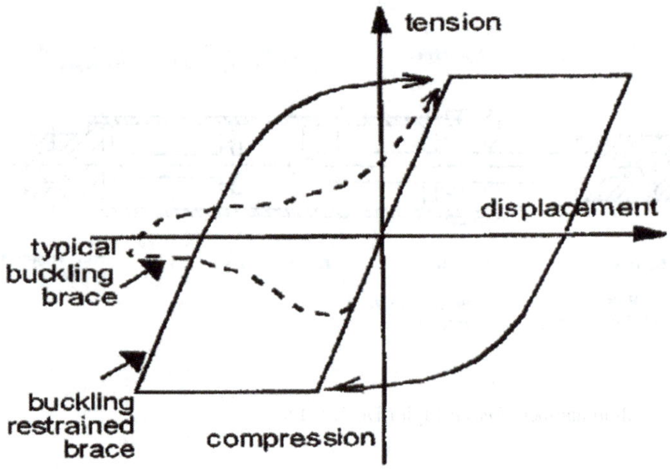

Fig. 4 Hysteretic behavior of B.R.B.

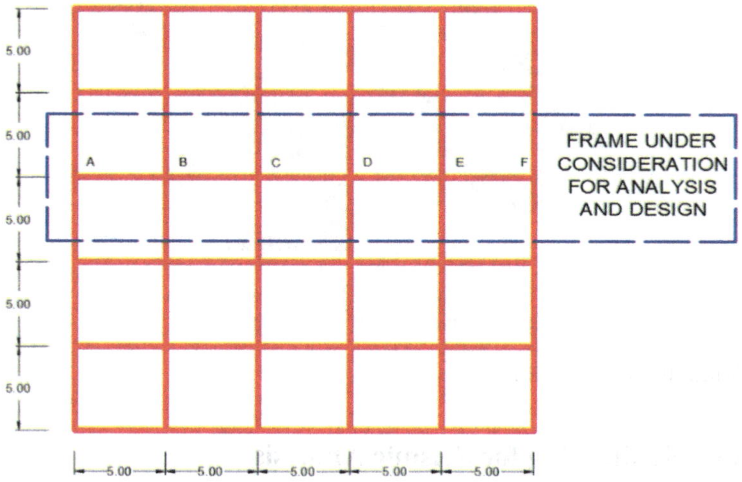

Fig. 5 Structural plan of building considered for analysis. All dimensions are in (m)

3.1 Seismic Data

The typical example building is assumed to be located in the Pune City of India. As per the current seismic code of India, the site seismic zone is categorized as Zone-III. The structural plan of the building was considered for analysis as shown in Fig. 5. To perform the seismic analysis through the time history solution, the plots for selected time history records for the three earthquake records and details of these earthquake records are given in Fig. 6 and Table 1, respectively.

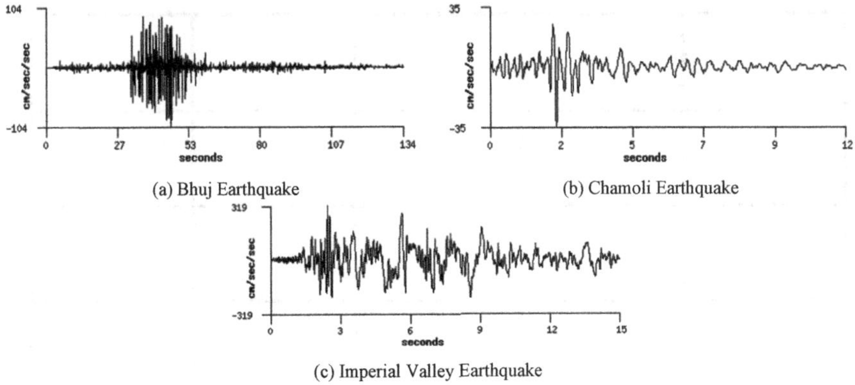

(a) Bhuj Earthquake

(b) Chamoli Earthquake

(c) Imperial Valley Earthquake

Fig. 6 Selected time history records

Table 1 Selected time history records

Sr. no.	Name of earthquake	Recording station	EQ component	PGA (m/s^2)
1	Bhuj earthquake EQ	Ahmedabad [AMD]	N 78 E	−1.038
2	Chamoli earthquake EQ	Gopeshwar [GOP]	N 20 E	−3.5283
3	Imperial valley EQ	UCSD station [US6616]	N 45 E	−3.2059

3.2 Different Frame Cases

For the comparative study, the considered frame of the example building is considered without any damping system, i.e., Bare frame; to evaluate the effectiveness of the passive hybrid damping system, the frame is installed with B.R.B. and V.F.D., and this damping system is termed as Hybrid Damping. Further, to verify the individual effects of the B.R.B. and V.F.D. on the frame, two cases with installed B.R.B. and V.F.D. only are considered for seismic analysis. The four different cases of frames as mentioned are shown in Figs. 7, 8, 9 and 10.

4 Results and Discussions

1. Buckling restrained braces (B.R.B.) and Viscous Fluid Dampers (V.F.D.) are modeled in the frame with built-in link elements in SAP200.
2. Figs. 11, 12, 13 and 14 show the yielding of these elements; thus, the full potential utilization of these dampers is achieved.
3. The results for column base shear, story displacements, and inter-story drifts are shown in the table below.

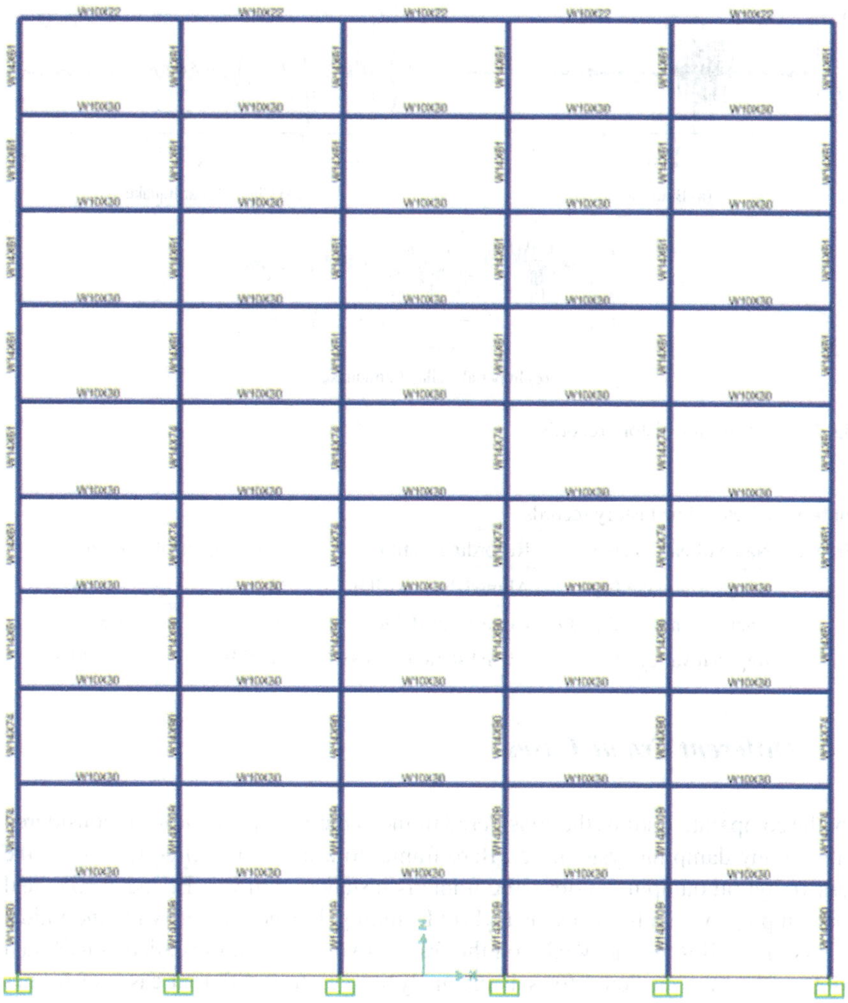

Fig. 7 Building frame without dampers (frame-1)

4.1 Hysteresis Behavior

4.2 Base Shear

The response results for the seismic base shear for the four cases of the frames with and without damping are shown in Table 2. The seismic base shear as per the Indian IS:1893-2016 code is also computed. It is seen that installing the hybrid damping system results in ductile behavior of the frame, and the resulting base shear is less than

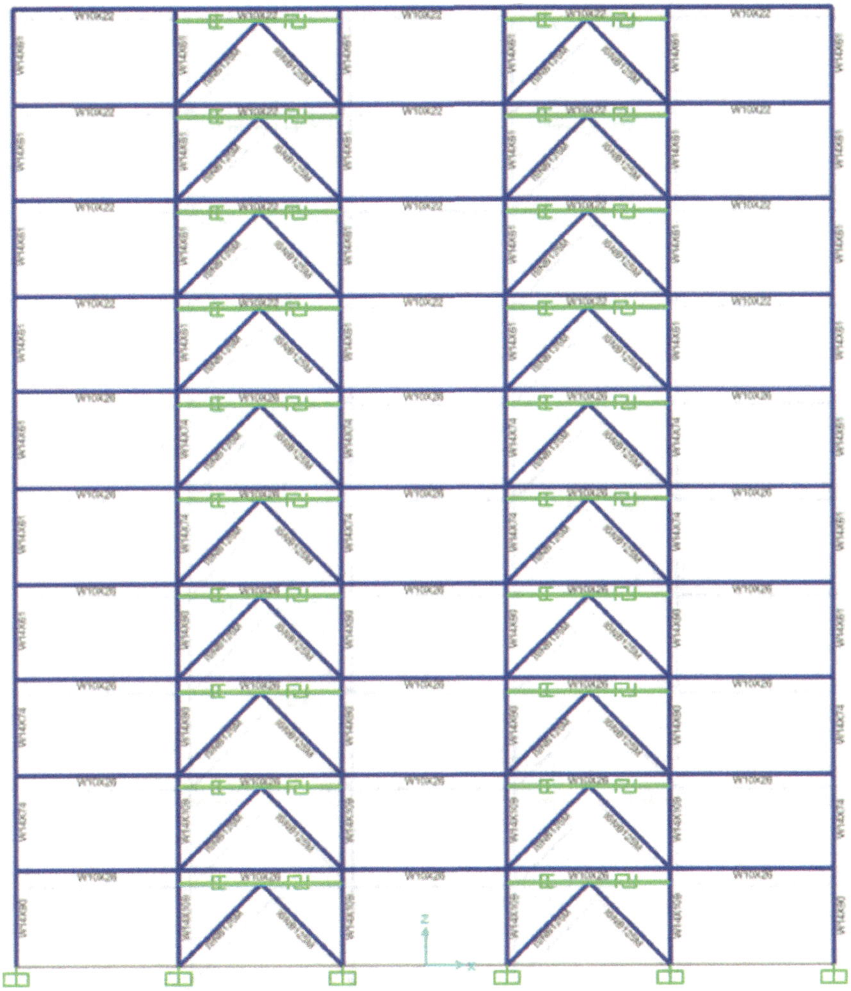

Fig. 8 Frame with hybrid damping system installed in 2nd and 4th bay (frame-2)

the frame without any damping systems. Installing B.R.B. made the frame highly stiff resulting in higher base shear.

4.3 Story Displacement

The calculated floor displacements are shown in Table 3 with plots for all four frames in Fig. 15. It is seen that the bare frame undergoes large deformation at the

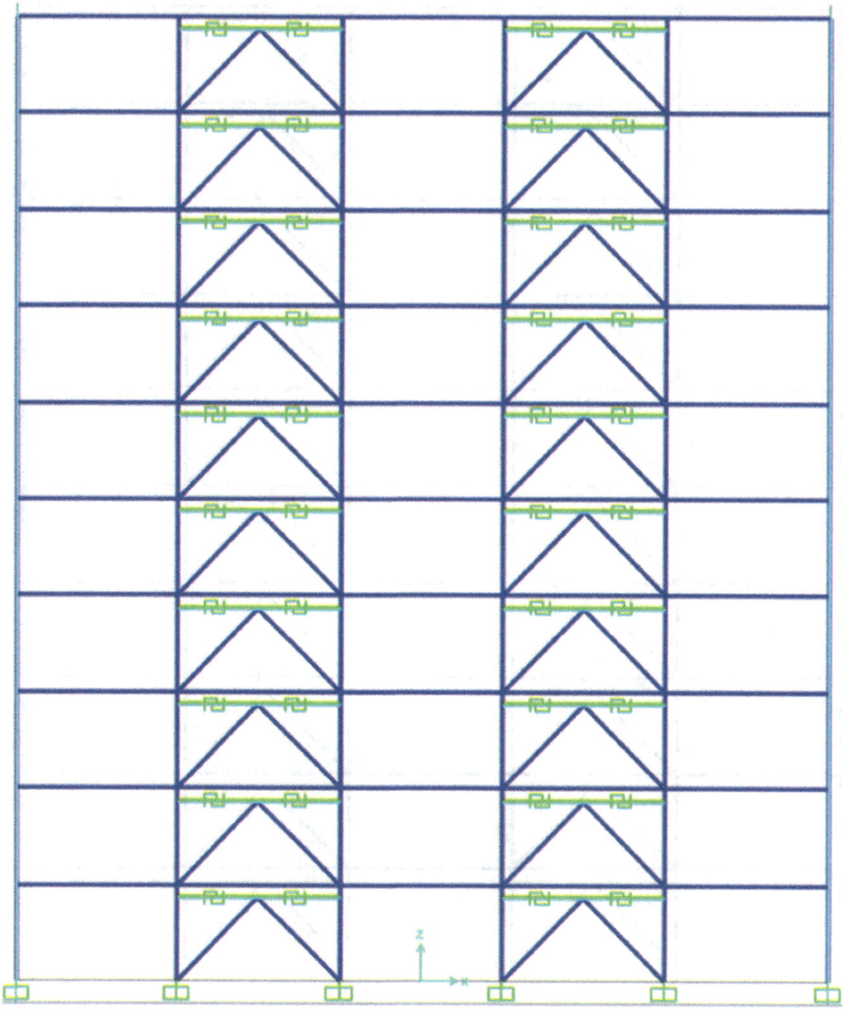

Fig. 9 Frame installed with B.R.B. in 2nd and 4th bay (frame-3)

roof. However, the other three frames with a damping system result in lower roof displacements.

4.4 Story Drift

The calculated story drifts along with the plots are shown in Table 4 and Fig. 16, respectively. The story drift for the damped model frames is significantly reduced when compared with the frames without a damping system.

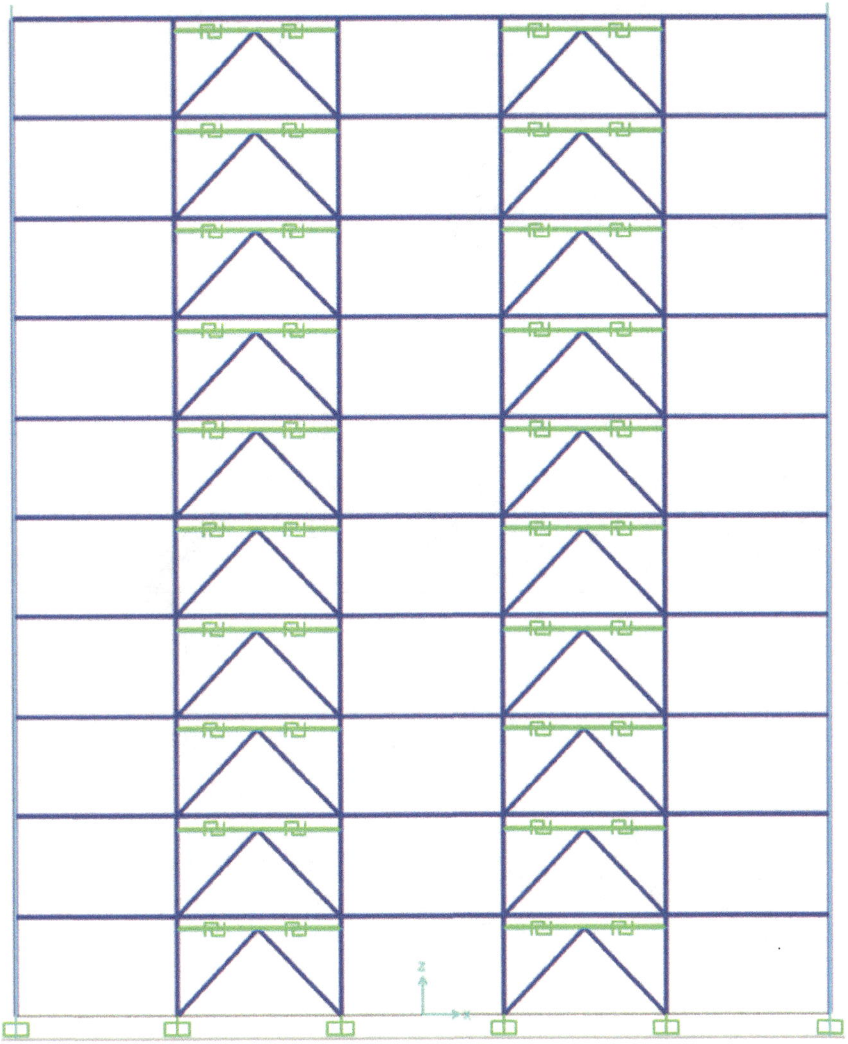

Fig. 10 Frame installed with V.F.D. in 2nd and 4th bay (frame-4)

5 Conclusions

The example building is analyzed and designed using the three earthquake records. For seismic analysis, the interior frames with four different cases with and without passive damping are considered, viz. Bare Frame, Frame with V.F.D., Frame with B.R.B., and frame with Hybrid damping (V.F.D. and B.R.B.). It is seen that the frame without any damping system performs poorly in terms of the large displacements at the roof story. There is also a slightly higher base shear force observed for the bare

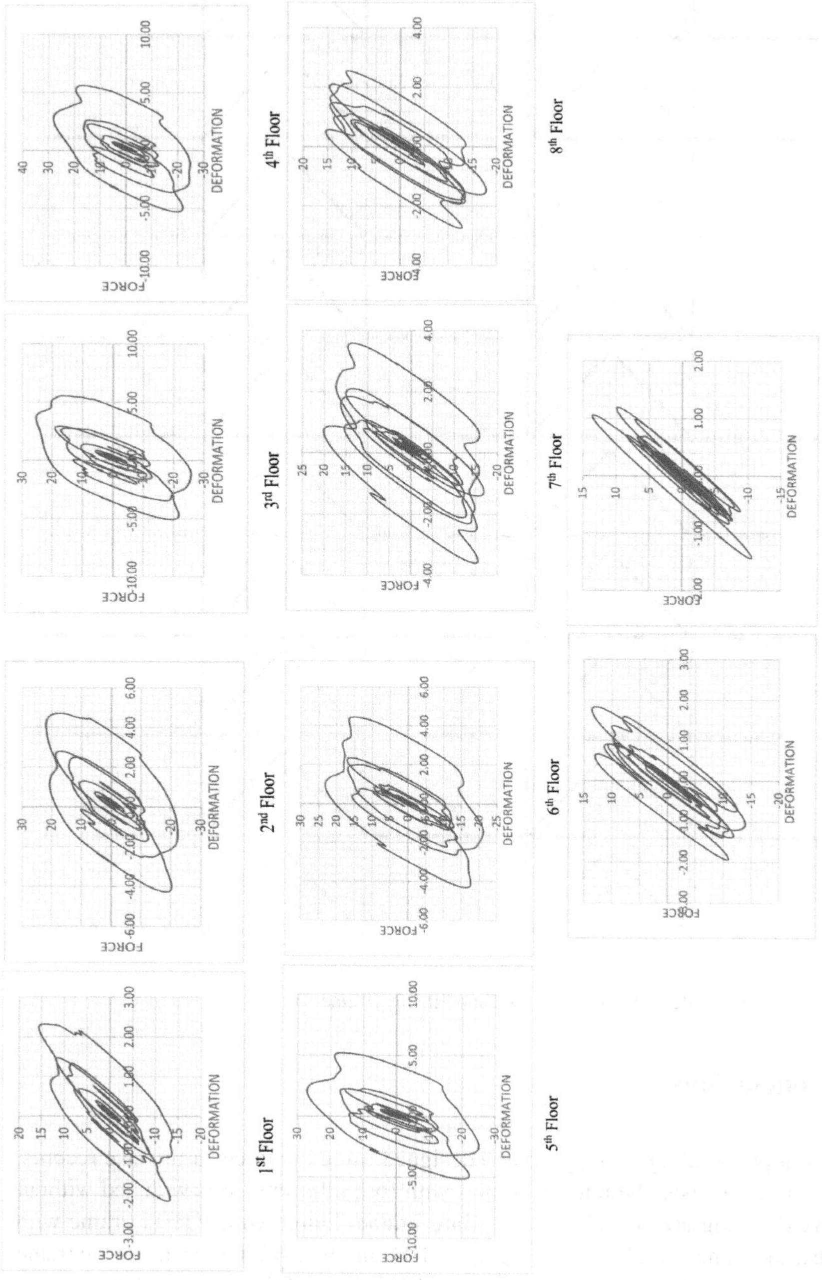

Fig. 11 Hysterstic plots for V.F.D. in hybrid damping frame

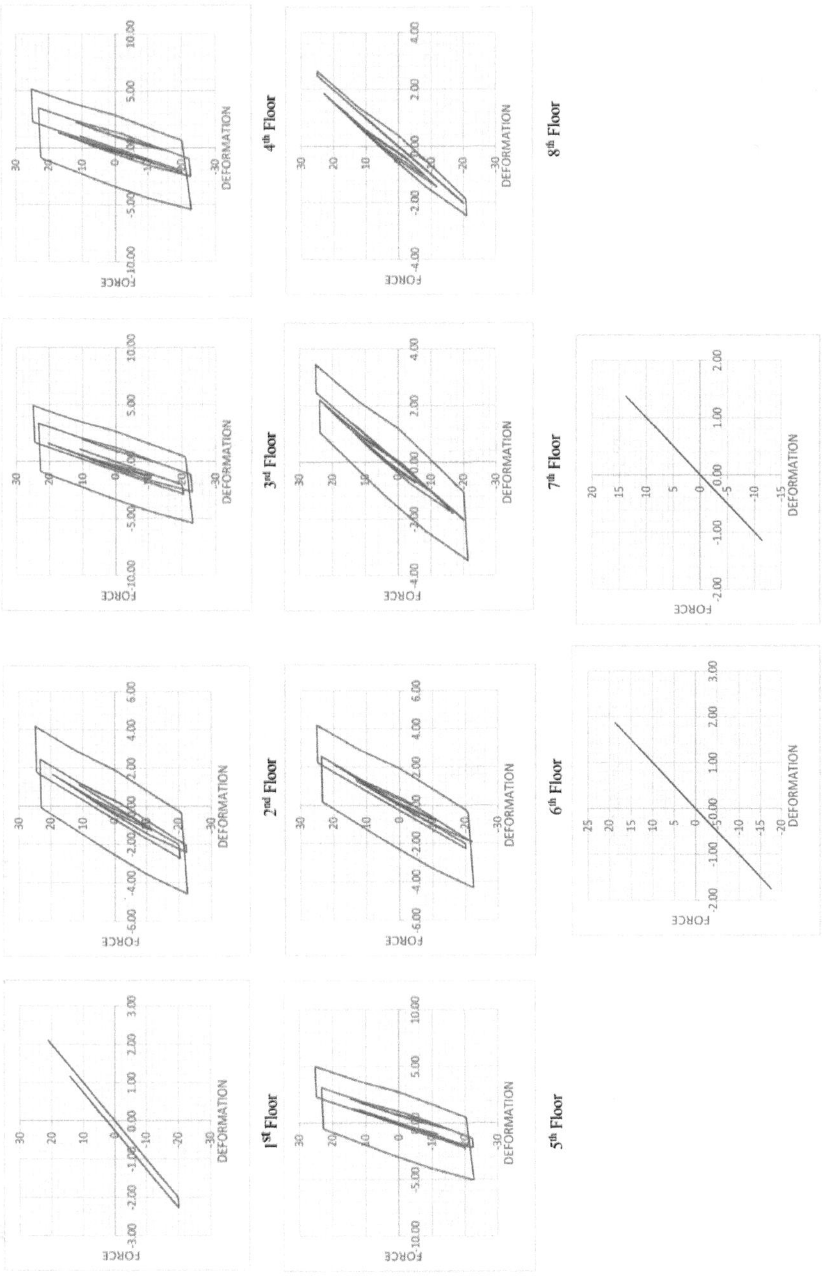

Fig. 12 Hysterstic plots for B.R.B. in hybrid damping frame

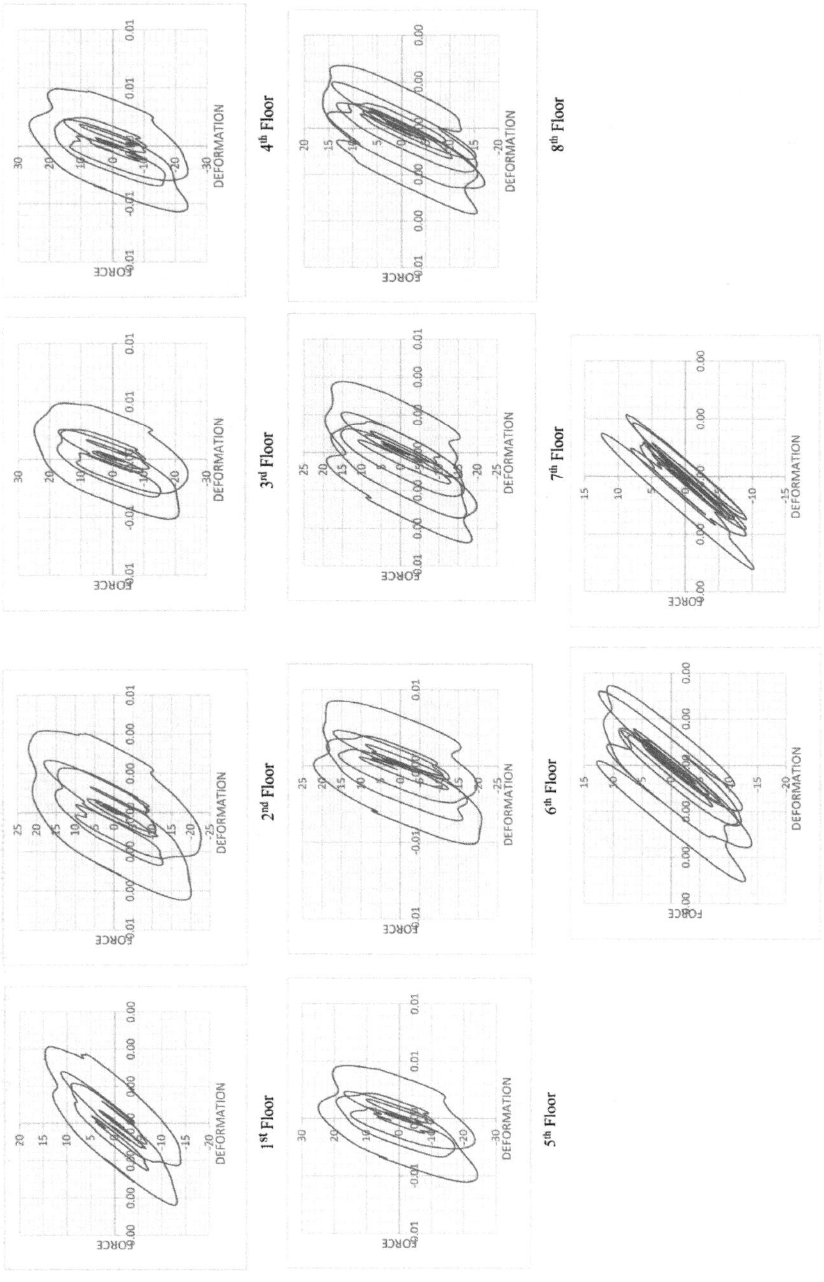

Fig. 13 Hysterstic plots for frame with V.F.D. only frame

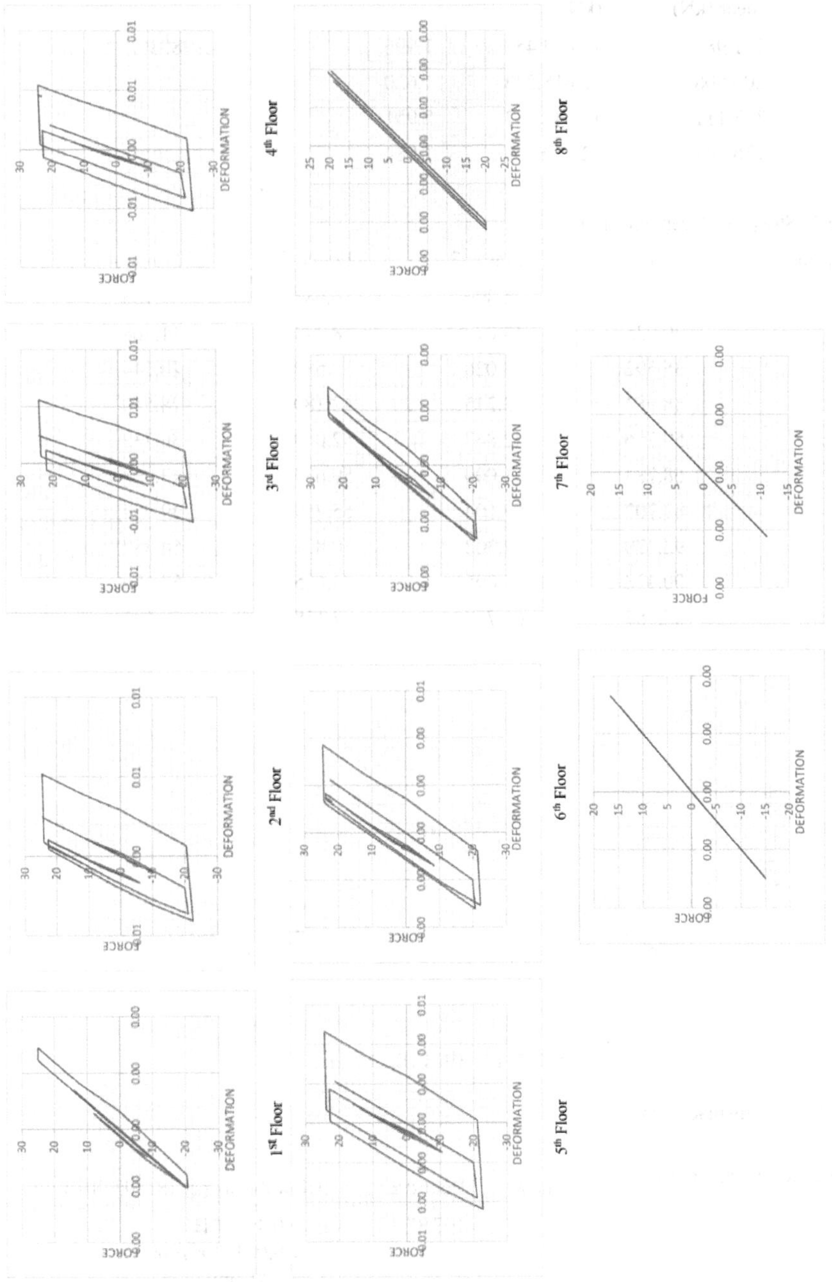

Fig. 14 Hysterstic plots for frame with B.R.B. only frame

Table 2 Base shear response

Frame #	Observed base shear (kN)	Seismic weight (kN)	% of seismic weight	As per IS1893:2016
1	251.986	12,624.453	1.996	1.99808%
2	204.508	12,618.243	1.620	
3	373.111	12,642.353	2.951	
4	136.947	12,629.709	1.084	

Table 3 Story displacement (mm)

Story no.	Frame-1	Frame-2	Frame-3	Frame-4
1	7.209	4.390	6.683	4.105
2	20.935	13.097	18.343	11.868
3	35.592	23.026	30.582	20.244
4	48.982	33.215	42.189	28.550
5	59.766	42.867	52.049	36.732
6	68.783	51.000	59.633	43.963
7	81.702	57.604	65.759	49.984
8	92.470	62.302	70.821	54.387
9	100.428	65.095	74.655	57.186
10	107.087	66.759	77.158	58.691

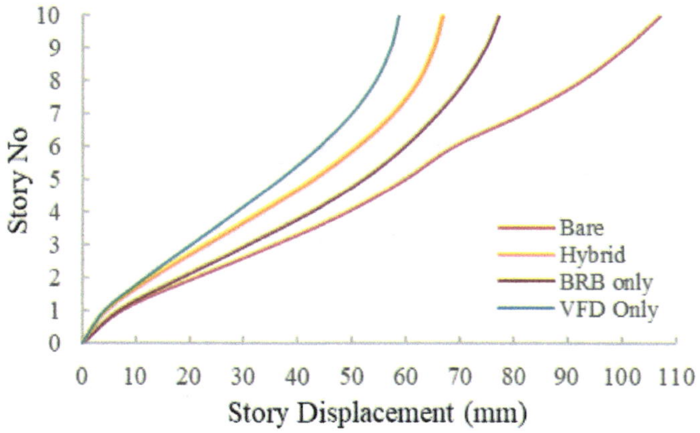

Fig. 15 Story displacements

Table 4 Roof drift (mm)

Frame #	Roof drift	Allowable as per IS1893:2016
1	107.87	$\Delta = 0.4\%$ of H $\Delta = 0.004*30*1000$ $\Delta = 120$ mm
2	66.759	
3	77.158	
4	58.691	

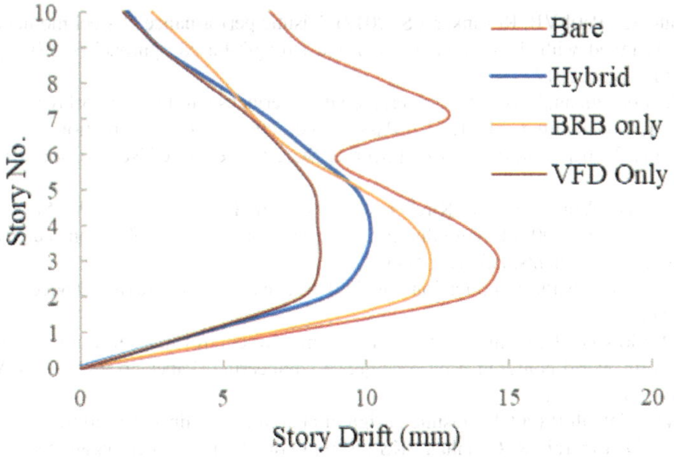

Fig. 16 Story drifts

frame. However, in the frames with damping systems, the lateral stiffness are lower except for the frame with B.R.B. For the frame with installed V.F.D., improvements with low roof displacements are observed. The base shear was also reduced to a large extent in the case of the V.F.D. only frame as compared to other frames. The joint displacement is reduced by 80% and the inter-story drift is decreased by 45% when compared with Frame-1. A roof displacement of 59 mm was seen when the damping and stiffness properties were changed accordingly. Finally, by reducing the displacement and drift, the effect for non-structural elements can be reduced to quite a large extent.

References

1. Gupta A, Krawinkler H (1999) Seismic demands for performance evaluation of steel moment resisting frame structures. Report 132, Stanford University
2. Taylor DP (1983) Seismic protection with fluid viscous dampers for the Torrey Mayor a 57-story office tower in Mexico City, Taylor Devices, Inc, 90 Taylor Drive, North Tonawanda
3. FEMA - 355 C (2000) State of the art report on systems performance of steel moment frames subject to earthquake ground shaking. SAC Joint Venture
4. Miyamoto HK, Gilani AS, Wada A (2008) State of the art design of steel moment frame buildings with dampers. WCEE 2008, Beijing, China
5. Miyamoto HK, Gilani ASJ, Wada A, Viscous damper limit states and collapse analysis of steel frame buildings with dampers. Paper no-146, July 25–29, Toronto, Ontario, Canada
6. Liu J (1986) Earthquake analysis of steel moment-resisting frames. South China University of Technology, China
7. Rajendran K, Rajendran CP, Thakkar M, Tuttle MP (2001) The 2001 Kutch (Bhuj) earthquake: seismic surface features and their significance. Curr Sci 80(11)
8. Magar Patil HR, Jangid RS (2013) Seismic vulnerability assessment of steel moment resisting frame due to infill masonry walls, variation in column size and horizontal buckling restrained braces. ACEE 2(1)

9. Kharmale SB, Patil VB, Revankar VS (2012) Seismic performance of steel moment-resisting frames designed with displacement-based and strength-based approaches. ISET paper no. D013, 20–21 Oct 2012
10. Kojic S, A design analysis of the Imperial country services building. Yugoslavia
11. Soong TT, Constantinou MC (2010) Passive and active structural vibration control in civil engineering. Department of Civil Engineering State University of New York at Buffalo, NY 14260
12. Symans MD, Whittaker AS, Kircher CA, Charney FA, Johnson MW, McNamara RJ, Constantinou MC (2008) Energy dissipation system for seismic applications current practice and recent developments. ASCE 134(1)
13. Taylor (2007) Structural applications of fluid viscous dampers. Taylor Devices Inc., North Tonawanda
14. Dogan T, Goodno BJ, Craig JI (2004) Hybrid passive control in steel moment frame buildings.In: 13th world conference on earthquake engineering, Paper no 2387, 1–6 Aug 2004, Vancouver B.C., Canada
15. López WA, Sabelli R (2004) Seismic design of buckling-restrained braced frames
16. Pandey Y, Dharmaraju R, Chauhan PKS (2001) Estimation of source parameters of Chamoli Earthquake, India. Indian Acad Sci (Earth Planet Sci) 110(2)

Thermal Buckling Analysis of Stiffened Composite Cutout Panels

K. S. Subash Chandra, T. Rajanna, and K. Venkata Rao

Abstract In aeroplanes, ships, and hypersonic space vehicles, stiffened plates with cutouts are often used. Aerodynamic heating produced by the interaction between the environment and the vehicle's surface quickly heats up this structural component. These activities affect the buckling characteristics of perforated plates and may result in an early failure due to the presence of cuts. This impact is avoided by properly stiffening the area surrounding the cutout. Thermal buckling effects must, thus, be taken into account when designing and analyzing thin-walled structural components. Using finite-element techniques, the effect of high temperatures on the buckling stresses of composite laminates with and without reinforced cuts is studied. This is accomplished via the use of a nine-noded heterosis element and a three-noded quadratic beam element. A transformation matrix ensures the displacement compatibility of the plate and stiffener. Thermal buckling of layered panels with cutouts is analyzed by employing eccentric stiffeners. It is witnessed from this investigation that various parameters significantly affect the buckling behavior of stiffened plates. These factors include cutout size, material anisotropy, thermal expansion ratio, modulus ratio, and boundary conditions.

Keywords Thermal buckling · Cutout · Stiffener · Laminates · Nine-noded heterosis

---

K. S. S. Chandra (✉)
Department of Technical Education, Government Polytechnic, Bagepalli, Karnataka, India
e-mail: subashks.17pm@bmsce.ac.in

T. Rajanna
Department of Civil Engineering, BMS College of Engineering, Bengaluru 560 019, India

K. V. Rao
Department of Mechanical Engineering, BMSE College of Engineering, Bengaluru 560 019, India

M. S. Ranadive et al. (eds.), *Recent Trends in Construction Technology and Management*, Lecture Notes in Civil Engineering 260,
https://doi.org/10.1007/978-981-19-2145-2_69

1 Introduction

The composite materials' qualities are known for their more excellent strength/weight ratio, high specific rigidity, and improved fatigue life. The inclusion of stiffeners to a panel, without significantly impacting its total weight, further enhances these features. There is an enormous increase in a minor weight penalty's strength and stability aspects of the constructions. Plates with a pattern of stiffeners are usually discovered in an aircraft's fuselage, including wings, a ship's hull and its deck, maritime drilling gears, pressure containers, roofing members, and the thrusting plinth of the missile. The extensive acceptance and utilization of the before-mentioned materials, under various incidents, are governed by severe environmental circumstances throughout their service life. The raised heat is one thing that dangerously influences the buckling properties of layered laminae, mainly in the geometrically ceased panels. These discontinuities may complicate the in-plane thermic stress concentration in the plates. Deficient stresses produce a deterioration of stability due to non-uniformity in the stress field. The fiber-reinforced composite constructions remain exposed to high-temperature loads throughout the operational cycle, leading to significant strength degeneration, including additional damaging consequences. Hence, the entire understanding of thin panels with cutouts is essential to prevent premature breakdown and utilize their full strength, which can be explored tremendously in thin-walled structures. Two kinds of buckling are customarily probable. They are local buckling due to cutout and the overall buckling essentially the another. The first is attributed to local and the global as the other. The local mode is observed across the cutout. However, panels are more cost-effective if they appear to buckle globally compared to those that buckle locally. Rigid reinforcing around the cutout removes the local mode. Thus, construction engineers need to consider composited laminates' stability behavior to achieve a practical design with reinforced cutouts.

In many literary works, the mechanical buckling of composites plates under in-plane compression is studied (for example, [1, 2]). Laminated composites cause thermal buckling due to extreme temperature changes. In some situations, the thermal load is the critical load for designing such plates. The three-dimensional layer-wise research analysis has helped predict composite laminated angles' thermal bending reaction [3]. Buckling of composite laminates plates in a thermal environment is carried out by many researchers, and a few studies are by References [4–12]. The maximum of the works was conducted without any discontinuities or cutouts. Panels with cutouts become a part of an integral component of complex structures. These perforations are present mainly to access the pipeline, electric cables, doors, or window structures. Research has been done for laminated panels considering the cutouts by References [13, 14]. Such perforated plates' thermal analysis can undergo thermal buckling mainly due to local buckling near the cutouts. Hence, various stiffeners strengthen the cutout panels [15] against thermal buckling.

In contrast, it is very much essential to consider weight addition due to stiffeners practically. Considering the total volume removed by the cutout portion being added as stiffeners around the cutouts, such work is done for various mechanical in-plane compressive loadings by Reference [16]. Nevertheless, thermal buckling for panels with cutouts and stiffeners considering the same is yet to be done. There are very few works available on the thermal buckling of laminated plates with cutouts. Studies with thermal buckling of laminates with stiffeners placed around the cutout, considering the total volume remaining the same throughout, are not detected within the open literature. While, in the current article, a nine-noded heterosis plate element is applied, the finite-element program is established in MATLAB to analyze the thermal buckling of laminates with and without stiffeners. The thermal buckling behavior for different lamination angles, cutout ratios, and modular ratios is analyzed only after verifying with the current formulation with previous studies. The effects of panels with an eccentrically placed stiffener around the circular cutouts are analyzed, investigated, and compared with the panels without stiffeners for various parameters under consideration.

2 Theoretical Formulation

In this study, the first-order shear deformation theory, including a shear rectification determinant of 5/6, is considered for the non-linear distribution of shear strains within the thickness. Reddy and Phan [2] express the equations for laminates. The displacement field is

$$\{\overline{u},\ \overline{v},\ \overline{w}\}^T = \{u(x, y),\ v(x, y),\ w(x, y)\}^T + \left\{z\theta_x,\ z\theta_y, 0\right\}^T \tag{1}$$

where $\overline{u}$, $\overline{v}$, and $\overline{w}$ are displacements ingredients in the x-, y-, and z-paths correspondingly. u, v, and w are associated mid-plane displacements in the x-, y-, and z-axis, sequentially, plus θ_x and θ_y perform rotations of normal regarding y- and x-axis, sequentially, being presented within Fig. 1a. A typical mesh is shown in Fig. 1b.

Green–Lagrange's strain–displacement relationship exists and is appropriated during the formulation. The strain–displacement association consists of (i) linear strain expressions applied to derive the elastic stiffness matrix also and (ii) non-linear strain concerning the geometrical stiffness matrix.

$$\{\varepsilon\}=\{\varepsilon_l\}+\{\varepsilon_{nl}\} \tag{2}$$

2.1 Constitutive Relations

Constitutive relations of the composite panel subjected to the present buckling analysis's temperature are described as follows. The stress–strain association for laminate is given by equation and is explained in detail in Subash Chandra et al. [18].

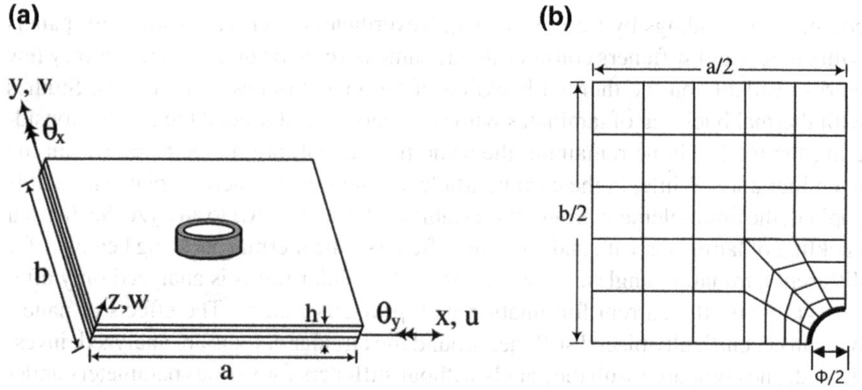

Fig. 1 **a** Plate geometry with cutout and eccentric stiffener **b** detailed mesh over quarter plate

$$\{F\}=[D]\{\varepsilon\}-\{F^N\} \tag{3}$$

The constitutive matrix in the above Eq. (3) is given by

$$[D]=\begin{bmatrix} [A_{ij}] & [B_{ij}] & 0 \\ [B_{ij}] & [D_{ij}] & 0 \\ 0 & 0 & [S_{ij}] \end{bmatrix} \tag{4}$$

The extension–extension, extension–bending, and bending–bending stiffness components exist presented in Eq. (4) as

$$\left(A_{ij}, B_{ij}, D_{ij}\right)=\sum_{k=1}^{n}\int_{z_{k-1}}^{z_k}\left(\overline{Q_{ij}}\right)_k\left(1, z, z^2\right)dz, \quad i, j=1,2,6 \tag{5}$$

and

$$\left[D_{ij}\right]=\alpha\sum_{k=1}^{n}\left(\overline{Q_{ij}}\right)_k(z_k-z_{k-1}); \quad i, j=4,5 \tag{6}$$

where $\left(\overline{Q_{ij}}\right)$ in the above equation is

$$\left(\overline{Q_{ij}}\right)=[T_1]^T\left(Q_{ij}\right)[T_1](i, j=1,26)$$
$$\left(\overline{Q_{ij}}\right)=[T_2]^T\left(Q_{ij}\right)[T_2](i, j=4,5)$$

in which,

$$[T_1] = \begin{bmatrix} \cos^2\theta & \sin^2\theta & \sin\theta\cos\theta \\ \sin^2\theta & \cos^2\theta & -\sin\theta\cos\theta \\ -2\sin\theta\cos\theta & 2\sin\theta\cos\theta & \cos^2\theta - \sin^2\theta \end{bmatrix} \text{ and}$$

$$[T_2] = \begin{bmatrix} \cos\theta & \sin\theta \\ -\sin\theta & \cos\theta \end{bmatrix}$$

$$\left[Q_{ij}\right]_k = \begin{bmatrix} Q_{11} & Q_{12} & 0 \\ Q_{12} & Q_{22} & 0 \\ 0 & 0 & Q_{66} \end{bmatrix} (i, j = 1, 2, 6) \text{ and}$$

$$\left[Q_{ij}\right]_k = \begin{bmatrix} Q_{44} & 0 \\ 0 & Q_{55} \end{bmatrix} (i, j = 4, 5)$$

in which,

$$Q_{11} = \frac{E_{11}}{1 - \upsilon_{12}\upsilon_{21}},\ Q_{12} = \frac{\upsilon_{12}E_{22}}{1 - \upsilon_{12}\upsilon_{21}},$$

$$\mathrm{Q}_{22} = \frac{\mathrm{E}_{22}}{1 - \upsilon_{12}\upsilon_{21}},\ \mathrm{Q}_{44} = \mathrm{G}_{13},\ \mathrm{Q}_{55} = \mathrm{G}_{23}$$

The non-mechanical strength and moment resultants concerning temperature implication are shown as

$$\{N_x^N, N_y^N, N_{xy}^N\}^T = \sum_{k=1}^{n} \left(\overline{Q_{ij}}\right)_k \{e\}_k (z_k - z_{k-1}).$$

$$\left\{M_x^N, M_y^N, M_{xy}^N\right\}^{\mathrm{T}} = \frac{1}{2}\sum_{k=1}^{n} \left(\overline{Q_{ij}}\right)_k \{e\}_k \left(z_k^2 - z_{k-1}^2\right).$$

Where

$$\{e\}_k = \left\{e_x e_y e_{xy}\right\}^T = [T]\{\alpha_1\alpha_2\}_k^T (\Delta T)$$

while $[T]$ = Transformation matrix fit to temperature and denotes

$$[T] = \begin{bmatrix} \cos^2\theta & \sin^2\theta \\ \sin^2\theta & \cos^2\theta \\ \sin 2\theta & -\sin 2\theta \end{bmatrix}$$

$e_x, e_y,$ and e_{xy} are the non-mechanical strains of the kth lamina, oriented at an arbitrary angle θ; $\alpha_1 and \alpha_2$ are the thermal coefficients in longitudinal and lateral directions.

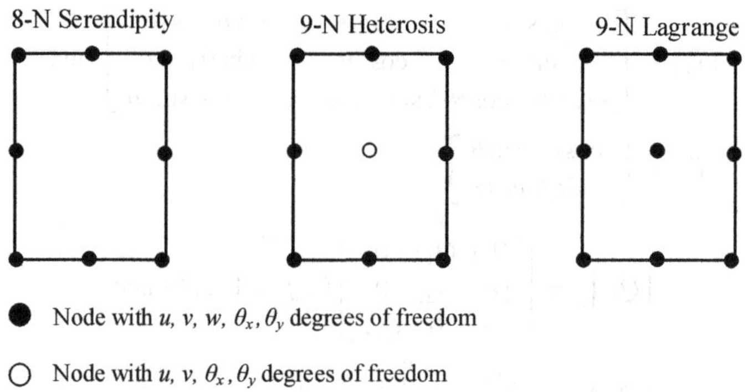

Fig. 2 Plate element types

2.2 Formulation of a Plate Element

In this plate element, there are nine nodes, and each has five degrees of freedom (three translational and two rotational). As shown in Figure 2, the heterosis element is a derivation of the Serendipity elements with eight nodes and the Lagrange elements with nine nodes. The full mathematical expression for the heterosis plate element and the corresponding stiffness matrices have been discussed by Subash Chandra [18] and have not been shown here for brevity.

2.3 Stiffener Element Formulation

Figure 3 shows the standard geometrical arrangement of an isoparametric three-nodal beam element with b_s in width and d_s in depth with a positive co-ordinate configuration. The stiffer is constructed as a laminated strip in the current formulation and is positioned at a different angle ψ to the x-axis in some arbitrary directions. Timoshenko beam principle is the basis of the stiffener formula. In the original hypothesis, only the consequence of shear deformation is considered. The principle is updated to handle the torsional effect into account as well. The displacement field "z" from the stiffener mid-plane is given according to a modified beam theory and is given as in [16].

Fig. 3 Geometry of stiffener with arbitrary orientation

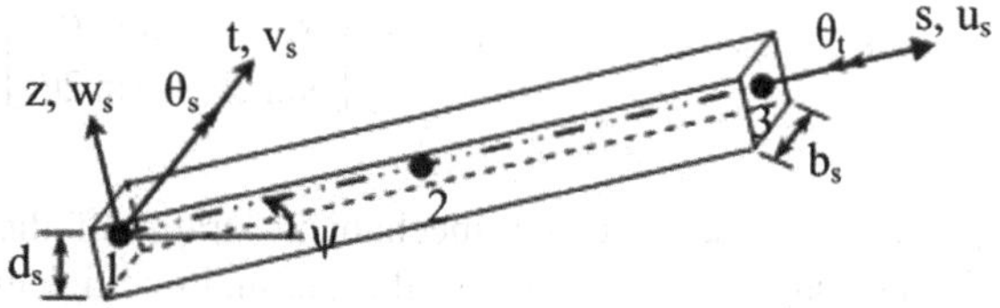

$$\overline{u}_s = u_s + z\theta_s$$
$$\overline{v}_s = z\theta_t$$
$$\overline{w}_s = w_s \tag{7}$$

where θ_s and θ_t imply bending and torsional rotations along each plane concerning t- and s-axis, and u_s and w_s of axial and transverse displacements. The relationship between the laminated stiffener is given as

$$\{N_s^1\} = [D_s]\{\varepsilon_s^1\} \tag{8}$$

where $\{N_s^1\} = [N_s\ M_s\ M_{st}\ Q_{sz}]^T$ and $\{\varepsilon_s^1\} = [\varepsilon_s\ \chi_s\ \chi_{st}\ \gamma_{sz}]^T$ indicate both stress and strain vectors, on sequence. $[D_s]$ is the stiffener, constitutive matrix comparable to Patel and Sheikh concept [17].

Applying Green–Lagrange's strain–displacement expression [17], elastic stiffness matrix $[k_{es}]$ and geometric stiffness matrix $[k_{Gs}]$ can be expressed as follows:

$$[k_{es}] = b_s \int_{-1}^{+1} [T_e]^T[T_o]^T[B_s]^T[D_s][B_s][T_o][T_e]d\xi \tag{9}$$

$$[k_{es}] = b_s \int_{-1}^{+1} [T_e]^T[T_o]^T[B_{Gs}]^T[D_s][B_{Gs}][T_o][T_e]|J_s|d\xi \tag{10}$$

where $[T_e]$ and $[T_o]$ are matrices of transformation, taking account of eccentricity and random orientation of stiffener, $[B_s]$ and $[B_{Gs}]$ are the linear and non-linear strain–displacement matrix, whose details are found in [16].

2.4 Governing Equations

The following eigenvalue problem may be used to describe the stability equation.

$$([K] - \lambda[K_g])\{q\} = 0 \tag{11}$$

where $[K]$ and $[K_g]$ indicate the global stiffness plus geometric matrix in order. Furthermore, $\{q\}$ is a generalized displacement vector.

Specific critical buckling temperature ΔT_{cr} exists and is collected from the lowest eigenvalue λ.

3 Results and Discussion

On a thin laminated squares panel, thickness ratio is considered to be $h/b = 0.01$. Numerical results are shown for stiffened panel with d_s/b_s is known as the ratio of the stiffener (d_s) depth to width (b_s), which is taken as 2 for all problems, except for where stiffener depth to width ratio is taken into consideration. The overall stiffener volume taken into account is proportional to the material quantity withdrawn by the cut; that is to say, the stiffeners form from the material extracted from the cut-over. The panel's lamination scheme varies from $\theta = (\pm 0^0)_{2s}$ to $(\pm 90^0)_{2s}$. In the stiffener's case, the fiber orientation remains impartial toward each stiffener's axis; nevertheless, the existing stiffener formulation is fit toward several assigned ply-orientation. Material properties analyzed toward both panel as well as stiffener denote $E_{11} = E_{22} = 25$; $G_{12} = G_{13} = 0.50\, E_{22}$; $G_{23} = 0.20\, E_{22}$ and $\nu_{12} = 0.25$, $\alpha_1 = 0.02 \times 10^{-6}\,{}^{o}C^{-1}$, $\alpha_2 = 22.52 \times 10^{-6}\,{}^{o}C^{-1}$, unless stated. The two sets of boundary conditions examined during the aforementioned study do the same as Rajanna et al. (2008).

3.1 Comparison and Convergence Study

The comparison investigations ensure the exactness of the outcomes obtained, and convergence studies for the present study with and without stiffener are done to meet the precise convergence during this current work and get relevant results. Comparison studies in Tables 1 and 2 show excellent rapport amidst this present finite-element formulation and previous studies. Convergence studies show the number of elements required to converge for the current formulation in Table 3. The value within the

Table 1 Comparison of thermal buckling with previous studies for laminate under simply supported boundary condition

Laminate [mesh size]	Shi et al. [8] [12 × 12]	Shi et al. [11] [12 × 12]	Ounis et al. [14] [12 × 12]	Zhao et al. (2016) [32 × 24]	Present [12 × 12]
$[0/90/90/0]_s$	12.26	12.2610	12.2612	12.25	12.261
$[0/45/\text{-}45/90]_s$	13.71	13.7519	13.7357	13.65	13.724

Table 2 Thermal buckling with ten-layered angle-ply laminate composite plates

Source/laminate $(\pm\theta)_{5s}$	Minimum critical buckling temperature (ΔT_{cr})			
	$\theta = 0^0$	$\theta = 15^0$	$\theta = 30^0$	$\theta = 45^0$
Present	0.7464×10^{-3}	0.1115×10^{-2}	0.1502×10^{-2}	0.1674×10^{-2}
Noor and Burton	0.7464×10^{-3}	0.1115×10^{-2}	0.1502×10^{-2}	0.1674×10^{-2}
Ounis et al.	0.7464×10^{-3}	0.1133×10^{-2}	0.1540×10^{-2}	0.1720×10^{-2}

Table 3 Mesh convergence for present investigations of ΔT_{cr} of the laminate including and without reinforcement around the circular cutout

Cutout ratio (ϕ/b)	No. of elements	Critical buckling temperature (ΔT_{cr})	
		Without stiffener	With stiffener
0.5 [laminate sequence $(45/-45)_{2s}$]	100	946.05	1492.02
	132	939.06	1487.03
	164	925.52	1478.21
	196	921.70	1477.22
	228	921.60 (921.70)	1477.19 (1477.20)

bracket in Table 3 indicates the thermal buckling value obtained through ABAQUS software.

3.2 *Parametric Studies on Thermal Buckling of Cutout Panels with and Without Stiffeners*

The various effects such as a change in cutout ratios, modulus ratio, thermal expansion ratios, and different boundary conditions are studied. Furthermore, analysis is done in detail for the square ($a/b = 1$) panels having centrally placed cutouts and is compared with the reinforced panel around the cutout with a stiffener depth (d_s) to width (b_s) ratio equal to 2.

3.2.1 Effect of Cutout Ratios on Panels with and Without Stiffener

The effect of different cutout ratios (ϕ/b) ranging from 0.1 to 0.7 is studied for the variations in the thermal buckling of square panels. A similar study is analyzed for panels with stiffener ($d_s/b_s = 2$) placed at the cutout surrounding, which amounts to equal to the volume of the cutout portion of the panel, as shown in Figure 4a and b, sequentially. It is found that $(\pm 45)_{2s}$ have the highest thermal buckling for cutout panels either with and without stiffeners. The Von-misses stress plot for $(\pm 45)_{2s}$ ply angle has been shown for cutout panel ϕ/b =0.5, with and without stiffener in Figure 5a and b sequentially. It signifies from Figure 5a that the panel without stiffener has a higher stress concentration near the hole, leading to local buckling. However, such a phenomenon is not seen for panels with reinforcement, as the stresses get redistributed, hence higher thermal buckling resistance.

The difference in thermal buckling between each cutout size increased with the rise in the cutout ratio. Larger the cutout size is found to have higher ΔT_{cr} irrespective of fiber orientation. However, the percentage change has increased between with and without stiffeners. For example, in $(\pm 45)_{2s}$ ply panel, with $\phi/b = 0.1$, percentage variations are 11.7%. In contrast, with $\phi/b = 0.7$, 169% change between the panels

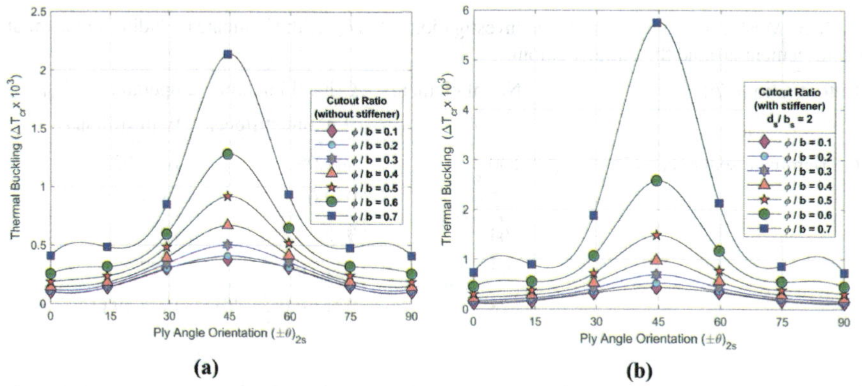

Fig. 4 Variation in the thermal buckling for different ply angles with increasing cutout ratios **a** without stiffener and **b** with stiffener

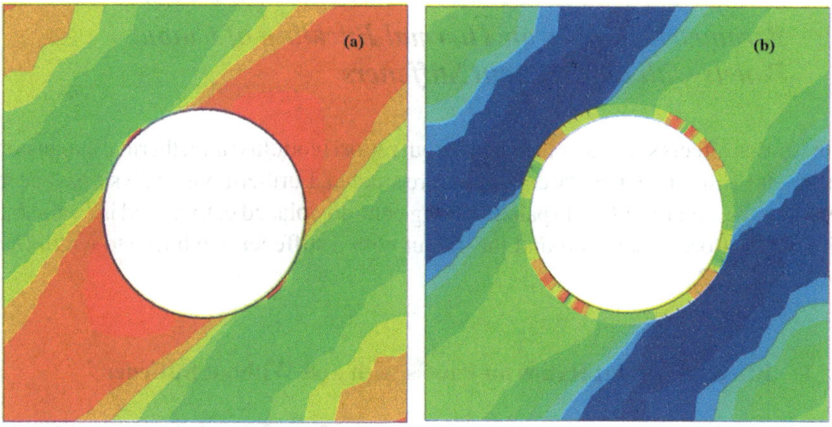

Fig. 5 Von-misses stress distribution for panels with cutout ratio $\phi/b = 0.5$, for ply angle $(\pm 45)_{2s}$ **a** without and **b** with stiffener, respectively

with and without stiffeners. Hence, adding stiffeners around the cutout for larger cutout sizes improved the thermal buckling resistance significantly.

3.2.2 Effect of Modulus Ratio on Cutout Panels with and Without Stiffener

The effect of change of modulus ratio observed for cutout panels (ϕ/b=0.5), with and without stiffener, is as seen in Figure 6a and b. It implies that as the modulus ratio increased, the ΔT_{cr} increased for both cutout panels with and without stiffeners.

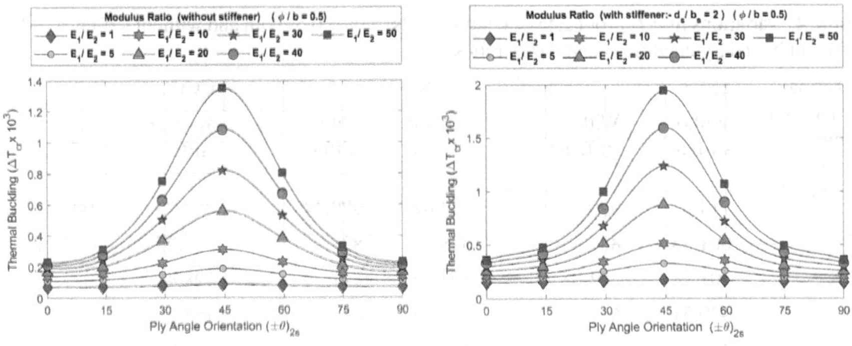

(a) (b)

Fig. 6 Variation in the thermal buckling for different ply angles with increasing modulus ratios with cutout ratio $\phi/b = 0.5$ **a** without stiffener and **b** with stiffener around the cutout

However, the percentage change in thermal buckling between stiffened and unstiffened panels decreased at higher modulus ratios. For example, in $(\pm 45)_{2s}$ ply-angle laminate, with modulus ratio = 1, the percentage change in ΔT_{cr} is 96%, while with modulus ratio = 50, it is 44.2%. It is also observed that ply-angle variations have no meaningful effect on ΔT_{cr} for modulus ratio = 1 (i.e., isotropic), which agrees with the [20] studies, which are also valid for cutout panels with and without stiffeners.

3.2.3 Effect of Boundary Conditions and Thermal Expansion Ratios (α_1/α_2) for Cutout Panels with and Without Stiffeners

The effect of change in boundary conditions with varying thermal expansion ratios (α_1/α_2) for $(\pm 45)_{2s}$ panels with and without stiffeners is analyzed with a cutout ratio = 0.5, as shown in Table 4. That denotes that as the α_1/α_2 ratio raised, the thermal buckling decreased irrespective of boundary restrictions. However, it was also observed that the higher the α_1/α_2 ratio, the lower the thermal buckling rate. It is found that for clamped conditions, the thermal buckling resistance is most elevated. For stiffened cutout, panels have higher ΔT_{cr} than without, irrespective of α_1/α_2. However, the percentage change increased with higher α_1/α_2 ratios. For example, in the supported boundary condition, for α_1/α_2 ratio = 1, the percentage change is found to be 36%, while for α_1/α_2 ratio = 35, it is 45%. Moreover, this change is found to decrease with boundary constraints.

3.2.4 Effect of Stiffener Depth Ratio (d_s/b_s) for Different Ply-Angle Orientations

The ply-angle orientation $(\pm\theta)_{2s}$ for different stiffener depth ratios (d_s/b_s) in the range of 1–10 is studied and analyzed for panel with $\phi/b = 0.5$ and is shown in

Table 4 Variations in ΔT_{cr} for $(\pm 45)_{2s}$ cutout panels ($\phi/b = 0.5$) with and without stiffener under different boundary conditions and α_1/α_2 ratios

Thermal expansion ratio (α_1/α_2)	SSSS		CCSS		CCCC	
	Without stiffener	With stiffener	Without stiffener	With stiffener	Without stiffener	With stiffener
1	714.8	970.8	723.07	979.44	752.73	1003.8
5	601.34	831.4	608.3	838.9	633.25	860.27
10	501.78	704.85	507.59	711.29	528.41	729.73
15	430.51	611.72	435.49	617.35	453.35	633.55
20	376.96	540.32	381.32	545.33	396.96	559.76
25	335.26	483.84	339.14	488.35	353.05	501.35
30	301.87	438.05	305.36	442.14	317.89	453.97
35	274.53	400.18	277.7	403.93	289.09	414.77

Table 5. The study is carried out for the same volume, equal to the cutout portion of the panel. It is understood that as the ply angle increases from $\theta = 0^0$ to $\theta = 45^0$, the percentage change in thermal buckling between any two successive d_s/b_s ratios decreases. It may be because the reduction in thickness of stiffener increases the thermal exposed area. It also signified that the minimum thermal buckling difference was observed at $\theta = 45^0$. Further, increase in ply angle from $\theta = 45^0$ to $\theta = 90^0$, the percentage change in thermal buckling increases and is determined to hold a maximum in $\theta = 90^0$.

Table 5 Variations in ΔT_{cr} for $(\pm\theta)_{2s}$ cutout panels ($\phi/b = 0.5$) with different stiffener d_s/b_s ratios

d_s/b_s ratio	$\theta = 0^0$	$\theta = 15^0$	$\theta = 30^0$	$\theta = 45^0$	$\theta = 60^0$	$\theta = 75^0$	$\theta = 90^0$
1	307.81	385.68	754.56	1477.7	755.51	386.04	307.96
2	304.38	382.71	751.04	1477.2	752.08	383.08	304.53
3	301.96	380.4	748.4	1475	749.51	380.77	302.11
4	300.42	378.9	746.62	1473.2	747.79	379.27	300.58
5	299.4	377.89	745.36	1471.8	746.5	378.27	299.56
6	298.69	377.19	744.43	1470.7	745.69	377.57	298.85
7	298.18	376.69	743.71	1469.9	745.01	377.07	298.35
8	297.81	376.32	743.15	1469.3	744.48	376.7	297.97
9	297.53	376.04	742.7	1468.7	744.05	376.43	297.7
10	297.33	375.84	742.34	1468.3	743.71	376.22	297.5

4 Conclusions

The comparison of thermal buckling has been made with the previous studies and holds good agreement for the laminated plates. The tasks are carried out with heterosis plate element and quadratic beam element in the present formulation. The square panels with centrally placed circular cutout are compared with the panels with stiffener (with its depth to width ratio equal to 2) placed encompassing the circular cutout corresponding to that volume separated of panel's cutout part. A few noteworthy points are extracted from the present study and are detailed as follows:

- The buckling variation within smaller cutouts with/without stiffener is not that significant, and hence, the smaller-sized cutouts may be neglected within the plate.
- The modular ratio and ply-orientation notably influence the thermal buckling resistance concerning plates with and without stiffeners. Toward any given ply angle, the thermal buckling detention progresses beside an increase in the modular ratio, which is significant for $(\pm 45)_{2s}$ ply-orientation. It is insignificant for $(\pm 0)_{2s}$ and $(\pm 90)_{2s}$ ply-orientations.
- An increase in boundary constraint significantly increases the thermal bucking resistance. Simply support boundary condition has the least thermal buckling, while clamped experienced the highest.
- Higher the value of α_1/α_2, the higher the thermal buckling resistance for stiffened panels.
- The panel with $(\pm 45)_{2s}$ has the highest thermal buckling resistance for either cutout panels with and without stiffener. However, for modulus ratio = 1, it does not affect either the ply angle or the panels with and without stiffener.
- The stiffener depth ratio notably influences the thermal buckling aspects of each panel. The plate's thermal bucking stability diminishes by increasing the stiffener depth ratio for a given cutout size.

References

1. Timoshenko SP, Gere JM (1961) Theory of elastic stability. New York, McGraw-Hill
2. Reddy JN, Phan ND (1985) Stability and vibration of isotropic, orthotropic and laminated plates according to a higher-order shear deformation theory. J Sound Vibration 98(2):157–170
3. Noor AK, Burton WS (1992) Three dimensional solutions for thermal buckling of multilayered anisotropic plates. J Eng Mech 18(4):683–701
4. Wu CH, Tauchert TR (1980) Thermoelastic analysis of laminated plates in symmetric specially orthotropic laminates. J Therm Stress 3(2):247–59
5. Chen L-W, Chen L-Y (1987) Thermal buckling of laminated composite plates. J Therm Stress 10(4):345–356
6. Chen WJ, Lin PD, Chen LW (1991) Thermal buckling behavior of thick composite laminated plates under non-uniform temperature distribution. Comput Struct 41(4):637–45
7. Prabhu M, Dhanaraj R (1994) Thermal buckling of laminated composite plates. Comput Struct 53(5):193–204

8. Shi YC et al (1999) Thermal postbuckling of composite plates using the finite element modal co-ordinate method. J Therm Stress 22(6):595–614. doi: https://doi.org/10.1080/014957399280779
9. Matsunaga H (2005) Thermal buckling of cross-ply laminated composite and sandwich plates according to a global higher-order deformation theory. Compos Struct 68(4):439–54. doi: https://doi.org/10.1016/j.compstruct.2004.04.010
10. Topal U, Uzman Ü (2008) Thermal buckling load optimization of laminated composite plates. Thin-Walled Struct 46(6):667–75. doi: https://doi.org/10.1016/j.tws.2007.11.005
11. Shiau L-C et al (2010) Thermal buckling behavior of composite laminated plates. Compos Struct 92(2):508–14. doi: https://doi.org/10.1016/j.compstruct.2009.08.035
12. Ounis H et al (2014) Thermal buckling behavior of laminated composite plates: a finite-element study. Front Mech Eng 9(1):41–49. doi: https://doi.org/10.1007/s11465-014-0284-z
13. Topal U, Uzman Ü (2010) Effect of rectangular/circular cutouts on thermal buckling load optimization of angle-ply laminated thin plates. Sci Eng Compos Mater 17(2). doi: https://doi.org/10.1515/SECM.2010.17.2.93
14. Ounis H, Belarbi MO (2017) Thermal buckling behaviour of laminated composite plates with cut-outs. J Appl Eng Sci Technol 3(2):63–69
15. Devarajan B, Kapania RK (2020) Thermal buckling of curvilinearly stiffened laminated composite plates with cutouts using isogeometric analysis. Compos Struct 238:111881. doi: https://doi.org/10.1016/j.compstruct.2020.111881
16. Rajanna T et al (2018) Effect of reinforced cutouts and ply-orientations on buckling behavior of composite panels subjected to non-uniform edge loads. Int J Struct Stab Dyn 18(04):1850058. https://doi.org/10.1142/S021945541850058X
17. Patel SN, Sheikh AH (2016) Buckling response of laminated composite stiffened plates subjected to partial in-plane edge loading. Int J Comput Meth Eng Sci Mech 17(5–6):322–38. doi:https://doi.org/10.1080/15502287.2016.1231235
18. Subash Chandra KS, Rao KV, Rajanna T (2021) Hygro-thermo-mechanical vibration and buckling analysis of composite laminates with elliptical cutouts under localized edge loads. Int J Struct Stab Dyn 2150150. doi:https://doi.org/10.1142/S0219455421501509
19. Subash Chandra KS, Rao KV, Rajanna T (2020) A parametric study on the effect of elliptical cutouts for buckling behavior of composite plates under non-uniform edge loads. Latin American J Solids Struct 17(8):e328. doi:https://doi.org/10.1590/1679-78256225
20. Jones RM (2005) Thermal buckling of uniformly heated unidirectional and symmetric cross-ply laminated fiber-reinforced composite uniaxial in-plane restrained simply supported rectangular plates. Composites. Part A, Appl Sci Manuf 36(10):1355–1367

Effect of Isolated Wind Incidence on Local Peak Pressure

Supriya Pal, Ritu Raj, and S. Anbukumar

Abstract This paper presents the results of the experimental investigations carried out on a Modified triangle shape tall building to evaluate its aerodynamic performance of such. The purpose is to study the associated environment arising due to wind forces in isolated conditions on a building model of scale 1:300 with a prototype having 180 m height. Local peak pressure coefficients for the isolated tall building are studied by using boundary layer wind tunnel experiments. Two wind incidences, 0° and 180°, are considered for the current isolated study. From the perspective of cladding as well as structural design, the experimental findings are analyzed and presented. A higher number of critical positions facing maximum peak pressure and maximum peak suction are observed at 1800 of wind incidence as compared to that at 00 wind incidence. The highest suction at 180° wind incidence is −1.25, which is nearly 36% stronger than the highest suction at 0° wind incidence (−0.92). The maximum fluctuation between maximum peak pressure (0.85) and minimum peak pressure (−0.85) is observed at Face-I of 180° of wind incidence. Due to the difference in the direction of wind incidence, the test results showed significant variations in the related wind environment around the building model. Isolated wind incidence condition at 180° is critical as compared to 0° wind incidence.

Keywords Tall buildings · Peak pressure coefficient · Turbulence · Suction · Vortex

1 Introduction

Building materials and practices have advanced recently, resulting in a new generation of taller and flexible constructions [1]. Tall buildings are more exposed to wind oscillations and dynamic loading [2], and that is why shape optimization is considered to increase wind resistance [3]. Wind-induced lateral movement of tall structures can agitate residents (especially those who are on the higher levels) and inflict structural

S. Pal (✉) · R. Raj · S. Anbukumar
Delhi Technological University, New Delhi, India
e-mail: supriya8788@gmail.com

M. S. Ranadive et al. (eds.), *Recent Trends in Construction Technology and Management*, Lecture Notes in Civil Engineering 260,
https://doi.org/10.1007/978-981-19-2145-2_70

damage [4] and, thus, is considered as one of the main serviceability criteria while designing tall buildings [5]. Underspecification can have massive impacts, necessitating extensive experimental and analytical research to better comprehend more realistic wind-induced situations [6].

With the development of structural systems and building materials, the possibility of the construction of unconventional shape tall buildings is very high. Wind-induced effects on various unconventional shape isolated tall building models have been investigated through a series of experimental and numerical work, including "+" [7–9], "H" [10–12], "Z" [13], "L" and "U" [14], "L" and "T" [15], "C" [16], "L" [17], "E" [18, 19] and "Triangle" shape building with various configurations [20–23] have investigated "Fish Shape" model; [24] have studied Remodel triangle shape building; [25] investigated two Fish-plan shape buildings, "Triangle" and "Y" [26],"Triangle", "Y", and "Circular" [27] shape buildings. In addition to studying the aforementioned unusually shaped tall building models, research has been conducted to better grasp the notion of aerodynamic shape optimization [28–31]. Despite the fact that many peculiar shaped structures are being investigated these days due to the versatile requirements, the majority of the studies are limited to the examination of mean coefficients of pressure and forces. Wind analysis is a project of tremendous unconventionality if the building's plan is unique because of the flow patterns originating from the exchanges of the wind with the elements. Because of the building's unique shape, peak pressure and suctions on the walls must be investigated, since they frequently cause damage in high-wind scenarios [21, 32, 33].

The formation of high turbulence in the wind at the principal building model causes the fluctuation in wind-induced load. The goal of researching wind flow characteristics surrounding any altered building plan is to compare the wind flow pattern on all of the building's walls to that of typical square or rectangular-shaped buildings [34]. The research concentrated on the experimental investigation of peak pressure and suctions coming at a face on an isolated Modified triangular shape building model for various isolated wind incidence directions of 0° and 180°. Isolated Modified triangle plan buildings are adapted triangular plan shape buildings that are typically erected on triangular plots; the best example being the Flatiron Building in New York (see Fig. 1). Small adjustments to the exterior building shape, according to [35], could result in force and moment diminution. Modified triangle-shaped buildings (Fig. 2) that are modified triangular plan shaped buildings provide better room availability as compared to triangular plan shaped buildings. The results demonstrate that understanding the position of high pressure and high suction zones is critical for cladding surface design since it creates considerably huge compression and strong tension on the external surface, resulting in cladding failure. Furthermore, the findings of this study can help structural designers choose from a variety of recent technologies in order to meet a structure's failure and serviceability standards in severe wind conditions.

Fig. 1 New York City's flatiron building

2 Experimental Work's Setup

Experiments were conducted in Roorkee's boundary layer wind tunnel in the Department of Civil Engineering at IIT Roorkee. The studies were conducted at a geometric scale of 1:300 in a simulated wind flow of terrain category II as per IS 875: part 3, where the mean wind velocity profile followed the power-law with a power exponent of 0.22. Figure 3 depicts the mean wind velocity profile. The average wind velocity at the building model's roof height was 10 m/s.

Measurements are made on a pressure model made of a transparent Perspex sheet for conditions of isolated wind incidence at 0° and 180° as shown in Fig. 4c. H/D = 3 was the height-to-breadth ratio. The models depicted full-size buildings with a height of 180 m at the specified geometric scale of 1:300. A total of 273 pressure taps, 39 on each level (Fig. 4b), are installed on the walls of the principal building. The pressure points are located at seven different heights levels of 10, 60, 180, 300, 420, 540, and 590 mm from the bottom as shown in Fig. 4a to obtain a proper distribution of wind pressure on all the faces of models.

Fig. 2 Downscale model in the wind tunnel

3 Results and Discussions

The peak pressure is the maximum and minimum pressure coefficients that come at a pressure point over the time of pressure observation. The peak maximum pressure coefficient ($C_{\mathrm{p}}^{\Delta}(i, \theta)$ is determined for each pressure tapping point using the formula Eq. (1):

$$C_{\mathrm{p}}^{\Delta}(i, \theta) = \frac{\text{Peak maximum } P_{\mathrm{a}} \text{ pressure point ``}i\text{'', at wind direction } \theta}{0.6\,\mathrm{V}^2} \tag{1}$$

The peak minimum pressure coefficient ($C_{\mathrm{p}}^{\nabla}(i, \theta)$) is determined for each pressure tapping point using the formula Eq. (2):

$$C_{\mathrm{p}}^{\nabla}(i, \theta) = \frac{\text{Peak minimum } P_{\mathrm{a}} \text{ pressure point ``}i\text{'', at wind direction } \theta}{0.6\,\mathrm{V}^2} \tag{2}$$

where P_{a} denotes the pressure at the corresponding pressure tapping point, and V is the mean wind velocity in m/s at the top of the building model, which in this case is 10 m/s.

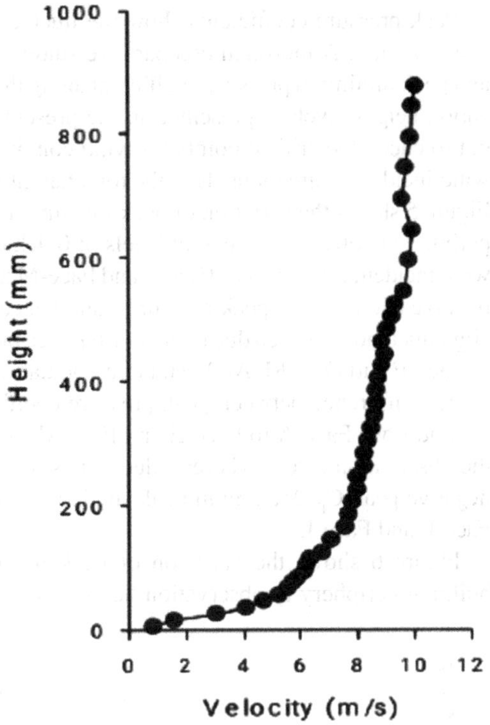

Fig. 3 Mean wind velocity profile

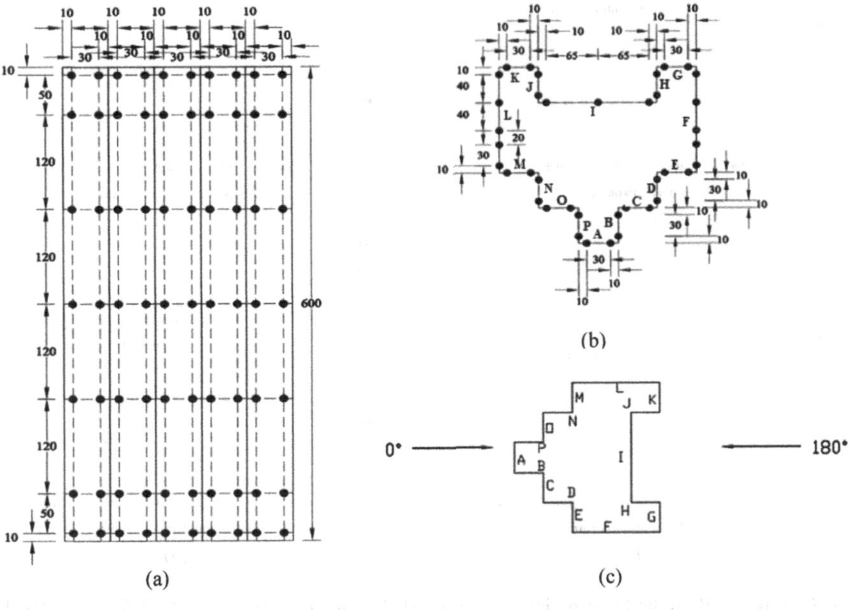

Fig. 4 Pressure taping layout of modified triangular-shaped model **a** elevation, **b** plan view, and **c** directions of wind incidence (all units: mm)

Peak pressure coefficients show the fluctuating component of induced pressure over a period. A thorough comparative study of the distribution of peak maximum and peak minimum pressure coefficient along the building periphery at certain observation height levels is presented in the present study. This study is mainly carried out to check the critical points for wind conditions at both the 0° and 180° isolated wind incidence directions. H is the total height of the building model, i.e., 600 mm. Figure 5 shows the variation of peak maximum and minimum C_p along the building periphery at observation height levels of 0.9H, 0.5H, 0.3H, and 0.1H for 0° isolated wind incidence conditions. Face-E and Face-M at all the observation levels show both positive and negative peak pressures, and hence, the face has maximum turbulence. High fluctuation is seen due to flow turbulence prior to flow separation from the edges of Face-E and Face-M. At 0.3H, large fluctuation is seen at Face-G and Face-K with a large difference between peak pressure coefficients due to the converging stream of fluid flow. Face-A to Face-E and Face-M to Face-P show positive peak C_p at all the observation levels, whereas, due to positioning of faces, Face-F to Face-L show negative peak C_p. Maximum peak suction at all the observation levels is shown by Face-F and Face-L.

Figure 6 shows the variation of peak maximum and minimum C_p along the building periphery at observation height levels of 0.9H, 0.5H, 0.3H, and 0.1H for

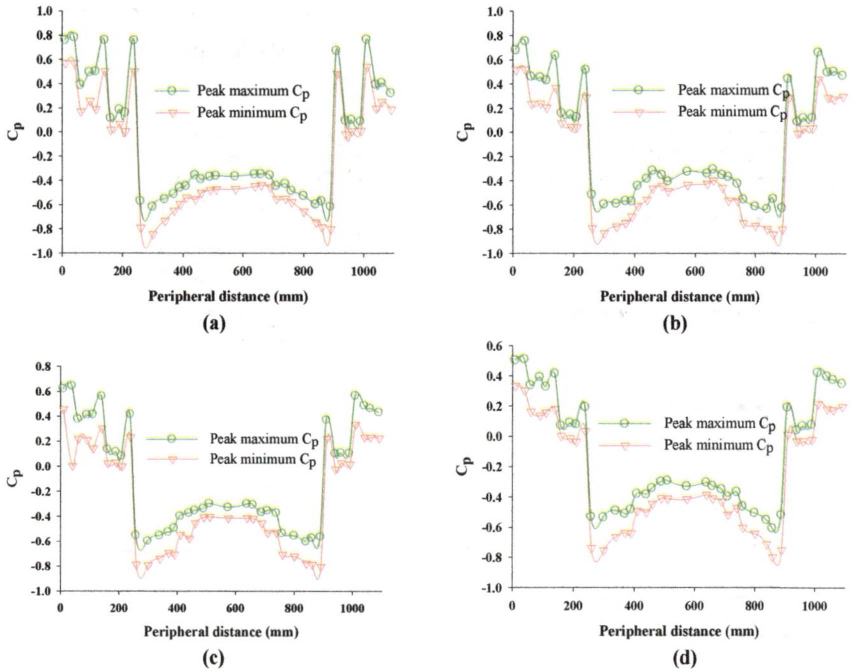

Fig. 5 Peak C_p along building periphery at **a** 0.9H, **b** 0.5H, **c** 0.3H, and **d** 0.1H for 0° isolated wind incidence

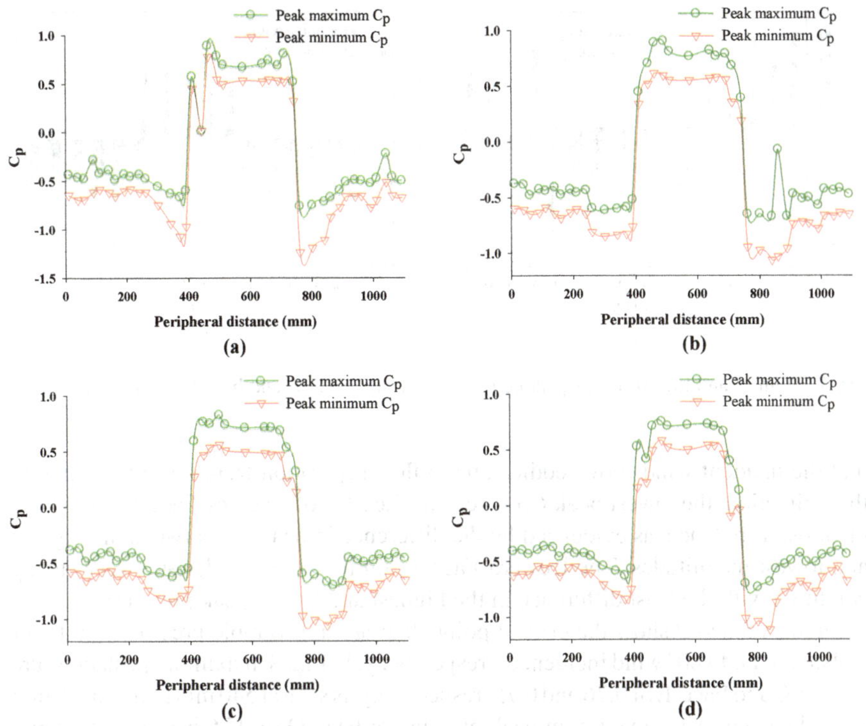

Fig. 6 Peak C_p along building periphery at **a** 0.9H, **b** 0.5H, **c** 0.3H, and **d** 0.1H for 180° isolated wind incidence

180° isolated wind incidence conditions. At this angle of wind incidence, no face shows transient behavior with both positive and negative peak C_p magnitudes due to the orientation of the model to the incoming wind. Due to the orientation of the model, Face-A to Face-F and Face-L to Face-P show negative magnitudes in peak maximum C_p and peak minimum C_p showing that the faces totally lie in the wake region. Maximum peak C_p is shown by Face-H at most of the observation levels as the wind is getting trapped between the adjacent edge of Face-H and Face-I, whereas, Peak minimum C_p of -1.25 is seen at Face-L at 0.9H.

For (a) 0° incident wind and (b) 180° incident wind circumstances, Fig. 7 displays the fluctuation of highest peak *Cp* and lowest peak *Cp* along the model's faces. Figure 7a shows a significant disparity between the highest and lowest peak Cp magnitudes at Face-N and Face-D, owing to the highest turbulence at the face due to the faces' depressed placement. Because Face-H and Face-J are on the leeward side and at depression, the difference in peak *Cp* between the highest and lowest magnitudes is the smallest. Because flow separation occurs from the faces, creating a suction area at the faces due to face orientation, Face-F and Face-L have the lowest peak *Cp*.

In Fig. 7b because of the windward orientation of the faces in the current circumstance, the highest peak *Cp* is seen at Face-H, followed by Face-I and Face-J. Given

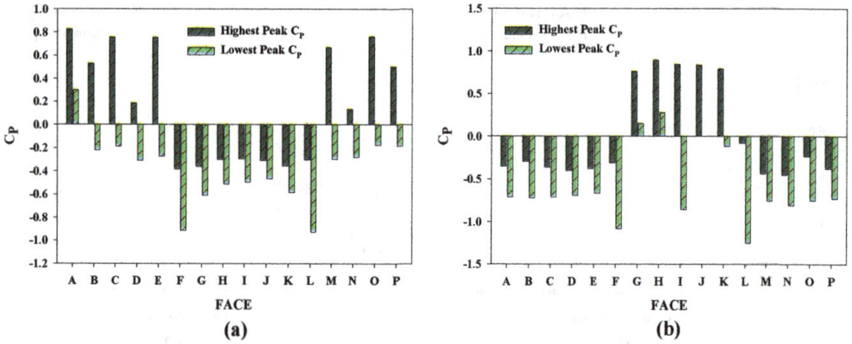

Fig. 7 Highest and lowest peak C_p along building façade for **a** 0° and **b** 180° wind incidences

that the incident wind forms eddies due to flow separation from these faces due to their direction, the lowest peak *Cp* is seen at Face-L, followed by Face-F. Face-I has the most turbulence, as evidenced by the difference in peak *Cp* between the highest and lowest magnitudes. The leeward Face-A, on the other hand, has a very steady condition with the least difference in the highest and lowest peak pressures.

Figures 8 and 9 show the critical points that are unfavorable pressure points for isolated 0° and 180° wind incidences, respectively. In Fig. 8, maximum peak pressure at Face-C and Face-E of 0.76 and 0.75, respectively, is seen at 540 mm height at 40 mm from the edge. Also, maximum peak pressure at Face-O of 0.76 is seen at 540 mm height at 40 mm from the edge. The highest peak maximum pressure is produced mainly due to the positioning of faces. The highest peak suction of −0.92 is produced at Face-F at 590 mm height and 10 mm from edge mainly due to top wash. Maximum fluctuation, i.e., the difference between peak maximum and minimum C_p is seen at Face-C at 540 mm height and 10 mm from the edge, at Face-B at 10 mm height and

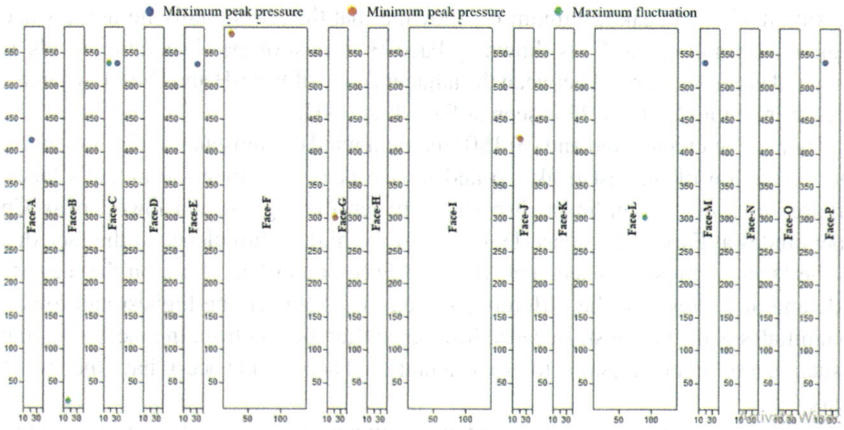

Fig. 8 Unfavorable positions on model for isolated 0° wind incidence condition

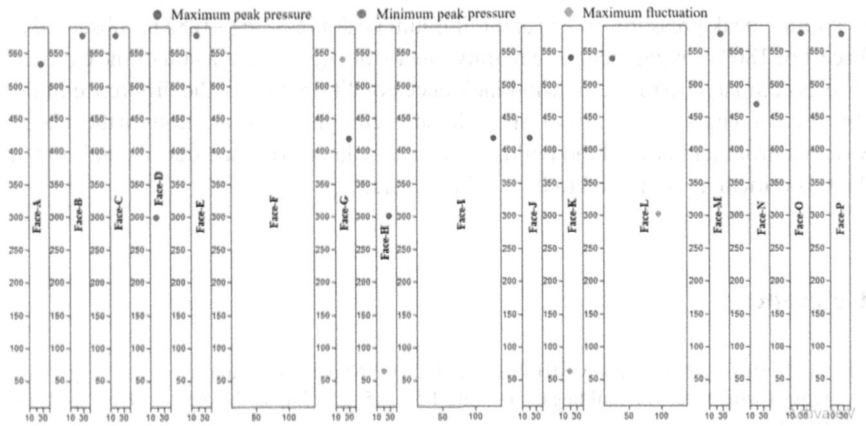

Fig. 9 Unfavorable positions on model for isolated 180° wind incidence condition

10 mm from the edge, and at Face-L at 300 mm height and 100 mm from the edge. The maximum difference between the peak values shows high turbulence areas in certain wind directions.

Figure 9 shows the critical points that are unfavorable pressure points for isolated 180° wind incidence conditions. Due to the orientation of the isolated model to the wind direction, higher peak pressure and peak suction are observed at the faces. The highest suction of −1.25 is manifested by Face-L at 540 mm height and 10 mm distance from the edge. Whereas, maximum peak pressure of 0.90 is manifested by Face-H at 300 mm height and 40 mm distance from the edge. The maximum fluctuation between peak pressures is seen at Face-F at 540 mm height and 10 mm distance from the edge.

4 Conclusion

Experiments are carried out to investigate the effects of varied wind directions on an isolated Modified triangle shape model. Peak pressure coefficients are calculated to find out the effect of fluctuating wind forces on various walls of the model at 0° and 180° wind incidence conditions. Due to the positioning of both faces and the fact that flow separation is prevalent at the faces, the smallest minimum peak pressure at both wind incidences is obtained at Face-F and Face-L. Critical positions in terms of maximum peak pressure and maximum peak suction are shown in order to highlight the unfavorable positions in both directions of wind incidences. A higher number of critical positions facing maximum peak pressure and maximum peak suction are observed at 180° of wind incidence as compared to that at 0° wind incidence. The highest suction at 180° wind incidence is −1.25, which is nearly 36% stronger than the highest suction at 0° wind incidence (−0.92). The maximum fluctuation between maximum peak pressures (0.85) and minimum peak pressure (−0.85) is observed at Face-I of 180° of wind incidence mainly due to the higher dimension of the exposed face as compared to that at 0° wind incidence condition. Due to the difference in the direction of wind incidence, the test results showed significant variations in the related wind environment around the building model. Isolated wind incidence condition at 180° is critical as compared to 0° wind incidence.

References

1. El-Heweity MM, Abdelnaby MH, Eshra EM (2019) Numerical simulation of buffeting longitudinal wind forces on buildings. Alex Eng J 58:225–236. https://doi.org/10.1016/j.aej.2018.08.001
2. Xu A, Xie ZN, Fu JY, Wu JR, Tuan A (2014) Evaluation of wind loads on super-tall buildings from field-measured wind-induced acceleration response. Struct Design Tall Spec Build 23:641–663. https://doi.org/10.1002/tal.1065
3. Zheng C, Xie Y, Khan M, Wu Y, Liu J (2018) Wind-induced responses of tall buildings under combined aerodynamic control. Eng Struct 175:86–100
4. Aly AM (2013) Pressure integration technique for predicting wind-induced response in high-rise buildings. Alex Eng J 52:717–773. https://doi.org/10.1016/j.aej.2013.08.006
5. Farouk MI (2016) Check the comfort of occupants in high rise building using CFD. Ain Shams Eng J 7:953–958
6. Raj R, Jha S, Singh S, Choudhary S (2020) Response analysis of plus shaped tall building with different bracing systems under wind load. Int J Adv Res Eng Technol (IJARET) 11(3):371–380
7. Chakraborty S, Dalui SK, Ahuja AK (2014) Wind load on irregular plan shaped tall building—a case study. Wind Struct 19(1):59–73
8. Chakraborty S, Dalui SK, Ahuja AK (2013) Experimental and numerical study of surface pressure on '+' plan shape tall building. J Civ Eng 8:251–262
9. Raj R, Ahuja AK (2013) Wind loads on cross shape tall buildings. J Acad Indus Res (JAIR) 2(2)
10. Cheng L, Kit ML, Wong SY (2015) POD analysis of crosswind forces on a tall building with square and H-shaped cross sections. Wind Struct an Int J 21(1):63–84
11. Nagar SK, Raj R, Dev N (2020) Experimental study of wind-induced pressures on tall buildings of different shapes. Wind Struct 31(5):441–453. doi: https://doi.org/10.12989/was.2020.31.5.441
12. Nagar SK, Raj R, Dev N (2021) Proximity effects between two plus-plan shaped high-rise buildings on mean and RMS pressure coefficients. Scientia Iranica. doi: https://doi.org/10.24200/SCI.2021.55928.4484
13. Paul R, Dalui SK (2016) Wind effects on 'Z' plan-shaped tall building: a case study. Int J Adv Struct Eng 8:319–335
14. Gomes MG, Rodrigues AM, Mendes P (2005) Experimental and numerical study of wind pressures on irregular-plan shapes. J Wind Eng Ind Aerodyn 93:741–756
15. Amin JA, Ahuja AK (2011) Experimental study of wind-induced pressures on buildings of various geometries. Int J Eng Sci Technol 3(5):1–19
16. Mallick M, Kumar A, Patra KC (2019) Experimental investigation on the wind-induced pressures on C-shaped buildings. KSCE J Civ Eng 23(8):3535–3546
17. Li Y, Li SQ, Chen F (2017) Wind tunnel study of wind-induced torques on L-shaped tall buildings. J Wind Eng Ind Aerodyn 167:41–50
18. Bhattacharyya B, Dalui SK, Ahuja AK (2014) Wind induced pressure on 'E' plan shaped tall buildings. Jordan J Civ Eng 8(2). https://doi.org/10.1016/j.engstruct.2018.08.031

19. Bhattacharyya B, Dalui SK (2020) Experimental and numerical study of wind-pressure distribution on irregular-plan-shaped building. J Struct Eng 146(7):04020137. https://doi.org/10.1061/(ASCE)ST.1943-541X.0002686
20. Bandi EK, Tamura Y, Yoshidaa A, Kim YC, Yang Q (2013) Experimental investigation on aerodynamic characteristics of various triangular-section high-rise buildings. J Wind Eng Ind Aerodyn 122:60–68
21. Bandi EK, Tanaka H, Kim Y, Ohtake K, Tamura Y (2013b) Peak pressures acting on tall buildings with various configurations. Int J High Rise Build 2:229–244
22. Pal S, Raj R, Anbukumar S (2021) Comparative study of wind induced mutual interference effects on square and Fish-plan-shape tall buildings. Sådhanå, Indian Acad Sci 46:86. https://doi.org/10.1007/s12046-021-01592-6
23. Tamura Y, Tanaka H, Ohtake K, Kim YC, Yoshida A, Bandi EK, Xu X, Yang Q, Aerodynamic control of wind-induced vibrations and flow around super-tall buildings. In: Proceedings of the 6th international conference advance experience strutural engineering international working advance smart mater, smart strutures technology, Urbana-Champaign
24. Pal S, Raj R, Anbukumar S (2021) Bilateral interference of wind loads induced on duplicate building models of various shapes. Latin American J Solids Struct 18(5):e386. https://doi.org/10.1590/1679-78256595
25. Pal S, Raj R (2021) Evaluation of wind induced interference effects on shape remodeled tall buildings. Arab J Sci Eng. https://doi.org/10.1007/s13369-021-05923-x
26. Ming G (2010) Wind-resistant studies on tall buildings and structures. Sci China Technol Sci 53(10):2630–2646. https://doi.org/10.1007/s11431-010-4016-2
27. Hayashida H, Iwasa Y (1990) Aerodynamic shape effects of tall buildings for vortex induced vibration. J Wind Eng Ind Aerodyn 33:237–242
28. Ahmed E, Bitsuamlak G, Damatty AE (2017) Enhancing wind performance of tall buildings using corner aerodynamic optimization. Eng Struct 136:133–148
29. Sharma A, Mittal H, Gairola A (2018) Mitigation of wind load on tall buildings through aerodynamic modifications: review. J Build Eng 18:180–194
30. Stathopoulos T, Zhou Y (1993) Numerical simulation of wind induced pressures on buildings of various geometries. J Wind Eng Ind Aerodyn 46–47:419–430
31. Xie J (2012) Aerodynamic optimization in super-tall building designs. In: The Seventh International Colloquium on Bluff Body Aerodynamics and its Applications (BBAA7) Shanghai, China (2012), pp 2–6
32. Stathopoulos T, Saathoff P (1991) Wind pressure on roofs of various geometries. J Wind Eng Ind Aerodyn 38(2–3):273–284. https://doi.org/10.1016/0167-6105(91)90047-z
33. Surry D, Djakovich D (1995) Fluctuating pressures on models of tall buildings. J Wind Eng Ind Aerodyn 58(1–2):81–112. https://doi.org/10.1016/0167-6105(95)00015-j
34. IS: 875-Part-3 (2015) Code of practice for design loads (other than earthquake loads) for buildings and structures-wind loads. India 2015
35. Gaur N, Raj R (2020) Wind load optimisation by aerodynamic mitigation techniques for tall buildings–a review. Solid State Technol 63(2s):5968–5986

Investigation of Performance of Perforated Core Steel Buckling Restrained Brace

Prajakta Shete, Suhasini Madhekar, and Ahmad Fayeq Ghowsi

Abstract Energy Dissipating Devices (EDDs) are used to control structural vibration. Buckling-Restrained-Braces (BRBs) are observed as one of the most effective structural techniques used for resisting lateral forces generated due to earthquakes as they exhibit symmetric load-deformation behavior. The use of all-steel BRB is getting widespread because of its advantages over the traditional concrete BRB. All-steel BRB is entirely made of steel by sandwiching the core plate with the outer restraining element. Perforated core plate BRB is a recently innovated EDD. The steel core segment is perforated in different configurations along the length to dissipate the seismic energy. In this study, a numerical investigation is carried out to predict all-steel perforated core BRB (PBRB) cyclic response. The numerical models are validated with the previous experimental study. Numerical finite element modeling is carried out using ABAQUS software. This validated model is further used to carry out the investigation of the cyclic performances of PBRBs. Various configurations are considered to study the cyclic performance of PBRBs. Axial strength, hysteresis response, and energy dissipation are observed to evaluate the cyclic performances of PBRBs. Results show that the configuration of PBRB has a substantial impact on the stability of perforated BRB.

Keywords Buckling restrained brace · Perforated buckling restrained brace · Finite element analysis · Axial strength · Hysteresis response

1 Introduction

Buckling restrained brace (BRB) is one of the most effective lateral load resisting systems. BRB is a structural brace component designed to permit the structure

P. Shete (✉) · S. Madhekar
Department of Civil Engineering, COEP, Pune, Maharashtra 411005, India
e-mail: ps15.civil@coep.ac.in

A. F. Ghowsi
Department of Civil Engineering, IIT Delhi, New Delhi 110 016, India

M. S. Ranadive et al. (eds.), *Recent Trends in Construction Technology and Management*, Lecture Notes in Civil Engineering 260,
https://doi.org/10.1007/978-981-19-2145-2_71

to survive earthquake-induced cyclic lateral loadings. Traditional concrete BRB involves a steel core, casing, and bond-preventing layer or unbonding layer. The unbonding layer separates the outer casing from the inner steel core. This permits the steel core to take the axial forces in the bracing component. The encasing member restrains the brace from buckling in compression. It comprises a square, rectangular, or circular hollow structural steel casing filled with mortar or concrete [1]. By preventing the buckling action of the brace, a symmetric and stable hysteresis curve is achieved that dissipates a high amount of energy in every single cycle. As the steel core is restrained from buckling action, it experiences nearly uniform axial strains. The core cross-sectional area can be considerably less than that of conventional braces as its behavior is not restricted by buckling. BRB is a displacement-dependent EDD that works after yielding. The middle portion of the core is designed to yield inelastically and rigid; the non-yielding lengths are provided on either end of the core plate. The enlarged cross-sectional area of the non-yielding segment confirms that it stays elastic, and thus plasticity is focused in the central portion of the steel core. The plastic hinges related to buckling do not form in properly designed and detailed BRBs. A significant numerical study has been done to reduce the seismic response using BRB frames [2, 3]. Celik and Bruneau [4] highlighted the potential of BRBs for the rehabilitation of existing buildings and bridges. The overall behavior of the BRB core is affected by its length; moderate length BRB can efficiently dissipate energy [5]. Khampanit et al. [6] show that the friction among the core and the restraining segment of BRB may cause overstrength in compression and may cause overall flexural buckling of the outer core. An extensive numerical and experimental study has been carried out in the past, which distinguishes the promising seismic performances of BRBs.

All-steel BRB is getting widespread because of its lightweight, easy maintenance, inspection, simple fabrication, and erection [7]. All-steel BRB is entirely made of steel by sandwiching the core plate with the outer restraining element. A gap is required between the steel core and the restraining element to cater Poisson's expansion. The stoppers are provided to keep the outer casing in its proper position. Jiang et al. [8] demonstrated that buckling load and buckling mode are affected by the core length of BRB. Comprehensive research has been carried out by Ghowsi and Sahoo [9] to study the influence of loading sequence and the restraining parameters of all-steel BRBs. They concluded that the loading history affects the energy dissipation demand. Shete et al. [10] carried out a detailed numerical analysis of all-steel BRBs on varying core lengths and gaps and concluded that the interaction between these influencing parameters must be considered to obtain a stable hysteretic performance.

The research work by Zhou et al. [11] reveals the requirement to keep the yielding of the core limited within the specific portion and keep the ends elastic and stable. This can be achieved by weakening the yielding portion, and weakening can be done by perforating the yielding steel core plate. [12] were the first to use the perforated core in BRB. Perforated core plate BRB is a recently innovated EDD. The core steel segment is perforated in different configurations along the length to dissipate the seismic energy. The previous study shows improper perforated plate configuration may cause premature buckling of the strong axis and cracks in the perforated zone.

Perforations in the BRB can be arranged as per the strength requirement, and hence there is a need for further research to investigate the best suitable arrangement and configuration.

In the present study, the research is carried out to evaluate the impact of the configuration of PBRB, location, and size of the perforated zone on the cyclic behavior of the PBRB. The evaluation compares the hysteresis response, axial strength, core fracture pattern, and energy dissipation.

2 Numerical Study

To explore the influence of perforated core and its configuration on the cyclic performance of BRB, finite element analysis has been carried out in ABAQUS software. In this study, a numerical investigation has been conducted to observe the cyclic behavior of all-steel perforated core BRB. The numerical models are calibrated with the previous experimental study, and the particulars of finite element simulation are discussed in the subsequent section. Three different configurations, hexagonal, circular, and rectangular, are considered to evaluate the cyclic effect of PBRB. The yielding core length of PBRB is kept constant for all cases. Further to study the effect of the size of perforation, two configurations, namely circular and rectangular, are considered.

2.1 Finite Element Analysis

To evaluate the impact of the perforations and different configurations on the cyclic behavior of the PBRB, rectangular, circular, and hexagonal shapes are considered to keep the cross-sectional area of the core plate constant. The numerical finite element model used for the parametric study is calibrated with the previous experimental work conducted by Wu et al. (2014). All-steel BRB with the core cross-section 160 mm × 20 mm is considered with a uniform core length of 1600 mm, as considered in the experimental work. The stoppers on either side of the steel core segment in the central portion along the length are provided, as shown in Fig. 1. The gap size between the steel core and the restrainers is provided as 1 mm along the strong surface and 2 mm along weak surfaces. The results of the experimental work done by Wu et al. [13] are used to validate the BRB model. Further, this BRB model is perforated in a different configuration to study the effect of the perforated BRB core and its configuration, as discussed earlier.

General-purpose eight-noded hexahedral first order (C3D8R) element with Hourglass control and reduced integration technique is used to model the brace in ABAQUS. Each node of C3D8R elements has three degrees of freedom. In case of the hexahedral element throughout, constant volumetric strain is considered. Hexahedral elements offer the solution of equal precision at less cost. The convergence rate is

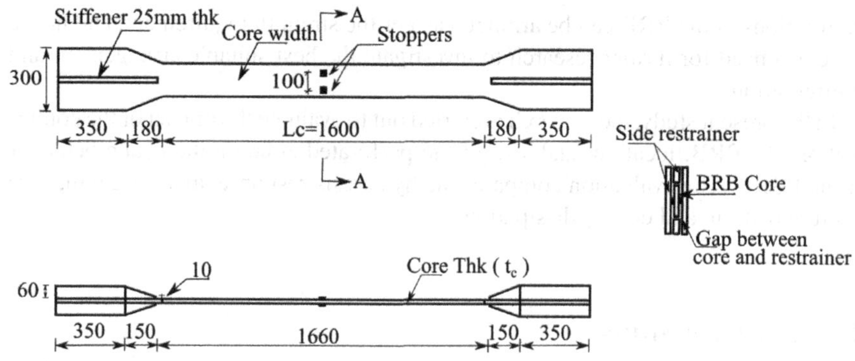

Fig. 1 Details of all-steel BRB

high, and their mesh sensitivity is less than the tetrahedral elements. The reduced integration technique decreases the computational time by using the lower order integration to form the element stiffness. Hourglass control prevents shear locking and volumetric control. Similar to past numerical modeling techniques [14, 15] are adopted in the present work.

The primary purpose of the restrainer segment is to offer adequate constraint to the buckling of the BRB core segment prior to compression yielding; hence at the restrainer interfaces, perfect contact with tie constraints has been assumed. In the experimental work by Wu et al. [13], the restrainers are connected with high-strength bolts. The fundamental research by [14] shows that the transverse flexural distortion of the restraining elements between the bolts is substantial compared to the axial distortion of bolts. Hence, they are not modeled in the finite element analysis. For surface-to-surface contact between the core and the restrainer segment, friction coefficient 0.15 in the tangential direction and the hard contact in the normal direction are considered. The BRB behavior under cyclic loading conditions is modeled using the damage plasticity model, including the nonlinear isotropic and kinematic hardening behavior. Combined isotropic and kinematic hardening behavior is applied to model the elastoplastic material properties of the steel core element. All other casing portion is modeled as perfectly elastic. The loading direction effect on the inelastic cyclic behavior of material is considered by the combined isotropic and kinematic hardening model. Kinematic hardening represents the yield surface translation in stress space through back stress α. The change in the size of the yielding surface is described through isotropic hardening. Back stress represents the constant kinematic shift of the yield surface and is beneficial to model the effect of residual stresses. The kinematic hardening component $\dot{\alpha}$ is expressed as the combination of a linear kinematic term and a nonlinear relaxation term [16]

$$\dot{\alpha} = C\frac{1}{\sigma^0}(\sigma - \alpha)\dot{\bar{\varepsilon}}^p = \gamma\alpha\dot{\bar{\varepsilon}}^p \quad (1)$$

Table 1 Calibrated kinematic hardening parameters of steel core

σ_Y (MPa)	C (GPa)	γ	Q_∞ (MPa)	b
330	8	52	25	10

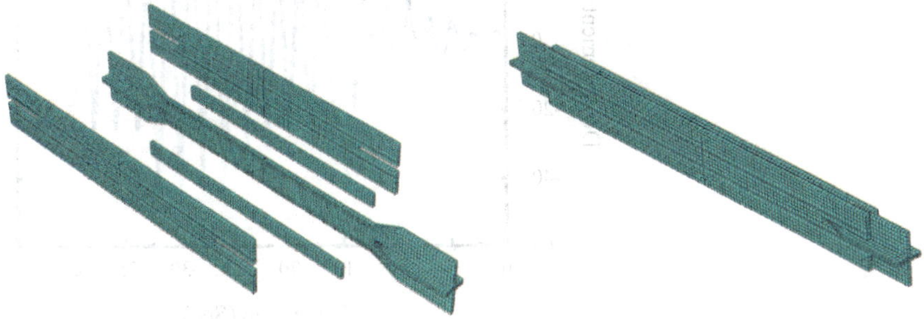

Fig. 2 All-steel BRB finite element model and assembled configuration

In Eq. (1), C represents the initial kinematic hardening modulus, and σ^0 is the initial yield stress, γ defines the rate at which the kinematic hardening modulus decreases with an increase in plastic deformation, $\dot{\bar{\varepsilon}}^p$ signifies the equivalent plastic strain rate, which considers the hardening and softening of the material as follows:

$$\sigma_0 = \sigma_Y + Q_\infty(1 - e^{-b\bar{\varepsilon}^p}) \tag{2}$$

where σ_Y is the yield stress when plastic strain is zero, Q_∞ and b are the material parameters. Q_∞ is the maximum change in the size of the yield surface, and b is the rate at which the size of the yield surface changes as plastic straining develops. The calibrated kinematic hardening parameters of the steel core used in the present research work are given in Table 1. Young's modulus of 200 GPa and Poisson's ratio of 0.3 is used. Based on the first buckling mode, the initial global imperfection factor of Length/1000 is considered [17]. The typical finite element model used for validation purpose and the perforated BRB model is shown in Fig. 2.

2.2 *Loading Protocol and Validation*

The loading protocol considered for the experimental work by Wu et al. [13] has been selected for the present work, as shown in Fig. 3. Loading protocol signifies the gradually rising core displacements till a maximum of 3.5% core strain in an interval of 0.5%, monitored by constant fatigue cycles equivalent to 2% core strain. The loading protocol includes all possible effects on BRBs, such as the gradually increasing, decreasing, or constant displacements during a seismic event. For validation of finite

Fig. 3 Loading protocol

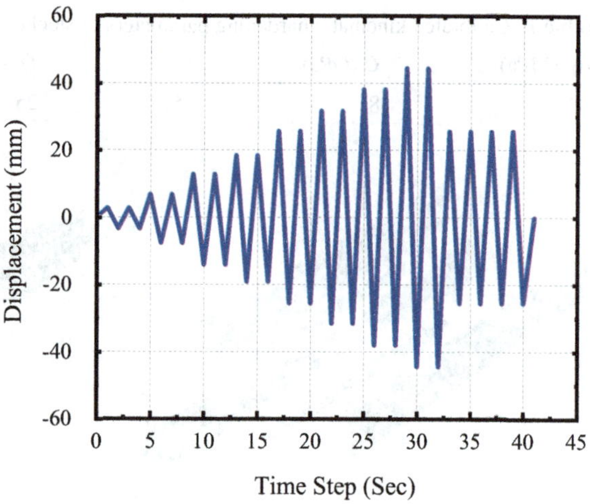

Fig. 4 Matching of hysteresis response

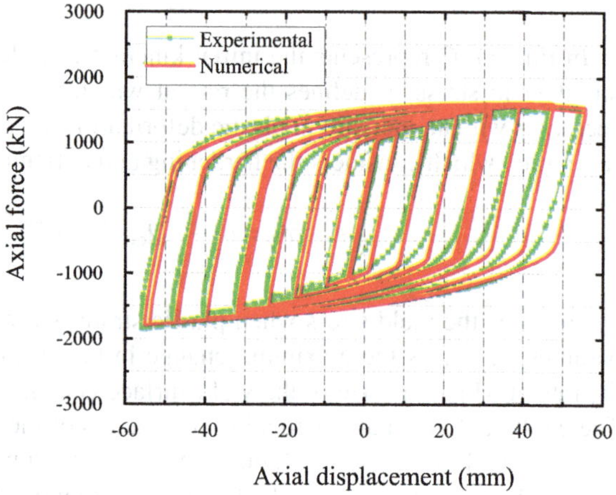

element models, the predicted hysteresis response of BRB is compared with the experimental work as discussed earlier. The hysteresis response, peak values of axial load, and hardening response at each cycle of the finite element model match the experimental results very well. 'Fig. 4'shows the hysteretic response of the validated numerical model and experimental result.

3 Results and Discussion

In the present research work, the influence of the configuration of PBRB is studied, and their seismic performances are evaluated with the hysteresis response, axial strength, fatigue pattern, and energy dissipation. In past studies by Zhou et al. [11], the influence of perforated core plate configuration on the strong axis stability is evaluated and determined the suitability of semi-circle configuration in enhancing the strong axis stability and performance during fatigue. A limited study is available on PBRB and its configuration. In this study, an attempt has been made to evaluate the various configuration to study their seismic response. Perforated cores with hexagonal, circular, and rectangular configurations are considered by keeping the yielding length of the core constant. The percentage openings are found to decide the performance of PBRB. Details of the perforated configuration are shown in Fig. 5. The

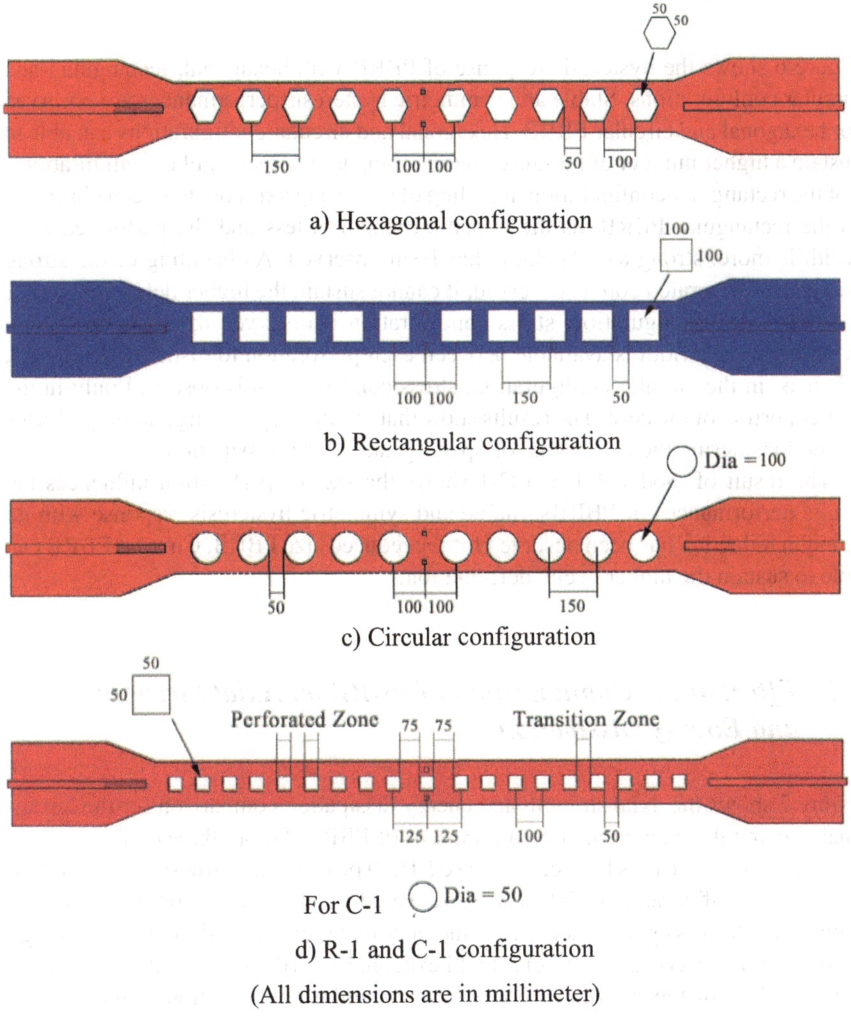

Fig. 5 Details of PBRB configuration

effect of reduced area perforation on BRB performance is studied by reducing the size of perforations. Model C-1 and R-1 are the PBRBs with reduced size of perforation. According to the configurations, the percentage of openings has been found. Hexagonal configuration has 25% area opening; rectangular configuration has 39% area opening; circular configuration has 31% area opening. R-1 and C-1 have 19% and 15% area of the opening, respectively.

3.1 Effect of the Configuration of PBRB on Hysteresis Response

Figure 6 shows the hysteresis response of PBRB with hexagonal, rectangular, and circular configurations. Stable and symmetric hysteresis performances are observed for hexagonal and circular PBRB. Hexagonal and circular configurations can able to sustain a higher number of fatigue cycles as compared to rectangular configurations. For the rectangular configuration, buckling of the strong axis core has been observed. In the rectangular PBRB, as the transition length is less and the perforated zone width is more, strong axis buckling has been observed. As buckling of the strong axis in the perforated core is observed, it cannot sustain the higher deformation. For the hexagonal configuration, stress concentration is observed in the end segment. Also, sufficient width is available between each perforation to distribute the stress, whereas, in the circular configuration, stress concentration is observed only in the center portion of the core. The results show that circular type configuration performs better considering the percentage of opening and energy dissipation.

The result of model R-1 and C-1 shows the size of perforation influences the cyclic performances of PBRBs. Stable and symmetric hysteresis response with no strength reduction has been observed for the reduced size PBRB. Circular PBRB can able to sustain the higher cyclic deformation.

3.2 Effect of the Configuration of PBRB on Axial Strength and Energy Dissipation

Figure 7 shows the axial strength and energy dissipation comparison for the hexagonal, rectangular, and circular configurations of PBRB. For all three configurations, the same elastic stiffness has been observed. High post elastic stiffness is observed for the circular configuration PBRB. The area covered under the corresponding hysteretic loops gives the dissipated energy by a member under cyclic loading. Higher energy dissipation is observed for circular and hexagonal PBRB. As a result of the strong axis buckling of the perforated core, the rectangular configuration cannot sustain all the given cyclic loading sequences. Hence low energy dissipation, as well as post-yield stiffness, are observed.

Table 2 summarises the results obtained for all described configurations. It is observed that the circular configuration PBRBs are the most promising PBRB. An overall improvement in the cyclic response has been observed. Enhancement in the strong axis stability, as well as fatigue performance, increases the energy dissipation capacity of circular configuration PBRB. To control the strong axis buckling of PBRB, the area of the perforated zone and the width of the transition segment must be designed carefully. In the case of hexagonal configuration, stress concentration at the edges is observed, hence as far as possible, make the edges round.

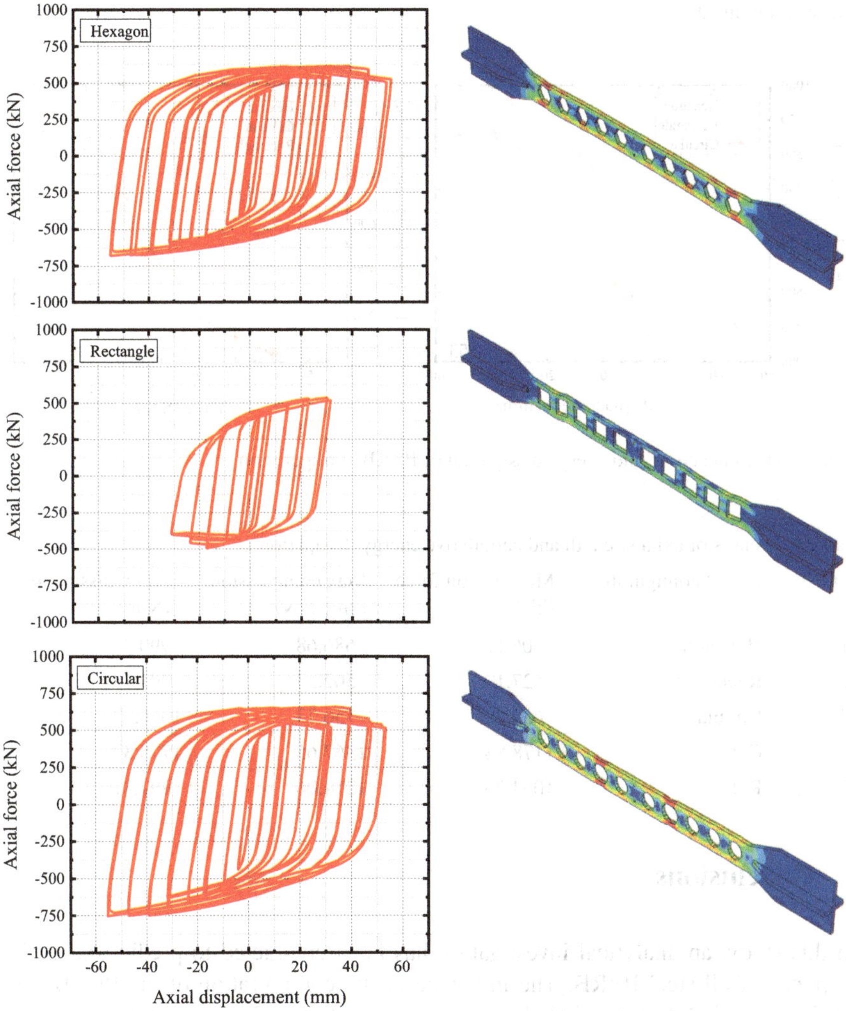

Fig. 6 Hysteresis response and PBRB yielding pattern

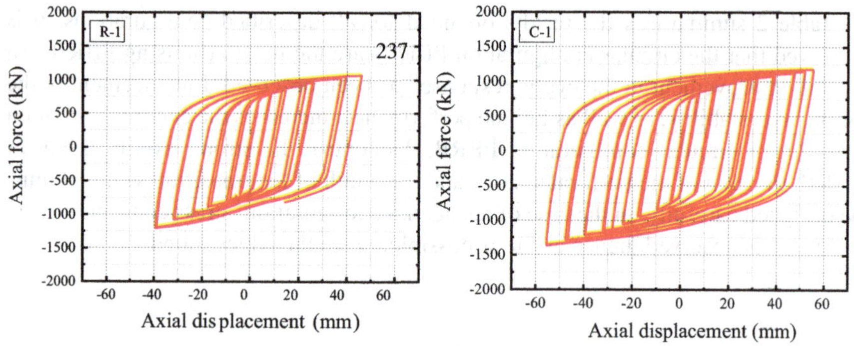

Fig. 6 (continued)

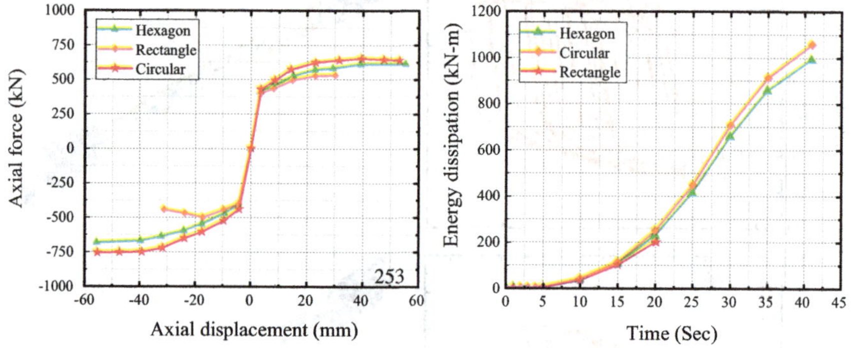

Fig. 7 Backbone curve and energy dissipation of PBRB configuration

Table 2 Details of axial strength and cumulative energy dissipation

Sr. No.	PBRB configuration	Max tension force (kN)	Max compression force (kN)	Energy dissipation (kN-m)
1	Hexagon	606.17	686.68	990.35
2	Rectangle	527.17	502.33	201.13
3	Circular	650.38	756.73	1056.73
4	C-1	1178.84	1367.65	1775.37
5	R-1	1063.76	1214.06	497.81

4 Conclusions

In this study, an analytical investigation has been conducted to predict the cyclic response of all-steel PBRB. The influence of the configuration of the PBRB core has been studied. It is concluded that the configuration and area of perforation have a significant impact on the overall performance of PBRB. Configuration, area, and the perforation location must be selected judiciously in order to achieve the required performance of PBRB. PBRB is a lightweight and economical option where less axial strength is required. In PBRB, post-yield stiffness changes with the change in the configuration, which significantly impacts strong axis stability, fatigue performance, and energy dissipation capacity. The strong axis buckling of the perforated core depends on the selection of the length of the transition zone. Further, the perforated yielding core length and the perforated configuration and their interaction effect can be studied. Hexagonal and circular PBRB configurations are the most suitable configurations.

References

1. Wu B, Mei Y (2015) Buckling mechanism of steel core of buckling-restrained braces. J Construct Steel Res 107:61–69
2. Sabelli R, Mahin S, Chang C (2003) Seismic demands on steel braced frame buildings with buckling-restrained braces. Eng Struct 25(5):655–666
3. Fahnestock LA, Sause R, Ricles JM (2007) Seismic response and performance of buckling-restrained braced frames. J struct Eng 133(9): 1195-1204
4. Celik OC, Bruneau M (2009) Seismic behavior of bidirectional-resistant ductile end diaphragms with buckling restrained braces in straight steel bridges. Eng Struct 31(2):380–393
5. Mirtaheri M et al (2011) Experimental optimization studies on steel core lengths in buckling restrained braces. J Construct Steel Res 67(8):1244–1253
6. Khampanit A, Leelataviwat S, Kochanin J, Warnitchai P (2014) Energy-based seismic strengthening design of non-ductile reinforced concrete frames using buckling-restrained braces. Eng Struct 81:110–122
7. Dusicka P, Tinker J (2013) Global restraint in ultra-lightweight buckling-restrained braces. J Compos Constr 17(1):139–150
8. Jiang Z, Guo Y, Zhang B, Zhang X (2015) Influence of design parameters of buckling-restrained brace on its performance. J Constr Steel Res 105:139–150
9. Ahmad Fayeq G, Ranjan Sahoo D (2019) Effect of loading history and restraining parameters on cyclic response of steel BRBs. Int J Steel Struct 19(4):1055–1069
10. Shete P, Madhekar S, Ghowsi F (2022) Numerical analysis of steel buckling restrained braces with varying length, gap, and stoppers. Pract Period Struct Des Construct 27(1):04021051. Doi: https://doi.org/10.1061/(ASCE)SC.1943-5576.0000629
11. Yun Z, Gong C, Zhao J, Zhong G, Tian S (2020) Strong-axis stability and seismic performance of perforated core plate buckling-restrained braces. Thin-Walled Structures 156:106997
12. Zhou Y, Qian H, Chu H (2009) A study on the design principle and performance of a new type of buckling-resistant brace. China Civil Engg J 42(4): 64-71
13. Wu A, Lin P, Tsai K (2014) High-mode buckling responses of buckling-restrained brace core plates. Earthquake Eng Struct Dyn 43(3):375-393
14. Korzekwa A, Tremblay R (2009) Numerical simulation of the cyclic inelastic behaviour of buckling restrained braces. In: International specialty conference on behaviour of steel structures in seismic area (STESSA), 2009
15. Hoveidae N, Tremblay R, Rafezy B, Davaran A (2015) Numerical investigation of seismic behavior of short-core all-steel buckling restrained braces. J Construct Steel Res 89-99

16. Chaboche JL, Dang Van K, Cordier G (1979) Modelization of the strain memory effect on the cyclic hardening of 316 stainless steel
17. Zhao J, Wu B, Ou J (2011) A novel type of angle steel buckling-restrained brace: cyclic behavior and failure mechanism. Earthquake Eng Struct Dyn 40(10):1083–1102
18. Prestandard FEMA (2000) Commentary for the seismic rehabilitation of buildings (FEMA356). Washington, DC: Federal Emergency Management Agency 7 (2000), 2

A Method for Evaluating Maximum Response in Multi-storied Buildings Due to Bi-directional Ground Motion

P. B. Kote, S. N. Madhekar, and I. D. Gupta

Abstract In the present paper, a method is proposed to evaluate the maximum response of multi-story buildings under the effect of simultaneous action of bi-directional components of earthquake ground motion using the concept of critical response spectrum. The critical response spectrum is computed using the resultant response of a bidirectional single degree of freedom system at each time step under the simultaneous action of the two horizontal components of ground motion. For an illustration of the proposed method, steel building asymmetric in the plan is analyzed using critical response spectra of the three different pairs of recorded earthquake ground motion, and the results obtained are validated by comparison with the exact time-history solution. The exact response is taken as the maximum of the responses estimated by applying the horizontal components of time-histories of ground acceleration at different angles from 0° to 180° at an interval of 10°. On the other hand, in the response spectrum method, two values of the desired response quantity are obtained by applying the critical response spectrum along the structural x and y directions and the multi-component seismic response is combined using the SRSS method. It has been found that the use of the critical spectrum provides a very convenient method for estimating the maximum response under bi-directional earthquake excitation without the need for computation of the critical incident angle of incidence.

Keywords Multi-component seismic excitation · Time-history · Critical earthquake response · Critical response spectrum

---

P. B. Kote (✉) · S. N. Madhekar
Department of Civil Engineering, College of Engineering Pune, Pune 411005, India
e-mail: kpb18.civil@coep.ac.in

I. D. Gupta
Central Water and Power Research Station (CWPRS), Pune 411024, India

M. S. Ranadive et al. (eds.), *Recent Trends in Construction Technology and Management*, Lecture Notes in Civil Engineering 260,
https://doi.org/10.1007/978-981-19-2145-2_72

1 Introduction

In seismic design practice, the translational components in the 3 orthogonal directions are considered and rotational components are usually ignored. The critical orientation of the earthquake components with the structural axes, as well as the multi-component response combination methods for combining individual effects, has been studied in the past by several researchers. In usual seismic design, structures are analyzed and designed for horizontal translational components of acceleration. However, for the seismic design of the bridges, the vertical component of ground motion is also considered. Penzien and Watabe [6] stated that recorded components of ground motion may not be necessarily uncorrelated, therefore based on the concept mechanics they proposed a procedure to compute three mutually perpendicular directions, resolving the recorded components in these principal directions results in the uncorrelated earthquake ground motion components. The axes to which these resolved components are uncorrelated are later termed principal axes. These idealized major and intermediate principal axes are horizontal. The major principal axis approximately gives the direction of the epicenter of the earthquake, the horizontal intermediate principal axis is perpendicular to the major principal component of ground motion, and the third minor principal axis is vertical. In general, when the principal components of ground motion are applied in the structural axes of the buildings the maximum seismic response of that building can be obtained.

In usual design practice, the seismic response of the structure is performed using the response spectrum method. In the response spectrum method, the spectra are applied along the structural axis of the buildings and multi-component seismic response is computed using different combination rules, the details of these combination rules are discussed in Sect. 2.2. Various researchers proposed rules for combining multicomponent seismic response, Newmark [5] was the first researcher to propose a simplified percentage rule, later Rosenblueth and Contreras [8] proposed a similar rule to estimate the approximate seismic response of the structure due to the application of the multi-component seismic ground motion. They assumed a linear combination of 100% of the response due to one seismic component and some percentage (β) of the responses resulting due to application of the other two components of ground motion. Newmark suggested $\beta = 40\%$ of the actual response and Rosenblueth and Contreras suggested $\beta = 30\%$ of the actual response. Smeby and Der Kiureghian [11] assumed the Penzien and Watabe idealized model of uncorrelated ground motions, and they proposed the CQC3 rule (the extension of the CQC) modal combination rule [1] for the combination of the multi-component seismic response. This rule was derived using elementary concepts of stationary random vibrations and accounts for the correlation between the components of ground motion and the structural response.

Many studies have investigated to relative performances of these rules, but to the best of the author's knowledge, none of them have compared the maximum response with the exact time-history solution. Wilson et al. [12] stated that the percentage combination rules in most of the cases underestimate the design forces in the structural members when compared with the SRSS rule of combining multi-component seismic response. Menun and Der Kiureghian [4] carried out a comparative study for the CQC3 response results with SRSS, the 30% ($\beta = 0.3$), and the 40% ($\beta = 0.4$) rules, assuming CQC3 results to provide a benchmark. The studies by Fernández-Dávila et al. [2] and Zaghlool et al. [13] have pointed out that percent rules, and the SRSS rules are inappropriate and do not take into account the orientation effects of horizontal orthogonal components of ground motion. Salazar et al. [10] showed that the percentage rule i.e. 30%, and the SRSS rule of combining multi-component responses underestimate when compared with the CQC3 rule. Reyes-Salazar et al. [7] studied the response of moment-resisting steel frames and observed that for the bi-directional seismic ground motion if combined with the SRSS rule, in most cases this method produces a lower estimate of response than the Max + 30%. They showed that maximum seismic response could be obtained for different orientations than that of the principal components of ground motion.

Sadhu and Gupta [9] formulated a new modal combination rule using the ordered peak seismic response of multi-storied buildings under the effects of the multi-component seismic ground motions. They have shown that their method performs better than the CQC3 rule for stiffer buildings. However, the structure considered for the analysis was not appropriate. The authors have also pointed out the method suggested by them underestimates the response. Secondly, this rule requires the response spectrums in principal directions, which may not be always available.

In the present study, an investigation of the various spatial combination rules for the computation of seismic response is performed, and a new method is proposed to estimate the maximum seismic response for multi-storied buildings under the simultaneous action of bi-directional components of seismic ground motion using the proposed concept of critical response spectrum. The critical response spectrum is computed using the resultant response of a single degree of freedom at each time step under the simultaneous action of the two horizontal components of motion. For illustration, a steel moment-resisting building, asymmetric in the plan, is analyzed using the proposed critical response spectrum of the three different pairs of recorded ground motion, and the results obtained are validated by comparison with the exact time-history solutions.

The exact response is taken as the maximum of the responses estimated by applying the two horizontal time histories of ground acceleration at different angles from 0° to 180° with respect to the structural axis. On the other hand, in the response spectrum method, two values of the desired response quantity are obtained by applying the critical response spectrum along the *x*- and *y*-directions of the structure and are combined using the SRSS method. It will be shown that the proposed critical spectrum provides a very good estimate of maximum response without the need for the computation of the critical incident angle. The proposed method evaluates the maximum response of multi-storied buildings subjected to bi-directional ground motions using the concept of the critical response spectrum, rather than the traditional principal components. The proposed method seismic response results are compared with the existing combination rules such as CQC3 rule, MAX + 30%, MAX +

40%, and SUM method of combining response for multi-component seismic ground motion.

2 Theoretical Background

2.1 Time History Method

The equation of motion for a linear, multi-degree of freedom system (MDF) due to the three input recorded components of ground motion is given by

$$M\ddot{U} + C\dot{U} + KU = -MI\ddot{X} \tag{1}$$

where M, C, and K are the lumped story mass matrix, modal damping matrix, and stiffness matrix respectively, U is the nodal relative displacements vector, $X = [X_1 X_2 X_3]^{\mathrm{T}}$ is components of ground motion in three translational directions, includes components in the horizontal $X_1(t)$ and $X_2(t)$ vertical $X_3(t)$, and $I = [I_1 I_2 I_3]^{\mathrm{T}}$ is the mass influence matrix. Usually, the response quantity of interest, $R(t)$, the nodal displacements are written in the generic form such that $R(t) = q^{\mathrm{T}}U$, where q constants vector. To obtain the response of internal forces in a member q represents the stiffness matrix of that member. Transformation to the normal coordinate system such that $U = \Phi Y$, and the ith uncoupled modal equation is given by

$$\ddot{Y}_i + 2\zeta_i\omega_i\dot{Y}_i + \omega_i^2 Y_i = -\Gamma_i^T\ddot{X} \tag{2}$$

where $Y = [Y_1 Y_2 \ldots Y_n]^{\mathrm{T}}$ is the vector of normal coordinates, $\Phi = [\Phi_1\Phi_2 \ldots \Phi_n]$ is the eigenvector matrix. ω_i and ζ_i are the structural natural frequency and corresponding modal damping ratio of ith mode, respectively, and $k = 1, 2, 3$ are the three input ground motion components. Time-history solution for multi-component excitation is as per Eq. (3)

$$U(t) = \sum_{k=x,y}\sum_{l=x,y}\sum_{i}\sum_{j}\psi_{ki}\psi_{lj}Y_i(t)Y_j(t) \tag{3}$$

where $Y_i(t)$ is the required modal solution of Eq. (2) for ith mode, which is like the equation of motion for a single degree of freedom oscillator.

The time history solution procedure given above is applicable when the earthquake ground motion is applied along the structural axis of the buildings, however, to find the maximum response the ground motion is required to be applied in the all-possible orientations. This process involves a large number of computations thus practically for design purposes it is not adopted.

2.2 Response Spectrum Method

For the seismic analysis and design of the buildings for a particular location, the actual site-specific earthquake time history is required, however, the availability of such records for every location may not be possible. Further, the seismic analysis of the structures depends on the amplitude of ground acceleration as well as the frequency content and the structure's dynamic properties. Therefore, in most cases, the response spectrum method is used for seismic analysis. This method provides a computational advantage as the prediction of the response involves the calculation of the maximum values of the response quantities in each mode of vibration. These maximum values of the response for the element forces in each mode are obtained from the smooth design spectrum which is based on the average of several earthquakes. The maximum modal response is obtained from Eq. (4).

$$Y_{i,\max} = |Y_i(t)|_{\max} = \Gamma_i S_d(\zeta_i, \omega_i) \quad (4)$$

where Γ_i is the modal participation factor for ith mode and $S_d(\zeta_i, \omega_i)$ is the ith mode nodal displacement. The final maximum displacement response $Y_{\max}$ is evaluated using the modal combination rules. The commonly used method for combining the modal maximum final response is given by Complete Quadratic Combination (CQC) method [1] given by

$$Y_{\max} = \sqrt{\sum_{i=1}^{n}\sum_{j=1}^{n} \rho_{ij} Y_i Y_j} \quad (5)$$

here Y_i and Y_j are the ith and jth modal responses respectively and ρ_{ij} is the modal correlation coefficient [1] given by Eq. (6)

$$\rho_{ij} = \frac{8(\zeta_i\zeta_j)^{1/2}(\zeta_i + \beta\zeta_j)\beta^{3/2}}{(1-\beta^2)^2 + 4\zeta_i\zeta_j\beta(1+\beta^2) + 4\left(\zeta_i^2 + \zeta_j^2\right)\beta^2} \quad (6)$$

The response due to multi-component seismic ground motion is determined using simplified combination methods such as the percentage rules (e.g. Max + 30%, Max + 40%), square root of the sum of squares (SRSS), and Complete Quadratic Combination (CQC3) rules the details of these methods are given below.

2.2.1 CQC-3 Rule

This method is proposed by Smeby and Der Kiureghian [11] and later by Menun and Der Kiureghian [4] for finding the critical response under the simultaneous action of ground motion. This method requires the ground motion in the principal directions.

The incident angle-dependent response is given by Eq. (7)

$$R_\theta = \left[\sum_{k=1,2,3} \sum_{i=1}^{n} \sum_{j=1}^{n} \rho_{ij} \psi_{ki} \psi_{kj} S_{ki} S_{kj} \right.$$
$$- \sum_{i=1}^{n} \sum_{j=1}^{n} \rho_{ij} [\psi_{1i}\psi_{1j} - \psi_{2i}\psi_{2j}][S_{1i}S_{1j} - S_{2i}S_{2j}] \sin^2\theta$$
$$\left. + 2 \sum_{i=1}^{n} \sum_{j=1}^{n} \rho_{ij} \psi_{1i}\psi_{2j} [S_{1i}S_{1j} - S_{2i}S_{2j}] \sin\theta \cos\theta \right]^{1/2} \tag{7}$$

where, ψ_{ki} is the effective participation factor in the kth structural axis, ρ_{ij} is the modal correlation coefficient [1], S_{1i} and S_{2i} are the horizontal principal components [6]. The θ in the equation represents the angle between the principal components w.r.t the structural axis. The critical angle for the response can be obtained by differentiating Eq. (7) w.r.t. θ. The critical angle is given by Eq. (8)

$$\theta_{\text{cr}} = \frac{1}{2} \tan^{-1} \left(\frac{2 \sum_{i=1}^{n} \sum_{j=1}^{n} \rho_{ij} \psi_{1i} \psi_{2j} [S_{1i}S_{1j} - S_{2i}S_{2j}]}{\sum_{i=1}^{n} \sum_{j=1}^{n} \rho_{ij} [\psi_{1i}\psi_{1j} - \psi_{2i}\psi_{2j}][S_{1i}S_{1j} - S_{2i}S_{2j}]} \right) \tag{8}$$

Equation (8) has two roots between 0° and 180° corresponding to those values of θ gives the maximum and minimum multi-component seismic response.

2.2.2 SRSS Rule

This method of combining response due to bi-directional seismic components of ground motion is given by Eq. (9)

$$R_{\text{crit}} = \left[\{R^x\}^2 + \{R^y\}^2 \right]^{1/2} \tag{9}$$

where R^x and R^y is the response due to application of the critical response spectrum in xth structural and yth structural axis respectively. The response quantities R^x and R^y are computed as given in Eq. (10)

$$\{R^x\} = \left\{ \sum_{i=1}^{n} \sum_{j=1}^{n} \rho_{ij} \psi_{xi} \psi_{xj} S_{xi} S_{xj} \right\}^{\frac{1}{2}}$$
$$\{R^y\} = \left\{ \sum_{i=1}^{n} \sum_{j=1}^{n} \rho_{ij} \psi_{yi} \psi_{yj} S_{yi} S_{yj} \right\}^{\frac{1}{2}} \tag{10}$$

where ψ_{xi}, ψ_{yi} are the effective modal participation factor in the xth and yth structural axis, ρ_{ij} is the modal correlation coefficient [1], S_{x_i} and S_{y_i} are the ith modal response spectral acceleration values of the xth and yth components.

2.2.3 Max + 30% Rule

As per the discussion in the introduction section [8] suggested the simplified percentage rule to estimate the approximate multi-component seismic response. This method is further recommended by ATC-3(1978), IS-1893:2016, and many other seismic codes. The response is given by Eq. (11)

$$R = \max\{R^x + 0.3R^y \text{ or } 0.3R^x + R^y\} \tag{11}$$

2.2.4 Max + 40% Rule

Newmark [5] proposed this method with a higher percentage of contribution from the other two components of ground motion. This method produces a higher estimate of response compared to the earlier rule. This method is recommended by ASCE (1986), ATC-32, for seismic analysis of nuclear structures. The response is given by Eq. (12)

$$R = \max\{R^x + 0.4R^y \text{ or } 0.4R^x + R^y\} \tag{12}$$

2.2.5 SUM Method

This method assumes the contribution of the response due to multi-component seismic ground motion occurs simultaneously hence, the response is combined as absolute sum of the response contribution as given in Eq. (13)

$$R = R^x + R^y \tag{13}$$

This method usually yields conservative results hence this method is not recommended by various design codes.

3 Response Spectrum Used for Analysis

The response spectra are the smooth curves of the absolute maximum response of an SDOF system (for a fixed damping ζ and natural frequency ω) due to earthquake ground motion to its natural time period (or frequency). The various types of response spectra considered for the present analysis for the structures are given below.

3.1 Response Spectra of Recorded Components [SAX, SAY]

The response spectra of the originally recorded components of the seismic ground motion for a given damping ratio (ζ) are considered as SAX and SAY.

3.2 Response Spectra of Principal Components [SAXP, SAYP]

The SAXP and SAYP are obtained by resolving the recorded earthquake components in the two horizontal major and intermediate principal direction θ_1, θ_2 respectively. The angles of the major and minor principal directions θ_1 and θ_2 are computed from [6] idealization.

3.3 Critical Response Spectra [SACRIT]

The proposed Critical Response Spectrum is obtained by the resultant response $R_{(t)}$ of an SDOF system obtained from SRSS of the response in the *x*-and *y*-directions are $R_{x(t)}$ and $R_{y(t)}$ respectively for each natural period (T) and damping ratio (ξ). The resultant response of the SDOF system is computed by Eq. (14)

$$SA_{\text{crit}(T,\xi)} = R_{(t)\max} \tag{14}$$

where

$$R_{(t)} = \sqrt{R_{x(t)}^2 + R_{y(t)}^2} \tag{15}$$

To perform a numerical study, three real ground motion time histories are considered. The details of the selected earthquake records are shown in Table 1. Time history components records and the corresponding response spectra of the three selected real earthquake records considered for analysis are shown in Figs. 1 and 2.

Table 1 Details of the selected earthquakes

REC#	Name of EQ	Date	M	Focal depth [km]	Epicenter distance [km]	Comp	PGA (g)	Recording site
1	Gorkha Nepal	25/04/2015	7.9	13.40	13.60	NORT	0.157	Kirtipur
						EAST	0.253	
2	Sikkim	18/09/2011	6.7	10.00	50.30	EAST	−0.145	Gangtok
						NORT	−0.158	
3	Uttarkashi	20/10/1991	6.9	13.20	73.90	N85E	0.253	Bhatwari
						N05W	0.246	

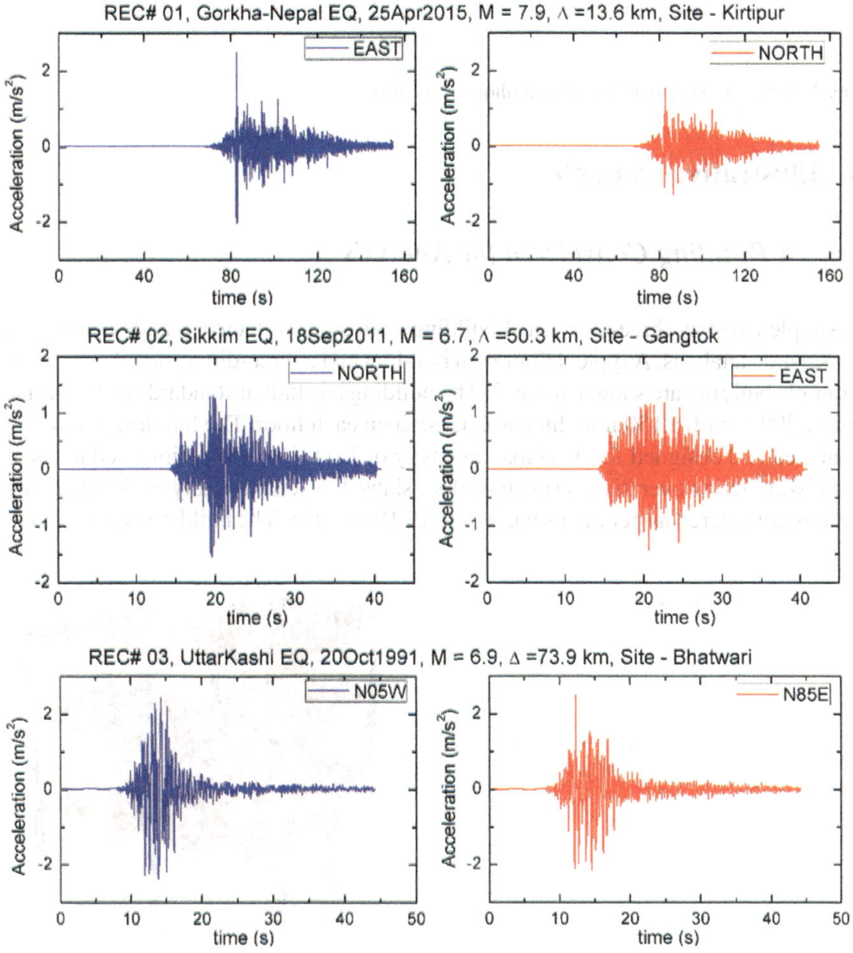

Fig. 1 Selected recorded earthquake time history

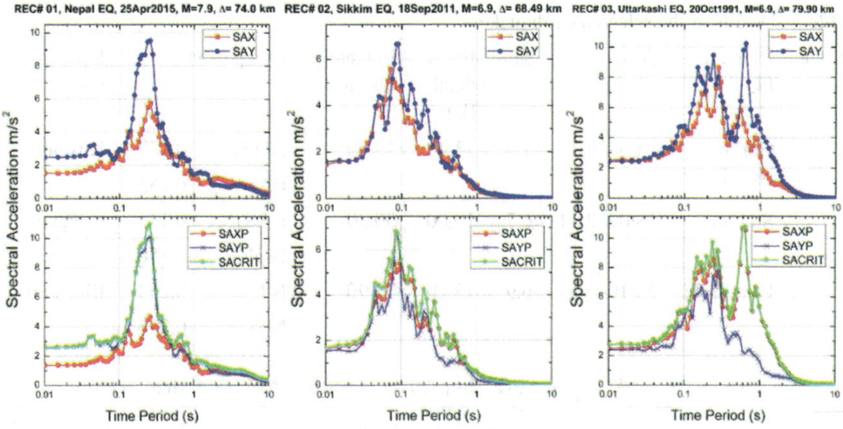

Fig. 2 Response spectra of selected earthquake records

4 Illustrative Example

4.1 A Building Considered for Analysis

A simple two-bay three-story steel building with an asymmetric plan is considered for seismic analysis. A typical floor plan building and a three-dimensional view of an example building are shown in Fig. 3. The building has Indian standard steel columns ISHB 200-1 with a 150 mm thick R.C.C. slab on each floor. The building is initially analyzed and designed for dead load intensity of 3.0 kN/m^2, Superimposed loads of 3.5 kN/m^2 acting over the intermediate floor slabs. For seismic analysis, 5% damping ratio is considered as per the provisions in IS 1893:2016. The building is assumed to

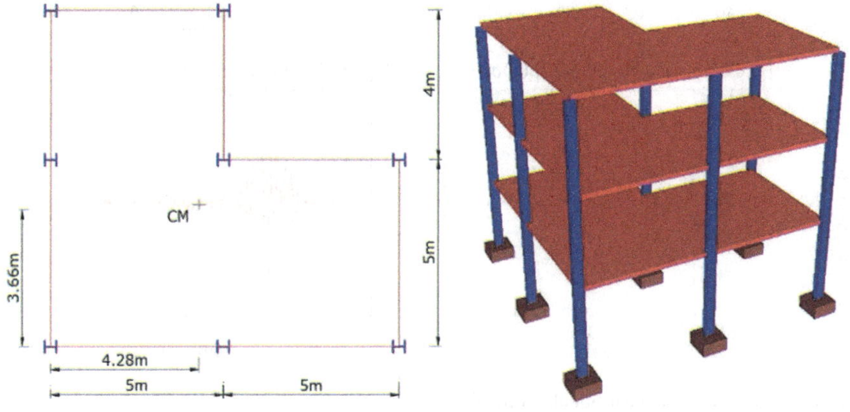

Fig. 3 Typical three-story steel building

Table 2 Natural modes and frequencies (ω_n) of example building

Mode No.	Time period [s]	Natural frequency (ω_n) [rad/s]	Cumulative mass participation (%)
1	1.34	4.70	0.87
2	0.69	9.08	80.13
3	0.63	9.91	91.87
4	0.49	12.95	91.87
5	0.37	18.22	91.87
6	0.25	25.02	98.13
7	0.23	27.30	99.04
8	0.18	35.20	99.88
9	0.16	38.41	100.00

have rigid diaphragms, mathematically modeled as 3 degrees-of-freedom (δ_x, δ_y, θ) per floor considering lumped mass at story levels. The computed Mass and Stiffness matrix for the example building is given in Eqs. (16) and (17). The modal analysis results for time period (s) for the example building for the 9 modes of vibration are given in Table 2.

$$M = 10^3 \begin{bmatrix} 60.65 & 0.00 & 0.00 & 0.00 & 0.00 & 0.00 & 0.00 & 0.00 & 0.00 \\ 0.00 & 60.65 & 0.00 & 0.00 & 0.00 & 0.00 & 0.00 & 0.00 & 0.00 \\ 0.00 & 0.00 & 48.17 & 0.00 & 0.00 & 0.00 & 0.00 & 0.00 & 0.00 \\ 0.00 & 0.00 & 0.00 & 60.65 & 0.00 & 0.00 & 0.00 & 0.00 & 0.00 \\ 0.00 & 0.00 & 0.00 & 0.00 & 60.65 & 0.00 & 0.00 & 0.00 & 0.00 \\ 0.00 & 0.00 & 0.00 & 0.00 & 0.00 & 48.17 & 0.00 & 0.00 & 0.00 \\ 0.00 & 0.00 & 0.00 & 0.00 & 0.00 & 0.00 & 838.51 & 0.00 & 0.00 \\ 0.00 & 0.00 & 0.00 & 0.00 & 0.00 & 0.00 & 0.00 & 838.51 & 0.00 \\ 0.00 & 0.00 & 0.00 & 0.00 & 0.00 & 0.00 & 0.00 & 0.00 & 665.88 \end{bmatrix} \quad (16)$$

$$K = 10^6 \begin{bmatrix} 51.32 & -25.66 & 0.00 & 0.00 & 0.00 & 0.00 & -17.41 & 8.71 & 0.00 \\ -25.66 & 51.32 & -25.66 & 0.00 & 0.00 & 0.00 & 8.71 & -17.41 & 8.71 \\ 0.00 & -25.66 & 25.66 & 0.00 & 0.00 & 0.00 & 0.00 & 8.71 & -8.71 \\ 0.00 & 0.00 & 0.00 & 13.75 & -6.88 & 0.00 & 1.23 & -0.61 & 0.00 \\ 0.00 & 0.00 & 0.00 & -6.88 & 13.75 & -6.88 & -0.61 & 1.23 & -0.61 \\ 0.00 & 0.00 & 0.00 & 0.00 & -6.88 & 6.88 & 0.00 & -0.61 & 0.61 \\ -17.41 & 8.71 & 0.00 & 1.23 & -0.61 & 0.00 & 862.71 & -431.35 & 0.00 \\ 8.71 & -17.41 & 8.71 & -0.61 & 1.23 & -0.61 & -431.35 & 862.71 & -431.35 \\ 0.00 & 8.71 & -8.71 & 0.00 & -0.61 & 0.61 & 0.00 & -431.35 & 431.35 \end{bmatrix} \quad (17)$$

4.2 Numerical Analysis and Results

A detailed numerical analysis is carried out for the example building as shown in Fig. 3. The dynamic analysis is performed for seven different analysis cases. Details of these analysis cases are given in Table 3. In the present study, time history analysis (Case-1) is considered as the benchmark method. The response obtained by the response spectrum superposition method for Case-2 to Case-7 are compared with Case-1 time history results. In the proposed critical spectrum method, the modal response due to simultaneous action of multi-component seismic excitation is obtained based on the critical spectra as defined in Sect. 3, and the SRSS rule is used to combine the multi-component modal response. The story response quantities such as story displacements (for diaphragm), base shear force, and bending moment (for all elements in a story) are tabulated in Table 4. The normalized displacement, shear force, and bending moments plots for all seven cases are shown in Fig. 4.

To illustrate the efficacy of the proposed method over the existing methods, plots with percentage error is shown considering the time history method (Case-1) as the benchmark method as shown in Fig. 5. As shown in Fig. 5, it is evident that the simplified percentage rules viz Max + 30%, Max + 40% rules, and Sum method underestimate the response. The main reason for lower estimates of the response is due to the non-consideration of the incident angle of the earthquake in the simplified percentage rule methods. The CQC3 rule underestimates the response when compared with the time history method although the critical angle of earthquake incidence is considered while evaluating the response. However, it is seen that the proposed critical spectrum method (Case-2) provides a very good estimate of the desired response without

Table 3 Cases considered for dynamic analysis

Case#	Case details
1	Time history analysis using time history of recorded components of ground motion for the angle of incidence from 0° to 180° with an interval of 10°
2	Response spectrum analysis using SACRIT Spectra [3] and combining response using SRSS rule (Present Study)
3	Response spectrum analysis using SAXP and SAYP spectra [6] and combining response using CQC3 Rule for the critical response [4, 11]
4	Response spectrum analysis using SAX, SAY spectra (considering real components as principal components) and combining response using CQC3 Rule for the critical response
5	Response spectrum analysis using SAX, SAY spectra and combining response using [Max + 30%] rule [8]
6	Response spectrum analysis using SAX, SAY spectra, and combining response using [Max + 40%] rule [5]
7	Response spectrum analysis using SAX, SAY spectra and combining response using SUM rule

Table 4 Story response results

REC#	Case #	Story displacement (mm)			Base shear force (kN)			Bending moment (kNm)		
		Story #			Story #			Story #		
		1	2	3	1	2	3	1	2	3
1	1	19.03	35.23	44.19	483.01	410.76	245.84	3418.83	1969.80	737.52
	2	19.43	33.41	40.72	487.55	376.08	230.21	3281.52	1818.87	690.63
	3	15.60	27.47	33.39	392.03	303.74	190.44	2658.63	1482.54	571.32
	4	15.14	25.78	31.46	379.59	292.59	188.52	2582.10	1443.33	565.56
	5	15.15	25.81	31.49	379.93	292.85	188.66	2584.32	1444.53	565.98
	6	15.16	25.81	31.50	380.04	292.94	188.71	2585.07	1444.95	566.13
	7	15.19	25.86	31.56	380.73	293.47	189.00	2589.60	1447.41	567.00
2	1	6.28	11.46	14.41	159.21	136.22	79.10	1123.59	645.96	237.30
	2	6.77	11.80	14.35	171.22	132.67	74.68	1135.71	622.05	224.04
	3	6.32	11.04	13.43	158.98	123.06	68.67	1052.13	575.19	206.01
	4	5.71	9.97	12.13	145.10	112.35	65.94	970.17	534.87	197.82
	5	5.72	9.98	12.14	145.24	112.46	66.00	971.10	535.38	198.00
	6	5.72	9.99	12.14	145.29	112.50	66.02	971.43	535.56	198.06
	7	5.73	10.01	12.17	145.57	112.72	66.15	973.32	536.61	198.45
3	1	59.14	105.20	128.13	1482.32	1153.08	573.77	9627.51	5180.55	1721.31
	2	59.54	105.49	128.14	1495.75	1160.96	583.03	9719.22	5231.97	1749.09
	3	58.23	103.22	125.39	1463.20	1135.92	568.23	9502.05	5112.45	1704.69
	4	56.52	100.13	121.63	1418.12	1100.65	553.42	9216.57	4962.21	1660.26
	5	56.55	100.17	121.69	1418.76	1101.14	553.67	9220.71	4964.43	1661.01
	6	56.56	100.19	121.71	1418.97	1101.30	553.76	9222.09	4965.18	1661.28
	7	56.61	100.27	121.81	1420.24	1102.29	554.26	9230.37	4969.65	1662.78

the need for consideration of any earthquake incident angle. The response quantities obtained by the proposed method are more consistent when compared to other methods of combining response due to multi-component seismic ground motion. The proposed method can be very easily implemented using commercial earthquake analysis programs.

5 Conclusion

The existing simplified percentage rule methods for combining the response of buildings under the simultaneous action of bi-directional seismic ground motion underestimate the desired response for all considered earthquake records. The main reason for underestimating response in percentage rule methods is due to the non-account

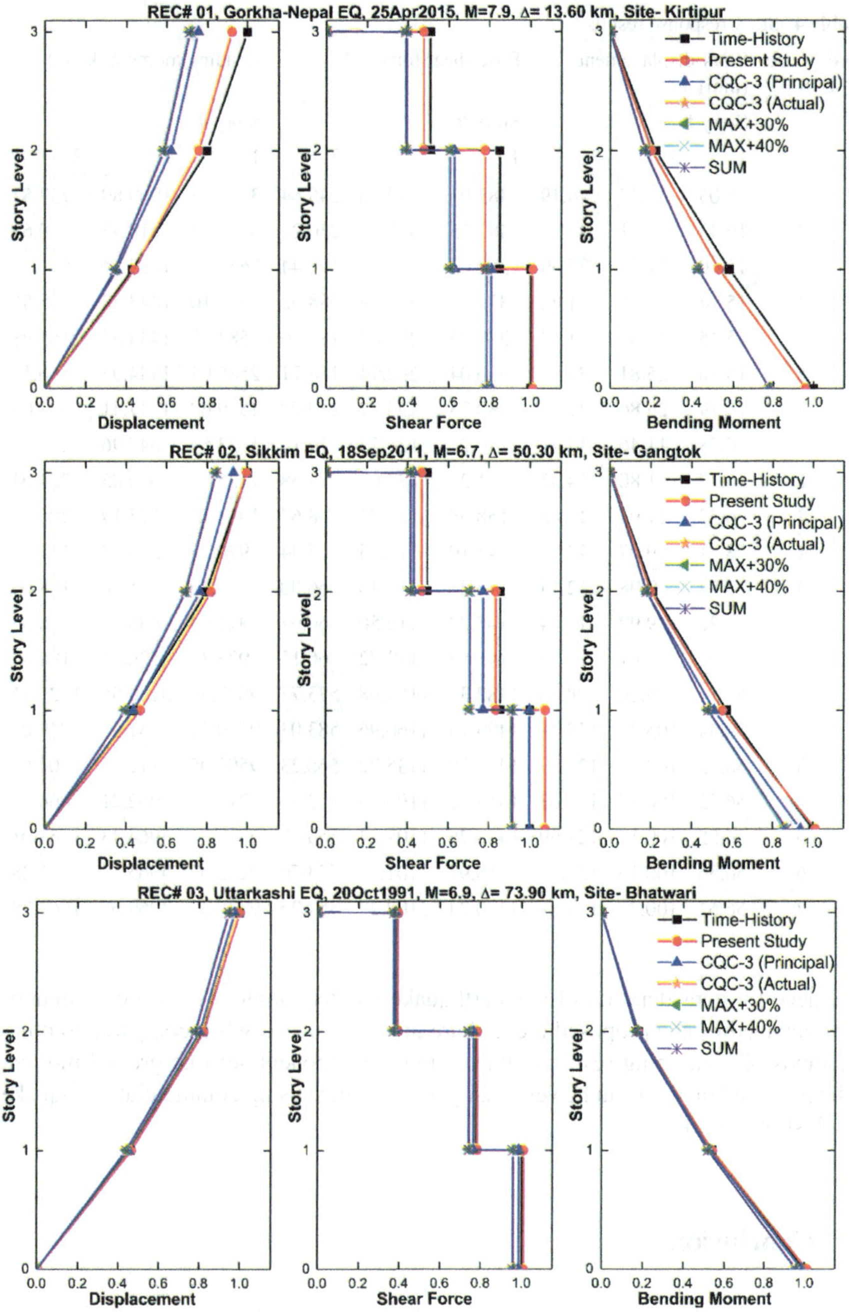

Fig. 4 Normalised response plot

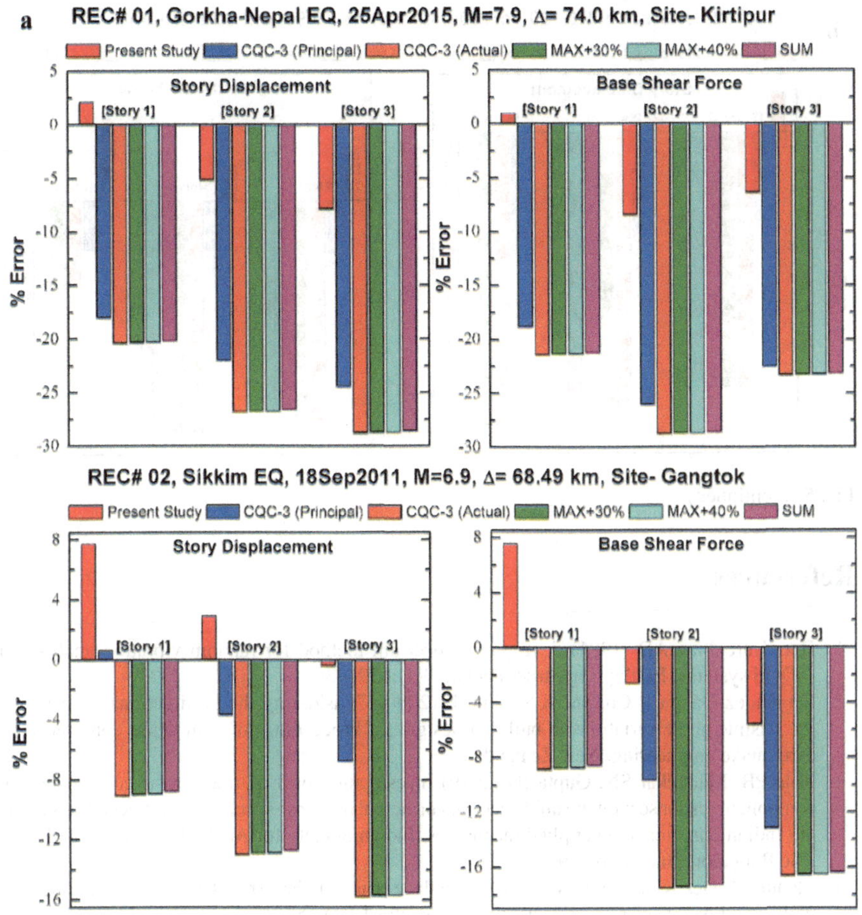

Fig. 5 % Error in story response

for the incident angle of the earthquake. The CQC-3 method considers the critical angle for a response, but in most cases, the response obtained by this method underestimates the response when compared with the exact time history solution.

From the illustrated analysis it is found that the proposed method results are consistent with the time history results and provide a very convenient approach to estimate the maximum seismic response of the buildings due to the simultaneous action of multi-component seismic ground motion without the need for computation of a critical earthquake angle. Therefore, the proposed method is computationally advantageous over the CQC-3, and thus the implementation of the proposed method for seismic analysis is easily possible using already available computer programs.

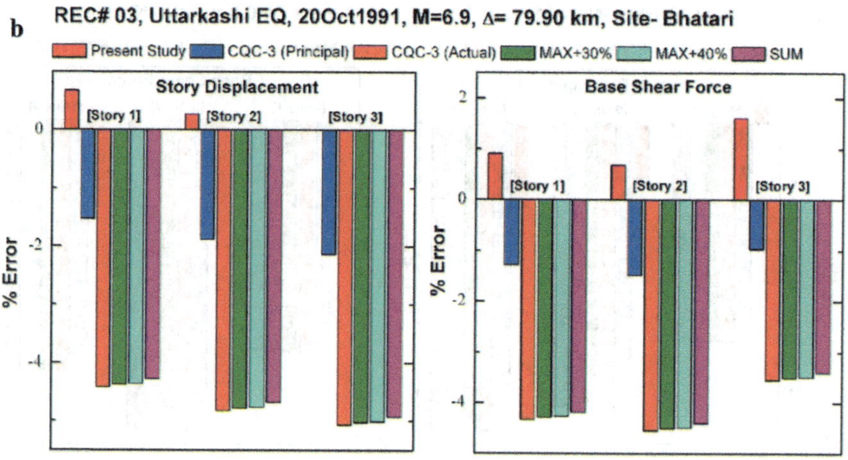

Fig. 5 (continued)

References

1. Der Kiureghian AD (1981) A response spectrum method for random vibration analysis of MDOF systems. Earthq Eng Struct Dynam 9(1):419–436
2. Fernández-Dávila I, Cominetti S, Cruz E (2000) Considering the bi-directional effects and the seismic angle variations in building design. In: Proceedings of 12th world conference on earthquake engineering, New Zealand
3. Kote PB, Madhekar SN, Gupta ID (2019) Investigation of 30% rule for the combination of component response with the unidirectional principal response spectrum. In: Proceedings of the 4th Indian conference on applied mechanics. Indian Society for Applied Mechanics (ISAM), IISc Bangalore, India, pp 209–212
4. Menun C, Der Kiureghian A (1998) A replacement for the 30%, 40% and SRSS rules for multi-component seismic analysis. Earthq Spectra 14(1):153–156
5. Newmark NM (1975) Seismic design criteria for structures and facilities, Trans-Alaska pipeline system. In: Proceedings of the U.S. national conference on earthquake engineering. Earthquake Engineering Institute, California, pp 94–103
6. Penzien J, Watabe M (1975) Characteristics of three-dimensional earthquake ground motion. Earthq Eng Struct Dynam 3:365–373
7. Reyes-Salazar A, Haldar A, Rivera-Salas JL, Bojórquez E (2015) Review of assumptions in simplified multi-component and codified seismic response evaluation procedures. KSCE J Civ Eng 19(5):1320–1335. https://doi.org/10.1007/s12205-015-0190-x
8. Rosenblueth E, Contreras H (1977) Approximate design for multi-component. J Eng Mech Div ASCE 881–893
9. Sadhu A, Gupta VK (2008) A modal combination rule for ordered peak response under multi-component ground motion. ISET J Earthq Technol 45(3–4):79–96
10. Salazar AR, Barraza AL, Lopez AL, Haldar A (2008) Multi-component seismic response analysis—a critical review. J Earthq Eng 12(1):779–799

11. Smeby W, Der Kiureghian AD (1985) Modal combination rules for multi-component earthquake excitation. Earthq Eng Struct Dyn 4(1):1–12
12. Wilson EL, Suharwardy I, Habibullah A (1995) A clarification of the orthogonal effects in a three-dimensional seismic analysis. Earthq Spectra 11(4):659–666
13. Zaghlool BS, Carr AJ, Moss PJ (2000) Inelastic behavior of three-dimensional structures under concurrent seismic excitations. In: Proceedings of 12th world conference on earthquake engineering, New Zealand

Finite Element Analysis of Piled Raft Foundation Using Plaxis 3D

Anupam Verma and Sunil K. Ahirwar

Abstract A piled raft foundation is a type of deep foundation capable for transferring the heavy load of superstructure into the soft soil by pile and raft support system. A numerical analysis of piled raft foundation is presented in this study. Various types of interactivity among components like pile-to-pile interactivity, raft-to-pile interactivity, pile-to-soil interactivity, and raft-to-soil interactivity are also being examined using a three-dimensional finite element software Plaxis 3D (Netherlands user manuals, [1]). The parameters taken in this study are raft thickness, pile spacing, pile cross-sectional shape, and pile length. These parameters were varied and compared with other available studies. The results obtained in the present studies are in good agreement with other research studies.

Keywords Piled raft · Soft soil · Foundation · Plaxis 3D

1 Introduction

The exponential increase in the number of high-rise structures of 150–300 m and more have presented a challenging situation among structural and geotechnical engineers in designing the foundation systems of such structures. Thus the mere application of conventional foundation design methods is insufficient for such structures; therefore, engineers are forced to follow more innovative and skillful designs. Instead of using piles and rafts alone, the concept of combination of the elements of foundations such as piles and raft can be applied to support a structure in which the role of piles is to reduce raft sinking and distinctive settlements and can also contribute to significant prudence without taking a trade-off between the safety and execution of the foundation. Such foundations can be called "piled enhanced raft" or "piled-raft foundations". A piled raft is a compounded geotechnical formation comprising foundation elements like piles, soil, and raft. It can be distinguished from the usual

A. Verma · S. K. Ahirwar (✉)
Department of Civil Engineering and Applied Mechanics, Shri G. S. Institute of Technology and Science, Indore, Madhya Pradesh, India
e-mail: skasgsits@gmail.com; sahirwar@sgsits.ac.in

M. S. Ranadive et al. (eds.), *Recent Trends in Construction Technology and Management*, Lecture Notes in Civil Engineering 260,
https://doi.org/10.1007/978-981-19-2145-2_73

design of the foundation, where either the raft or piles transport the loads. The first to introduce the concept and design approach for piles under a raft foundation were Burland et al. [2] and called the piles "settlement reducing piles". The Combined Piled Raft Foundation (CPRF) has been successfully implemented in various parts of the world over the last four decades to optimize foundations for structures in civil engineering. During 1994–97, one of the International Society of Soil Mechanics and Foundation Engineering's (ISSMFE) technical committees based their efforts on piled raft foundations and provided detailed reports on collective knowledge on various design methods and case history.

First, the piled raft foundations were considered an alternative for high-rise building foundations on cohesive active settlement soils such as the Frankfurt clay, but as a result of extensive researches on its performance, pile-raft now has been preferred as a foundation for other soils too.

Clancy and Randolph [3] studied the spring model plate in which a plate element was taken in place of the raft and supported by the number of spring elements taken instead of the pile group as shown in Fig. 1 and described the interactivity between different elements. Poulos [4] conducted a similar study on the plate-spring model by performing 2D numerical analysis by examining the impact on load sharing of CPRF. The development of a numerical method carried out the study of piled raft bearing behavior by Reul [5]. The findings of variation in foundation geometry on differential and total settlements were studied by Prakoso and Kulhawy [6]. Sinha and Hanna [7] stimulated a 3D finite element analysis of a piled raft foundation and analyzed in ABAQUS software using modified Drucker-Prager Constitutive Law.

This paper compares current studies and research done by Sinha and Hanna [7] by Plaxis 3D [1] software.

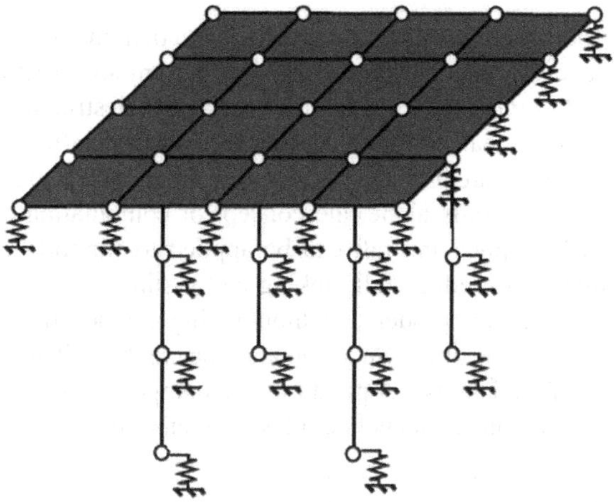

Fig. 1 Plate on spring model [3]

2 Numerical Model

The development of a three-dimensional numerical model was performed for the stimulation of the combined piled raft foundation. The model consisted of a soil block, the foundation elements, zone of contact, and prescribed displacements. The software program PLAXIS 3D [1] was used in the development of the model. Figure 2 presents the structural model of the foundation bed, and the deformed mesh of CPRF is shown in Fig. 3. Because of the symmetrical conditions, only the quarter part of the foundation is modeled and analyzed.

In this study, the soil is taken as a homogenous, isotropic, and single-phase medium. Tables 1 and 2 list the soil and other component parameters.

The water table effect was not taken into consideration. A comparison of study has been made between Sinha et al. [7] and the current study by taking the similar properties in later cases and modeling the parameters in Plaxis 3D [1].

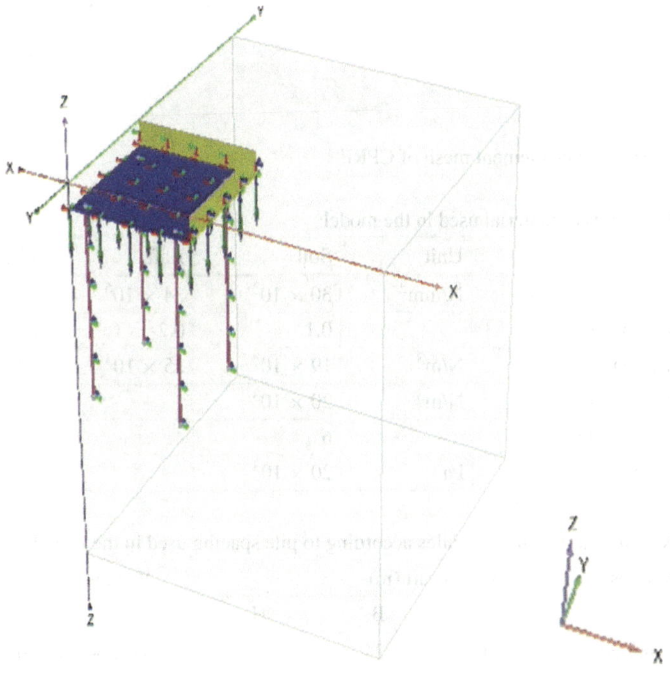

Fig. 2 Structural model of CPRF

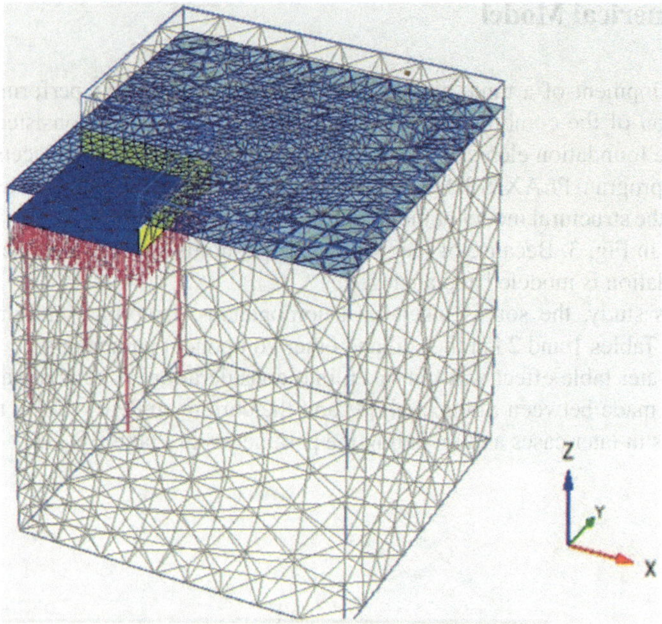

Fig. 3 Deformed finite element mesh of CPRF

Table 1 Properties of material used in the model

Parameter	Unit	Soil	Raft	Pile
Modulus of elasticity (*E*)	N/mm^2	30×10^3	34×10^6	25×10^6
Poisson's ratio (υ)		0.1	0.2	0.2
Dry density (γ)	N/m^3	19×10^3	25×10^3	25×10^3
Saturated density (γ')	N/m^3	20×10^3		
Internal friction angle (°)		6		
Soil cohesion (C')	Pa	20×10^3		

Table 2 Raft size and number of piles according to pile spacing used in the model

Spacing of piles	Size of raft (m)			Pile number
	L	B	H	
2d, 3d, 4d and 6d	24	24	2	144, 64, 36 and 16
7d	28	28	2	16
8d	32	32	2	16
10d	40	40	2	16

3 Study of Parameters

The parameters studied for the load and displacement characteristics of combined piled raft foundations are examined in the corresponding section. The results of variation of structural models are compared with the previous study using different software, and the variation in results is discussed.

3.1 Variation in Raft Thickness

In this case, a 24 m square raft, 6d pile spacing, and 15 m pile length are examined. The variation in raft thickness was examined over the thickness as 0.5 m, 1 m, 1.5 m, 2 m, and 2.5 m. These variations are also compared with the behavior of an unpiled raft 0.5 m thick. The raft was subjected to the prescribed displacement of 0.5 m which was applied on the raft surface, and corresponding loads are obtained from the Load versus Settlement plot of the model. Figure 4 represents the results obtained by the analysis of the model in the form of load settlement curves. It was observed that for smaller thickness raft, the load-bearing capacity was higher as compared to thicker raft for the same pile spacing. These values were obtained for a given raft size, pile spacing and loading conditions; thus, the optimization in design can be done to obtain more economical and safe construction.

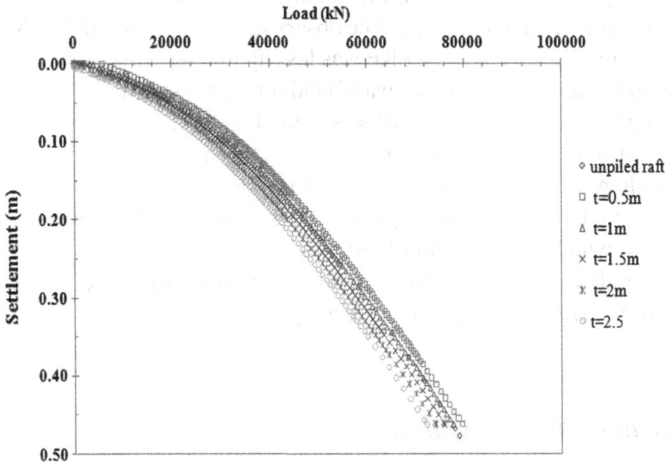

Fig. 4 Load versus settlement for variation in raft thickness

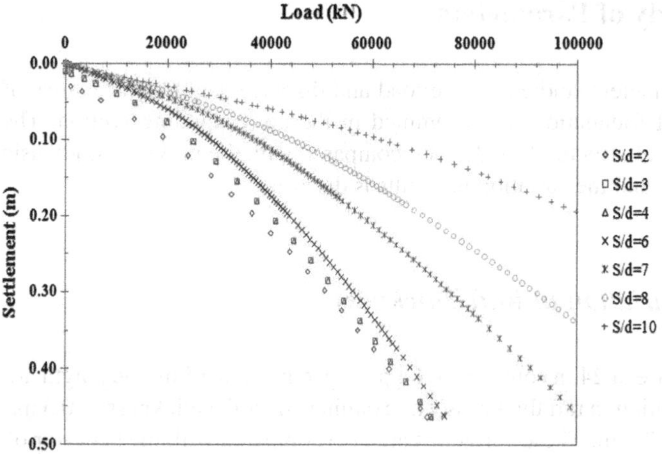

Fig. 5 Load versus settlement for variation in pile spacing

3.2 Variation in Pile Spacing

In this case, a raft of 2 m thickness and piles of 1 m diameter and 15 m length were examined. A prescribed displacement of 0.5 m was given to the raft, and analysis was performed for pile spacing varying from 2d, 3d, 4d, 6d, 7d, 8d, and 10d, where d represents the pile diameter taken into account. Figure 5 represents the load settlement relation for variation in pile spacing. The observation was recorded that the increment in load-carrying capacity of the CPRF was less up to spacing 6d, beyond that drastic increment in load carrying capacity was found for higher spacing as it can be believed that the contribution of larger size raft was more dominant, resulting in compensation for loss of capacity of the system. In addition, in the studies conducted by Sinha and Hanna [7], there was a decrease in load-carrying capacity up to 6d beyond which the authors observed a similar type of pattern. This contradicts the conventional design philosophies, which show a limit of the maximum pile spacing to 3.5d as it was an observation indicate that a decrement in pile interactions was observed beyond 3.5d, which in turn decreases the system's capacity.

3.3 Variation in Pile Length

In this case, analysis was conducted on a pile length of 5 m, 10 m, and 15 m. The spacing adopted was 6d, and 2 m thick raft was taken into consideration. The pile diameter was kept as 1.0 m, and the size of the raft was taken as 24 m × 24 m. A prescribed displacement of 0.5 m was given in Plaxis 3D, and the corresponding load was obtained by a load-settlement curve. Since the continuum of raft and soil is symmetrical, only a quarter part has been modeled to save computation and model

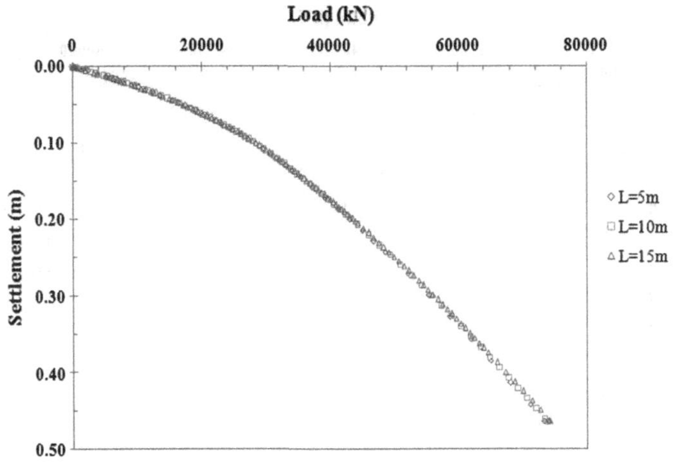

Fig. 6 Load versus settlement for variation in pile length

time. The results obtained by the analysis in the form of the load versus settlement curve are represented in Fig. 6. In the load-carrying capacity of the system, a slight increase has been observed, so there can be compensation in the design between pile length and pile spacing for the more economical design of the foundation.

3.4 Variation in Pile Cross-Section

In this case, the square cross-section was analyzed; its sides varied as 0.4 m, 0.8 m, and 1.2 m. 6d pile spacing and a 24 m and 2 m thick square raft were adopted. A prescribed displacement of 0.5 m was given in Plaxis 3D, and a corresponding load was obtained from the load settlement curve. As a symmetrical raft was adopted, so the only quarter part was modeled to save computational time. Figure 7 represents the results obtained from analysis in the form of load versus settlement curve. It was interpreted from the analysis that there was no significant effect observed by the variation in the cross-sectional size of piles. Hence it is up-to-the structural and geotechnical designers to adopt a suitable cross-section and size of the pile for the economical and safe design of the foundation.

4 Conclusions

In order to examine the effects of various parameters, FE analysis was conducted to evaluate the performance of piled raft foundations. The PLAXIS 3D [1] software program is used for successful problem stimulation. From this study, the following

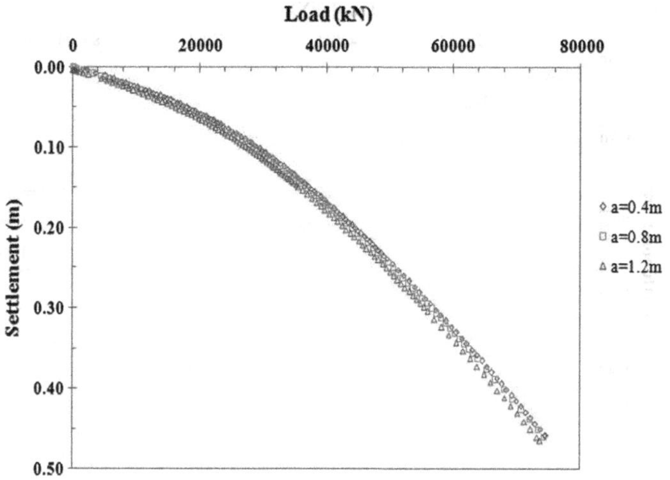

Fig. 7 Load versus settlement for variation in pile cross-section

conclusions can be made from the examination of different parameters of piled raft foundations:

1. Raft settlement increases as pile spacing increases and decreases as pile size and length increase. The system acts as a raft when the spacing between piles exceeds six times the pile diameter.
2. Increased spacing between piles diminishes the aid of increased size and length of the piles in settlement reduction. A swap between spacing, size, and length of piles should be considered to create an affordable design.
3. For the variation in raft thickness, it was noted that the system capacity increased with an increase in raft thickness of up to 1.5 m, beyond which the settlements increased due to raft self-weight, resulting in system failure and less load-carrying capacity. A thinner raft results in unequal load sharing among components of CPRF, and a thicker raft will result in excessive settlements due to more load on piles.

References

1. Plaxis BV (2013) Netherlands user manuals. Plaxis 3D
2. Burland JB, Broms BB, de Mello VFB (1977) Behaviour of foundation and structures. In: Proceedings of 9th ICSMFE, Tokyo, vol 2, pp 495–546
3. Clancy P, Randolph MF (1993) An approximate analysis procedure for piled raft foundations. Int J Numer Anal Meth Geomech 17:849–869
4. Poulos HG (2001) Piled raft foundations: design and applications. Geotechnique 51(2):95–113
5. Reul O, Randolph MF (2003) Piled rafts in over consolidated clay: comparison of in situ measurements and numerical analyses. Géotechnique 53(3):301–315

6. Prakoso WA, Kulhawy FH (2001) Contribution to piled raft foundation design. J Geotech Geoenviron Eng (ASCE) 127:1(17):17–24
7. Sinha A, Hanna AM (2017) 3D numerical model for piled raft foundation. Int J Geomech 17(2):040160551–040160559

FRP Strengthened Reinforced Concrete Beams Under Impact Loading: A State of Art

Swapnil B. Gorade, Deepa A. Joshi, and Radhika Menon

Abstract Reinforced Concrete structures are subjected to impact loading like collision due to vehicles, ships, and airplanes, impact due to rock-fall or debris, impact due to fall of heavy machines or object on structural member and terrorist activities such as blast or missile impact. The impact loading create severe damage to various structural members. It is not feasible to replace the structure due to high cost, hence it is desirable to repair or strengthen the members by retrofitting techniques. For retrofitting of damaged structural members under static loading the fiber reinforced polymer materials are highly preferred and used rapidly. These materials are being popular because of advantages like high strength and high fatigue resistant, higher resistant to corrosion, lighter in weight and higher stiffness, durable and easy to apply. Lot of investigation was already done on performance of retrofitted RC beams under the static loading but very limited investigation has been carried out on performance of retrofitted RC beams using FRP under impact load. The impact loading characterized by a high intensity load over a short period of time and categorized into low velocity impact or high velocity impact. Longitudinally bonded FRP sheet improve the flexural capacity where FRP bonded in transverse direction provides resistant to shear failure. It was found that the most common method used by researcher is drop weight hammer test. FEA analysis has been carried out previously to simulate the reinforced concrete beams under impact load. The aim of this paper is to review the literature on behavior of reinforced concrete beams strengthen with FRP under impact load. This paper also reviews the retrofitting materials, their properties and impact testing methods for RC beams under impact loading. It is suggested that FRP bonding to the RC beams shows improvements in strength recovery against impact loading.

S. B. Gorade (✉)
Department of Civil Engineering, Dr. D. Y. Patil Institute of Technology, Savitribai Phule Pune University, Pimpri, Pune, India
e-mail: sbgorade7@gmail.com

D. A. Joshi
Department of Civil Engineering, Dr. D. Y. Patil Institute of Technology, Pimpri, Pune, India

R. Menon
Department of Mathematics, Dr. D. Y. Patil Institute of Technology, Pimpri, Pune, India

M. S. Ranadive et al. (eds.), *Recent Trends in Construction Technology and Management*, Lecture Notes in Civil Engineering 260,
https://doi.org/10.1007/978-981-19-2145-2_74

Keywords FRP · Retrofitting · Strengthening · Impact loading · RC beams

1 Introduction

RC structures are subjected to static as well as dynamic loading. Considering the military application and recent terrorism activities to public infrastructure the study of performance of RC beams strengthened using FRP against impact loading is of great interest. Due to increased traffic density the possibility of impact of vehicle against RC structure like bridges has also increased. Many RC structures are subjected to damaging impact load due to vehicle, ship, airplane, debris, rock fall, missile impact, collapse of crane, and unregulated motions of heavy machines [1–3]. The retrofitting of existing damaged RC beams, done using conventional techniques like steel plate bonding, concrete or steel jacketing increases dead load, it is difficult to apply and tends to corrosion. However in recent years to prolong the life of damaged RC structure the most effective repairing technique is use of externally bonded FRP.

FRP material got popularity for strengthening existing RC structure because it has high strength to weight ratio, high tensile strength, non-corrosive, easy to handle and install without interruption and good resistant to chemical attack [4–11]. A considerable amount of research has been carried out to show the increase in strength, stiffness and ductility of reinforced concrete beams retrofitted with FRP under static loading. However studies on reinforced concrete beams strengthened with FRPs under impact are still limited [1, 12]. Pham and Hao [1] mentioned in their literature that under impact loading the performance of retrofitted reinforced concrete beams also includes the study of FRP-concrete bonding and energy absorption capacity of beam. In such cases size, mass and velocity of impactor are also important parameters. Design guidelines for RC beams strengthened with FRP were also developed by many codes for example guidelines given by ACI 440-2R-17 code. A review of literature was carried out in this paper on FRP retrofitted RC beams under impact loading. It was observed that research on FRC strengthen reinforced concrete beam under impact load is still limited and need more recommendations from research community.

2 FRP Materials and Strengthening Techniques for Impact

An inappropriate selection of FRP material and strengthening technique cannot solve the purpose of repair and strengthening of RC structure. To upgrade the strength requirement and increase the service life of structure, selection of proper strengthening materials and methods needed [13]. The traditional strengthening techniques may include use of concrete or steel jacketing, textile fiber sheet, wire mesh, post tension cables, metallic plate etc. need much space, and they are difficult to apply [14, 15]. For different structural members like columns, beams, slabs and walls the objectives of FRP strengthening techniques are one or more as follows: (1) to increase the load carrying capacities under axial, flexural or shear, (2) to increase the stiffness, (3) to improve ductility, and (4) to improve durability against external environmental agents [16].

2.1 Types of FRP Materials

FRP composites are manufactured by adding two or more materials to form compound which is superior to its parent material. These composites are manufactured by embedding high strength continuous fibers in a polymer matrix also called resin. There are four FRP composites available in industry, and they are carbon (CFRP), glass (GFRP), basalt (BFRP) and aramid (AFRP) [17, 18].

Fibers are the main reinforcing elements in the polymer matrix which is act as a binder. Polymer matrices (resin) divided into two groups thermosetting (polyesters epoxies and vinyl esters) and thermoplastics (polypropylene, polyvinyl chloride, polyethylene and polyurethane). Thermoplastics are sensitive to environmental conditions, low creep and their production is costly. Thermosetting resigns are most popular in construction industry due to their improved mechanical properties [19].

According to ACI 440-2R-17, the unit weight of structural steel is 7.9 g/cm^3, unit weight of CFRP is 1.5–1.6 g/cm^3, unit weight of BFRP is 2.2–2.7 g/cm^3, unit weight of GFRP is 2–2.1 g/cm^3, and unit weight of AFRP 1.2–1.5 g/cm^3. Carbon FRP has very high tensile strength (8–10 times of steel) and possesses high strength to weight ratio. It is five times lighter than steel, and it shows very high resistant to creep and fatigue than all others materials.

Epoxy resins are used for excellent bonding between FRP composites and concrete surface. They have 0.9% elongation at break, tensile strength of 30–90 MPa, and elastic modulus of 1.1–6 GPa. CFRP shows better performance than GFRP after crack propagation till failure. When CFRP and GFRP combined together named as hybrid, FRP is the novel retrofitting material shows high ductility and deformability [4].

Figure 1 shows a relation between the stress strain curves for different FRP materials and structural steel. It clearly indicate that FRP material have high strength compare to other conventional materials.

The compressive strength of FRP composite is less than tensile strength and hence is not suitable as compression side reinforcement. BFRP is costlier than GFRP due to lack of producer but have better strength and alkalis resistant. GFRP has the lowest cost, resistant to chemical attack, good thermal insulation and heat resistant and consumed mostly for strengthening of secondary structures. AFRP has heat resistant and is good option in high alkaline environment [17].

As per ACI 440 part 6, CFRP have low specific gravity, density, and high tensile strength than structural steel and GFRP. AFRP shows good tensile strength, lower density than GFRP. FRP shows brittle character and low modulus of elasticity. Table 1 shows the mechanical properties of FRP materials and structural steel.

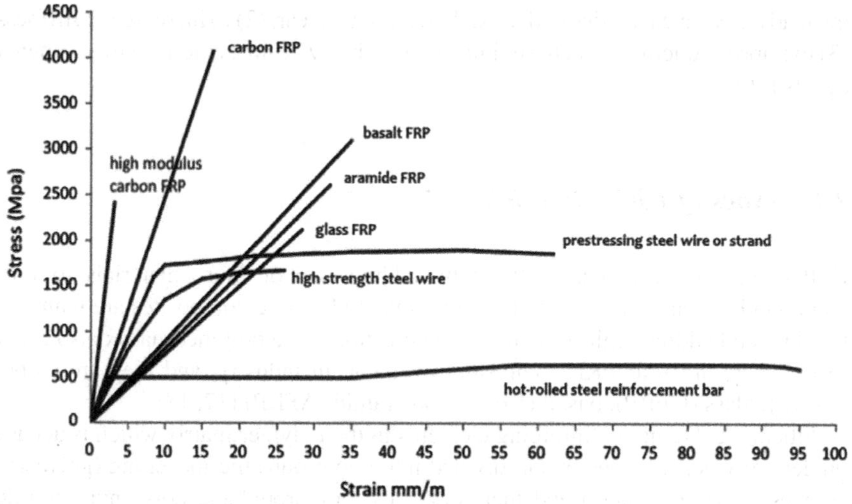

Fig. 1 Comparison of structural steel and various FRP materials [20].

Table 1 Mechanical properties of FRP materials (ACI 440 part 6)

FRP material	Specific gravity	Yield strength (MPa)	Tensile strength (MPa)	Density (g/cm^3)	Elastic modulus (GPa)	Strain at break, in %
CFRP	1.0–1.1	1755-3600	1720–3690	1.55–1.76	120–580	0.5–1.9
GFRP	1.5–2.5	600–1400	480–1600	2.11–2.70	35–51	1.2–3.1
AFRP	1.38–1.39	1700–2500	1720–2540	1.28–2.6	41–125	1.9–4.4
BFRP	2.7–2.89	1000–1600	1035–1650	2.15–2.70	45–59	1.6–3.0
Steel	7.8	500	–	7.75–8.05	200	–

CFRP shows 50–60% mass reduction but higher cost compare to structural steel. GFRP has low weight and high strength. It shows high resistant to chemical attack. Use of AFRP is restricted in construction industry up to light loaded structures due to its extreme low compressive strength. BFRP have good thermal resistivity and ductility than CFRP, and it will be used to strengthen the beams subjected to high temperature. BFRP composite shows improvement in all mechanical properties but its production is costly. While selecting the retrofitting materials, the strength requirement, availability, cost, and environmental conditions need to be considered. The alignment and orientation of fibers in FRP are also important factors and should consider while selecting the retrofitting material [4].

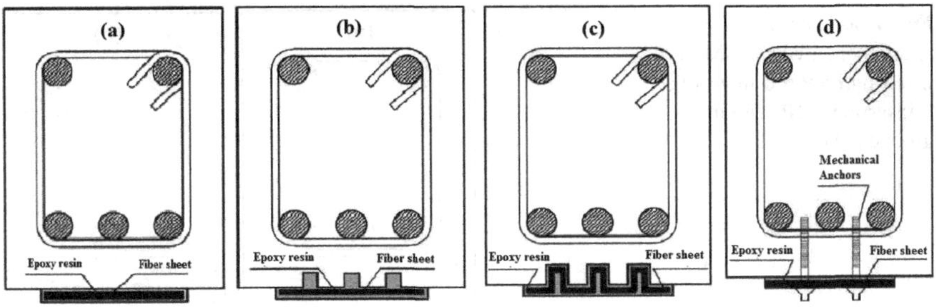

Fig. 2 Beams strengthened with EB technique, **a** using only FRP, **b** FRP on grooves, **c** FRP in grooves, **d** using mechanical anchors [21]

2.2 Strengthening System with FRP for RC Beams

The FRPs are available in the form of laminates, sheets, plates, and rods. The different strengthening systems of RC beams using FRP are as follows:

1. Externally Bonded technique (EB): with grooves, without grooves and mechanically anchored EBR
2. Near Surface Mounted (NSM) bar/strips.

2.2.1 Externally Bonded Technique (EB)

Externally bonded technique is the most common technique used to strengthen RC beam. The main deficiency of this method is possibility of failure due to early debonding of FRP laminate from the concrete surface. Surface preparation is necessary in this method, and it was done by roughening surface by sandblast technique or use of air jet and water jet which can slightly postponed the premature debonding. In this technique, number of FRP layers applied on the beam soffit with groove, without groove and using mechanical anchors as shown in Fig. 2. To strengthen the beams in shear and flexure, FRP is applied in different styles like full wrap, U wrap, side wrap, inclined U wrapping, cross wrap, etc. [1, 4, 21].

2.2.2 Near Surface Mounted (NSM) Bar/Strips

In externally bonded technique, the tensile capacity of FRP is not fully utilized due to early debonding. This deficiency is removed in NSM method. In Near surface mounted (NSM) method FRP bar or FRP strip inserted in the grooves which are provided along the bottom face of the beam (tension face) and then filling the groove completely with epoxy as shown in Fig. 3. The probability of FRP strip deboning from concrete can be reduced by providing anchors, fasteners at the end of strips. It is costly and time consuming process [1, 4, 21].

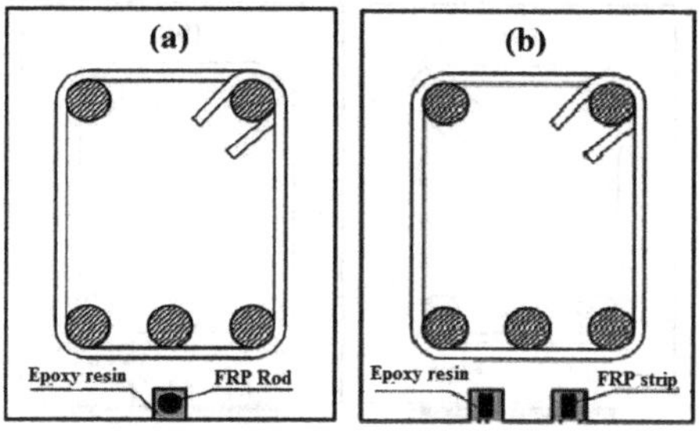

Fig. 3 RC beam strengthened with NSM, **a** inserting FRP rod in groove, **b** inserting FRP strip in groove [21]

Table 2 Studies on FRP material used for strengthening of RC beams under impact

Researcher	Year	Section mm	Span in m	Samples nos	Type of FRP used
Erki and Meier	1999	400 × 300	8.15	4	CFRP laminates and steel plates
Tang and Saadatmanesh	2003	203 × 95	1.83	5	Carbon and kevlar FRP
Tang and Saadatmanesh	2005	203 × 95	1.98–2.9	27	Carbon and kevlar FRP
Soleimani et al	2007	150 × 150	0.8	10	Sprayed GFRP
Thong M. Pham et al	2016	150 × 250	1.5	13	FRP wrap
Fujikakea et al	2017	160 × 170	1.7	20	CFRP sheet and laminate
T. Liu et al	2017	200 × 400	2.4	11	CFRP wrap
A. Remennikov et al	2017	100 × 150	1.2	12	CFRP wrap
Soleimani and Roudsari	2019	150 × 150	1.0	17	Sprayed GFRP
Kishi et al	2020	200 × 250	3.0	12	AFRP and CFRP sheet

Table 2 shows the various studies on FRP material used for strengthening of RC beam under impact loading.

3 Testing Methods Under Impact Loading

Testing methods for reinforced concrete beams strengthened with FRP under impact load should be such that it will simulate the mechanism, loading conditions and failure modes likely to occur. Impact test methods divided into two types (1) impact due to large mass with low velocity and (2) impact due to small mass with high velocity. The most popular method is impact due to large mass with low velocity which includes

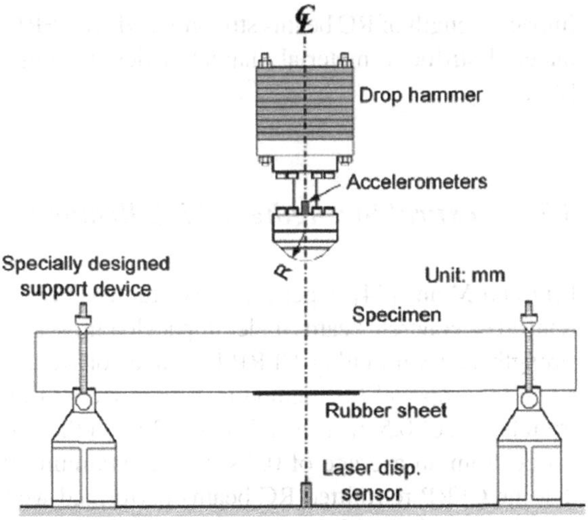

Fig. 4 Impact loading setup [2]

drop-weight tests, Izod and Charpy test, hydraulic test machines in which impact velocities designed upto 10 m/s [1]. Izod and Charpy pendulum test shows number of difficulties for researcher hence most common impact testing method adopted by researcher is dropping a weight on RC beam specimen from certain height. In this test the impact velocity of the hammer can be determined by accelerometer or an optical sensor or using high speed camera [1]. Generally a shape of impactor is this test is semispherical. The dynamic capacities of RC beam specimen were calculated by single drop weight test whereas fracture energy can be determined by multi drop test where specimen failed due to number of blows. Main advantage of the drop hammer test is that specimen of large range geometries can be tested. The impact loading applied on beam specimen using drop weight hammer was shown in Fig. 4.

3.1 RC Beams Under Impact Load

RC structures are subjected under impact loading during various instances like crashing vehicles, ships, airplanes, rock fall, failure of crane etc. [2]. An RC beam under impact load was described by its high amplitude in a short duration which causes inertia effects in the beams and strain rate effects in the materials. Hence compared to static loading, beams under impact load behave differently. Impact load have low loading rate of 10 s^{-1}. Impact loads are concentrated on RC beams as point loads for very short duration with higher strain rates than static loading [1, 22].

Externally bonded FRP system is decreases the crack width, decreases the deflection of RC beam and increase the impact strength. It was observed that shear deficient RC beams when retrofitted with FRP and tested under impact load, the impact strength and shear strength were increased by 15% and 96% respectively [22, 23].

Impact strength of RC beams strengthened with FRP is depends upon impact energy, material stiffness, material characteristics, loading rate and strengthening system [22].

3.2 Flexural Strengthened RC Beams Under Impact

Erki and Meier [24] experimentally studied the behavior of flexural strengthened reinforced concrete beams under impact loading. They tested four 8.15 m long beams strengthened with either CFRP laminates or steel plates. The impact loading was given by drop and lift method as lifting one end of a test beam and then dropped from height of 0.5 m, 1 m, 1.5 m and 2 m on the support. The strain rate of loading varied from an average of 0.7 s^{-1} to a maximum of 0.84 s^{-1}. The results showed that the CFRP retrofitted RC beams performed well under impact loading. Beams retrofitted with CFRP laminates could not deliver same energy absorption as the beams strengthened with steel plates. The first occurrence of rupture of FRP or deboning of FRP was not explained in the paper. However from this work is difficult to study the performance of RC beams strengthened with FRP impacted in the mid-span of beam.

Tang and Saadatmanesh [22] used five RC beams of size 203 × 95 × 1980 mm, without stirrups in their experimental work. The beams were strengthening with Carbon and Kevlar FRP laminates at top and bottom face of beam under drop weight impact hammer. A steel cylinder was used to apply impact load on top face of the beams. They observed that improvement in the strength gain depends on the type, weight, material properties and thickness of the FRP laminates. They concluded that the composite laminates increases resistant to impact load and reduce maximum deflection. Flexural cracks occurred at bottom face of beam and then propagated upward. After increase in the impact load, the transverse shear cracks were observed. These cracks were extended to the interface between FRP laminate and concrete. The stiffer laminate showed improved capacities and less deflection of the beam under impact loading. Cracks were not found between the interface of concrete and FRP because the compression strain was smaller than the tension strain. The rebound and vibrations response need to study using load cell to calculate reaction forces on both compression and tension forces.

Tang and Saadatmanesh [25] were done experimental study on twenty seven RC beams retrofitted with FRP laminates under impact loading. A steel cylinder was used as a drop weight hammer for impact loading. They observed that composite laminates perform better against impact loading. FRP laminates bonded to RC beam, increased flexural and cracking strength, stiffness of beams, reduce number of cracks. They found that FRP laminates were reduced maximum deflection of strengthen beams by 30–40% than unstrengthen beams under impact loading. During the impact loading, the maximum reaction force was three times higher than the beams under static loading. They suggested an equation for predicting the deflection of beams under impact load based on flexural wave theory. FRP delimitation was reported by same author in their previous work [22] but it was not recorded by same author in current work. Further investigation will be needed to understand the failure mechanism and structural capacities. The negative moments caused by the structural vibrations due to impact loading was not considered in the work.

Kantar and Anil [26] experimentally investigated ten retrofitted RC beams in flexure with CFRP strips. They conducted experimental work by varying the concrete compressive strength and drop height. Five different drop heights were selected for drop hammer. They found that the dissipated energy and number of drops required for failure were increased. Yilmaz and Anil [27] experimentally tested shear deficient RC beams under impact load which are strengthened with CFRP U-wraps. CFRP strips spacing and impact velocities were the variables in this experimental work. ANSYS software was used to model the RC beams. They concluded that impact behavior of shear deficient RC beams was changes due to CFRP strips attached to beams. They found that, decrease in CFRP strip spacing reduces the acceleration, velocity and displacement.

Pham and Hao [28] experimentally investigated the performance of RC beams retrofitted with FRP under static and impact loading. They tested total thirteen RC beams out of that six RC beams in static loading where seven beams under impact loading. Beams were wrapped with U-wrap system and FRP strips in longitudinal direction. The peeling stress in the normal section of RC beams was not reduced by the tensile force in FRP. To prevent the deboning of longitudinal FRP, the cross-section of four beams soffit was modified and curved shape was given such that the confining stresses from the U-wrap FRP and the stresses in the adhesive resist the peeling stress concentration. They found that due to modification in RC beam section with the same amount of FRP material, delayed the deboning and improved the capacity of longitudinal FRP. Due to the modified RC section, the U-wrap FRP provides confinement effect on longitudinal FRP system. This also reduced the stress concentration at corners of the beam.

Kishi et al. [6] conducted experiments on reinforced concrete beams strengthen with FRP sheets on tension side. They were tested using drop weight impact test under low velocity. Total twelve specimens of 3 m length were tested. AFRP and CFRP sheets having equal axial stiffness were used. This work was carried out to study the improvements in impact resistant, its characteristics and failure pattern of AFRP, CFRP strengthen beam. Out of twelve, ten specimens of RC beams were strengthened with AFRP and CFRP sheets. These specimens were tested by dropping a 300 kg drop hammer from various heights 1, 2, 2.5 and 3 m.

3.3 Shear Strengthened RC Beams Under Impact

Huo et al. [29] investigated that the performance of bond between the concrete and CFRP under drop weight impact test from dynamic loading point of view. They also studied the result of impact loading rate on bond strength, bond length and strain distribution along CFRP strip bonded to RC beams. They found that the impact loading rate considerably influences the bond strength. This paper focused only on impact loading effects on the bond length and strength between FRP and concrete.

Pham and Hao [28] experimentally studied the impact behavior of FRP retrofitted RC beams and contribution of CFRP to improve the shear strength. They tested thirteen beams of size 150 × 250 × 1500 mm without stirrups and strengthened with CFRP U-wraps and CFRP 45° wraps to study effect of different wrapping schemes. They found that using the same amount of CFRP strips, CFRP at 45° wrapping gave better results than U-wrapping w.r.to the load carrying capacity and deflection of the beams.

Remennikov et al. [30] investigated the performance of RC beams strengthen externally with different CFRP wrapping schemes under static and impact loading. Twelve shear deficient RC beams of size 100 mm × 150 mm × 1200 mm were strengthen with one or two layers of U-wraps and also one layer of CFRP was fully wrapped. To study the load-deflection, energy absorption capacity and failure modes the six beams were tested under static loading. Remaining six beams were tested under free falling drop weight impact hammer to analyze failure modes, crack patterns and mid span deflection. They concluded that the use of CFRP wrapping as an external reinforcement for shear deficient beams showed improvements in the energy absorption capacity. The failure modes of beams transformed from shear failure to flexural failure.

Fujikake [2] conducted experimental work on RC beams strengthen in shear and weaker in flexural. Twenty RC beams of cross-section 160 × 170 mm size and 1700 mm length were strengthened with unidirectional CFRP sheets and CFRP laminates. CFRP strengthening was applied to RC beams in four different ways. The CFRP sheets having 150 mm width and 0.222 mm thickness were bonded to the soffit of the beams. For few RC beams, CFRP sheet of width 250 mm and 0.111 mm thickness was installed at the ends of the soffit perpendicular to axis of beam. Some RC beams were bonded with 1.0 mm thick and 50 mm wide CFRP laminates to the soffit of the RC beams. A 300 kg drop hammer was dropped freely from the height of 100, 200, 400 mm to produce single impact on beams. It was observed that crack width is significantly decreased for CFRP strengthened RC beams compare to unstrengthen RC beam.

Liu and Xiao [12] done experimental work on eleven RC beams without stirrups, of cross section 200 mm × 400 mm and length 2400 mm under static and impact loading. CFRP strips were used to retrofit RC beams and wrapped fully around it such that the strips were perpendicular to the axis of the beam. In this experimental work the effect of impact mass, impact velocity and impact energy was mainly considered. They also studied the failure modes, crack patterns, mid span deflection, strains in CFRP and dynamic behavior including time history. They concluded that for shear deficient RC beams without stirrups, full wrapping with CFRP strips found effective. To prevent local punching shear near impact area the increased width of CFRP strips was effective. The strain of CFRP strips which was far away from impact area found low. They also found that the high impact velocity under same energy decreases the reaction force and deflection but increases the impact force.

Table 3 Studies on impact technique used for FRP strengthened RC beams

Researcher	Year	Type of FRP used	Hammer weight (kg)	Drop height in m
Erki and Meier	1999	CFRP laminates and steel plates	Lift and drop	0.5 m, 1 m, 1.5 m and 2 m
Tang and Saadatmanesh	2003	Carbon and kevlar FRP	22.6	0.305, 0.61, 0.91, 1.22, 1.53, 1.83, 2.14, 2.44, 2.74 m
Tang and Saadatmanesh	2005	Carbon and kevlar FRP	22.6	0.3–1.7 m
Soleimani et al	2007	Sprayed GFRP	591	0.8 m
Thong M. Pham et al	2016	FRP wrap	230.5	2.5 m and 3 m
Fujikakea et al	2017	CFRP sheet and laminate	300	0.1, 0.2 and 0.4 m
T. Liu et al	2017	CFRP wrap	383, 513, 773	1.5 m and 3 m
A. Remennikov et al	2017	CFRP wrap	580	50 mm
Soleimani and Roudsari	2019	Sprayed GFRP	591	2.5 m
Kishi et al	2020	AFRP and CFRP sheet	300	1, 2, 2.5 and 3 m

Huo et al. [23] studied experimentally the performance of CFRP wrapping system bonded with RC beams without stirrups under impact loading and failure mechanism was observed. They also studied FRP contribution to shear strength. Total fifteen beams casted and out of that five beams for static and 10 beams for impact loading were tested. From experiments, they found that the strengthening with single 45° CFRP wraps was very effective than that of CFRP U-wraps. Test result showed that amount of CFRP, its arrangement and impact energy was played an important role. Table 3 shows the summery of studies on impact technique used for FRP strengthened RC beams.

4 Discussion and Conclusion

The proper selection of FRP material and strengthening technique achieve the objectives of strengthening such as improvements in strength, stiffness, ductility. The above literature review provides a clear understanding of the performance of reinforced concrete strengthened with FRP under impact. It is concluded that FRP bonding to the RC beams shows improvements in strength recovery and protect the RC beams against impact load. The outcome of this review listed below:

1. RC beams shows better flexural, shear resistant capacities under impact loading when wrapped externally bonded FRP.
2. The important contribution of FRP is the improvement of service life of existing RC beams.
3. The rupture and deboning mechanism of EB FRP require more investigation. If the bond strength between reinforced concrete beam and FRP improves then it gives high stiffness and resistant to impact.
4. Low velocity impact test with drop weight hammers is generally observed in most of the research work.
5. Use of end anchorage control the FRP debonding and depends on loading condition and strengthening configuration.
6. Due to change of loading from static to impact loading the failure pattern of strengthened RC beam changes from flexural to flexural shear.
7. Further investigation is needed on performance of RC beams strengthened with FRP under impact loading.

References

1. Pham TM, Hao H (2016) Review of concrete structures strengthened with FRP against impact loading. Structures 7:59–70
2. Fujikake K et al (2017) CFRP strengthened RC beams subjected to impact loading. Procedia Eng 210:173–181
3. Soleimani, Roudsari (2019) Analytical study of reinforced concrete beams tested under quasi-static and impact loading. J Appl Sci 9(2838):2–16
4. Siddika A et al (2019) Strengthening of reinforced concrete beams by using fiber-reinforced polymer composites: a review. J Build Eng 25(100798):1–12
5. Nasrin S et al (2017) Behavior of retrofitted UHPC beams using carbon fiber composites under impact loads. J Struct Congr 392–402
6. Kishi N et al (2020) Low-velocity impact load testing of RC beams strengthened in flexure with bonded FRP sheets. J Compos Constr 24(5):04020036, 1–10
7. Snehal K, Das BB (2019) Mechanical and permeability properties of hybrid fibre reinforced porous concrete. Indian Concr J 93(1):54–59
8. Srikumar R, Das BB, Goudar SK (2019) Durability studies of polypropylene fibre reinforced concrete. In: Sustainable construction and building materials. Springer, Singapore, pp 727–736
9. Yadav S, Das BB, Goudar SK (2019) Durability studies of steel fibre reinforced concrete. In: Sustainable construction and building materials. Springer, Singapore, pp 737–745
10. Goudar SK, Shivaprasad KN, Das BB (2019) Mechanical properties of fiber-reinforced concrete using coal-bottom ash as replacement of fine aggregate. In: Sustainable construction and building materials. Springer, Singapore, pp 863–872
11. George RM, Das BB, Goudar SK (2019) Durability studies on glass fiber reinforced concrete. In: Sustainable construction and building materials. Springer, Singapore, pp 747–756
12. Liu T, Xiao Y (2017) Impact behavior of CFRP-strip–wrapped RC beams without stirrups. J Compos Constr 21(5):04017035, 1–14
13. Sumukh EP, Goudar SK, Das BB (2021) Predicting the service life of reinforced concrete by incorporating the experimentally determined properties of steel–concrete interface and corrosion. In: Recent trends in civil engineering. Springer, Singapore, pp 399–417

14. Al-Mahaidi R, Kalfat R (2018) Methods of structural rehabilitation and strengthening. In: Rehabilitation concrete structure with fiber-reinforced polymer, chap 2. Elsevier, Amsterdam, pp 7–13
15. Siddika A et al (2019) Flexural performance of wire mesh and geotextile-strengthened reinforced concrete beam. SN Appl Sci 1:1324
16. Buyukozturk O, Gunes O, Karaca E (2004) Progress on understanding debonding problems in reinforced concrete and steel members strengthened using FRP composites. Constr Build Mater 18(1):9–19
17. Amran YHM et al (2018) Properties and applications of FRP in strengthening RC structures: a review. Structures 16:208–238
18. Shakir Abbood I et al (2021) Properties evaluation of fiber reinforced polymers and their constituent materials used in structures: a review. Mater Today Proc 43:1003–1008. https://doi.org/10.1016/j.matpr.2020.07.636
19. Naser MZ et al (2019) Fiber-reinforced polymer composites in strengthening reinforced concrete structures: a critical review. Eng Struct 198:109542
20. Rousakis T et al (2010) Retrofitting and strengthening of contemporary structures: materials used, Encycl. Earthq. Eng., Berlin, Heidelberg: Springer Berlin Heidelberg (eds) pp. 1–15 https://doi.org/10.1007/978-3-642-36197-5-303-1
21. Seyyed MB et al (2015) Reviewing the FRP strengthening systems. Am J Civ Eng 3(2):38–43. Special issue: research and practices of civil engineering in developing countries
22. Tang T, Saadatmanesh TH (2003) Behavior of concrete beams strengthened with fiber-reinforced polymer laminates under impact loading. J Compos Constr 7(3):209–218
23. Huo J et al (2018) Dynamic behavior of CFRP-strengthened reinforced concrete beams without stirrups under impact loading. ACI Struct J 115:775–787
24. Erki M, Meier U (1999) Impact loading of concrete beams externally strengthened with CFRP laminates. J Compos Constr 3(3):117–124
25. Tang T, Saadatmanesh TH (2005) Analytical and experimental studies of fiber-reinforced polymer-strengthened concrete beams under impact loading. ACI Struct J 102(1):139–149
26. Kantar E, Anil Ö (2012) Low velocity impact behavior of concrete beam strengthened with CFRP strip. Steel Compos Struct 12(3):207–230
27. Anil Ö, Yilmaz T (2015) Low velocity impact behavior of shear deficient RC beam strengthened with CFRP strips. Steel Compos Struct 19(2):417–439
28. Pham TM, Hao H (2016) Impact behavior of FRP-strengthened RC beams without stirrups. J Compos Constr ASCE 20(4):04016011, 429–440
29. Huo J et al (2016) "Experimental study on dynamic behavior of CFRP to concrete interface", beams without stirrups under impact loading. J Compos Constr 20(5):1–12
30. Remennikov A et al (2017) Impact performance of concrete beams externally bonded with carbon FRP sheets. In: Mechanics of structures and materials: advancements and challenges, pp 1695–1699. ISBN 978-1-138-02993-4

Effect of Lateral Stiffness on Structural Framing Systems of Tall Buildings with Different Heights

A. U. Rao, Sradha Remakanth, and Aditya Karve

Abstract Even though the gravity loads are the primary loading on a building, as the building height increases, it will be exposed to the lateral loads due to wind and earthquakes. Hence, the building must be designed to have sufficient stiffness and strength in order to resist the lateral loadings. In order to limit the lateral displacement, additional structural materials will be required for tall buildings, and to achieve economy it is always desirable to use an appropriate structural system. In this paper, three-dimensional structural systems are studied and compared using six different building heights. The parameters such as lateral displacement, inter storey drifts, time period and base shear are compared between these systems with respect to the height. From the study, it has been observed that outrigger structural systems are not only proficient in controlling the top displacements but also play a substantial role in reducing the inter storey drifts.

Keywords Core walls · Lateral loads · Lateral displacement · Structural system · Tall building

1 Introduction

A building is said to be tall or high rise when it has different aspects of design, construction and use compared to those of common buildings in certain regions and periods. It is a matter of human perception and hence the definition of tall buildings cannot be universally applied. When a tall building is subjected to appreciable lateral loading, such as earthquake and high wind pressure, special structural arrangements such as moment frame, shear wall, tube structure and outrigger must be adopted to resist these forces. The selection of a structural system for a particular building

A. U. Rao (✉) · S. Remakanth
Department of Civil Engineering, Manipal Institute of Technology, Manipal Academy of Higher Education, Manipal 576104, Karnataka, India
e-mail: asha.prabhu@manipal.edu

A. Karve
Walter P Moore Engineering India Pvt. Ltd., Pune, India

M. S. Ranadive et al. (eds.), *Recent Trends in Construction Technology and Management*, Lecture Notes in Civil Engineering 260,
https://doi.org/10.1007/978-981-19-2145-2_75

depends upon several factors such as architectural requirements, functionality of building, loading, occupancy and aspect ratio. Various structural systems available for tall building constructions are as follows.

1.1 Moment Frame System

It is the structural system which consists of frames, i.e. beams and columns in order to resist the gravity and lateral loading. The frames can be parallel or inclined in which the beams and columns are connected by moment-resistant joints. These structural systems are convenient due to their rectangular frame which is clear of structural forms, thereby providing an unobstructed layout. Moment frame systems are found to be efficient up to a height of 60 m, above which the sizing of the members has to be increased in order to limit the lateral displacement. The effectiveness of the building improves when the moment frame systems are adopted along with shear walls.

1.2 Structural Wall System

In this system, the walls are rigidly connected with the floor diaphragm which helps in proper load path when the building is subjected to lateral loadings. Buildings with structural walls are found to add more stiffness to the building compared with frame systems as they resist the vertical as well as lateral loading.

1.3 Moment Frame and Structural Wall System

This system is a combination of moment frame and structural wall system, where gravity loading is mainly taken by the moment frame and lateral loading is resisted by the structural wall.

1.4 Tube System

Tube system consists of a perimeter frame with closely spaced columns connected by a deep spandrel. It acts as a hollow vertical cantilever. The lateral resistance for this system is provided by the perimeter frame. A special type of tube system includes a braced tube, which allows greater spacing between the columns with diagonal bracing connected at the exterior. Another variation is the bundled tube which consists of two or more tubes tied together to form a single multi-cell tube.

1.5 Structural Wall with Outrigger System

In this system, a core wall is connected to these perimeter columns using beam elements called outriggers at discrete floor locations. It is a lateral load-resisting system where the outriggers tie the central core with the external peripheral columns. The lateral loads, due to wind and earthquake, acting on the central core will be transferred to the peripheral columns through the outriggers and hence reduce the overturning moment. Outrigger systems can be further classified into three:

1.5.1 Conventional Outrigger

In this, the outriggers are directly connected with the central core with the columns at the periphery.

1.5.2 Offset Outrigger System

Outriggers are located elsewhere than in the plane of the core wall.

1.5.3 Virtual Outrigger System

In this system, the direct load is transferred from the core to the outrigger without a direct connection. This is achieved with the help of the floor diaphragm which converts the overturning moment into a horizontal couple. Belt trusses are also added for this system.

2 Literature Review

Badami and Suresh [1] have studied the behaviour of various structural systems (moment frame system, core wall system and outrigger) subjected to lateral loading in the case of tall buildings. ETABS was used to model the buildings, wherein parameters such as lateral displacement, storey drifts, storey displacement and base shear values were checked in order to find the structural efficiency. Wind loading considered was not uniform throughout the structure, nor it was constant with time, and it varied on different sides of the building. It was observed that this creates a lot of problems for tall buildings and is the major concern of engineers. In this paper, it was concluded that outrigger systems are the most efficient in controlling top displacement as well as inter storey drifts compared to rigid frame systems which have relatively high lateral stability.

Baraskar and Kawade [2] made a comparison between the structural wall system and the conventional system for a 15-storied building using ETABS software. Wind loading and seismic loading were considered for the analysis of both structural systems. In this paper, it was concluded that the time period of the beam-column system is higher compared to the structural wall system. Also, the lateral displacement and the storey drifts of the conventional systems are found to be more than the structural wall system. The structural wall system is found to provide better performance than the conventional system. It was also concluded from the study that the structural framing of the structural wall system is cheaper than the moment frame system.

Titiksha and Gupta [3] studied various structural framing systems subjected to lateral loadings. The study was conducted for a 4-storied building with three structural systems—ordinary moment resisting frames (OMRF), special moment resisting frames (SMRF) and braced steel frames (BSF). The seismic Zone V was adopted in the analysis, and efficiencies of the building were determined by checking the storey drifts and base shear. From this study, it was concluded that the SMRF showed better results than OMRF. In BSF, lateral loading is resisted effectively and their service life is more even though they are costly. They concluded that the storey drifts are effectively resisted in BSF compared to that in OMRF and SMRF.

Jose et al. [4] have studied the effect of lateral stiffness on different structural systems (frame system, core wall system and dual system) which were subjected to lateral loadings. Staad Pro software was used for the analysis of three-dimensional models to assess all the lateral load-resisting systems under the effect of wind loads and seismic loading. It was concluded that the lateral displacement of bare frame systems is high compared to systems with external and internal infills. It was observed that there is considerable variation in the lateral displacement of the frame with core and staircase compared to that of the bare frame system.

Hemavati and Ramya [5] studied lateral load-resisting systems for high-rise buildings. They incorporated a rigid frame system, shear wall system with opening and framed tube system, and studied their behaviour by comparing the parameters such as lateral displacement, lateral drifts and base shear. They concluded that the displacement was significantly reduced by using a braced frame system than a rigid frame or frame tube system. It was observed that the stiffness parameter is achieved by the frame tube system compared to the rigid frame or braced frame system. Also, the braced frame system helped in minimizing the storey drifts.

Yadhav and Rai [6] conducted a comparison study of moment-resisting frames with regular as well as irregular structures for different seismic zones using Staad Pro software to find out their behaviour when subjected to lateral loadings. They observed that maximum bending moment, shear force and displacement occurs in irregular plaza building, whereas it is minimum in regular frame building. Thus, it was concluded that the performance of a regular frame is better than an irregular frame considering parameters such as displacement, shear force and bending moment.

Baygi and Khazaee [7] investigated the optimal number of outriggers in a tall structure to reduce the top storey displacement. They considered three types of lateral loading—uniform, triangular and centralized, and also studied the effects of variation in structural elements such as column, outriggers and core on the optimal number of outriggers. They observed that four or more outriggers did not improve the reduction in structural displacements.

Hafner et al. [8] have studied the impact of the negative and positive shear lag effect on a tall tube structure. They conducted a parametric analysis with parameters including the ratio of column spacing and column dimensions, the ratio of storey height and the height of connecting beams, the ratio of the layout dimensions of the entire building and the type (profile) of horizontal load being used. They conducted a parametric analysis with the main purpose of detecting the critical height of the tall building where the transfer from positive to negative shear lag occurs. Based on the study, they suggested the transformation from the positive to the negative shear lag is conditioned by the layout dimensions ratio in tall tube structural systems.

Sawant and Bogar [9] have studied the seismic behaviour of tall structures with outriggers and the belt truss system using linear dynamic and non-linear time history analysis along with the V and X bracing systems. They concluded that X bracing systems are better, and time history analysis gave a better response for tall buildings under earthquake loading.

Nassani and Ali [10] compared the structural response of different types of lateral load-resisting systems (moment-resisting frame system, shear wall system, dual system and framed tube system) under the effect of seismic and wind loads using ETABS. They analysed a building having 28 storeys each with an area of 625 m^2 (25 m $\times$ 25 m), and the storey displacements were evaluated for all lateral load-resisting systems. They found that of all the types of lateral load-resisting systems, the dual system showed a very suitable structural response for this height without exceeding the limitation.

Kavyashree et al. [11] have reviewed the outrigger structural systems from conventional outriggers as a rigid connection to damped outriggers with a passive, active, semi-active and hybrid control system. They have briefly reviewed the outriggers with respect to types, analysis of tall buildings without outriggers, formulation of equations for outrigger structure to simplify the analysis, approaches to locate optimum positioning of outriggers and the damped outrigger system. The authors particularly stressed outrigger structure with semi-active control and performance enhancement of the outrigger system with the effective devices. Based on their review, they have stressed the need for precise semi-active and hybrid control techniques which would make outrigger structural systems perform better and be economically feasible.

Based on the above brief literature review, in order to find out the effective structural systems for buildings of heights 60–250 m, it is decided to conduct a comparative study. Six different heights are considered and the lateral displacements are compared for buildings consisting of different systems such as moment frame system, structural wall system, tube system and outrigger system.

In the present paper, various lateral force-resisting structural systems such as moment frame system, structural wall system, tube system and outrigger system are compared with conventional structures for six different heights (60 m, 80 m, 120 m, 180 m, 200 m and 250 m), to conclude on an efficient system that can be adopted.

3 Modelling of Different Structures

3.1 Model Specifications

Plan dimension: 45 m × 30 m.
Storey height: 4 m.
Seismic zone: III.
Location: Pune (Wind speed is 39 m/s).
Building type: Commercial building with LL 2.5 kN/m^2 and SDL 2.5 kN/m^2.
Grade of beam: M30.
Grade of column: M40.

The building is modelled for 4 different structural systems, such as moment frame, structural wall system, tube system and outrigger system using ETABS. Also, a comparison of these systems is done for 6 different heights such as 60 m, 80 m, 120 m, 180 m, 200 m and 250 m. Loads acting on the building are taken to be the same for all the structural systems.

3.2 Modelling of Moment Frame System

A moment frame building was modelled in ETABS with a plan dimension of 45 m × 30 m which can be seen in Fig. 1. The beam size adopted is 300 mm × 600 mm and the column size adopted is 800 mm × 800 mm. The thickness of the slab provided is 175 mm. The spacing of columns is between 5 and 5.75 m.

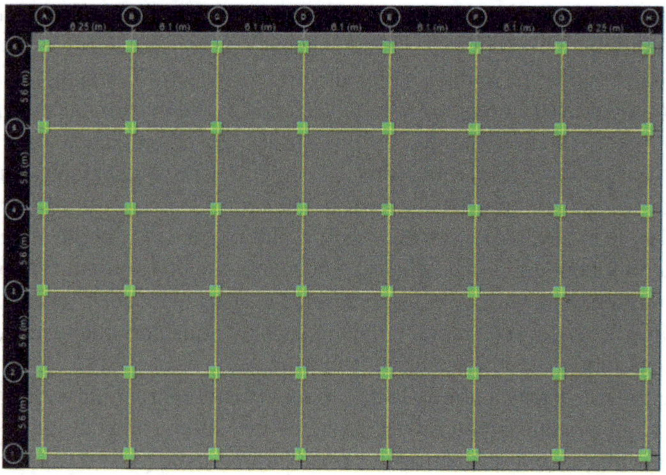

Fig. 1 Plan of moment frame building

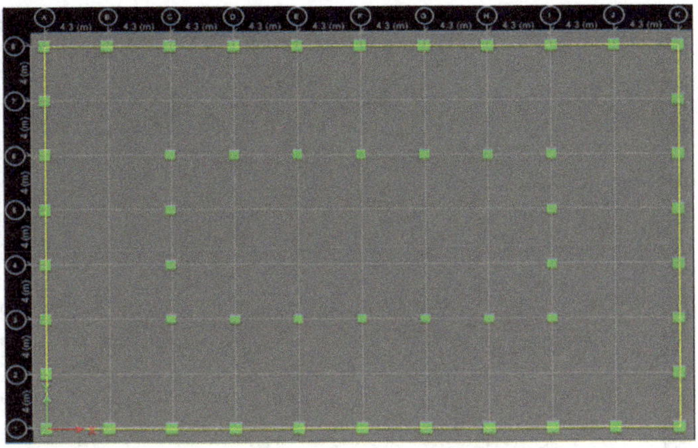

Fig. 2 Plan of tube structure

3.3 Modelling of Tube System

The columns of the tube system having sizes 800 mm × 800 mm are spaced between 4 and 4.5 m, and are connected by deep beams of 600 mm × 1200 mm, as shown in Fig. 2. The perimeter tube frame mainly resists the lateral loads acting on the building. Interior columns of 650 mm × 650 mm size are provided which resist mostly the gravity loading acting on the building. Loads acting on the building are taken to be the same for all the structural systems. The slab thickness provided is 275 mm and the stiffness modifier applied to them is 0.01. The diaphragm assigned is rigid, as there are no irregularities in the plan.

3.4 Modelling of Structural Wall System

In this system, the core walls are provided in the form of three I sections, and the thickness of the wall is 300 mm, as shown in Fig. 3. Perimeter beams of size 300 mm × 600 mm and columns of dimensions 800 mm × 800 mm are provided around the building. Perimeter beams and columns take care of the gravity loading. The slab thickness provided is 275 mm and the slab is modelled as a shell. Hence, the lateral loading acting on the building will be transferred to the wall through the meshing of the slab. The core walls take part in resisting lateral loading.

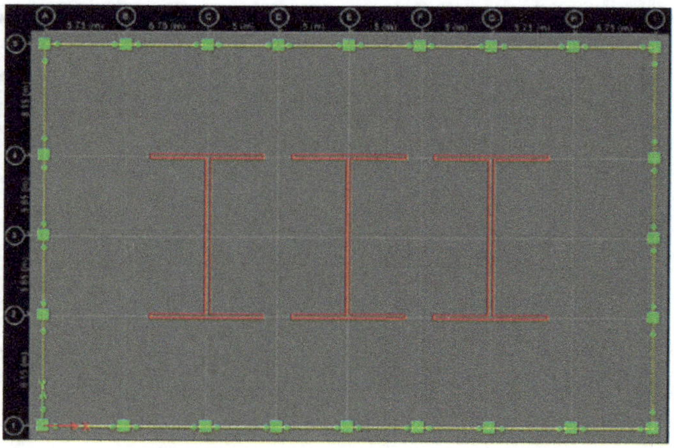

Fig. 3 Plan of structural wall system

3.5 *Modelling of Outrigger System*

The ETABS model is the same as that of the structural wall with a perimeter frame, where the structural core wall is connected to the exterior columns using outriggers, which are modelled as a spandrel beam having one storey height, as shown in Fig. 4. Outrigger is provided in one storey for a building with a height of 200 m, as shown in Fig. 5. The location of the outrigger is taken as mid-height of the building which is found to be more efficient.

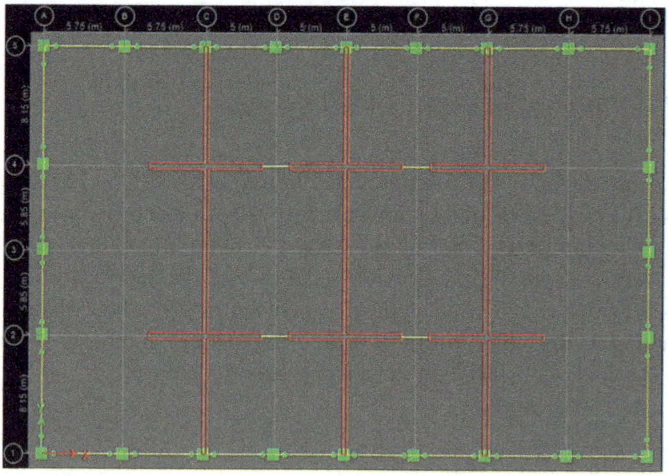

Fig. 4 Plan of outrigger system

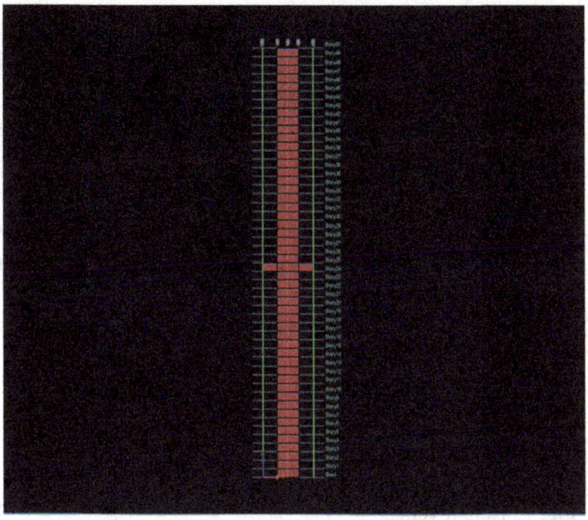

Fig. 5 Elevation of outrigger system

4 Results and Discussions

4.1 Moment Frame System

For moment frame building of 60 m, the displacement in the *X* direction and *Y* direction obtained are 64.336 mm and 93.882 mm, respectively. As the limit of displacement for a height of 60 m is 120 mm (H/500), the building displacements are within the limit at this particular height for a bare frame system.

When the building height is increased to 80 m, the displacement values obtained in the *X* and *Y* directions are 127.794 mm and 187.238 mm, respectively. The limit of displacement for a building height of 80 m is 160 mm, indicating that displacement in the *Y* direction is more than the limited value.

- Figure 6 shows that the displacement of the building increases with the building height. It is seen that the increase in displacement in the *Y* direction is around 40% more compared to displacement in the *X* direction.

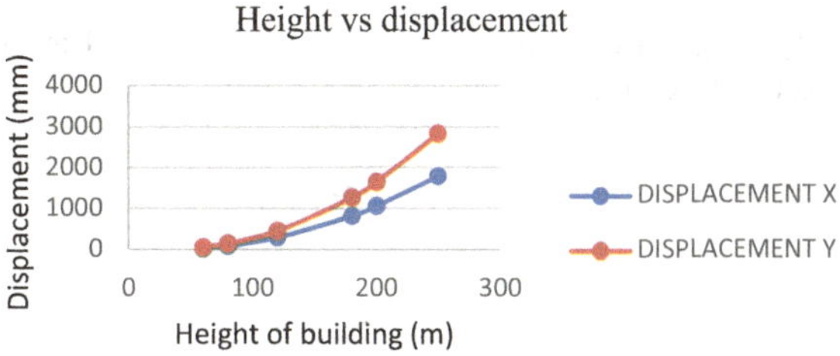

Fig. 6 Height versus displacement plot for moment frame

- For moment frame building of 60 m, the drifts in the *X* direction and *Y* direction obtained are 0.0016 and 0.00157, respectively. The limit of inter storey drift is H/250 which is 0.004, which is satisfied.
- When the height is increased to 180 m, the drift values obtained are 0.00594 and 0.00581 in the *X* and *Y* directions, respectively. Hence, the building is not safe for seismic drift when it exceeds a height of 120 m (Fig. 7).
- The depth of the perimeter beam was changed from 1200 to 2000 mm and the results were checked. By increasing the depth of the spandrel beam, the displacement in the *Y* direction was brought down to 342 mm which is within the limit (400 mm) which can be observed in Fig. 8.
- From Fig. 9, it is observed that the drift of the building increases as the building height increases. The percentage difference of drift values in the *X* and *Y* directions is around 36%.

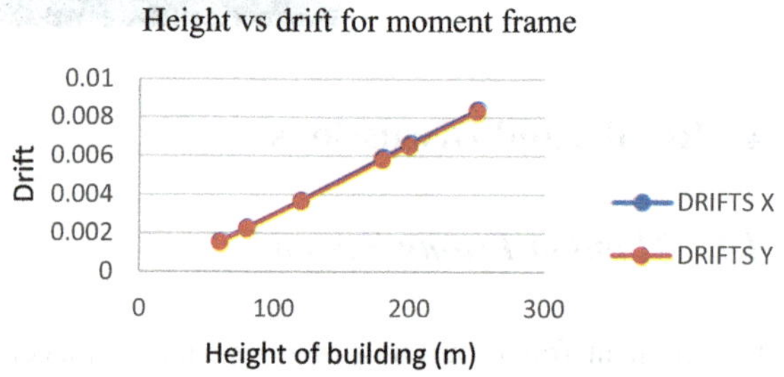

Fig. 7 Height versus displacement for tube structure with spandrel depth 1200 mm

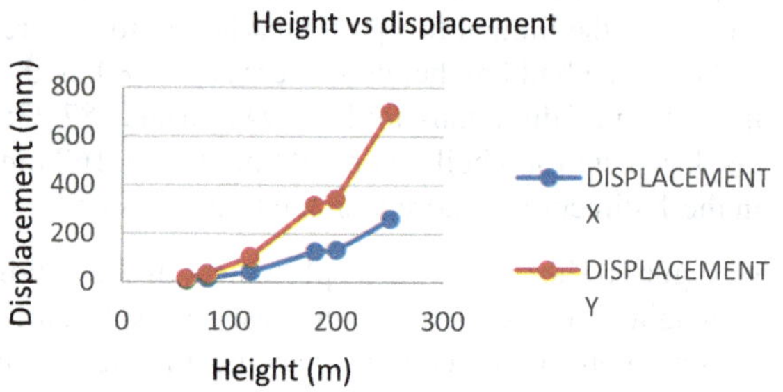

Fig. 8 Height versus displacement for tube structure with spandrel depth 2000 mm

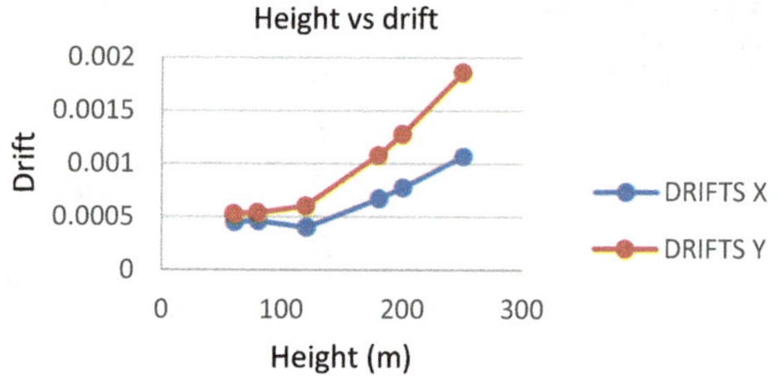

Fig. 9 Height versus drift plot for tube structure

4.2 Structural Wall with Perimeter Frame System

- From Fig. 10, it is seen that displacement is 480 mm at a height of 180 m, which exceeds the limit of H/500, (i.e. 400 mm).
- In order to reduce these displacements, link beams are provided to connect the structural walls, as shown in Fig. 11.
- For building height after 180 m, link beams of size 300 mm × 600 mm are used at each floor level. Coupling beams/link beams are used to connect shear wall or other elements and are used to withstand the lateral loading. Also, the horizontal wall thickness was changed from 300 to 450 mm to bring the displacements within the limit. By doing so, the displacement is found to be off the limit from a height of 200 m.

Fig. 10 Height versus displacement of building with structural wall and frame

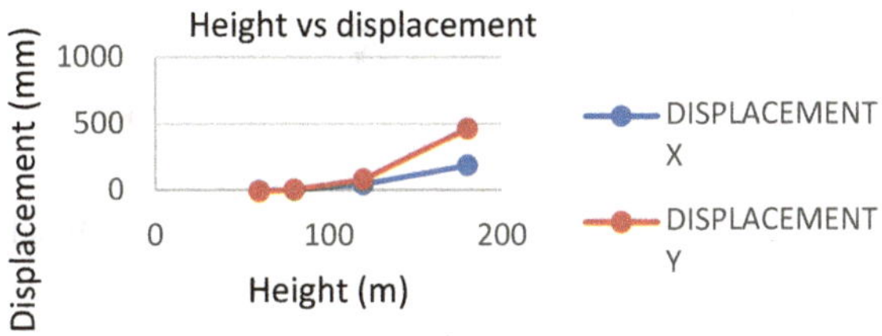

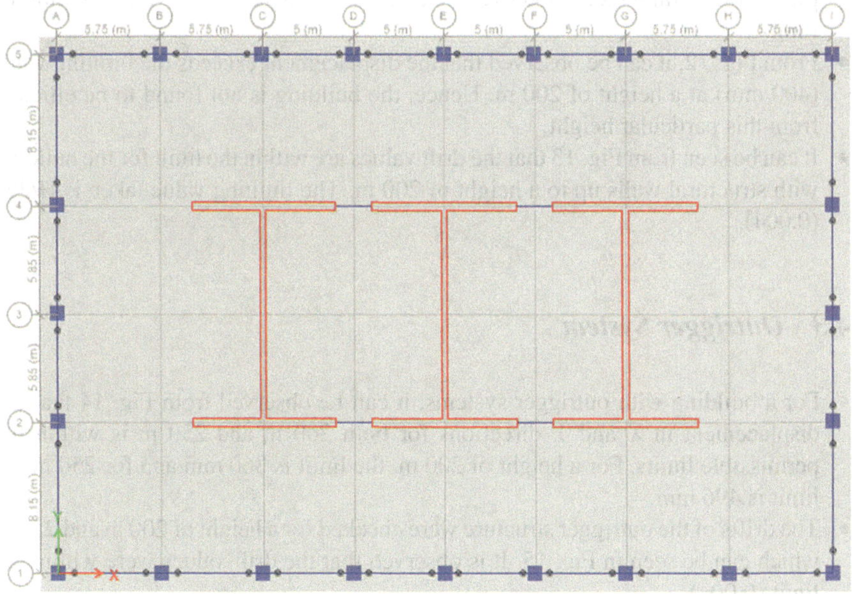

Fig. 11 Plan of structural wall system connected with link beams

Fig. 12 Height versus displacement plot of structural wall connected with link beams

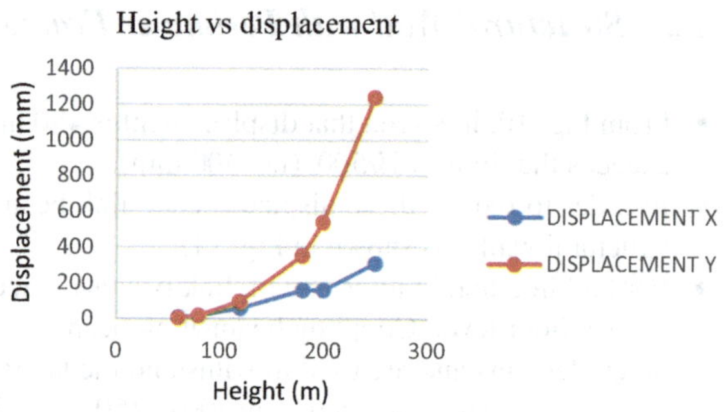

Fig. 13 Height versus drift plot for structural wall system

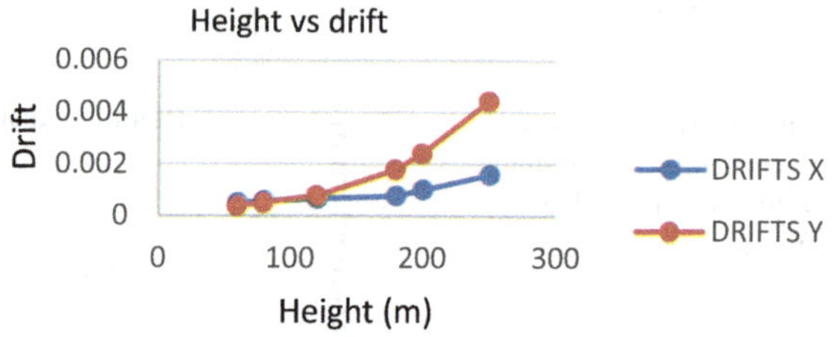

- Outriggers are introduced in the structural wall building which helps in resisting the lateral loading more effectively and bring the displacement values within the limit.
- From Fig. 12, it can be observed that the displacement exceeds the limiting value (400 mm) at a height of 200 m. Hence, the building is not found to be efficient from this particular height.
- It can be seen from Fig. 13 that the drift values are within the limit for the building with structural walls up to a height of 200 m. The limiting value taken is H/250 (0.004).

4.3 *Outrigger System*

- For a building with outrigger systems, it can be observed from Fig. 14 that the displacement in *X* and *Y* directions for both 200 m and 250 m is within the permissible limits. For a height of 200 m, the limit is 360 mm and for 250 m the limit is 496 mm.
- The drifts of the outrigger structure were checked for a height of 200 m and 250 m which can be seen in Fig. 15. It is observed that the drift values were within the limit (0.004).

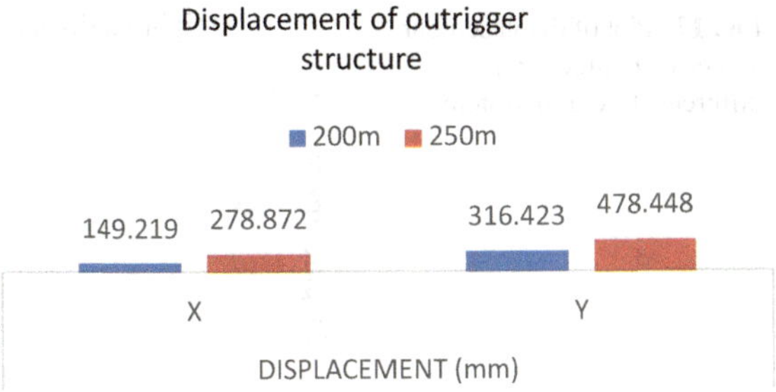

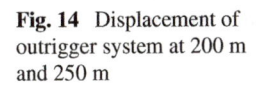

Fig. 14 Displacement of outrigger system at 200 m and 250 m

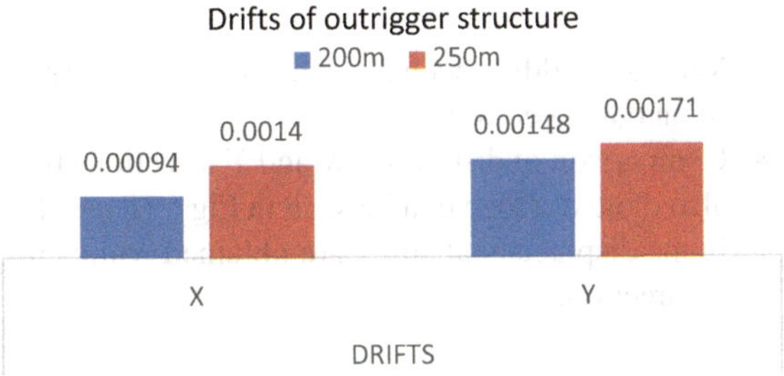

Fig. 15 Drifts in X and Y of outrigger system with heights 200 and 250 m

5 Comparison of Results

- In the present study, four different structural systems (moment frame, tube system, structural wall with perimeter frame and outrigger system) are compared for parameters such as displacement and drift, in buildings with six different heights (60 m, 80 m, 120 m, 180 m, 200 m and 250 m).
- By comparing the displacements from Figs. 16 and 17, it is seen that the outrigger is the most efficient for very tall buildings above a height of 200 m.
- The conventional moment frame system displaces highly as the height increases. This displacement can be brought down by the introduction of a structural wall system and also with tube structure.

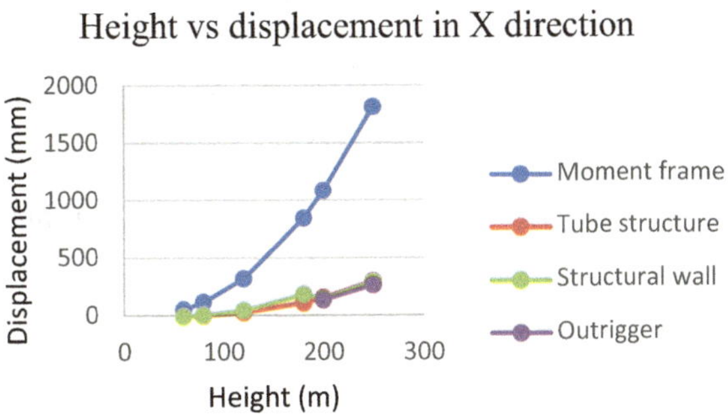

Fig. 16 Plot of displacement values in *X* direction for different structural systems

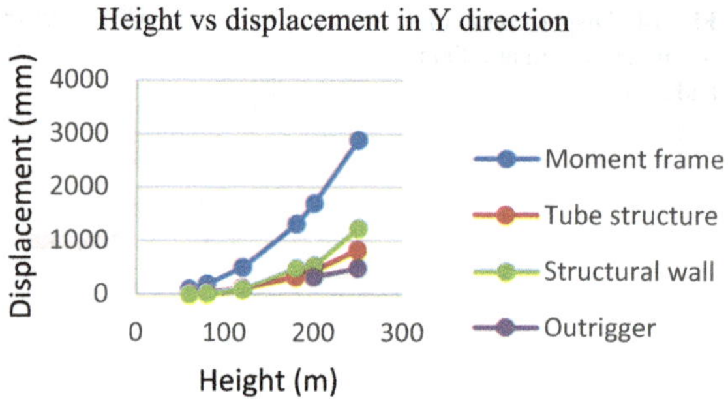

Fig. 17 Plot of displacement values in *Y* direction for different structural systems

- Very tall building such as with a height of 250 m is effective with the introduction of an outrigger system.
- Comparison of drift along *X* and *Y* directions for different structural systems was also plotted which can be seen in Figs. 18 and 19, respectively. Similar results as in the displacement study are obtained while checking the drift values along the *X* direction.

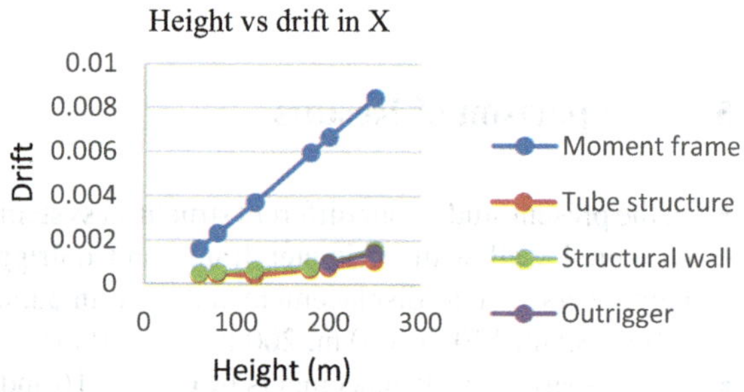

Fig. 18 Plot of drifts in *X* direction for different structural systems

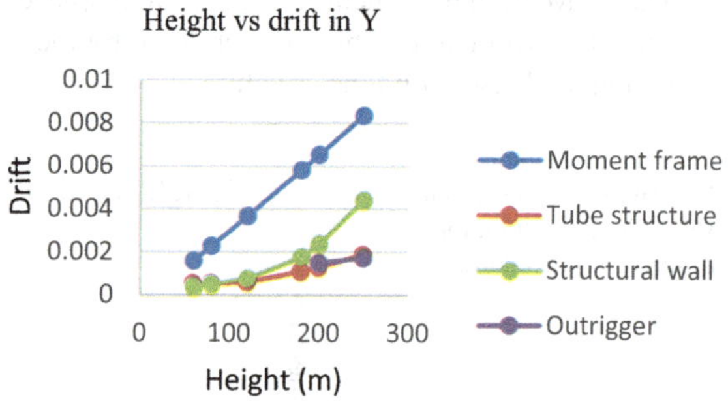

Fig. 19 Plot of drifts in *Y* direction for different structural systems

6 Conclusions

- From the study on the conventional system, it is observed that the lateral displacements (both X and Y) are within permissible limits for smaller heights up to 60 m. As the building height increases, the lateral displacement also increases.
- It is observed that increasing the member sizes for moment frame systems in order to control the deflection tends to be uneconomical, and hence other structural systems must be adopted.
- Structural wall systems and tube systems can be adopted for buildings from 60 to 200 m.
- In the case of the structural wall frame systems beyond 200 m, outriggers can be used to connect the core to the exterior columns so that the lateral displacements can be controlled effectively.
- For the max height for a tall building of 250 m (according to IS 16700: 2017), the outrigger system is found to be more suitable. The outrigger structural systems are not only proficient in controlling the top displacements but also play a substantial role in reducing the inter storey drifts.

Thus, from the study conducted and comparison made using six different building heights for three-dimensional structural systems with parameters such as lateral displacement, inter storey drifts, time period and base shear, it has been observed that outrigger structural systems are very efficient not only in controlling the top displacements but also reducing the inter storey drifts.

References

1. Badami S, Suresh MR (2014) A study on behavior of structural systems for tall buildings subjected to lateral loads. Int J Eng Res Technol 3(7):2278–3181
2. Baraskar NB, Kawade UR (2015) Structural performance of RC structural wall system over conventional beam-column system in G+15 storey building. Int J Eng Res Gen Sci 3(4):2091–2730
3. Titiksha A, Gupta MK (2015) A study of various structural framing systems subjected to seismic loads. Int J Civ Eng 2(4):2348–8352
4. Jose SM, Rao AU, Abubaker KA (2017) Comparative study on the effect of lateral stiffness on different structural framing systems subjected to lateral loads. Int J Civ Eng Technol 8(6):398–410
5. Hemavati S, Ramya K (2017) The structural behavior of lateral load resisting system induced in tall buildings—a comparative study. J Chem Pharm Chem 83:0974–2115
6. Yadav AK, Rai A (2017) A seismic comparison of RC special moment resisting frame considering regular and irregular structures. Int J Eng Sci Res Technol 6(2):2277–9655
7. Baygi, Khazaee (2019) The optimal number of outriggers in a structure under different lateral loadings. J Inst Eng (India) Ser A 100:753–761. http://doi.org/10.1007/s40030-019-00379-7
8. Hafner I, Vlašić A, Kišiček T, Renić T (2019) Parametric analysis of the shear lag effect in tube structural systems of tall buildings. Appl Sci 11:278. https://doi.org/10.3390/app11010278
9. Sawant VD, Bogar VM (2019) Parameters comparison of high rise RCC structure with steel outrigger and belt truss by linear and non-linear analysis. Int J Eng Trends Technol 67(7):15–23

10. Nassani DE, Ali K (2020) Lateral load resisting systems in high-rise reinforced concrete buildings. Eur J Sci Technol. http://doi.org/10.31590/ejosat.808269
11. Kavyashree BG, Patil S, Rao VS (2021) Evolution of outrigger structural system: a state-of-the-art review. Arab J Sci Eng. https://doi.org/10.1007/s13369-021-06074-9

Free Vibration Response of Functionally Graded Cylindrical Shells Using a Four-Node Flat Shell Element

R. B. Dahale, S. D. Kulkarni, and V. A. Dagade

Abstract In the present work, a flat shell quadrilateral element with four nodes is developed for the free vibration analysis of Functionally Graded Cylindrical Shells by modifying the 'discrete Kirchhoff quadrilateral plate element' based on Reddy's third-order theory developed earlier by the second author. This four-node shell element has seven degrees of freedom (DOF) per node, namely three displacements, two rotations of mid-plane, and two transverse shear strain components. In the present work, Functionally Graded Cylindrical shell panels with simply supported and clamped boundary conditions, with various R/a ratios and with different volume fraction indices, are analysed for free vibration response. The property variation through the thickness is according to the power law. To assess the performance of the developed element, the results, of non-dimensionalised frequencies, are compared with the results presented in the literature. A comparison of results shows that the developed element yields quite accurate results with a coarser mesh, which leads to computational efficiency.

Keywords Functionally graded material · Free vibration · Cylindrical shell · Transformation matrix

1 Introduction

Functionally graded materials (FGM) are used in the aerospace industry as well as in other fields of modern technology, especially where structures are subjected to high differential temperatures. FG materials are composite material that is microscopically heterogeneous and made from metals and ceramic combination. Ceramic plays a role of a thermal barrier and the metal gives the required ductility. Unlike laminated composites, these are free from de-bonding, as the transition from one phase to another is smooth. Many researchers have tried to understand this material

R. B. Dahale (✉) · S. D. Kulkarni · V. A. Dagade
Department of Civil Engineering, College of Engineering Pune, Pune, Maharashtra 411005, India
e-mail: rajendra2299@gmail.com

M. S. Ranadive et al. (eds.), *Recent Trends in Construction Technology and Management*, Lecture Notes in Civil Engineering 260,
https://doi.org/10.1007/978-981-19-2145-2_76

with different geometrical variations, supporting condition variations, material variations, and volume fraction of those material variations. The plethora of mathematical techniques of problem-solving including the 3D elasticity approach, meshless technique, 2D higher-order shear deformation theories with an analytical approach, and 2D theories with finite element solutions have been tried to get its dynamic and static study. The finite element having optimum nodes (4 nodes) and degrees of freedom (7 DOF) is the one used here with the third-order theory.

2 Material Properties of FGM

FG Material has two phases, namely the ceramic phase and the metal phase. The properties change gradually from one phase to another. The material property variation, through the thickness, is predicted by Voigt's rule of mixtures (ROM).

$$P(_z) = P_b + (P_t - P_b)(0.5 + / - z/h)^n \tag{1}$$

The material property variation profile through the thickness of the shell is characterised by power-law index n ($0 \leq n \leq \infty$). P_b is the material properties at the bottom layer, whereas P_t is for the top layer of the shell. The shell thickness is 'h'. Poisson's ratio is the same for both materials.

It is attainable to generate an infinite number of composition arrangements in the metal-ceramic phase by differing the value of 'n'. The variations in the volume fraction of the ceramic phase and metal phase by non-dimensional thickness coordinate (z/h) are set out in Fig. 1 for five values of the power-law indices. 'n_c' for 'ceramic top configured' shell and 'n_m' for 'metal top configured' shell show the volume fraction index in the graphs.

It is understandable from the graphs that as volume fraction index 'n_c' increases in the 'ceramic top configured FGM' shell, the ceramic content drops, and metal

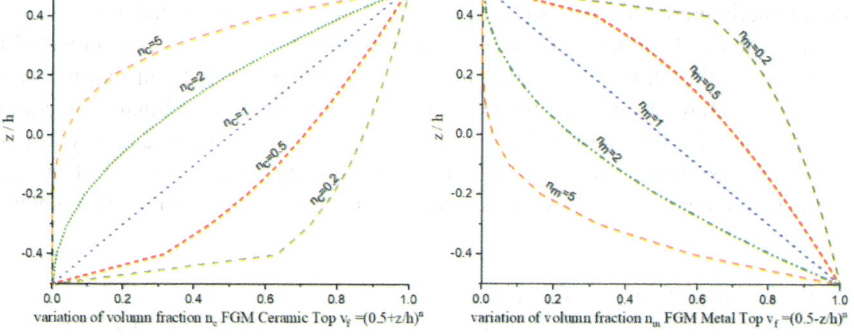

Fig. 1 Effect of variation of the volume fraction through the thickness of a shell

content increases. It means at 'n_c' = zero, the shell turns to ceramic whereas for 'n_c' = ∞, it turns to metal and vice versa in the 'Metal top configured' shell.

2.1 Displacement Field Approximation

Consider a cylindrical shell (Fig. 2) with a thickness 'h' with a ceramic-layered top and metal-layered top configuration. The shell's middle surface is known to be a reference plane where z is zero and the top surface is at z = '$h/2$' and the bottom surface is at z = '$-h/2$'. These shells are analysed using the plate element developed by the second author and his co-workers and presented in Deshpande [1] and Deshpande [2], with necessary modifications. The plate element is based on Reddy's Third-Order Theory.

The displacement field approximation is as follows:

$$w(x, y, z, t) = wo(x, y, t), \tag{2}$$

$$u(x, y, z, t) = uo(x, y, t) - zwo_d + R(z)\psi_0(x, y, t) \tag{3}$$

where

$$u = \begin{bmatrix} u_x \\ u_y \end{bmatrix}, wo_d = \begin{bmatrix} w_{o,x} \\ w_{o,y} \end{bmatrix}, u_o = \begin{bmatrix} u_{ox} \\ u_{oy} \end{bmatrix}, \psi_o = \begin{bmatrix} \psi_{ox} \\ \psi_{oy} \end{bmatrix}$$

and

$$R(z) = \left[z - \left\{4z^2/\left(3h^2\right)\right\}\right] \tag{4}$$

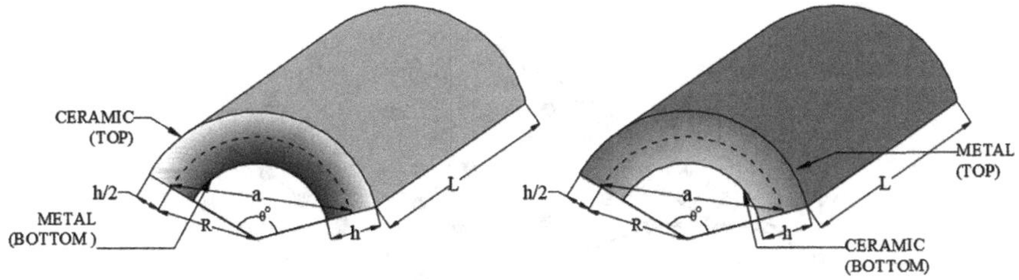

Fig. 2 Geometry of FGM ceramic top and FGM metal top configured cylindrical shells

2.2 Finite Element Formulation of Flat Shell Element

The cylindrical shell shown in Fig. 2 is discretised by using the quadrilateral elements as shown in Fig. 3. Before assembling these elements, it is required to transform the actions and displacements from local directions to global directions. The concept of two coordinate systems used for obtaining transformation matrices given by Clough [3] is extended here for the free vibration analysis of moderately a thick FG cylindrical shell. The orientations of local, global, and surface coordinate axes are shown in Fig. 3.

'A fixed set of surface coordinates (ξ_1, ξ_2, ξ_3) in which ξ_3 is taken normal to the shell surface at every point is shown in Fig. 3. Cartesian coordinates (*x*, *y*, *z*) are shown in the same figure. It is assumed that rotation about the normal to the tangent plane would be negligible in the actual shell. Accordingly, rotational degrees of freedom referred to as the surface coordinates (ξ_1, ξ_2) per nodal point are only included in the base system describing the assembled structure by Clough [3]'. Thus in this approach, the local translational degrees $u_{0x}^{'i}$, $u_{0y}^{'i}$, $w_0^{'i}$ are transformed to a base coordinate system with reference to global Cartesian coordinates (*x*, *y*, *z*), and the rotational degrees $\theta_{0x}^{'i}$, $\theta_{0y}^{'i}$, $\psi_{0x}^{'i}$, $\psi_{0y}^{'i}$ are transformed to a surface coordinate system (ξ_1, ξ_2, ξ_3), neglecting the contribution to (ξ_3).

$U_i^{e'}$ is local and U_i^e is the global displacement vector for *i*th node shown in the following expressions.

$$U^{e'} = \left[u_{0x}^{'i} \; u_{0y}^{'i} \; w_0^{'i} \; \theta_{0x}^{'i} \; \theta_{0y}^{'i} \; \psi_{0x}^{'i} \; \psi_{0y}^{'i} \right]^{\mathrm{T}} \tag{5}$$

$$U^{e} = \left[u_{0x}^{i} \; u_{0y}^{i} \; w_0^{i} \; \theta_{0\xi 1}^{i} \; \theta_{0\xi 2}^{i} \; \psi_{0\xi 1}^{i} \; \psi_{0\xi 2}^{i} \right]^{\mathrm{T}} \tag{6}$$

Transformation matrix T^i for transforming the local translational DOF to global translational DOF for an *i*th node is given by the following expression.

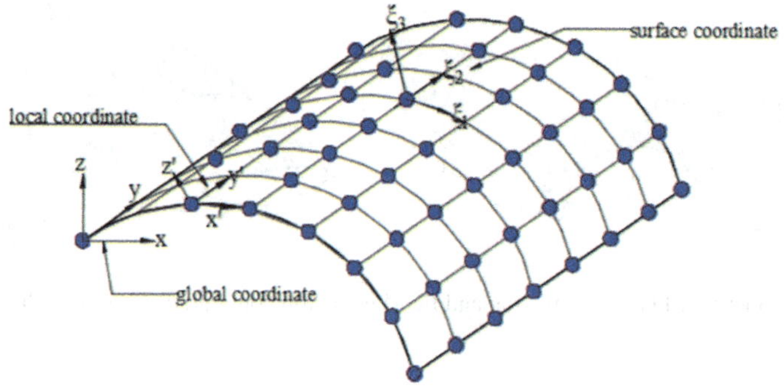

Fig. 3 Local, global, and surface coordinate systems in discretised cylindrical shell

$$\begin{bmatrix} u'^{i}_{0x} \\ u'^{i}_{0y} \\ w'^{i}_{0} \end{bmatrix} = \begin{bmatrix} \lambda_{11} & \lambda_{12} & \lambda_{13} \\ \lambda_{21} & \lambda_{22} & \lambda_{23} \\ \lambda_{31} & \lambda_{32} & \lambda_{33} \end{bmatrix} \begin{bmatrix} u^{i}_{0x} \\ u^{i}_{0y} \\ w^{i}_{0} \end{bmatrix} \quad \text{i.e.} \quad u'^{i}_{0} = T_i u^{i}_{0} \tag{7}$$

where λ_{11}, λ_{12}, etc. are the direction cosines of the local (primed) axes with respect to global (unprimed) axes.

The relationship between the local rotations and global rotations for an *i*th node is given by the following expression.

$$\begin{bmatrix} \theta'^{i}_{0x} \\ \theta'^{i}_{0y} \\ \theta'^{i}_{0z} \end{bmatrix} = \begin{bmatrix} \lambda_{22} & \lambda_{21} & \lambda_{23} \\ \lambda_{12} & \lambda_{11} & \lambda_{13} \\ \lambda_{32} & \lambda_{31} & \lambda_{33} \end{bmatrix} \begin{bmatrix} \theta^{i}_{0x} \\ \theta^{i}_{0y} \\ \theta^{i}_{0z} \end{bmatrix} \quad \text{i.e.} \quad \theta'^{i}_{0} = \overline{T}_i \theta^{i}_{0} \tag{8}$$

The relationship between the surface rotations $\theta^{i}_{0\xi}$ and θ^{i}_{0} for the *i*th node can be obtained as follows, For a circular cylindrical shell, the angle between the global *x*-axis and tangent ξ_1 can be obtained easily as this tangent is actually tangent to the circle $x^2 + z^2 = R^2$ in the *x*–*z* plane. Differentiating this expression for the circle yields $\tan\alpha = -x/z$. α is the angle between the axes ξ_1 and *x*. In addition, as the axis ξ_2 is along the global *y*-axis, the angle between them is always zero. Using this, the following relationship between $\theta^{i}_{0\xi}$ and θ^{i}_{0} for the *i*th node can be obtained as given below

$$\begin{bmatrix} \theta^{i}_{0\xi_1} \\ \theta^{i}_{0\xi_2} \\ \theta^{i}_{0\xi_3} \end{bmatrix} = \begin{bmatrix} 1 & 0 & 0 \\ 0 & \cos(\alpha) & \cos\left(\frac{\pi}{2} - \alpha\right) \\ 0 & \cos\left(\frac{\pi}{2} + \alpha\right) & \cos(\alpha) \end{bmatrix} \begin{bmatrix} \theta^{i}_{0x} \\ \theta^{i}_{0y} \\ \theta^{i}_{0z} \end{bmatrix} \quad \text{i.e.} \quad \theta^{i}_{0\xi} = T^{i}_{\xi} \theta^{i}_{0} \tag{9}$$

Using Eqs. (8) and (9), relationship between the θ'^{i}_{0} and $\theta'^{i}_{0\xi}$ is given as

$$\theta^{i}_{0\xi} = T^{i}_{\xi} \overline{T}^{T^{i}}_{\xi} \theta'^{i}_{0} = \overline{T}^{\xi}_{i} \theta'^{i}_{0} \tag{10}$$

The transformation matrix T^{ξ}_{i} of size 2 × 2, which relates the local rotations θ'^{i}_{0x}, θ'^{i}_{0y} to tangential rotations $\theta^{i}_{0\xi 1}$, $\theta^{i}_{0\xi 2}$, are obtained by removing the third row and third column of the transformation matrix $\overline{T_{\xi}}^{Ti}$. T^{ξ}_{i} is also used for transforming ψ'^{i}_{0x}, ψ'^{i}_{0y}, to tangential shell directions. Finally, the element transformation matrix of size 28 × 28 T^{e} is given as

$$T^e = \begin{bmatrix} T_1 & 0 & 0 & 0 & 0 & 0 & 0 & 0 & 0 & 0 & 0 & 0 \\ 0 & T_1^\xi & 0 & 0 & 0 & 0 & 0 & 0 & 0 & 0 & 0 & 0 \\ 0 & 0 & T_1^\xi & 0 & 0 & 0 & 0 & 0 & 0 & 0 & 0 & 0 \\ 0 & 0 & 0 & T_2 & 0 & 0 & 0 & 0 & 0 & 0 & 0 & 0 \\ 0 & 0 & 0 & 0 & T_2^\xi & 0 & 0 & 0 & 0 & 0 & 0 & 0 \\ 0 & 0 & 0 & 0 & 0 & T_2^\xi & 0 & 0 & 0 & 0 & 0 & 0 \\ 0 & 0 & 0 & 0 & 0 & 0 & T_3 & 0 & 0 & 0 & 0 & 0 \\ 0 & 0 & 0 & 0 & 0 & 0 & 0 & T_3^\xi & 0 & 0 & 0 & 0 \\ 0 & 0 & 0 & 0 & 0 & 0 & 0 & 0 & T_3^\xi & 0 & 0 & 0 \\ 0 & 0 & 0 & 0 & 0 & 0 & 0 & 0 & 0 & T_4 & 0 & 0 \\ 0 & 0 & 0 & 0 & 0 & 0 & 0 & 0 & 0 & 0 & T_4^\xi & 0 \\ 0 & 0 & 0 & 0 & 0 & 0 & 0 & 0 & 0 & 0 & 0 & T_4^\xi \end{bmatrix} \tag{11}$$

After deriving the expressions for transformation matrix T^e, local stiffness matrix $K^{e'}$, local mass matrix $M^{e'}$, and local load vector $P^{e'}$, finally the element stiffness matrix K^e of size 28 × 28, element mass matrix M^e of size 28 × 28, and element load vector P^e of size 28 × 1 with reference to global axes are given as

$$\begin{aligned} M^e &= T^{e^T} M^{e'} T^e, \\ K^e &= T^{e^T} K^{e'} T^e, \\ P^e &= T^{e^T} P^{e'} \end{aligned} \tag{12}$$

2.3 *Assembly of Element Matrices*

Summing up the contributions of all elements, Hamilton's principle yields

$$\overline{M}\ddot{U} + \overline{K}\ddot{U} = \overline{P} \tag{13}$$

where $\overline{K}$, $\overline{M}$, and $\overline{P}$ are the assembled counterparts of matrices K^e, M^e, and P^e. For synchronous vibration, the $\overline{P}$ (load vector) is set to zero. The undamped natural frequencies ω_n are obtained by solving the eigenvalue problem.

3 Numerical Results

Using the above formulation, a computer program in FORTRAN compiler 95 is developed for the free vibration analysis. The formulation is evaluated by comparing the results of non-dimensionalised natural frequencies with the results available in the literature of Pradyumna and Bandyopadhyay [4], Aragh and Hedayati [5], and Zhu et al. [6]. In all the problems, except that presented in Table 2, the full shell 'ceramic top' configured panel shown in Fig. 2 is modelled with an $M \times M$ mesh of equal size elements.

The material properties considered for metal and ceramic in all the problems unless specified are:

Aluminium: Modulus of Elasticity $E_\mathrm{m} = 70$ GPa; density $\rho_\mathrm{m} = 2707$ kg/m^3; Poisson's ratio $\upsilon_\mathrm{m} = 0.3$ and
Zirconia: Modulus of Elasticity $E_\mathrm{c} = 380$ GPa; density $\rho_\mathrm{c} = 3000$ kg/m^3; Poisson's ratio $\upsilon_\mathrm{c} = 0.3$.

The obtained natural frequencies are non-dimensionalised as

$$\overline{\omega} = \omega a^2 \sqrt{\rho_m h / D_m}, \quad \text{where } D_m = E_m h^3/(12(1 - \upsilon_m^2))$$

3.1 Simply Supported Cylindrical Shell Panels

The non-dimensionalised fundamental frequencies of the FGM cylindrical shell with all edges simply supported are presented in Table 1. The table highlights the radius to span ratio variation from deep to shallow shell for a moderately thick ($a/h = 10$) shell. When the 'R/a' ratio is 0.5, the shell is deep and when the 'R/a' is 10 it becomes a shallow shell. The results are obtained using 12 × 12, 16 × 16, and 24 × 24 mesh and are compared with those presented in Pradyumna [4] obtained using 8 node elements and 2D analytical results in Aragh [5], based on the higher-order shear deformation theories.

From Table 1, it is observed that the frequencies are inversely proportional to the 'R/a' ratio. Stiffness is higher for the deep shell, whereas it reduces in the shallow shell and so the frequencies decrease. It is also important to note that with an increase in volume fraction, the shell becomes metallic and frequencies get reduced. With the increase in volume fraction index, the mass of the shell decreases, but it has less effect on frequencies than the stiffness.

Table 2 shows the results of non-dimensionalised natural frequencies for FGM 'ceramic top configured' and 'metal top configured' rectangular cylindrical shell panels with volume fractions $n_\mathrm{c} = 1$ and $n_\mathrm{m} = 1$, respectively. The panels with a subtended angle of 30° and 90°, $L/R = 1$, $h/R = 0.1$, and $R = 1$ m are solved with the developed element considering various mesh sizes of $M \times M$ with unequal size elements.

The material used as metal is Aluminium and Ceramic is Zirconia, the properties of which are as follows:

Aluminium: Modulus of Elasticity $E_\mathrm{m} = 70$ GPa; density $\rho_\mathrm{m} = 2707$ kg/m^3, Poisson's ratio $\upsilon_\mathrm{m} = 0.3$,

Table 1 Non-dimensionalised fundamental frequency parameter for (SSSS) cylindrical FG shell panels with different *R*/*a* ratios

n_c	*R*/*a* →	0.5	1	5	10
0	Pradyumna [4]	68.8645	51.5216	42.2543	41.9080
	Aragh [5]	69.9700	52.1003	42.7160	42.3677
	Present (12 × 12)	70.2200	52.9231	42.0741	41.7270
	Present (16 × 16)	68.5123	50.6418	42.0952	41.7570
	Present (24 × 24)	66.7074	50.3306	42.1069	41.7570
0.2	Pradyumna [4]	64.4001	47.5968	40.1621	39.8472
	Aragh [5]	65.4304	48.1341	39.0836	38.7568
	Present (12 × 12)	67.7320	49.0650	40.3388	39.9896
	Present (16 × 16)	66.0990	48.8372	40.3596	40.0186
	Present (24 × 24)	64.3677	48.5362	40.3716	40.0390
0.5	Pradyumna [4]	59.4396	43.3019	37.2870	36.9995
	Aragh [5]	60.3574	43.7689	36.0944	35.7891
	Present (12 × 12)	61.8813	44.7951	36.2649	35.9135
	Present (16 × 16)	60.4155	44.5774	36.2844	35.9392
	Present (24 × 24)	58.8568	44.2988	36.2964	35.9577
1	Pradyumna [4]	53.9296	38.7715	33.2268	32.9585
	Aragh [5]	54.7141	39.1621	32.0401	31.7608
	Present (12 × 12)	55.6641	40.2597	32.1270	31.7834
	Present (16 × 16)	54.3597	40.0517	32.1445	31.8053
	Present (24 × 24)	52.9788	39.7949	32.1559	31.8213
2	Pradyumna [4]	47.8259	34.3338	27.4449	27.1789
	Aragh [5]	48.5250	34.6852	27.5614	27.3238
	Present (12 × 12)	49.4001	35.7563	28.5729	28.2627
	Present (16 × 16)	48.2350	35.5578	28.5886	28.2813
	Present (24 × 24)	47.8259	35.3223	28.5994	28.2952

Zirconia: Modulus of Elasticity $E_c = 151$ GPa, density $\rho_c = 3000\ \text{kg/m}^3$, Poisson's ratio $\upsilon_c = 0.3$.

The obtained natural frequencies are non-dimensionalised as

$$\overline{\omega} = \omega L^2 / h\sqrt{\rho_c / E_c}$$

It can be perceived from the tables that the results obtained using the present element with a rectangular profile are close to the 2D analytical results of Zhu (2014). The maximum percentage difference as compared to the 2D analytical results of Zhu (2014) is less than 1.6%. The table also gives the insight that for subtended angle 30°, the frequencies are more though it is a shallow shell ('*R*/*a*' = 1.92) in comparison

Table 2 Convergence of frequencies for functionally graded all-round simply supported (SSSS) rectangular cylindrical shells

FGM config.	θ	Results	$\overline{\omega}_1$	$\overline{\omega}_2$	$\overline{\omega}_3$	$\overline{\omega}_4$	$\overline{\omega}_5$	$\overline{\omega}_6$
'Ceramic Top'	30°	Zhu [6]	1208.65	1939.38	1982.04	3100.41	3593.17	3683.06
		Present (12 × 12)	1213.65	1968.95	1984.76	3097.74	3598.06	3689.05
		Present (16 × 16)	1211.72	1967.54	1982.92	3097.77	3597.30	3686.88
		Present (24 × 24)	1210.32	1966.07	1981.33	3097.28	3596.71	3685.71
	90°	Zhu [6]	763.86	766.02	1208.79	1228.35	1464.65	1604.52
		Present (12 × 12)	771.79	773.03	1212.47	1228.62	1464.68	1607.40
		Present (16 × 16)	769.94	770.13	1210.27	1227.93	1464.09	1605.29
		Present (24 × 24)	768.10	768.48	1208.77	1227.55	1463.53	1603.64
'Metal Top'	30°	Zhu [6]	1227.60	1919.12	2001.33	3120.25	3649.26	3683.06
		Present (12 × 12)	1229.98	1936.20	2000.02	3113.17	3649.58	3689.14
		Present (16 × 16)	1228.77	1932.00	1999.18	3114.51	3650.24	3686.95
		Present (24 × 24)	1227.89	1928.70	1998.36	3114.97	3650.57	3685.77
	90°	Zhu [6]	767.33	774.87	1227.74	1228.35	1470.40	1615.29
		Present (12 × 12)	772.80	779.70	1230.37	1230.80	1469.95	1616.76
		Present (16 × 16)	770.44	777.21	1229.11	1229.92	1469.65	1615.19
		Present (24 × 24)	768.71	775.47	1228.22	1229.45	1469.28	1613.92

with 90° ('*R*/*a*' = 0.639) shells. With a 30° subtended angle, the shell becomes thick as '*a*/*h*' ratio is 5.29, whereas for the 90° subtended angle '*a*/*h*' ratio is 15.708, and the shell becomes very thin. The span to thickness ratio has affected the frequency values. Metal top configured cylindrical shell gives slightly higher values of the non-dimensionalised frequencies for both the subtended angles.

In Figs. 4, 5, 6 and 7, the graphs show the variations of the first six non-dimensionalised frequencies of FGM Aluminium/Zirconia all-round simply supported (SSSS) moderately thick ('*a*/*h*' = 10) cylindrical shells with volume fraction index for different '*R*/*a*' ratios. Figures 4 and 5 present the variation for deep shell '*R*/*a*' = 0.5 and 1, whereas Figs. 6 and 7 present for a shallow shell with '*R*/*a*' = 5 and 10. The results were obtained using a mesh size of 24 × 24.

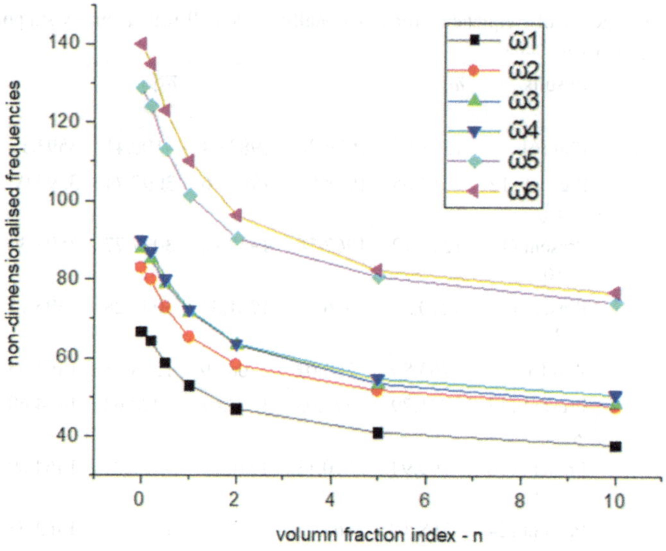

Fig. 4 Variations of the first six consecutive frequencies with volume fraction for ('R/a' = 0.5)

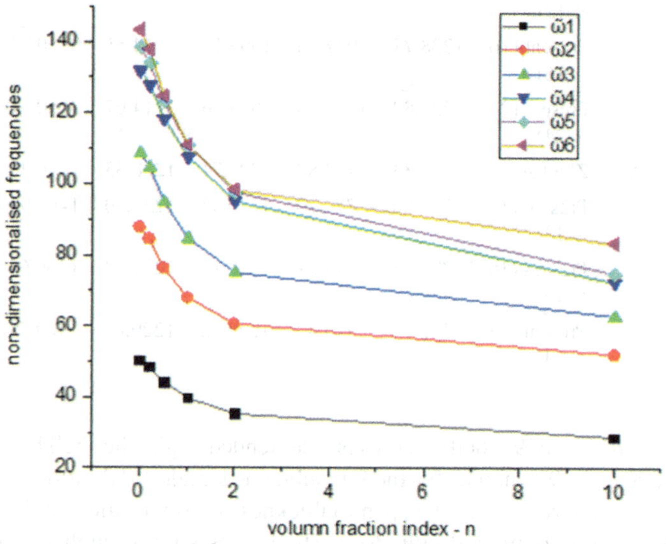

Fig. 5 Variations of the first consecutive frequencies with volume fraction for ('R/a' = 1)

With the increase in volume fraction as the shell becomes metallic, frequencies get reduced. From the graph, it is also observed that as the 'R/a' ratio increases, the shell becomes shallow thereby resulting in reduced values of frequencies.

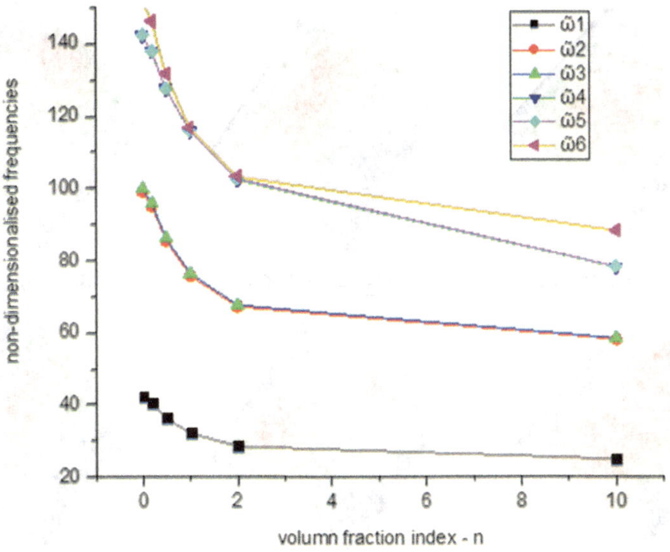

Fig. 6 Variations of the first consecutive frequencies with volume fraction for (‘R/a’ = 5)

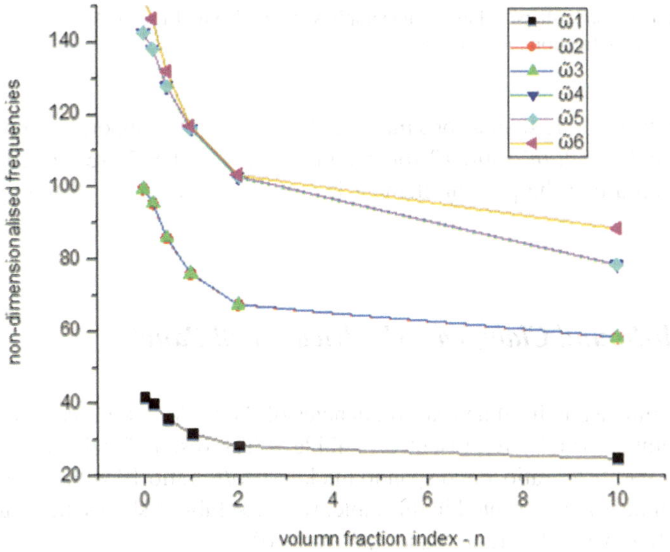

Fig. 7 Variations of the first consecutive frequencies with volume fraction for (‘R/a’ = 10)

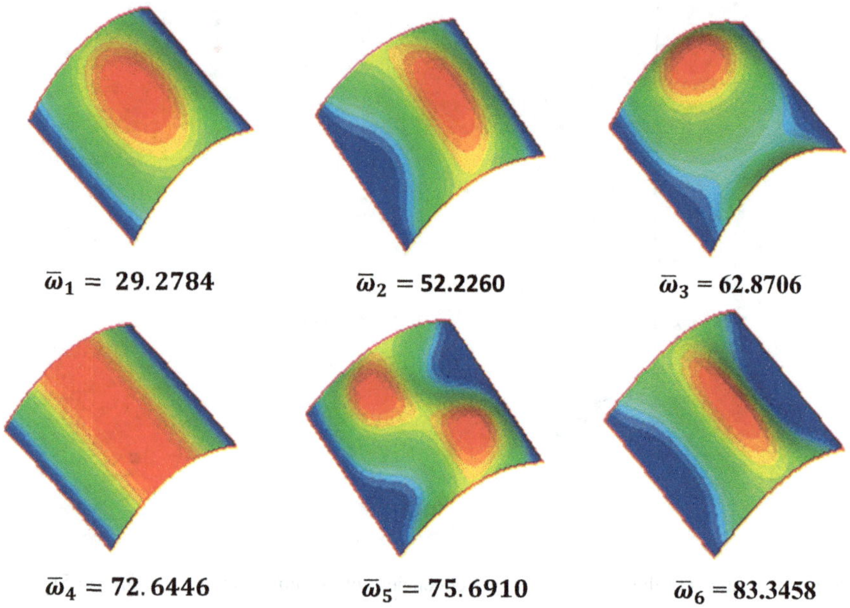

Fig. 8 First six mode shapes of all-round simply supported cylindrical shells for ($R/a = 1$ and $a/h = 10$) for volume fraction index $n = 5$

The first six mode shapes for a moderately thick simply supported deep shell with '*R*/*a*' ratio 1 and '*a*/*h*' ratio 10 for volume fraction index 5 are plotted in Fig. 8. It is observed that the present element is capable of capturing all the modes very precisely.

3.2 All-Round Clamped Cylindrical Shell Panel

The non-dimensionalised natural frequencies of the FGM cylindrical shell with all-round clamped panels are presented in Table 3. The results for the deep shell with radius to span '*R*/*a*' ratio 1 and span to thickness '*a*/*h*' ratio 10 with various volume fraction indices are presented in this table, whereas Table 4 shows the results for the shallow shell with '*R*/*a*' ratio 5 and '*a*/*h*' ratio 10.

From both the tables, it is clear that the frequencies converge with finer mesh. It is confirmed from the results presented in Tables 3 and 4 that an increased volume fraction index leads to reduced frequency values. With volume fraction zero, the frequencies are around 43% more than the frequencies for volume fraction 10. Comparing both the tables, considering geometry, it is confirmed that the deep shell has more stiffness and has increased frequency values.

Table 3 Convergence of frequencies for FGM 'ceramic top configured' (CCCC) deep cylindrical shells ($R/a = 1$)

n	Results	$\overline{\omega}_1$	$\overline{\omega}_2$	$\overline{\omega}_3$	$\overline{\omega}_4$	$\overline{\omega}_5$	$\overline{\omega}_6$
0	Pradyumna [4]	94.497	–	–	–	–	–
	Present (12 × 12)	93.481	129.726	150.783	191.401	216.895	240.024
	Present (16 × 16)	93.522	129.360	150.626	190.854	215.764	239.081
	Present (24 × 24)	93.525	128.977	150.413	190.250	214.704	238.169
0.2	Pradyumna [4]	87.393	–	–	–	–	–
	Present (12 × 12)	86.756	119.962	139.340	176.948	200.868	221.730
	Present (16 × 16)	86.794	119.630	139.225	176.469	199.851	220.925
	Present (24 × 24)	86.802	119.287	139.055	175.936	198.897	220.140
0.5	Pradyumna [4]	79.569	–	–	–	–	–
	Present (12 × 12)	79.267	109.000	126.560	160.686	182.804	201.222
	Present (16 × 16)	79.300	108.671	126.450	160.221	181.798	200.449
	Present (24 × 24)	79.308	108.346	126.298	159.724	180.882	199.719
1	Pradyumna [4]	71.245	–	–	–	–	–
	Present (12 × 12)	71.265	97.470	113.084	143.485	163.631	179.534
	Present (16 × 16)	71.286	97.113	112.943	142.981	162.543	178.692
	Present (24 × 24)	71.287	96.781	112.781	142.483	161.599	177.943
2	Pradyumna [4]	62.975	–	–	–	–	–
	Present (12 × 12)	63.255	86.657	100.201	127.122	145.194	158.688
	Present (16 × 16)	63.259	86.263	100.009	126.554	144.016	157.746
	Present (24 × 24)	63.250	85.912	99.817	126.025	143.026	156.944
10	Pradyumna [4]	51.380	–	–	–	–	–
	Present (12 × 12)	51.437	72.406	82.924	105.390	119.791	130.629
	Present (16 × 16)	51.427	72.055	82.711	104.847	118.815	129.798
	Present (24 × 24)	51.405	71.728	82.495	104.327	117.959	129.058

4 Conclusions

The present element gives sufficiently accurate results for natural frequencies for all cases of functionally graded shells including the shallow and deep shells considered in this study. The natural frequency parameters decrease for cylindrical shell panels as the radius to span ratio increases for all boundary conditions. It is observed from the table that the results obtained are converging as the mesh becomes fine, for equal as well as unequal sizes of the elements. This proves the suitability of the developed element for the functionally graded cylindrical shells.

Table 4 Convergence of frequencies for FGM 'ceramic top configured' (CCCC) shallow cylindrical shells ($R/a = 5$)

n	Results	$\overline{\omega}_1$	$\overline{\omega}_2$	$\overline{\omega}_3$	$\overline{\omega}_4$	$\overline{\omega}_5$	$\overline{\omega}_6$
0	Pradyumna [4]	71.8861	–	–	–	–	–
	Present (12 × 12)	72.7577	137.7135	138.4774	194.0111	230.3940	232.2535
	Present (16 × 16)	72.6719	137.3134	138.0906	193.4305	229.0914	231.0594
	Present (24 × 24)	72.5741	136.9046	137.6921	192.7794	227.8841	229.9528
0.2	Pradyumna [4]	68.1152	–	–	–	–	–
	Present (12 × 12)	69.8543	132.3580	133.0561	186.5449	221.6653	223.4163
	Present (16 × 16)	69.7719	131.9721	132.6844	185.9895	220.4056	222.2610
	Present (24 × 24)	69.6788	131.5805	132.3038	185.3694	219.2444	221.1966
0.5	Pradyumna [4]	63.1896	–	–	–	–	–
	Present (12 × 12)	63.0075	119.6566	120.2121	168.7720	200.8786	202.3988
	Present (16 × 16)	62.9234	119.2702	119.8418	168.2247	199.6302	201.2439
	Present (24 × 24)	62.8342	118.8949	119.4791	167.6418	198.5142	200.2145
1	Pradyumna [4]	56.5546	–	–	–	–	–
	Present (12 × 12)	55.9603	106.4272	106.8503	150.1176	178.9560	180.2746
	Present (16 × 16)	55.8637	106.0034	106.4443	149.5180	177.6236	179.0237
	Present (24 × 24)	55.7701	105.6174	106.0719	148.9283	176.4842	177.9598
2	Pradyumna [4]	36.2487	–	–	–	–	–
	Present (12 × 12)	49.6639	94.2795	94.5904	132.6947	158.1199	159.2932
	Present (16 × 16)	49.5510	93.8111	94.1398	132.0170	156.6989	157.9466
	Present (24 × 24)	49.4575	93.4250	93.7674	131.4272	155.5595	156.8827
10	Pradyumna [4]	33.6611	–	–	–	–	–
	Present (12 × 12)	43.0428	79.1237	79.1789	110.2672	130.1092	131.3297
	Present (16 × 16)	42.3387	78.9925	79.2662	110.1433	129.6637	130.7659
	Present (24 × 24)	42.2368	78.6148	78.8973	109.5411	128.6415	129.8151

References

1. Deshpande GA, Kulkarni SD (2016) Free vibration analysis of functionally graded sandwich plates using discrete Kirchhoff quadrilateral element based on Reddy's third-order shear deformation theory. In: ICCMS2016, IIT Bombay, Mumbai, India, June 27–July 1, 2016
2. Deshpande GA, Kulkarni SD (2019) Free vibration analysis of functionally graded plates under uniform and linear thermal environment. Acta Mech 230:1347–1354
3. Clough RW, Johnson CP (1968) A finite element approximation for the analysis of thin shells. Int J Solids Struct 4(1):43–60
4. Pradyumna S, Bandyopadhyay JN (2008) Free vibration analysis of functionally graded panels using higher-order finite-element formulation. J Sound Vib 318:176–192

5. Aragh BS, Hedayati H (2012) Static response and free vibration of two-dimensional functionally graded metal/ceramic open cylindrical shells under various boundary conditions. Acta Mech 223:309–330
6. Zhu S, Guoyong J, Tiangui Y (2014) Free vibration analysis of moderately thick functionally graded open shells with general boundary conditions. Compos Struct 117:169–186

The Behaviour of Transmission Towers Subjected to Different Combinations of Loads Due to Natural Phenomenon

Devashri N. Varhade and R. R. Joshi

Abstract Many Transmission towers have collapsed during cyclones, tornadoes, thunderstorms, downbursts, lightning, ice disasters and natural hazards. These failures have been attributed mostly to wind loading. The research work to date emphasizes the consideration of rain load, lightning effect, snow load, etc. The present study is as an attempt to understand the behaviour of tower structures under strong wind and rainfall excitation taking lightning damage and snow load into consideration. The Finite element model of the tower structure was developed using STAAD-PRO Software. Then the computation of wind load for the six zones as per the provisions of IS 802-1995 was done and the rain load was also computed as per the theory proposed in previous research work. The effect of lightning was taken into account in the form of temperature-induced load. The snow load was computed as per the clause mentioned in IS 875(Part 4)-1987. Several load combinations were made which included dead, wind, rain, snow, lightning-induced loads. The finite element model of the vulnerable tower hit by the loads was analysed for different load combinations mentioned to estimate its ultimate strength and to find out the failure modes after analysis. Collapse modes of the transmission tower for various load combinations were compared. Lightning-induced load being instantaneous in nature was not included in any load combination and was applied as the only load acting on the structure and it was found to be most critical and hazardous. Out of the load combinations made, it was found that the load combination which included the combined effect of dead load, wind load and rain load was critical for all the zones, and failure was observed in the bracings from the bottom part.

Keywords Rainfall excitation · Lightning · Transmission tower · FEM analysis

D. N. Varhade (✉) · R. R. Joshi
Department of Civil Engineering (Applied Mechanics Division), College of Engineering Pune, Pune, India
e-mail: varhade.devashri@gmail.com

M. S. Ranadive et al. (eds.), *Recent Trends in Construction Technology and Management*, Lecture Notes in Civil Engineering 260,
https://doi.org/10.1007/978-981-19-2145-2_77

1 Introduction

Many transmission towers have failed during cyclones, tornadoes, thunderstorms, downbursts, lightning, ice disasters and other natural hazards. These failures have been generally attributed simply to wind loading. But the research work to date emphasizes the consideration of rain load, lightning effect, snow load, etc. and also on the structure along with wind load. The present study is an attempt to understand the behaviour of the tower structures under strong wind and rainfall excitation taking lightning damage and snow load into consideration for the transmission towers built in different zones of the Indian region.

Up till now, many researchers have focused on the structural properties of the transmission towers under stochastic winds. A few research works have given due importance to other natural loads such as rain, snow and lightning. Alipour et al. [1] deduced an analytical approach to characterize tornado-induced loads on the lattice structures. The wind risk of electric transmission power lines due to hurricane hazards was assessed by Reinoso et al. [2]. Huang et al. [3] provided the Bayesian approach for typhoon-induced fragility analysis of real overhead transmission lines. Tian et al. [4] in their study derived the velocity ratio of wind-driven rain and also studied its application on a transmission tower subjected to wind and rain loads. The failure analysis of transmission towers struck by tropical storms was carried out by Zhang et al. [5]. Fu et al. [6] studied the wind resistance of a lightning-damaged transmission tower. The impact of extreme weather on the transmission line structures was studied by Hathout et al. [7]. Yang et al. [8] studied the tower destruction mechanics of the Overhead transmission lines and prevention technologies in ice disasters. A case study of the failure of a 220 kV Double circuit transmission line tower due to lightning was presented by Nair et al. [9]. Many such research works till date present the importance of consideration of rain, snow and lightning-induced loads along with wind load acting on the transmission towers.

In the present study, the different combinations of the mentioned loads were made, and the most critical combination was found. The wind load acting on the structure was computed according to the specifications mentioned in the Indian Standard (IS) code 802-1995 for all the six zones given in the code. The rain load was computed as per the equation approach provided by Li et al. [4] in their research work. Computation of snow load acting on the conductors and cross arms was made according to the clause mentioned in IS 875-1987 (part 4). The considerations for lightning-induced loads were taken from the work of Nair et al. [9]. Different load combinations were made for the loads mentioned above taking the Indian topography, meteorology and physiography into consideration and the most critical load combination was found.

2 Overview of the Study Performed

In this study, different natural loads that are supposed to act on the transmission tower were computed as per various theories proposed, and various load combinations were made depending upon the zone in which the tower was supposed to be built taking the weather conditions and climate of that area into consideration. First of all, a finite element model of the tower was developed. The different regions of the country were categorized into six zones depending on the basic wind speeds. The wind loads were computed for all the six zones for prescribed basic wind speeds as per the equation approach taking different empirical factors such as risk factor, terrain category and topography into consideration. The equation approach proposed to compute the rain load was used which in turn was dependent on an empirical parameter called the velocity ratio. The lightning-induced load was computed and applied as temperature load. The snow load was computed for the areas where snowfall was expected; as per the IS code. Depending on the region in which the tower was located, only the expected loads in those regions were applied to the tower and corresponding combinations were made and the most critical combination for each zone was found.

3 Establishment of the Finite Element Model

The finite element (FE) model of the steel transmission tower was established in STAAD-PRO Software. The tower height was taken to be 36 m. The vertical spacing between the conductors was 4 m. The vertical spacing between the top conductor and ground wire was 7 m. Cross arm length was 4.5 m. The base width was 6.6 m and the span length was 100 m. The sections used for the tower in the first zone were Indian Standard Angle (ISA) 70 × 70 × 10 mm, ISA 75 × 75 × 10 mm, ISA 80 × 80 × 12 mm and ISA 120 × 120 × 8 mm LD (Long Leg Back-to-Back Double Angle), and they were further modified for different zones to make the tower safe against wind load. The yield stress was taken as 250 MPa (Fig. 1).

The Modulus of Elasticity of steel was taken as 200 GPa and the thermal coefficient of expansion was taken as 12×10^{-6}/°C.

4 Computation of Loads

4.1 Wind Load

The wind load calculations were made according to the specifications mentioned in IS 802-1995. The overall region is divided into six zones. Basic wind speeds for the six wind zones are 33, 39, 44, 47, 50 and 55 m/s, respectively. The sample

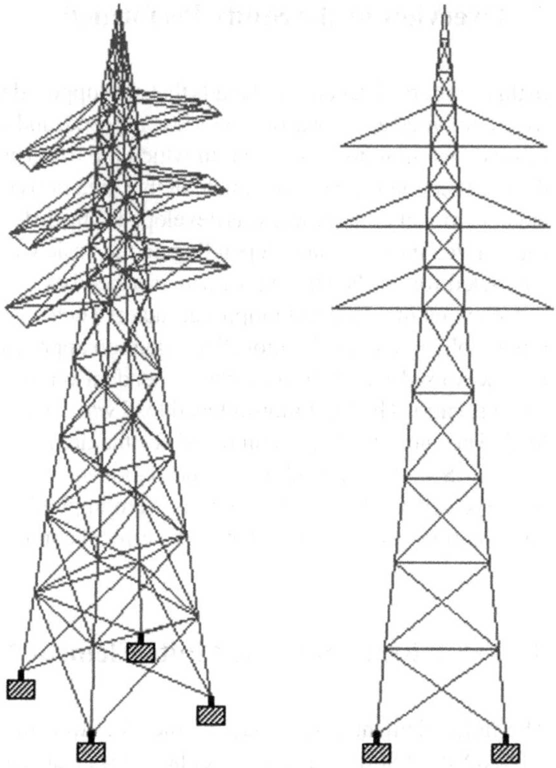

Fig. 1 Finite element model of the transmission tower

calculations for Wind Load for Zone 1 are as presented below in Table 1. For the Equations for Load, computation reference could be made to IS 802-1995 (Part 1).

Table 2 shows the wind forces acting on the tower for different zones. All the forces are in kN.

4.2 Rain Load

The rain load acting on the transmission tower was computed according to the equation approach mentioned by Li et al. [4] in their research work where they have derived the velocity ratio for wind-driven rain. The rain load was computed using the expressions mentioned as follows:

$$
\begin{aligned}
V_{\mathrm{m}} &= u^{*} \times \ln(z/z_{0})/k \quad z < 30\,\mathrm{m} \\
V_{\mathrm{m}} &= u^{*} \times \left(\ln(z/z_{0}) + 34.5pzu^{*}\right)/k \quad 300\,\mathrm{m} > z \geq 30\,\mathrm{m}
\end{aligned} \tag{1}
$$

Table 1 Wind load calculations for Zone 1 (Basic wind speed = 33 m/s)

P_d	Panel	$(S.R)_x$	C_{dtx}	$(S.R)_z$	C_{dtz}	G_t	F_{wx} (kN)	F_{wz} (kN)
403	1	0.080	3.60	0.080	3.60	1.70	5.147	5.147
N/m^2	2	0.103	3.40	0.103	3.40	1.70	5.329	5.329
	3	0.114	3.35	0.114	3.35	1.74	4.930	4.930
	4	0.131	3.25	0.131	3.25	1.80	4.531	4.531
	5	0.161	3.10	0.161	3.10	1.87	4.085	4.085
	6	0.244	2.66	0.163	3.10	1.89	1.992	5.021
	7	0.258	2.66	0.258	2.66	1.92	1.941	1.941
	8	0.275	2.60	0.203	2.60	1.94	1.846	5.076
	9	0.297	2.51	0.297	2.51	1.96	1.735	1.735
	10	0.325	2.50	0.171	3.10	1.98	1.675	4.902
	11	0.410	2.20	0.410	2.20	2.00	1.230	1.230
	12	0.615	2.00	0.615	2.00	2.02	1.016	1.016
	13	1.034	2.00	1.034	2.00	2.04	0.595	0.595

$(S.R)_x$, $(S.R)_z$ = Solidity Ratio; C_{dtx}, C_{dtz} = Drag Coefficients; G_t = Gust Response Factor; F_{wx}, F_{wz} = Wind Forces in *X* and *Z* directions, respectively

Table 2 Wind load for different zones

Zone 2		Zone 3		Zone 4		Zone 5		Zone 6	
P_d = 563 N/m^2		P_d = 717 N/m^2		P_d = 818 N/m^2		P_d = 925 N/m^2		P_d = 1120 N/m^2	
F_{wx}	F_{wz}	F_{wx}	F_{wz}	F_{wx}	F_{wz}	F_{wx}	F_{wz}	F_{wx}	F_{wz}
7.19	7.19	9.16	9.16	10.45	10.45	11.81	11.81	14.31	14.31
7.45	7.45	9.48	9.48	10.82	10.82	12.23	12.23	14.81	14.81
6.89	6.89	8.77	8.77	10.01	10.01	11.31	11.31	13.70	13.70
6.33	6.33	8.06	8.06	9.19	9.19	10.40	10.40	12.60	12.60
5.71	5.71	7.27	7.27	8.29	8.29	9.38	9.38	11.36	11.36
2.78	7.02	3.55	8.93	4.04	10.19	4.57	11.53	5.54	13.96
2.71	2.71	3.45	3.45	3.94	3.94	4.45	4.45	5.40	5.40
2.58	7.09	3.29	9.03	3.74	10.30	4.24	11.65	5.13	11.57
2.42	2.42	3.08	3.08	3.52	3.52	3.98	3.98	4.82	4.82
2.34	6.85	2.98	8.72	3.40	9.95	3.85	11.25	4.66	13.63
1.72	1.72	2.19	2.19	2.50	2.50	2.82	2.82	3.42	3.42
1.42	1.42	1.81	1.81	2.06	2.06	2.33	2.33	2.82	2.82
0.83	0.83	1.06	1.06	1.21	1.21	1.37	1.37	1.65	1.65

where V_m = mean wind speed of height z; k = Karman constant, equalling 0.4; z_0 = surface roughness length (m); $u^* = [kV_m(z_r)/\ln(z/z_0)]$ is the friction velocity based on the mean wind velocity V_m at the reference height of z_r; and $p = 10^{-4}\ s^{-1}$ is the Coriolis parameter.

$$\gamma = (0.5117\alpha^{2.792}\ln(d) + \beta)z^{-\alpha} + (-0.0442\alpha^{-0.2} + 0.0497)d^{1.5} + (0.0493/\alpha) + 0.8928 \quad z < 30\,\text{m}$$
$$\gamma = \chi(0.9492z^{-0.9} + 0.0037)(1 - 0.9541^{d}) + 1 \quad 150 \geq z \geq 30\,\text{m} \tag{2}$$

where α, β and χ = parameters associated with the surface roughness length, γ = velocity ratio and d = raindrop diameter.

$$\gamma = V_r/V_w \tag{3}$$

$$F = (\rho\Pi^2 d^3 n V_r^2 A)/12 \tag{4}$$

where n = total number of raindrops with average diameter in a unit volume, ρ = density of water and V_r = rain velocity. Table 3 gives the Rain Load acting on the transmission tower for different zones. All the forces in Table 3 are in kN.

Table 3 Rain load acting on the tower for different zones

Zone 1		Zone 2		Zone 3		Zone 4		Zone 5		Zone 6	
F_{wx}	F_{wz}	F_{wx}	F_{wz}	F_{wx}	F_{wz}	F_{wx}	F_{wz}	F_{wx}	F_{wz}	F_{wx}	F_{wz}
3.39	3.39	4.74	4.74	6.03	6.03	6.88	6.88	7.78	7.78	9.43	9.43
5.80	5.80	8.11	8.11	10.30	10.30	11.77	11.77	13.31	13.31	16.12	16.12
6.16	6.16	8.60	8.60	10.93	10.93	12.48	12.48	14.12	14.12	17.09	17.09
6.07	6.07	8.50	8.50	10.77	10.77	12.30	12.30	13.91	13.91	16.84	16.84
5.78	5.78	8.09	8.09	10.29	10.29	11.75	11.75	13.30	13.30	16.10	16.10
1.93	9.01	2.69	12.60	3.42	16.00	3.91	18.27	4.42	20.67	5.35	25.03
1.86	1.86	2.61	2.61	3.32	3.32	3.80	3.80	4.28	4.28	5.19	5.19
1.79	13.60	2.51	18.97	3.20	24.12	3.65	27.55	4.12	31.16	5.00	37.73
1.73	1.73	2.40	2.40	3.06	3.06	3.50	3.50	3.95	3.95	4.79	4.79
2.15	12.00	3.15	17.52	4.15	23.11	4.85	26.97	5.60	31.18	7.02	39.11
1.52	1.52	2.24	2.24	2.95	2.95	3.45	3.45	3.98	3.98	5.01	5.01
1.28	1.28	1.87	1.87	2.50	2.50	2.90	2.90	3.36	3.36	4.22	4.22
0.45	0.45	0.65	0.65	0.86	0.86	1.02	1.02	1.17	1.17	1.47	1.47

4.3 Snow Load

The snow load was calculated as per the specifications mentioned in the IS 875(part 4)-1987. The thickness of the ice deposit all around the wire was taken to be 10 mm. The mass density of ice was assumed to be equal to 0.9 g/cm^2. By converting the snow load into a uniformly distributed load and applying it to the conducting wire, the analysis of the conducting wire was done and the reactions in the form of the nodal load were then transferred to the nodes of the tower where they were connected to it.

The same approach was used to compute the load on the cross arms of the tower except that the load acting on the members was applied as area load which was uniformly distributed on them.

4.4 Lightning-Induced Load

The lightning-induced load was computed as per the theory proposed by Nair et al. [9]. The lightning-induced load was applied as temperature load, and hence it led to the development of thermal stresses in the members of the transmission tower and the nature of stresses was compressive. The temperature rise was taken to be 260 °C as per the results obtained by Nair et al. [9]. As the lightning striked the structure, there was a tremendous rise in the temperature which in turn induced thermal stresses in the members which were compressive in nature. The temperature load is governed by the following expression, where σ_t = thermal stress, α = coefficient of thermal expansion, Δt = rise in temperature and E = Modulus of Elasticity.

$$\sigma_t = \alpha \times \Delta t \times E \quad (5)$$

5 Load Combinations

The following load combinations were chosen to record the observations depending upon the zone in which the structure will lie. The zones were classified on the basis of physiography and topography of the country and so the load combinations were made relevant to the location of the structures in those regions. The combinations were made taking into consideration the probability of the loads acting together.

1. DL + WL 2. DL + RL 3. DL + WL + SL 4. DL + SL 5. DL + LIGHTNING.

6 Results and Conclusions

The research to date shows that rain load has a non-negligible effect on the structures and must be given due consideration. As the wind velocity was increased, the rain velocity which was based on the velocity ratio specified based on empirical considerations and based on wind velocity also increased. The maximum relative top displacements of the transmission tower due to rain load compared with the wind load for all the six zones were 13.18%, 15.50%, 17.54%, 18.80%, 20.16% and 22.31%, respectively. Figures 2 and 3 show the comparison of the top displacement of the transmission tower for the wind load and the rain load acting one at a time on the tower in *X* and *Z* directions, respectively. As the wind velocity was increased, the rain velocity also increased and the relative top displacement also increased significantly. The Transmission tower which was safe for wind load failed when it was subjected to rain load. Table 4 shows the sections that were used for the transmission tower design in the respective zone for wind load and the sections that failed when it was subjected to the rain load.

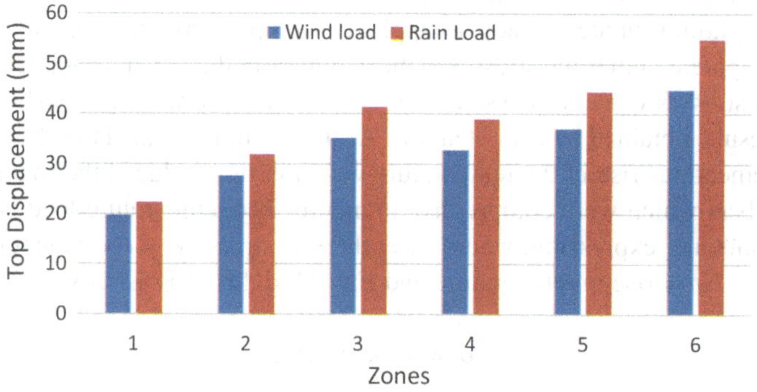

Fig. 2 Top displacement of the transmission tower for wind and rain load in *X* direction

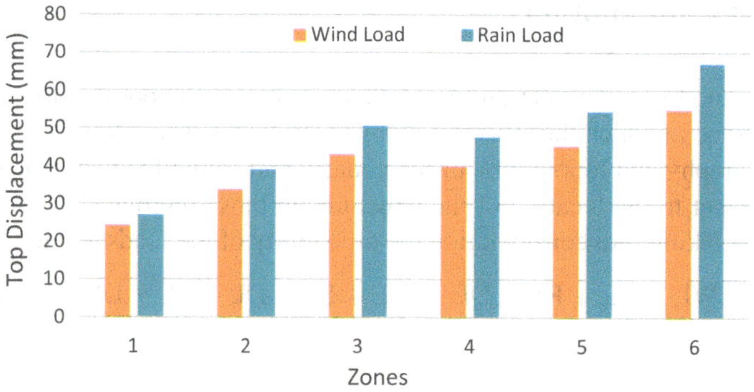

Fig. 3 Top displacement of the transmission tower for wind and rain load in *Z* direction

Table 4 Members safe for wind load but failed when subjected to rain load

Zone	Sections used in design for wind load	Failed members for rain load
1	ISA 70 × 70 × 10, ISA 75 × 75 × 10, ISA 80 × 80 × 12, ISA 120 × 120 × 8 LD	ISA 70 × 70 × 10
2	ISA 75 × 75 × 10, ISA80 × 80 × 12, ISA 90 × 90 × 10, ISA 120 × 120 × 8 LD	ISA 75 × 75 × 10
3	ISA 80 × 80 × 10, ISA 90 × 90 × 10, ISA 90 × 90 × 12, ISA 100 × 100 × 10, ISA 120 × 120 × 8 LD	ISA 80 × 80 × 10
4	ISA 80 × 80 × 10, ISA90 × 90 × 10, ISA100 × 100 × 10, ISA 120 × 120 × 10 LD	ISA 80 × 80 × 10
5	ISA 90 × 90 × 10, ISA 100 × 100 × 10, ISA 120 × 120 × 10 LD	ISA 90 × 90 × 10
6	ISA 90 × 90 × 10, ISA 100 × 100 × 10, ISA 110 × 110 × 10, ISA 120 × 120 × 10 LD	ISA 100 × 100 × 10

Figure 4 shows the failure of the transmission tower when subjected to the rain load. The failure has initiated from the bottom part bracings up to the bottom cross arms. The members highlighted in red are those which will fail first in compression followed by the blue ones.

The Lightning induced load is an electrical discharge caused by imbalances between storms, clouds and the ground or within the clouds themselves. When lightning strikes a structure the, temperature rise is very high which can induce thermal stresses in the tower. According to the results obtained in the analysis, the stress developed was about 642 MPa which was much higher than the yield stress of steel members which is about 250 MPa, and the ultimate stress which is about 430 MPa. The failure of the tower members due to thermal stresses is shown in Fig. 5.

The snow load on the transmission wires was computed according to the IS code 875 (part 4) for Zone 5 as this was the zone which includes the regions of northeast and hence snowfall was expected in this region. By taking the ice density to be 0.9 g/cm^2, the snow load acting on the wires as well as the cross-arm members were computed. The relative top node displacement of the transmission tower for the load combination which included dead, wind, snow altogether was only 0.35% as compared to the displacement for the load combination which included dead and wind load only. Hence, in this zone as well, the most critical combination was the dead and rain loads. Figure 6 shows the failure of the transmission tower when it is subjected to the load combination that includes dead, wind and snow acting on the structure simultaneously. Figure 7 gives the maximum displacements of the tower for different load combinations for Zone 5.

The conclusions from the above study are as follows:

- The combination DL + RL was found to be the critical combination for all the six zones.

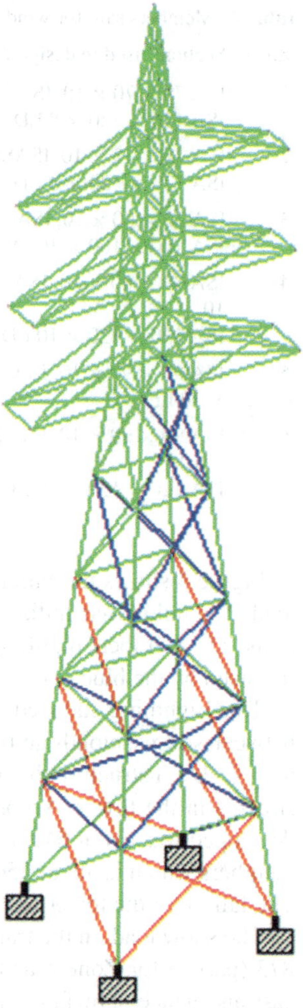

Fig. 4 Failure of the bracing due to the rain load for Zone 3

- Lightning-induced load was not included in the load combination as it was instantaneous and was not going to act continuously on the structure. As per the analysis results, the tower will fail due to lightning itself even in the absence of other prominent design loads. It led to tremendous temperature rise which in turn caused high thermal stresses in the members of the structure.
- For the snow-prone region, the top displacement of the member due to snow load was not found as significant as that for the rain load. Hence, for this region also the DL + RL load combination was the most critical one.

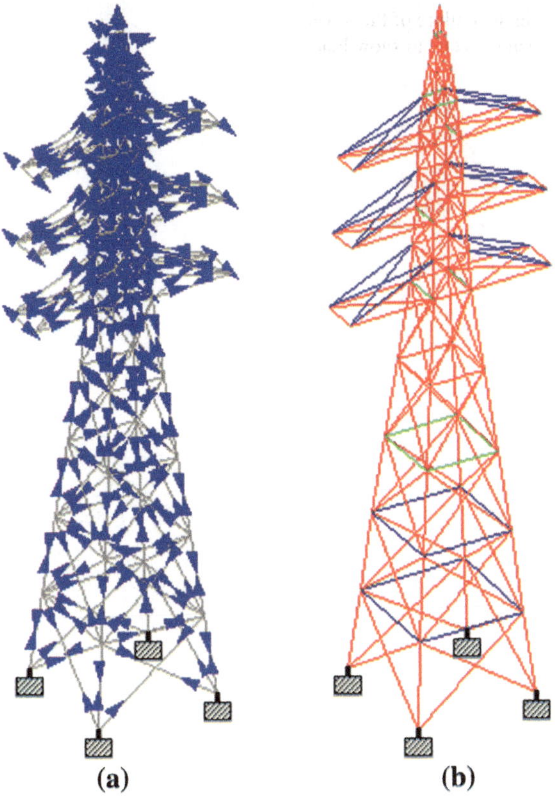

Fig. 5 **a** Application of temperature load, **b** failure of the transmission tower due to lightning

Fig. 6 Failure of the tower members due to snow load

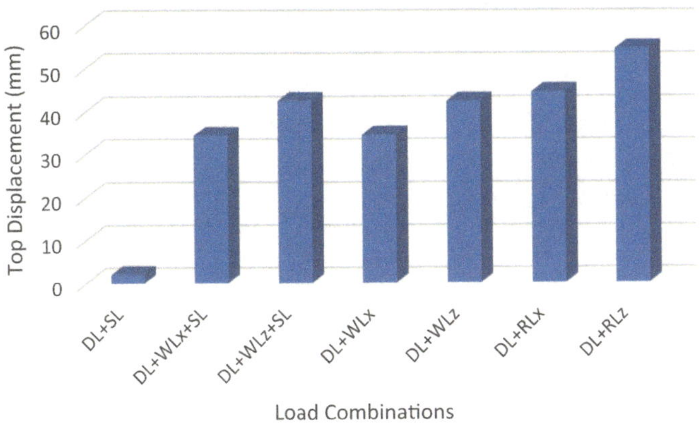

Fig. 7 Bar diagram showing top displacement of the transmission tower for Zone 5

References

1. Alipour A, Sarkar P, Dikshit S, Razavi A, Jafari M (2020) Analytical approach to characterize tornado-induced loads on lattice structures. J Struct Eng 146(6):04020108. http://doi.org/10.1061/(asce)st.1943-541x.0002660
2. Reinoso E, Niño M, Berny E, Inzunza I (2020) Wind risk assessment of electric power lines due to hurricane hazard. Nat Hazards Rev 21(2):04020010. http://doi.org/10.1061/(asce)nh.1527-6996.0000363
7 Mingfeng Huang, Lieyang Wu, Qing Xu, Yifan Wang (2020) Bayesian approach for typhoon-induced fragility analysis of real overhead transmission lines.https://ascelibrary.org/doi/abs/10.1061/%28ASCE%29EM.1943-7889.0001816
4. Tian L, Zeng Y-J, Fu X (2018) Velocity ratio of wind-driven rain and its application on a transmission tower subjected to wind and rain loads. J Perform Constructed Facil 32(5):04018065. http://doi.org/10.1061/(asce)cf.1943-5509.0001210
8 Jian Zhang, Qiang Xie (2018) Failure analysis on transmission tower struck by tropical storms.https://ascelibrary.org/doi/10.1061/9780784481837.010
6. Fu X, Li H-N, Li J-X (2018) Wind-resistance and failure analyses of a lightning-damaged transmission tower. J Perform Constructed Facil 32(1):04017127. http://doi.org/10.1061/(asce)cf.1943-5509.0001121
9 Ibrahim Hathout, Karen Callery, M.Sc.E (2015) Impact of extreme weather on transmission line structures.https://ascelibrary.org/doi/10.1061/9780784479414.044
8. Yang F, Yang J, Han J, Zhang Z (2014) Tower destruction mechanics of overhead transmission lines and prevention technologies in ice disasters. In: Sustainable development of critical infrastructure—proceedings of the 2014 international conference on sustainable development of critical infrastructure, pp 355–367. http://doi.org/10.1061/9780784413470.038
9. Nair Z, Aparna KM, Khandagale RS, Gopalan TV (2005) Failure of 220 kV double circuit transmission line tower due to lightning. J Perform Constructed Facil 19(2):132–137. http://doi.org/10.1061/(asce)0887-3828(2005)19:2(132)

Fig. 7 [illegible]

References

[illegible]

Fragility Assessment of Mid-Rise Flat Slab Structures

B. P. Dhumal and V. B. Dawari

Abstract Flat slab structures have advantages over conventional RCC framed structures in terms of construction time as no beams are required to transfer loads from slabs to columns and thus the faster rate of construction is achieved, with less floor height demand, larger clearance of slab to floor, etc. These structures have disadvantages like higher flexibility, high punching shear at slab column junction, and poor performance under seismic loads which cannot be ignored. This study presents fragility curves for mid-rise flat slab RCC structures without masonry infill walls. A nonlinear static pushover analysis is performed to capture the nonlinear seismic response of a sample of structures while monitoring the performance limit states. The HAZUS methodology was used for developing the fragility curves given by the HAZUS technical manual. This study evaluates the losses in flat slab RCC structures due to earthquakes for specific seismic zones in India and compares them with the conventional frame structures. FEM-based software CSI ETABS 2018 has been used as a tool. The present study contributes to quantifying seismic fragility. The fragility curves are interconnected and planned to provide a logical and consistent probability treatment of damage or loss. The fragility curves are found to evaluate and interpret the results.

Keywords Fragility curves · Flat slabs · HAZUS methodology · Nonlinear static pushover analysis

1 Introduction

Prior to an earthquake, vulnerability assessments of buildings are usually made to judge the requirements for improving facilities and important buildings counter to incoming earthquakes. The best way to achieve such evaluations is the fragility curves. The significant loss to society and the economy due to various earthquakes on various steel and RCC structures has increased awareness of assessing the seismic

B. P. Dhumal (✉) · V. B. Dawari
Department of Civil Engineering, College of Engineering Pune, Shivajinagar, Pune, India
e-mail: dhumal1613@gmail.com

M. S. Ranadive et al. (eds.), *Recent Trends in Construction Technology and Management*, Lecture Notes in Civil Engineering 260,
https://doi.org/10.1007/978-981-19-2145-2_78

hazard analysis and vulnerability studies of existing buildings. The damage and losses in buildings caused due to earthquakes should be anticipated with an agreeable degree of confidence to mitigate the losses and post-disaster repair works. Fragility curves are a tool for assessing the likelihood of structural damage due to earthquakes as a function of ground motion indicators and otherwise design parameters.

A flat slab system is a type of reinforced concrete system that has several benefits over traditional moment-resistant frame systems. Flat slab system has a good architectural look, unobstructed space, requirement of less building height, and less construction time with easy formwork systems. However, it causes issues which demand inspection with flat slab construction systems. One of the major disadvantages of the system is large transverse displacement under seismic excitations due to the absence of beams thus resulting in lower lateral stiffness. This in turn causes damage to even non-structural members. Another issue is punching shear at slab column junctions due to the transfer of shear and destabilizing moments from slab to directly columns; these destabilizing moments generate high shear stresses under seismic excitations. A typical schematic representation of a flat slab system is shown in Fig. 1.

Therefore, flat slabs are not recommended in high seismic hazard regions such as earthquake zones 4 and 5 of the Indian seismic region. Thus, they are used only as vertical load-carrying systems combined with shear walls to carry and resist lateral loads. The design of flat slab systems is conducted in the same approach as ordinary frames. Thus, response under moderate earthquake shows extensive damage even when the codal conditions for drift limits are satisfied. This highlights the requirement to assess the susceptibility of a flat plate structure for which the fragility curves available are lower in comparison to traditional moment resistant frames.

Ahanthem et al. [1] modelled RC frames in compliance and not in compliance with strong column-weak beam (SCWB) and derived the fragility curves using the

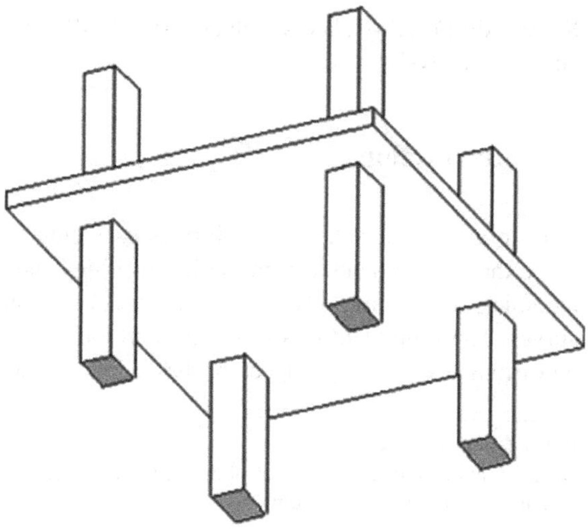

Fig. 1 Typical flat slab system

instructions specified in the HAZUS manual and observed that frames confirming to SCWB had a lower probability of damage than frames non confirming to SCWB. Schotanus [5] applied a general method for seismic fragility analysis of systems proposed by Veneziano et al. to an RCC frame. Response surface was used to switch the capacity part in an analytical limit state function (g-function), with a categorical functional relationship which fits a second-order polynomial and is used as input for SORM analysis.

2 Methodology

There is no absolute strategy to develop fragility curves intended for a specific structure, wide uncertainty is involved in all the steps of precision; this is due to variability in analytical modelling materials defined design process ground motion characteristics and also defining limits states current study uses accepted procedures from his manual for vulnerability assessment ensuring that rational decisions are taken by deriving fragility curves for a structural system. The assumed approach used is outlined in Fig. 2.

2.1 Pushover Analysis

Pushover analysis is a nonlinear static method in which a structure is exposed to gravity loads and monotonous lateral load pattern that increases in elastic and non-elastic behaviour until an ultimate stage is attained. Pushover curve calculates the capacity of the structure and presents in the form of a capacity curve, and the performance point is calculated by intersecting the capacity and demand curve of the structure Guidelines to perform nonlinear static pushover analysis are given in FEMA 356, FEMA 273, and ATC-40. The force deformation curve for pushover analysis is shown in Fig. 3; the points marked as A, B, C, D, and E represent the force–deformation behaviour of hinges formed. The segment BC is again subdivided into Intermediate

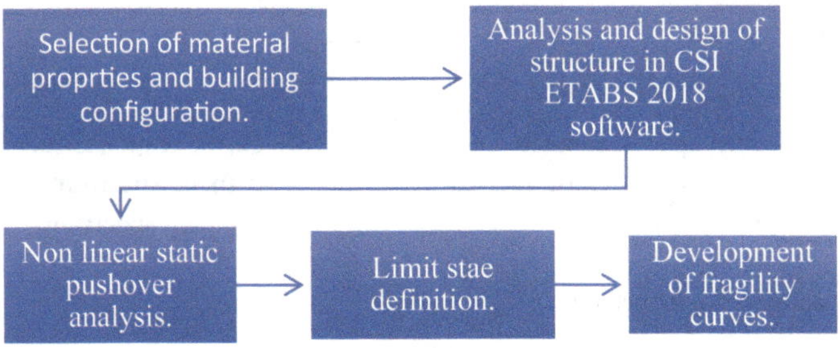

Fig. 2 Methodology used in the derivation of fragility curves

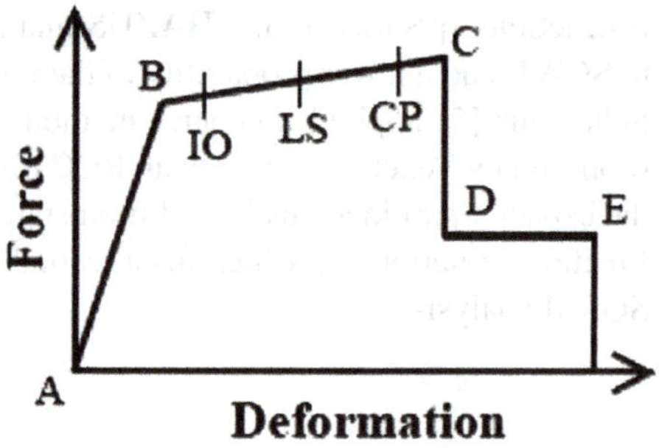

Fig. 3 Force deformation curve

Occupancy, Life Safety, and Collapse Prevention phases which are receiving criteria of hinges formed.

2.2 *Capacity Spectrum Method (CSM)*

The Capacity Spectrum Method is an approximate method of analysing the structure with a nonlinear static analysis (pushover), the seismic response of a structure. CSM helps to investigate the seismic response of the structure in relation to force and displacements and describes the seismic performance of structures. The concept of CSM assumes that the greatest lateral storey drifts illustrate the seismic building response and the maximum lateral storey drifts are controlled by deformations of the fundamental mode of the original flexible system. Performance point is the point where the capacity curve and demand curve meet, i.e. capacity and demand of structure equals. Figure 4 shows the performance point obtained by the capacity spectrum method.

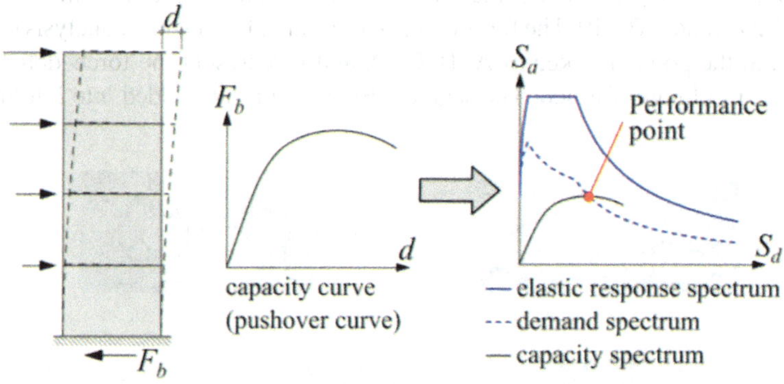

Fig. 4 Capacity spectrum method

2.3 *HAZUS Methodology*

2.3.1 Fragility Function

HAZUS technical manual FEMA HAZUS-MH MR4 Flood [4] gives a methodology to obtain fragility curves for several types of structures. HAZUS is a compilation of parameters that operate integrally to quote fatalities, loss of function, and financial crashes in a province owing to earthquakes. For a particular building category, various sets of curves are constructed in HAZUS separated by height as low, mid, and high rise, and level of seismic design as low, moderate, and high and seismic performance level. Figure 5 shows fragility curves given by the HAZUS manual.

HAZUS fragility curves are constructed based on damped elastic spectral displacement at the intersection of the pushover curve and the earthquake response spectrum, $S_{d(Tc)}$. Building fragility curves are log-normal functions describing the probability of reaching or exceeding structural and non-structural damage states, at certain mean estimates of the spectral response, for example, the spectral displacement. These curves consider the differences and uncertainties associated with capacity curve properties, damage states, and ground vibration.

The conditional probability (P) of being in or exceeding a given damage condition, ds, is defined given the spectral displacement, S_d, (or other seismic demand parameters) by Eq. (1)

$$P[\mathrm{ds}/S_d] = \phi\left[\frac{\ln}{\beta_{\mathrm{ds}}}\left(\frac{S_d}{S_{d,\mathrm{ds}}}\right)\right] \tag{1}$$

where

S_d Spectral displacement that defines the threshold of a particular damage state.

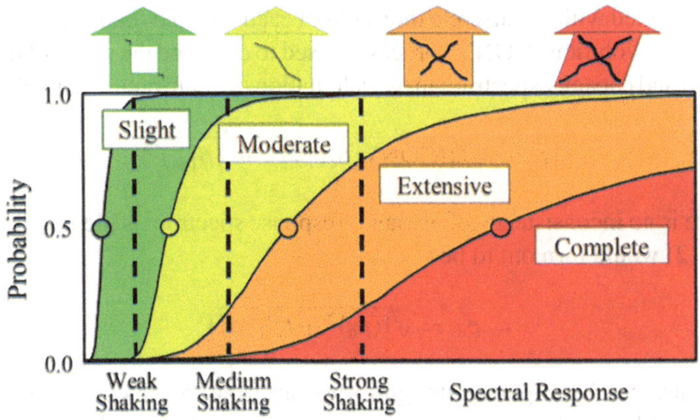

Fig. 5 HAZUS fragility curves

Table 1 Damage state model

Damage state	Spectral displacement (S_d)
Slight	$0.7\ S_{dy}$
Moderate	$1.5\ S_{dy}$
Extensive	$0.5(S_{dy} + S_{du})$
Complete	S_{du}

$S_{d,ds}$ Median value of spectral displacement for which the building reaches the threshold of damage states.
β_{ds} Standard deviation of the natural logarithm of spectral displacement for damage state.
ϕ Standard normal cumulative distribution function.

In this article, fragility curves are constructed based on the capacity spectrum that is achieved from pushover analysis. These capacity spectrums are utilized to get the yield spectral displacement (S_{dy}) and ultimate spectral displacement (S_{du}). Capacity spectra obtained from pushover analysis are used to construct fragility curves in this paper. Values of medians at several damage states are calculated using yield spectral displacement (S_{dy}) and ultimate spectral displacement (S_{du}). Values given by Giovinnazi et al. (2006) for the damage state model for median values of spectral displacement ($S_{d,ds}$) are incorporated in this paper. The values are presented in Table 1.

2.3.2 Development of Damage State Variability

Overall variability of fragility curves is depicted by log-normal standard deviation (Beta). It is contributed by three primary sources, namely the variability linked with the capacity curve, β_C, the variability linked with the demand spectrum, β_D, and the variability linked with the discrete threshold of each damage state, $\beta_{T,ds}$, as described in Eq. (2). Convolution "CONV" process is used to combine the effect of dependence of demand and capacity spectrum in the calculation of contribution in total variability.

$$\beta_{ds} = \sqrt{\{(\text{CONV}[\beta_c, \beta_d])^2 + (\beta_{Tds})^2\}} \quad (2)$$

If there is no inconsistency of demand (response spectrum is identified precisely), then Eq. (2) would turn out to be

$$\beta_{ds} = \sqrt{[(\beta_c)^2 + (\beta_{Tds})^2]} \quad (3)$$

β_{ds} values can be imported straightforwardly from the HAZUS manual as they are already calculated considering appropriate values of kappa factors (k), β_C, and β_D values for distinct kinds of buildings. Table 2 illustrates the variability numbers

Table 2 Variability values for five-storey model

Degradation values				
Damage state	Kappa factor (k)	β_{Tds}	β_c	β_{ds}
Slight	0.9 (Minor degradation)	0.4 (Moderate)	0.3 (Moderate)	0.8
Moderate	0.5 (Major degradation)	0.4 (Moderate)	0.3 (Moderate)	0.95
Extensive	0.1 (Extreme degradation)	0.4 (Moderate)	0.3 (Moderate)	1.05
Complete	0.1 (Extreme degradation)	0.4 (Moderate)	0.3 (Moderate)	1.05

applied to the five-storied flat slab RCC model taking into account the moderate cases of degradation devoid of infill walls.

3 Structural Modelling

A hypothetical (G + 4) RCC flat slab building frame has been considered in the present study. There are two reasons for considering a mid-rise building. It might not be feasible to fulfil the drift requirements in high-rise structures because of the inherent flexibility of flat slab buildings. On the other side, low-rise buildings would be much stiffer and may not require particular attention. For simplicity, a flat slab RCC building is taken as a bare frame omitting the infill walls. The geometric specification of chosen RCC frame is shown in Table 3. The frame has the same floor plan which is displayed in Fig. 6.

Table 3 Details of geometry of flat slab RCC frame

Item	Dimensions
Plan dimension	30 m (X-Dir) × 20 m (Y-Dir)
Beam size	300 mm × 450 mm
Column size	550 mm × 550 mm
Slab thickness	250 mm
Floor height	Ground storey 3.5 m and 3 m typical storey
Drop thickness and size	375 mm of 2000 mm × 2000 mm
Live load	4 kN/m^2
Seismic zone	Zone III
Response reduction factor	3
Importance factor	1.2
Type of soil	Medium type soil
Material grades	Fe 500, Fe 415 and M30

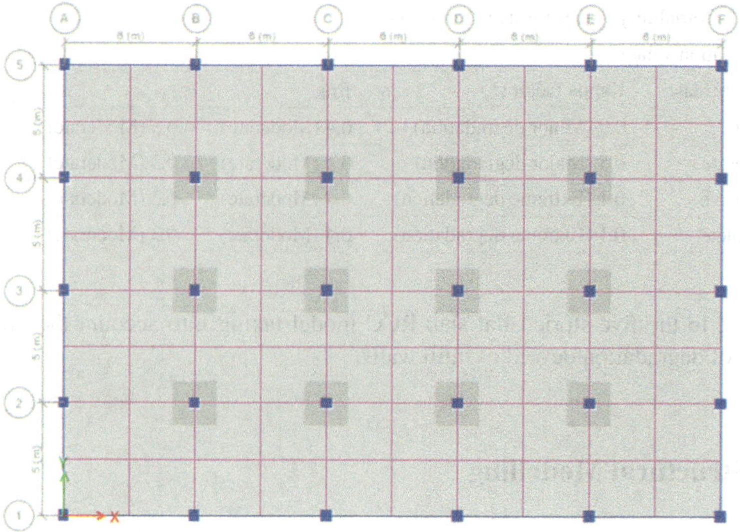

Fig. 6 Plan

3D modelling of the RCC structure conforming to the needs of IS 456:2000, IS 1893 (Part1):2016 was created in CSI ETABS 2018 software. The hinge properties provided for modelling nonlinear frame elements in ETABS 2018 are as follows. P-M2-M3 hinges for columns and M3 hinges for beams are assigned as per FEMA 356. Live load of 4 and 1 kN/m^2 of floor finish have been applied on each floor excluding the roof. A linear analysis was performed using the load combinations specified in IS 456:2000 and the design of RCC members was carried out for the critical load combinations. Since the most severe challenge of a flat slab system is the punching shear failure, safety measures are to be observed in the design stage to avoid this unattractive behaviour. The slab depth of 250 mm is provided which passed the punching shear check of IS 456:2000 BIS IS 456 [2].

The seismic design is carried out conforming to IS 1893:2016, IS 13920:2016. The flat slab building is supposed to be in Pune, Maharashtra, India, i.e. seismic zone 3. After the successful linear analysis and design of RC members, a nonlinear static pushover analysis was performed in ETABS 2018 software, and the required results were extracted such as pushover curve, capacity curve, and performance point.

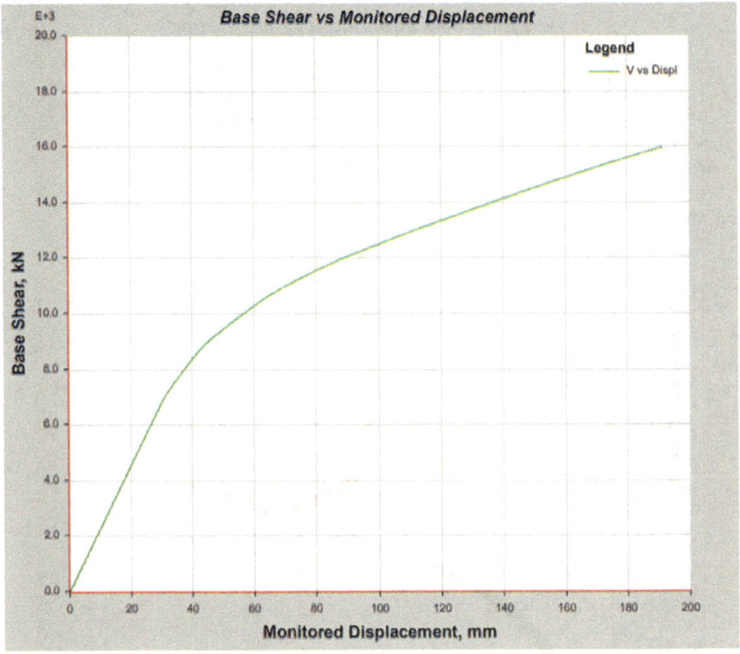

Fig. 7 Pushover curve

4 Results and Discussion

4.1 Pushover Analysis Results

To monitor the overall ductility of the structure, displacement-controlled pushover analysis was carried out. Figure 7 represents the pushover curve obtained for the flat slab structure. The structure was observed to be at the Life safety limit after the development of hinges. The plastic hinges were found to be clustered at the first three storeys, consistent with the sizeable drifts experienced by the same storeys. The performance point of the frame was found at 14,408 kN at a displacement of 147.43 mm. The performance point obtained from the Capacity Spectrum Method is shown in Fig. 8.

4.2 Fragility Curves

The hazard parameter is chosen as spectral displacement (S_d) for creating vulnerability curves. To define the damage limit states, control points were obtained from the bilinear curve obtained from pushover analysis. Table 4 shows the control points

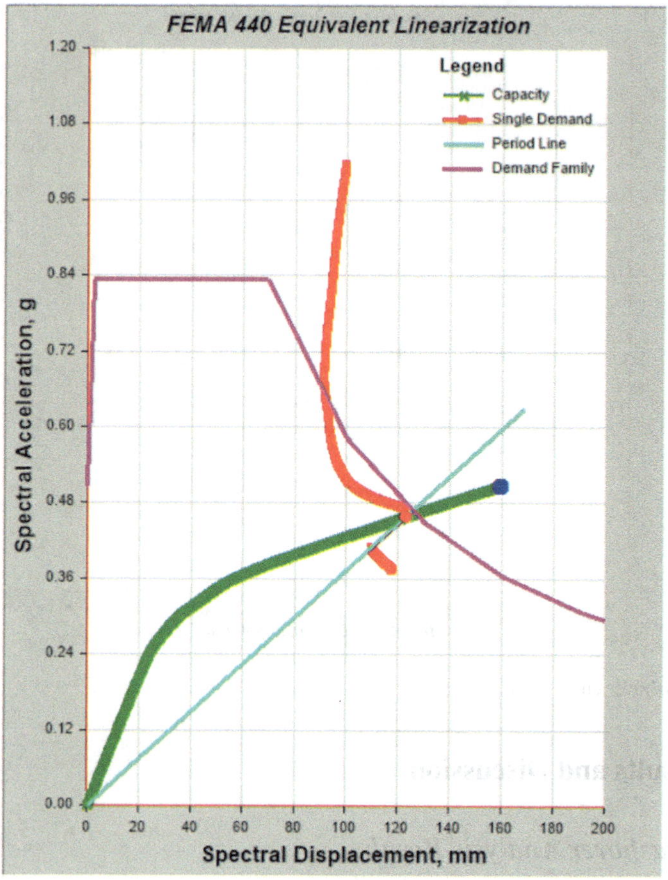

Fig. 8 Capacity spectrum

Table 4 Control points

Control points	Value
A_y	0.9
D_y	0.75
A_u	0.65
D_u	0.44

obtained and Table 5 shows the values for damage limit states in terms of spectral displacement. After calculating the probabilities of various damage states, a curve was plotted between damage hazard and probability corresponding to that damage state.

Table 5 Damage states

Damage state	Ds (mm)
Slight	28.98
Moderate	62.11
Extensive	88.43
Complete	135.45

The fragility curves for the RCC structure have also been plotted as shown in Fig. 9 as per the HAZUS methodology. Table 6 shows the probabilities of various damage states at a spectral displacement of 50 mm.

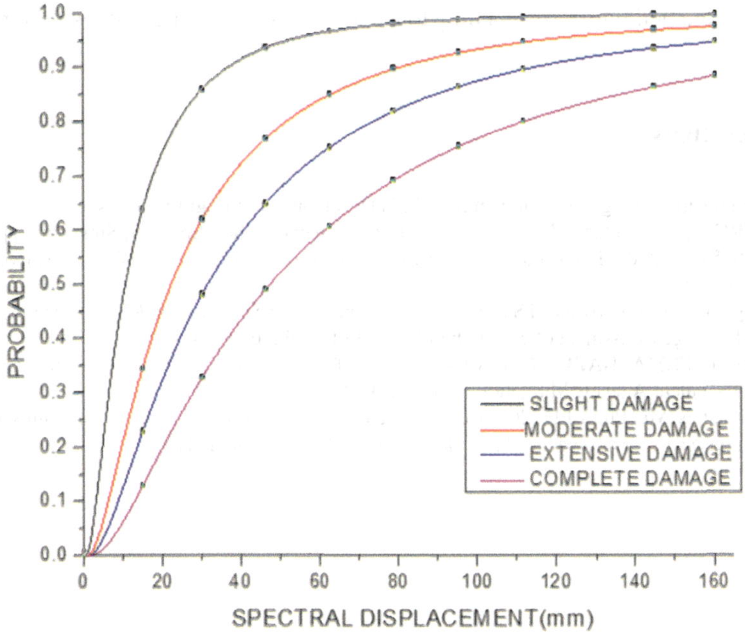

Fig. 9 Fragility curves

Table 6 Probabilities

Damage state	Probability
Slight	0.9
Moderate	0.75
Extensive	0.65
Complete	0.44

5 Conclusion

In this work, seismic fragility curves are obtained for mid-rise flat slab RCC building. It is found that the flat slab structure is more flexible than traditional moment-resisting frames owing to the absence of horizontal elements such as beams and lateral load-resisting elements like shear walls. Further, the HAZUS methodology for the development of fragility curves has been discussed for the flat slab RCC frame structure. From fragility curves, it is seen that for the same spectral displacement the probability of damage decreases as the damage level increases. The method for fragility assessment given by HAZUS technical manual is well suited for flat slab structures also to assess the damage intensities under seismic excitations. From the results obtained, it is determined that the HAZUS methodology is well suited to predict the damage level of the building corresponding to a particular value of spectral displacement.

References

1. Ahanthem N, Ningthoukhongjam SS (2021) Development of fragility curves for different types of RC frame structures. In: Advances in structural technologies. Springer, Singapore, pp 71–87
2. BIS IS 456 (2000) Plain and reinforced concrete code of practice, 4th revision. Bureau of Indian Standards
3. Lagomarsino S, Giovinazzi S (2006) Macroseismic and mechanical models for the vulnerability and damage assessment of current buildings. Bull Earthq Eng 4(4):415–443
4. Model, FEMA HAZUS-MH MR4 Flood (2009) Technical manual. Federal Emergency Management Agency, Mitigation Division, Washington, DC
5. Schotanus MI, Pinto PE (2002) Fragility analysis of reinforced concrete structures using a response surface approach. Master degree thesis, IUSS, Pavia, Italy

Seismic Response of RC Elevated Liquid Storage Tanks Using Semi-active Magneto-rheological Dampers

Manisha V. Waghmare, Suhasini N. Madhekar, and Vasant A. Matsagar

Abstract The paper presents the use of semi-active magneto-rheological (MR) dampers for the structural response reduction of the reinforced concrete (RC) elevated liquid storage tanks. The effectiveness of MR dampers is investigated based on the control strategies and the placement of the dampers in the staging. The RC elevated liquid storage tank is modeled as a multi-degree freedom system for the staging with a two-mass model for the container with liquid. Two control systems, viz., open-loop and closed-loop control systems are considered. The control algorithms employed are (1) Passive-OFF, (2) Passive-ON, (3) Clipped-Optimal Control (COC), and (4) Simple Semi-active Control (SSC). The study is also focused on the effect of change of voltage on the response quantities. For the COC algorithm, the feedback gain is obtained by considering velocity feedback. The present study proposes a SSC algorithm which is an effective way of controlling the response of RC elevated tank, which uses the ratio of the damper force and capacity of the damper. For numerical simulation, a code is developed in MATLAB. The coupled differential equations of motion for the system are solved using the state-space method. The response of the broad and slender tanks is studied by taking the ratio of the height of the liquid to the radius of the container (S) as 0.5 and 2.0, respectively. The controlled response of the tank under eight different ground motions comprising of near-field and far-field components of earthquakes is evaluated and compared with that of the uncontrolled response. Base shear, displacement, and damper force are obtained. The results showed that the structural response is effectively controlled using semi-active MR dampers with the proposed SSC strategy. The simple semi-active control algorithm proposed in this study could be considered as a proficient algorithm for the seismic response reduction of RC elevated liquid storage tanks using SAMRD.

M. V. Waghmare (✉)
Department of Civil Engineering, AISSMS College of Engineering, Pune, India
e-mail: mvwaghmare@aissmscoe.com

S. N. Madhekar
Department of Civil Engineering, College of Engineering Pune, Shivajinagar, Pune 411005, India

V. A. Matsagar
Department of Civil Engineering, Indian Institute of Technology (IIT) Delhi, Hauz Khas, New Delhi 110016, India

M. S. Ranadive et al. (eds.), *Recent Trends in Construction Technology and Management*, Lecture Notes in Civil Engineering 260,
https://doi.org/10.1007/978-981-19-2145-2_79

Keywords RC elevated liquid storage tank · Semi-active magneto-rheological damper (MR) · Semi-active control algorithms · Optimal linear control · Open-loop control · Closed-loop control

1 Introduction

Recently, a class of semi-active dampers with smart materials whose properties can be changed by the application of a magnetic field has become very popular in the field of structural control. These materials respond to the change in the magnetic field and are thus called magneto-rheological (MR) materials [2]. These materials are mainly in liquid form and are used in the devices for structural vibration control. Due to the mechanical simplicity, high dynamic range, low power requirements, large force capacity, and robustness, this device offers an attractive means of protecting civil infrastructure systems against severe earthquakes and wind loading [13].

The major objective of the present study is to study the effectiveness of semi-active MR dampers and their control algorithms in reducing the seismic response of the RC elevated liquid storage tanks subjected to different earthquakes. The detailed objectives of the present study are: (i) to investigate uncontrolled and semi-actively controlled seismic response of the RC elevated liquid storage tanks installed with MR dampers at different levels of staging and operated with Passive-OFF, Passive-ON, Clipped-Optimal Control (COC), and Simple Semi-active Control (SSC) algorithm, (ii) to examine the effect of command voltage on the efficiency of semi-active magneto-rheological dampers (SAMRDs), and (iii) to explore the effect of placement of SAMRDs on response reduction.

The focus of this paper is to investigate the effectiveness of proposed control strategies for MR dampers for installation in RC elevated liquid storage tanks. The control algorithms are evaluated for their ability to reduce the peak base shear and displacements. The controlled responses are compared with the corresponding uncontrolled responses of the RC elevated liquid storage tank.

2 Modeling of Elevated Liquid Storage Tank

In the present study, to facilitate the installation of dampers and to study the effect of placement of dampers, multi-degree freedom staging for the RC elevated liquid storage tank with a two-mass model of liquid is presented. The two-mass model for the rigid concrete container is used as per the guidelines specified by the ACI 350.3 [1] and EN 1998-4 [4]. The assumptions and formulation of the mathematical modeling RC tank are described in detail by Waghmare et al. [11, 12].

Figure 1a shows the idealized mathematical model of the RC elevated liquid storage tank considered in the study. Figure 1b, c show the placement of SAMRDs, Configuration I and Configuration II, respectively, in the staging of the RC elevated

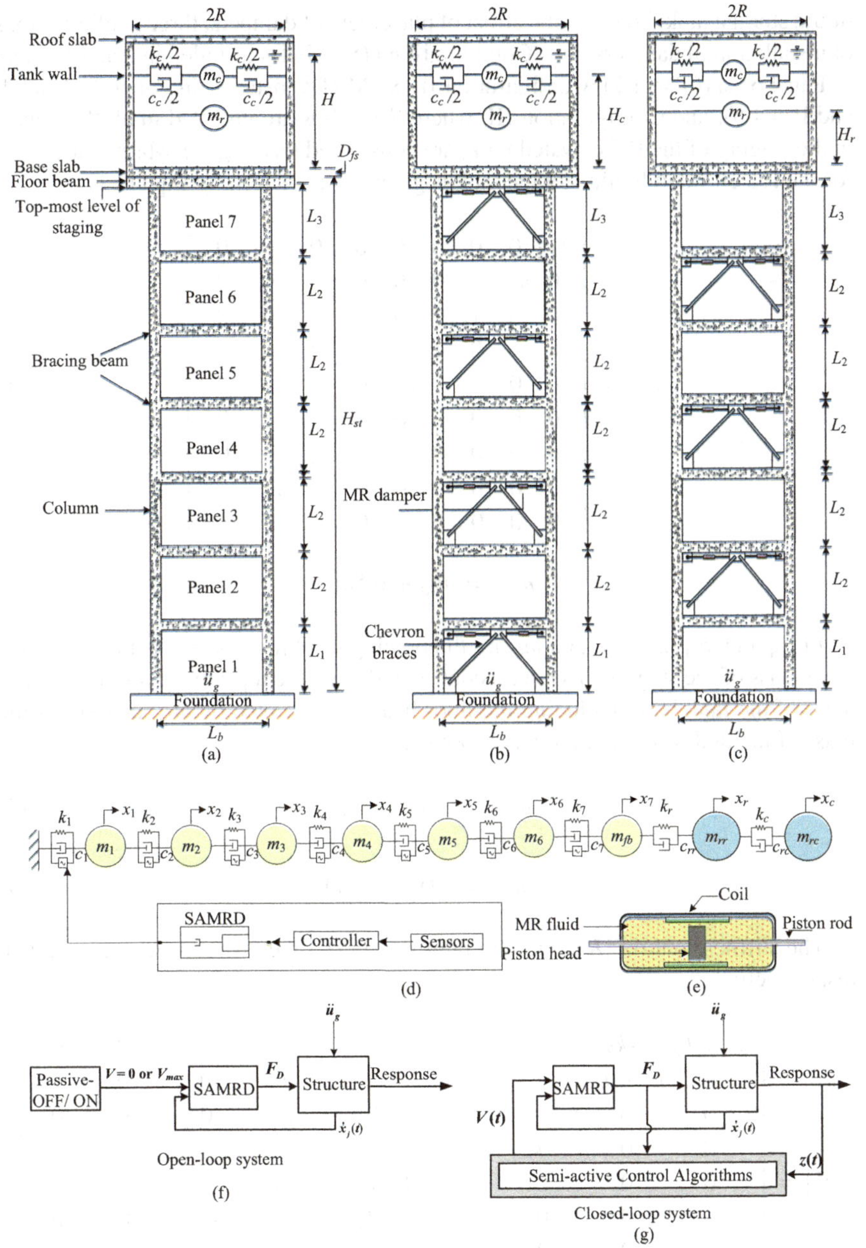

Fig. 1 Schematic diagram of the RC elevated liquid storage tank, **a** the idealized mathematical model, **b** configuration I and **c** configuration II of placement of semi-active MR dampers (SAMRDs), **d** mathematical model of RC elevated liquid storage tank with SAMRDs, **e** schematic diagram of the construction of SAMRD, **f** block diagram of open-loop control system, and **g** closed-loop control system

liquid storage tank. To study the effect of placement of dampers, three configurations of installation of dampers, viz., Configuration I (SAMRDs installed in Panel 1, Panel 3, Panel 5, and Panel 7), Configuration II (SAMRDs installed in Panel 2, Panel 4, and Panel 6); and Configuration III, where SAMRDs are installed in all the Panels in the staging of the RC elevated tank, are considered (see Fig. 1). Mass matrix $[\boldsymbol{M}]$ for the RC elevated liquid storage tank is given, respectively, by Eq. (1) as

$$[\boldsymbol{M}] = \begin{bmatrix} m_1 & 0 & 0 & 0 & 0 & 0 & 0 & 0 & 0 \\ 0 & m_2 & 0 & 0 & 0 & 0 & 0 & 0 & 0 \\ 0 & 0 & m_3 & 0 & 0 & 0 & 0 & 0 & 0 \\ 0 & 0 & 0 & m_4 & 0 & 0 & 0 & 0 & 0 \\ 0 & 0 & 0 & 0 & m_5 & 0 & 0 & 0 & 0 \\ 0 & 0 & 0 & 0 & 0 & m_6 & 0 & 0 & 0 \\ 0 & 0 & 0 & 0 & 0 & 0 & m_{fb} & 0 & 0 \\ 0 & 0 & 0 & 0 & 0 & 0 & 0 & m_{rr} & 0 \\ 0 & 0 & 0 & 0 & 0 & 0 & 0 & 0 & m_{rc} \end{bmatrix} \tag{1}$$

$$m_{fb} = m_{tb} + 0.5m_{ts} \tag{2}$$

In Eq. (1), m_1, m_2, …, and m_6 are the bracing level masses, m_{fb} is the top-most-level mass of the staging. The mass contributed from the top-most *panel* is m_{ts}. The mass of the tank base, including the floor beams, is denoted by m_{tb}; and m_{tw} is the mass of the tank wall, including the roof slab.

$$m_{rc} = \gamma_c(m + m_{tw}) \tag{3}$$

$$m_{rr} = \gamma_r(m + m_{tw}) \tag{4}$$

The stiffness matrix $[\boldsymbol{K}]$, and damping matrix $[\boldsymbol{C}]$ are given in Eqs. (5) and (6), respectively.

$$[\boldsymbol{K}] = \begin{bmatrix} k_1+k_2 & -k_2 & 0 & 0 & 0 & 0 & 0 & 0 & 0 \\ -k_2 & k_2+k_3 & -k_3 & 0 & 0 & 0 & 0 & 0 & 0 \\ 0 & -k_3 & k_3+k_4 & -k_4 & 0 & 0 & 0 & 0 & 0 \\ 0 & 0 & -k_4 & k_4+k_5 & -k_5 & 0 & 0 & 0 & 0 \\ 0 & 0 & 0 & -k_5 & k_5+k_6 & -k_6 & 0 & 0 & 0 \\ 0 & 0 & 0 & 0 & -k_6 & k_6+k_7 & -k_7 & 0 & 0 \\ 0 & 0 & 0 & 0 & 0 & -k_7 & k_7+k_r & -k_r & 0 \\ 0 & 0 & 0 & 0 & 0 & 0 & -k_r & k_r+k_c & -k_c \\ 0 & 0 & 0 & 0 & 0 & 0 & 0 & -k_c & k_c \end{bmatrix} \tag{5}$$

$$[C] = \begin{bmatrix} c_1+c_2 & -c_2 & 0 & 0 & 0 & 0 & 0 & 0 & 0 \\ -c_2 & c_2+c_3 & -c_3 & 0 & 0 & 0 & 0 & 0 & 0 \\ 0 & -c_3 & c_3+c_4 & -c_4 & 0 & 0 & 0 & 0 & 0 \\ 0 & 0 & -c_4 & c_4+c_5 & -c_5 & 0 & 0 & 0 & 0 \\ 0 & 0 & 0 & -c_5 & c_5+c_6 & -c_6 & 0 & 0 & 0 \\ 0 & 0 & 0 & 0 & -c_6 & c_6+c_7 & -c_7 & 0 & 0 \\ 0 & 0 & 0 & 0 & 0 & -c_7 & c_7+c_{rr} & -c_{rr} & 0 \\ 0 & 0 & 0 & 0 & 0 & 0 & -c_{rr} & c_{rr}+c_{rc} & -c_{rc} \\ 0 & 0 & 0 & 0 & 0 & 0 & 0 & -c_{rc} & c_{rc} \end{bmatrix} \tag{6}$$

The damping associated with the column members of the staging for Panel 1, Panel 2, Panel 3, Panel 4, Panel 5, Panel 6, and Panel 7 (see Fig. 1a), denoted by c_1, c_2, c_3, c_4, c_5, c_6, and c_7 is given by

$$c_j = 2\xi_t\sqrt{k_j m_j} \tag{7}$$

For $j = 1$ to p, where p is the number of bracing levels, and ξ_t is the damping ratio of the staging in the elevated tank, c_{rr} is the damping associated with k_r and m_{rr}, while c_{rc} is the damping associated with m_{rc} and k_c. The stiffness associated with the Panel 1, Panel 2, Panel 3, Panel 4, Panel 5, Panel 6, and Panel 7 is denoted by k_1, k_2, k_3, k_4, k_5, k_6, and k_7.

3 Equation of Motion

In this section, governing equations of motion for n degree of freedom system and their solution for time history analysis using the state-space method; for the RC elevated liquid storage tank installed with the SAMRDs is presented. The mathematical model of the RC liquid storage tank installed with SAMRDs is presented in Fig. 1d. Further, Fig. 1e shows the construction of SAMRD. Figure 1f, g show a block diagram representation of the open-loop and closed-loop control systems, respectively. The governing differential equation of motion under earthquake excitation, with semi-active dampers, is given by

$$[\boldsymbol{M}]\left\{\ddot{\boldsymbol{X}}(t)\right\} + [\boldsymbol{C}]\left\{\dot{\boldsymbol{X}}(t)\right\} + [\boldsymbol{K}]\{\boldsymbol{X}(t)\} = \{\boldsymbol{F}_e(t)\} + [\boldsymbol{L}]\{\boldsymbol{f}(t)\} \tag{8}$$

where $[\boldsymbol{L}]_{n\times n}$ is the control force distribution matrix.

In Eq. (8), $\{\boldsymbol{F}_e(t)\} = \left\{ F_1(t)\ F_2(t)\ \ldots\ F_n(t) \right\}^{\mathrm{T}}$ is a vector such that $\{\boldsymbol{F}_e(t)\} = -[\boldsymbol{M}]\{\boldsymbol{r}\}\ddot{u}_g(t)$ where, $\{\boldsymbol{r}\} = \left\{ 1\ 1\ \ldots\ 1 \right\}^{\mathrm{T}}$ is the influence coefficient vector. Here, the degrees of freedom (n) is nine. Also, $\{\boldsymbol{f}(t)\} = \left\{ F_{D1}(t)\ F_{D2}(t)\ \ldots\ F_{Dn}(t) \right\}^{\mathrm{T}}$ is a vector containing the control force exerted by the dampers corresponding to each degree of freedom, n.

The base shear, V_b, in the staging at the foundation level is expressed as

$$V_b = m_{rc}a_c + m_{rr}a_r + \cdots + m_3a_3 + m_2a_2 + m_1a_1 \tag{9}$$

3.1 State-Space Solution

In the state-space method, the response of the system is analyzed using both displacement and velocity as independent variables, called as states [6]. Vector, $z(t)$ is defined in Eq. (10) to represent both states, viz., displacement ($\boldsymbol{X}(t)$) and velocity ($\dot{\boldsymbol{X}}(t)$) of the system.

$$z(t) = \begin{Bmatrix} \boldsymbol{X}(t) \\ \dot{\boldsymbol{X}}(t) \end{Bmatrix} \tag{10}$$

Every degree of freedom is associated with two states, viz., displacement and velocity. Thus, if the degree of freedom of a structure is n, then there will be $2n$ states, first n for the displacement and the remaining n for the velocity. Equation (8) can be written in the following equivalent state-space form:

$$\dot{z}(t) = \begin{Bmatrix} \dot{\boldsymbol{X}}(t) \\ \ddot{\boldsymbol{X}}(t) \end{Bmatrix} = \begin{bmatrix} \boldsymbol{0} & \boldsymbol{I} \\ -\boldsymbol{M}^{-1}\boldsymbol{K} & -\boldsymbol{M}^{-1}\boldsymbol{C} \end{bmatrix} \begin{Bmatrix} \boldsymbol{X}(t) \\ \dot{\boldsymbol{X}}(t) \end{Bmatrix} + \begin{Bmatrix} \boldsymbol{0} \\ \boldsymbol{M}^{-1}\boldsymbol{F}_e(t) \end{Bmatrix} + \begin{Bmatrix} \boldsymbol{0} \\ \boldsymbol{M}^{-1}\boldsymbol{L}\boldsymbol{f}(t) \end{Bmatrix} \tag{11}$$

where $\boldsymbol{0}$ and $\boldsymbol{I}$ are, respectively, null matrix with all zero elements and identity matrix. Time history analysis is carried out, and response quantities, viz., displacements, base shear, and damper forces are obtained.

4 Modeling of Semi-active MR Damper

The magneto-rheological (MR) fluid is a class of smart materials whose rheological properties may be controlled by the application of an external magnetic field. Further, standard electrical connectors, wires, and feed-through can be reliably used, and MR dampers can be powered directly from conventional and low voltage sources [2]. Dyke et al. [3] and Spencer et al. [10] developed a phenomenological model for MR dampers. Figure 1d shows the details of the damper.

4.1 Modified Bouc–Wen Model

The Bouc–Wen model and the modified Bouc–Wen models are computationally more tractable than the Bingham model [7]. In the present study, the modified Bouc–Wen model proposed by Spencer et al. [10] for the magneto-rheological dampers, as shown in Fig. 2a, is used. The nonlinear relationship between the damper force and velocity of the damper is shown in Fig. 2b.

The equation governing the force predicted by this model is

$$F_{Dj} = c_1\dot{y} + k_1(u_d - x_0) \tag{12}$$

$$Z_{mr} = -\gamma|\dot{u}_d - \dot{y}|Z_{mr}|Z_{mr}|^{nmr-1} - \beta(\dot{u}_d - \dot{y})|Z_{mr}|^{nmr} + A_{mr}(\dot{u}_d - \dot{y}) \tag{13}$$

$$\dot{y} = \frac{1}{c_0 + c_1}\{\alpha Z_{mr} + c_0\dot{u}_d + k_0(u_d - y) \tag{14}$$

where Z_{mr} is an evolutionary variable that accounts for the history dependence of the response; u_d is the damper displacement; $\dot{u}_d$ is the velocity across the damper; y is the internal pseudo-displacement of the damper. k_1 is the accumulator stiffness; c_1 is viscous damping at lower velocity in the model to produce roll-off; c_0 is viscous damping at larger velocity; x_0 is the initial displacement in the spring and α, β, γ, nmr and A_{mr} are the characteristic parameters of the model. The model parameters α, $c_{0,}$ and c_1 depend on the input voltage to the current driver such that $\alpha = \alpha_a + \alpha_b U$, $c_0 = c_{0a} + c_{0b}U$, and $c_1 = c_{1a} + c_{1b}U$, where the dynamics involved in the MR fluid reaching rheological equilibrium and in driving the electromagnet in the MR damper are accounted through the first-order filter $U = -\eta(U - V_j)$. V_j is the voltage applied to the current driver and η is the constant.

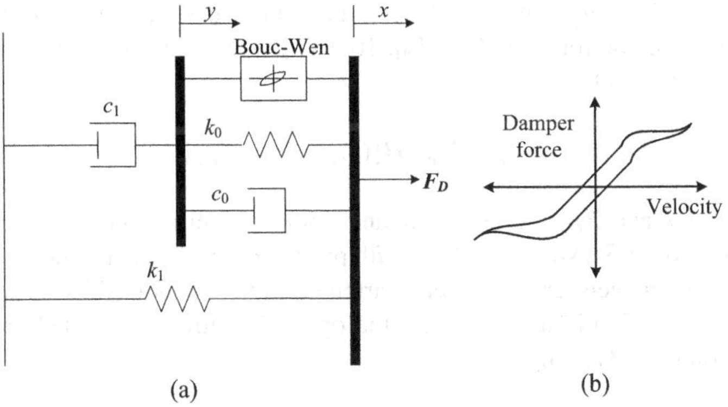

Fig. 2 **a** Mechanical model modified Bouc–Wen model [10], **b** Damper force–velocity relationship of magneto-rheological damper

5 Semi-active Control Algorithms

Figure 1f, g show the flow chart for the selection of control systems. In the present study, four types of controls using open-loop and closed-loop control are designed to achieve the desired response. These include, viz., (1) Passive-OFF, (2) Passive-ON, (3) Clipped-optimal control (COC), and (4) Simple semi-active control (SSC) as described in the following sections. The open-loop control system where the command voltage $V(t)$ is fed to the MR damper is shown in Fig. 1f for Passive-OFF and Passive-ON algorithms. In the open-loop Passive-OFF mode of SAMR damper, the command voltage is held at 0 V, while in the Passive-ON case, it is held at the maximum level $V_{\max}$.

5.1 *Clipped-Optimal Control (COC)*

The clipped-optimal control algorithm, referred to herein as a classical Linear Quadratic Regulator (LQR), is widely used for active as well as for semi-active control [9]. In this algorithm, the quadratic expression for the cost function (J) (Eq. 15) is minimized over the duration of the excitation to obtain the control force $\boldsymbol{f}_c$.

$$J = \int_0^{t_f} [z^{\mathrm{T}}(t)\boldsymbol{Q}z(t) + \boldsymbol{f}_c^T(t)\boldsymbol{R}_r\boldsymbol{f}_c(t)]\mathrm{d}t \tag{15}$$

For the closed-loop system, minimizing the cost function in Eq. (15) results in a control force vector $\boldsymbol{f}_c(t)$ controlled only by the state vector $z(t)$ [5], The force generated by the SAMRD can be regulated by changing the command voltage, thereby changing the damping constant. Hence, the command voltage $V(t)$ is obtained by substituting the control force $\boldsymbol{f}_c$ in Eq. 16, using a clipped control algorithm based on the velocity feedback

$$V_j = V_{\max} H[(f_{cj} - F_{Dj})F_{Dj}] \tag{16}$$

where V_j, $f_{cj,}$ and F_{Dj} are the command voltage, control force, and the damper force required for SAMRD installed at jth panel. $H(\cdot)$ is the Heaviside step function which varies between zero for negative argument and one for positive argument. The clipped-control algorithm along with the optimal controller is called the clipped-optimal control (COC) algorithm [3].

5.2 *Simple Semi-active Control (SSC)*

A simple semi-active control (SSC) is proposed in the present study, which contains the force induced in the damper (F_{Dj}) and the maximum damper force ($F_{D\max}$). The significant advantage of this control algorithm is that it does not require any control parameter to be decided. SSC is defined as

$$V_j = \begin{cases} \frac{V_{\max}}{F_{D\max}} \left|F_{Dj}\right| & \text{for } \left|F_{Dj}\right| \leq F_{D\max} \\ V_{\max} & \text{for } \left|F_{Dj}\right| \geq F_{D\max} \end{cases} \tag{17}$$

6 Numerical Study

The numerical study is conducted to investigate seismic response reduction of the RC elevated liquid storage tanks installed with semi-active MR dampers. A code has been developed in MATLAB [8] for conducting time history analyses of the RC elevated liquid storage tank. The particulars of the ground motion time histories are, Imperial Valley, 1940 (PGA = 0.34 g), Tabas, 1978 (PGA = 0.93 g), Loma Prieta, 1989 (PGA = 0.57 g), Turkey, 1992 (PGA = 0.5 g), Northridge, 1994 (PGA = 0.84 g), Kobe, 1995 (PGA = 0.59 g), Chi-Chi, 1999 (PGA = 0.21 g), and Bhuj, 2001 (PGA = 0.11 g). The geometric properties of the tank considered in the numerical study are the same as in Waghmare et al. [11, 12].

The parameters of the RC tank considered for numerical simulation are presented in Table 1. The volume of the liquid is considered as 100% and 25% ($S = 2.0$ and $S = 0.5$) of the tank capacity excluding freeboard. The dynamic properties of the tank modeled herein are obtained by conducting free vibration analysis. The fundamental natural time period for Mode 1 is 5.06 s and 3.68 s; for Mode 2 is 1.12 s and 2.18 s; and for Mode 3 is 0.57 s and 0.58 s for $S = 0.5$ and $S = 2.0$, respectively.

The details of the control systems and the algorithms used in the numerical study are described in Table 2. To study the effect of voltage on open-loop and closed-

Table 1 Parameters of the RC elevated tank considered in the numerical study

Parameter	Numerical values	Parameter	Numerical values
Height of the liquid	$H = 5.05$ m	Damping ratio for convective mass	$\xi_c = 0.5\%$
Inner diameter of container	$2R = 5.06$ m	Modulus of elasticity of RC	$E = 25 \times 10^6$ kN/m^2
Aspect ratio, $S = H/R$	0.5 and 2.0	Damping ratio of RC	$\xi_t = 5.0\%$
Mass density of the contained liquid (water)	1000 kg/m^3	Number of columns	4

Table 2 Details of the control algorithms

S. No	Classification of the control system	Semi-active control algorithms	Voltage	Semi-active device/controller used for response reduction
1	Open-loop control system	Passive-OFF	$V_{\max} = 0$ V	SAMRD
2	Open-loop control system	Passive-ON	$V_{\max} = 2.75$ V	SAMRD
3	Closed-loop feedback control system	Clipped-optimal control (COC)	$V_{\max} = 2.75$ $V_{\max} = 10$ V	SAMRD + controller
				$\boldsymbol{Q_d} = \boldsymbol{K}$ and $\boldsymbol{Q_v} = \boldsymbol{M}$
				$\boldsymbol{R_r} = 10^{-6}(\boldsymbol{I})_{n \times n}$
				$q = 1 \times 10^5$
4	Closed-loop feedback control system	Simple semi-active control (SSC)	$V_{\max} = 2.75$ V $V_{\max} = 10$ V	SAMRD + controller
				$F_{D\max} = 40$ kN

loop systems response, the command voltage $\boldsymbol{V}$ is varied from 1 to 10 V (Passive-ON, COC, and SSC). For the Passive-OFF algorithm, the command voltage is kept constant at 0 V. In case of optimal control theory, the control force parameter, $\boldsymbol{R_r}$, and the response reduction parameter, q, are considered as 10^{-6} and 10^5, respectively.

To study the effect of placement of dampers, three configurations of placement of dampers, viz., Configuration I (SAMRDs installed in Panel 1, Panel 3, Panel 5, and Panel 7), Configuration II (SAMRDs installed in Panel 2, Panel 4, and Panel 6); and Configuration III, where SAMRDs are installed in all the Panels in the staging of the RC elevated tank, are considered (see Fig. 1).

The seismic response quantities, viz., normalized base shear, $F_b = V_b/W$, shear at the level of lumped masses, and damper forces are obtained. Here, $W = (M_t)\, g$ is the total weight of the tank, where M_t is the total mass, considering bracing level masses, rigid mass, and convective mass.

The modified Bouc–Wen model is used to model nonlinear hysteretic behavior of SAMR damper. Bouc–Wen parameters used in the present study are listed in selected as $\alpha_a = 102$ N/m, $\alpha_b = 496$ N/m/V, $k_0 = 5000$ N/m, $k_1 = 1200$ N/m, $\gamma = 300$/m, $\beta = 300$/m, $x_0 = 0$, $\boldsymbol{c_{0a}} = 440$ N-s/m, $\boldsymbol{c_{0b}} = 4400$ N-s/m/V, $\boldsymbol{c_{1a}} = 29{,}000$ N-s/m, $\boldsymbol{c_{1b}} = 500$ N-s/m, $\eta = 50$ s^{-1}, $A_{mr} = 1900$, $n_{mr} = 1$. The parameters selected are on the basis to suit the damper force–velocity behavior (see Fig. 2).

6.1 *Effect of Voltage*

The base shear response of tanks with the open-loop systems (Passive-OFF and Passive-ON) and closed-loop systems (COC and SSC) is considered when the elevated tank is installed with SAMR dampers (Configuration III). The effect of change in command voltage is studied when the command voltage, $\boldsymbol{V}(t)$, is changed from 0 to 10 V. Figure 3 shows the normalized base shear response of the tanks installed with SAMR dampers (Configuration III) for $S = 0.5$ and $S = 2.0$ for four earthquakes. Additionally Fig. 3 presents the comparison of the Passive-OFF ($V = 0$), Passive-ON, COC ($q = 10^5$, $R_r = 10^{-6}$), and SSC ($F_{D\max} = 40$ kN) algorithms.

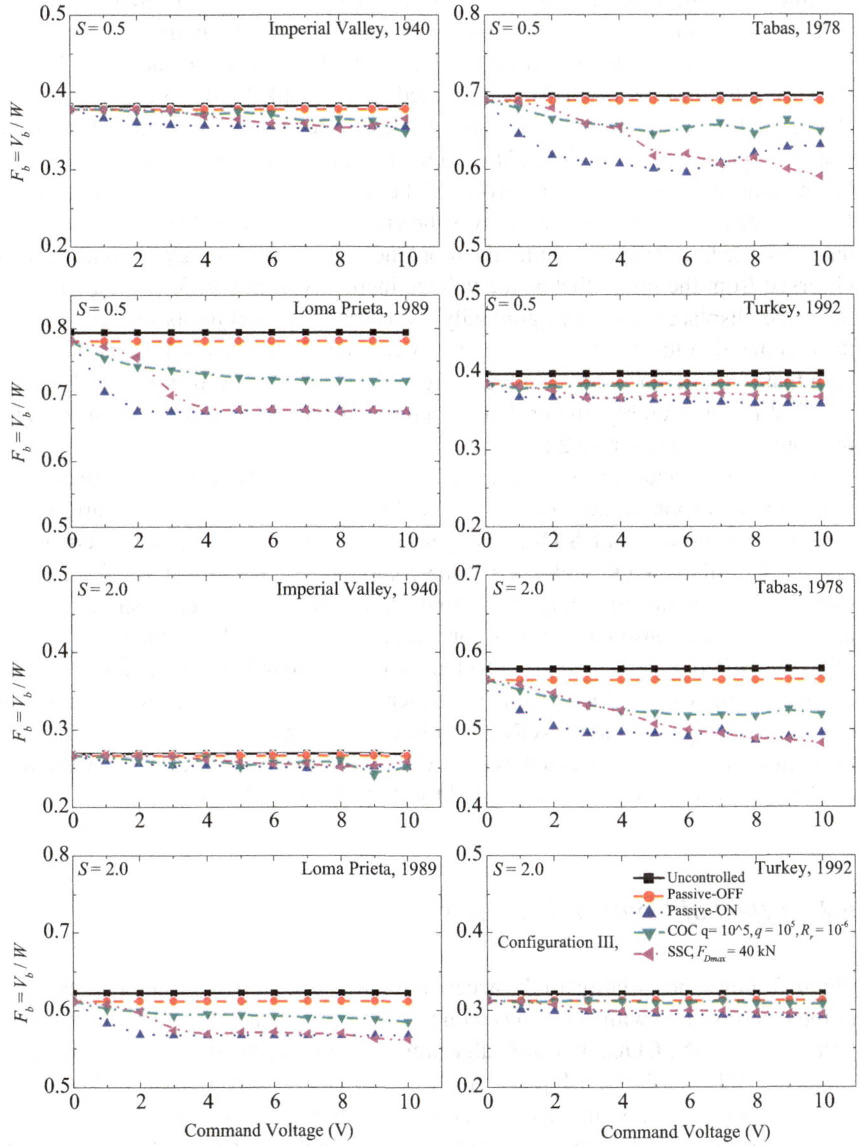

Fig. 3 Variation of base shear response with the command voltage when the RC elevated tank is installed with SAMRDs and operated with Passive-OFF, Passive-ON, COC, and SSC algorithms

From Fig. 3, it is observed that SAMR dampers reduce the base shear response of the tanks under all considered earthquakes. The increase in voltage increases the response reduction. It is the specialty of magneto-rheological dampers as the voltage is applied, the MR fluid is converted into the semisolid state, increasing the resistance motion and leading to increased response reduction. Remarkable response reduction is achieved as the voltage is increased from 0 to 3 V; however, a further increase in voltage has resulted in negligible or no additional response reduction.

Figure 4 describes the effect of voltage on the Force–displacement and Force–velocity relationship for SAMRD installed in RC tank ($S = 2.0$, Configuration III) under Loma Prieta, 1989 earthquake. The figure shows the relationship of the response quantities of the SAMRD installed in Panel 4, which portrays the peculiar nonlinear force–velocity behavior of the MR damper. The significance of the force–displacement loop is that it gives the energy dissipation characteristics of the dampers; the larger the area under the loop, the higher is the energy dissipation. It is observed from the figure that as the voltage increases from 0 to 3 V, the area under the force–displacement curve gradually increases. This highlights the fact that, as the command voltage increases, the energy dissipation increases. Further, it is also seen that the difference between the energy dissipation achieved from 0 to 1 V; and 1 to 2 V is considerably higher than the energy dissipation achieved by increasing the command voltage from 2 to 3 V.

From Figs. 3 and 4, it is seen that although Passive-ON strategy is effective in reducing the response, the saturation of the MR effect seems to have occurred after 3 V. The performance of SAMR damper can be enhanced by applying advanced control algorithms (COC and SSC) to the command voltage. It is noted from Fig. 3 that at higher command voltage SCC algorithm is the most effective strategy as far as base shear response reduction is concerned. It is observed that the effectiveness of the SAMR damper is greatly affected by the command voltage and the control algorithm employed. Moreover, it is further noted that there exists the optimum command voltage for Passive-ON algorithm. Accordingly, for further simulations, the command voltage is selected as 2.75 V for Passive-ON algorithm; and COC and SSC algorithms are further investigated for 2.75 V and 10 V.

6.2 *Effect of Control Strategies*

Figure 5 shows the variation of the acceleration, velocity, and displacement response of the tank installed with SAMRDs with Configuration I and operated with Passive-OFF, Passive-ON, COC, and SSC algorithms with designated command voltages, subjected Tabas, 1978 earthquake. It is observed that acceleration and velocity response is profoundly affected by the control strategies employed. Further, it is exciting to know that though there is a reduction in acceleration and velocity response, the response at the convective level remains unaffected. Moreover, displacement response is not much affected by the installation of the SAMRDs. It is observed from

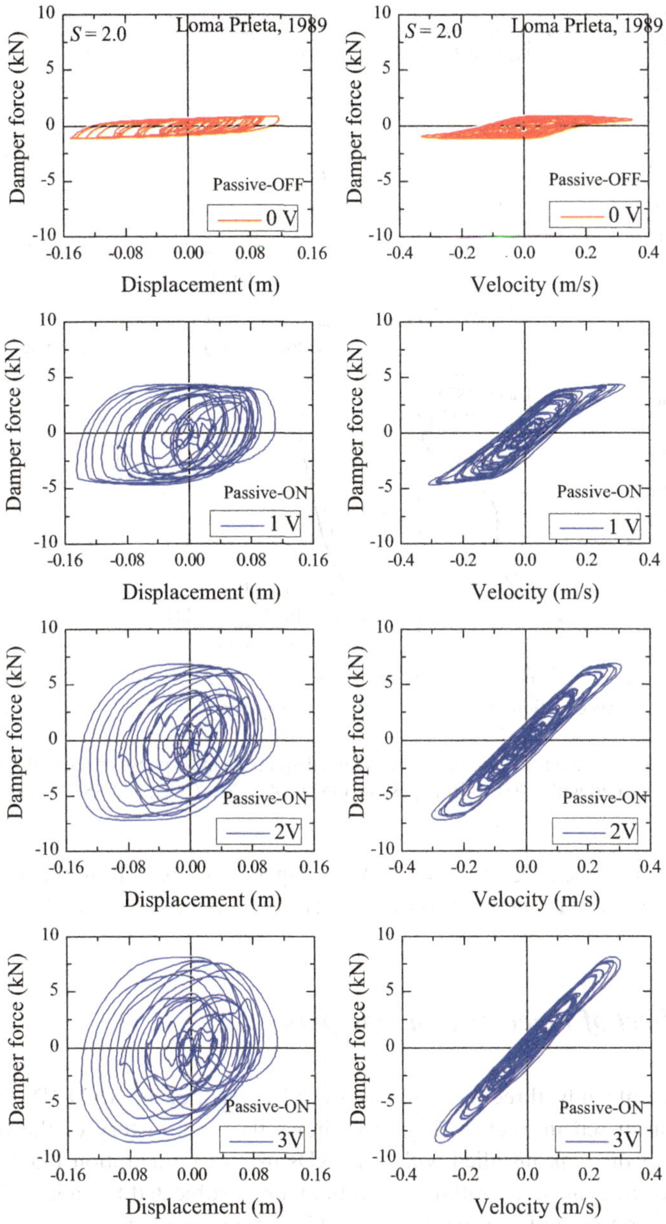

Fig. 4 Effect of voltage on the force–displacement and force–velocity relationship for SAMRDs installed at the panel 4, in RC tank ($S = 2.0$, Configuration III), under Loma Prieta, 1989 earthquake

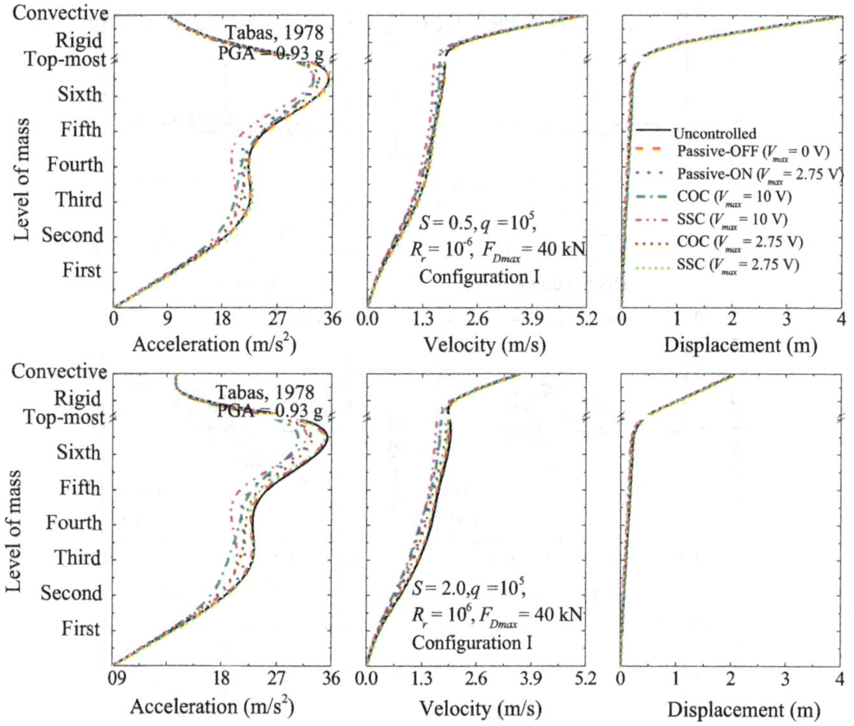

Fig. 5 Variation of the acceleration, velocity, and displacement response of the tank installed with SAMRDs operated with passive-OFF, passive-ON, COC, SSC, COC, and SSC control algorithms

the figures that the SSC ($V_{\max} = 10$ V) is superior in controlling the acceleration, velocity, and displacement response of an elevated liquid storage tank.

6.3 *Effect of Placement of Dampers*

In the present study, three configurations of placement of the SAMRDs are considered as discussed in Sect. 5. Figure 6 shows the time history of the base shear response of the tank installed with SAMRDs using Configuration I and employed with different control algorithms when the tank is subjected to Tabas, 1978 earthquake. It is observed from the figure that Configuration I is the suitable placement configuration of SAMRDs for Tabas, 1978 earthquake. Further, the suitability of configuration may be different for different earthquakes and different control algorithms. Though SSC ($V_{\max} = 10$ V) algorithm is found to be outstanding (30% reduction), the other algorithms such as Passive-OFF, Passive-ON, COC ($V_{\max} =$ 10 V), COC ($V_{\max} = 2.75$ V) and SSC ($V_{\max} = 2.75$ V), also controlled the base

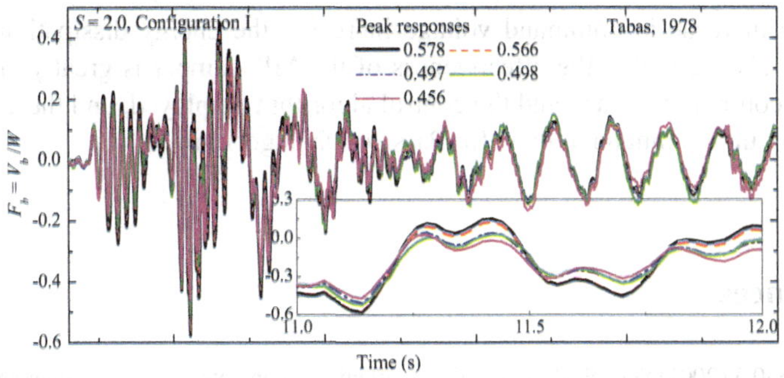

Fig. 6 Time variation of peak normalized base shear (F_b) for RC elevated liquid storage tank (S = 2.0), installed with SAMR dampers in configuration I, operated with passive-OFF, passive-ON, COC, and SSC algorithms and subjected to Tabas, 1978 earthquake

shear response with maximum reduction as 5%, 29%, 18%, 10%, and 16%, respectively, for broad tank; and for slender tanks it is 9%, 29%, 24%, 12%, and 12%, respectively. COC (V_{max} = 10 V) algorithm proved to be more capable than all the other algorithms in case of Imperial Valley, 1940, (Configuration III, $S = 0.5$ and $S = 2.0$), Tabas, 1978 (Configuration II, $S = 0.5$ and $S = 2.0$), Northridge, 1994 (Configuration I, $S = 0.5$), Kobe, 1995 (Configuration II and Configuration III, $S = 2.0$), and Bhuj, 2001 (Configuration I and Configuration III, $S = 0.5$). Whereas better results for SSC (V_{max} = 2.75 V) are obtained under Loma Prieta, 1989, Turkey, 1992, and Kobe, 1995 (Configuration II, $S = 0.5$ and $S = 2.0$), earthquakes. Passive-OFF and COC (V_{max} = 2.75 V) algorithms are not that much striking as far as base shear response reduction of the elevated liquid storage tanks is concerned.

7 Conclusions

In this paper, the effectiveness of semi-active magneto-rheological dampers (SAMRDs) for seismic response reduction of the RC elevated liquid storage tank has been demonstrated. The SAMRDS are installed in three placement configurations within the staging of the tank and are employed with four control algorithms as Passive-OFF, Passive-ON, Clipped Optimal Control (COC), and Simple Semi-active Control (SSC).

1. A simple semi-active control (SSC) is proposed in the present study which shows the outstanding performance in reducing the response of the tank. The significant advantage of this control algorithm is that it does not require any control parameter to be decided.
2. Remarkable response reduction is achieved as the voltage is increased from 0 to 3 V; however, a further increase in voltage has resulted in negligible or no additional response reduction.

3. The increase in command voltage increases the energy dissipation. Further, it is also seen that the effectiveness of the MR damper is greatly affected by the command voltage and the control algorithm employed, and there exists the optimum command voltage for Passive-ON algorithm

References

1. ACI 350.3 (2001) Seismic design of liquid containing concrete structures. American Concrete Institute, Farmington Hill, Michigan (MI), USA
2. Carlson JD, Jolly MR (2000) MR fluid, foam and elastomer devices. Mechatronics 10:555–569
3. Dyke SJ, Spencer BF, Sain MK, Carlson JD (1996) Modeling and control of magnetorheological dampers for seismic response reduction. Smart Mater Struct 5(5):565–575
4. EN 1998-4 (2006) Euro-code 8—design of structures for earthquake resistance. Part 4: silos, tanks and pipelines. European Committee for Standardization, Brussels
5. Gluck N, Reinhorn AM, Gluck J, Levy R (1996) Design of supplemental dampers for control of structures. J Struct Eng ASCE 122(12):1394–1399
6. Hart GC, Wong K (2000) Structural dynamics for structural engineers. Wiley, New York
7. Jung HJ, Spencer BF Jr, Lee IW (2003) Control of seismically excited cable-stayed bridge employing magnetorheological fluid dampers. J Struct Eng 129(7):873–883
8. Math Works, MATLAB (2018) The language of technical computing. The Math Works Inc., USA
9. Sadek F, Mohraz B (1998) Semi-active control algorithms for structures with variable dampers. J Eng Mech ASCE 124(9):981–990
10. Spencer BF Jr, Dyke SJ, Sain MK, Carlson JD (1997) Phenomenological model for magnetorheological dampers. J Eng Mech ASCE 123(3):230–238
11. Waghmare MV, Madhekar SN, Matsagar VA (2019) Semi-active fluid viscous dampers for seismic mitigation of RC elevated liquid storage tanks. Int J Struct Stab Dyn 19(03):1950020. https://doi.org/10.1142/s0219455419500202
12. Waghmare MV, Madhekar SN, Matsagar VA (2020) Influence of nonlinear fluid viscous dampers on seismic response of RC elevated storage tanks. Civ Eng J 6: 98–118 (Special issue Emerging materials in civil engineering)
13. Yang G, Spencer BF Jr, Carlson JD, Sain MK (2002) Large-scale MR fluid dampers: modeling and dynamic performance considerations. Eng Struct 24:309–323

Virtual Testing of Prototypes Using Test Frame Designed for Lateral Load

Suyog Nikam and I. P. Sonar

Abstract Structures such as Retaining Walls, Sheet piles, Frames, Masonry Walls, and Poles are subjected to lateral load. They are designed considering lateral load. lateral load-carrying capacity is an important factor for these structures. For verification of lateral load-carrying capacity of structures, lateral load testing of the actual prototype of such structures is needed. A vertical load testing facility is mostly available in laboratories but a lateral load testing facility is very rarely available. The lateral load testing facility is available in some of the engineering institutions in India, such as IIT Kanpur, SERC Chennai, and IIT Guwahati. During the testing of prototypes, an adequate reaction must be provided by the reaction wall specially manufactured for the lateral load test setup. The reaction wall should act as rigid lateral support and should not deflect during testing. This work includes the development and design of a reaction frame for Lateral load testing. The reaction frame is designed in such a way that prototypes of different dimensions can be tested. The designed reaction frame is used for virtual testing of the shear wall, retaining wall, and steel frame by using finite element software STAAD PRO. Obtained deflection results of the reaction frame and prototypes are analyzed. The results showed that the designed reaction frame is rigid enough to test different prototypes for lateral loading. As lateral load test setup is rarely available, it will have great demand.

Keywords Lateral load · Virtual testing · Retaining wall · Shear wall · Frame

S. Nikam (✉)
College of Engineering Pune, Pune, India
e-mail: suyognikam1765@gmail.com

I. P. Sonar
Civil Engineering Department, College of Engineering Pune, Pune, India

M. S. Ranadive et al. (eds.), *Recent Trends in Construction Technology and Management*, Lecture Notes in Civil Engineering 260,
https://doi.org/10.1007/978-981-19-2145-2_80

1 Introduction

Structures like masonry walls, shear walls, and frames are subjected to lateral loading due to wind and earthquake actions. The retaining wall is under lateral loading due to soil pressure. Electric poles are under significant lateral loading due to the action of wires. These structures are designed considering lateral load. The lateral load is more critical for structures than vertical loads. Construction materials like concrete and brick are very strong in compression but very weak in tension. Lateral loading produces significant tension in these structures. Hence, lateral load-carrying capacity is an important factor for these structures. In ancient construction, resistance to lateral load is provided by using thick walls. However, in modern construction, thick wall is not preferred due to size constrain. For verification of lateral load-carrying capacity of structures, lateral load testing of the actual prototype of such structures is needed. Detail study of the behavior of structures under the application of lateral load is essential. For testing of actual prototypes, a reaction frame is required to provide adequate support during testing. As the sizes differ from specimen to specimen, it is difficult to use the same reaction frame for testing different types of test specimens [1–3]. In earlier studies, reaction frames are designed and constructed to conduct an experimental study of a single type of specimen. In this type of test, setup is very expensive and not feasible. Earlier test setups can only check the specimens of specific height and type. This test setup can check specimens of different heights and types [4]. The lateral load testing facility is available in some of the engineering institutions in India, such as IIT Kanpur, SERC Chennai, and IIT Guwahati [5]. Testing of lateral load behavior of Assam-type houses by testing full-scale frame specimens was done at IIT Guwahati. The objective of this study is to design such a reaction frame to conduct virtual testing of a different specimen such as retaining wall, shear wall, and frame using a designed reaction frame on STAAD Pro. software and also to check the performance of the reaction frame during testing.

2 Reaction Frame

2.1 Description of Reaction Frame

Lateral load testing requires a strong reaction frame while imposing lateral loads employing a hydraulic jack. For this purpose, the steel frame is designed to be built behind the test specimen. This Steel frame has a dimension of 3.6 m in height and a plan area of 2.4 m * 2.4 m. Prototypes up to height 3.4 m and 2.4 m wide can be tested using this reaction frame. Normally, the height of the shear wall, masonry wall, retaining wall, and frame is up to 3.4 m. Therefore, this height is selected for the reaction frame. For out-of-plane testing of a masonry wall and retaining wall, 2.4 m width is selected. The capacity of the test frame is 300 kN. That is frame can resist the total summation of the lateral load of 300 kN. Lateral loads can be applied by

Fig. 1 3D rendered view of reaction frame

employing horizontal members attached with jacks. The 3D shape of the test frame is shown in Fig. 1.

The Front Horizontal members are simply supported by the vertical column members. Bolted simple connection is used. These front horizontal members are movable along the vertical plane at different heights from 0.2 to 3.4 m with an interval of 0.1 m. This movement of front horizontal members is important for testing specimens with different heights. The lateral load should be applied using a hydraulic jack attached to the front horizontal members. The connection of the hydraulic jack to these members is made in such a way that hydraulic jacks are movable along the horizontal axis.

2.2 STAAD Pro. Modeling of Reaction Frame

Rolled steel sections are considered for the design of the reaction frame. Properties of steel are shown in Table 1. A Reaction frame of a steel structure presented in Fig. 3 has been analyzed using STAAD Pro. Software. Three Column Reaction frame with dimensions as given in Fig. 2. All connections are fixed connections

Table 1 Material property

Material	Elastic modulus (E) (N/mm^2)	Poisson's ratio (μ)	Density (kg/m^3)	Alpha (/°C)	Yield strength (Fy) (N/mm^2)	Ultimate strength (Fu) (N/mm^2)
Steel	200,000	0.3	7833.409	0.000012	250	407.8

except the connection of the front horizontal member (M3) to the vertical column (M1). Connections have been designed and size parameters such as the thickness of the plate, diameter of bolts, gauge distance of bolts, weld size, and weld length have been evaluated from connection details using IS800-2007 provisions. Details of loading considered for frame analysis have been as follows. 300 kN of point load was applied at each point after a 0.1 m interval separately along both vertical and horizontal axis. So, during in-plane lateral load testing, an adequate reaction is provided. Then distributed load along the horizontal beam with a summation of 300 kN is applied on the same points. This type of loading will provide adequate reaction during out-of-plane lateral load testing for specimens. Finally, distributed load with a summation of 300 kN is applied along the vertical axis so that testing of the retaining wall is possible by converting soil pressure into an equivalent point load system.

Rolled Channel sections and equal angle sections are used for designing the reaction frame. Steel sections are selected and designed considering arrangements for lateral load application. Steel sections of the reaction frame are designed considering IS 800-2007 provisions. Steel sections selected for design are as shown in Fig. 2 and Table 2.

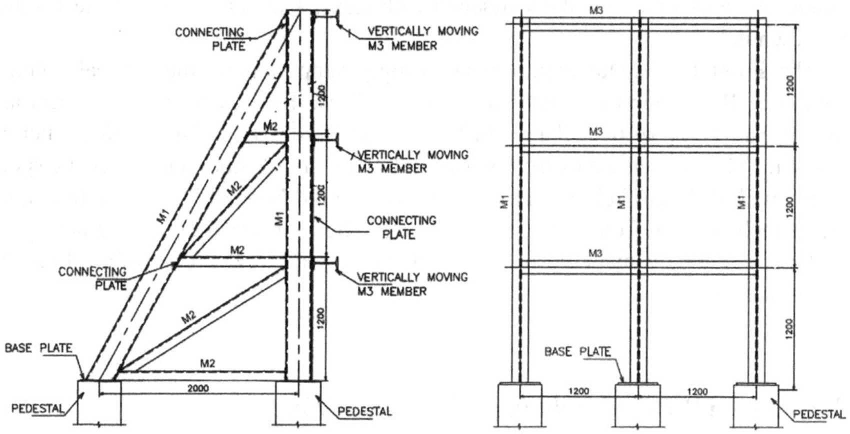

Fig. 2 A side view and front view of reaction frame

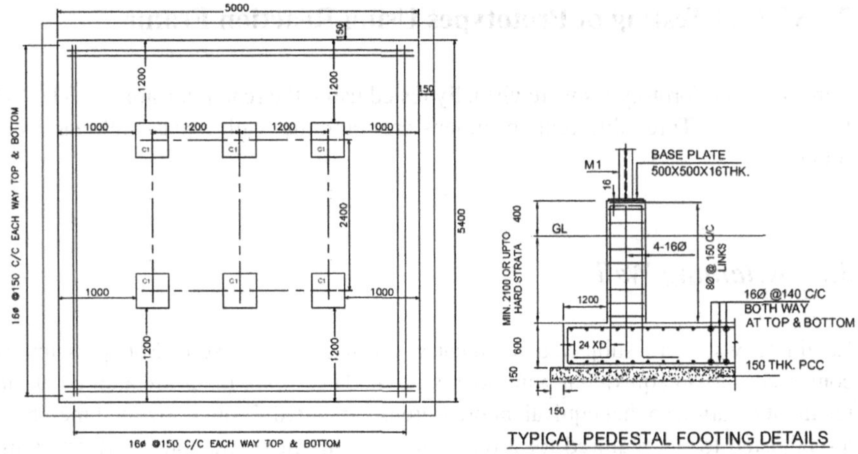

Fig. 3 Footing layout and details

Table 2 Steel sections

Member	Description	Selected section
M1	Vertical and inclined column	2ISMC 350 Back to back channel section
M2	Inclined bracing members	2 ISA 100 * 100 * 10 Back to back equal angle section
M3	Front horizontal members	2ISMC 250 Back to back channel section

2.3 Foundation Design for Reaction Frame

The foundation design of the reaction frame is critical. The reaction frame is subjected to 300 kN of lateral force. Due to the application of lateral load, the overturning and sliding effect is very significant for the reaction frame. For countering, the effect of overturning and sliding, mat footing must be provided at least at the depth of 2.1 m below the ground level. As the spacing between the columns is very less, mat footing is an appropriate option. For the design of the foundation, M30—the grade of concrete and Fe 500—grade of steel reinforcement are considered. The required depth of Mat footing is 600 mm. Details of foundation design are shown in Fig. 3. For designing the foundation, provisions of IS 456-2000 are used.

3 Virtual Testing of Prototypes Using Reaction Frame

Three different prototypes were virtually tested using the reaction frame on STAAD Pro. Software. Three different specimens are retaining wall, Shear wall, and steel frame.

3.1 Retaining Wall

Retaining walls are subjected to linearly varying soil pressure. Soil pressure is converted into an equivalent point load system. Then point loads are applied at the resultant location so that equivalent pressure is generated. Front horizontal members are adjusted to the required location. Using the designed reaction frame retaining walls up to a height of 3.4 m and a width of 2.4 m can be tested. For the different heights of the retaining walls, the locations of front horizontal members are shown in Table 3.

For virtual testing retaining wall, 3 m high retaining wall is modeled in front of designed reaction frame in STAAD Pro. An appropriate model of reaction frame is selected as per Table 3. As shown in Fig. 4, unit weight of soil is considered as 19 kN/m^3 and the angle of internal friction is taken as 30°. Concrete of grade M30 and steel of grade Fe500 are considered for designing the retaining wall. Provisions of IS 456:2000 are considered while designing the retaining wall.

Surcharge load is increased from 15 to 150 kN/m^2. Deflection results of reaction frame and retaining wall during the virtual testing using STAAD Pro. Software, are as shown in Table 4 and Fig. 7.

Table 3 Location of horizontal members along vertical axis

Height of retaining wall (m)	Location of horizontal member 1 (m)	Location of horizontal member 2 (m)	Location of horizontal member 3 (m)
3.4	0.5	1.5	2.7
3.2	0.5	1.5	2.6
3.0	0.5	1.5	2.5
2.8	0.5	1.5	2.4
2.6	0.5	1.5	2.3
2.4	0.4	1.2	2.0
2.2	0.4	1.2	1.9
2.0	0.4	1.1	1.7
1.8	0.3	0.9	1.5
1.6	0.3	0.9	1.4

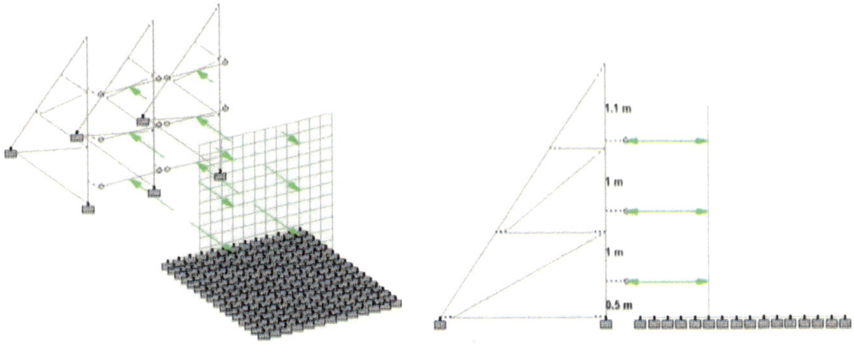

Fig. 4 3D and side view of virtual testing of retaining wall

Table 4 Test results for retaining wall

Surcharge (kN/m^2)	Deflection of retaining wall (mm)	Deflection of reaction frame (mm)
0	2.572	0.080
15	3.415	0.110
30	4.258	0.140
45	5.101	0.160
60	5.945	0.195
75	6.787	0.224
90	7.629	0.253
105	8.473	0.281
120	9.316	0.310
135	10.159	0.331
150	11.002	0.368

3.2 *Shear Wall*

A shear wall is a vertical element of a structural system that is designed to resist lateral forces. Virtual testing of the shear wall for out-of-plane lateral forces is performed using an earlier designed reaction frame. A shear wall of 3.0 m height and 250 mm thickness is modeled in front of an earlier designed reaction frame in STAAD Pro. As shown in Fig. 5, concrete of grade M30 is considered for designing the shear wall. Out-of-plane, lateral load from 45 to 225 kN is applied at three different locations on the shear wall.

Provisions of IS 456:2000 are considered while designing the retaining wall. Deflection Results of reaction frame and shear wall during the virtual testing using STAAD Pro. Software is shown in Table 4 and Fig. 5.

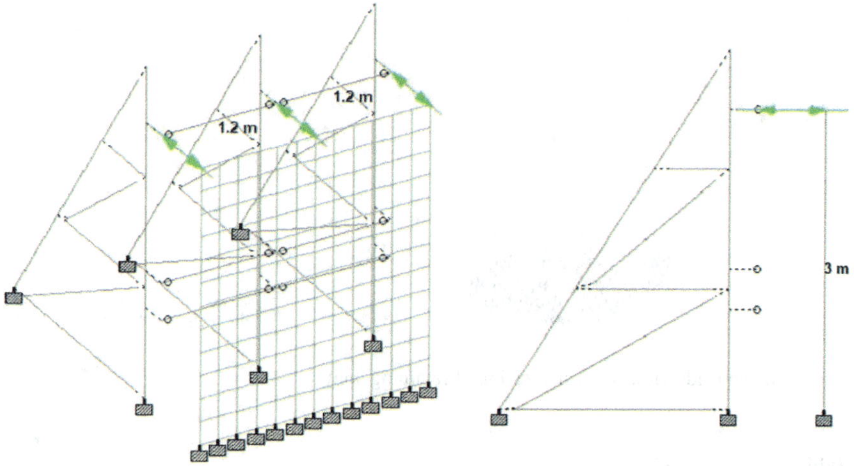

Fig. 5 3D and side view of virtual testing of shear wall

3.3 Steel Frame

The last two virtual tests are performed for out-of-plane lateral load action. This virtual test is performed for the in-plane lateral load. The steel frame of ISMB 250 and ISA 60 * 60 * 10 is modeled in front of the reaction frame and lateral point load is applied until failure (300 kN) of the steel frame prototype. The height of the steel frame prototype is 2.4 m and the length is 2.0 m (Fig. 6).

Deflection Results of reaction frame and shear wall during the virtual testing using STAAD Pro. Software is shown in Table 5 and Fig. 7.

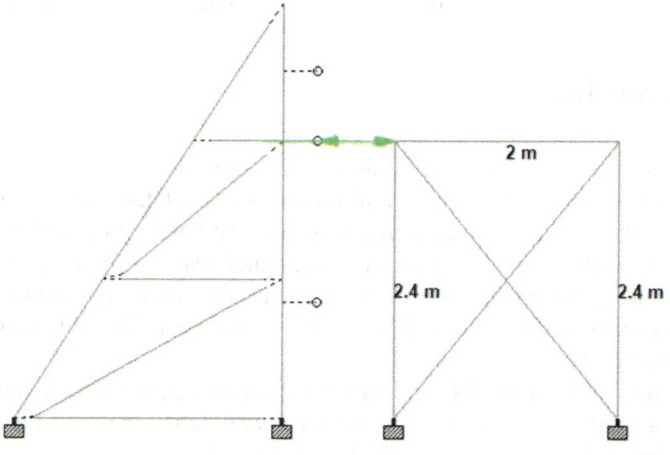

Fig. 6 3D and side view of virtual testing of the shear wall

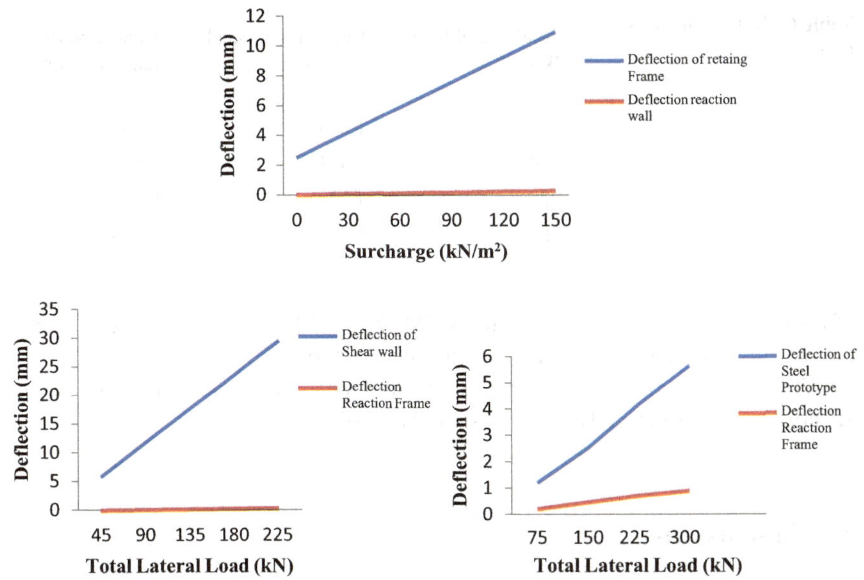

Fig. 7 Deflection of reaction wall and different prototypes

4 Results and Discussions

The results obtained from the FEM analysis using STAAD Pro. are summarized in Tables 4, 5, and 6. The comparison graph between the deflection of different prototypes and reaction frames is shown in Fig. 7. Maximum deflection on test prototypes is observed at the top end as fixed support is only assumed at the bottom of the reaction frame. The maximum deflection for retaining is 11.022 mm, for the shear wall, it is 29.741 mm, and for the steel test frame, it is 7.429 mm after members started failing. While for all prototypes, deflection of the reaction frame is very low. It shows that the reaction frame is rigid enough for testing prototypes of different dimensions up to a height of 3.4 m and a width of 2.4 m. The maximum deflection

Table 5 Test results for shear wall

Total lateral load (kN)	Deflection of shear wall (mm)	Deflection of reaction frame (mm)
45	5.948	0.086
90	11.896	0.172
135	17.844	0.259
180	23.793	0.355
225	29.741	0.444

Table 6 Test results for steel frame

Total lateral load (kN)	Deflection of steel frame (mm)	Deflection of reaction frame (mm)
75	1.225	0.254
150	2.574	0.509
225	4.227	0.763
300	5.703	0.938

of the reaction frame is 0.938 mm for a 300 kN point load. Due to Covid-19, actual construction of the reaction frame is currently postponed.

While performing an actual experiment, this deflection error should be considered. As no structure is perfectly rigid, this error is considerable.

5 Conclusions

Virtual testing using finite element (STAAD Pro. software) was performed under different lateral loading using a reaction frame specially designed for the lateral load test. Three different prototypes, namely, retaining wall, shear wall, and steel frame prototypes were tested. The performance of the reaction was checked during virtual testing and it showed good results. As the designed reaction frame is cost-efficient and compact in size, it is capable of testing various types of prototypes. Following are a few conclusions that are drawn from the study:

1. Maximum deflection of the reaction frame is 0.938 mm for 300 kN of lateral point load. This shows that the reaction frame is strong and capable testing of different prototypes.
2. Total steel requirement for assembly of the reaction frame is only 33.586 kN and the area required for the total lateral load test setup is 5.5 m by 9 m. Hence, it is cost-efficient and compact.
3. Lateral load testing of various types of prototypes with different heights from 0.2 to 3.4 m is possible using the single reaction frame.
4. Experimental setup suggested that the lateral load testing of prototypes will help for understanding the true behavior of the large specimens and this facility will create an opportunity for research work as well a revenue generation through testing of large prototypes, which is needed for the Engineering Institute.

References

1. Choudhury T et al (2020) Experimental and numerical analyses of unreinforced masonry wall components and building. Constr Build Mater 257. http://doi.org/10.1016/j.conbuildmat.2020.119599
2. Arslan ME et al (2017) Structural behavior of rammed earth walls under lateral cyclic loading: a comparative experimental study. Constr Build Mater 133:433–442. http://doi.org/10.1016/j.conbuildmat.2016.12.093
3. Graham DA et al (2010) Performance of log shear walls subjected to monotonic and reverse-cyclic loading. J Struct Eng 136(1):37–45. http://doi.org/10.1061/(asce)st.1943-541x.0000035
4. Morandi P et al (2014) In-plane experimental response of strong masonry infills. In: 9th international masonry conference, July 7, 8, 9, 2014, Guimarães, pp 1–12
5. Chand B et al (2019) Lateral load behavior of traditional Assam-type wooden house. J Struct Eng 145(8):04019072. http://doi.org/10.1061/(asce)st.1943-541x.0002359

Crack Simulation and Monitoring of Beam-Column Joint by EMI Technique Using ANSYS

Tejas Shelgaonkar and Suraj Khante

Abstract Health monitoring of structures has a great contribution to the growing interest of researchers in damage detection. The Structural Engineering Community is nowadays employing Electro-Mechanical Impedance (EMI) technique on a wide range for monitoring the health of structures. EMI technique involves the use of Piezoelectric (PZT) patches bonded to the host structure and also instruments such as Impedance analyzer/LCR meter, Wired/Wireless connections, etc. The change in conductance/impedance/susceptance signatures in comparison with the healthy signatures defines the presence of damage. The experimental arrangement in this regard resembles the impedimental system approach. So, substitutable modeling and analysis with the help of simulation are preferable. In this research, a beam-column joint has been modeled in ANSYS 19.2 software. After analysis, the joint has been monitored for baseline (healthy) results on the basis of the simulated EMI technique for health inspection. After getting the baseline conductance and susceptance signatures, the beam-column joint has been formed with a crack modeled on the joint. The fracture analyzed model was then monitored in the same way as a healthy model with the help of the PZT patch. The conductance signature so obtained was then compared with the baseline (healthy) signature, the change in which identified the presence of a crack on the joint. Root mean squared deviation (RMSD) values have been calculated for computing the change in conductance values for respective frequency ranges.

Keywords ANSYS · Damaged/cracked beam-column joint · EMI technique · SHM

T. Shelgaonkar (✉)
Structural Engineering, Government College of Engineering, Amravati, India
e-mail: shelgaonkartejas@gmail.com

S. Khante
Applied Mechanics, Government College of Engineering, Amravati, India

M. S. Ranadive et al. (eds.), *Recent Trends in Construction Technology and Management*, Lecture Notes in Civil Engineering 260,
https://doi.org/10.1007/978-981-19-2145-2_81

1 Introduction

The important units of structures are expected to fulfill safety considerations all through the existence period. That is why damages and defects ranging from incipient stage to greater stage due to fatigue, overburden, and environmental impacts are necessary to be detected [1]. Around the whole world, health monitoring of structures has been continuously playing an important role in the fine development of civil, aerospace, and mechanical engineering [2]. Structural Health Monitoring (SHM) is the process of checking the present condition of structures. For reducing the repairing and maintenance cost, Non-Destructive Testing (NDT) is preferably used. Perceptible inspection is usually executed on the anticipated locations of damages by the inspectors with fine experience. Moreover, health monitoring is a work-intensive program incorporating risk factors about safety. This is the reason behind SHM works leaning toward real-time sensing systems [3]. One of the most popular methods in this regard include Electro-Mechanical Impedance Technique (EMI technique) for health monitoring [4]. Over the past years, SHM has got a prime part in improving the reliability and safety of structures, which reduces the cost of operation [5]. Structural Health Monitoring (damage detection) contains various methods according to the use, type of damage, condition of structure, etc. From all those methods, Electro-Mechanical Impedance based damage detection has gained popularity these days. Middle-sized to the greater-sized damages are easily visible, but holding the severe damages back in the incipient stage itself by detecting the damage earlier can head toward the security and well-being of the infrastructure [6]. EMI technique stands ahead in detecting the incipient level damages in the structure. It involves the use of Piezoelectric (PZT) patches. In the EMI technique, a PZT can serve as both actuator and sensor when attached to a structure. When a PZT is provided with an alternating electric field, deformation is produced in the PZT patch which then excites the parent structure to which it is glued. This excitation response is again transferred back to the piezoelectric patch as an electrical response. When any change in the response (mechanical) due to some damage is generated, the effect of that is seen there at the LCR meter (Fig. 1)/impedance analyzer in the electrical response of the piezoelectric (PZT) patch [7]. Beam-Column joint is one of the most crucial elements of a

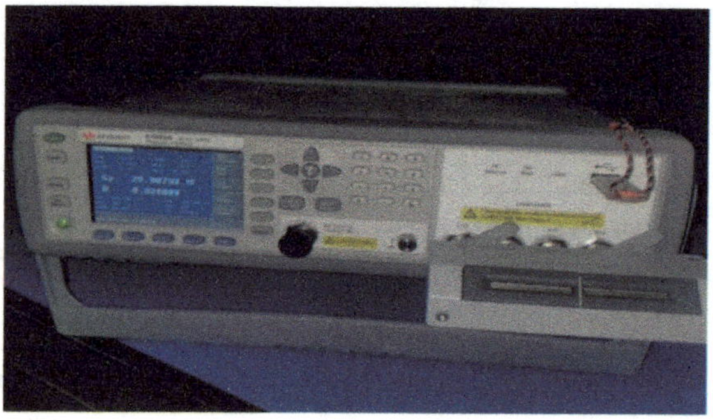

Fig. 1 LCR meter

structure, the damage of which may lead to large-scale destruction of a structure. As soon as lateral loads act, critical problems start to arise at a joint and so in the whole structure. In the absence of an experimental setup, the simplest way of working on a system for specific results to be obtained after analysis is finite element simulation. Generation of finite element model to evaluate the performance of the system can get the required results in an efficient manner.

Divsholi et al. [8] took four differently detailed concrete structures with PZT sensors, and the impedance extracted from PZT was compared with admittance. A relation between load step and damage index was obtained. Yan and Chen [5] correlated the signature changes to properties of structures and damage detection was ensured. The thematic experimental setup is shown in Fig. 1. Baluch et al. [9] presented the finite element model of a beam-column junction with concrete having low strength, retrofitted with CFRP sheets by using 'DIANA' software. Hamzeloo et al. [10] performed the EMI technique on aluminum and steel hollow cylinders. FEM results were also compared with experimental data. The effects of the damage on damage metrics were explained. For the same, experimental and FEM results of the system were used. Lim and Soh [11] investigated fatigue crack detection and characterization by the EMI technique. Bharathi Priya et al. [7] experimentally analyzed differentially cured concrete beam embedded with PZT patches of varying thickness connected in series. Using the 2nd-day signature as baseline signature, the variation in frequency peaks was quantified using RMSD as strength index. Results show that the serial sensing method using PZT is a quick and efficient method. Venkatesan et al. [12] analyzed the seismic performance of beam-column joint with new detailing of reinforcement. Group-wise samples with various joint detailing formats along with their finite element models for un-strengthened and ferro-cement strengthened specimens were compared. Lal [13] considered beam-column joint model from a STAAD.Pro designed building. The joint was further modeled in NX-Cad and then imported to ANSYS and analyzed for corresponding shear stress and deformation. Gedam and Khante [2] performed experiments on 2 concrete beams. In the first experiment, different methods of embedment of PZT by subjecting the model to damage were implemented. In the second experiment, an embedded smart aggregate with varying orientations was implemented to get the signatures (susceptance and conductance) of various piezoelectric transducers. Khante and Chikte [14] worked for defect detection of CFST (Concrete filled steel tubes) by using the EMI technique. Four CFST columns with different properties in terms of bonding and compaction and imparted debonding were taken to detect the defect in terms of RMSD index. Na and Baek [15] reviewed studies on the EMI technique in past decades and new ideas proposed by authors were also surveyed. Zhou et al. [1] proposed a modern damage index based on EMI-based principle and a double stepped method of inspecting the damage with respect to the damage indicator. A number of piezoelectric sensors attached to the steel beam were implemented to get the output responses. Kaur et al. [16] presented damage and retrofitting monitoring in RC structures with PZT patches. They carried out both experimental and FEM analysis for checking the suitability of use of 2nd order derivative along with other derivatives with high order for mode shapes of displacement and detection of damage. It was concluded that the response of sensors based on curvature mode shape can take out the damaged location successfully. Taha et al. [3] has taken out electromechanical responses calculated from PZT sensors for damaged and repaired concrete elements. The responses in the form of admittance signatures for damaged and repaired cases were compared and analyzed. Jayachitra and Priyadarshini [4] represented the detection of damage in concrete structures using a PZT patch. The conductance, admittance, impedance, and susceptance graphs were generated at various frequency levels. And damage index was determined by RMSD values. It was concluded that the PZT patch was found to be accurate for damage detection of structure. Su et al. [17] studied the use of 2 types of coatings with polymer on sensors as protection. The effectiveness of the polymer coated piezoelectric sensor was evaluated using an impedance analyzer. For simulation of different coating configuration sensors, finite element analysis was conducted. Kaur and Singla [6] reviewed the Electromechanical Impedance technique and its development to date for the evaluation of structures.

The review of literature gives a brief overview about the research work progressed related to the implementation of EMI technique to beam-column joints. It can be seen that simulation of finite element model for health monitoring using EMI technique is to be primely focused yet along with the experimental works to get another choice for efficient inspection. Present work consists of finite element modeling of beam-column joint in ANSYS 19.2 software with piezoelectric patch modeled for health monitoring of the system. The conductance and susceptance signatures, which are said to be the real and imaginary parts of admittance, respectively, were expected from the responses recorded from the PZT patch model. To implement the same, a healthy beam-column joint along with a fracture analyzed (crack introduced) model were created and both the models were monitored with the help of PZT patch connected to the joint for conductance/susceptance signatures to be obtained. The change in signatures of the damaged (crack introduced) model in comparison with the healthy model was to be acquired for justification of the methodology. After this qualitative analysis of both the models on the basis of signatures, to analyze the results quantitatively, Root Mean Squared Deviation (RMSD) values were to be calculated with respect to the different values of conductance at various frequency substeps. The results in the form of an RMSD index for various frequency ranges showed the significance of variation in signatures of a healthy and damaged case.

2 Structural Health Monitoring Using EMI Technique

Structural engineering accompanying SHM forms a valued subject in the view of the economy. The performance of structures is directly linked to the strength of the economy [16]. For the safety of various types of structural systems, the development and modification of health monitoring techniques have got increasing interest in researches nowadays. Recently, piezoelectric materials, such as Lead Zirconate Titanate (PZT), have shown good performance in health monitoring [17]. Due to its various advantages, the EMI technique which is a non-destructive evaluation technique of health monitoring; is becoming a successful health inspection strategy. The development of EMI technique was introduced initially by Liang et al. [15]. In the EMI technique, impedance/admittance can be used as an indicator of damage [8]. EMI technique is said to be a cost-effective option for SHM and non-destructive testing of engineering systems [7]. When an electric field is applied to PZT patch, the patch gets deformed [14]. Impedance/Admittance/Conductance/Susceptance curves are formed in LCR meter/impedance analyzer because of piezo sensors to detect changes in structural response, i.e., presence of structural damage. Electro-mechanical Impedance technique involves the use of piezoelectric (PZT) patches attached to the host structure which excite the host structure with provided frequency and detect the changes in electromechanical impedance curves (signatures). Getting the impedance for the damaged/defected case and comparing the same with the baseline impedance; from the observed change in impedance signature, it can be determined that the damage has taken place. In order to perform the operation with high damage sensitivity, the measurement should be taken at high frequencies. At such a high frequency, the excitation wavelength is small and it is very much sensitive for detecting the structural changes. The usage of material sensing with smart material like EMI method has provided an optional choice to the conventional techniques of health inspection, because of which, some drawbacks have overcome. The Electro-Mechanical Impedance method, with PZT as sensor/actuator is applicable with its well-known ability in the detection and characterization of damages [11].

3 Beam-Column Joint Dominance

In RC building frames, junctions can be said to be the most critical elements. Under earthquake/wind loads, flexural and shear stresses simultaneously act at the junction. As the included materials have limited capacities, joints also have limited strengths. Joints get damaged severely when forces larger than those strengths are applied to the structure during wind/earthquake action [9]. Past earthquakes have shown that failure of beam-column joint leads to the collapse of a structure partially or fully. The columns and beams that are near to the joints are most critical, and when strong waves of earthquake hit the structure, they get large shear forces and bending moments [12]. That is why; beam-column junctions are the most crucial elements of a structure, especially under the action of lateral loads. Over the past years, researches have been conducted on beam-column joints and till now, a clear picture has not been derived [13].

4 Finite Element Modeling

Finite element modeling of the system has been carried out in the software ANSYS. FE Modeling gives the numerical computation of reaction charges on each sensor node for generating the corresponding impedance/admittance/susceptance/conductance signatures [10]. A beam-column joint with steel reinforcement was modeled in ANSYS and primarily analyzed for shear stress, total deformations, etc. The dimensions and reinforcement details for beam-column joint were as shown in Table 1.

Modeling process in ANSYS involves the stepwise provision of material properties, geometry formation, model creation, and setup for meshing before analysis and result generation. In ANSYS, the analysis can be performed in both workbench and APDL tools. Some properties are in built and provided in the workbench for general analysis and some are to be given manually as per the analysis requirement. As far as reinforced concrete structure is considered, concrete and steel general material properties can be selected as they are in the workbench tool. But for modeling PZT patch, piezoelectric properties are to be given as input to the tool. In the present work, PZT patch geometry was modeled initially as steel material in the workbench tool itself. The whole model was then imported to APDL tool and properties of the PZT patch were changed to piezoelectric properties thereafter in APDL for getting solution in the form of conductance signatures. The properties of steel, concrete, and PZT materials included in the work were taken as given in Tables 2, 3, and 4, respectively.

Table 1 Dimensions, reinforcement details of beam-column joint

Beam specifications		Column specifications	
Parameter	Value (mm)	Parameter	Value (mm)
Width	150	Width	150
Depth	200	Depth	200
Span	800	Height	1000
Cover	25	Cover	25
Top steel	2#10	Longitudinal steel	4#10
Bottom steel	2#10		
Transverse steel dia	6	Transverse steel dia	6
Transverse steel spacing	120	Transverse steel spacing	150

Table 2 Steel material properties used for modeling

Density	7850	Kg/m^3
Young's modulus	2×10^{11}	Pa
Poisson's ratio	0.3	
Bulk modulus	1.67×10^{11}	Pa
Shear modulus	7.69×10^{10}	Pa
Tensile yield strength	2.5×10^{8}	Pa
Comp. yield strength	2.5×10^{8}	Pa
Tensile ultimate strength	4.6×10^{8}	Pa

Table 3 Properties of concrete material used for modeling

Density	2.3×10^{3}	Kg/m^3
Young's modulus	3×10^{10}	Pa
Poisson's ratio	1.8×10^{-1}	
Bulk modulus	1.56×10^{10}	Pa
Shear modulus	1.27×10^{10}	Pa
Tensile ultimate strength	5×10^{6}	Pa
Comp. ultimate strength	4.1×10^{7}	Pa

Table 4 Properties of PZT patch for modeling

Parameters	Symbols	Values	Unit
Density	ρ	7800	Kg/m^3
Dielectric loss factor	$\tan\delta$	0.02	
Compliance coefficients	S_{11}	15.0	10^{-12} m^2/N
	$S_{22} = S_{33}$	19.0	
	$S_{12} = S_{21}$	−4.50	
	$S_{13} = S_{31}$	−5.70	
	$S_{23} = S_{32}$	−5.70	
	$S_{44} = S_{55}$	39.0	
	S_{66}	49.4	
Electrical permittivity	ε_{11}^{T}	1.75	10^{-8} F/m
	ε_{22}^{T}	1.75	
	ε_{33}^{T}	2.12	
Piezoelectric strain coefficients	d_{31}	−2.10	10^{-10} m/V
	d_{32}	−2.10	
	d_{33}	5.0	
	d_{24}	5.80	
	d_{15}	5.80	

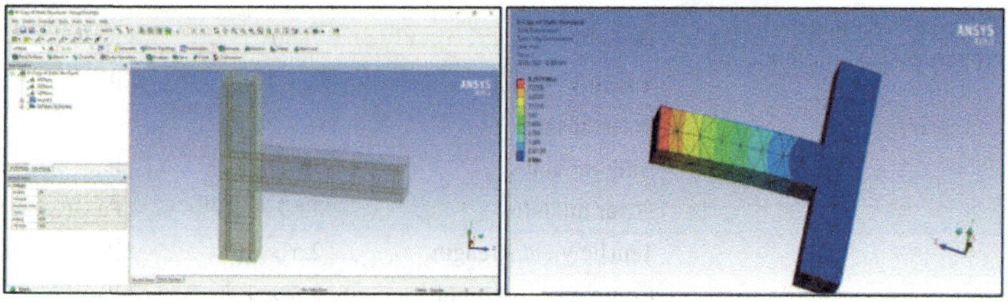

(a) (b)

Fig. 2 **a** Geometry creation and **b** analysis results (total deformation) in ANSYS workbench

5 System Development

After selecting static structural analysis, the next step in ANSYS is there to select the engineering data (properties of materials). Then, for creating geometry, two tools, i.e., design modeler and space claim can be used. Creating geometry (Fig. 2a) is the previous step to the prime mechanical model formation (Fig. 2b), which includes meshing, defining loads, analysis, and required results generation.

5.1 Crack Introduction

For fracture analysis of the beam-column joint model, a small crack was introduced near the junction of the model (Fig. 3). The portion targeted for the development of crack was extruded along the width of the beam and the setup involved in fracture analysis was prepared. After generating the mesh, analysis similar to the healthy model along with fracture analysis for the crack was performed. Finally, whole systems of both the healthy (Fig. 4) and the damaged models were imported to APDL tool for monitoring and PZT analysis was performed in order to get the conductance/susceptance signatures.

6 Results and Discussion

As the PZT patch was initially modeled as a steel element in a mechanical workbench, after importing the whole geometry in APDL, firstly the piezoelectric element was added to the element list and after that, the specified piezoelectric properties were provided to the element by removing the earlier steel material properties. In the PZT patch, top nodes were provided with a voltage of 1 V and bottom nodes were provided with a voltage of 0 V. Both the top nodes and bottom nodes were coupled by volt degree of freedom (volt DOF). For the same frequency ranges, the conductance and

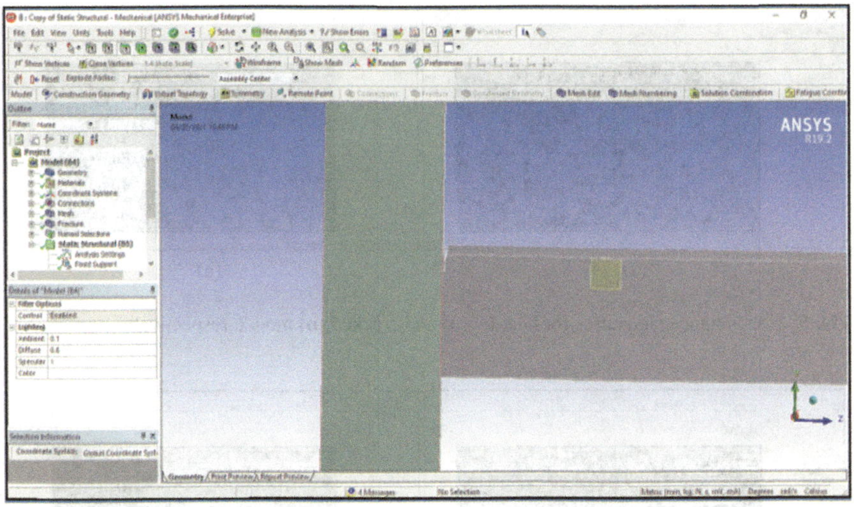

Fig. 3 Introducing crack in the model

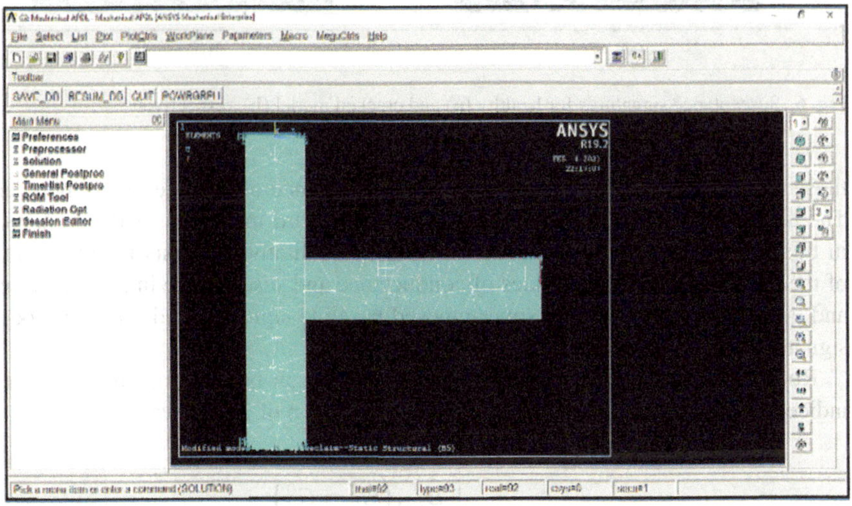

Fig. 4 Importing geometry to APDL tool for PZT patch specification

susceptance values for the healthy model and cracked model were recorded (Figs. 5 and 6).

The introduction of damage in the structural member was to be examined with the change in conductance values at the same frequency ranges. That is why the results were calculated by accounting same frequency ranges provided to the healthy and the damaged models, respectively. Consequently, the signatures (curves showing values of conductance for various frequency substeps) were observed to be varying from

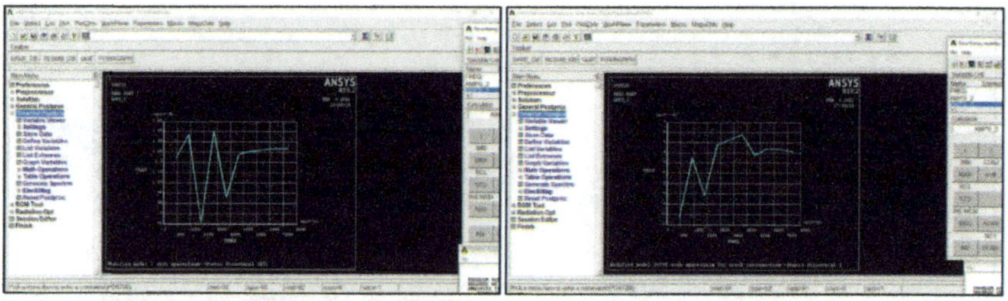

(a) (b)

Fig. 5 Conductance signatures for healthy (**a**) and cracked (**b**) model, respectively

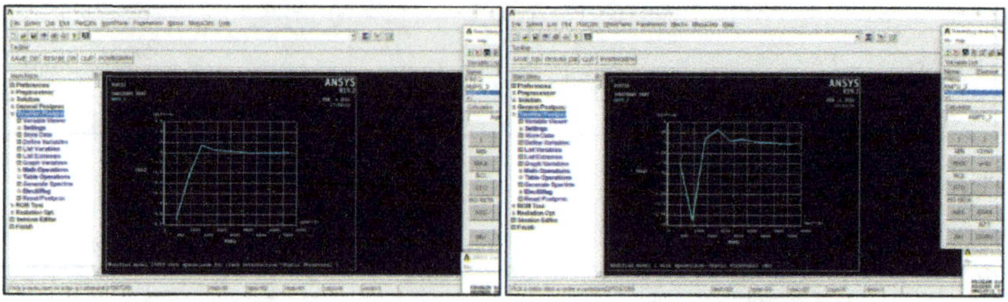

(a) (b)

Fig. 6 Susceptance signatures for healthy (**a**) and cracked model (**b**), respectively

those of healthy signatures which justified the presence of damage in the member. The upward/downward and also the forward/backward migration of the signature in this regard can be marked to be the base of qualitative analysis for inspection of the damage. Getting the values of conductance and susceptance in a spreadsheet and generating the curve, the superimposed baseline signature and cracked model signature were obtained as shown in Figs. 7 and 8.

Root Mean Squared Deviation (RMSD) values can be determined as damage indices for various ranges. The formula for calculation of RMSD values is

$$\mathrm{RMSD} = \sqrt{\frac{\Sigma_{j=1}^{N}\left(G_j^1 - G_j^0\right)^2}{\Sigma_{j=1}^{N}\left(G_j^0\right)^2}}$$

where

G_j^1 = Conductance at *j*th measurement point after damage.

G_j^0 = Conductance at *j*th measurement point before damage.

RMSD values were calculated for different ranges of frequencies, i.e., 75–225 kHz, 300–450 kHz, and 525–750 kHz (Fig. 9).

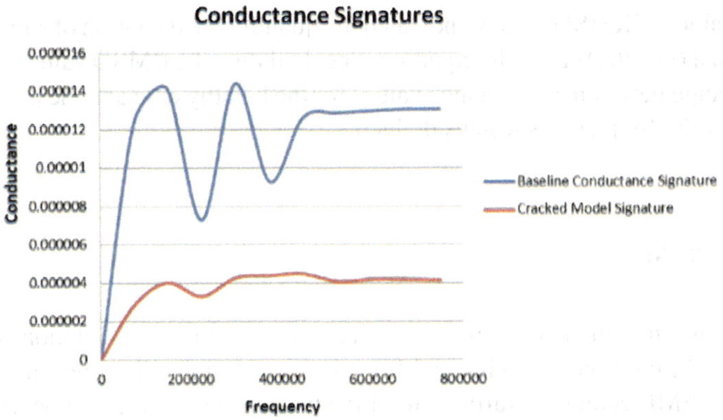

Fig. 7 Conductance signatures for healthy and cracked model combination

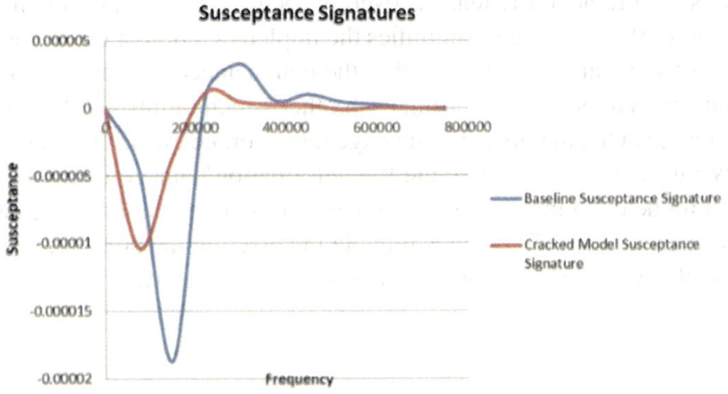

Fig. 8 Susceptance signatures for healthy and cracked model combination

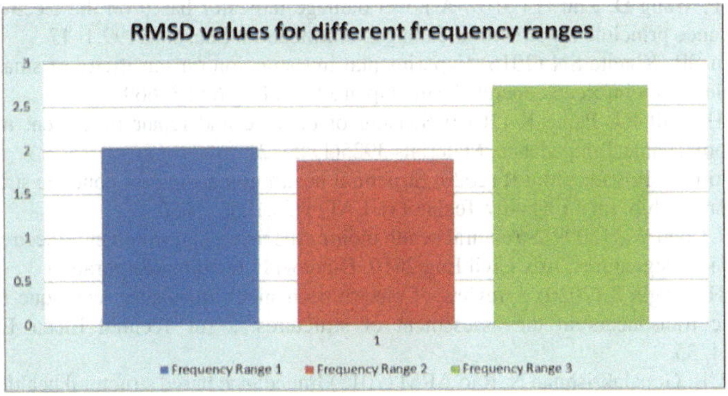

Fig. 9 RMSD values for various frequency ranges

The values of RMSD indices indicated the quantitative deviation of signature from the original (healthy) case. The equation for calculating the RMSD value corresponds to the change between conductance values for the healthy case and the damaged case at the specific frequency substeps defined.

7 Conclusion

For avoiding strength deduction of structures, non-destructive evaluation is preferred under which, the Electromechanical Impedance technique gives promising results. Under the EMI technique, further simplification of the process can be achieved by using finite element analysis of the structural element to be monitored. ANSYS tool achieves greater efficiency in modeling and analyzing structural system according to the analysis requirement. FE analysis overcomes the problems faced in the complex arrangement of the system and simplifies the implementation of the method. Results generated after the analysis showed that the conductance signature for the healthy model and cracked model differ from each other which ultimately defined the presence of damage in the member. The damage indicator, i.e., RMSD values for various frequency ranges also confirmed that the deformation/damage is monitored by the PZT patch modeled in the system by authenticating the variation in signatures. The results synthesize that PZT patch is reliable in detecting the presence of damage in a structure also with the use of a simulated system efficiently.

References

1. Zhou P, Wang D, Zhu H (2018) A novel damage indicator based on the electromechanical impedance principle for structural damage identification. Sensors 2199:1–17
2. Gedam SR, Khante SN (2016) Experimental investigation on sensitivity of smart aggregate embedded in reinforced concrete beam. Open J Civil Eng 6:653–669
3. Taha H, Ball RJ, Paine K (2019) Sensing of damage and repair of cement mortar using electromechanical impedance. Materials 3925(12):1–21
4. Jayachitra T, Priyadarshini R (2020) Structural health monitoring for concrete structure using impedance chip. Int J Eng Adv Technol (IJEAT) 9(3):1058–1060
5. Yan W, Chen WQ (2009) Structural health monitoring using high-frequency electromechanical impedance signatures. Adv Civil Eng 2010. Hindawi Publishing Corporation
6. Kaur H, Singla S (2020) A review of electromechanical impedance technique using piezoelectric transducers in the assessment of structures. J Inf Technol Electr Engi (ITEE) 9(5):24–35
7. Priya CB, Gopalakrishnan N, Rao ARM (2015) Impedance based structural health monitoring using serially connected piezoelectric sensors. J Inst Smart Struct Syst (JISSS) 4(1):38–45
8. Divsholi BS, Yaowen Yang, Bing L (2009) Monitoring beam-column joint in concrete structures using piezo-impedance sensors. Adv Mater Res 79–82:59–62
9. Baluch MH, Ahmed D, Rahman MK, Ilki A (2012) Finite element simulation of low concrete strength beam-column joint strengthened with CFRP. In: Proceedings of 15th World conference on earthquake engineering

10. Hamzeloo SR, Shamshirsaz M, Rezaei SM (2012) Damage detection on hollow cylinders by electro-mechanical impedance method: experiments and finite element modelling. Comptes Rendus Mecanique 340:668–677
11. Lim YY, Soh CK (2014) Electro-mechanical impedance (EMI)-based incipient crack monitoring and critical crack identification of beam structures. Res Nondestr Eval Am Soc Nondestr Testing 25(2):82–98
12. Venkatesan B, Ilangovan R, Jayabalan P, Mahendran N, Sakthieswaran N (2016) Finite element analysis (FEA) for the beam-column joint subjected to cyclic loading was performed using ANSYS. Circuits Syst 7:1581–1597
13. Lal A (2016) Analysis of exterior beam column joint using ANSYS. Int J Sci Res (IJSR) 5(7):947–950
14. Khante SN, Chikte SS (2017) Defect detection of concrete filled steel tubes with PZT based technique. Int J Innovative Res Sci Eng Technol 6(1):673–678
15. Na WS, Baek J (2018) A review of the piezoelectric electromechanical impedance based structural health monitoring technique for engineering structures. Sensors 1307:1–18
16. Kaur N, Bhalla S, Maddu SCG (2019) Damage and retrofitting monitoring in reinforced concrete structures along with long-term strength and fatigue monitoring using embedded Lead Zirconate Titanate patches. J Intell Mater Syst Struct 30(1):100–115
17. Su Y-F, Han G, Kong Z, Nantung T, Lu N (2020) Embeddable piezoelectric sensors for strength gain monitoring of cementitious materials: the influence of coating materials. Eng Sci 11:66–75

The Impact of Perforation Orientation on Buckling Behaviour of Storage Rack Uprights

Kadeeja Sensy, Ashish Gupta, K. Swaminathan, and J. Vijaya Vengadesh Kumar

Abstract Industrial storage rack uprights are the most widely used framed structures of thin-walled cold-formed steel members which are meant for the depot of palletized goods. The sensibility of uprights towards local buckling, distortional buckling, global buckling or interaction between these buckling modes in the presence of perforations results in complex behaviour of the uprights. The sustainability of the uprights in terms of the perforations, cross-section geometry, buckling and complexity might be difficult to design through the existing analytical methods alone. The test-based design being expensive leads to the necessity of an analysis-based design approach. The limited analytical methods available in the literature focus on orthogonal perforations and there is no systematic study available in terms of perforation orientation which would imply a minimum net section. This article sheds light effect of increased perforated width due to orientation which directly influences the elastic local critical buckling load calculations. The parametric analysis using finite element software is systematically done for the simple rack section with orthogonal cross-section elements having two idealized transverse web perforations of square and rectangular shape oriented at an angle other than orthogonal angles. The pre-validated FE model is used in this study, and thus the critical elastic buckling loads are procured for various perforation orientations. The CUFSM buckling analysis is done by using the reduced thickness expression recommended in the literature and compared with the FE results. The applicability of reduced thickness equations in accounting perforations in the view of perforation orientation is compared and summarized.

Keywords Elastic buckling · Perforation orientation · Storage rack uprights · CUFSM · Reduced thickness equations

---

K. Sensy · A. Gupta · K. Swaminathan · J. Vijaya Vengadesh Kumar (✉)
National Institute of Technology Karnataka, Surathkal, Mangalore 575025, India
e-mail: vj@nitk.edu.in

M. S. Ranadive et al. (eds.), *Recent Trends in Construction Technology and Management*, Lecture Notes in Civil Engineering 260,
https://doi.org/10.1007/978-981-19-2145-2_82

1 Introduction

Storage rack uprights are the framed structure that are assimilated from cold-formed steel columns having multiple perforations along longitudinal and/or transverse directions to accommodate the beams or bracings employed for storing industrial goods. These perforations lower the load-carrying capacity of the uprights. Thus, buckling analysis is vital in the investigation and design of cold-formed members. Particularly, this article is focused on the buckling analysis aspects of industrial storage rack uprights, which is the most significant storage structure in industries and thus demands the most reliable and simple design methods. To overwhelm the failure of the racks, the research on the characteristics of industrial storage uprights with incorporating safe design is performed.

The methods for buckling analysis comprise of the Finite Element Method (FEM), the Generalized Beam Theory (GBT) [2] and the Finite Strip Method (FSM) [9]. FEM is the most flexible method adopted for the buckling analysis of complex geometries and variable load and boundary conditions using software such as ANSYS and ABAQUS. Despite that, FSM and GBT are used with the free available software CUFSM and GBTUL, respectively, as a robust tool. The design of the uprights is done using the Direct Strength Method (DSM) [11] once the buckling results are obtained [1].

Analytical-based approach through ANSYS and GBT in accordance with the European design standards for the perforated members are introduced by Davies [5] and inferred that GBT analysis has provided an adequate analytical solution for the complex problem on account of perforations and a safe design approach.

The elastic buckling analysis of cold-formed steel members having general boundary conditions by combining conventional FSM with the constrained finite strip method (*c*FSM) was proposed by Li and Schafer [7]. FSM solutions together with DSM were adopted in member design through a signature curve for S–S boundary conditions, while the application of DSM for other boundary conditions was not developed completely. However, the global buckling mode for other boundary conditions can be determined using the effective length factor in elastic critical buckling equations.

The determination of elastic buckling loads of rectangular perforated cold-formed storage rack columns with the help of FSM using the concept of reduced thickness expressions was demonstrated by Casafont et al. [3] and the proposed reduced thickness equations can be adopted in practical design even if it is not accurate for local buckling [4].

The determination of the critical elastic buckling loads of five cold-formed steel pallet racks upright sections, with sharp edge cross-sections and pinned-warping free boundary conditions, having web perforations was proposed by Smith and Moen [10] with a focus on the AISI's Direct Strength Method [8].

FSM and GBT are adopted for the buckling analysis of perforated members using the incorporation of reduced thickness expressions [3, 10]. Most of this literature focused on the analysis of idealized perforations oriented at orthogonal angles.

In this article, the buckling analysis of the simple rack section with orthogonal cross-section elements having two idealized transverse web perforations of square and rectangular shape oriented in an angle other than orthogonal angles are done using FE software ABAQUS. The pre-validated FE model for gross sections and idealized perforations [6] are further accounted for in this study. The CUFSM buckling analysis is done using the reduced thickness expression recommended in the literature and compared with the FE results. The aptness of reduced thickness expressions in accounting perforations in the terms of perforation orientation is compared using the analysis results and summarized.

2 Cross-Section Geometry

As a preliminary study, only one cross-section is considered in the initial phase of the investigation. The considered cross-section is finalized based on the elastic critical buckling behaviour of the cross-section in which the minimum elastic critical local and distortional modes are nearly equal. This study further limits the cross-section as sharp cornered. The dimensions of rack section are 80 × 40 × 20 × 15, in which the numeral values represent the height, flange width, rear flange width and flange stiffener length in mm (Fig. 1). The rack section of this particular dimension is referred to as R. The length of the specimen considered is three times the maximum dimension of the cross-section, i.e., 240 mm. The base metal sheet thicknesses are 1.0 mm, 1.6 mm, and 2.0 mm and the two idealized transverse web perforations of rectangular shape referred to as P, having the size of 10 × 20 mm, oriented in angles such as 0, 30, 45, 60, and 90 degrees are deemed (Fig. 2). The yield stress considered is 255.91 MPa. The model name R_t()_P() is used for depicting the rack section, thickness, and perforation orientation as in Table 1. The model name for 1.6 mm thickness and web perforation oriented at 30 degree is R_t1.6_P30. Thus, a total of 18 combinations, including gross sections that are unperforated sections (G), are studied.

3 FE Analysis

The software ABAQUS was used to perform the Finite Element (FE) eigenbuckling analyses to predict the perforation pattern effects on elastic local buckling load and mode shapes by the use of shell elements. The centre lines of the upright sections are modelled by means of 3D shell extrusion for the geometric modelling (Fig. 3). The material properties of the upright are considered as isotropic and homogeneous with Young's modulus, $E = 2 \times 10^5$ MPa and Poisson's ratio, $\mu = 0.30$. The elements of the end cross-section are constrained with the centroid of cross-section using Multi-Point Constraint (MPC), which is called a reference point at top and bottom end and the unit load is applied at the top end of the specimen (Fig. 4). Boundary conditions were pinned at the ends with allowing axial deformations.

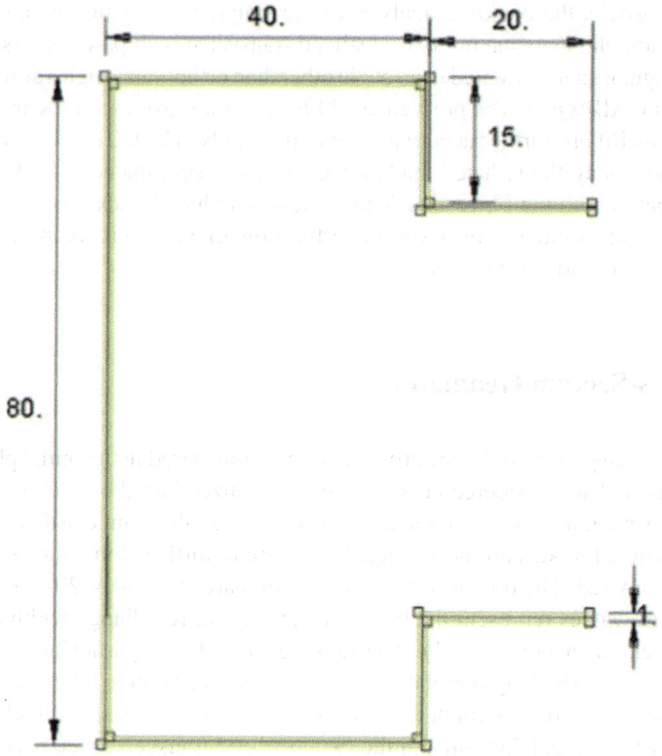

Fig. 1 Cross-section dimensions

As per the detailed convergence study done for gross sections and selected perforated sections, four-node shell elements with reduced integration are considered here to mesh the model with the mesh size equal to the thickness of the base sheet (Fig. 5). In total, the buckling analysis using FE analysis is performed for 3 gross sections and 15 perforated sections. The critical elastic local buckling loads (P_{crl}_FE) for all the sections considered are obtained with their corresponding mode shapes directly from the Eigen value since the unit load is applied (Fig. 6).

4 FSM Analysis Using Reduced Thickness Equations

4.1 Reduced Thickness Equations

The determination of critical elastic buckling load of the perforated section can be done using both CUFSM and GBTUL software by incorporating the reduced thickness equation. The element with perforations can be divided into perforated

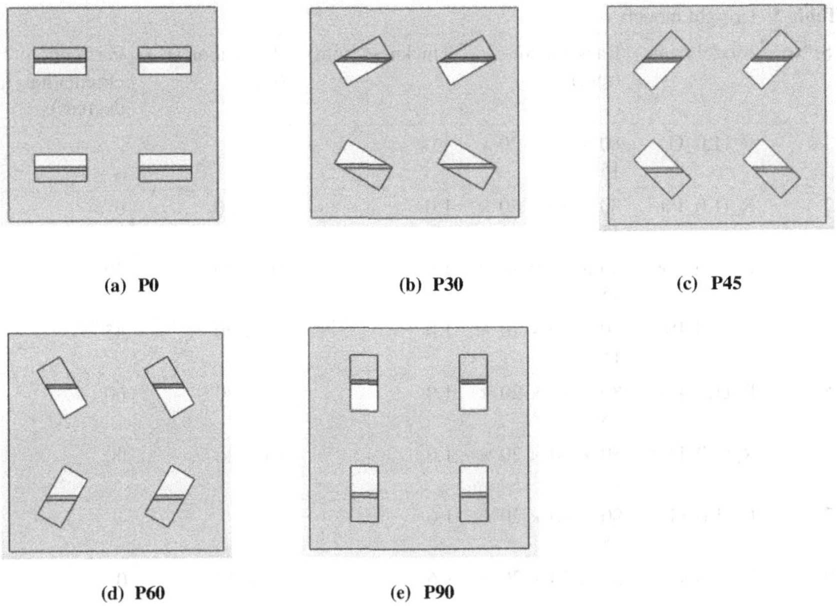

(a) **P0** (b) **P30** (c) **P45**

(d) **P60** (e) **P90**

Fig. 2 Perforation orientation is one-third length of the specimen

and unperforated strips, in which the reduced thickness equation is applied for the perforated strip and the actual thickness is applied for all unperforated strips and elements. The reduced thickness expressions accounting for the thickness on the perforated strip are demonstrated in literature [3, 10].

Casafont's reduced thickness equation for local buckling is

$$t_r = 0.61t\frac{L_{np}B_{np}}{LH} + 0.18t\frac{B_p}{L_p} + 0.11 \quad (1)$$

where

- L pitch length
- L_{np} non-perforated strip length between perforations
- B_{np} width of the non-perforated strip
- B width of the cross-section
- H height of the cross-section
- B_p width of the perforation
- L_p length of the perforation

Smith and Moen's reduced thickness equation for local buckling is

$$t_r = t\left[1 - \frac{n_l n_t (L_h d_h + \mu d_h \alpha + \mu L_h \beta + \alpha\beta)}{Lb}\right]^{1/2} \quad (2)$$

Table 1 Upright models

S. No.	Model name	Rack section (mm)	Thickness (mm)	Perforation size (mm)	Perforation Orientation (in degrees)
1	R_t1.0_G	80 × 40 × 20 × 15	1.0	–	–
2	R_t1.0_P0	80 × 40 × 20 × 15	1.0	10 × 20	0
3	R_t1.0_P30	80 × 40 × 20 × 15	1.0	10 × 20	30
4	R_t1.0_P45	80 × 40 × 20 × 15	1.0	10 × 20	45
5	R_t1.0_P60	80 × 40 × 20 × 15	1.0	10 × 20	60
6	R_t1.0_P90	80 × 40 × 20 × 15	1.0	10 × 20	90
7	R_t1.6_G	80 × 40 × 20 × 15	1.6	–	–
8	R_t1.6_P0	80 × 40 × 20 × 15	1.6	10 × 20	0
9	R_t1.6_P30	80 × 40 × 20 × 15	1.6	10 × 20	30
10	R_t1.6_P45	80 × 40 × 20 × 15	1.6	10 × 20	45
11	R_t1.6_P60	80 × 40 × 20 × 15	1.6	10 × 20	60
12	R_t1.6_P90	80 × 40 × 20 × 15	1.6	10 × 20	90
13	R_t2.0_G	80 × 40 × 20 × 15	2.0	–	–
14	R_t2.0_P0	80 × 40 × 20 × 15	2.0	10 × 20	0
15	R_t2.0_P30	80 × 40 × 20 × 15	2.0	10 × 20	30
16	R_t2.0_P45	80 × 40 × 20 × 15	2.0	10 × 20	45
17	R_t2.0_P60	80 × 40 × 20 × 15	2.0	10 × 20	60
18	R_t2.0_P90	80 × 40 × 20 × 15	2.0	10 × 20	90

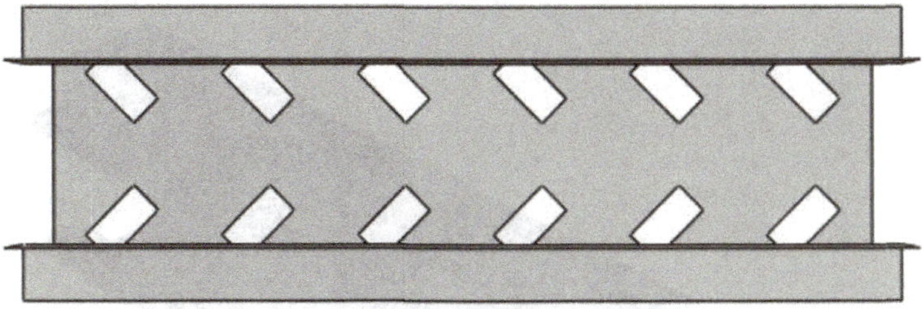

Fig. 3 Geometric modelling

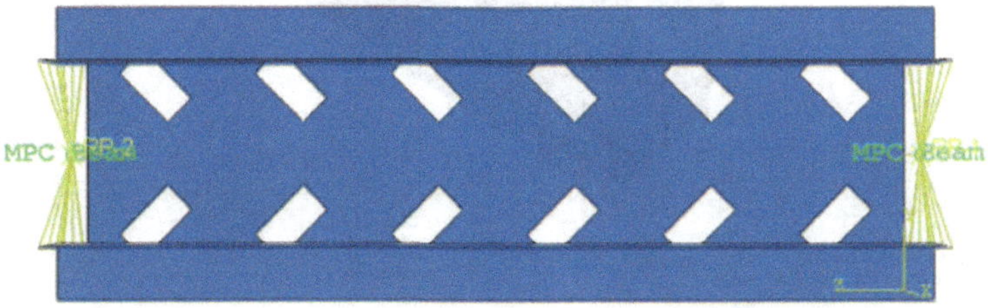

Fig. 4 MPC

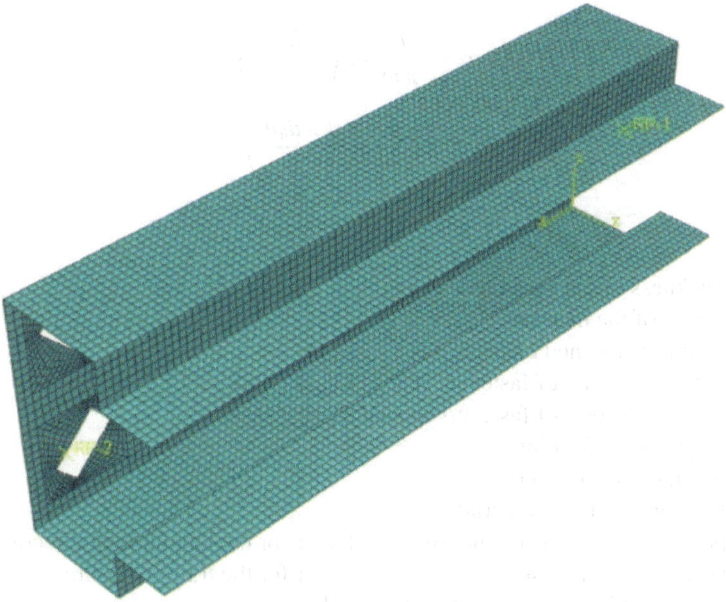

Fig. 5 Shell elements

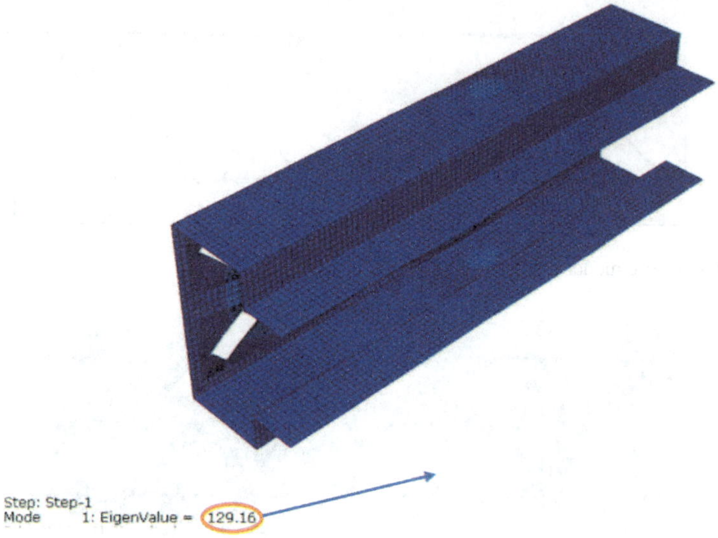

Fig. 6 Buckling load from FE analysis

$$\alpha = \frac{L}{\pi m}\sin\left(\frac{\pi L_h m}{L}\right) \tag{3}$$

$$\beta = \frac{b}{\pi n}\sin\left(\frac{\pi d_h n}{b}\right) \tag{4}$$

where

t	thickness of the base metal
L	length of the member
B	width of stiffened element
n_l	number of rows of fasteners spaced longitudinally
n_t	number of rows of fasteners spaced transversely
L_h	length of perforation
d_h	width of perforation
μ	Poisson's ratio of material
α	perforation dimension modification factor for the longitudinal direction
β	perforation dimension modification factor for the transverse direction
m	number of longitudinal half-wavelengths
n	number of transverse half-wavelengths

α and β are the modification factors which are determined from Eqs (3) and (4), by inputting $m = L/b$ and $n = 1$.

Using these reduced thickness equations, the elastic critical local buckling load can be determined using CUFSM or GBTUL.

4.2 Finite Strip Analysis

As both [3, 10] literatures are referred for the tr equations, the concerning remarks made by them should be noted. Casafont mentioned that the stress should not be applied at the perforated strip, which is not possible in GBTUL because the user has no freedom to apply the load as stress, rather it is applied as a point load at the centroid. For that reason, CUFSM analysis is chosen for the buckling analysis.

Once the reduced thickness expressions are calculated for all 15 sections, finite strip analysis was done using CUFSM to get the elastic critical local buckling stress from the signature curve. Thus, the elastic critical local buckling load (Pcrl_trS as the elastic critical local buckling load from Smith's tr equation and Pcrl_trC as the elastic critical local buckling load from Casafont's tr equation) can be obtained from the first minimum of signature curve.

5 Results and Discussion

The FE results of perforated stub columns are compared with the finite strip analysis results using reduced thickness equations proposed by Smith and Moen [10] and Casafont et al. [3] in Table 2. It should be noted that the boundary condition considered in the FE analysis is flexurally simply supported but twisting and warping are restrained. In other words, simply supported condition restraint to warping deformation is considered in FE analysis. This is to mimic the experimental behaviour to validate the FE model. By the way, in most practical cases, the warping deformations are restrained. The results from Casafont's reduced thickness equation are deviating from the FE results, however, the values are lower than FE results and hence conservative. The Moen's reduced thickness equation is relatively better than Casafont's method for local buckling. The Smith's method is best suitable for the idealized perforations, whereas it is unduly conservative for other perforation orientations which are handled as idealized perforations. The deviation of results is more pronounced for larger thickness sections with perforations oriented in angles other than 0 and 90°. Therefore, further investigation is required to improve the reduced thickness equation by including the perforation orientation in the finite strip analysis. For the slender plate elements, that is for lower thickness the perforation resulting in minimum net section (P90 orientation) exhibits a higher buckling load than other orientations in FE analysis. Although the same trend is followed in Casafont method, the Smith method leads to a different conclusion as the highest ultimate load for P0 orientation having maximum net section area. For other orientations, the buckling load is relatively in between P0 and P90 orientations and this effect is more pronounced in lower slender web elements, that is for higher thickness. As shown in Table 2, the perforation orientation significantly affects the buckling load and both analytical methods are inconsistent. Therefore, more refined research is required in terms of element slenderness as well as the perforation orientations.

Table 2 Analysis results

S. No.	Model name	P_y (kN)	P_{crl}_FE (kN)	P_{crl}_FE/P_y	P_{crl}_$t_r S$ (kN)	P_{crl}_$t_r S/P_y$	P_{crl}_$t_r C$ (kN)	P_{crl}_$t_r C/P_y$
1	R_t1.0_G	58.35	38.85	0.67	36.36	0.62	36.36	0.62
2	R_t1.0_P0	58.35	31.84	0.55	34.19	0.59	31.54	0.54
3	R_t1.0_P30	58.35	31.59	0.54	27.09	0.46	23.11	0.40
4	R_t1.0_P45	58.35	31.80	0.55	24.53	0.42	22.80	0.39
5	R_t1.0_P60	58.35	32.19	0.55	25.08	0.43	25.72	0.44
6	R_t1.0_P90	58.35	32.86	0.56	31.91	0.55	39.27	0.67
7	R_t1.6_G	92.86	159.14	1.71	149.94	1.61	149.94	1.61
8	R_t1.6_P0	92.86	129.33	1.39	140.86	1.52	124.88	1.34
9	R_t1.6_P30	92.86	128.34	1.38	110.99	1.20	84.90	0.91
10	R_t1.6_P45	92.86	129.16	1.39	100.16	1.08	82.98	0.89
11	R_t1.6_P60	92.86	130.83	1.41	102.58	1.10	93.24	1.00
12	R_t1.6_P90	92.86	133.76	1.44	131.34	1.41	151.90	1.64
13	R_t2.0_G	115.67	310.16	2.68	293.93	2.44	281.92	2.44
14	R_t2.0_P0	115.67	251.01	2.17	275.89	2.39	237.06	2.05
15	R_t2.0_P30	115.67	248.89	2.15	216.65	1.87	156.49	1.35
16	R_t2.0_P45	115.67	250.51	2.17	195.11	1.69	153.45	1.33
17	R_t2.0_P60	115.67	253.86	2.19	199.98	1.73	174.32	1.51
18	R_t2.0_P90	115.67	259.79	2.25	257.11	2.22	230.02	1.99

6 Conclusions

In this study, the impact of perforation orientation in the elastic critical local buckling behaviour of storage rack upright is focused. Numerical analysis results were generated using FE analysis. The minimum elastic critical buckling load was determined using Smith's reduced thickness approach and Casafont's reduced thickness approach for local buckling. Comparing the results, the perforation orientation significantly influences the elastic critical local buckling load compared to FE results. Although most of the results fall on the conservative side, few results are showing unconservative behaviour for the same perforations but oriented in angles. This nature is significant for the larger thickness sections. Therefore, the reduced thickness equations currently available in the literature for local buckling can be further improved by including the element slenderness and perforation orientation as one of the key parameters to determine the reduced thickness for perforated strip in the cross-section. Detailed research in this domain is under process.

Acknowledgements The authors highly acknowledge and appreciate Prof. S Arul Jayachandran's (IIT Madras) help in various stages of this current research, in particular the numerical analysis.

References

1. American Iron And Steel Institute (2016) AISI S100: North American specification for the design of cold-formed steel structural members. p 505
2. Bebiano R, Silvestre N, Camotim D (2008). GBTUL—a code for the buckling analysis of cold-formed steel members. In: 19th International specialty conference on recent research and developments in cold-formed steel design and construction, pp 61–79
3. Casafont M, Pastor M, Bonada J, Roure F, Peköz T (2012) Linear buckling analysis of perforated steel storage rack columns with the finite strip method. Thin-Walled Struct 61:71–85. https://doi.org/10.1016/j.tws.2012.07.010
4. Casafont M, Pastor MM, Roure F, Bonada J, Peköz T (2013) Design of steel storage rack columns via the direct strength method. J Struct Eng (United States) 139(5):669–679. https://doi.org/10.1061/(ASCE)ST.1943-541X.0000620
5. Davies JM, Leach P, Taylor A (1997) The design of perforated cold-formed steel sections subject to axial load and bending. Thin-Walled Struct 29(1–4):141–157. https://doi.org/10.1016/s0263-8231(97)00024-4
6. Harisanth KS (2019) Buckling behaviour of cold-formed steel storage racks. MTech Thesis, Indian institute of technology madras
7. Li Z, Schafer BW (2010) Buckling analysis of cold-formed steel members with general boundary conditions using CUFSM: conventional and constrained finite strip methods. In: 20th International specialty conference on cold-formed steel structures—recent research and developments in cold-formed steel design and construction, pp 17–31
8. Moen CD, Schafer BW (2011) Direct strength method for design of cold-formed steel columns with holes. J Struct Eng 137(5):559–570. https://doi.org/10.1061/(ASCE)ST.1943-541X.0000310
9. Schafer B, Adany S (2006) Buckling analysis of cold-formed steel members using CUFSM. In: 18th International specialty conference on cold-formed steel structures

10. Smith FH, Moen CD (2014) Finite strip elastic buckling solutions for thin-walled metal columns with perforation patterns. Thin-Walled Struct 79:187–201. https://doi.org/10.1016/j.tws.2014.02.009
11. Yu W, Laboube RA, Wiley J (2010) Cold-Formed steel design

Modelling Interfacial Behaviour of Cement Stabilized Rammed Earth Using Cohesive Contact Approach

T. Pavan Kumar Reddy and G. S. Pavan

Abstract A monolithic construction formed by compacting processed soil in progressive layers in a formwork is called a rammed earth wall. A lot of applications of using rammed earth walls for both load bearing and non-load bearing can be seen across the world and carbon content is low in this building material. The construction of rammed earth structures involves layered compaction, thus forming an interface between two layers. Modelling of interface plays an important role in the strength and durability of these structures. The interface is modelled using a cohesive contact approach and the response of the rammed earth triplet is obtained. The slope of load vs displacement curve of the rammed earth triplet is 81% accurate with the experimental slope and the peak load of the triplet is 82% accurate with the experimental peak load. Thus, the comparison of load versus displacement obtained from the finite element method with experimental data of load versus displacement yields almost similar results.

Keywords Rammed earth · Interface · Cohesive contact · Triplet

1 Introduction

A monolithic construction formed by compacting processed soil in progressive layers in a formwork is called a rammed earth wall. A lot of applications of using rammed earth walls for both load bearing and non-load bearing can be seen across the world. Rammed earth constructions can be classified into two categories: stabilized rammed earth and unstabilized rammed earth. A mixture of soil, sand and gravel can be defined as unstabilized rammed earth. There are no additional stabilizers that are

T. Pavan Kumar Reddy
NITK Surathkal, Mangaluru, India

G. S. Pavan (✉)
Department of Civil Engineering, NITK Surathkal, Mangaluru, India
e-mail: pavan.gs@nitk.edu.in

M. S. Ranadive et al. (eds.), *Recent Trends in Construction Technology and Management*, Lecture Notes in Civil Engineering 260,
https://doi.org/10.1007/978-981-19-2145-2_83

added to this. In addition to the materials used in unstabilized rammed earth, stabilizers (cement, lime, etc.) are added for rammed earth to improve the strength and durability of the structure and this is called stabilized rammed earth. An interface is defined as a common layer or a surface that is present between two bodies. The construction of rammed earth structures involves layered compaction, thus forming an interface between two layers. This interface plays an important role in the stability and durability of these structures. During earthquakes, there will be in-plane shear forces which makes the behaviour of the interface important. Also, when there is an uneven settlement of foundations, in-plane shear forces play a major role in the behaviour of the interface. Hence, it is important to model the interface such that it resists the loads due to structural loads or an earthquake, etc. making it more durable. This presentation shows how an interface of a rammed earth triplet is modelled using a cohesive contact approach. The main objectives of our present study are to model the interface of the rammed earth triplet using the cohesive contact approach present in commercial approaches, to obtain the values of applied loads and corresponding displacements till the failure of the rammed earth triplet and to plot load vs displacement curve and compare with the curve obtained from experimental values. Reddy and Kumar [1] conducted an experimental study on the physical properties and compaction characteristics of compacted cement stabilized rammed earth (CSRE) and cement stabilized soil mixtures. Standard Proctor test is conducted to determine the values of OMC and MDD of the samples tested. Results show that the addition of ordinary Portland cement (OPC) does not show any effect on OMC and MDD values of soil. Also, as the clay fraction of the cement mixed soil increases, MDD decreases and OMC increases. As the time lag increases, OMC increases and MDD decreases irrespective of soil type and cement content. Reddy and Kumar [2] explained the elastic properties of CSRE and stress–strain relationships considering the effects of density and moisture. Compressive strength tests both wet and dry are conducted on the rammed earth specimens. Results show that the optimum clay fraction required is about 16% for yielding maximum compressive strength for CSRE. Strength is directly proportional to the density, i.e., there is an increase in strength as the density increases and the relationships are linear. There is a significant increase in the strength as the density is changed from 1600 to 2000 kg/m^3. An increase in 20% of density resulted in increasing the strength by 300–500%. Lepakshi and Reddy [3] explained that the objective of this paper is to establish shear strength parameters and Mohr–Coulomb failure envelopes for CSRE. Tri-axial shear tests are conducted to determine the shear strength parameters. The results show that the shear strength of the specimen in wet condition is almost half of the shear strength of the specimen in dry condition. Irrespective of the confining pressure and the testing condition, the shear strength and cohesion of CSRE were greatly influenced by cement content. The tri-axial compressive strength and the shear strength increase as the confining pressure increases. Pavan et al. [4] explained enhancing the interface shear strength of CSRE and the overall shear behaviour of CSRE by bonding techniques. Three bonding techniques: (i) using dents, (ii) applying a coating of cement slurry and iii) a combination of dents and cement slurry are formed in between the interfaces of rammed earth specimens and are tested under loading. A 90% increase in shear bond strength of CSRE specimens with bonding techniques is observed when compared to triplets without any bonding technique, under dry condition. Also, CSRE specimens with cement slurry as a bonding technique exhibits higher strength when compared to CSRE specimens with bonding techniques of dents or a combination of both dents and cement slurry. Some of the other studies focused on rammed earth have been conducted by Lepakshi and Reddy [5], Shrestha et al. [6], El-Nabouch et al. [7] and Hiroyuki et al. [8] (Figs. 1 and 2).

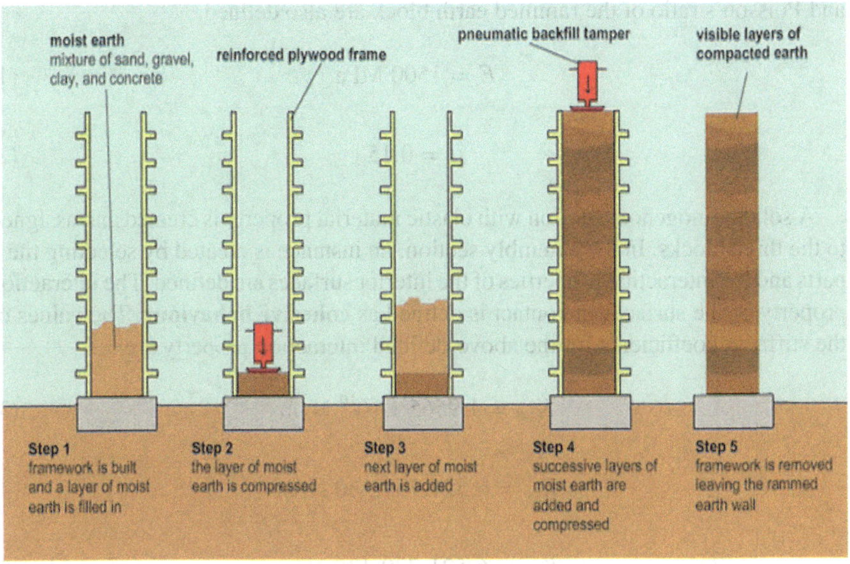

Fig. 1 Steps involved in rammed earth construction [2]

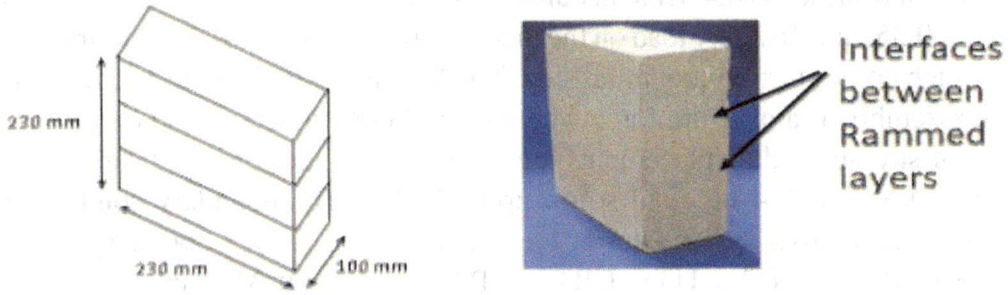

(a) Schematic representation of triplet specimen

(b) Actual triplet specimen in dry condition

Fig. 2 Rammed earth triplet showing the interface [3]

2 Methodology

Modelling is done by using Abaqus 6.14, Finite Element Method (FEM) based software. This section shows step by step procedure showing the modelling of the rammed earth triplet adopted in the software package. In the parts section, three rammed earth blocks of dimensions 230 mm × 230 mm × 100 mm are developed. In the materials section, the blocks are defined with material properties and are named as elastic. The density of the material is defined as $\rho = 1800\ \text{kg/m}^3$. Young's modulus and Poisson's ratio of the rammed earth block are also defined.

$$E = 1500\ \text{MPa} \tag{1}$$

$$\mu = 0.15 \tag{2}$$

A solid homogenous section with elastic material property is created and assigned to the three blocks. In the assembly section, an instance is created by selecting the 3 parts and the interaction properties of the interior surfaces are defined. The interaction property of the surfaces in contact is defined as **cohesive behaviour**. The values of the stiffness coefficients for the above-defined interaction property are

$$K_{nn} = 1.9565 * 10^{10}\ \text{N/m} \tag{3}$$

$$K_{tt} = 6{,}521{,}739{,}130\ \text{N/m} \tag{4}$$

$$K_{ss} = 6{,}521{,}739{,}130\ \text{N/m} \tag{5}$$

The max nominal stress of damage initiation criteria in normal mode is 7 MPa and in shear 1 and shear 2 is 0.4 MPa. In damage evolution, the maximum displacement at failure is 0.35 mm. To apply load on the member, we need to create a step named apply load which comes next to the initial step. The load is applied on the middle member of the assembly as a pressure force which is gradually increased and corresponding values of stresses and displacements are noted. The values of load taken are 20 N, 200 N, 2 kN, 3 kN, 4 kN, and so on up to 20 kN. The boundary conditions are applied at the bottom of the exterior members and it is a fixed boundary condition (ENCASTRE U1 = U2 = U3 = UR1 = UR2 = UR3 = 0). Meshing the member is an important step in modelling and analysis of the member. The approximate global size of the seed is taken as 0.0115 mm. The elements of the mesh are defined as an 8-node linear brick element (C3D8R). By selecting the mesh part instance, the members are meshed with a seed size of 0.0115 mm and an element type of C3D8R. A job is created by giving a name to the file and model as the source. The job created is submitted and the results are observed from the visualization section.

3 Results and Discussion

The rammed earth triplet is loaded and for each load, corresponding stresses and displacements are calculated. At a peak load of 16.25 kN, the contour plot of shear stresses (τ_{xy}) is shown in Fig. 3. It can be observed that the shear stresses are symmetric about both the interfaces. The values of the shear stress also attain peak value at the interface and gradually reduce as we move away from the interface towards the ends.

Further, Fig. 4 shows the vertical displacement contour plot of the triplet corresponding to a load of 16.5 kN. We can observe the sliding behaviour of the middle layer of the triplet relative to the outer layers of the rammed earth triplet. The middle layer was displaced to a vertical displacement of about 0.26 mm. The results obtained from the numerical simulation compare well with the results obtained from experiments conducted in the study by Pavan et al. [4].

Upon reaching a load of 16.5 kN, the simulation failed to converge and was unable to simulate the post–peak deformation behaviour of the triplet.

A graph is plotted between the numerical values of loads and displacements obtained from the numerical model and experimental values of loads and displacements and is shown in Fig. 5.

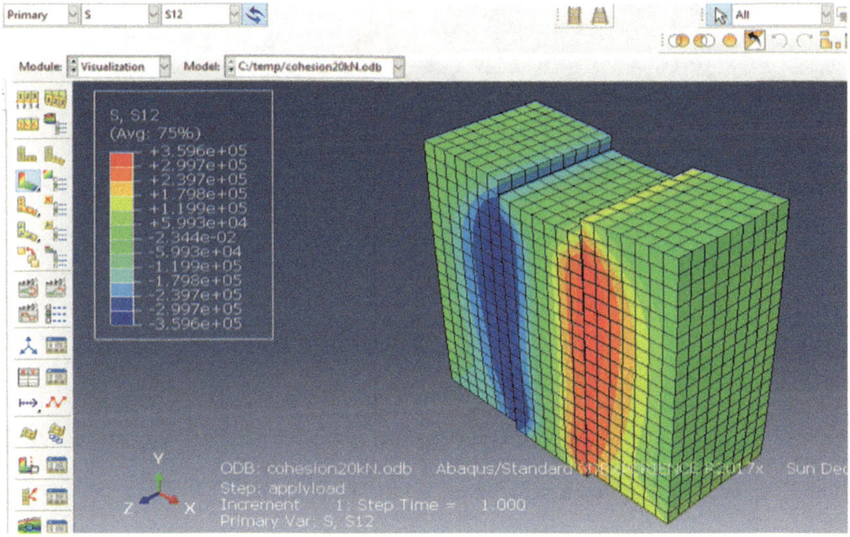

Fig. 3 Stress contour plot of rammed earth triplet

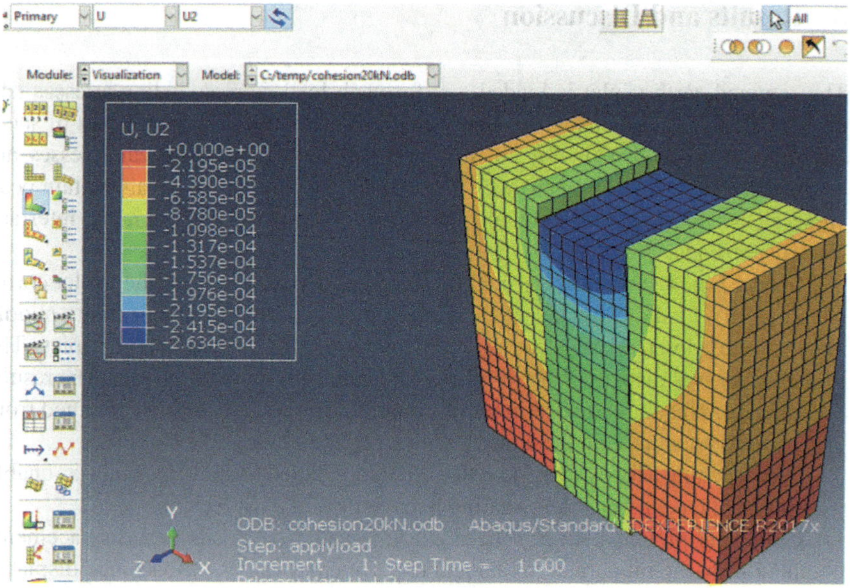

Fig. 4 Displacement contour plot of rammed earth triplet

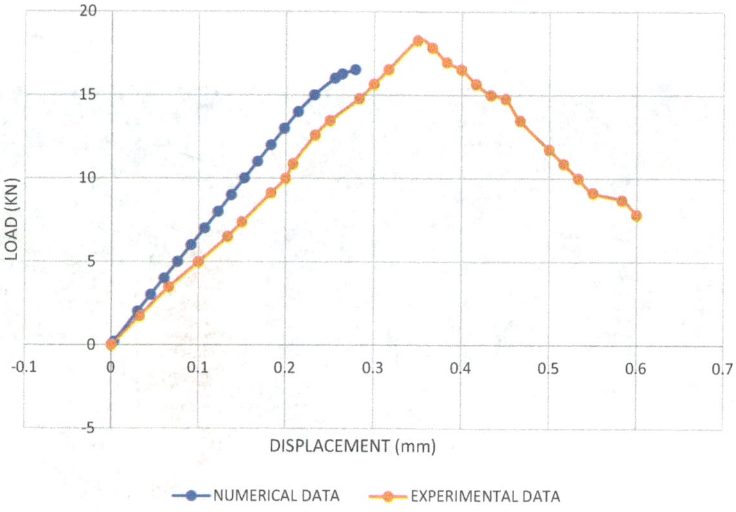

Fig. 5 graph of numerical and experimental data [4]

4 Conclusion

From the numerical modelling done above, we can conclude that the simulation was successful in producing the sliding action between successive layers of rammed earth triplet when it is subjected to in-plane shear loading. The slope of the load–displacement diagram as per the numerical simulation was found to be equal to 59.6 kN/mm and was found to be 81% accurate in comparison to experimental values. The peak load as captured by numerical simulation was found to be 16.5 kN and is 82% accurate in comparison to the peak load witnessed during experiments available in the literature. The elastic response of the triplet, along with a part of the inelastic response of the rammed earth triplet under in-plane shear loading was captured by the numerical model in this study. Further studies can adopt advanced contact modelling strategies to capture the entire inelastic response of the rammed earth triplet under in-plane shear loading.

References

1. Venkatarama Reddy BV, Prasanna Kumar P (2010) Cement stabilised rammed earth. Part A: compaction characteristics and physical properties of compacted cement stabilised soils. Mater Struct (RILEM). https://doi.org/10.1617/s11527-010-9658-9
2. Venkatarama Reddy BV, Prasanna Kumar P (2010) Cement stabilised rammed earth. Part B: compressive strength and stress–strain characteristics. Mater Struct (RILEM). https://doi.org/10.1617/s11527-010-9659-8
3. Lepakshi R, Venkatarama Reddy BV (2020) Shear strength parameters and Mohr-Coulomb failure envelopes for cement stabilised rammed earth. Constr Build Mater 249:118708
4. Pavan GS, Ullas SN, Nanjunda Rao KS (2020) Shear behavior of cement stabilized rammed earth assemblages. J Build Eng 27(2020):100966
5. Lepakshi Raju BV, Reddy V (2018) Influence of layer thickness and plasticizers on the characteristics of cement-stabilized rammed earth. Am Soc Civil Eng. https://doi.org/10.1061/(ASCE)MT.1943-5533.0002539
6. Shrestha KC, Aoki T, Miyamoto M, Wangmo P (2020) In-Plane shear resistance between the rammed earth blocks with simple interventions: experimentation and finite element study. Buildings 10(3):57
7. El-Nabouch R, Bui QB, Perrotin P, Plé O (2018) Shear parameters of rammed earth material: results from different approaches. Adv Mater Sci Eng 2018. https://doi.org/10.1155/2018/8214604
8. Arakia H, Kosekia J, Satob T (2015) Tensile strength of compacted rammed earth materials. Jpn Geotech Soc 56:189–204

Four Node Flat Shell Quadrilateral Finite Element for Analysis of Composite Cylindrical Shells

V. A. Dagade and S. D. Kulkarni

Abstract This work presents the free vibration and static analysis of laminated composite cylindrical shells with a 2D computationally efficient facet-shell finite element that is formed by combining the plate bending and membrane element. The element is based on zigzag theory and is augmented using the discrete Kirchhoff quadrilateral element developed earlier by the second author for the analysis of composite and sandwich plates. It is vital to convert the actions and displacements from local to global direction using a transformation matrix. DKZigTS is developed by adding two fictitious degrees of freedom and has nine local and global degrees of freedom per node. It does not face the problem of the ill-conditioned stiffness matrix. The performance of the developed element is assessed for its exactitude by comparing the results obtained with analytical 2D, 3D and finite element solutions available in the literature. Various boundary conditions, geometrical shapes and material properties are used in this study. The element exhibits a very satisfactory performance for moderately thick and thick shell panels.

Keywords Finite element · Cylindrical shell · Composite · Free vibration · Static

1 Introduction

Composite laminated shell characteristics can be modified, by tailoring the orientation of fiber and stacking sequence of the layer, which results in the use of laminated shells in high-performance and weight-sensitive engineering applications. It is showing good performance in the medical field applications such as bone fixation plates, bone grafts and hip joint replacement by virtue of its dimensional stability, high specific strength, thermal stability and tribological properties. Shell is a much stronger structural form than others have; hence understanding the static and dynamic response of shell structure with composite material is essential. Deflections, stresses,

V. A. Dagade (✉) · S. D. Kulkarni
College of Engineering Pune, Pune, Maharashtra 411005, India
e-mail: dagadeva14.civil@coep.ac.in

M. S. Ranadive et al. (eds.), *Recent Trends in Construction Technology and Management*, Lecture Notes in Civil Engineering 260,
https://doi.org/10.1007/978-981-19-2145-2_84

natural frequencies and their associated mode shapes are studied in this work for the symmetric and anti-symmetric composite cylindrical shell.

2 Displacement Field Approximation

Consider composite material cylindrical shell of thickness h and radius R_x at mid-plane as shown in Fig. 1. The mid-plane of the shell is chosen as the reference plane $z =$ zero. The top surface is at $z = \frac{h}{2}$, and bottom surface is at $z = \frac{-h}{2}$.

"The zigzag theory being 2D layer-wise theory has approximations, which are listed here:

1. Vertical Deflection 'w' is Independent of 'z'.

$$w(x, y, z, t) = wo(x, y, t). \tag{1}$$

2. The in-plane displacements 'u_x' and 'u_y' are a combination of global third-order variation in 'z' across the thickness and layer-wise piecewise linear variation, i.e. with discontinuity in slopes $u_{x,z}$, $u_{y,z}$ at the layer interfaces. Accordingly, u_x, u_y for the kth layer can be expressed as

$$u(x, y, z, t) = u_0(x, y, t) - zw_{0d}(x, y, t) + R^k(z)\psi_0(x, y, t)$$
$$u = \begin{bmatrix} u_x \\ u_y \end{bmatrix}, wo_d = \begin{bmatrix} w_{o,x} \\ w_{o,y} \end{bmatrix}, u_o = \begin{bmatrix} u_{ox} \\ u_{oy} \end{bmatrix}, \psi_o = \begin{bmatrix} \psi_{ox} \\ \psi_{oy} \end{bmatrix} \tag{2}$$

$R^k(z)$ is a 2 × 2 matrix of layer-wise functions of z of the form

$$R^k(z) = \widehat{R}_1^k + z\widehat{R}_2^k + z^2\widehat{R}_3 + z^3\widehat{R}_4 \tag{3}$$

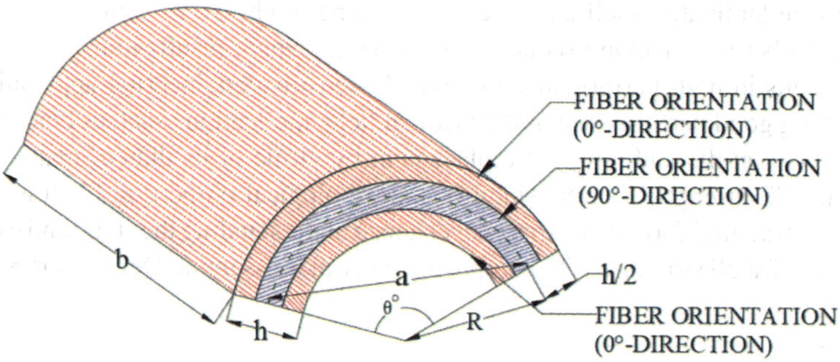

Fig. 1 Geometry of composite (0°/90°/0° stacking sequence configured) cylindrical shell

where $\widehat{R}_1^k$, $\widehat{R}_2^k$, $\widehat{R}_3$ and $\widehat{R}_4$ coefficient matrices depend on the material properties and lay-ups. Expressions for these matrices are given in the Kapuria and Kulkarni." [1].

2.1 Finite Element Formulation

Figure 2 shows the four-node quadrilateral element. *x*, *y*, *z* are the global and *x*', *y*', *z*' are the local axes.

The plate element initially developed by Kapuria and Kulkarni [1] has seven local DOF per node but to transform it globally, two additional fictitious degrees of freedom need to be added per node as mentioned in Zienkiewicz et al. [2]. After the considerations of these two additional fictitious DOF, namely, $\theta'_{oz}{}^{i}$ and $\psi'_{oz}{}^{i}$, $U_i^{e\prime}$ and U_i^e which are the local and the global nodal displacement vectors for the *i*th node are given in the following expressions.

$$U^{e'} = \left[u'^i_{0x} \; u'^i_{0y} \; w'^i_0 \; \theta'^i_{0x} \; \theta'^i_{0y} \; \theta'^i_{0z} \;\; \psi'^i_{0x} \; \psi'^i_{0y} \;\; \psi'^i_{0z} \right]^T$$

$$U^e = \left[u^i_{0x} \; u^i_{0y} \; w^i_0 \; \theta^i_{0x} \; \theta^i_{0y} \; \theta^i_{0z} \;\; \psi^i_{0x} \; \psi^i_{0y} \;\; \psi^i_{0z} \right]^T. \quad (4)$$

It is crucial to transform actions and displacements from local to global directions, to use plate elements for analysing cylindrical shells. The transformation is achieved using the following expression.

$$U_i^{e'} = T_i^e \, U_i^e \, ; \qquad i = 1, 2, 3, 4$$

T_i^e = the transformation matrix for *i*th node given as

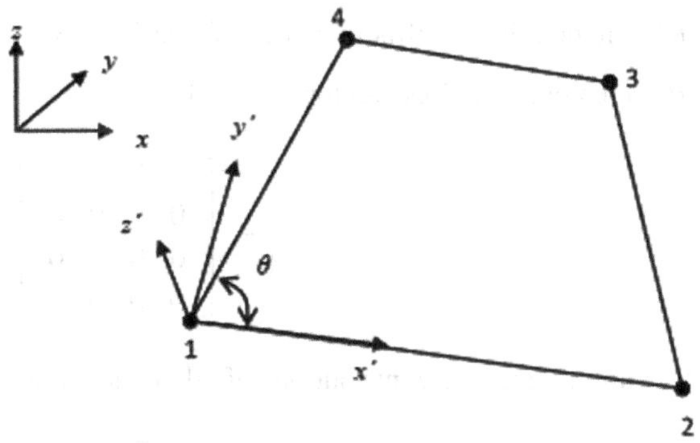

Fig. 2 Geometry of the element

$$T_i = \begin{bmatrix} \lambda_{11} & \lambda_{12} & \lambda_{13} & 0 & 0 & 0 & 0 & 0 & 0 \\ \lambda_{21} & \lambda_{22} & \lambda_{23} & 0 & 0 & 0 & 0 & 0 & 0 \\ \lambda_{31} & \lambda_{32} & \lambda_{33} & 0 & 0 & 0 & 0 & 0 & 0 \\ 0 & 0 & 0 & \lambda_{22} & \lambda_{21} & \lambda_{23} & 0 & 0 & 0 \\ 0 & 0 & 0 & \lambda_{12} & \lambda_{11} & \lambda_{13} & 0 & 0 & 0 \\ 0 & 0 & 0 & \lambda_{32} & \lambda_{31} & \lambda_{33} & 0 & 0 & 0 \\ 0 & 0 & 0 & 0 & 0 & 0 & \lambda_{22} & \lambda_{21} & \lambda_{23} \\ 0 & 0 & 0 & 0 & 0 & 0 & \lambda_{12} & \lambda_{11} & \lambda_{13} \\ 0 & 0 & 0 & 0 & 0 & 0 & \lambda_{32} & \lambda_{31} & \lambda_{33} \end{bmatrix} \quad (5)$$

where λ_{11}, λ_{12}, etc. are the direction cosines of the local axes with respect to the global axes.

Stiffness matrix becomes singular, due to zero stiffness corresponding to $\theta_{0z}^{\prime i}$ and $\psi_{0z}^{\prime i}$. This problem of singularity is overcome by placing a small stiffness coefficient corresponding to the degrees of freedom $\theta'_{oz}{}^{i}$ and $\psi'_{oz}{}^{i}$ equal to an arbitrary number in the local stiffness matrix of size 36 × 36, where the arbitrary number can be taken equal to the minimum value of stiffness coefficient in the diagonal of local stiffness matrix × 10^{-2}. The same treatment is given to the mass matrix.

The global element mass matrix M^e of size 36 × 36, global element stiffness matrix K^e of size 36 × 36, and global element load vector P^e of size 36 × 1 are given in Eq. (6).

$$M^e = T^{e^T} M^{e'} T^e, \quad K^e = T^{e^T} K^{e'} T^e, \quad P^e = T^{e^T} P^{e'} \quad (6)$$

where

T transformation matrix of size 36 × 36,

$M^{e'}$ element local mass matrix of size 36 × 36,

$K^{e'}$ element local stiffness matrix of size 36 × 36,

$P^{e'}$ element local load vector of size 36 × 1

$$T^e = \begin{bmatrix} T_1 & 0 & 0 & 0 \\ 0 & T_2 & 0 & 0 \\ 0 & 0 & T_3 & 0 \\ 0 & 0 & 0 & T_4 \end{bmatrix}. \quad (7)$$

Summing up the contributions of all elements, we get

$$\overline{M}\ddot{U} + \overline{K}U = \overline{P} \quad (8)$$

where $\overline{K}$, $\overline{M}$ and $\overline{P}$ are the assembled counterparts of matrices K^e, M^e and P^e, respectively. For free vibration, analysis right-hand side is set to zero and natural frequencies are obtained using subspace iteration technique. For static analysis, $\overline{M}$ is set to null matrix.

3 Numerical Results

The formulation of the four node flat shell element (DKZigTS) and the computer program developed are assessed by comparing the results of non-dimensionalised frequencies, deflection and stresses. The composite cylindrical shells of different configurations are modeled with $M \times N$ mesh of equal size elements in all the problems considered in this study. In the static analysis, shells are subjected to the transverse load acting on the top surface of the shell.

Uniformly distributed load of intensity q_0 and Sinusoidal load $q(x, y) = q_0 \sin(\frac{x\pi}{a}) \sin(\frac{y\pi}{b})$ are considered in this study.

The results are non-dimensionalised as follows:

- Natural frequencies

$$\overline{\omega} = \omega \times \frac{a^2}{h} \times \sqrt{\frac{\rho}{E_2}}; \tag{9}$$

- In-plane displacement and vertical deflection:

$$\overline{u}_x(0, \frac{b}{2}, \frac{z}{h}) = u_x \times E_2 \times \frac{h^2}{a^3 \times q_0};$$
$$\overline{w}(\frac{a}{2}, \frac{b}{2}, \frac{z}{h}) = 100 \times w \times E_2 \times \frac{h^3}{a^4 \times q_0}; \tag{10}$$

- In-plane stresses:

$$\overline{\sigma_x}, \overline{\sigma_y}(\frac{a}{2}, \frac{b}{2}, \frac{z}{h}) = \sigma_x, \sigma_y \times \frac{h^2}{a^2 \times q_0};$$

- Transverse stresses:

$$\overline{\tau_{xy}}(0, 0, \frac{z}{h}) = \tau_{xy} \times \frac{h^2}{a^2 \times q_0}; \quad \overline{\tau_{yz}}(\frac{a}{2}, 0, \frac{z}{h})$$
$$= \tau_{yz} \times \frac{h}{a \times q_0}; \overline{\tau_{zx}}(0, \frac{b}{2}, \frac{z}{h}) = \tau_{zx} \times \frac{h}{a \times q_0}; \tag{11}$$

The material properties used for analysis is:

$$E_1 = 172.5\text{ GPa}, E_2 = E_3 = 6.9\text{ GPa};$$
$$G_{12} = G_{31} = 3.45\text{ GPa}, G_{23} = 1.38\text{ GPa};$$
$$\mu_{12} = \mu_{13} = \mu_{23} = 0.25, \rho = 1580\text{ kg/m}^3.$$

3.1 Free Vibration Response

3.1.1 Symmetric Cross-Ply Composite Shell

In Table 1, the natural frequencies for the first seven modes obtained using the developed elements with the mesh sizes 12 × 12, 16 × 16 and 24 × 24 are presented. The results are compared for all-round simply supported (0°/90°/0°) configured, shallow shell panels of square planform with the results of 2 D higher-order shear deformation theories given by various researchers and 3D FE results of [3] for the various radius to span ratios ranging from 'R/a' = 5 to 'R/a' = 20. The 'h/a' ratio considered is 0.1. 3D FE results of (ANSYS 2019) are using mesh size of (24 × 24 × 6).

It is observed for the moderately thick shells, that the results of present elements are quite close to the 3D finite element results obtained with 20 node brick elements

Table 1 Non-dimensionalised seven consecutive frequencies for (SSSS) composite cylindrical shell panels (0°/90°/0°) with varying 'R/a' ratios

R/a	Theory	$\overline{\omega}_1$	$\overline{\omega}_2$	$\overline{\omega}_3$	$\overline{\omega}_4$	$\overline{\omega}_5$	$\overline{\omega}_6$	$\overline{\omega}_7$
	3D FE [3]	11.545	18.242	22.211	22.180	28.170	30.656	31.886
	HSDT [4]	11.850	–	–	–	–	–	–
5	HSDT [5]	11.746	–	–	–	–	–	–
	HSDT [6]	11.851	–	–	–	–	–	–
	HSDT [7]	11.846	18.489	–	–	29.399	30.856	32.961
	DKZigTS (12 × 12)	11.424	18.418	22.249	22.309	28.236	31.488	31.838
	DKZigTS (16 × 16)	11.429	18.438	22.222	22.283	28.245	31.506	31.909
	DKZigTS (24 × 24)	11.433	18.457	22.202	22.263	28.254	31.531	31.968
	3D FE [3]	11.434	18.225	22.200	22.205	28.088	30.598	31.822
10	DKZigTS (12 × 12)	11.413	18.311	22.277	22.285	28.294	31.363	31.861
	DKZigTS (16 × 16)	11.419	18.332	22.249	22.257	28.305	31.380	31.934
	DKZigTS (24 × 24)	11.425	18.350	22.230	22.238	28.314	31.402	31.992
	3D FE [3]	11.452	18.222	22.132	22.216	28.157	30.586	31.574
20	DKZigTS (12 × 12)	11.444	18.342	22.28	22.285	28.342	31.38	31.948
	DKZigTS (16 × 16)	11.453	18.368	22.253	22.257	28.354	31.401	32.027
	DKZigTS (24 × 24)	11.460	18.389	22.233	22.237	28.365	31.427	32.089

of [3], even for higher modes with maximum percentage difference is less than 2.85. The present element has captured the third and fourth shear modes perfectly, which are not reported by Fiorenzo et al. [7], as their results are based on 2D analytical theory. The fundamental frequencies for the case 'R/a' = 5 and 'h/a' = 0.1 are presented by Mantari et al. [5], Ferreira et.al. [6] and Reddy [4] using 2D shear deformation theories are also presented in the table.

Table 2 shows the comparison of non-dimensionalised fundamental frequencies obtained by the present element DKZigTS and 3D FE results of [3] for composite cylindrical shell with (0°/90°/90°/0°) stacking sequence. The radius to span ratios varying from 1 to 10 is studied in this table for thickness to span ratio 'h/a' = 0.1 and 'h/a' = 0.25. The 2D results are obtained by solving closed-form formulations by Garg et al. [8], and 2D higher-order shear deformation theories results of [4] are also presented in the table.

From the table, it can be easily understood that the augmentation of the radius span ratio leads to the decrease of frequencies in the case of the moderately thick shell, whereas it is reversed in the case of the thick shell. It is seen that results of DKZigTS are close to 3D FE results of [3] obtained using 20 node brick element with mesh size of (24 × 24 × 8), for the deep and shallow shell as well as for the moderately thick and thick shell. The maximum percentage difference is less than 2.10 for moderately thick shell and 2.14 for thick shell.

Figure 3 shows the first 5 mode shapes obtained using present elements of mesh size (12 × 12) for a cylindrical shell panel (0°/90°/0°) of square planform, with one curved edges clamped and all other three edges free. The mode shapes obtained using 3D FE 20 node elements of ANSYS [3] with a mesh size of (24 × 24 × 6) are also

Table 2 Dimensionless natural frequencies for (0°/90°/90°/0°) cylindrical shell (SSSS) panel with varying 'R/a' ratio and 'h/a' ratio

'R/a'	'h/a'	3D FE [3]	HSDT [4]	HOST [8]	DKZigTS (12 × 12)	DKZigTS (16 × 16)	DKZigTS (24 × 24)
1	0.1	11.991	–	12.698	12.096	12.041	11.963
	0.25	6.805	–	7.057	6.966	6.902	6.833
2	0.1	11.513	–	11.999	11.641	11.634	11.625
	0.25	6.905	–	7.150	7.068	7.062	7.053
3	0.1	11.463	–	11.856	11.586	11.610	11.610
	0.25	7.052	–	7.171	7.099	7.100	7.100
4	0.1	11.383	–	11.805	11.584	11.592	11.596
	0.25	7.053	–	7.178	7.130	7.133	7.135
5	0.1	11.367	11.830	11.781	11.578	11.584	11.589
	0.25	7.072	–	7.182	7.140	7.144	7.148
10	0.1	11.346	11.790	11.749	11.571	11.579	11.585
	0.25	7.152	–	7.187	7.157	7.157	7.157

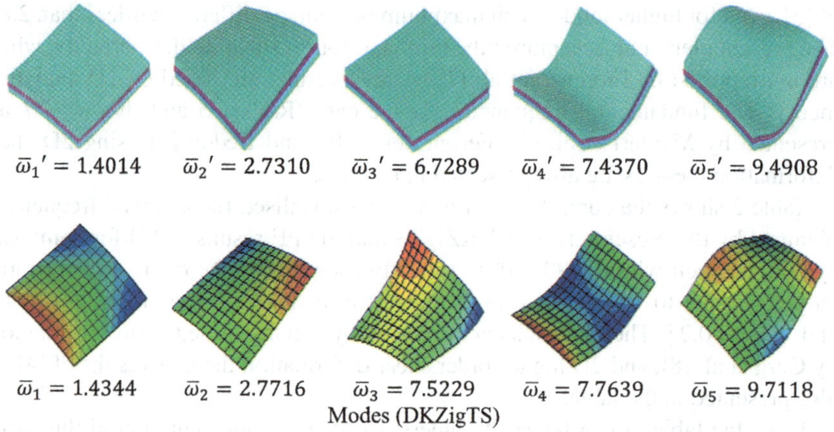

Fig. 3 Comparison of first five mode shapes for cantilever composite cylindrical shell with staking sequence (0°/90°/0°), 'R/a' = 5 and 'h/a' = 0.15

shown for comparison. The present element has correctly depicted all the five modes indicating the effectiveness of the present element based on zigzag theory.

3.1.2 Anti-Symmetric Cross-Ply Composite Shell

Table 3 shows the developed DKZigTS element is assessed for moderately thick ('h/a' = 0.1) and thick ('h/a' = 0.15) anti-symmetric cross-ply composite shell panel with (0°/90°) stacking sequence and all-round simply supported boundary conditions. The results are compared with the 3D elasticity results of Bhimaraddi [9], Ye and Soldatos

Table 3 Dimensionless natural frequencies for (0°/90°) cylindrical shell (SSSS) panel with varying 'R/a' ratio and 'h/a' ratio

'R/a'	'h/a'	3D Elast. [9]	3D Elast. [10]	HSZT [11]	DKZigTS (12 × 12)	DKZigTS (16 × 16)	DKZigTS (24 × 24)
1	0.1	10.408	10.697	10.655	10.482	10.388	10.258
	0.15	8.606	8.969	8.911	8.704	8.585	8.453
2	0.1	9.362	9.495	9.489	9.270	9.258	9.241
	0.15	8.358	8.507	8.513	8.343	8.328	8.304
4	0.1	9.061	9.116	9.137	9.036	9.031	9.026
	0.15	8.273	8.331	8.378	8.337	8.333	8.328
5	0.1	9.020	9.062	9.089	9.017	9.013	9.010
	0.15	8.256	8.257	8.357	8.347	8.343	8.341
10	0.1	8.956	8.978	9.018	8.981	8.980	8.979
	0.15	8.225	8.247	8.319	8.353	8.351	8.351

[10] and Higher-order zigzag theory (HSZT) results of Garg [11]. Garg et al. [11] used nine node finite elements having seven degrees of freedom per node with 18 × 18 mesh size resulting in 9583 degrees of freedom. It is seen that the present results with 24 × 24 mesh size and 5625 degrees of freedom, even for deep shell panels are closer to 3D elasticity results of Alavandi (1992), than the 2D HSZT results of Garg [11]. It is clear from the comparison that the results of present elements are in good agreement with 3D elasticity results with a maximum percentage difference is less than 1.44 for moderately thick shells and 1.77 for a thick shell.

3.2 *Static Response*

Table 4 working is for computation of deflection and stresses of all-round simply supported (0°/90°/0°) configured composite cylindrical shells under sinusoidal distributed load. The shell has rectangular planform with 'b/a' = 3 and radius to span ratio 'R/a' = 4. The results for thick 'h/a' = 0.2 and moderately thick 'h/a' = 0.1 shells are presented in the table. The results obtained using present formulation are compared with result published by Exact 3D of Huang et al. [12]. Equivalent single layer theory results using finite element solution of Dau et al. [13] having triangular element with 6 nodes with 18 DOF at a corner node and 9 DOF at a mid-side node based on sinus model SIN-C are also presented in the Table. The mesh of size (4 × 4) with 32 triangular element, which gives total 954 DOF on full shell, is used by Dau et al. [13] for analysing the results. The results of the present element with mesh size (24 × 72) give 16,425 DOF, but give results that are more accurate, with the maximum percentage difference for the transverse deflection is 1.16, for normal stresses it is 3.99.

Table 4 Deflection and stresses of (0°/90°/0°) configured rectangular cylindrical shells

'h/a'	Theory	$\overline{w}$	$\overline{\sigma_x}$	$\overline{\sigma_y}$	$\overline{\tau_{xy}}$	$\overline{\tau_{zx}}$	$\overline{\tau_{yz}}$
0.2	Exact 3D [12]	2.1180	1.0220	1.1161	0.2588	0.3867	0.2729
	ESL FE [13]	1.9371	0.9230	1.0240	0.2358	0.2930	0.2858
	DKZigTS (12 × 36)	2.3504	1.0501	1.1345	0.2321	0.4152	0.2507
	DKZigTS (16 × 48)	2.2596	1.0289	1.1295	0.2332	0.404	0.2591
	DKZigTS (24 × 72)	2.1425	1.0177	1.1245	0.2532	0.3928	0.2631
	(% Diff.)	1.16	0.42	0.75	2.16	1.58	3.59
0.1	Exact 3D [12]	0.9396	0.7463	0.6468	0.151	0.4271	0.1555
	ESL FE [13]	0.8763	0.7026	0.6076	0.1412	0.3029	0.1734
	DKZigTS (12 × 36)	0.9515	0.7095	0.6402	0.1252	0.4041	0.1558
	DKZigTS (16 × 48)	0.9308	0.7207	0.6614	0.1464	0.4053	0.1670
	DKZigTS (24 × 72)	0.9305	0.7319	0.6726	0.1576	0.4064	0.1682
	(% Diff.)	0.97	1.93	3.99	4.37	4.85	8.17

Table 5 Deflection and stresses of (0°/90°/0°) and (0°/90°) configured cylindrical shells

Type	'R/a'	Mesh Size	$\overline{u}_x$	$\overline{w}$	$\overline{\sigma_x}$	$\overline{\sigma_y}$	$\overline{\tau_{xy}}$	$\overline{\tau_{zx}}$	$\overline{\tau_{yz}}$
0°/90°/0°	5	(12 × 12)	0.0193	1.1529	0.8696	0.0531	0.0469	0.4920	0.6355
		(16 × 16)	0.0193	1.1534	0.8696	0.0531	0.0479	0.5029	0.6353
		(24 × 24)	0.0193	1.1537	0.8695	0.0531	0.0490	0.5086	0.6346
	10	(12 × 12)	0.0157	1.1565	0.8753	0.0526	0.0526	0.4960	0.6427
		(16 × 16)	0.0157	1.1568	0.8752	0.0526	0.0530	0.5061	0.6424
		(24 × 24)	0.0156	1.1570	0.8751	0.0525	0.0534	0.5115	0.6417
	20	(12 × 12)	0.0139	1.1574	0.8757	0.0523	0.0559	0.4969	0.6445
		(16 × 16)	0.0139	1.1577	0.8756	0.0523	0.0561	0.5069	0.6442
		(24 × 24)	0.0139	1.1578	0.8755	0.0523	0.0563	0.5123	0.6440
0°/90°	5	(12 × 12)	0.0527	1.8634	1.1015	0.1260	0.0466	0.1902	0.3471
		(16 × 16)	0.0527	1.8653	1.1019	0.1261	0.0506	0.1900	0.3487
		(24 × 24)	0.0527	1.8665	1.1022	0.1261	0.0530	0.1898	0.3494
	10	(12 × 12)	0.0476	1.8634	1.1147	0.1261	0.0460	0.1898	0.3598
		(16 × 16)	0.0476	1.8653	1.1149	0.1262	0.0485	0.1896	0.3613
		(24 × 24)	0.0476	1.8665	1.1150	0.1262	0.0508	0.1893	0.3620
	20	(12 × 12)	0.0449	1.8874	1.1150	0.1257	0.0466	0.1888	0.3651
		(16 × 16)	0.0449	1.8883	1.1152	0.1258	0.0477	0.1885	0.3666
		(24 × 24)	0.0448	1.8888	1.1153	0.1258	0.0488	0.1883	0.3672

Table 5 presents computation of normal and transverse stresses along with vertical and in-plane deflections of all-round simply supported, symmetrical (0°/90°/0°) and anti-symmetrical (0°/90°) layered composite cylindrical shells under uniformly distributed load. The results for the moderately thick (h/a = 0.1) having square plan-form shell panels with 'R/a' = 5, 10 and 20 are presented in this table. From the table, it can be concluded that deflection is proportional, whereas in-plain displacement is inversely proportional to the 'R/a' ratio.

From Figs. 4, 5, 6 and 7 the through thickness distribution of in-plane normal stresses $\overline{\sigma_x}$, $\overline{\sigma_y}$, and transverse shear stress $\overline{\tau_{zx}}$, $\overline{\tau_{yz}}$ are plotted, respectively, for (0°/90°/0°) laminated composite shell Panel, with all-round simply supported boundary conditions and subjected to uniformly distributed load. All the figures are plotted for two 'R/a' ratios, namely, 5 and 10 with 'h/a' ratio is equal to 0.1. Full shell is discretised with a mesh size of (24 × 24) of present element.

From Figs. 4 and 5, it is clear that there is discontinuity at layer interface of 0° and 90°. In Figs. 6 and 7, it is clear that the transverse shear stresses are zero at top and maximum value at 'z/h' = 0 at bottom.

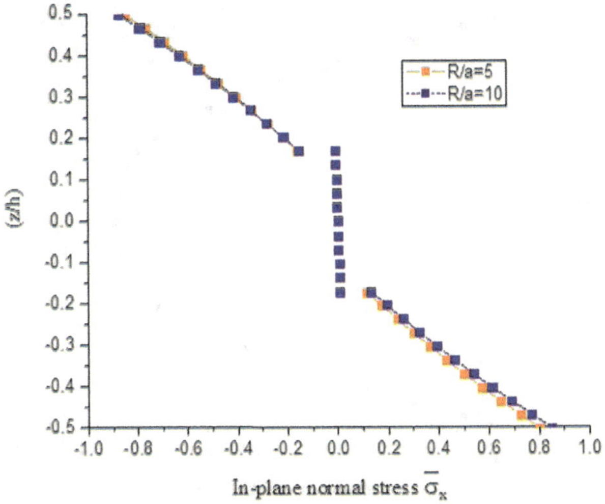

Fig. 4 In-plane normal stress $\overline{\sigma_x}$ variation for (0°/90°/0°) configured shell

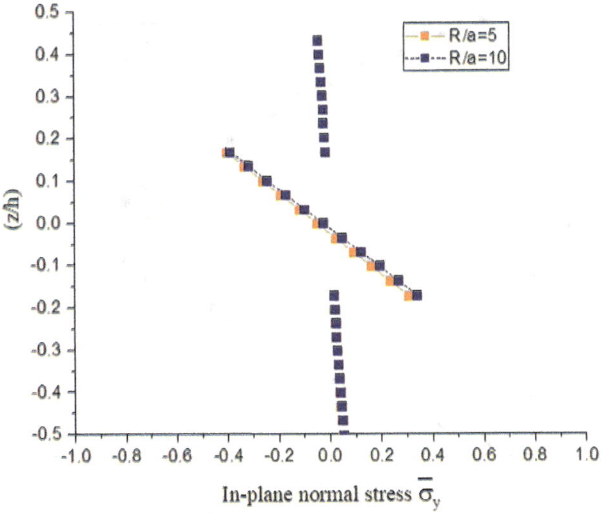

Fig. 5 In-plane normal stress $\overline{\sigma_y}$ variation for (0°/90°/0°) configured shell

4 Conclusions

The developed DKZigTS element gives sufficiently accurate results for static and free vibration analysis of composite cylindrical shell panels for moderately thick as well as thick shells considered in the study. The element results are very close to 3D exact results from literature for moderately thick and thick shells. The present

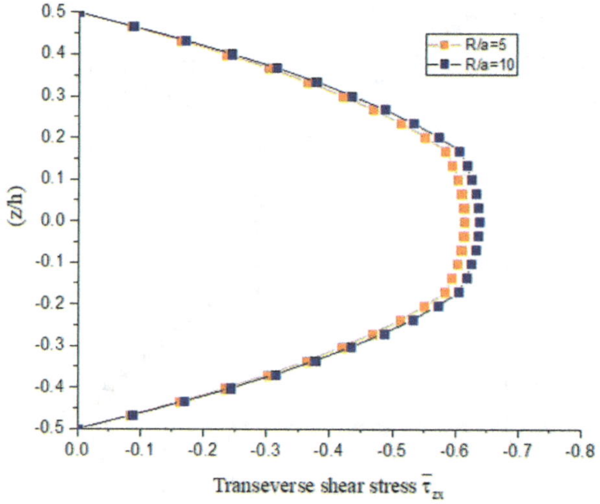

Fig. 6 The transverse shear stress $\overline{\tau_{zx}}$ variation for (0°/90°/0°) configured shell

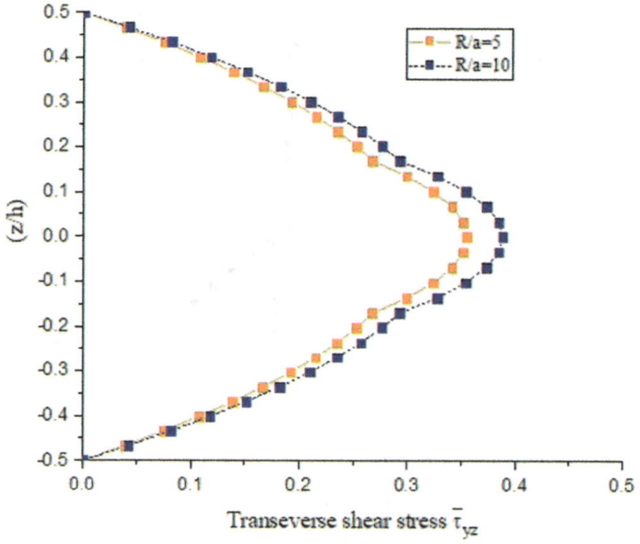

Fig. 7 The transverse shear stress $\overline{\tau_{yz}}$ variation for (0°/90°/0°) configured shell

element results are in close agreement with the results reported in the literature when tested for symmetric as well as anti-symmetric laminates. The present formulation is quite capable of depicting the mode shapes and in depicting the through thickness variation of transverse shear stresses. The formulation is suitable for general-purpose finite element programming as it is unpretentious. The novelty of this element is in its potential to analyse any curved pattern shell, without any modification.

Referencess

1. Kapuria S, Kulkarni SD (2007) An improved discrete Kirchhoff quadrilateral element based on third-order zigzag theory for static analysis of composite and sandwich plates. Int J Numer Meth Eng 69:1948–1981
2. Zienkiewicz OC, Taylor RL (2000) The finite element method: solid mechanics, vol 2. Butterworth-Heinemann, Oxford
3. ANSYS Mechanical APDL (2019) ANSYS User Manual
4. Reddy JN, Liu CF (1985) A higher-order shear deformation theory of laminated elastic shells. Int J Eng Sci 23(3):319–330
5. Mantari JL, Oktem AS, Soares CG (2011) Static and dynamic analysis of laminated composite and sandwich plates and shells by using a new higher-order shear deformation theory. Compos Struct 94:37–49
6. Ferreira AJM, Castro LM, Bertoluzza S (2011) A wavelet collocation approach for the analysis of laminated shells. Compos B 42:99–104
7. Fiorenzo AF (2014) A refined dynamic stiffness element for free vibration analysis of cross-ply laminated composite cylindrical and spherical shallow shells. Compos B 62:143–158
8. Garg AK, Khare RK, Kant T (2006) Higher-order closed-form solutions for free vibration of laminated composite and sandwich shells. J Sandwich Struct Mater 8(3):205–235
9. Bhimaraddi A (1991) Free vibration analysis of doubly curved shallow shells on rectangular plane-form using three–dimensional elasticity theory. Int J Solids Struct 27:897–913
10. Ye J, Soldatos KP (1994) Three-dimensional vibration of laminated cylinders and cylindrical panels with symmetric or antisymmetric cross-ply lay-up. Composs Eng 4:429–444
11. Garg AK, Chakrabarti A, Bhargava P (2013) Vibration of laminated composites and sandwich shells based on higher order zigzag theory. Eng Struct 56:880–888
12. Huang NN (1994) Influence of shear correction factors in the higher order shear deformation laminated shell theory. Int J Solids Struct 31:1263–1277
13. Dau F, Polit O, Touratier M (2004) An efficient C1 finite element with continuity requirements for multi-layered/sandwich shell structures. Comput Struct 82:1889–1899

Exact Elasticity Analysis of Sandwich Beam with Orthotropic Core

Ganesh B. Irkar and Y. T. LomtePatil

Abstract Sandwich beam consists of three layers such that two face sheets at top and bottom are separated by a comparatively soft core between them. Due to its specific advantages over a conventional structure, it is been widely used in areas like aerospace, automobiles, and many more. As the use of Sandwich structure is increasing nowadays, it becomes necessary to explore this area in more depth and provide new solution technique for the analysis of such structures. In this study, the sandwich beam with orthotropic core is analyzed as plane stress problem. With the help of elasticity equations, governing differential equations (GDEs) are developed in terms of displacements. Series expansion approach is used to solve GDEs and unknown stresses and displacements within the sandwich laminate are evaluated. Obtained results concur with available analytical solution ensuring correctness of adopted methodology.

Keywords Displacement based formulation · Exact elasticity analysis · Sandwich structure · Series expansion approach

1 Introduction

Sandwich is type of composite structure. As sandwich structure outplays conventional structure in terms of structural advantages, it is used in many industries such as aerospace, marine, and automobiles. Structures are constructed to serve certain purpose throughout their life span. Safety is the most important parameter in structural engineering field. When stress induced inside the structural element due to external loads exceeds strength of that element then that member fails. This kind of local failure can result in complete collapse of structure which is not acceptable by safety point of view. So, it is necessary to study the distribution of stresses inside the structural element when subjected to external loading in the same sense as actual loading in working life. After analyzing distribution of stresses and considering the

---

G. B. Irkar · Y. T. LomtePatil (✉)
Civil Engineering Department, College of Engineering Pune, Pune 411005, India
e-mail: lomteyt@gmail.com

M. S. Ranadive et al. (eds.), *Recent Trends in Construction Technology and Management*, Lecture Notes in Civil Engineering 260,
https://doi.org/10.1007/978-981-19-2145-2_85

material strength, safety can be ensured by keeping lemniscates in appropriate direction. There are several methods using which analysis of sandwich structure can be done. The literatures which provided elasticity solution to sandwich beam has been discussed in the next paragraph.

Saint–Venant [1] considered cantilever beam with rectangular cross section having anisotropy of special form and obtained the exact elasticity solution. Silverman [2], Hashin [3], Gerstner [4], Rao and Ghosh [5], and Cheng et al. [6] used Airy's stress functions and provided exact solution to shear deformable laminated composite beam. Holt and Webber [7] presented exact solution for honeycomb sandwich beams. Pagano [8, 10] analyzed laminated composite plate under cylindrical bending and obtained exact elasticity solution. Srinivas et al. [9] developed 3D linear, small deformation theory of elasticity for the free vibration of simply supported homogeneous isotropic thick rectangular plate. Venkataraman [12] considered simply supported sandwich beam for analysis. In this study, the face sheets of sandwich beam are modeled as Euler–Bernoulli beam theory. Displacement-based formulation is used for obtaining the GDEs for core and then solved using Pagano's approach. This methodology to analyze sandwich beam is explained in Sankar [11]. Kant et al. [13] provided the semi-analytical elasticity solution to simply supported sandwich beam. The GDEs developed using mixed formulation are solved using numerical integration technique. The exact elasticity solution is always used as a basis to check the accuracy of any refined shear deformation theory.

In this study, exact 2D elasticity solution to sandwich beam with orthotropic core is provided. To achieve this displacement-based formulation is employed and the governing differential equations are expressed in terms displacements. The solution to these equations is obtained using series expansion approach. Series expansion approach is adopted because it gives the solution to differential equations with constant coefficients as well as with variable coefficients without making any modification in the formulations. MATLAB R2019a software is used for programming. The results obtained by this solution technique are thoroughly compared with available literature. The through thickness variation of displacements and stresses is also provided in the tabular and graphical form for better understanding.

2 Theoretical Formulation

Simply (diaphragm) supported sandwich beam having span L along x-axis and total depth H along z-axis subjected to sinusoidal loading is considered as plane stress problem (Fig. 1). The loading is applied on top surface in positive z-direction, whereas bottom surface is free from any traction. The x-axis is along the mid-depth of the beam. The depth of each face sheet is $H_f = 0.1H$ and that of core is $2H_c$. For the simplicity of analysis, the sandwich beam is divided into four layers. Layer no. 1 corresponds to bottom face sheet, whereas layer nos. 2 and 3 indicate the lower half and upper half part of the core, respectively, and layer 4 represent the top face sheet.

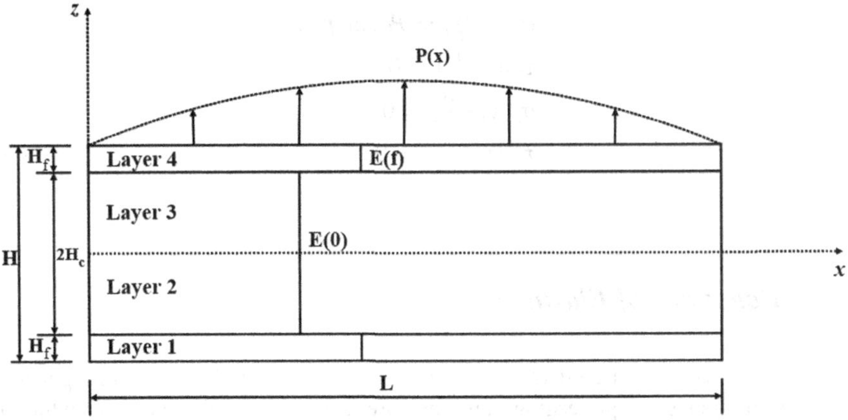

Fig. 1 Sandwich beam

If the properties of face sheets and core are matched, the same formulation can be adopted for homogeneous isotropic/orthotropic beam.

The sandwich beam shown in Fig. 1 $E(0)$ represents the modulus of elasticity for the core, whereas $E(f)$ represents the modulus of elasticity for face sheet. From this, it can be seen that the material properties for the sandwich beam change abruptly at the interface which may lead to the delamination between core and face sheets.

The loading acting on the beam is expressed in terms of Fourier series as given by Eq. (1) so that any kind of loading can be handled. In this case, only sinusoidal load is considered so only one term in the series is sufficient.

$$P(x) = \sum_{m=1,3}^{\infty} P_0 \sin(px) \tag{1}$$

where $p = \frac{m\pi}{L}$.

The simple support condition yields boundary conditions (BCs) at ends of the beam as given by Eq. (2),

At $x = 0$ and $x = L$

$$\sigma_x = w = 0 \tag{2a}$$

At $x = \frac{L}{2}$

$$u = 0 \tag{2b}$$

Also the BCs at top and bottom of the beam are

$$\sigma_z(x, \tfrac{H}{2}) = P_0 \sin(px)$$
$$\tau_{zx}(x, \tfrac{H}{2}) = 0$$
$$\sigma_z(x, \tfrac{-H}{2}) = 0$$
$$\tau_{zx}(x, \tfrac{-H}{2}) = 0. \tag{2c}$$

2.1 *Equations of Elasticity*

In this case as the material for both core as well as face sheets is homogeneous orthotropic hence the governing equations are derived for only one layer which is equally applicable to the remaining layers. Only we need to match the displacements and transverse stresses at the interface to maintain the continuity at the interface.

2.1.1 Differential Equations of Equilibrium

In the absence of body force, differential equation of equilibrium becomes

$$\frac{\partial \sigma_x}{\partial x} + \frac{\partial \tau_{zx}}{\partial z} = 0$$
$$\frac{\partial \sigma_z}{\partial z} + \frac{\partial \tau_{xz}}{\partial x} = 0 \tag{3}$$

where $\tau_{zx} = \tau_{xz}$

2.1.2 Material Constitutive Relationship

For 2D elasticity plane stress problem material constitutive relationship can be represented as

$$\begin{Bmatrix} \sigma_x \\ \sigma_z \\ \tau_{xz} \end{Bmatrix} = \begin{bmatrix} C_{11} & C_{12} & 0 \\ C_{21} & C_{22} & 0 \\ 0 & 0 & C_{33} \end{bmatrix} \begin{Bmatrix} \varepsilon_x \\ \varepsilon_z \\ \gamma_{xz} \end{Bmatrix} \tag{4}$$

2.1.3 Strain–displacement Relation

Strain can be written in terms of displacement as

$$\begin{Bmatrix} \varepsilon_x \\ \varepsilon_z \\ \gamma_{xz} \end{Bmatrix} = \begin{Bmatrix} \frac{\partial u}{\partial x} \\ \frac{\partial w}{\partial z} \\ \frac{\partial u}{\partial z} + \frac{\partial w}{\partial x} \end{Bmatrix} \tag{5}$$

2.2 *Displacement Based Formulation*

In displacement-based formulation, elastic displacements are considered as primary variables and differential equations of equilibrium are written in terms of displacements using material constitutive relation (Eq. 4) and strain–displacement relationships (Eq. 5). This gives a set of two coupled partial differential equations (PDEs) (Eq. 6).

Substituting strain–displacement relations Eq. (5) and constitutive relations Eq. (4) in Equilibrium equations Eq. (3) set of equation is obtained

$$\begin{aligned} C_{11}\frac{\partial^2 u}{\partial x^2} + C_{33}\frac{\partial^2 u}{\partial z^2} + (C_{12} + C_{33})\frac{\partial^2 w}{\partial x.\partial z} &= 0 \\ (C_{21} + C_{33})\frac{\partial^2 u}{\partial x.\partial z} + C_{33}\frac{\partial^2 w}{\partial x^2} + C_{22}\frac{\partial^2 w}{\partial z^2} &= 0. \end{aligned} \tag{6}$$

PDEs are quite difficult to tackle as it contains more than one variable. To convert PDEs into a set of ordinary differential equations (ODEs), Fourier series expansion technique is used. Primary variables are expressed in the following manner such that it satisfies end boundary conditions (2a) and (2b) exactly.

$$\begin{aligned} u(x, z) &= U(z)\cos(px) \\ w(x, z) &= W(z)\sin(px). \end{aligned} \tag{7}$$

Substituting Eq. (7) in Eq. (6) and applying orthogonality condition following set of ODEs is obtained

$$\begin{aligned} &C_{33}U''(z) - p^2C_{11}U(z) + p(C_{12} + C_{33})W'(z) = 0 \\ &- p(C_{12} + C_{33})U'(z) + C_{22}W''(z) - p^2C_{33}W(z) = 0 \end{aligned} \tag{8}$$

where $(') = \frac{\mathrm{d}}{\mathrm{d}z}$

2.2.1 Solution Technique

Series expansion approach is solution technique for displacement-based formulation. In this technique, the solution obtained ODEs Eq. (8) is assumed in the form of power series in transverse direction of the beam.

$$U(z) = \sum_{r=0}^{\infty} A_r z^r$$
$$W(z) = \sum_{r=0}^{\infty} B_r z^r \tag{9}$$

where A_r and B_r are the coefficients in the power series.

The differentiation of two displacement components after simple algebraic manipulation is

$$U'(z) = \sum_{r=0}^{\infty} A_{r+1}(r+1)z^r,$$
$$U''(z) = \sum_{r=0}^{\infty} A_{r+2}(r+2)(r+1)z^r,$$
$$W'(z) = \sum_{r=0}^{\infty} B_{r+1}(r+1)z^r,$$
$$W''(z) = \sum_{r=0}^{\infty} B_{r+2}(r+2)(r+1)z^r. \tag{10}$$

Substituting Eq. (9) and Eq. (10) in Eq. (8),

$$\begin{Bmatrix} C_{33}A_{r+2}(r+2)(r+1) - p^2C_{11}A_r \\ +p(C_{12}+C_{33})B_{r+1}(r+1) \end{Bmatrix} \sum_{r=0}^{\infty} z^r = 0$$
$$\begin{Bmatrix} -p(C_{21}+C_{33})A_{r+1}(r+1) \\ +C_{22}B_{r+2}(r+2)(r+1) - p^2C_{33}B_r \end{Bmatrix} \sum_{r=0}^{\infty} z^r = 0 \tag{11}$$

Equation (11) is obtained. For solution to be nontrivial, the term inside curly bracket must be zero. Thus, recurrence relationship occurring in power series is obtained.

$$A_{r+2}(r+2)(r+1) = I_1A_r + I_2B_{r+1}(r+1)$$
$$B_{r+2}(r+2)(r+1) = I_3A_{r+1}(r+1) + I_4B_r \tag{12}$$

where

$$I_1 = \frac{p^2C_{11}}{C_{33}}, I_2 = \frac{-p(C_{12}+C_{33})}{C_{33}},$$

$$I_3 = \frac{p(C_{12} + C_{33})}{C_{22}}, \ I_4 = \frac{p^2 C_{33}}{C_{22}}.$$

The successive evaluation of Eq. (12) for $r = 0, 1, 2 \ldots$ gives the coefficients $A_2, B_2, A_3, B_3, \ldots$ of the two power series in terms of only four independent coefficients A_0, B_0, A_1, B_1. Hence the primary variables $U(z)$ and $W(z)$ can also be expressed in terms of the same four coefficients. These unknown coefficients are determined using the BCs given by Eq. (2c). Once four unknown coefficients of the power series are obtained the complete solution for the primary variables $u(x, z)$ and $w(x, z)$ are also known and hence the stresses at any point in the beams can be evaluated in a straightforward manner.

3 Results

3.1 Validation

To check the accuracy of the present approach, the results obtained by series expansion approach are compared with Kant et al. [13]. For validation purpose, same material properties (Table 1) and normalization (Eq. 13) as used by the author are followed.

Table 2 shows the convergence study for the aspect ratio ($\eta = {}^1/_4$) from which it can be seen that the results are converging for 9 terms of series.

$$\eta = \frac{H}{L}$$

Table 3 gives the comparison of results for homogeneous isotropic beam on the

Table 1 Material properties

Sr. No.	Material		Property	Source
1	Homogeneous Isotropic		$E = 1.0$ GPa, $\nu = 0.3$, $G =$ 0.3846 GPa	–
2	Homogeneous Orthotropic		$E_x = 172.4$ GPa; $E_z = 6.89$ GPa $\upsilon_{xz} = 0.25$ $G_{xz} = 3.45$ GPa	Pagano [14]
3	Sandwich	Face sheets- Homogeneous orthotropic	*For face sheet* $E_x = 172.4$ GPa; $E_z = 6.89$ GPa $\upsilon_{xz} = 0.25$ $G_{xz} = 3.45$ GPa	Pagano [14]
		Core- Homogeneous orthotropic	*For core* $E_z = 3.450$ GPa; $E_x = 0.276$ GPa $\upsilon_{xz} = 0.25$ $G_{xz} = 0.414$ GPa	

Table 2 Convergence study for Homogeneous Isotropic beam ($\eta = \frac{1}{4}$)

	$\overline{u}$		$\overline{w}$	$\overline{\sigma}_x$		$\overline{\tau}_{xz}$
No of terms retained in series (r)	$x = 0$	$x = 0$	$x = L/2$	$x = L/2$	$x = L/2$	$x = 0$
	$z = H/2$	$z = -H/2$	$z = 0$	$z = H/2$	$z = -H/2$	$z = 0$
3	−0.4157	0.4337	7.6666	0.3452	−0.3406	0.2524
4	−0.7838	0.8036	14.3489	0.6344	−0.6312	0.4756
5	−0.7591	0.7790	13.9368	0.6150	−0.6118	0.4692
6	−0.7686	0.7884	14.1085	0.6224	−0.6192	0.4750
7	−0.7684	0.7883	14.1058	0.6223	−0.6191	0.4749
8	−0.7685	0.7883	14.1067	0.6223	−0.6192	0.4750
9	0.7685	0.7883	14.1067	0.6223	−0.6192	0.4750

Table 3 Validation of results for Homogeneous Isotropic beam

		$\overline{\sigma}_x$		$\overline{\tau}_{xz}$	$\overline{w}$
Aspect ratio (η)	Source	$x = L/2$	$x = L/2$	$x = 0$	$x = L/2$
		$z = H/2$	$z = -H/2$	$z = 0$	$z = 0$
1/4	Present work	0.6223	−0.6192	0.4750	14.1067
	Kant et. al. [12]	0.6223	−0.6192	0.4750	14.1076
1/10	Present work	0.6100	−0.6099	0.4771	12.6075
	Kant et. al. [12]	0.6100	−0.6099	0.4771	12.6086
1/20	Present work	0.6084	−0.6084	0.4774	12.3913
	Kant et. al. [12]	0.6084	−0.6084	0.4774	12.3925
1/30	Present work	0.6081	−0.6081	0.4774	12.3513
	Kant et. al. [12]	0.6082	−0.6082	0.4774	12.3524
1/40	Present work	0.6081	−0.6081	0.4774	12.3372
	Kant et. al. [12]	0.6081	−0.6081	0.4774	12.3383
1/50	Present work	0.6080	−0.6080	0.4774	12.3307
	Kant et. al. [12]	0.6080	−0.6080	0.4774	12.3318

contrary Tables 4 and 5 presents the results for homogeneous orthotropic and sandwich beam with orthotropic core, respectively. From these results, it can be concluded that the results obtained by the present formulation are in well agreement with the considered literature.

$$\overline{w} = \frac{100 E_z \eta^3 w(L/2, 0)}{p_0 L}, \ \overline{\sigma}_{\mathrm{x}} = \frac{\eta^2 \sigma_x(L/2, 0)}{p_0}, \ \overline{\tau}_{xz} = \frac{\eta \tau_{xz}(0, z)}{p_0}. \tag{13}$$

Table 4 Validation of results for Homogeneous Orthotropic beam

		$\overline{\sigma}_x$		$\overline{\tau}_{xz}$	$\overline{w}$
Aspect ratio (η)	Source	$x = L/2$	$x = L/2$	$x = 0$	$x = L/2$
		$z = H/2$	$z = -H/2$	$z = 0$	$z = 0$
1/4	Present work	0.9026	−0.8468	0.4328	1.9483
	Kant et. al. [12]	0.9028	−0.8469	0.4328	1.9509
1/10	Present work	0.6570	−0.6551	0.4683	0.7325
	Kant et. al. [12]	0.6570	−0.6551	0.4683	0.7333
1/20	Present work	0.6203	−0.6201	0.4751	0.5527
	Kant et. al. [12]	0.6203	−0.6201	0.4751	0.5532
1/30	Present work	0.6134	−0.6134	0.4764	0.5192
	Kant et. al. [12]	0.6134	−0.6134	0.4764	0.5197
1/40	Present work	0.6110	−0.6110	0.4769	0.5075
	Kant et. al. [12]	0.6110	−0.6110	0.4769	0.5079
1/50	Present work	0.6099	−0.6099	0.4771	0.5020
	Kant et. al. [12]	0.6099	−0.6099	0.4771	0.5024

Table 5 Validation of results for sandwich beam

		$\overline{\sigma}_x$		$\overline{\tau}_{xz}$	$\overline{w}$
Aspect ratio	Source	$x = L/2$	$x = L/2$	$x = 0$	$x = L/2$
(η)		$z = H/2$	$z = -H/2$	$z = 0$	$z = 0$
1/4	Present work	2.3319	−2.3775	0.3408	11.0675
	Kant et. al. [12]	2.3839	−2.3470	0.3396	11.0600
1/10	Present work	1.4252	−1.4347	0.3506	2.6658
	Kant et. al. [12]	1.4317	−1.4314	0.3504	2.6688
1/20	Present work	1.2893	−1.2918	0.3521	1.4237
	Kant et. al. [12]	1.2910	−1.2910	0.3521	1.4252
1/30	Present work	1.2640	−1.2651	0.3524	1.1925
	Kant et. al. [12]	1.2647	−1.2647	0.3524	1.1936
1/40	Present work	1.2551	−1.2557	0.3525	1.1114
	Kant et. al. [12]	1.2555	−1.2555	0.3525	1.1125
1/50	Present work	1.2510	−1.2514	0.3526	1.0739
	Kant et. al. [12]	1.2512	−1.2512	0.3525	1.0750

3.2 *Results and Discussion for the Present Work*

In the present work, an attempt has been made to check the applicability of the assumptions made in Euler–Bernoulli beam theory, for this the variation of displacements and stresses for various aspect ratios through the depth of the beam are plotted which will also help to give the demarcation between thick and thin beam. Also the properties of face sheets and core are varied in the ratio (β) of 1/1, 1/25, 1/50, 1/75, and 1/100 to study the effect of core softening on the behavior of sandwich beam. Equation (14) represents the non-dimensional form used. The properties of the face sheets used for the present work are same as used by Pagano [14] and mentioned in Table 1 sr. no. 3.

$$\overline{u} = \frac{E_z\sqrt{\eta^5}u(0,z)}{p_0 H}, \overline{w} = \frac{100E_z\eta^3 w(L/2,0)}{p_0 L},$$
$$\overline{\sigma}_x = \frac{\sqrt{\eta^5}\sigma_x(L/2,0)}{p_0}, \overline{\tau}_{xz} = \frac{\eta^2\tau_{xz}(0,z)}{p_0}, \overline{z} = \frac{z}{H}. \quad (14)$$

Table 6 gives results for homogeneous isotropic beam with various aspect ratios, whereas the through thickness variation of displacements and stresses for aspect ratio 1/4, 1/10, 1/20, and 1/50 are given in Tables 7, 8, 9, and 10, respectively. The same results are shown graphically in Fig. 2. Similarly, Tables 11, 12, 13, 14 and 15, and Fig. 3 correspond to homogeneous orthotropic material.

Table 6 Results of homogeneous isotropic beam for various aspect ratios

	$\overline{u}$		$\overline{w}$		$\overline{\sigma}_x$		$\overline{\tau}_{xz}$
Aspect ratio (η)	$x=0$	$x=0$	$x=L/2$	$x=L/2$	$x=L/2$	$x=L/2$	$x=0$
	$z=H/2$	$z=-H/2$	$z=H/2$	$z=-H/2$	$z=H/2$	$z=-H/2$	$z=0$
1/4	−0.3842	0.3942	13.9818	13.7865	0.3112	−0.3096	0.1187
1/10	−0.6110	0.6139	12.5660	12.5610	0.1929	−0.1929	0.0477
1/20	−0.8650	0.8661	12.3802	12.3799	0.1360	−0.1360	0.0239
1/50	−1.3682	1.3685	12.3289	12.3289	0.0860	−0.0860	0.0095

Table 7 Variation of displacements and stresses thorough the depth of homogeneous isotropic beam for $\eta = 1/4$

$\overline{z}$	$\overline{u}(x=0)$	$\overline{w}(x=L/2)$	$\overline{\sigma}_x(x=L/2)$	$\overline{\tau}_{xz}(x=0)$
−0.5	0.3942	13.7865	−0.3096	0
−0.4	0.3101	13.8905	−0.2433	0.0434
−0.3	0.2305	13.9722	−0.1801	0.0766
−0.2	0.1541	14.0344	−0.1190	0.1001
−0.1	0.0798	14.0789	−0.0594	0.1141
0	0.0065	14.1067	−0.0004	0.1187
0.1	−0.0670	14.1182	0.0587	0.1142
0.2	−0.1416	14.1130	0.1186	0.1003
0.3	−0.2186	14.0899	0.1801	0.0768
0.4	−0.2991	14.0471	0.2440	0.0436
0.5	−0.3842	13.9818	0.3112	0

Table 8 Variation of displacements and stresses thorough the depth of homogeneous isotropic beam for $\eta = 1/10$

$\bar{z}$	$\bar{u}$ $(x = 0)$	$\bar{w}$ $(x = L/2)$	$\bar{\sigma}_x$ $(x = L/2)$	$\bar{\tau}_{xz}$ $(x = 0)$
−0.5	0.6139	12.5610	−0.1929	0
−0.4	0.4898	12.5775	−0.1538	0.0172
−0.3	0.3667	12.5903	−0.1151	0.0306
−0.2	0.2445	12.5995	−0.0766	0.0401
−0.1	0.1229	12.6053	−0.0383	0.0458
0	0.0015	12.6075	0.0000	0.0477
0.1	−0.1198	12.6063	0.0383	0.0458
0.2	−0.2415	12.6015	0.0766	0.0401
0.3	−0.3637	12.5933	0.1151	0.0306
0.4	−0.4868	12.5815	0.1539	0.0172
0.5	−0.6110	12.5660	0.1929	0

Table 9 Variation of displacements and stresses thorough the depth of homogeneous isotropic beam for $\eta = 1/20$

$\bar{z}$	$\bar{u}$ $(x = 0)$	$\bar{w}(x = L/2)$	$\bar{\sigma}_x$ $(x = L/2)$	$\bar{\tau}_{xz}(x = 0)$
-0.5	0.8661	12.3799	−0.1360	0
-0.4	0.6924	12.3840	−0.1088	0.0086
-0.3	0.5191	12.3872	−0.0815	0.0153
-0.2	0.3461	12.3895	−0.0543	0.0201
-0.1	0.1733	12.3909	−0.0272	0.0229
0	0.0005	12.3913	0.0000	0.0239
0.1	−0.1722	12.3909	0.0272	0.0229
0.2	−0.3450	12.3896	0.0543	0.0201
0.3	−0.5180	12.3874	0.0815	0.0153
0.4	−0.6913	12.3842	0.1088	0.0086
0.5	−0.8650	12.3802	0.1360	0

From these tables and figures, it can be clearly seen that for higher aspect ratios greater than 1/10 significant warping of section is observed hence plane section remains plane this assumption is not applicable and the transverse displacement w is also not constant which means that thickness of the beam does not remain constant. It is observed that for thicker beam the distribution of the transverse shear stress is not symmetric. This may be because the load is applied on the top surface of the beam which is nonsymmetric. The in-plane normal stress σ_x distribution is also non-linear.

For laminated sandwich beam with orthotropic core, the results are presented for the material property used for validation purpose because in that problem very soft core and stiff face sheets are considered. Table 16 presents summarized results of

Table 10 Variation of displacements and stresses thorough the depth of homogeneous isotropic beam for $\eta = 1/50$

$\bar{z}$	$\bar{u}(x = 0)$	$\bar{w}(x = L/2)$	$\bar{\sigma}_x(x = L/2)$	$\bar{\tau}_{xz}(x = 0)$
−0.5	1.3685	12.3289	−0.0860	0
−0.4	1.0947	12.3296	−0.0688	0.0034
−0.3	0.8210	12.3301	−0.0516	0.0061
−0.2	0.5473	12.3304	−0.0344	0.0080
−0.1	0.2737	12.3307	−0.0172	0.0092
0	0.0001	12.3307	0.0000	0.0095
0.1	−0.2734	12.3307	0.0172	0.0092
0.2	−0.5470	12.3304	0.0344	0.0080
0.3	−0.8207	12.3301	0.0516	0.0061
0.4	−1.0944	12.3296	0.0688	0.0034
0.5	−1.3682	12.3289	0.0860	0

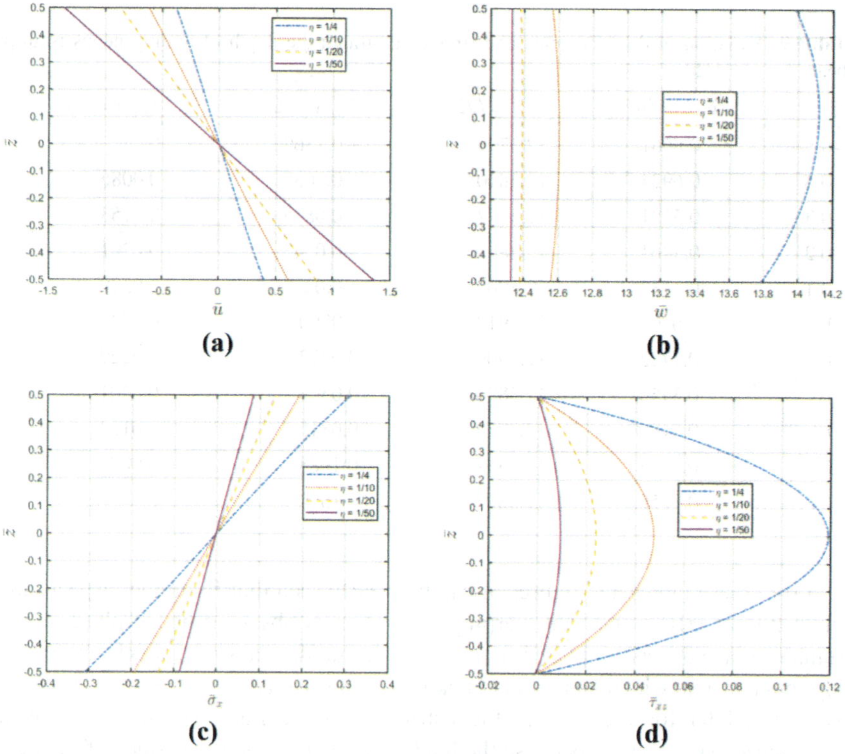

Fig. 2 Variation of normalized (**a**) in-plane displacement $\bar{u}$ (**b**) transverse displacement $\bar{w}$ (**c**) in-plane normal stress $\bar{\sigma}_x$ (**d**) transverse shear stress $\bar{\tau}_{xz}$ through the thickness of a homogeneous isotropic beam

Table 11 Results of homogeneous orthotropic beam for various aspect ratios

	$\overline{u}$		$\overline{w}$		$\overline{\sigma}_x$		$\overline{\tau}_{xz}$
Aspect ratio (η)	$x=0$	$x=0$	$x=L/2$	$x=L/2$	$x=L/2$	$x=L/2$	$x=0$
	z = H/2	$z=-H/2$	$z=H/2$	$z=-H/2$	$z=H/2$	$z=-H/2$	$z=0$
1/4	−0.0226	0.0215	2.0953	1.9015	0.4513	−0.4234	0.1082
1/10	−0.0263	0.0264	0.7351	0.7301	0.2078	−0.2072	0.0468
1/20	−0.0353	0.0353	0.5526	0.5523	0.1387	−0.1387	0.0238
1/50	−0.0549	0.0549	0.5020	0.5019	0.0862	−0.0862	0.0095

Table 12 Variation of displacements and stresses thorough the depth of homogeneous orthotropic beam for $\eta = 1/4$

$\overline{z}$	$\overline{u}(x=0)$	$\overline{w}(x=L/2)$	$\overline{\sigma}_x(x=L/2)$	$\overline{\tau}_{xz}(x=0)$
−0.5	0.0215	1.9015	−0.4234	0
−0.4	0.0124	1.9060	−0.2427	0.0510
−0.3	0.0071	1.9112	−0.1378	0.0802
−0.2	0.0039	1.9193	−0.0755	0.0965
−0.1	0.0020	1.9315	−0.0356	0.1051
0	0.0005	1.9483	−0.0051	0.1082
0.1	−0.0011	1.9701	0.0263	0.1066
0.2	−0.0032	1.9967	0.0690	0.0993
0.3	−0.0066	2.0274	0.1371	0.0836
0.4	−0.0125	2.0610	0.2523	0.0538
0.5	−0.0226	2.0953	0.4513	0

Table 13 Variation of displacements and stresses thorough the depth of homogeneous orthotropic beam for $\eta = 1/10$

$\overline{z}$	$\overline{u}(x=0)$	$\overline{w}(x=L/2)$	$\overline{\sigma}_x(x=L/2)$	$\overline{\tau}_{xz}(x=0)$
−0.5	0.0264	0.7301	−0.2071	0
−0.4	0.0197	0.7307	−0.1547	0.0179
−0.3	0.0140	0.7312	−0.1099	0.0310
−0.2	0.0090	0.7316	−0.0705	0.0399
−0.1	0.0044	0.7321	−0.0345	0.0451
0	0.0001	0.7325	−0.0001	0.0468
0.1	−0.0043	0.7331	0.0342	0.0451
0.2	−0.0089	0.7336	0.0703	0.0400
0.3	−0.0139	0.7342	0.1099	0.0311
0.4	−0.0196	0.7347	0.1550	0.0179
0.5	−0.0263	0.7351	0.2078	0

Table 14 Variation of displacements and stresses thorough the depth of homogeneous orthotropic beam for $\eta = 1/20$

$\bar{z}$	$\bar{u}(x=0)$	$\bar{w}(x=L/2)$	$\bar{\sigma}_x(x=L/2)$	$\bar{\tau}_{xz}(x=0)$
−0.5	0.0353	0.5523	−0.1387	0
−0.4	0.0277	0.5524	−0.1090	0.0087
−0.3	0.0205	0.5525	−0.0806	0.0153
−0.2	0.0135	0.5526	−0.0532	0.0200
−0.1	0.0067	0.5527	−0.0264	0.0228
0	0.0000	0.5527	0.0000	0.0238
0.1	−0.0067	0.5527	0.0264	0.0228
0.2	−0.0135	0.5527	0.0532	0.0200
0.3	−0.0205	0.5527	0.0806	0.0153
0.4	−0.0277	0.5527	0.1090	0.0087
0.5	−0.0353	0.5526	0.1387	0

Table 15 Variation of displacements and stresses thorough the depth of homogeneous orthotropic beam for $\eta = 1/50$

$\bar{z}$	$\bar{u}(x=0)$	$\bar{w}(x=L/2)$	$\bar{\sigma}_x(x=L/2)$	$\bar{\tau}_{xz}(x=0)$
−0.5	0.0549	0.5019	−0.0863	0
−0.4	0.0438	0.5020	−0.0688	0.0034
−0.3	0.0327	0.5020	−0.0515	0.0061
−0.2	0.0218	0.5020	−0.0343	0.0080
−0.1	0.0109	0.5020	−0.0171	0.0092
0	0.0000	0.5020	0.0000	0.0095
0.1	−0.0109	0.5020	0.0171	0.0092
0.2	−0.0218	0.5020	0.0343	0.0080
0.3	−0.0327	0.5020	0.0515	0.0061
0.4	−0.0438	0.5020	0.0688	0.0034
0.5	−0.0549	0.5020	0.0863	0

displacements and stresses for various aspect ratios on the other hand the detailed through thickness variation of these quantities is shown in Tables 17, 18, 19, and 20 for aspect ratio 1/4, 1/10, 1/20, and 1/50, respectively, and the graphical representation is given in Fig. 4. For aspect ratios greater than 1/10, the variation of in plane displacement u is different for face sheets and core but its value is same at the interface, whereas for aspect ratios lower than 1/10 relatively smooth variation is observed. On the other hand for transverse displacement w, no significant change is observed. In this case as the core is very soft, the in-plane normal stress σ_x (bending stress in case of beam) is completely resisted by the face sheets. Besides this a sudden change is observed in σ_x at the interface of face sheets and core. In case of transverse

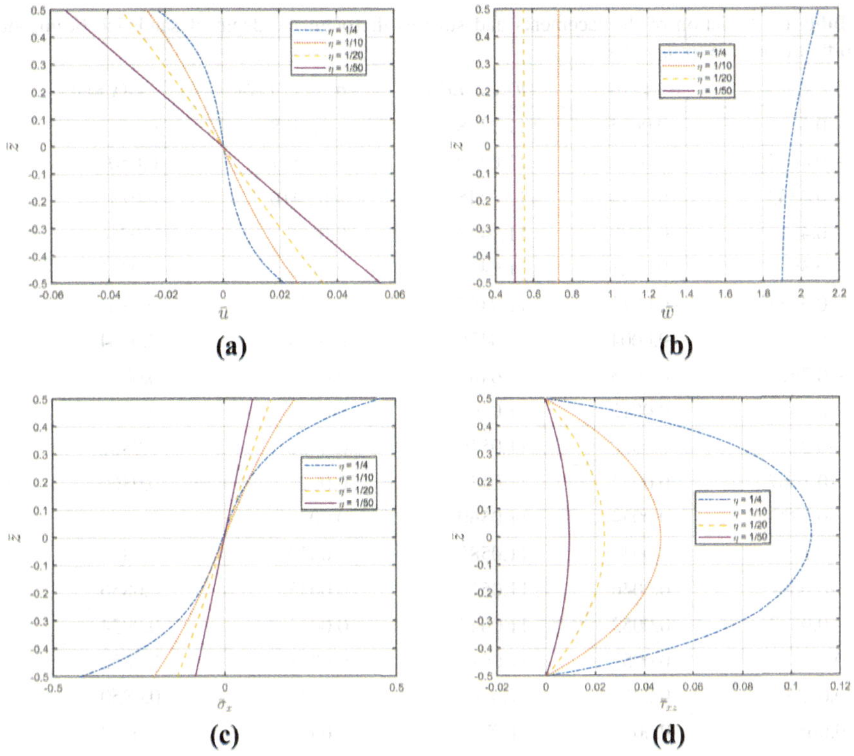

Fig. 3 Variation of normalized (**a**) in-plane displacement $\overline{u}$ (**b**) transverse displacement $\overline{w}$ (**c**) in-plane normal stress $\overline{\sigma}_x$ (**d**) transverse shear stress $\overline{\tau}_{xz}$ through the thickness of a homogeneous orthotropic beam

Table 16 Results of sandwich beam for various aspect ratios

	$\overline{u}$		$\overline{w}$		$\overline{\sigma}_x$		$\overline{\tau}_{xz}$
Aspect ratio (η)	$x = 0$	$x = 0$	$x = L/2$	$x = L/2$	$x = L/2$	$x = L/2$	$x = 0$
	$z = H/2$	$z = -H/2$	$z = H/2$	$z = -H/2$	$z = H/2$	$z = -H/2$	$z = 0$
1/4	−0.0589	0.0605	11.1545	11.0378	1.1660	−1.1888	0.0852
1/10	−0.0572	0.0577	2.6654	2.6626	0.4507	−0.4537	0.0351
1/20	−0.0733	0.0735	1.4231	1.4229	0.2883	−0.2889	0.0176
1/50	−0.1125	0.1126	1.0738	1.0738	0.1769	−0.1770	0.0071

shear stress, a non-symmetric variation is observed for higher aspect ratio and also a significant change is observed in the distribution for face sheets and core but the values are matching at the interface.

Table 17 Variation of displacements and stresses thorough the depth of sandwich beam with orthotropic core for $\eta = 1/4$

$\overline{z}$	$\overline{u}(x=0)$	$\overline{w}(x=L/2)$	$\overline{\sigma}_x(x=L/2)$	$\overline{\tau}_{xz}(x=0)$
−0.5	0.0605	11.0378	−1.1888	0
−0.4667	0.0383	11.0419	−0.7521	0.0507
−0.4333	0.0174	11.0445	−0.3410	0.0792
−0.4	−0.0029	11.0458	0.0585	0.0866
−0.3667	−0.0021	11.0461	0.0009	0.0865
−0.3333	−0.0012	11.0467	0.0012	0.0865
−0.3	−0.0004	11.0476	0.0014	0.0864
−0.2667	0.0005	11.0487	0.0017	0.0863
−0.2333	0.0012	11.0502	0.0020	0.0862
−0.2	0.0020	11.0519	0.0022	0.0861
−0.1667	0.0027	11.0538	0.0025	0.0860
−0.1333	0.0034	11.0560	0.0027	0.0859
−0.1	0.0041	11.0585	0.0030	0.0857
−0.0667	0.0046	11.0612	0.0033	0.0856
−0.0333	0.0052	11.0642	0.0035	0.0854
0	0.0057	11.0675	0.0038	0.0852
0.0333	0.0061	11.0709	0.0041	0.0850
0.0667	0.0064	11.0746	0.0043	0.0848
0.1	0.0067	11.0786	0.0046	0.0845
0.1333	0.0069	11.0827	0.0049	0.0843
0.1667	0.0070	11.0871	0.0051	0.0840
0.2	0.0071	11.0916	0.0054	0.0837
0.2333	0.0070	11.0964	0.0057	0.0835
0.2667	0.0069	11.1014	0.0060	0.0832
0.3	0.0067	11.1065	0.0062	0.0828
0.3333	0.0063	11.1119	0.0065	0.0825
0.3667	0.0059	11.1173	0.0068	0.0822
0.4	0.0053	11.1230	0.0071	0.0818
0.4333	−0.0153	11.1348	0.3079	0.0763
0.4667	−0.0365	11.1454	0.7244	0.0494
0.5	−0.0589	11.1545	1.1660	0

4 Conclusions

Exact elasticity analysis of sandwich beam with orthotropic core is presented using series expansion approach. As this approach is exact analytical approach, it can be used for thick as well as thin beams. In case of series expansion approach, same

Table 18 Variation of displacements and stresses thorough the depth of sandwich beam with orthotropic core for $\eta = 1/10$

$\overline{z}$	$\overline{u}(x=0)$	$\overline{w}(x=L/2)$	$\overline{\sigma}_x(x=L/2)$	$\overline{\tau}_{xz}(x=0)$
−0.5	0.0577	2.6626	−0.4537	0.0000
−0.4667	0.0490	2.6630	−0.3855	0.0139
−0.4333	0.0406	2.6634	−0.3195	0.0256
−0.4	0.0325	2.6637	−0.2552	0.0351
−0.3667	0.0298	2.6640	−0.0003	0.0351
−0.3333	0.0271	2.6642	−0.0002	0.0351
−0.3	0.0245	2.6645	−0.0002	0.0351
−0.2667	0.0218	2.6647	−0.0001	0.0351
−0.2333	0.0191	2.6649	0.0000	0.0351
−0.2	0.0165	2.6651	0.0000	0.0351
−0.1667	0.0138	2.6652	0.0001	0.0351
−0.1333	0.0111	2.6654	0.0001	0.0351
−0.1	0.0085	2.6655	0.0002	0.0351
−0.0667	0.0058	2.6656	0.0003	0.0351
−0.0333	0.0031	2.6657	0.0003	0.0351
0	0.0004	2.6658	0.0004	0.0351
0.0333	−0.0022	2.6659	0.0005	0.0350
0.0667	−0.0049	2.6659	0.0005	0.0350
0.1	−0.0076	2.6660	0.0006	0.0350
0.1333	−0.0103	2.6660	0.0006	0.0350
0.1667	−0.0130	2.6660	0.0007	0.0350
0.2	−0.0157	2.6660	0.0008	0.0349
0.2333	−0.0184	2.6659	0.0008	0.0349
0.2667	−0.0211	2.6659	0.0009	0.0349
0.3	−0.0238	2.6658	0.0010	0.0349
0.3333	−0.0265	2.6657	0.0010	0.0348
0.3667	−0.0292	2.6656	0.0011	0.0348
0.4	−0.0320	2.6655	0.0011	0.0348
0.4333	−0.0401	2.6655	0.3163	0.0254
0.4667	−0.0486	2.6655	0.3825	0.0138
0.5	−0.0572	2.6654	0.4507	0.0000

formulation can be adopted for differential equations with constant coefficients as well as variable coefficients. The through thickness variation of displacements as well as stresses presented here can work as benchmark for other beam theories and finite element solution. Typical design guideline of aspect ratio being less than 1/20 can be considered appropriate for thin beam designs as shown by the present analysis.

Table 19 Variation of displacements and stresses thorough the depth of sandwich beam with orthotropic core for $\eta = 1/20$

$\overline{z}$	$\overline{u}(x = 0)$	$\overline{w}(x = L/2)$	$\overline{\sigma}_x(x = L/2)$	$\overline{\tau}_{xz}(x = 0)$
−0.5	0.0735	1.4229	−0.2889	0.0000
−0.4667	0.0669	1.4230	−0.2628	0.0065
−0.4333	0.0604	1.4231	−0.2372	0.0123
−0.4	0.0539	1.4232	−0.2119	0.0176
−0.3667	0.0494	1.4233	−0.0003	0.0176
−0.3333	0.0449	1.4233	−0.0003	0.0176
−0.3	0.0405	1.4234	−0.0002	0.0176
−0.2667	0.0360	1.4235	−0.0002	0.0176
−0.2333	0.0315	1.4235	−0.0002	0.0176
−0.2	0.0270	1.4236	−0.0001	0.0176
−0.1667	0.0225	1.4236	−0.0001	0.0176
−0.1333	0.0180	1.4236	−0.0001	0.0176
−0.1	0.0136	1.4237	0.0000	0.0176
−0.0667	0.0091	1.4237	0.0000	0.0176
−0.0333	0.0046	1.4237	0.0000	0.0176
0	0.0001	1.4237	0.0001	0.0176
0.0333	−0.0044	1.4237	0.0001	0.0176
0.0667	−0.0089	1.4237	0.0001	0.0176
0.1	−0.0133	1.4237	0.0002	0.0176
0.1333	−0.0178	1.4237	0.0002	0.0176
0.1667	−0.0223	1.4237	0.0002	0.0176
0.2	−0.0268	1.4236	0.0003	0.0176
0.2333	−0.0313	1.4236	0.0003	0.0176
0.2667	−0.0358	1.4235	0.0003	0.0176
0.3	−0.0403	1.4235	0.0004	0.0176
0.3333	−0.0448	1.4234	0.0004	0.0176
0.3667	−0.0492	1.4234	0.0004	0.0175
0.4	−0.0537	1.4233	0.0005	0.0175
0.4333	−0.0602	1.4232	0.2366	0.0123
0.4667	−0.0667	1.4232	0.2623	0.0064
0.5	−0.0733	1.4231	0.2883	0.0000

Table 20 Variation of displacements and stresses thorough the depth of sandwich beam with orthotropic core for $\eta = 1/50$

$\overline{z}$	$\overline{u}(x = 0)$	$\overline{w}(x = L/2)$	$\overline{\sigma}_x(x = L/2)$	$\overline{\tau}_{xz}(x = 0)$
−0.5	0.1126	1.0738	−0.1770	0.0000
−0.4667	0.1046	1.0738	−0.1645	0.0025
−0.4333	0.0967	1.0738	−0.1520	0.0049
−0.4	0.0888	1.0738	−0.1396	0.0070
−0.3667	0.0814	1.0739	−0.0002	0.0070
−0.3333	0.0740	1.0739	−0.0002	0.0070
−0.3	0.0666	1.0739	−0.0002	0.0070
−0.2667	0.0592	1.0739	−0.0001	0.0070
−0.2333	0.0518	1.0739	−0.0001	0.0070
−0.2	0.0444	1.0739	−0.0001	0.0070
−0.1667	0.0370	1.0739	−0.0001	0.0070
−0.1333	0.0296	1.0739	−0.0001	0.0070
−0.1	0.0222	1.0739	−0.0001	0.0071
−0.0667	0.0148	1.0739	0.0000	0.0071
−0.0333	0.0074	1.0739	0.0000	0.0071
0	0.0000	1.0739	0.0000	0.0071
0.0333	−0.0074	1.0739	0.0000	0.0071
0.0667	−0.0148	1.0739	0.0000	0.0071
0.1	−0.0222	1.0739	0.0001	0.0070
0.1333	−0.0296	1.0739	0.0001	0.0070
0.1667	−0.0370	1.0739	0.0001	0.0070
0.2	−0.0444	1.0739	0.0001	0.0070
0.2333	−0.0518	1.0739	0.0001	0.0070
0.2667	−0.0592	1.0739	0.0002	0.0070
0.3	−0.0666	1.0739	0.0002	0.0070
0.3333	−0.0740	1.0739	0.0002	0.0070
0.3667	−0.0814	1.0739	0.0002	0.0070
0.4	−0.0888	1.0738	0.0002	0.0070
0.4333	−0.0967	1.0738	0.1520	0.0049
0.4667	−0.1046	1.0738	0.1644	0.0025
0.5	−0.1125	1.0738	0.1769	0.0000

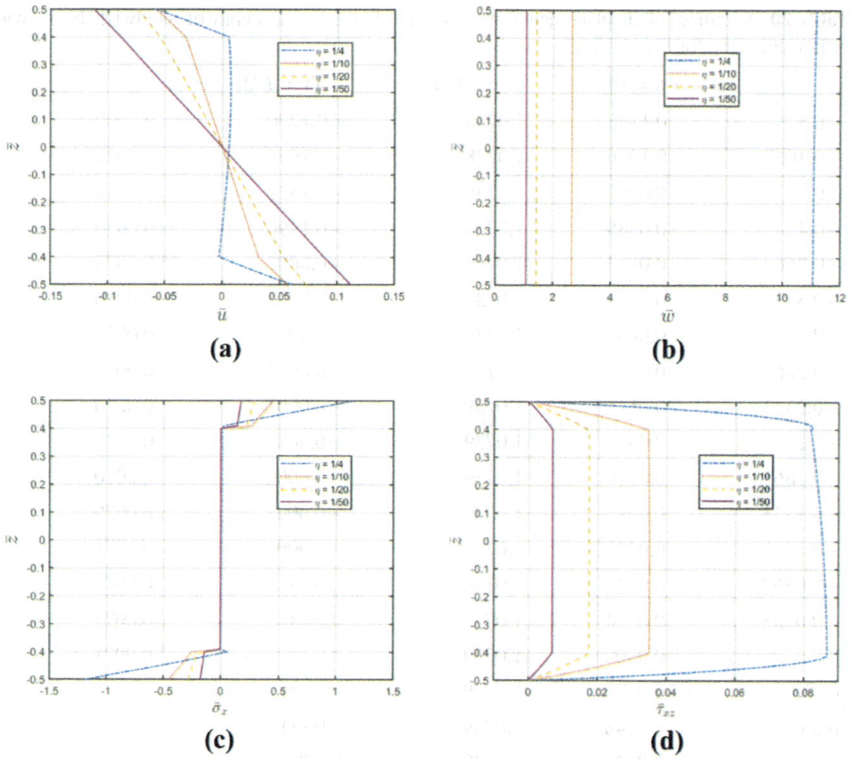

Fig. 4 Variation of normalized (**a**) in-plane displacement $\overline{u}$ (**b**) transverse displacement $\overline{w}$ (**c**) in-plane normal stress $\overline{\sigma}_x$ (**d**) transverse shear stress $\overline{\tau}_{xz}$ through the thickness of a sandwich beam

References

1. Saint Venant A and Barre de (1856) Memoire sur la felxion des prismes. J De Math Pures et appl 2:89–189
2. Silverman IK (1964) Orthotropic beams under polynomial loads. ASCE J Eng Mech 90:293–319
3. Hashin Z (1967) Plane an isotropic beam. ASME J Appl Mech 34:257–262
4. Gerstner RW (1968) Stresses in composite cantilever. J Compos Mater 2:498–501
5. Rao KM, Ghosh BG (1979) Exact analysis of unsymmetric laminated beam. ASCE J structural 105:2313–2325
6. Cheng S, Wei X, Jiang T (1989) Stress distribution and deformation of adhesive- bonded laminated composite beams. ASCE J Eng Mech 115:1150–1162
7. Holt PJ, Weber JBH (1982) Exact solution to some honeycomb sandwich beam, plate and shell problem. J Strain Anal Eng 17:1–8
8. Pagano NJ (1969) Exact solution for composite laminates in cylindrical bending. J Compos Mater 3:398–411
9. Srinivas S, Rao G, Rao AK (1970) An exact analysis for vibration of simply-supported homogeneous and laminated thick rectangular plates. J Sound Vib 12:187–199
10. Pagano NJ (1970) Influence of shear coupling in cylindrical bending of anisotropic laminates. J Compos Mater 4:330–343

11. Sankar BV (2001) An elasticity solution for functionally graded beams. Compos Sci Technol 61:689–696
12. Venkataraman S, Sankar BV (2003) Elasticity solution for stresses in sandwich beam with functionally graded core. AIAA J 41:2501–2505
13. Kant T, Pendhari SS, Desai YM (2007) On accurate stress analysis of composite and narrow sandwich beams. Int J Comput Methods Eng Sci Mech 8:165–177
14. Pagano NJ (1970) Exact solution for rectangular bidirectional composites and sandwich plates. J Compos Mater 4:20–34

Flexural Fatigue Analysis of Cross Ply and Angle Ply Laminates

Sammed Patil and Y. T. LomtePatil

Abstract Composite materials are believed to be materials of future with potential application in high performance structure. Composites being light weight and having higher strength to weight ratio proves very efficient and economical. The static strength of laminated composite materials can be calculated easily but it's analysis under cyclic loading is a quite complicated phenomenon. Continuous degradation of material occurs under cyclic loading which causes redistribution of the stresses within the laminate hence failure of single lamina does not indicate failure of entire laminate which in turn increases the complexity of the analysis. Also laminated composite materials being nonhomogeneous and anisotropic its fatigue analysis involves a lot of calculations; this can be reduced with the help of finite element-based software. Plenty of work has been carried out for fatigue analysis of laminated composite materials subjected to axial loading but the flexural fatigue analysis needs some attention. Here, the attempt has been made to perform the flexural fatigue analysis of laminated composite material using finite element-based software ANSYS. The results are presented for cross ply and angle ply laminates with two material combinations, namely, E-Glass epoxy and Carbon epoxy for different aspect ratios.

Keywords Laminated composite · Flexural fatigue analysis · ANSYS · E-glass epoxy

1 Introduction

Flexural fatigue analysis of a cantilever laminated composite beam (formed by stacking different plies together) is subjected to a point load at the free end is carried out. for this purpose, unidirectional, cross ply, and angle ply laminates with different span to depth ratios and stacking sequences are considered. The S–N curves from the literatures are given as an input data while performing fatigue analysis. Constant amplitude fully reversible cyclic loading is applied and stress life-based approach

S. Patil · Y. T. LomtePatil (✉)
Department of Civil Engineering, College of Engineering Pune, Pune, 411005, India
e-mail: lomteyt@gmail.com

M. S. Ranadive et al. (eds.), *Recent Trends in Construction Technology and Management*, Lecture Notes in Civil Engineering 260,
https://doi.org/10.1007/978-981-19-2145-2_86

is used to predict the fatigue life of laminated composite material. These results are obtained for material, namely, E-Glass epoxy. Results are tabulated for each stacking sequence by defining maximum stress in beam and corresponding fatigue life and fatigue life contour plots are shown. Graphs are plotted by varying parameters to understand the behavior under the fatigue for different cases.

Laminated composite materials are believed to be materials of future with potential application in high-performance structures. Laminated composite materials being lightweight and having high directional stiffness as well as high strength to weight ratio are gaining more attention. Fatigue loads for many composite structures are unavoidable so recent designs of composite materials without fatigue analysis are not possible. The heterogeneous and anisotropic nature of laminated composites leads to the formation of different stress levels within the material, which may cause damage modes such as matrix cracking, fiber breakage, delamination, debonding, and ply failure. Laminated composite material under fatigue loading is subjected to continuous degradation of material properties which are responsible for redistribution of the stresses within the laminate so failure of single lamina does not mean failure of entire laminate. This involves a lot of calculation work to make the calculation simple finite element-based softwares helpful. Under fatigue action failure of the material occurs way before the load reaching its ultimate value. So, the fatigue strength of material should be taken into consideration while designing any member subjected to cyclic loading.

The work done by various researchers in the field of fatigue analysis of laminated composite materials is discussed here [1]. The damage mechanism of laminated composite materials has been thoroughly explained by taking different sets of examples and entire idea of fatigue failure is explained [2]. Here, authors have studied flexural fatigue behavior of glass and Kevlar fiber using cross ply laminate. They have also considered the effect of reduction in stiffness for 5% and 10% reduction and studied the effect of the stacking sequence and the reinforcement type on the behavior of cross ply laminates in cyclic loading [3]. In this paper, 3D fatigue progressive damage models are developed for damage accumulation and life of CFRP laminate is studied. Fatigue failure analysis was performed by using a set of Hasin's failure criteria and Ye delamination criteria [4]. In this paper, author prepared FEA model to evaluate damage in composite laminate and predicted fatigue life of laminates with different layup sequence based on fatigue characteristics of longitudinal, transverse, and in plane shear direction [5]. Here authors discussed fatigue life estimation for multidirectional laminates and proposed mathematical models for the improvement of fatigue life of multidirectional laminates. But it is important to note that those models have not considered progressive damage and stiffness degradation occurring throughout the life of the laminate [6]. In this work, the axial fatigue analysis has been done by considering reduction in stiffness after every cycle of loading, and Tsai-Hill failure criteria are used to predict fatigue life of cross and angle ply laminates [7]. This paper discusses the prediction of fatigue response of composites using an empirical strength and stiffness degradation scheme coupled to a cumulative damage accumulation approach. The fatigue analysis was performed using ABAQUS™ finite element software using a user-defined material subroutine UMAT developed for the material response.

From the literature, it can be seen that most of the work is done on axial fatigue analysis, whereas flexural fatigue analysis needs more attention. Hence in the present paper flexural fatigue analysis of laminated composite material is performed using finite element-based software ANSYS. The cantilever beam of 1 m span and different stacking sequences, subjected to a point load at the free end is considered. The results are presented for the aspect ratios ranging from 5 to 25. A comprehensive comparison of results for unidirectional, cross ply and angle ply laminates with E-Glass epoxy and carbon epoxy material combination is done.

2 Theoretical Formulation

In the present paper, flexural fatigue analysis of the laminated composite member having different stacking sequences and aspect ratio (span to depth ratio) is presented. One end of the flexural member is assumed to be fixed and other is completely free. A point load is applied at the free end of the member. The span and width of the member is held constant, whereas its depth is varied to get the aspect ratios ranging from 5 to 25. The details are given in Fig. 1. The flexural member is assumed to be made up of two materials, namely, E-glass epoxy and carbon epoxy with various laminate arrangements like unidirectional laminate, symmetric cross ply laminate, and angle ply laminate. The details are mentioned in Table 1. A constant amplitude fully reversible cyclic loading is applied to know the flexural fatigue life of the member (as shown in Fig. 2).

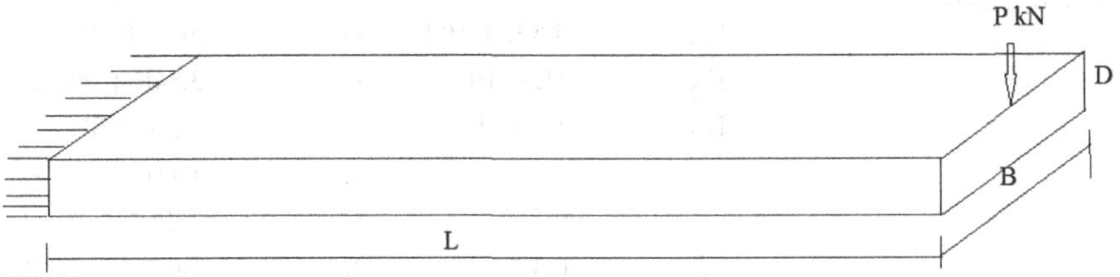

Fig. 1 Flexural member with point load at free end

Table 1 Details of stacking sequence and aspect ratios

Sr. No.	Stacking sequence	Aspect ratio	Actual depth (mm)
1	$[0]_S$	5	200
2	$[90]_S$	10	100
3	$[0/90/90/0]_S$	15	66
4	$[+45/-45]_S$	20	50
5	[0/30/60/90]	25	40

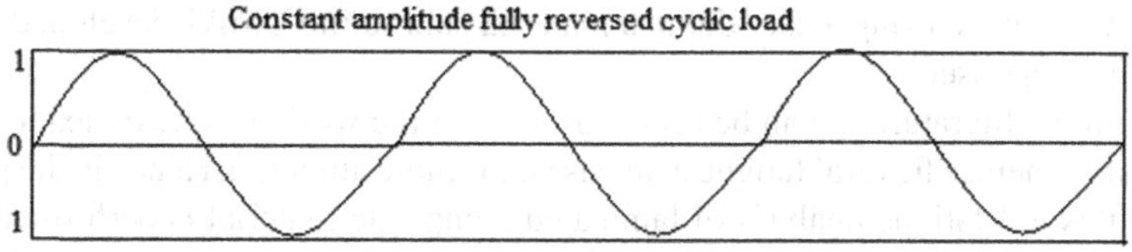

Fig. 2 Type of cyclic loading applied

Table 2 Material properties of E-Glass epoxy

Parameter	Property	Parameter	property
E_{xx}	1.21E + 11 Pa	G_{xy}	4.7E + 09 Pa
E_{yy}	8.6E + 9 Pa	G_{yz}	3.1E + 09 Pa
E_{zz}	8.6E + 9 Pa	G_{xz}	4.7E + 09 Pa
υ_{xy}	0.27	X_T	2.31E + 09 Pa
υ_{yz}	0.4	Y_T/Z_T	2.9E + 07
υ_{xz}	0.27	X_C	−1.08E + 09
		Y_C/Z_C	−1E + 08

Here, the S–N curve-based approach is used to find out flexural fatigue life of laminated composite material under out of plane loading. The S–N curves for laminated composite materials are given as input in the ANSYS software, and flexural fatigue life is determined. The material properties for E-glass epoxy and carbon epoxy are given in Tables 2 and 3, respectively.

Table 3 Material properties of carbon epoxy

Parameter	Property	Parameter	Property
E_{xx}	4.5 E + 10 Pa	G_{xy}	5E + 09 Pa
E_{yy}	1E + 10 Pa	G_{yz}	3.84E + 09 Pa
E_{zz}	1E + 10 Pa	G_{xz}	5E + 09 Pa
υ_{xy}	0.3	X_T	1.11E + 09 Pa
υ_{yz}	0.4	Y_T, Z_T	35E + 07 Pa
υ_{xz}	0.3	X_C	−6.75E + 08 Pa
		Y_C, Z_C	−1.2E + 08 Pa

3 Results

3.1 Validation of Results

In the reference [8], fatigue analysis of E-glass epoxy material is carried out experimentally and the results are presented for different stacking sequences such as [90] 4 s, [0/90/90/0]s, [45/0/0/−45]s, [45/90/−45/0], [0/45/−45/90], and [0/45/90/−45]s. The material properties used by the author are mentioned in Table 2. For this purpose, an axially loaded member having dimensions 500 × 127 × 8 mm and subjected to a tensile load of 70 kN (Fig. 3) is considered. The same problem is analyzed for the validation purpose using ANSYS software and for modeling 8 noded 3D solid 46 element with 3D.O.F is used. Table 4 shows the comparison of results from which it can be seen that the results obtained by ANSYS are in good agreement with the literature.

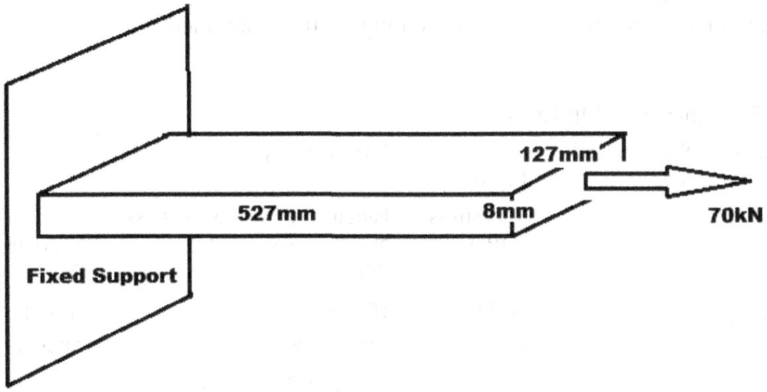

Fig. 3 Fixed at one end and axially loaded

Table 4 Validation of results with results

Sr. No.	Stacking sequence	No. of cycles (N) By reference [8]	ANSYS results (N)	Percentage error
1	[90]4 s	1.2944E + 6	1.1949E + 6	−7.68
2	[0/90/90/0]s	1.936E + 6	1.998E + 6	3.2
3	[45/0/0/−45]s	7.6674E + 5	7.8449E + 5	2.41
4	[45/90/−45/0]	9.4883E + 5	9.1883E + 5	−3.161
5	[0/45/−45/90]	2.1943E + 6	2.1089E + 6	−3.71
6	[0/45/90/−45]s	1.275E + 6	1.2068E + 6	−5.34

3.2 Present Work Results and Discussion

In the present work, the results are presented for E-glass epoxy and carbon epoxy material with the aspect ratio ranging from 5 to 25 with an increment of 5. The results are given in the form of maximum stress developed in the laminate and fatigue life in terms of number of cycles. Table 5 presents the results for unidirectional laminate where all the fibers are oriented along longitudinal direction $[0]_s$, whereas Table 6 shows the results for unidirectional laminate with fibers oriented along transverse direction $[90]_s$.

Table 7 gives the results for cross ply laminate $[0/90]_s$. Tables 8 and 9 present the results for angle ply laminates with symmetric $[45/-45]_s$ and asymmetric [0/30/60/90] stacking sequences, respectively.

From the results obtained it can be seen that the fatigue life for the laminate where fibers are oriented along longitudinal direction is maximum compared to all other stacking sequences, as the fibers are oriented along the longitudinal direction they help in resisting flexural action in a better manner. Contrarily for laminate with fibers oriented in a transverse direction least fatigue life is obtained.

Table 5 Fatigue life of [0]s laminate

Stacking sequence	Span to depth ratio	E-glass Epoxy	Carbon epoxy		
		Max stress (MPa)	Fatigue life in No. of cycles (N)	Max stress (MPa)	Fatigue life in No. of cycles (N)
$[0]_{200s}$	5	30.54	1E + 10	48.79	1E + 10
$[0]_{100s}$	10	46.04	4.91E + 09	74.24	7.09E + 09
$[0]_{66s}$	15	76.11	6.887E + 7	100.72	5.71E + 8
$[0]_{50s}$	20	130.67	8.57E + 5	166.71	3.84E + 5
$[0]_{40s}$	25	205.58	43,336	275.02	47,463

Table 6 Fatigue life of [90]s laminate

Stacking sequence	Span to depth ratio	E-glass Epoxy		Carbon epoxy	
		Max stress (MPa)	Fatigue life in No. of cycles (N)	Max stress (MPa)	Fatigue life in No. of cycles (N)
$[90]_{200s}$	5	38.82	1E + 10	56.79	1E + 10
$[90]_{100s}$	10	56.51	8.64E + 8	87.42	5.94E + 8
$[90]_{66s}$	15	86.04	1.495E + 6	115.67	1.675E + 7
$[90]_{50s}$	20	142.88	2.39E + 5	183.44	9.22E + 4
$[90]_{40s}$	25	218.63	30,099	296.27	30,274

Table 7 Fatigue life of [0/90]s

Stacking sequence	Span to depth ratio	E-glass Epoxy		Carbon epoxy	
		Max stress (MPa)	Fatigue life in No. of cycles (N)	Max stress (MPa)	Fatigue life in No. of cycles (N)
$[0/90]_{100s}$	5	36.73	1E + 10	53.44	1E + 10
$[0/90]_{50s}$	10	52.19	1.221E + 9	83.58	9.06E + 8
$[0/90]_{33s}$	15	83.47	6.96E + 6	111.94	3.19E + 7
$[0/90]_{25s}$	20	140.51	3.9737E + 5	179.09	1.33E + 5
$[0/90]_{20s}$	25	215.72	35,889	291.78	34,704

Table 8 Fatigue life of [+45/−45]s

Stacking sequence	Span to depth ratio	E-glass Epoxy		Carbon epoxy	
		Max stress (MPa)	Fatigue life in No. of cycles (N)	Max stress (MPa)	Fatigue life in No. of cycles (N)
$[45/-45]_{100s}$	5	32.67	1E + 10	52.79	1E + 10
$[45/-45]_{50s}$	10	49.93	2.617E + 9	79.21	1.05E + 9
$[45/-45]_{33s}$	15	80.37	1.4774E + 7	107.38	6.65E + 7
$[45/-45]_{25s}$	20	136.21	5.778E + 5	174.23	4.18E + 5
$[45/-45]_{20s}$	25	210.96	41,230	286.93	39,044

Table 9 Fatigue life of [0/30/60/90]s

Stacking sequence	Span to depth ratio	E-glass Epoxy		Carbon epoxy	
		Max stress (MPa)	Fatigue life in No. of cycles (N)	Max stress (MPa)	Fatigue life in No. of cycles (N)
[0/30/60/90]	5	31.82	1E + 10	53.44	1E + 10
	10	47.24	4.04E + 9	76.27	4.55E + 9
	15	78.18	2.6732E + 7	103.37	9.67E + 7
	20	133.23	7.389E + 5	170.24	8.47E + 5
	25	207.37	42,085	280.96	43,296

The cross ply laminate will have less fatigue life than that of angle ply laminate as it contains alternate ply's oriented along transverse direction that makes the cross ply laminate susceptible to the fiber matrix debonding. Among the angle ply laminate, the laminate with gradual introduction of angle leads to better fatigue resistance compared to laminate where sudden change in angle is provided. In all stacking sequences, it is observed that as the span to depth ratio increases fatigue life of the flexural member decreases which is quite obvious.

In Figs. 4 and 5, numbers 1, 2, 3, 4, 5 indicate stacking sequences [0]s, [90]s, [0/90]s, [45/−45]s, [0/30/60/90], respectively.

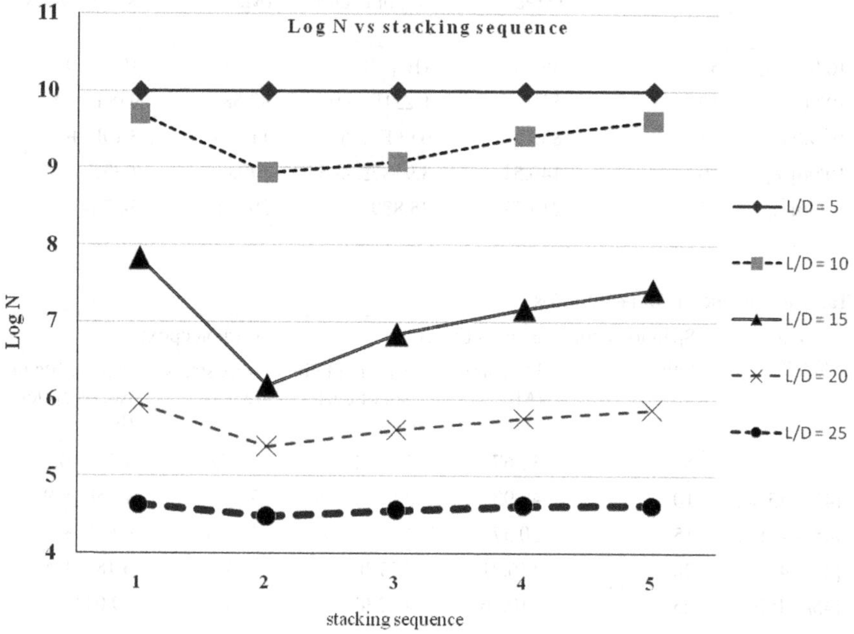

Fig. 4 Log N versus stacking sequence for different Span to depth ratio for E-glass epoxy material

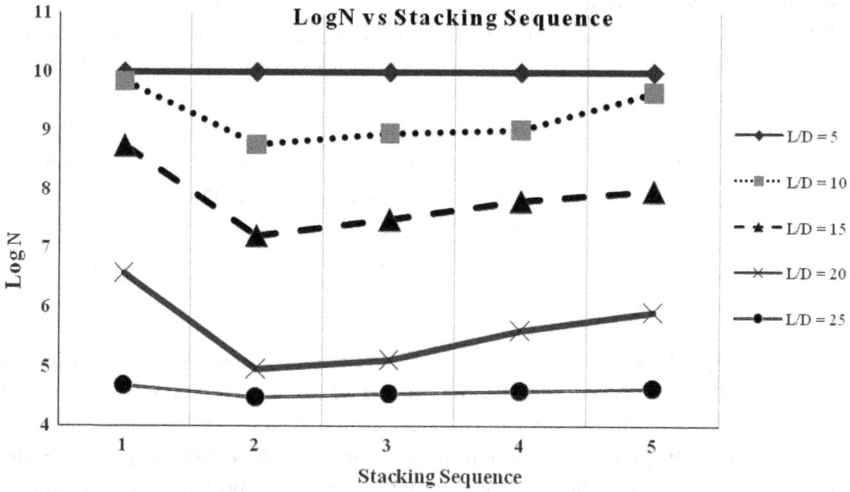

Fig. 5 Log N versus stacking sequence for different Span to depth ratio for carbon epoxy material

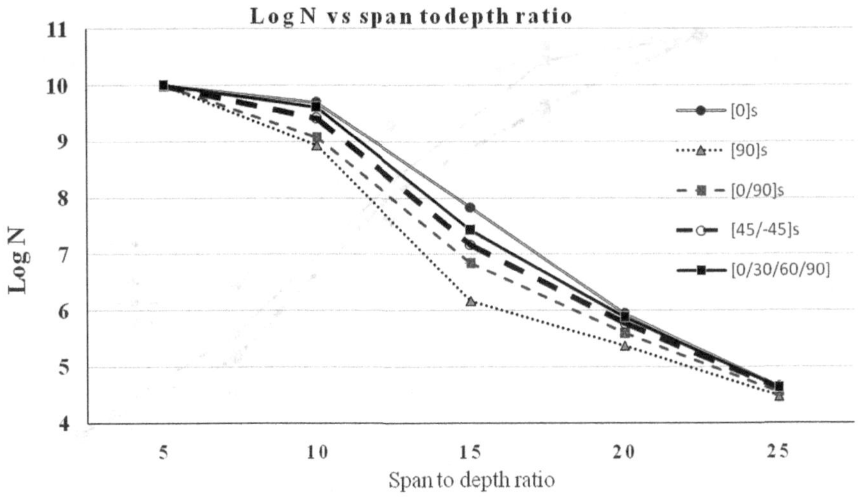

Fig. 6 Log N versus Span to depth ratio for different stacking sequence for E-glass epoxy material

Above results are also expressed in the graphical form for better understanding. Figure 4 shows the variation of fatigue life with different stacking sequences for various aspect ratios for E-glass epoxy material, whereas Fig. 5 gives these results for carbon epoxy material.

From the graph, it can be clearly seen that for aspect ratio 25 there is no significant difference between the fatigue life for all stacking sequences. From this, it can be concluded that for thin flexural members fatigue life is not affected by stacking sequences. Figures 6 and 7 show the variation of fatigue life with different span to depth ratio.

4 Conclusion

Flexural fatigue life is obtained for 5 different stacking sequence and 5 different span to depth ratios for two materials, namely, E-glass epoxy and carbon epoxy. The Laminated composite beam with point load at free end is modeled in ANSYS software and results for fatigue life are obtained. The following conclusions are drawn from the above study.

1. Unidirectional laminate with fiber orientation along longitudinal direction has the highest fatigue life among all stacking sequences. On the other hand, unidirectional laminate with fiber orientation in the transverse direction has the lowest fatigue life. So when the fibers are oriented along the longitudinal direction which is also the direction along which fiber bends, they help to resist the flexural action in a better manner.

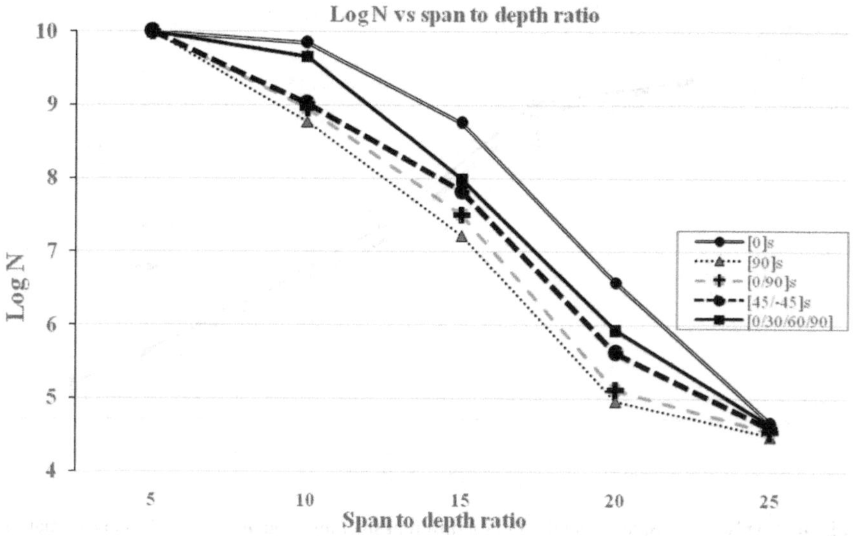

Fig. 7 Log N versus Span to depth ratio for different stacking sequence for carbon epoxy material

2. Among the angle ply and cross ply laminate the [0/30/60/90] configuration will have better fatigue strength than that of [45/−45]s and [0/90]s laminate because of the gradual introduction of angle.
3. In all stacking sequences, it is observed that as the span to depth ratio increases fatigue life of the flexural member decreases which is quite obvious.

References

1. Talreja R (1981) Fatigue of composite material: damage mechanism and fatigue life diagram. Pro Royal Soc London A 378:461–47
2. Bezazi AR, El Mahi A, Berthelot JM, Bezzazic B (2003) Flexural fatigue behavior of cross ply laminate an experimental approach. Strength Mater 35(2)
3. Toumi RB, Renard J et al (2013) Fatigue damage modelling of continuous E-glass fibre/epoxy composite. Procedia Eng 66:723—736
4. Lian W, Yao W (2010) Fatigue life prediction of composite laminates by FEA simulation method. Int J Fatigue 32:123–133
5. Arda Deveci H, Secil Artem H (2018) On the estimation and optimization capabilities of the fatigue life prediction models in composite laminates. J Reinf Plast Compos 0(0):1–18
6. Alireza AA, Sayid A (2011) Experimental and finite element analysis approach for fatigue of unidirectional fibrous composites. Mech Mater 87:106–112

7. Vasantha Kumar K, Ram Reddy P (2014) Flexural fatigue analysis of carbon/Epoxy angle Ply laminates. Int J Eng Res Dev 10(7):52–62
8. Plumtree H, Cheng GX (1999) Fatigue parameter for off-axis unidirectional fiber reinforced composites. Int J Fatigue 21:844–856

Recent Trends in Transportation and Traffic Engineering

Review on Mechanisms of Bitumen Modification: Process and Variables

N. T. Bhagat and M. S. Ranadive

Abstract The use of bitumen in road construction has been done since over a century, and the modification of bitumen has started in the 1950s. Bitumen plays a very substantial role in various aspects of road construction quality such as adhesion, friction, durability, ageing resistance and strength of pavement. It has been the subject of research to modify the bitumen for increasing these benefits and decreasing the damages in pavements like rutting, loss of strength, cracking, etc. Many modifiers like polymers, plastics, nano materials, organic wastes, oils and other waste materials have been used to modify bitumen over the years. However, the mechanisms of bitumen modification include certain processing conditions and variables like mixing temperature, mixing time, shear rate, etc. that affects the quality and performance of resulting modified bitumen. This paper presents an overview of the processes used and the processing conditions that are adopted for bitumen modification. It also reviews the effects on the performance of the resulting modified bitumen caused due to variations in the processing conditions and factors.

Keywords Bitumen · Modification process · Modifiers · Processing conditions

1 Introduction

Bitumen is the oldest engineering material know to be used as sealant, adhesive, preservative, waterproofing agent, and paving material for road construction. Earlier natural bitumen was used directly as obtained from the Earth's surface. Later on, the use of refined bitumen started in early 1900s in USA that was obtained from the refinement of crude oil [25]. After that, world utilization of bitumen increased rapidly over the years, especially as a binder in pavement construction. Bitumen binders have

N. T. Bhagat (✉) · M. S. Ranadive
Department of Civil Engineering, College of Engineering Pune, Pune 411005, India
e-mail: bhagat.nikita14@gmail.com

M. S. Ranadive
e-mail: msr.civil@coep.ac.in

M. S. Ranadive et al. (eds.), *Recent Trends in Construction Technology and Management*, Lecture Notes in Civil Engineering 260,
https://doi.org/10.1007/978-981-19-2145-2_87

a varying range of thermal and mechanical demands for road construction. Defects in pavements like rutting due to high temperatures, fatigue, and propagation of cracks due to low temperatures are the results of the thermal susceptibility of bitumen in addition to repeated traffic loading [11]. In order to reduce these defects and increase the performance characteristics of pavement such as strength, durability, moisture resistance and resistance to rutting and fatigue cracking, the enhancement of binder and bituminous mixtures has become a necessity in this era of changing weather conditions and increased traffic on roads. Also, addition of modifiers in the form of municipal solid wastes, plastic wastes, organic wastes, fibres and rubber wastes in bituminous binders and mixtures helps to reduce the solid waste pollution from the environment and proves to be an effective way of waste management [12]. Ranadive et al. [18] proposed the use of fibres extracted from refrigerator door panels and waste plastic in the stone mastic asphalt and asphaltic concrete mixtures. They found the Marshall stability and indirect tensile strength of the mixtures increased by addition of fibres from the refrigerator door panels. Similarly, lot of research is being done to improve the properties of bitumen by modifying it with various modifying agents such as polymers, recycled oils, organic waste, nanomaterials, etc. to improve the moisture damage resistance, adhesion, rutting resistance, fatigue resistance and strength of binders. Kulkarni and Ranadive [11], modified VG10 cutback bitumen using pyrolytic oil obtained from low density polyethylene (LDPE) to improve the shear strength of the conventional cutback.

The pristine characteristics of bitumen depend majorly on the production and processing mechanisms as well as the crude oil characteristics [2, 17]. Similarly, the improved performance characteristics of the modified bitumen is dependent on the compatibility between modifier and base bitumen, which majorly depends on the circumstances of their polarity, particle size, morphology, surface energy, interfacial stability, etc. Nowadays, the factors such as procedure of modification, preparation temperature, shear rates of mixing, etc. are found to be responsible for the improved or enhanced characteristics of the modified bitumen as well [3]. This paper focuses on the review of the bitumen modification process and factors of modification procedures that are responsible for the behaviour of the resulting modified bitumen.

2 Process

The susceptibility to temperature changes is a virtue of bitumen on which its performance depends, and the performance characteristics of bitumen are affected by the rate of the temperature changes and its range. Indian road applications need a range of 0–50 °C for material of pavement, whereas a 60–150 °C range of temperature is necessary for mixing, compaction, pumping and other intermediate operations. Hence, the characteristics of bitumen that help to withstand the deformation occurring because of the thermal shocks during the process of applying and operations, determine its performance. Bitumen can be qualified as high-performance grade bitumen depending on the rheological behaviour with respect to temperature changes

[21]. Chemical composition of bitumen consists an intricate linking of majorly polar and non-polar compounds. The feasibility of producing high-performance modified bituminous binder, having correctly balance chemical composition and physical characteristics, depends on the proper selection of crude source and processing conditions.

For stable formulation and obtaining desirable mechanical properties of the bitumen, the chemical compatibility and processing conditions like time, shear and temperature are necessary. The binder composition and mechanical performance of the modified bitumen may alter due to effect of processing variables and in turn may sometimes lead to undesirable in service properties [5]. It is not new to blend, mix or alloy two or more materials to get a new material product with different and enhanced physical and mechanical properties that those of the constituent materials. According to McNally [14], the heat and entropy of mixing are the major driving factors for homogeneity of resulting mixtures. He states that when the components in the mixture are polymers, the entropy change is very small and the heat of mixing alone will be responsible for the homogeneity of the mixture.

Bitumen also contains polymeric compounds that require proper range of temperature for mixing with any type of modifier. Many researchers have used similar procedure of mixing modifier in bitumen using a shear mixer at a temperature range for a specific time. McNally [14], explained why high shear and high processing temperatures were necessary for polymer modification of bitumen. According to the author the polymer modifiers that are most frequent in use for modification such as styrene–butadiene–styrene (SBS), styrene-butadiene rubber (SBR), crumb tire rubber (CR), ethylene vinyl acetate (EVA), low as well as high density polyethylene (LDPE, HDPE), and some other waste polymers, may have a significant effect of the mixing conditions on the resulting blend as well as the cost of production. Hence, high processing temperatures (170–180 °C) and high shear rates are required for reducing the difference of viscosity between polymers and bitumen, and for obtaining suitable polymer dispersions withing the bitumen matrix. He also showed that, the addition of reactive polymers such as isocyanate based reactive polymers require gentle processing conditions as they are in liquid form and they react with the functional groups of bitumen to form chemical bonds. The blends of these reactive polymers with bitumen are prepared using a low shear stirring device at temperatures between 60 and 180 °C.

Guilani et al. [4] used a high shear mixer to blend the bitumen and polyvinyl acetate at 180 °C at a shear mixing rate of 4000 revolutions per minute for quarter of an hour. The bitumen was pre heated to 180 °C before adding the modifier. Rossi et al. [19] prepared organosilicon based surfactant modified bitumen using a shear mixing homogeniser (high speed shear mixer). Bitumen was heated to 150 °C until it was fully flowing and then the modifier was added at a mixing speed of 800–1000 rpm for the next 10 min. Naskar et al. [16] followed the similar procedure of preparing polymer modified bitumen blends, by mixing at a high shear rate of 3500 rpm for 45 min at a temperature of 180 °C with a laboratory mechanical stirrer. They state that the mixing heat energy should not exceed 185 °C and there should be enough mixing time for homogeneous distribution of plastic molecules in the bitumen. Munera and

Ossa [15] prepared SBS polymer, crumb rubber, and polyethylene wax modified bitumen blends by using a conventional mechanical mixer. The binder mixes were mixed at a shear rate of 1200 ± 10 rpm for two hours in case of SBS and CR, while for PW 45 min of shearing time was required. The mixing temperature was kept between 180 and 190 °C for all the blends. Yu et al. [24] prepared asphalt rubber (AR) binder by mixing the crumb rubber with the base binder with the help of a high shear mixer at a shear rate of 4000 rpm at 176 °C for one hour. There is another method of adding polymeric modifiers in the bitumen, which is done by using the oil obtained from the pyrolysis of waste polymers as modifiers. Hadole and Ranadive [7] prepared HDPE pyrolytic oil modified bitumen at a rate of 3000 rpm at 160 °C for 2 h. Al-Sabaeei et al. [1] used the tyre pyrolysis oil (TPO) for modifying the conventional 60/70 penetration Grade bitumen. TPO was mixed in the binder using a multi-mix high shear mixer in laboratory. First the binder was heated to 160 °C in oven to achieve sufficient fluidity, then the tyre pyrolysis oil was added and blend was mixed manually for two minutes followed by mixing at 100 rpm for one hour to obtain a homogenous bled of TPO and bitumen. Hadole et al. [6] prepared pyro-oil modified VG30 bitumen at a temperature of 150 °C at 5000 rpm for first 15 min and then at 3000 rpm for the next 5 min to ensure proper mixing of the oil in bitumen. You et al. [23] fabricated asphalt nanocomposite binder, by adding nano clay in base asphalt using a high shear mixer. The base binder was first heated to 160 °C till it was liquid, then the nano clay was blended in it at a rate of 2500 rpm for 3 h so that the inserted montmorillonite (MMT) nano clay is dispersed properly. Saltan et al. [20] used high shear mixer to prepare ZnO nano particles modified bitumen at a rate of 4000 rpm for 2 h at 160 °C. Kavussi and Barghabany [9] used organo-montmorillonite nano clay to prepare modified bitumen at high shear rate of 5200 rpm, for 30 min at a temperature of 155 °C. Khapne et al. [10] used a high shear mixer for blending HDPE pyro-oil into base bitumen VG30 at a shear rate of 3000 rpm at 160 °C for 15 min.

3 Processing Variables

The important processing variables that determine the properties and behaviour of the modified binder are as follows.

3.1 Mixing Temperature

Shaffie et al. [22] examined the effect of varying processing variables on the physical properties of modified bituminous binder. The authors used an 80/100 penetration grade bitumen and modified it with natural rubber latex at two different processing temperatures of 140 and 160 °C. The resulting blend was tested for physical properties such as softening point, penetration and penetration index. Both softening point and penetration index values (PI) have found to be increased with increase in mixing temperature. The higher PI value indicates more resistance to temperature susceptibility. Hence the higher temperature of 160 °C was found more suitable for the modification process. Martin-Alfonso et al. [13] studied the effect of temperature at the time of mixing on the reactive polymer modified bitumen. They recorded rheokinetics curves at different processing temperatures for the blending of bitumen and polymer. A varying temperature range of 60–180 °C was adopted for preparation of modified bitumen. The results showed, that low processing temperature led to normalized torque increase, while higher processing temperature led to a low increase in the torque and faster modification process. When the mixing temperature is low, larger asphaltene particles will react with other higher molecules, resulting an increase in the viscosity and torque of mixing. While with high processing temperature (180 °C), the rate of reaction increases and the viscosity lowers, with a low increase in torque. Hence high processing temperature is recommended. Gonzales et al. [5] investigated the effect of processing variables on the rubber and PMB. They found that increase in temperature reduces the amount of insoluble material in the binder and also allows more rubber and polymer digestion. However, no effect was observed on the softening point and penetration of the binder due to increased process temperature, signifying very little influence over ageing. They observed that processing temperature increase between 180 and 200 °C didn't modify linear viscoelastic behaviour of the modified binder and a significant lowering in the viscosity had been observed up to 200 °C corresponding to the amount of polymer digestion increase. There is an increase in viscosity and viscoelastic behaviour due to ageing phenomenon, whereas the digestion of rubber/polymer have a reverse effect of decreasing the same, which in turn means that the effect of ageing in these conditions is less significant than that of digestion. However, further increase in processing temperature up to 210 °C causes the modified binder to have an increased viscosity as the effect of ageing dominates. It was observed that the highest amount of polymer digestion reached up to 200 °C, and any further increase in processing temperatures will not improve the binder.

3.2 *Mixing Time*

The effects of two different mixing times 30 min and 60 min, keeping the mixing speed constant, on the physical properties of modified bitumen were observed [22]. They found that for longer mixing time from 30 to 60 min, the softening point and penetration value increased. This was due to better dispersion of natural rubber latex modifier within the bitumen for longer mixing time. They found that the bitumen showed improved physical properties at higher temperature of 160 °C for more processing time of 60 min. Gonzales et al. [5] examined the results of using different mixing times ranging from 15 to 120 min on the characteristics of polymer and rubber blended bitumen. It was observed that at low temperatures, the viscosity value remains practically similar for low mixing time. However, more than 30 min processing times, show an increase in the viscosity exponentially, which is less marked at 135 °C, than at 60 °C. In other words, viscosity remains unchanged for a minimum processing time of 45 min, whereas it increases with an increase in a further mixing time, at low temperatures. Whereas, for higher temperatures (180 °C), viscosity does not show a much higher increase for processing times up to 45 min. But high mixing temperature may cause the binder to undergo primary ageing.

3.3 Shear Rate/Speed

Shaffie et al. [22] observed the effect of four different mixing shear rates of 500, 1000, 1500, and 2000 rpm on the physical properties of the modified bitumen. They found that mixing speeds have considerable effect on the properties. Softening point value increased significantly after a specified mixing speed, whereas the shear rate of blending did not show any considerable effect on the penetration value which showed up and down trend during the test. They determined the optimum shear rate of mixing using the penetration index values form softening point and penetration values, and found the optimum mixing shear rate to be 1270 rpm. Jamal and Giustozzi [8] studied the effect of two different shear rates of 700 rpm and 3500 rpm on the rheological performance of the crumb rubber modified binder. It was found that there was almost no effect on the complex shear modulus when the binder was mixed at low shear rate of 700 rpm. However, the complex shear modulus was found to be similar to that of neat binder when the higher shear rate of 3500 rpm was used for mixing. The ageing analysis of different shear rates indicated that higher shear rate enhances the ageing. They concluded that shear rate/ mixing speed have a significant influence on the performance of the binder, and the complex shear modulus of the modified binder increases along with ageing, and the phase angle decreases at higher mixing rates.

4 Summary and Conclusions

This study was mainly focussed on the common methods or processes adopted for the modification of standard base bitumen with different modifiers, and the processing conditions that are necessary for the preparation of modified bitumen. We reached at following conclusions;

- The mixing temperature, mixing time, and shear rate of mixing are the major processing conditions that are significant for the process or mechanism of modification.
- The first and foremost important thing to know about these three processing conditions is that the performance of the resulting modified bitumen, physical and rheological, is dependent on the interdependence of the variations in mixing time, temperature and shear rate.

- Different types of modifiers require different mixing temperatures, mixing time, and mixing speeds depending upon the chemical reaction, dispersion and reactivity of the modifier with the base bitumen.
- High mixing temperatures ensure proper modifier digestion in the bitumen resulting in lowering of viscosity and making the blend more resistant to temperature susceptibility.
- More mixing time increases the viscosity at low processing temperatures, whereas more mixing time up to 45 min does not have a considerable effect on viscosity with high processing temperatures.
- Shear rate imparts a considerable influence on the properties of modified bitumen. Higher shear rate mixing gives a similar complex modulus as that of base bitumen. Ageing of the bitumen increases at higher shear rates along with the rutting characteristics.
- High mixing temperatures and high shearing rates are usually required for decreasing the difference in viscosities of the bitumen and modifiers with optimum processing time for proper dispersion and reaction of the modifier within the bitumen matrix.

References

1. Al-Sabaeei AM, Napiah MB, Sutanto MH, Alaloul WS, Yusoff NIM, Khairuddin FH, Memon AM (2021) Evaluation of the high-temperature rheological performance of tire pyrolysis oil-modified bio-asphalt. Int Jof Pavem Eng 1–16
2. Carrera V, Partal P, García-Morales M, Gallegos C, Pérez-Lepe A (2010) Effect of processing on the rheological properties of poly-urethane/urea bituminous products. Fuel Process Technol 91(9):1139–1145
3. Fang C, Liu P, Yu R, Liu X (2014) Preparation process to affect stability in waste polyethylene-modified bitumen. Constr Build Mater 54:320–325
4. Giuliani F, Merusi F, Filippi S, Biondi D, Finocchiaro ML, Polacco G (2009) Effects of polymer modification on the fuel resistance of asphalt binders. Fuel 88(9):1539–1546
5. González V, Martínez-Boza FJ, Gallegos C, Pérez-Lepe A, Páez A (2012) A study into the processing of bitumen modified with tire crumb rubber and polymeric additives. Fuel Process Technol 95:137–143
6. Hadole HP, Suryawanshi SD, Khapne VA, Ranadive MS (2021) Moisture damage resistance of short-term aged pyro-oil–modified bitumen using rolling thin film oven by surface free energy approach. J Mater Civ Eng 33(10):04021268
7. Hadole HP, Ranadive MS (2019), Analysis of ageing mechanism of HDPE pyro-oil modified bitumen compared to VG30 based on fourier transform infrared spectrum. In: 5th national conference of transportation research group India. CTRG, Bhopal
8. Jamal M, Giustozzi F (2020) Low-content crumb rubber modified bitumen for improving Australian local roads condition. J Clean Prod 271:122484
9. Kavussi A, Barghabany P (2016) Investigating fatigue behavior of nanoclay and nano hydrated lime modified bitumen using LAS test. J Mater Civ Eng 28(3):04015136
10. Khapne V, Hadole H, Ranadive M (2020) Assessment of anti-stripping property of pyro-oil modified bituminous mixes using surface free energy approach. In: International conference on transportation and development 2020, pp 127–137. American Society of Civil Engineers, Reston, VA

11. Kulkarni SB, Ranadive MS (2020) Modified cutback as tack coat by application of pyro-oil obtained from municipal plastic waste: experimental approach. J Mater Civ Eng 32(5):04020100
12. Kulkarni SB, Ranadive MS (2021) A feasibility study towards the application of municipal waste pyrolysis oil in bituminous pavement. In: Civil infrastructures confronting severe weathers and climate changes conference, pp 130–147. Springer, Cham
13. Martín-Alfonso MJ, Partal P, Navarro FJ, García-Morales M, Bordado JCM, Diogo AC (2009) Effect of processing temperature on the bitumen/MDI-PEG reactivity. Fuel Process Technol 90(4):525–530
14. McNally T (ed) (2011) Polymer modified bitumen: properties and characterisation. Elsevier
15. Munera JC, Ossa EA (2014) Polymer modified bitumen: optimization and selection. Mater Des 1980–2015(62):91–97
16. Naskar M, Chaki TK, Reddy KS (2010) Effect of waste plastic as modifier on thermal stability and degradation kinetics of bitumen/waste plastics blend. Thermochim Acta 509(1–2):128–134
17. Porto M, Caputo P, Loise V, Eskandarsefat S, Teltayev B, Oliviero Rossi C (2019) Bitumen and bitumen modification: A review on latest advances. Appl Sci 9(4):742
18. Ranadive MS, Hadole HP, Padamwar SV (2018) Performance of stone matrix asphalt and asphaltic concrete using modifiers. J Mater Civ Eng 30(1):04017250
19. Rossi CO, Caputo P, Baldino N, Szerb EI, Teltayev B (2017) Quantitative evaluation of organosilane-based adhesion promoter effect on bitumen-aggregate bond by contact angle test. Int J Adhes Adhes 72:117–122
20. Saltan M, Terzi S, Karahancer S (2019) Mechanical behavior of bitumen and hot-mix asphalt modified with zinc oxide nanoparticle. J Mater Civ Eng 31(3):04018399
21. Selvavathi V, Sekar VA, Sriram V, Sairam B (2002) Modifications of bitumen by elastomer and reactive polymer—a comparative study. Pet Sci Technol 20(5–6):535–547
22. Shaffie E, Arshad AK, Alisibramulisi A, Ahmad J, Hashim W, Abd Rahman Z, Jaya RP (2018) Effect of mixing variables on physical properties of modified bitumen using natural rubber latex. Int J Civil Eng Technol 9(7):1812–1821
23. You Z, Mills-Beale J, Foley JM, Roy S, Odegard GM, Dai Q, Goh SW (2011) Nanoclay-modified asphalt materials: preparation and characterization. Constr Build Mater 25(2):1072–1078
24. Yu H, Leng Z, Zhou Z, Shih K, Xiao F, Gao Z (2017) Optimization of preparation procedure of liquid warm mix additive modified asphalt rubber. J Clean Prod 141:336–345
25. Zhu J, Birgisson B, Kringos N (2014) Polymer modification of bitumen: advances and challenges. Eur Polymer J 54:18–38

Alkali Activated Black Cotton Soil with Partial Replacement of Class F Fly Ash and Areca Nut Fiber Reinforcement

B. A. Chethan, A. U. Ravi Shankar, Raghuram K. Chinnabhandar, and Doma Hemanth Kumar

Abstract Alkali activation has received great attention for improving the soil properties with suitable precursor materials. Industrial byproduct class F fly ash was suitably utilized to improve Black Cotton (BC) soil properties along with ordinary Portland cement by various researchers. However, the CO_2 emission associated with cement production has enforced the evaluation of alternative binders. Laboratory investigations were conducted on BC soil by admixing various fly ash dosages (0–50%) and reinforcing the mix with 0.5% areca nut fiber. Alkali activator solution prepared using 8 molar sodium hydroxide solution (SH) and sodium silicate solution (SS) at 1.5 SS/SH ratio showed significant improvement in Unconfined Compressive Strength (UCS) of stabilized BC soil on 7 and 28 days curing. The reinforcement was effective in improving the flexural strength of stabilized mixes. Exorbitant unsoaked California Bearing Ratio (CBR) values were observed on 28 days of curing. However, the samples could retain low soaked CBR values despite reinforcement. Scanning Electron Microscope (SEM) images showed the reduction of shrinkage cracks and strong bonding of fibers in the stabilized mix. X-Ray Diffraction (XRD) patterns evidenced the formation of various hydration products due to the alkali reaction, which resulted in the high strength gain of mixes at ambient temperature curing. The leaching of mineral constituents from the set mix lead to the failure of durability samples. Due to nondurability, the alkali activation with a selected precursor cannot suit pavement materials requirements.

Keywords Alkali solution · Black cotton soil · Strength · Durability · SEM · XRD

1 Introduction

The stable and durable foundation soils are essential for the construction of any civil engineering structure. Soil stabilization is widely used to improve desirable properties of soil with various cementitious and marginal materials. In the recent

B. A. Chethan (✉) · A. U. R. Shankar · R. K. Chinnabhandar · D. H. Kumar
Department of Civil Engineering, NITK, Surathkal, India
e-mail: chethanba@gmail.com

M. S. Ranadive et al. (eds.), *Recent Trends in Construction Technology and Management*, Lecture Notes in Civil Engineering 260,
https://doi.org/10.1007/978-981-19-2145-2_88

era, alkali-activated industrial byproduct fly ash is considered as one of the green construction materials. Portland cement utilization can be minimized by the use of fly ash along with a suitable alkali additive. The proper utilization of fly ash can reduce environmental problems due to its reduced disposal [6]. Alkali-activated precursor materials are clinker-free binders, which can greatly reduce CO_2 emissions by 70–90% resulting from cement production [4]. Alkali-activated composites can provide high early-age strength [9]. A low-ordered crystalline structure of sodium aluminum silicate hydrate (NASH) gel is a common product resulting from the alkali activation process [17]. The use of fly ash in the black cotton soil can control its swelling [5]. BC soil's inherent volume change behavior due to moisture content variations reduces its workability, strength, and suitability for construction, which can be suitably improved with fly ash geopolymer [11]. The use of 7.5 molar NaOH solution for alkali-activated cement using fly ash for improvement of silty sand was considered to give better strength at a lesser cost [12]. An optimum SS/SH ratio of 1.5 was suitable for a fly ash geopolymer [11]. When the fly ash was admixed to BC soil, the rise in stiffness and reduction in the free swell index were observed due to reduced water affinity by particle flocculation [13].

As a ground improvement technique, natural coir fibers are effectively utilized with different cementitious compounds due to their low cost and abundant availability [14]. Cellulose percentage governs natural fiber's tensile strength, and lignin percentage governs biodegradation resistance [7]. The fiber reinforcement can greatly reduce bare soil's crack propagation potential [8]. Along with improved strength, stabilized soil's brittleness can be decreased by using natural coir fiber reinforcement [2]. The use of fibers is an economical and eco-friendly means of soil improvement. The formation of the spatial thread bridge network enhances the particle holding capacity around the fiber surface and results in strength improvement of alkali-activated BC soil with class F fly ash and slag mixes [15]. Areca nut fiber reinforcement effectively improved the strength, durability, and fatigue life of cement-treated lateritic soil due to enhanced bonding of the mix [10]. Linseed oil treatment was found to increase the interface friction and tensile strength of the treated fibers, which can further contribute to improvement in the strength of stabilized soil [16].

Numerous studies stated the benefit of various fibers for soil reinforcement along with different binders. However, limited studies were reported on areca fiber reinforcement's suitability with fly ash geopolymer to improve BC soils strength and durability. Hence comprehensive research was conducted to evaluate the strength, durability, and fatigue performance of stabilized BC soil. The bonding effect of alkali-activated soil with fiber was studied using flexure, CBR, and durability tests.

2 Materials and Methods

Expansive BC soil was procured from Chikmaglur District, Karnataka. A representative disturbed soil sample was obtained at 0.3 m depth below the ground surface.

Class F fly ash was procured from M/s Udupi thermal power plant, a rich aluminosilicate material suitable for alkali activation. Coarser areca fibers were extracted from mature areca nut shells. Preliminary investigations showed that 0.5% of coarser areca fibers with 25 mm length give a better reinforcing effect. Areca fiber has a density of 1.09 g/cc, an average diameter of 0.35 mm, and tensile strength of 2.2 kN/m^2. These fibers consist of 35–64.8% hemicellulose and 13–14.8% lignin content [10]. Areca fibers were coated with linseed oil and dried in the oven at 120 °C to obtain rougher morphology. Alkali solution was prepared by using commercially available sodium hydroxide pellets and sodium silicate solution. NaOH pellets were dissolved in a calculated quantity of potable water to prepare 8 molar solution. NaOH pellets reaction with water is exothermic, liberating enormous heat; therefore, it is allowed to cool down. After that, the NaOH solution and Na_2SiO_3 solutions were mixed to obtain a SS/SH ratio of 1.5. Alkali solution prepared was used for stabilization after 24 h. The dry ingredients, namely, BC soil, class F fly ash, and areca fibers, were thoroughly mixed. To this dry mixture, a calculated quantity of alkali solution was added, and mixing is continued. The prepared homogeneous mixture was immediately used for the preparation of specimens without any delay. Compacted samples were cured at ambient room temperature. Stabilized specimens after 3, 7, and 28 days of curing were tested for UCS. The flexural strength of beam specimens was determined after 3, 7, and 28 days of curing by a two-point loading method. Evaluation of CBR was done on 28 days cured specimens. Durability of specimens cured for 7 days was evaluated by conducting wetting–drying and freezing–thawing tests. Fatigue behavior of 28 days cured UCS samples was evaluated using a repeated load testing machine. Changes in the morphology of the stabilized sample were studied using Scanning Electron Microscope (SEM) images. The end products responsible for strength improvement were identified using X-Ray Diffraction (XRD) micrograms (Tables 1 and 2).

3 Experimental Outcomes

3.1 Compaction

Modified Proctor compaction characteristics of BC soil replaced with 0–50% of fly ash along with 0.5% areca fiber reinforcement compacted using water are depicted in Fig. 1. The use of water for compaction instead of alkali solution resulted in insignificant changes in compaction characteristics. The compact packing of the mix at 10% fly ash replacement resulted in the highest density of 18.5 kN/m^3. The addition of a higher dosage of fly ash reduced the mix density due to low specific gravity fly ash particles. BC soil with areca fibers has achieved the MDD of 18.1 kN/m^3 at the OMC of 16.8%. The OMC of the mix has reduced to 15.3% at 15% fly ash replacement. Further addition of fly ash resulted in an increase in OMC due to the

Table 1 Index properties of BC soil

Test properties	Unit	Test method	Average value
Specific gravity, Gs	–	IS 2720 (Part 3), 1980	2.56
γ_d	kN/m^3	IS 2720 (Part 7), 1980	16.1
OMC	%	IS 2720 (Part 7), 1980	22.2
Liquid limit (LL)	%	IS 2720 (Part 5), 1985	60
Plastic limit (PL)	%	IS 2720 (Part 5), 1985	33
Plasticity index (PI)	%	IS 2720 (Part 5), 1985	27
Shrinkage limit (SL)	%	IS 2720 (Part 6), 1972	23
Gravel (%)	%	IS 2720 (Part 4), 1987	2
Sand (%)	%	IS 2720 (Part 4), 1987	26
Silt (%)	%	IS 2720 (Part 4), 1987	68
Clay	%	IS 2720 (Part 4), 1987	4
Free swell index (FSI)	%	IS 2720 (Part 40), 1977	58
Organic content	%	IS 1498 (1970)	2.25
Indian standard soil classification system	–	IS 1498 (1970)	MH

Table 2 Chemical properties of BC soil and class F fly ash

Test properties	Unit	BC soil	Class F fly ash
pH	–	8.24	10.7
SiO_2	%	75.4	70.5
Al_2O_3	%	7.06	10.98
Fe_2O_3	%	2.64	1.84
CaO	%	0.007	0.003
SO_3	%	0.16	0.15
LOI	%	2.30	1.70

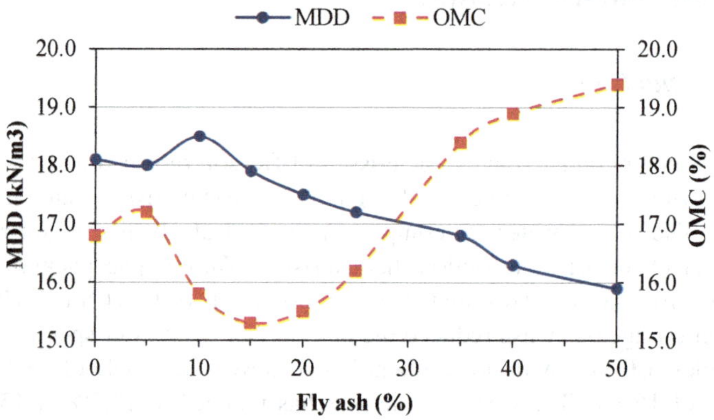

Fig. 1 Variation of modified proctor compaction characteristics of fly ash admixed BC soil

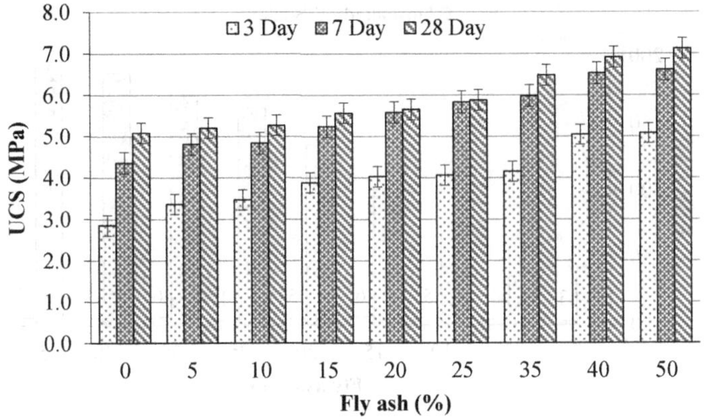

Fig. 2 Variation of UCS of stabilized BC soil

higher demand for water required for particle coating. The higher specific surface area of fly ash particles has contributed to a rise in the OMC.

3.2 UCS

It is revealed that the alkali activation of fiber-reinforced BC soil with fly ash precursor significantly influenced UCS strength gain. On a low curing period of 3 days, the dissolution of aluminosilicate precursor and subsequent geopolymerization was not enough; therefore, low UCS improvements were observed. The use of alkali solution significantly increased the UCS of stabilized samples on 7 days of curing. On further curing, a slight improvement in UCS values was observed for most of the mixes. Higher UCS values were observed at the maximum fly ash replacement level due to the availability of a high quantity of precursor material. On 28 days of curing, the alkali-activated BC soil constituting 50% fly ash along with 0.5% areca fiber reinforcement has achieved a maximum UCS of 7.1 MPa. The reinforcement has made the specimens more ductile when compared to unreinforced ones (Fig. 2).

3.3 CBR

Based on the UCS test results, it was observed that 28 days of curing promotes the UCS of stabilized soil significantly. Hence, CBR samples were prepared and cured for 28 days. The samples became very hard; therefore, excellent plunger penetration resistance was observed for the unsoaked test. Samples with more than 15% fly ash exhibited unsoaked CBR values of more than 100%. Areca fibers can take tensile load; hence the reinforcement formed a strong matrix, due to which resistance to

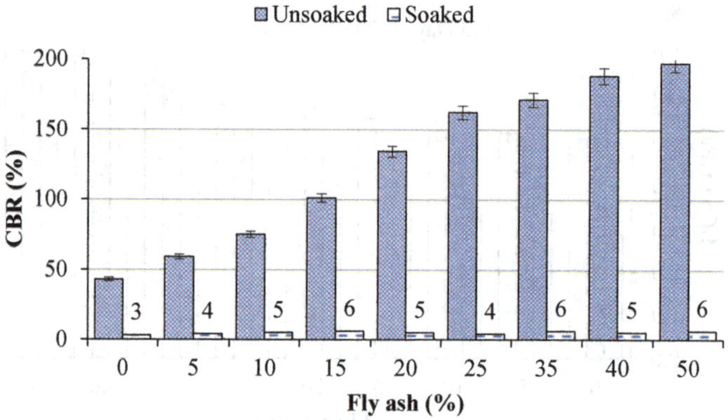

Fig. 3 Variation of CBR of stabilized BC soil

the plunger penetration was increased. When immersed in water, due to the delectation of structure by minerals leaching from the set mix, the samples lost most of their strength. The soaked samples exhibited low resistance to plunger penetration. However, a small improvement in the soaked CBR of the mix was observed when compared to untreated soil (Fig. 3).

3.4 *Flexural Strength*

Beam specimens of size 76 by 290 mm [1] cured for 3, 7, and 28 days showed significant flexural strength improvement. A high fly ash dosage of 50% has resulted in maximum flexural strength of 1.47 MPa on 28 days of curing. Marginal improvement in flexural strength was noticed after 7 days of curing for all the mixes. BC soil at modified Proctor density exhibited a low flexural strength of <0.02 MPa. The presence of fiber reinforcement along with improvement in flexural strength also reduced the development of shrinkage cracks on curing. Fibers strongly bonded to the stabilized soil and resisted failure during flexural load applications due to their high tensile capacity. Interfacial friction between fibers and bonded soil restricted the pullout of fibers from the set soil. Frictional resistance to the shrinkage was offered due to soil fiber interaction and interlocking mechanism [3] (Fig. 4).

3.5 *Fatigue*

The fatigue life is defined as the number of repetitions a stabilized material can withstand at the specified load stress. The 28 days cured samples were tested using a fatigue testing machine to evaluate the resistance under repeated load application.

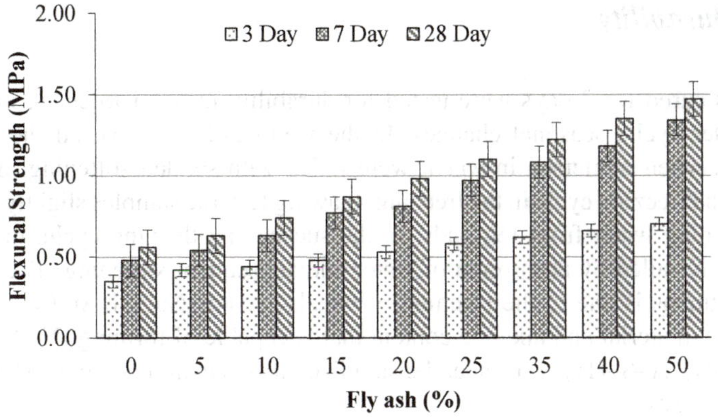

Fig. 4 Variation of flexural strength of stabilized BC soil

A frequency of 1 Hz and a rest period of 0.1 s was selected for testing. The samples were subjected to fatigue tests under different loads, which is considered as a certain percentage of UCS load. A fraction of load stress, which is equivalent to 0.35, 0.5, and 0.65 times that of UCS of each specimen, was applied, and the number of repetitions to failure was recorded. At applied stress of 0.35, the specimens failed after taking >47,000 repetitions, as shown in Fig. 5. All the samples at an applied stress of 0.2 times UCS have sustained >2 lakh fatigue repetitions, and the test was not continued further.

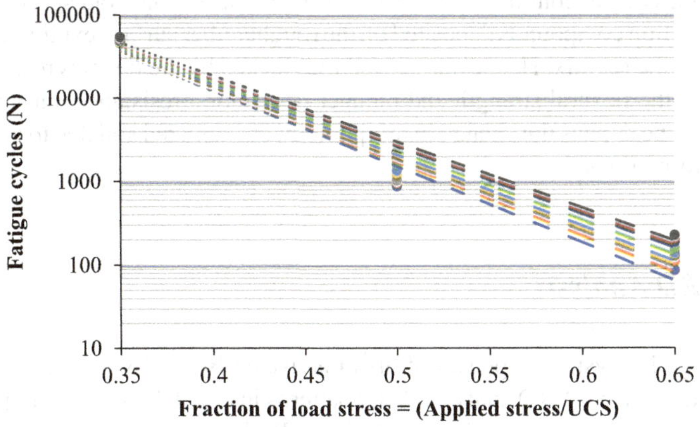

Fig. 5 Fatigue cycles to failure at different applied stress values

3.6 Durability

Samples cured for 7 days were tested for durability in order to check their retention under cyclic seasonal changes. In the wetting–drying test, all the stabilized samples, when immersed in water, were failed with sudden softening of the mix. During the freezing cycle in the freezing–thawing test, the samples slightly expanded by absorbing water from felt pads. In the succeeding thawing cycle, the samples completely failed by absorption of water and subsequent softening. The failure of the specimens is due to the leaching of metals such as Na, Al, Si, Ca, Fe present in the raw material and alkaline content that is capable of forming (N–A–S–H) gel and (C–(N)–A–S–H) gel. Hence, the stabilized mixes exhibited no resistance to the durability tests.

3.7 SEM Analysis

The effect on the morphology of the stabilized mix using fly ash precursor for improving BC properties using alkali solution is depicted in Fig. 6. Figure 6a depicts the discontinues and loose structure of the BC soil with more pores. The shrinkage cracks developed due to dehydration of soil are clearly visible. Due to shrinkage, the debonding of areca fibers with soil was observed. Due to polymerization compound formations by alkali-activated fly ash precursor, the pore spaces were sealed; thus, no cracks and discontinuities are observed, as depicted in Fig. 6b. Dense packing of soil particles due to continuous geopolymer matrix sealed the void spaces. Also, the polymerization compounds deposited on the rougher fiber surface exhibited excellent bonding of the set mix. Hence, the set soil interlocked by the fibers is responsible for improving the flexural strength and plunger penetration resistance during the CBR test. Figure 6c depicts the change in morphology of stabilized soil due to geopolymer product formations.

3.8 XRD Analysis

The change in crystalline phases of alkali-activated BC soil with precursor fly ash was identified using XRD analysis. XRD patterns justified the strength improvement by the formation of many hydration products. Peaks observed at different 2θ angle revealed the involvement of dissolvable alumina and silica in the formation of varying hydration products. XRD microgram depicted below shows the formation of these hydration products at the highest intensity. However, the mentioned hydration products were spread over multiple 2θ angles. Gismondine ($CaAl_2Si_2O_8{\cdot}4H_2O$), Liottite ($(Ca, Na)_4$ $(Si, Al)_6$ O_{12} $(SO_4, OH, Cl, CO_3)_2{\cdot}xH_2O$), Millosevichite ($Al_2$ $(SO_4)_3$), Foshagite (Ca_4 $(SiO_3)_3$ $(OH)_2$), Nordstrandite (Al $(OH)_3$), Xonotlite ($Ca_6Si_6O_{17}$

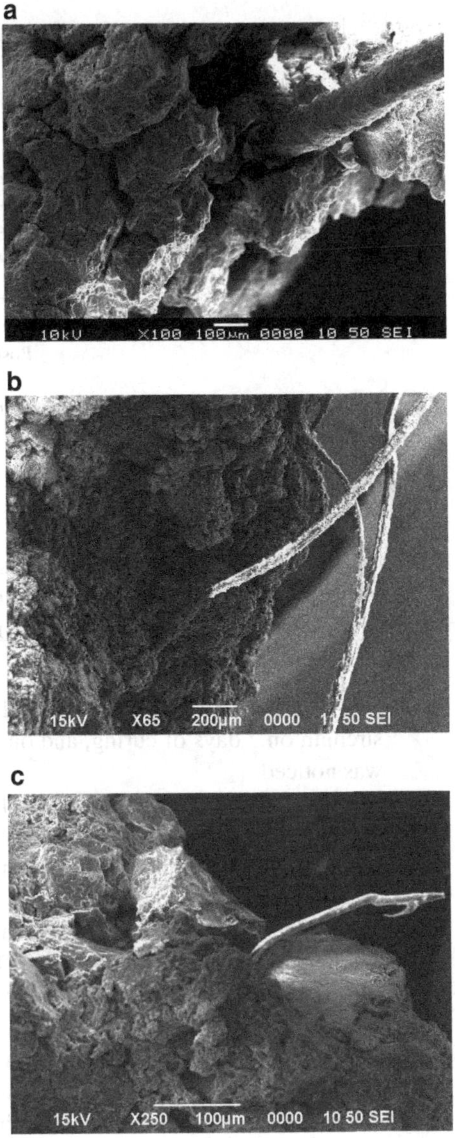

Fig. 6 **a** Black cotton soil exhibiting shrinkage cracks along with debonded areca fiber, **b** areca nut fibers bonded in alkali-activated mix, **c** mix showing geopolymer compounds

$(OH)_2$), Calcium Sulfate Hydrate ($CaSO_4 \cdot 0.15H_2O$), and Sodium Hydrogen Sulfate ($NaHSO_4$) are the common phases observed in the set stabilized mix. Along with this, Quartz low (SiO_2) and Pigeonite ((Ca, Mg, Fe) (Mg, Fe) Si_2O_6), which is a mineral of the clinopyroxene subgroup of the pyroxene group, were also found in the stabilized mix (Fig. 7).

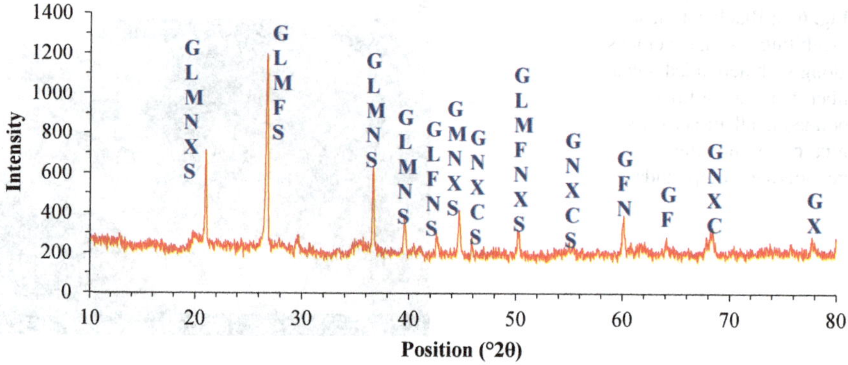

Fig. 7 XRD peaks showing the hydration products

4 Conclusions

Effect of using alkali-activated class F fly ash binder with areca nut fiber as reinforcement was reported in this investigation. The effect of stabilization was studied in terms of strength development, durability, and microstructural changes. The following outcomes are highlighted.

(1) Regardless of fly ash dosage, all alkali-activated specimens attained maximum strength on 7 days of curing, and on further curing slight increase in strength was noticed.
(2) The use of areca fiber reinforcement has made the specimens more ductile and reduced the density of mixes. However, the reinforcement has a significant influence on improving flexural strength and the crack propagation resistance of specimens.
(3) Under the dry condition, the specimens showed a noticeable improvement in plunger penetration resistance. The stabilized mixes exhibited excellent performance under the fatigue load repetitions.
(4) The alkali-activated BC soil with fly ash and areca fibers reinforcement was highly susceptible to weathering due to seasonal changes associated with wetting, drying, and frost actions like freezing and thawing. The leaching of minerals from the set mix weakens the soil.

The tested mix combinations are performing better at low moisture content conditions. The weakness of mixes lies in softening whenever they come in contact with water. As the pavement layers are seasonally subjected to moisture changes, the damages may be foreseen due to mix softening and the entire pavement deformation. Hence, these stabilized mix combinations cannot be used for pavements.

References

1. ASTM (2000) Standard test method for flexural strength of soil-cement using simple beam with third-point loading. D1635
2. Bai Y, Liu J, Song Z, Chen Z, Jiang C, Lan X, ... Kanungo DP (2019) Unconfined compressive properties of composite sand stabilized with organic polymers and natural fibers. Polymers 11(10):1576
3. Dang LC, Fatahi B, Khabbaz H (2016) Behaviour of expansive soils stabilized with hydrated lime and bagasse fibres. Procedia Eng
4. Davidovits J (1993) Geopolymer cements to minimize carbon dioxide greenhouse warming. Ceram Trans 37:165–182
5. Gajghate V, Patil P (n.d.) Experimental study of use of flyash as retarding agent in black cotton soil. In: Advances in civil engineering and infrastructural development, pp 655–664. Springer
6. Gupta S, GuhaRay A, Kar A, Komaravolu VP (2018) Performance of alkali-activated binder-treated jute geotextile as reinforcement for subgrade stabilization. Int J Geotech Eng 1–15
7. Hearle JWS, Morton WE (2008) Physical properties of textile fibres. Elsevier
8. Hussain R, Bordoloi S, Gadi VK, Garg A, Ravi K, Sreedeep S (2020) Effect of filament type and biochemical composition of lignocellulose fiber in vegetation growth in early plant establishment period. In: Problematic soils and geoenvironmental concerns, pp 201–213. Springer
9. Kong DLY, Sanjayan JG (2010) Effect of elevated temperatures on geopolymer paste, mortar and concrete. Cem Concr Res 40(2):334–339
10. Lekha BM, Goutham S, Shankar AUR (2015) Evaluation of lateritic soil stabilized with Arecanut coir for low volume pavements. Trans Geotech 2:20–29. https://doi.org/10.1016/j.trgeo.2014.09.001
11. Murmu AL, Patel A (2020) Studies on the properties of fly ash-rice husk ash-based geopolymer for use in black cotton soils. Int J Geosynth Ground Eng 6(3):1–14
12. Rios S, Ramos C, Viana da Fonseca A, Cruz N, Rodrigues C (2019) Mechanical and durability properties of a soil stabilised with an alkali-activated cement. Eur J Environ Civ Eng 23(2):245–267
13. Saride S, Dutta TT (2016) Effect of fly-ash stabilization on stiffness modulus degradation of expansive clays. J Mater Civ Eng 28(12):4016166
14. Shenal Jayawardane V, Anggraini V, Emmanuel E, Yong LL, Mirzababaei M (2020) Expansive and compressibility behavior of lime stabilized fiber-reinforced marine clay. J Mater Civ Eng 32(11):4020328
15. Syed M, GuhaRay A (2020) Effect of fiber reinforcement on mechanical behavior of alkali-activated binder-treated expansive soil: reliability-based approach. Int J Geomech 20(12):04020225. https://doi.org/10.1061/(asce)gm.1943-5622.0001871
16. Tan T, Huat BBK, Anggraini V, Shukla SK, Nahazanan H (2019) Strength behavior of fly ash-stabilized soil reinforced with coir fibers in Alkaline environment. J Nat Fibers 1–14
17. Zhang Y, Zhang J, Jiang J, Hou D, Zhang J (2018) The effect of water molecules on the structure, dynamics, and mechanical properties of sodium aluminosilicate hydrate (NASH) gel: a molecular dynamics study. Constr Build Mater 193:491–500

Development of Road Safety Models by Using Linear and Logistic Regression Modeling Techniques

Krantikumar V. Mhetre and Aruna D. Thube

Abstract Road accidents are caused by many factors like alcohol consumption, uncontrolled vehicle speed, poor road surface conditions, bad weather, inadequate traffic signs, and vehicle defects. A single solution cannot address these factors. Therefore, various engineering departments are engaged in road accident studies and their minimization. One of the solutions to study road accidents is to make use of accident prediction models. Regression is one of the essential techniques of predictive analytics in which a lot of prediction problems are modeled. Regression is a kind of supervised learning algorithms since a regression model requires both the dependent as well as the independent variables in the data set. Four simple linear regression models are developed, and two of them are best fitted. The fitted models are logarithmic-linear. The model's output is to find the number of fatalities based on the total number of accidents. The special cases considered here are accidents on a T-junction and accidents due to the intake of alcohol. A binary logistic regression model is developed for the accidents from the year 2014–2019, and the prediction of causing fatality is computed by using a cut-off probability value of 0.33. The overall model is accepted.

Keywords Road accidents · Road safety · Road safety modeling · Accident prediction modeling · Regression

1 Introduction

The third global road safety conference on was held in February 2020 at Stockholm, Sweden. In this conference, all the countries promised to reduce the deaths due to road accidents by 50% for 2030. China ranks 2nd, USA ranks 3rd, whereas India ranks 1st in deaths due to road accidents among the 199 countries involved in the World Road Statistics, 2018. World Health Organization's global report on road safety 2018 says India has a contribution of 11% in the world's accident-related deaths. In 2019,

K. V. Mhetre (✉) · A. D. Thube
Department of Civil Engineering, College of Engineering Pune, Pune, Maharashtra, India
e-mail: kvm18.civil@coep.ac.in

M. S. Ranadive et al. (eds.), *Recent Trends in Construction Technology and Management*, Lecture Notes in Civil Engineering 260,
https://doi.org/10.1007/978-981-19-2145-2_89

4,49,002 road accidents were recorded by India, out of which 1,51,113 are deaths and 4,51,361 are injuries. Savolainen and Mannering [17] developed models for motorcycle injuries in single as well as multiple vehicle crashes for Indiana, USA, using multinomial and nested logit modeling techniques. The authors concluded that motorcyclist age was one of the critical parameters causing severe injuries. Candappa et al. [3] proved the importance of developing the accident prediction models and the data required for road safety analysis. The data related to road geometry, accidents, traffic volume, and road users are useful for developing accident prediction models. Tay et al. [19] studied the factors contributing to hit and run cases in California, USA, by developing a logistic regression model. The parameters like traffic control devices, speed limit, and road profile are associated with "hit and run" cases for fatal crashes. Chen and Chen [5] collected ten years of accident data for a rural highway and developed accident prediction models using a mixed logit-modeling technique and studied the severity of injury in case of truck accidents.

Hu and Donnell [9] collected 5-years of crash data and developed injury-severity models for rural divided highways in Pennsylvania for crossover and rollover crash types. They considered influential factors related to vehicle, driver, median, and roadway in developing models. Altwaijri et al. [1] used mixed logit and multinomial logit-modeling techniques to perform statistical analysis on the crash data of Riyadh city in Saudi Arabia. The authors concluded that single-vehicle crashes were more severe than multiple vehicle crashes. The involvement of non-Saudi and illiterate people was found significant with the severity of crashes. Theofilatos et al. [20] developed separate models for urban areas of Greece to determine factors responsible for severity of accident. They have used the binary logistic regression technique to compute the probability of fatality/severe injury versus slight injury. Landge and Sharma [10] conducted a study to identify the accident-prone locations on NH-58 in India, connecting New Delhi to Mana. They have used 3 years of accident data to develop accident prediction models by using multiple regression techniques. Martensen and Dupont [11] concluded about the single and multi-vehicle accidents by using logistic multiple regression techniques. They have used the crash data from six countries of Europe and identified the variables like traffic flow, junction, and the median are critical. Yannis and Dragomanovits [23] prepared a questionnaire to understand the use of accident prediction models worldwide. They conclude that the accident prediction models are essential for reducing accidents and must be used by all organizations. The study reports that many organizations worldwide are not using accident prediction modeling in achieving road safety.

Atipathi et al. [2] developed an accident prediction model for a national highway in India using linear regression analysis. The authors have used four years of accident data from the year 2009–2012 and studied the impact of crashes on the Indian economy. Naqvi et al. [15] developed accident prediction models for three national highways in India using the binomial logistic regression modeling technique and identified the factors responsible for fatal motorcycle crashes. The influential variables were type of collision, number of lanes, and vehicles. Authors recommended that riders should use properly fastened helmets along with the reflective tapes.

Wegman [22] stated that most accidents in the world happened in low and middle-income countries and observed road safety implementation in high-income countries. The author concludes that road accidents will stay an alarming issue in the future if road safety is not improved. Sing [18] analyzed the Indian road accidents at a micro-level and commented on the variation of road accidents. The interpretation depends upon the date and time of accidents, age and sex of the road user. The author predicts that the fatalities will be around 2,50,000 for the year 2025. Chandra and Mohan [4] studied the behavior of drivers on roads from the USA and India. The authors made a comparison and concluded that driver's fault is the primary reason responsible for accidents in India. Goel [7] established a model by considering the modal split as the explanatory variable and the number of accidents as the response variable. The author concludes that 2-wheelers, cars, and heavy vehicles have a higher risk of fatalities, whereas walking, bicycle and public transport use has a lower risk of fatalities.

Haghighattalab et al. [8] concluded about the importance of implementing engineering ethics, which can prevent the accidents resulting from the decisions of engineers. Valen et al. [21] examined associations of accidents due to intake of alcohol or drugs with the risk factors of driver, which significantly contribute to fatal road traffic crashes. The important factors like absence of driving licence, over speeding, no use of helmets or seat-belts were responsible for fatal injury cases. Fountas et al. [6] studied the combined result of atmospheric and lighting conditions on the severity of injury for accidents involving single-vehicles in Scotland, United Kingdom, using injury-severity data of the years 2016 and 2017. The model shows that all the conditions have the variable impacts on the severity of injury.

Qureshi et al. [16] studied the impact of mandated societal lockdown on the road accidents. Authors have found a significant minimization in road accidents leading to minor injuries. For accidents connected with fatal or severe injuries after mandated societal lockdown.

Road accidents occur by many factors like improper geometric design of roads, defects in the vehicle, driver behavior, and vehicle's speed. There is a need to reduce the number of accidents since it produces fatalities, which we cannot ignore. No doubt, as the road accidents are uncertain and cannot be zero, the minimization of road accidents can be achieved and must be studied to reduce road accidents. The use of road safety models is one of the essential tools for achieving road safety worldwide.

2 Case Study

Two case studies have the consideration for developing simple linear regression and binomial logistic regression models. For simple linear regression models, the data is taken from the report "road accidents in India-2016" which is published by the 'ministry of road transport and highways [13], government of India. This information is available for entire India, consisting of 36 states for the year 2016. The accident data selected for developing models from the report is as shown in Table 1.

Table 1 Accident data for developing models 1–4

Sr. No.	Location	Type of classification	Categorization	Actual records
1	36 States of India	Type of junctions	T-junction	Total number of accidents Total number of fatalities Total number of injuries
2		Responsibility of driver	Intake of alcohol	

For developing a binomial logistic regression model, the accident data is collected from the local police stations of Pune city, Maharashtra, India, over a stretch of 19 km.

The accident data from the year 2014–2019 is collected, which has classified accidents like date and time, age of complainer, number of accidents, and number of fatalities. The complainer's age is the dependent variable, and the model predicts the likelihood of fatality.

3 Methodology

Regression is one of the essential techniques of predictive analytics in which a lot of prediction problems are modeled. Regression is a kind of supervised learning algorithms since a regression model requires both the dependent as well as the independent variables in the data set [12].

3.1 Simple Linear Regression

Regression is used to find the relationship a tool for finding an association relationship between a response variable (Y) and one or more explanatory variables ($X_1, X_2 \ldots X_n$) of the study. The relationship can be linear or non-linear. The Eq. (1) is not exact; due to the error term ε.

$$Y = \beta_0 + \beta_1 X + \varepsilon \tag{1}$$

where Y = response variable, X = explanatory variable, β_0 = Y-intercept, and β_1 = slope of the line.

The intercept is the mean value of the dependent variable when the independent variable is 0. The slope is the change in the dependent variable's value for a unit change in the independent variable.

3.1.1 Assumptions in Linear Regression Models

- The error term is normally distributed.
- For various values of explanatory variable, the variance of the error remains same.
- There is no auto-correlation between the two error values.

3.1.2 Coefficient of Determination (R-square)

R-square measures the fitness of the regression line with the data. Its estimate ranges from 0 to 1, and it explains the percentage of variation of the regression model. R-square is equal to the ratio of the sum of squares due to regression to the total sum of squares.

3.1.3 Standard Error and T-test

Standard error estimates the regression errors' standard deviation, which measures the observed values' variability or scatters around the regression line. A smaller standard error of the estimate indicates a better fit. If the standard error is zero, then there is a perfect fit. If the significance value for the T-test is less than 0.05, the model is accepted.

3.2 Logistic Regression

Logistic regression estimates the conditional probability of an event. It evaluates how the likelihood of an event gets affected by explanatory variables. The logistic function is as shown in Eq. (2).

$$P(Y = 1) = \pi = e^z/\left(1 + e^z\right) \tag{2}$$

3.2.1 Binomial Logistic Regression

In this method, the response variable is binary. The response variable can have only two values. The logistic function (sigmoidal function) is as shown in Eq. (3).

$$\pi(z) = e^z/\left(1 + e^z\right) \tag{3}$$

where, $z = \beta_0 + \beta_1 x_1 + \ldots. + \beta_n x_n$.

3.2.2 Interpretation of Logistic Regression Coefficients

Wald's test checks the importance of explanatory variables, which is identical to the t-test in simple linear regression. When the p-value is below 0.05, then the explanatory variable is significant.

The omnibus test checks the significance of the model coefficients. When the p-value is less than 0.05, then the explanatory variable is significant. Hosmer–Lemeshow goodness of fit test checks the model's overall fitness for a binary logistic regression. When the p-value is greater than 0.05, then there is no difference between the predicted and observed frequencies; that is, the model is accepted.

3.2.3 Classification Table

It is the table, which classifies all the probabilities into outcomes depending on the cut-off value used.

$$\text{Sensitivity} = [P/(P + Q)] \tag{4}$$

$$\text{Specificity} = [R/(R + S)] \tag{5}$$

where P = true positive; Q = false negative; R = true negative; S = false positive.

3.2.4 Receiver Operating Characteristic (ROC) Curve

It plots the graph of false-positive ratio and the true positive ratio. The higher the area covered under the ROC curve, the better the prediction ability of the model.

4 Results and Discussion

The results with discussion for all developed models are as follows, showing separate computations for both the case studies considered here.

4.1 Prediction of Fatal Accidents Based on the Total Number of Accidents on T-junction

SLR predicts the total number of fatalities with the help of a total number of accidents as the explanatory variable, as shown in Table 2. The r-square value is 0.826,

Table 2 Model-1 summary and coefficients

ANOVA						
Model-1	*Sum of squares*	*Df*		*Mean square*	*F*	*Significance*
Regression	20,411,348.17	1		20,411,348.17	160.993	0.000
Residual	4,310,669.827	34		126,784.407		
Total	24,722,018.00	35				
Model summary						
Model-1	*R*	*R-square*		*Adjusted R-square*	*Standard error*	
	0.909	0.826		0.821	356.0680	
Coefficients						
Model-1	*Unstandardized coefficients*			*Standardized coefficients*	*t*	*significance*
	B		*Standard error*	*Beta*		
(constant)	75.945		70.224		1.081	0.287
Total number of accidents	0.271		0.021	0.909	12.688	0.000

which means the developed model explains 82.6% of the variation. The SLR model developed is given by Eq. (6).

$$Y = 75.945 + 0.271 * X \tag{6}$$

where Y = Total number of fatalities, X = Total number of accidents.

The independent variable's significance value is less than 0.05; the model is accepted. The significance value for F-test is less than 0.05, which implies that the model is accepted. The histogram with the residuals' distribution in Fig. 1 shows the curve's approximate normal distribution. The p-p plot of the dependent variable indicating the expected and observed cumulative probabilities, as shown in Fig. 2, ensures the model's fitting. This p-p plot can be more close concerning the diagonal line, which implies developing a more accurate model than the model-1.

Model-2 has a better fit as shown in Table 3. The dependent variable is the log natural of the total number of fatalities, and the independent variable is the total number of accidents. The r-square value is 0.455, which implies that the model explains 45.5% of the variation, and the standard error is 0.7583, which is significantly less than model-1.

The SLR model developed is given by Eq. (7).

$$\text{Log}(Y) = 1.645 + 0.0003 * X \tag{7}$$

where Y = Total number of fatalities, X = Total number of accidents.

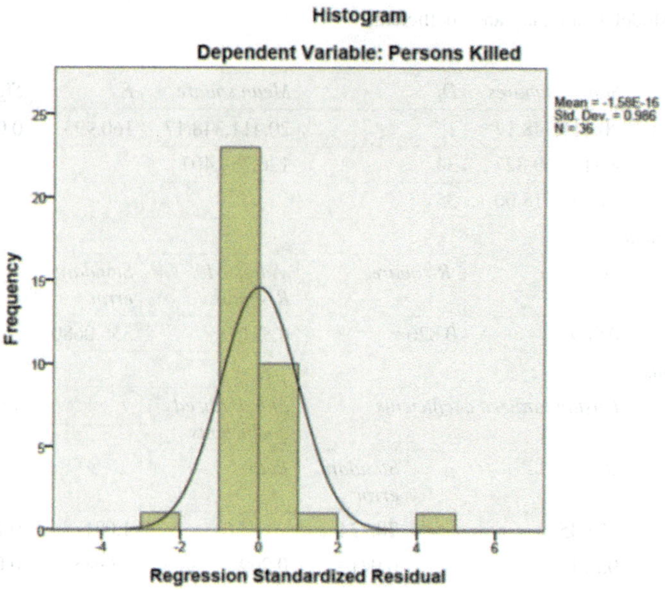

Fig. 1 Histogram, model-1

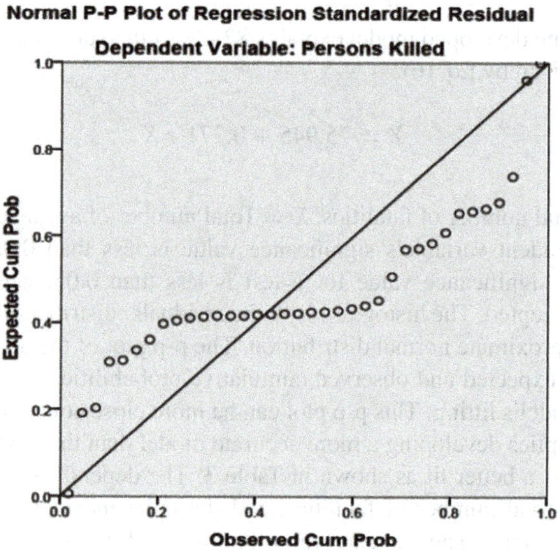

Fig. 2 p-p plot, model-1

Table 3 Model-2 summary and coefficients

ANOVA						
Model-1	*Sum of squares*	*Df*		*Mean square*	*F*	*Significance*
Regression	15.357	1		15.357	26.700	0.000
Residual	18.405	32		0.575		
Total	33.762	33				
Model summary						
Model-1	*R*	*R-square*		*Adjusted R-square*	*Standard error*	
	0.674	0.455		0.438	0.758397	
Coefficients						
Model-1	*Unstandardized coefficients*			*Standardized coefficients*	*t*	*Significance*
	B		*Standard error*	*Beta*		
(constant)	1.645		0.156		10.560	0.000
Total number of accidents	0.0003		0.000	0.674	5.167	0.000

The significance value is also less than 0.05; therefore, the model is accepted.

The histogram with the residuals' distribution in Fig. 3 shows the normal distribution of the curve.

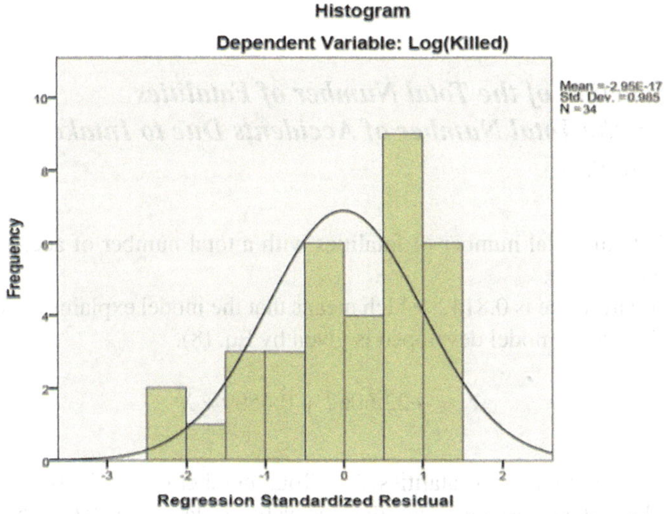

Fig. 3 Histogram, Model-2

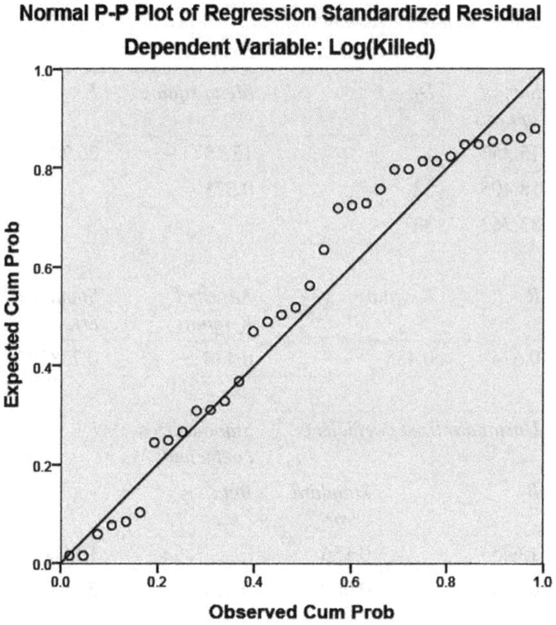

Fig. 4 p-p plot, Model-2

The dependent variable's p-p plot shows the expected and observed cumulative probabilities given in Fig. 4, which ensures the model's fitting. This p-p plot is more close concerning the diagonal line, and hence model-2 is more accurate than model-1.

4.2 *Prediction of the Total Number of Fatalities from the Total Number of Accidents Due to Intake of Alcohol*

SLR predicts the total number of fatalities with a total number of accidents as the explanatory variable.

The r-square value is 0.8142, which means that the model explains 81.42% of the variation. The SLR model developed is given by Eq. (8).

$$Y = -22.6062 + 0.4599 * X \tag{8}$$

where Y = Total number of fatalities, X = Total number of accidents

The independent variable's significance value is less than 0.05; the model is accepted, as shown in Table 4. Also, the F-value's significance is less than 0.05, which implies that the model is accepted. The histogram showing the residuals'

Table 4 Model-3 summary and coefficients

Summary Output					
Regression statistics					
Multiple R	0.902377083				
R-square	0.8142844				
Adjusted R-square	0.807880414				
Standard error	215.3632453				
Observations	31				
ANOVA					
	df	*SS*	*MS*	*F*	*Significance F*
Regression	1	5,897,512.924	5,897,512.924	127.1527412	4.04361E-12
Residual	29	1,345,058.495	46,381.32743		
Total	30	7,242,571.419			
	Coefficients	*Standard error*	*t*	*P-value*	
Intercept	−22.60615674	43.3374	−0.52163	0.605888	
X Variable 1	0.459898409	0.04078	11.27620	4.043E-12	

distribution in Fig. 5 indicates the curve's approximate normal distribution. The dependent variable's p-p plot shows the expected and observed cumulative probabilities are shown in Fig. 6, ensuring the model's fitting. This p-p plot can be more

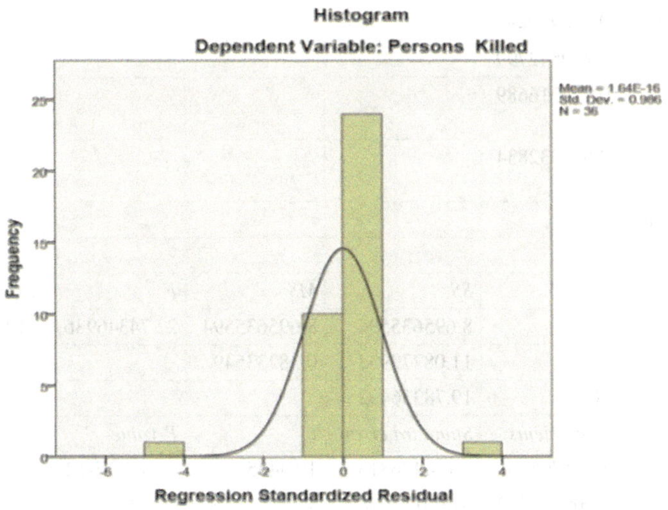

Fig. 5 Histogram, Model-3

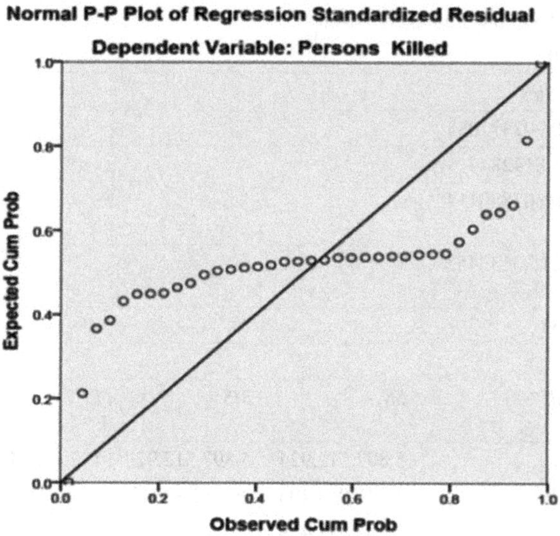

Fig. 6 p-p plot, Model-3

close concerning the diagonal line, which implies developing a more accurate model than model-3.

Table 5 Model-4 summary and coefficients

Summary output					
Regression statistics					
Multiple R	0.66298024				
R-square	0.439542799				
Adjusted R-square	0.420216689				
Standard error	0.618332834				
Observations	31				
ANOVA					
	df	*SS*	*MS*	*F*	*Significance F*
Regression	1	8.695635594	8.695635594	22.74346936	4.81901E-05
Residual	29	11.08772933	0.38233549		
Total	30	19.78336492			
	Coefficients	*Standard error*	*t*	*P-value*	
Intercept	1.36179556	0.124426813	10.9445828	8.19322E-12	
X Variable 1	0.000558442	0.000117098	4.769011361	4.81901E-05	

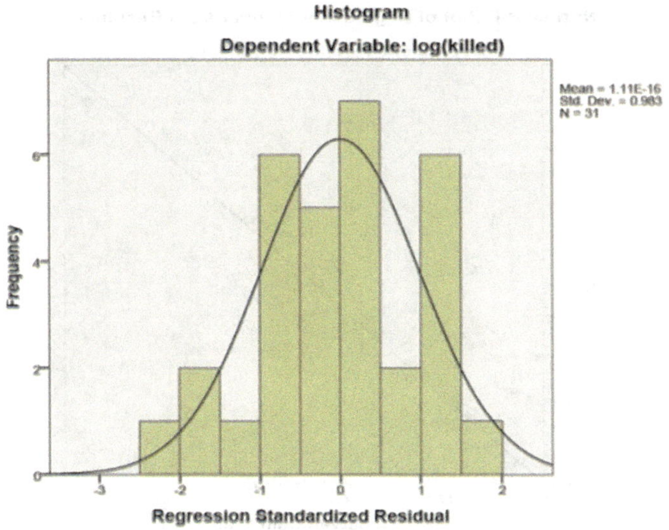

Fig. 7 Histogram, Model-4

Model-4 has a better fit, shown in Table 5. The dependent variable is the log natural of the total number of fatalities, and the independent variable is the total number of accidents. The r-square value is 0.4395, which implies that the model explains 43.95% of the variation, and the standard error is 0.6183, which is significantly less than the model-3.

The SLR model developed is given by Eq. (9).

$$\text{Log}(Y) = 1.3618 + 0.00056 * X \tag{9}$$

where Y = Total number of fatalities, X = Total number of accidents.

The histogram and p-p plot of the logarithmic model is as shown in Figs. 7 and 8. The fitting of the logarithmic model-4 is perfect as compared to the original model-3.

4.3 *Binomial Logistic Regression*

This model helps to predict whether the occurrence of an individual accident results in fatal or non-fatal injury. The fatality has an output of 1, and the non-fatality has an output of 0. The explanatory variable is the age of the complainer. The model output is as shown in Table 6. The total observations are 118, and the data for all 118 observations have consideration in the model building.

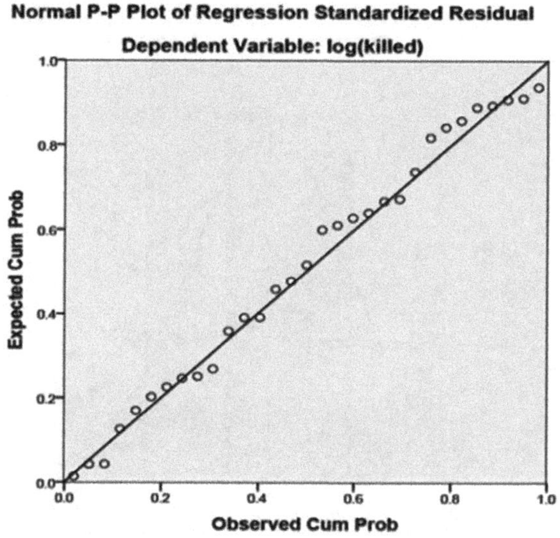

Fig. 8 p-p plot, Model-4

The omnibus test result shows that the significance values are less than 0.05; therefore, the model is accepted. Similarly, the Hosmer–Lemeshow test's output reveals a significance value greater than 0.05; therefore, the model is accepted.

The classification table automatically uses a cut-off value of 0.5, which means that if the probability of the outcome is less than 0.5, it takes the value 0 as an output. If the outcome's likelihood is higher than 0.5, it has 1 as an output value. Here, as shown in Table 6, all cases are classified correctly out of eighty-five numbers of 0-cases. The specificity value is 100%. Similarly, Out of thirty-three numbers of 1-cases, all are classified incorrectly. The sensitivity value is 0%, and the overall percentage is 72%. This value implies that the cut-off should be clearly defined to increase the sensitivity of the model.

The variables in the equation are as shown in Table 6, and the model built is as given in Eq. (10).

$$\pi(z) = e^{z}/\left(1 + e^{z}\right) \tag{10}$$

where z = 0.627–0.048* age of complainer.

The explanatory variable's significance value is less than 0.05, and the confidence interval does not contain the value of 1; therefore, the model is accepted. Now, changing the cut-off value to 0.33, the results are as shown in Table 6. The omnibus test result shows that the significance values are less than 0.05; therefore, the model is accepted. Similarly, the Hosmer–Lemeshow test's output shows a significance value greater than 0.05; therefore, the model is accepted. The cut-off value of 0.33 means that if the probability of the outcome is less than 0.33, it takes the value 0

Table 6 Model-5 summary and coefficients

Summary							
Selected cases	Included	*N*	*Percent*				
		118	100				
	Missing	0	0				
	Total	118	100				
Omnibus test of model coefficients							
		Chi-square	*Df*	*Significance*			
	Step	7.542	1	0.006			
	Block	7.542	1	0.006			
	Model	7.542	1	0.006			
Hosmer and Lemeshow Test							
		Chi-square	*Df*	*Significance*			
	1	7.324	8	0.502			
Classification Table with cut-off value 0.5							
	Observed		*Predicted*				
			Fatal/Non-fatal		*% correct*		
			0	*1*			
1	Fatal/Non-fatal	0	85	0	100		
		1	33	0	0		
	Overall %				72		
Variables							
	B	*S.E*	*Wald*	*df*	*Significance*	*Exp (B)*	*95% C. I for Exp (B)*

(continued)

Table 6 (continued)

Summary

							Lower	*Upper*
Age of complainer	- 0.048	0.019	6.080	1	0.014	0.953	0.917	0.990
Constant	0.627	0.638	0.968	1	0.325	1.873		

Classification Table with cut-off value 0.33

Step	*Observed*		*Predicted*					
			Fatal/Non-fatal		*% correct*			
			0	*1*				
1	Fatal/Non-fatal	0	59	26	69.4			
		1	13	20	60.6			
	Overall %				66.9			

as an output. If the outcome's likelihood is greater than 0.33, it takes the value 1 as an output. Here, as shown in Table 6; Out of eighty-five numbers of 0-cases, fifty-nine cases are classified correctly, and twenty-six instances have been classified incorrectly. The specificity value is 69.40%. Similarly, Out of thirty-three numbers of 1-cases, twenty cases are classified correctly, and thirteen cases are classified incorrectly.

The sensitivity value is 60.60%, and the overall percentage is 66.90%. Now, these computations look good as compared to the cut-off value of 0.5.

Finally, the ROC curve explains the model's overall effect, as shown in Table 7 and Fig. 9.

The area covered under the ROC curve is 0.678, which means that with 67.8% accuracy, the model quantifies the output as 1. Therefore, the model is acceptable and can predict fatality or non-fatality when a particular accident occurs, depending upon the complainer's age.

Table 7 Area under the ROC curve

ROC summary (model-5)				
Area	Standard error	Asymptotic significance	Asymptotic 95% confidence interval	
			Lower bound	Upper bound
0.678	0.056	0.003	0.567	0.788

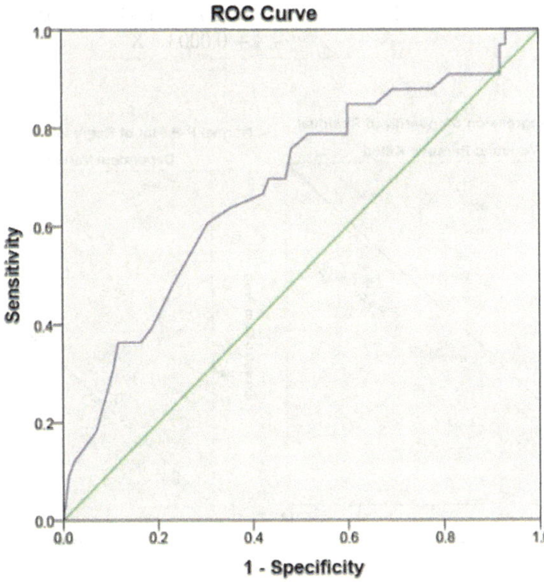

Fig. 9 ROC curve

5 Conclusions

The four simple linear regression models predict the number of fatalities with the total number of accidents that happened in various conditions.

5.1 Prediction of Fatalities Based on the Number of Accidents on T-junction

Out of Models 1 and 2, model-2 is the best-fitted model as the standard error value is significantly less than model-1, as shown in Table 8. Also, the comparison of the p-p plot between model-1 and model-2, as shown in Fig. 10, suggests using model-2 for predicting the number of fatalities.

Since the variables in model-1 are number of accidents and number of fatalities, they have a perfect correlation and hence the r-square value is high. In model-2, the correlation between the number of accidents and log of the number of fatalities is not as perfect as model-1 and hence the r-square for model-2 is less than model-1.

Table 8 Model 1 and 2 summary

Model number	Model	R-square	Standard error
1	$Y = 75.945 + 0.271 * X$	0.826	356.06
2	$\text{Log}(Y) = 1.645 + 0.0003 * X$	0.455	0.7584

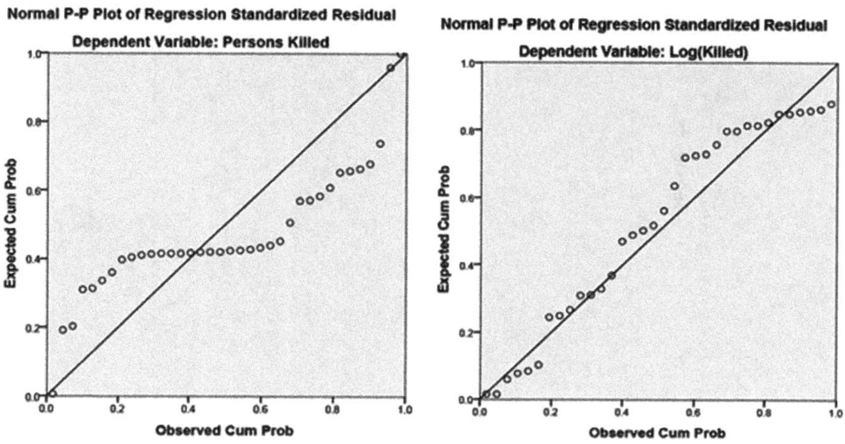

Fig. 10 Comparison of the p-p plot between model-1 and model-2

Table 9 Model 3 and 4 summary

Model number	Model	R-square	Standard error
3	$Y = -22.6062 + 0.4599 * X$	0.8143	215.36
4	$\text{Log}(Y) = 1.3618 + 0.00056 * X$	0.4395	0.6183

5.2 Prediction of Fatalities Based on the Total Number of Accidents Due to Intake of Alcohol

Out of models 3 and 4, model-4 is the best-fitted model as the standard error value is significantly less than model-3, as shown in Table 9.

Also, the comparison of the p-p plot between model-3 and model-4, as shown in Fig. 11, suggests using model-4 for predicting the number of fatalities.

Four simple linear regression models are developed, and two of them are the best fit. The fitted models are logarithmic-linear models. The model's output is to find the number of fatalities based on the total number of accidents. The special cases considered here are accidents on a T-junction and accidents due to intake of alcohol. The binary logistic regression model is developed for 118 number accidents from the year 2014 to 2019, and the prediction of causing fatality is computed by using a cut-off probability value of 0.33. The overall model is accepted. The developed models are valid.

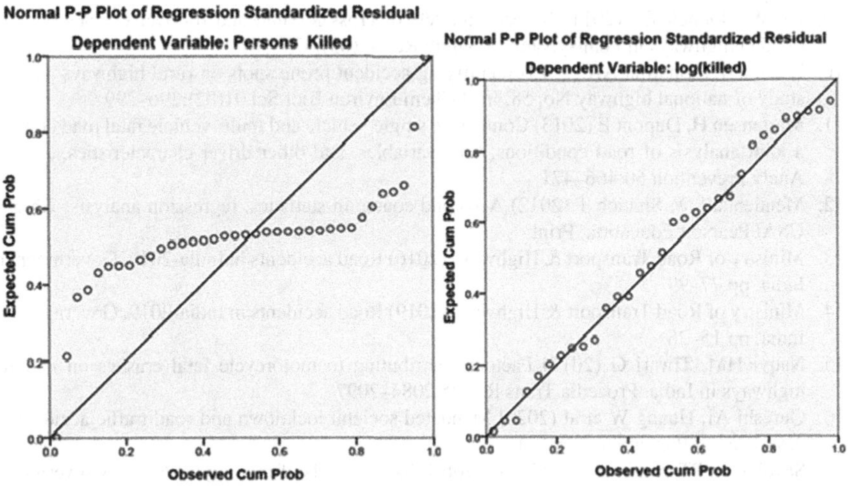

Fig. 11 Comparison of the p-p plot between model-3 and model-4

Acknowledgements The authors wish to thank the Civil Engineering Department, College of Engineering Pune, for giving the opportunity to do research. Also, extremely grateful to Local Police stations Pune City, Ministry of Road Transport and Highways for providing data used in this study.

References

1. Altwaijri S, Quddus M, Bristow A (2012) Analysing the severity and frequency of traffic crashes in Riyadh city using statistical models. Int J Trans Sci Technol 01(04):351–364
2. Athipathi G, Nagan S, Baskaran T (2017) Accident prediction on national highways in India. Int J Theor Appl Mech 12(01):153–165
3. Candappa NL, Schermers G et al (2014) Data requirements for road network inventory studies and road safety evaluations- guidelines and specifications. Road Infrast Saf Manage Evaluat Tools 03:1–87
4. Chandra S, Mohan M (2018) Analysis of driver behavior at unsignalized intersections. J Indian Road Congress 79(02):05–10
5. Chen F, Chen S (2011) Injury severities of truck drivers in single and multi-vehicle accidents on rural highways. J Accid Analy Prevent 43(05):1677–1688
6. Fountas G, Fonzone A et al (2020) The joint effect of weather and lighting conditions on injury severities of single-vehicle accidents. J Accid Analyt Methods Accid Res 27:1–19
7. Goel R (2018) Modeling of road traffic fatalities in India. J Accid Analy Prevention 112:105–115
8. Haghighattalab S, Chen A, Fan Y, Mohammadi R (2019) Engineering ethics within accident analysis models. J Accide Analy Prevention 129:119–125
9. Hu W, Donnell ET (2011) Severity models of cross-Median and rollover crashes on rural divided highways in Pennsylvania. J Safety Res 42(05):375–382
10. Landge VS, Sharma AK (2013) Identifying accident prone spots on rural highways—a case study of national highway No. 58. Int J Chem Environ Biol Sci 01(02):296–299
11. Martensen H, Dupont E (2013) Comparing single vehicle and multi-vehicle fatal road crashes: a joint analysis of road conditions, time variables, and other driver characteristics. J Accid Analy Prevention 60:466–471
12. Mendenhall W, Sincich T (2012) A second course in statistics: regression analysis, 7th edn. USA: Pearson Education, Print
13. Ministry of Road Transport & Highways (2016) Road accidents in India-2019. Government of India, pp 77–99
14. Ministry of Road Transport & Highways (2019) Road accidents in India-2019. Government of India, pp 15–25
15. Naqvi HM, Tiwari G (2017) Factors contributing to motorcycle fatal crashes on national highways in India. Procedia Trans Res 25:2084–2097
16. Qureshi AI, Huang W et al (2020) Mandated societal lockdown and road traffic accidents. J Accid Analy Prevention 146:1–4
17. Savolainen PT, Mannering F (2007) Probabilistic models of motorcyclists injury severities in single and multi-vehicle crashes. J Accid Analy Prevention 39(05):955–963
18. Singh SK (2017) Road traffic accidents in India: issues and challenges. Trans Res Procedia 25:4708–4719
19. Tay R, Barua U, Kattan L (2009) Factors contributing to hit and run in fatal crashes. J Accid Analy Prevention 41(02):227–233
20. Theofilatos A, Graham D, Yannis G (2012) Factors affecting accident severity inside and outside urban areas in Greece. J Traffic Injury Prevention 13(05):458–467

21. Valen A, Bogstrand ST, Vindenes V, Frost J, Larsson M, Holtan A, and Gjerde H (2019) Driver-related risk factors of fatal road traffic crashes associated with alcohol or drug impairment. J Accid Analy Prevention 131:191–199
22. Wegman F (2017) The future of road safety: a worldwide perspective. J IATSS Res 40:66–71
23. Yannis G, Dragomanovits A et al (2016) Use of accident prediction models in road safety management: an international inquiry. Trans Res Procedia 14:4257–4266

Finite Element Analysis for Parametric Study of Mega Tunnels

Shilpa Kulkarni and M. S. Ranadive

Abstract The development of any country has a positive relation to the infrastructural development, and the tunnels are a very much important part of it. The increased need for tunnel infrastructure in urban areas demands tunnel excavation using Tunnel Boring Machine (TBM) to avoid surface traffic disturbance during the project execution period. The construction of Mega Tunnels provides a successful and profitable solution for urban development. Mega Tunnel is a large-diameter tunnel in which the diameter is greater than 10 m. The construction of Mega Tunnels involves more cost of construction for the station, as we must go deeper for stability purposes. Face Stability of such types of tunnels has been a hot topic for researchers, as this issue gets more critical with the increase in tunnel diameter. The collapse pressure plays an important role in the face stability of tunnels with increased overburden. Therefore, with the advantage of mega tunnel construction, ground-induced settlement and deformation should be taken into account. This paper focuses on the ground-induced settlement at various cover to diameter (C/D) ratios. The parametric study had been done by three-dimensional finite element analysis using Midas Gtx Nx software with different C/D ratios and pressure parameters for soft soil. The diameter of the tunnel is considered as 15 m. The analysis showed that the application of requisite pressures plays a significant role in the face stability of mega tunnels. Conclusions were drawn based upon analysis, and approximate preliminary lining thickness was suggested for different C/D ratios, which will be beneficial to ascertain preliminary tunnel design in actual practice.

Keywords Finite element analysis · Mega tunnel · TBM · Face stability · Collapse pressure

S. Kulkarni (✉)
College of Engineering Pune, Pune, India
e-mail: shilpakul10@gmail.com

M. S. Ranadive
Department of Civil Engineering, College of Engineering Pune, Pune, India
e-mail: msr.civil@coep.ac.in

M. S. Ranadive et al. (eds.), *Recent Trends in Construction Technology and Management*, Lecture Notes in Civil Engineering 260,
https://doi.org/10.1007/978-981-19-2145-2_90

1 Introduction

Urban infrastructure development is a key to progress in the livelihood of any country. It saves time as well as the per capita cost of transport. Due to urbanization, the city population growth rate increased rapidly in the last few decades. It minimizes the usable surface area required for public utility purposes. Therefore, as the need for urbanization increases, the importance of tunnel construction for infrastructure development also rises. The single-bore, large-diameter tunnels are a feasible solution for time and cost-saving in the construction of tunnel infrastructure projects in urban areas. It proves beneficiary as it can be facilitated with a multi-operating transport system.

Tunnel excavation using TBM gives various major benefits for such projects, as it gives more structural stability and safety at the face and working area which has more prime concern while tunneling work. It also gives more speedy construction due to continuous operations. Though some challenges are intercrop while accessing the face stability, Wang et al. [1] proposed DEM model to study the face stability, and impact of depth and friction angle was clarified. Khezri et al. [2] assessed the stability of tunnel face in layered soil by using an upper bound theorem of limit analysis. A numerical approach to study tunnel face stability with the experimental and analytical approach presented and found in good agreement using Midas was developed by many researchers [3]. Mollon et al. [4] presented a probabilistic approach to determine the range of pressures that can be applied to tunnel face to avoid face collapse or blow-out. Authors have also suggested that the failure probability is very high due to the application of insufficient pressures. Pan and Dias [5, 6] present the need for consideration of pore water pressure in face stability analysis of circular tunnel and conclude that water table exerts significant unfavorable impact on it.

The face stability of the tunnel was affected by two main aspects such as geotechnical parameters of surrounded ground strata and the application of various pressures in the excavation and construction stages. Previous research showed the impact of one of the above aspects with consideration of only one pressure parameter. The objective of this study is to analyze the face stability of the mega tunnels by analyzing the effect of various pressures such as earth pressure, Jack thrust, and skin friction between ground strata and TBM. The effect of pore water pressure and tunnel lining thickness on face stability was also analyzed. This integrated approach of study benefited to tunnel industry for preliminary tunnel design.

2 Literature Review

As per Mollon et al. [7], the determination of minimal pressure and ground deformation to prevent the collapse of tunnel face was very important for the determination of tunnel face stability issue. The C/D ratio has more importance in the failure mechanism. Mollon et al. [8] suggests that the soil mass verging, the tunnel face can collapse into the tunnel if the applied face pressure drops below a critical value. It also states that if C > D, the collapse mechanism never offshoots. Chen et al. [9] studied the face instability of shield tunnels in the sand by proposing an experimental model at various C/D ratios. Soil arching during failure was monitored. Souza et al. [10] gives various critical pressure parameters to be considered for EPB TBM while tunnel advancing and its effect on structural stability. It also gives required calculation models for various pressures like thrust pressure and grout pressure. Zeng et al. [11] concludes that the face pressure is an important parameter to avoid face collapse. This also confirmed that the reliability of tunnel face stability increases significantly as face support increases or tunnel diameter decreases. Zhang et al. [12] concluded that the face stability had become a more challenging issue with increment in the tunnel diameter. Hernandez et al. [13] analyzed the face stability by numerical method at higher tunnel depth and suggest more pressure needed to support excavation face in soil. Shiau and Al-Asadi [14] describes the combined effect of surcharge load, self-weight, and internal supporting pressure and studies stability numbers for the circular tunnel in cohesive soil.

3 Data Collection

Data were collected for various geotechnical and pressure parameters required for tunnel modeling for soft computing analysis, as shown in Tables 1 and 2.

Table 1 Geotechnical parameter

No.	Design parameter	Soil	Segment	Steel	Grout
1	Modulus of elasticity (E) (MPa)	$1.3e^6$	$2.1e^7$	$2.5e^8$	$1e^7$
2	Poisson's ration	0.3	0.3	0.2	…
3	Unit weight (kN/m^3)	19	24(saturated)	78	22.5
4	Cohesion (Mpa)	15	…	…	…
5	Friction angel (°)	30	…	…	…
6	K_0	0.5	1	…	…

Table 2 Pressure parameters (set 1)

No.	Type of pressure	Pressure QTY (kN/m^2)
		Pressure set 1
1	Drilling/Horizontal/Face pressure	200
2	Jack thrust	−4500
3	Pressure due to contraction over the body of the shield TBM	50
4	Pressure due to gap between shield TBM and the segment diameter	1000

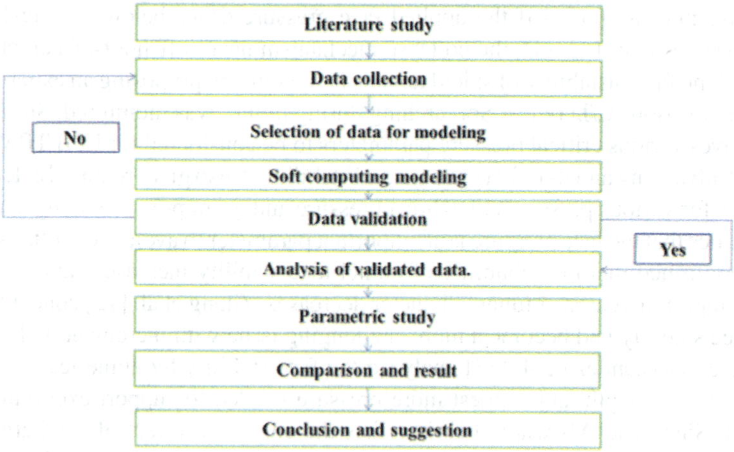

Fig. 1 Research methodology flow chart

4 Problem Statement

To perform a face stability analysis of a large single-bore tunnel of 15 m. diameter in soft soil.

5 Research Methodology

The adopted research methodology is given in the flow chart. At first, the geotechnical data and primary set of pressure parameter to be applied to the tunnel model was collected from various literature. These data were analyzed by the soft computing method for validation purposes. It was observed that collected pressure parameters were inadequate for the face stability of large-diameter tunnels. Hence, again data were collected. Once it validates, a parametric study had been done for drawing conclusions, and suggestions were made according to it (Fig. 1 flow chart).

6 Data Analysis

The three-dimensional models were prepared for analysis at different C/D ratios by keeping longitudinal margin and vertical side margin below tunnel invert. Cover depth (C) variate from C = 1D to C = 12D. Geotechnical data of soft soil and pressure parameters (shown in Tables 1 and 2) were taken from the collected data. Mohr's coulomb model was used. Five main cases were considered for study as follows and they are discussed in detail in the next section.

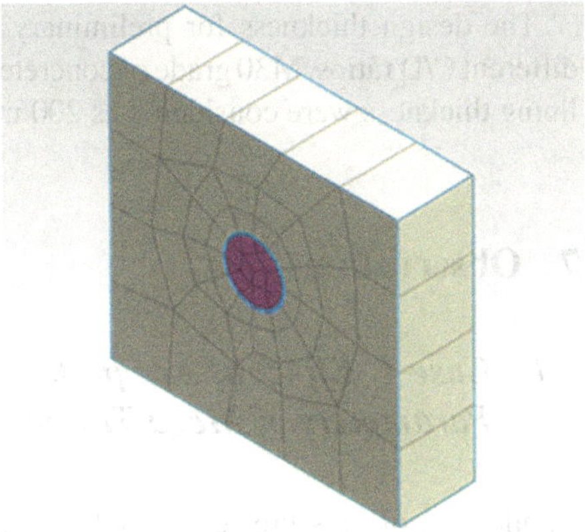

Fig. 2 Typical modeling and meshing for analysis

- Effect of adequate pressure on tunnel face.
- Effect of surcharge load on tunnel face.
- Effect of surcharge load with 10 m water table on tunnel face.
- Effect of surcharge load with 10 m submerged water level on tunnel face.
- Effect of lining thickness on tunnel face.

Figure 2 shows the typical tunnel modeling used for analysis. 13 sets of analysis, for each main case, were performed for observation purposes

Initially, to validate the collected data, models were prepared for C/D = 1, C/D = 2, C/D = 4, and C/D = 6, respectively, and the displacement profile was observed for the tunnel face at key locations, i.e., at the crown, at the invert, and at the right and left vertical side walls. The analysis and obtained profile showed that the applied pressure parameters were inadequate. Hence, required face pressure was calculated from modeling, i.e., by numerical method. From empirical formulas which were based upon the relation of jack thrust, TBM contraction pressure, and TBM gap pressure with respect to pipe diameter of TBM, the weight of pipe, length of pipe, the weight of Shield, and segment thickness were calculated by considering 2% volume loss. Meshing was done at the closed interval for accuracy. EPB TBM was considered, and construction stages for simulation were given according to it. The parametric study was done at the crown, at the invert, and at the right and left vertical walls of the tunnel for each model.

Maximum displacement and Total translation due to three directional loads at four key locations were observed. Self-weight was considered for analysis. After the application of calculated pressures, get an acceptable displacement profile. To observe it closely, again analysis was done by modeling at C/D = 1to C/D = 4 at every 0.5 m interval, C/D = 4 to C/D = 8 at every 1 m. interval, and C/D = 10 and C/D = 12 at 2 m. interval. Analysis result was observed for all cases. From observed results, conclusions were drawn for tunnel face stability at the preferred C/D ratio on structural stability and on economical basis.

The design thickness for preliminary lining purposes was suggested for three different C/D ratios. M30 grade of concrete was considered. Minimum and maximum lining thickness were considered as 200 mm and 450 mm, respectively.

7 Observations

7.1 Case 1. Effect of Adequate Pressure on Various Parameters of Mega Tunnel

Application of pressure set 1 (Table 2) on TBM while tunnel advancing gives the minimum displacement of the crown and the invert, comparatively with both vertical walls. Vertical walls displace more because of inadequate face pressure and jack thrust. Figure 3 shows the displacement profile. It refers that while advancing of mega tunnels adequate application of pressure plays an important role.

Mollon et al. [8] suggested the collapse pressure should be based on the work equation, to avoid tunnel face collapse into the tunnel which are dependent on the following:

- The weight of soil.
- The collapse face pressure.
- A possible surcharge loading on the ground.

After calculating the adequate pressure parameter (Table 3), maximum displacement of the crown and the invert of the tunnel was observed, and comparatively less displacement of both vertical walls was found, as shown by the graph in Fig. 4.

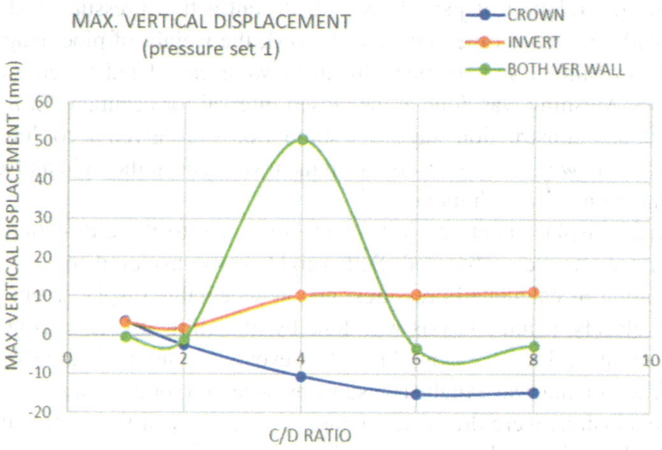

Fig. 3 Graph showing vertical displacement profile due to application of pressure set 1

Table 3 Pressure parameter (set 2)

No.	Type of pressure	Pressure QTY (kN/m^2)
		Pressure set 2
1	Drilling/Horizontal/Face pressure	232.5
2	Jack thrust	−75670
3	Pressure due to contraction over the body of the shield TBM	2120
4	Pressure due to gap between shield TBM and the segment diameter	5510

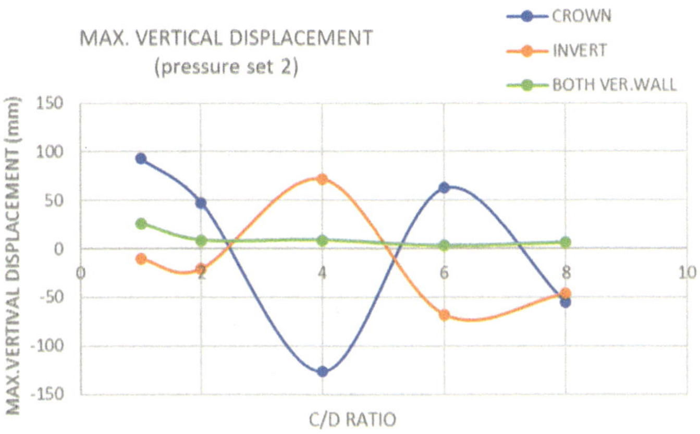

Fig. 4 Graph showing vertical displacement profile due to application of pressure set 2

7.2 *Case 2. Effect of Surcharge Load on Various Parameters of Mega Tunnel*

The arching effect was observed at C/D = 1 and C/D = 1.5 because of unequal distribution of stresses due to sufficiently lesser critical cover respective to the tunnel diameter. After C/D = 2, the displacement of nodes starts to change, and the crown of the tunnel starts to accept a particular trend of displacement till C/D = 4. The tunnel showed stabilized behavior in C/D = 4 to C/D = 8. The tunnel crown gives stable displacement but the invert showed a considerable rise in displacement in this region.

Due to the application of uniformly distributed pressure and static load, both side walls, i.e., right and left vertical walls give equal displacement. But it is negligible as compared to the crown and the invert (Fig. 6 for displacement profile) (Fig. 5).

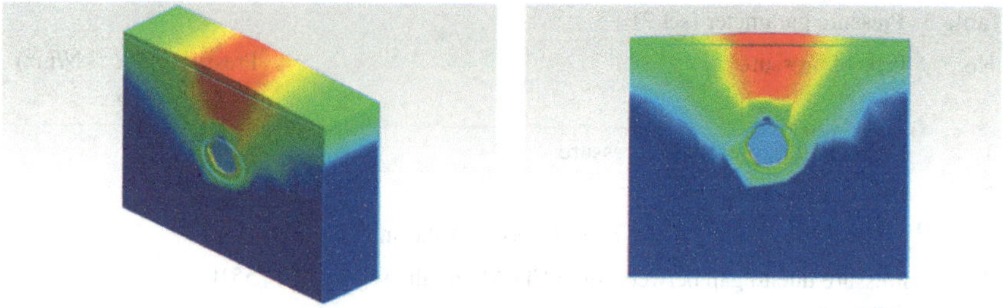

Fig. 5 Figure showing front and 3D view of the arching effect of the ground

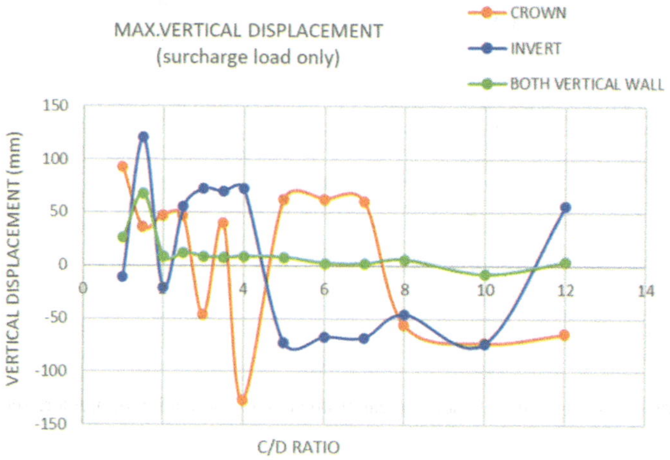

Fig. 6 Graph showing vertical displacement profile for the effect of surcharge load on tunnel key locations

7.3 Case 3: Effect of Surcharge Load with 10 m. Groundwater Table on Various Parameters of Mega Tunnel

The crown and the invert show the same displacement profile at C/D = 1 to C/D = 2, as that of the previous case in the application of only surcharge load. But, from C/D = 3, the downward displacement will be decreased up to C/D = 4. The displacement of the crown increases from C/D = 5 and C/D = 7 and shows a slight decrement at C/D = 6. The displacement of the crown at C/D = 8 decreased considerably. As stated above, the crown displacement decreased rapidly but from C/D = 8 to C/D = 12, it goes on increasing continuously. The invert also showed the same displacement as that of the previous case. From C/D = 3 to C/D = 12, the only change was the direction.

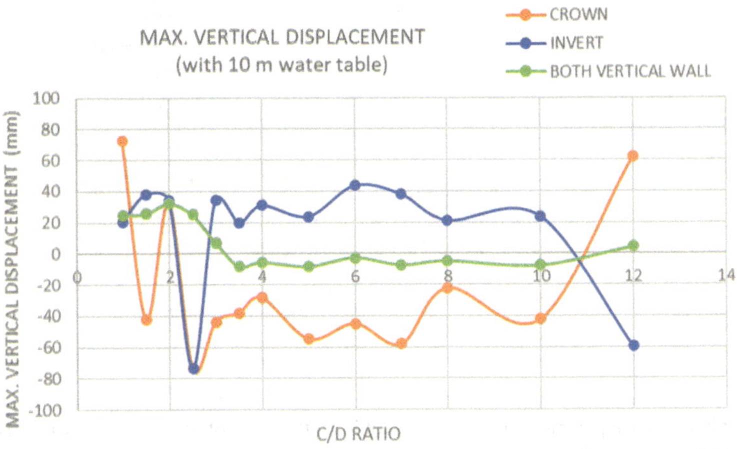

Fig. 7 Graph showing vertical displacement profile for the effect of surcharge load with 10 m groundwater table on tunnel key location

In this case, each vertical wall shows the same displacement profile as that of the invert and the crown though it was negligible in its comparison (Fig. 7 for displacement profile).

7.4 *Case 4: Effect of Surcharge Load with 10 m. Submerged Water Level on Various Parameters of Mega Tunnel*

The crown was displacing considerably in the downward direction from C/D = 1–2, but at C/D = 2, the crown displaced in the upward direction. The displacement gets increased slightly up to C/D = 3 followed by the change in the profile after it. At C/D = 3.5, the sudden change was observed in the displacement profile of the crown and the invert in opposite direction followed by a sudden change at C/D = 4.

The crown gets displaced in the downward direction up to C/D = 6. After it continued to be displaced in the upward direction up to C/D = 8. The invert showed the same displacement profile as that on the crown but in the opposite direction. Vertical displacement profile starts to be stable from C/D = 8, and it showed a stable profile till C/D = 12. The important behavior shown by the crown and the invert, in this case, is that the amount of displacement from C/D = 3.5 to C/D = 12 was the same but in the opposite direction.

In this case, each vertical wall shows a negligible displacement as compared to the crown and the invert (Fig. 8 for displacement profile).

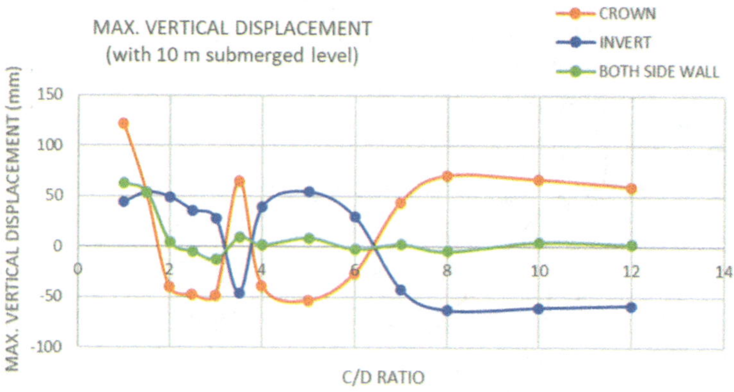

Fig. 8 Graph showing vertical displacement profile for the effect of surcharge load with 10 m water table on tunnel key locations

7.5 Total Translation of Mega Tunnel

Total translation of the tunnel is the resultant displacement that occurs of the node due to the application of the pressures on that node in three directions. For tunnel face stability purposes, the total translation of the tunnel due to pressure in all three direction is important to observed, with maximum displacement of the crown, the invert, and the vertical walls. From analysis, it was seen that between C/D = 2 and C/D = 4, case 2 (with only surcharge load) and case 4 (surcharge with10 m submerged level) show the nearly equal translation. Case 3 (surcharge load with 10 m water table) had a maximum total translation than the other two cases. Case 2 and case 4 show consequently low total translation.

At C/D = 3.5, C/D = 7 and C/D = 8 give a nearly equal total translation of tunnel. At C/D = 5, case 2 and case 4 show the equal total translation (Fig. 9).

8 Discussion on Cases

From all above-observed parameters, a comparison was made based on maximum vertical displacement and total translation of all four nodes, i.e., the crown, the invert, and both vertical walls.

It was seen that the displacement and the translation of mega tunnel were more due to the effect of surcharge load with respect to surcharge load with the water table and submerge level. The total translation of the tunnel was the same in all cases at C/D = 3.5, C/D = 7 from C/D = 8. Though the effect of the water table and water level was more on vertical displacement, the translation was minimum due to forces in the X- and Y-directions.

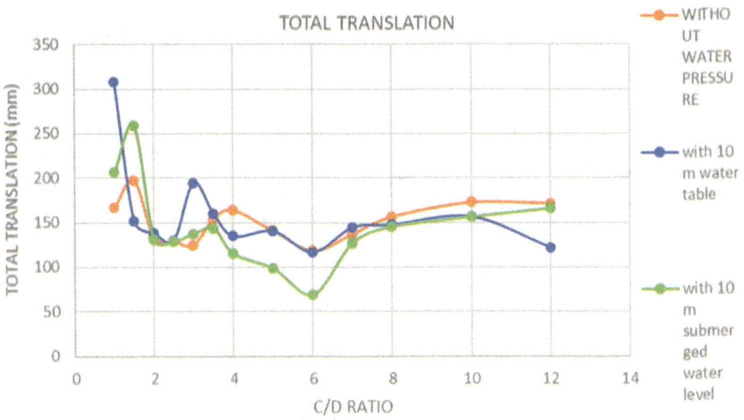

Fig. 9 Graph showing the total translation of mega tunnel

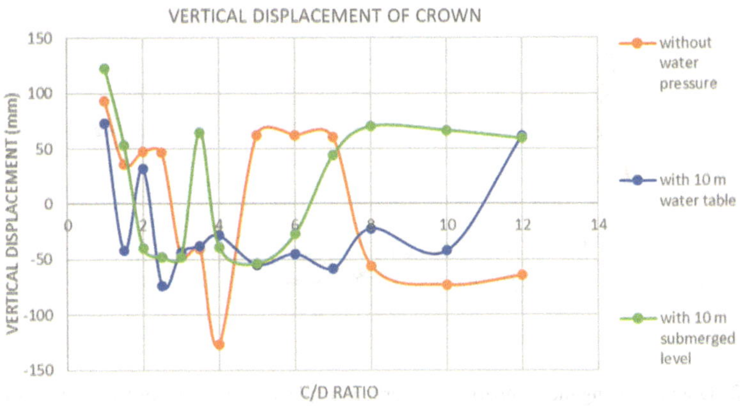

Fig. 10 Graph showing the comparison of vertical displacement profile of the crown

Previous research concludes that pore water pressure due to the presence of water table is affected more on face stability of tunnel [5, 6], but this study proposes the overburden or earth pressure above tunnel possesses more impact on tunnel face stability in soft soil (Figs. 10, 11 and 12).

9 Effect of Lining Thickness

From the above parametric study, it was observed that the tunnel starts to stabilize from C/D = 3.5, and it will show equal displacement and the total translation of the crown, the invert, and both vertical walls up to C/D = 7. Therefore, for observing the effect of lining thickness, C/D = 3.5, C/D = 5, and C/D = 7 were selected

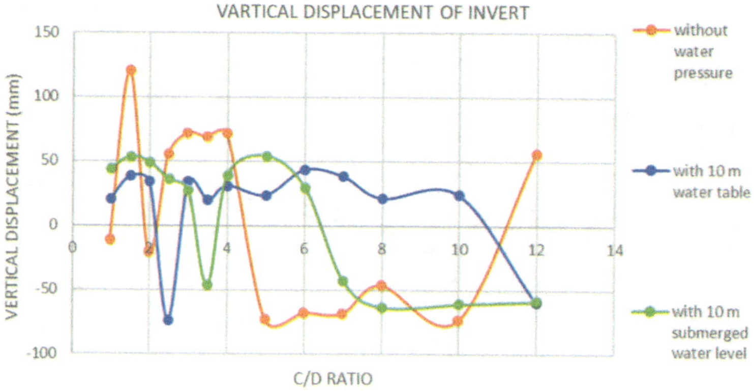

Fig. 11 Graph showing the comparison of vertical displacement profile of the invert

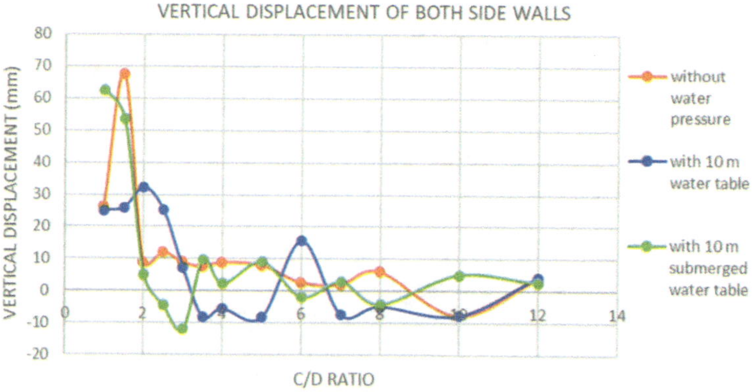

Fig. 12 Graph showing the comparison of vertical displacement profile of the vertical walls

for observation from a case surcharge with a 10 m submerged level. M30 grade of in-situ lining was considered. Maximum displacement and maximum solid stresses were observed at the crown and at the invert only. From the above study, it was observed that the displacement and translation at both vertical walls were negligible; hence, it was omitted. The range of lining thickness is in which the displacement of the node and the invert changes were observed (Figs. 13, 14, 15, 16, 17, and 18). Similarly, solid stresses were also observed. Lining thickness for preliminary design was selected. In the selected range, 50 mm thickness was added for safety purposes. The graph was plotted for the required C/D ratio vs lining thickness (Fig. 19).

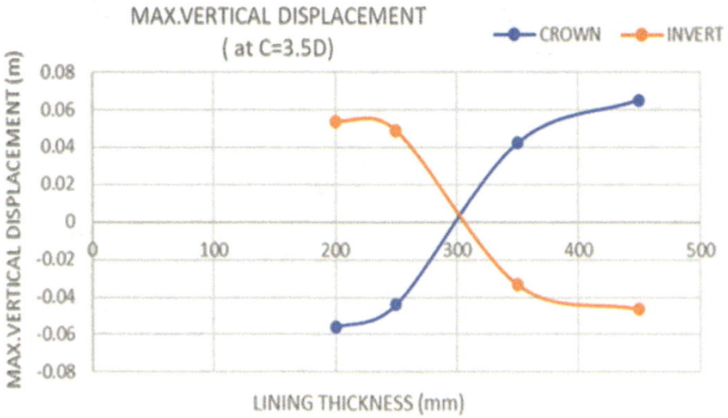

Fig. 13 Graph showing the effect of lining thickness on vertical displacement profile at C = 3.5D

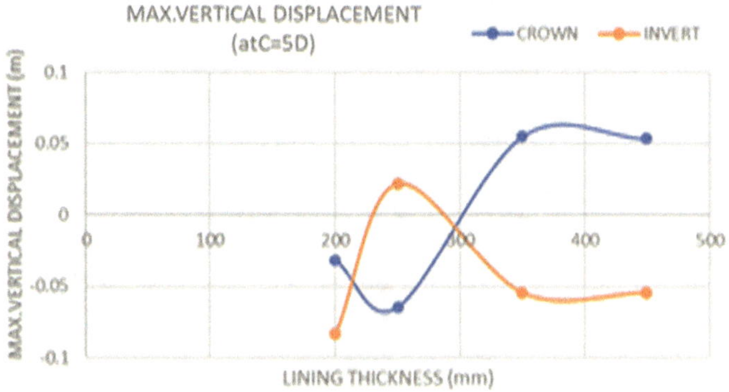

Fig. 14 Graph showing maximum solid stresses at C = 3.5D

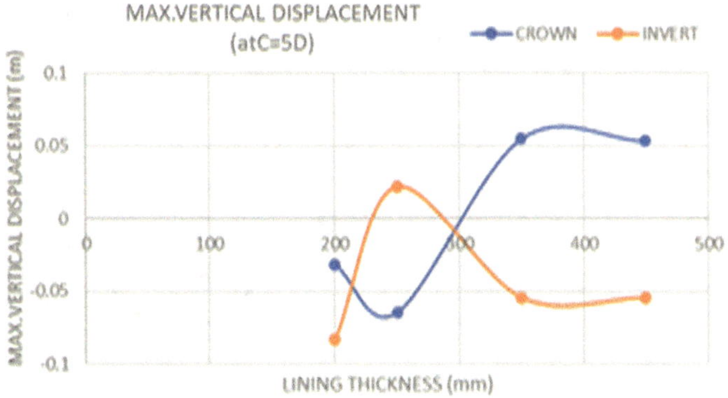

Fig. 15 Graph showing the effect of lining thickness on vertical displacement at C = 5D

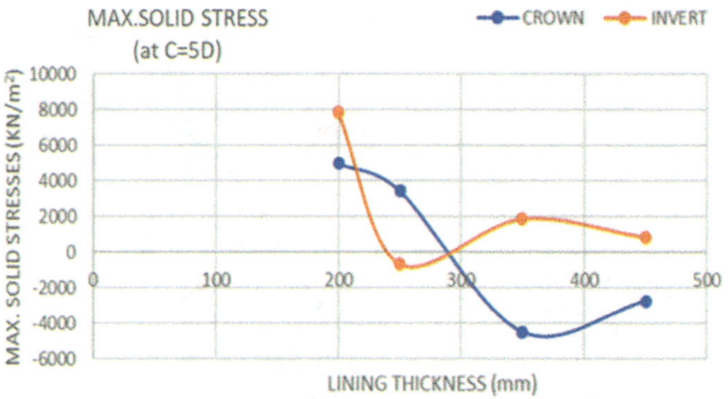

Fig. 16 Graph showing maximum solid stresses at C = 5D

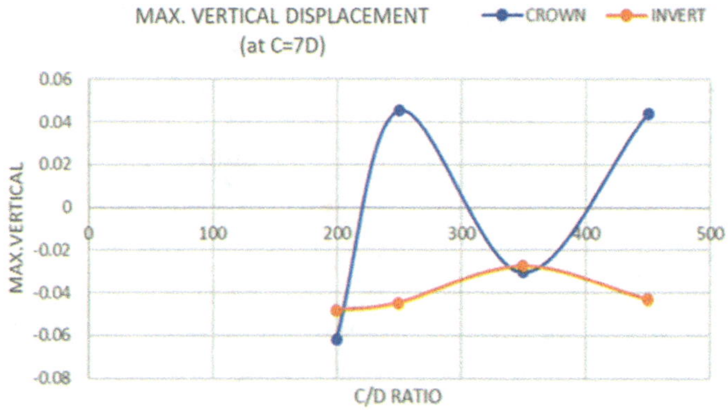

Fig. 17 Graph showing the effect of lining thickness on vertical displacement at C = 7D

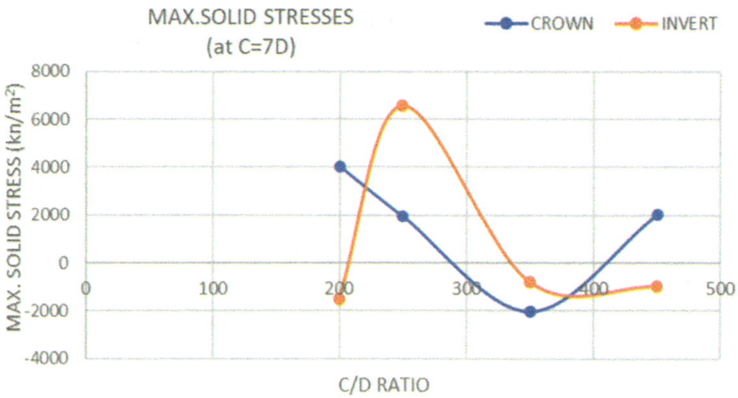

Fig. 18 Graph showing maximum solid stresses at C = 7D

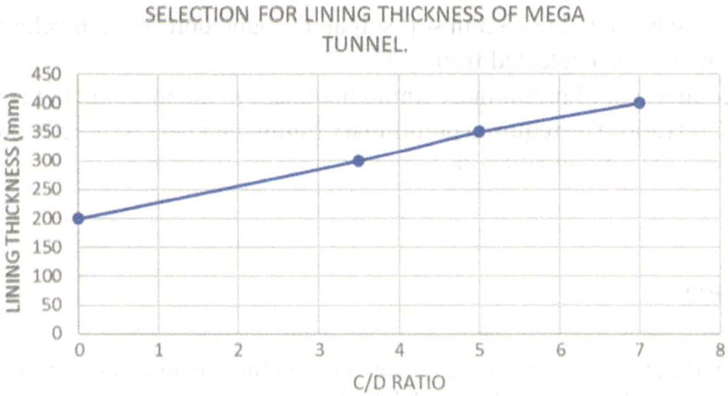

Fig. 19 Graph showing suggested preliminary lining thickness for required C/D ratio of mega tunnels

10 Suggestions

As per the above study, it can be concluded that the tunnel face can be stable at C/D = 3.5, C/D = 5, C/D = 7, and C/D = 8. From an economic point of view, preference will be given to C/D = 3.5. The required preliminary lining for the required C/D ratio is suggested and can be selected by referred graphical representation (Fig. 19).

11 Conclusions

Large-diameter tunnels more than 10 m in diameter are known as Mega Tunnel. It is beneficial to the society as it can facilitate a multi-transport system; hence, it is a profitable solution for urban infrastructure.

Face stability of Tunnel concerns with two aspects such as geotechnical strata and an adequate amount of pressure application while tunnel excavation. The elastic Modulus of strata is an important factor to decide the application of horizontal thrust. Earth pressure above the crown shows agreed behavior with horizontal thrust when adequately investigated.

The jack thrust application also varies with soil strength parameters such as unit weight and cohesion. TBM specification is one more prime concern while calculating jack thrust. It was observed that significant application of jack thrust is necessary to support excavating strata. The verge of collapse of ground is more in the initial three excavation stages.

Earth pressure or overburden is critical pressure for tunnel face stability in the case of Mega Tunnel rather than water pressure; hence, face stability may be analyzed by considering it. Hence, it can be concluded that a significant amount of applied pressure plays an important role in tunnel face stability; at this pressure, the tunnel face can be stable at C/D = 3.5 in soft soil and ground deformation behavior will be nullified and can be neglected from C/D = 7.

The requirement of preliminary lining thickness in the mega tunnel is depending upon the C/D ratio. The required preliminary lining thickness is directly proportional to the given C/D ratio in soft soil.

References

1. Wang J et al (2019) Face stability analysis of EPB shield tunnels in dry granular soils considering dynamic excavation process. J Geotech Eng, ASCE
2. Khezri N, Mohamad H et al (2016) Stability assessment of tunnel face in a layered soil using upper bound theorem of limit analysis. Geomech Eng
3. Ibrahim E, Soubra AH et al (2015) Three—dimensional face stability analysis of pressurized tunnel driven in multilayered purely frictional medium. Elsevier, Tunneling and Underground Space Technology
4. Mollon G, Dias D et al (2017) Range of the safe retaining pressures of a pressurized tunnel face by probabilistic approach. J Geotech Geoenviron Eng, ASCE
5. Pan Q, Dias D (2016) The effect of pore water pressure on tunnel face stability. Int J Numer Analy Methods Geomech
6. Pan Q, Dias D (2017) Three-dimensional face stability of tunnel in weak rock masses subjected to seepage forces. Elsevier, Tunnelling and Underground Space Technology incorporating to Trenchless technology
7. Mollon G, Dias D, Soubra A-H (2010) Face stability analysis of circular tunnels driven by a pressurized shield. J Geotech Geoenviron Eng https://doi.org/10.1061/(ASCE)GT.1943-5606.0000194
8. Mollon G, Phoon KK, Dias D, Soubra A-H (2011) Validation of a new 2D failure mechanism for the stability analysis of a pressurized tunnel face in spatially varying sand. J Eng Mech10.1061/(ASCE)EM.1943-7889.0000196
9. Chen R-P et al (2013) Experimental study on face instability of shield tunnel in sand. Elsevier, Tunnelling and Underground Space Technology
10. Souza TG et al (2015) TBM pressure models—observations. Theory and Practice, Geotechnical Synergy in Buenos Aires
11. Zeng P et al (2014) Reliability analysis of circular tunnel face stability obeying Hoek-Brown failure criterion considering different distribution types and correlation structure ASCE
12. Zhang ZX, Liu C, Huang X (2017) Numerical analysis of volume loss caused by tunnel face stability in soft soil. Springer
13. Hernandez YZ et al (2019) Three-dimensional analysis of excavation face stability of shallow tunnels. Elsevier, Tunnelling and Underground Space Technology
14. Shiau J, Al-Asadi F (2020) Three-dimensional analysis of circular tunnel heading using broms and bennermark's original stability number. Int J Geomech 20(7):06020015, ASCE
15. Broere W (1998) Face stability calculation for a slurry shield in heterogeneous soft soil. Tunnels Metropol 215–218
16. Kawadas MJ (2005) Monitoring ground deformation in tunnelling: current practices in transportation tunnels. Elsevier, Engineering Geology
17. Kim S-H et al (2006), Evaluation of shield tunnel face stability in soft ground. In: International symposium on underground excavation and tunneling
18. Kirsch A (2010) Experimental investigation of the face stability of shallow tunnel in sand. Springer
19. Senent S, Mollon G et al (2013) Tunnel face stability in heavily fractured rock masses that follows the Hoek-Brown failure criterion. Int J Rock Mech Min Sci 60:440–451

20. Zeng P, Senent S, Jimenez R (2014) Reliability analysis of circular tunnel face stability obeying Hoek-Brown failure criterion considering different distribution types and correlation structures, (ASCE) https://doi.org/10.1061/(ASCE)CP.1943-5487.0000464
21. Elarabi H, Mustafa A (2014) Comparison of numerical and analytical method of analysis of tunnel. In: Conference paper
22. Vu MN et al (2015) The impact of shallow cover on stability when tunnelling in soft soils. Elsevier, Tunnelling and Underground Space Technology incorporating to Trenchless technology
23. Xizang CL (2016) Three dimensional FEA on ground responses during twin tunnel construction using the URUP method. Elsevier, Tunneling and Underground Space Technology
24. Shiau J et al (2017) Stability charts for unsupported plane strain tunnel heading in homogeneous undrained clay. Int J GEOMATE
25. Dias D (2018) Three-dimensional face stability analysis of circular tunnels by numerical simulation, ASCE
26. Liu C, Peng Z et al (2020) Influence of TBM advance on adjacent tunnel during URUP tunneling: a case study and numerical investigation. Appl Sci
27. Neuner M et al (2020) On discrepancies between time—dependent nonlinear 3D and 2D finite element simulation of deep tunnel advance: a numerical study on Brenner Base tunnel. Comput Geotech 119:103355

Development of Financial Model for Hybrid Annuity Model Road Project

Pratiksha B. Gilbile and Gayatri S. Vyas

Abstract Public-Private-Partnership (PPP) models are in demand for infrastructure development all over the world. Road transport is an important infrastructure for the economic development of the country. Commonly adopted PPP models by the National Highways Authority of India in road projects are Build-Operate-Transfer (BOT) toll, BOT—annuity, Hybrid Annuity Model (HAM), toll-operate-transfer, operate–maintain–transfer, Engineering Procurement Construction (EPC). HAM is a combination of the traditional EPC and BOT—annuity model. HAM considers the involvement of both the parties that is the public and private sector. Mixed initial investment and different risk allocations make HAM distinct from other PPP models. PPP models are often long-term in nature, which makes both the parties check project's financial viability. This study aims to develop a financial risk model using Net Present Worth (NPW)-at-risk method. From the literature, financial key risk parameters are identified. A case study is selected and cash flows from the public's perspective are prepared. Probability distributions are allocated to risk parameters and Monte Carlo simulation is carried out using @Risk tool. The study also compares central and state policies of HAM. The findings from the developed financial model show variation of NPW as well as critical risk parameters. This financial model can be used as a decision tool for public authority and certain mitigation techniques can be applied to optimize project characteristics for the identified critical risk parameter.

Keywords Financial risk model · Hybrid annuity model · Net present worth-at-risk · Public-private-partnership

P. B. Gilbile (✉)
Construction Management, College of Engineering Pune, Pune, India
e-mail: gilbilepb19.civil@coep.ac.in

G. S. Vyas
Department of Civil Engineering, College of Engineering Pune, Pune, India
e-mail: gsv.civil@coep.ac.in

M. S. Ranadive et al. (eds.), *Recent Trends in Construction Technology and Management*, Lecture Notes in Civil Engineering 260,
https://doi.org/10.1007/978-981-19-2145-2_91

1 Introduction

Road transport is a critical infrastructure for the country's economic growth. India has the world's second-largest road network of around 58.98 lakh kilometers [1]. Road networks in India comprise expressways, national highways, state highways, major district roads, village roads. Historically, the government has been making investments in the transport sector. The Engineering, Procurement, and Construction (EPC) contracting method was in practice before the introduction of Public-Private Participation (PPP) in the early 1990s. However, to encourage private sector participation, the ministry has laid down comprehensive policy guidelines for private sector participation in the development of National Highways. In the last two decades, PPP models have been used in road infrastructure. Build-Operate-Transfer (BOT) is the most popular used PPP model. Variants of PPP are evolved to encourage the private sector's participation. Hybrid Annuity Model (HAM) is one of the variants of PPP introduced by the National Highways Authority of India (NHAI) in 2016. This introduced model is a combination of traditional EPC and BOT. One of the main risks associated with infrastructure projects is financial risk. HAM involves initial investment by both the sectors—public and private. So risk associated with the finance is altered in this model as it incorporated the concept of the time value of money and life cycle cost by introducing new bidding parameter-lowest project life cycle cost [2].

In this study, a financial risk model is developed by using Net Present Worth (NPW) @Risk concept. For developing a financial model following objectives are defined (i) To identify financial risk parameters, (ii) to define cash flows, (iii) to perform MCS in @Risk tool, (iv) to compare between state and government policies of HAM model. A case study is taken for comparison between state and government policies about HAM. Financial risk parameters affecting the Bid Project Cost (BPC) are identified and risks associated with those parameters are assigned. Monte Carlo Simulation (MCS) is done in @Risk tool to obtain the results. This developed model gives variable NPW thus helping the public sector to understand variations in financial statements. It also gives an idea about the critical risk parameter involved in cash flows.

2 Literature Review

2.1 Overview of PPP in India

The contracting method used earlier for the road network is EPC. The Government of India (GoI) has initiated economic liberalization and globalization since 1991. Inadequacies in the national road network, both in terms of magnitude and quality, have been identified as one of the hurdles to economic growth [3]. This necessitates the introduction of PPP models in India. Various PPP models used by NHAI

Table 1 Risk allocations of EPC, BOT, and HAM

Risks	EPC	BOT	HAM
Financing risk	Public sector	Private sector	Public sector
Demand risk	Public sector	Private sector	Public sector
O and M risk	Public sector	Private sector	Private sector
Construction risk	Public sector	Private sector	Private sector

are Build-Operate-Transfer (BOT) toll, BOT—annuity, HAM, toll-operate-transfer, operate–maintain–transfer. Among PPP models, BOT is one of the most popular models in roads [4]. In a typical BOT highway project, a private sector is authorized to construct and operate a transportation infrastructure facility and in return, is allowed to collect tolls for a specified period to recover all the costs and earn a reasonable profit. The contractor has financial risk in a project under the BOT model. So government authority introduced HAM which is with different risk allocations. Table 1 summarizes financing, demand, O and M, construction risk in EPC, BOT, and HAM.

2.2 Key Features of HAM

HAM is one of the PPP models in which initial investment is done by both the sectors—public and private. Following are the key features of HAM [1, 2].

- The public sector contributes to 40% of the BPC which is to be given in equal installments according to project milestones.
- 60% of BPC initially invested by the private sector, which designs, construct and maintain the project for 15 years.
- These investment ratios can be different for state and central government policies.
- Public Sector returns this (60% of project cost) in biannual annuity payments (inflation-indexed) over 15 years. Along with annual payments, interest shall be paid on reducing the balance of the cost.
- Bid parameter in HAM is the life cycle cost that is NPW of the BPC and first-year NPW of Operation and Maintenance (O and M) should be quoted. Equation 1 represents project life cycle cost.

$$\text{Lowest Project Life cycle cost} = \sum_{k=1}^{n} (\text{NPW of BPC} + \text{NPW of 1st O and M}) \tag{1}$$

- For the calculation of NPW, discounting rate is taken as bank rate + 3% are used.
- The private sector will be responsible for O and M throughout the concession period.
- Toll collection will be the responsibility of the public sector.

2.3 Financial Risk Parameters

The following 4 financial risk parameters are considered as uncertain parameters while developing a financial model:

- **BPC**

BPC is the most uncertain parameter as it involves construction cost. The minimum cost is favorable. For the analysis, a lognormal probability distribution is taken with Coefficient of Variation (CoV) 0.1 [4–6].

- **O and M cost**

O and M cost is the cost for all the expenditures during the concession period. It is the responsibility of the private sector. For this project, a normal probability distribution is considered with CoV 0.1 [6].

- **Inflation rate**

The inflation rate may vary during the project life and it can majorly affect project cash flows. It is taken as normal probability distribution with CoV 0.1 [4, 5].

- **Discounting rate**

For the calculation of NPW, discounting rate is taken as bank rate + 3%. A probability distribution is taken as a normal distribution with CoV 0.1 [4, 5].

2.4 Financial Risk and Evaluation Methods

Various risks are associated with infrastructure projects like construction risk, O and M risk, financial risk, demand risk, etc. Projects for road networks are mostly long-term in nature and require large initial expenditure. Financial management becomes necessary to determine the feasibility of the project. Various researchers have focused on different evaluation methods. These methods can be categorized into discounted cash flows and non-discounted cash flows [5]. These two methods can be differentiated based on the Time Value of Money (TVM) concept. TVM is nothing but the value of money changes with time. Discounted cash flows determine the present value of the future cash flows by considering the time value of money. Methods used in discounted cash flows are—NPW, Internal Rate of Return (IRR). Methods that are used in non-discounted cash flows do not consider the time value of money. For example—payback period, accounting rate of return. However, all these methods are based on assumption that cash flows are certain but the project's cash flows may differ from forecasted cash flows. Tiong [6] has developed an NPW @Risk method that incorporates NPW and MCS to account for uncertainty in the cash flow diagram. MCS is used to generate the possible distribution of NPW. Kumar et al.

[5] applied the NPW @Risk concept to 30 real-world BOT projects and developed a financial model. This NPW @Risk method considers probability associated with risk parameters and calculates variable NPW.

3 Research Methodology

From the literature review, financial risk parameters affecting project life cycle cost are identified. Various financial methods are studied from the literature survey and the NPW @Risk method is selected for this study. For the development of the financial risk model, a case study of "improvement of Belhe-Satara road, Maharashtra, India" is selected. Differences in-between state and central policies for HAM are studied and their effect on cash flows is comparatively analyzed. For that cash flows are developed for both cases. MCS is performed using the @Risk tool on both case studies. Comparative analysis of central and state policies is done. Figure 1 shows the research methodology of the study.

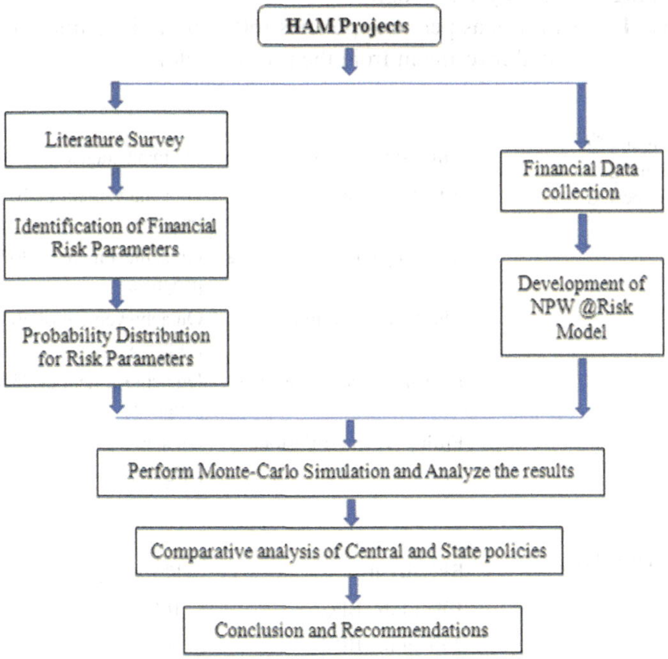

Fig. 1 Research methodology

4 Development of Financial Risk Model

A case study of improvement of Belhe–Satara road, Maharashtra India is selected. The BPC for the case study is 211 crore having a construction period of 2 years. First-year O and M quoted is 1.2 crore. Bank rate is assumed as 6.25%. For calculation of NPW, it is taken as bank Rate + 3% that is 9.25%. The inflation rate is taken as 5%. Bid due date is 23-04-2018.

Details for calculating NPW of BPC and first NPW of O and M payment are as follows:

- BPC—This is the cost of construction of the project which is payable by the public sector to the private sector. Initial Investment (60% of the BPC) given by the public sector is given in 5 equal installments during the construction period.
  Table 2 shows details of payments during the construction period.
- Adjusted BPC—It is the revised BPC accounted for variation in price index multiple. Revised values are calculated for NPW calculation. Table 3 shows details of adjusted BPC for the price index multiple confirming to respective physical progress. In Table 3, 10% of the BPC is adjusted for the price index multiple confirming to 10% physical progress.
  Case 1: Cash flows as per state policies (60% initial investment from public sector + 40% Initial investment from the private sector).

Table 2 Details for payments during the construction period

Milestone details	Payment stages
First payment milestone	On achievement of 10% physical progress
Second payment milestone	On achievement of 30% physical progress
Third payment milestone	On achievement of 50% physical progress
Fourth payment milestone	On achievement of 75% physical progress
Fifth payment milestone	On achievement of 90% physical progress

Table 3 Adjusted BPC details

BPC details	Physical progress
10% of the BPC	10%
20% of the BPC	30%
20% of the BPC	50%
25% of the BPC	70%
15% of the BPC	90%
10% of the BPC	On completion

Table 4 Calculation of construction support and adjusted BPC for case 1 and case 2

Date	Construction support for case 1	Construction support for case 2	Adjusted BPC for case 1 and case 2
23-04-2018	–	–	–
05-06-2019	26.74	17.82	22.28
28-01-2020	27.56	18.37	45.93
24-06-2020	28.07	18.71	46.78
27-10-2020	28.50	19.00	59.38
13-01-2021	28.77	19.18	35.97
05-03-2021	–	–	24.13

All values are in crore

Case 2: Cash flows as per central policies (40% initial investment from public sector + 60% Initial investment from the private sector).

- Construction support is calculated as 20% of each payment (inflation-indexed). Table 4 shows the calculation of construction support and adjusted BPC for case 1 and case 2. On a date 05-06-2019, first payment should be released which is 20% of the BPC subjected to inflation on achievement of 10% physical progress is 26.74 crore. Similarly for case 2 17.82 crore is calculated. In both the cases, adjusted BPC is calculated as 10% of the BPC on achievement of 10% physical progress.

The completion cost to be paid out in O and M period is 94.83 crore in case 1 and 141.38 crore in case 2.

O and M (inflation-indexed) calculations as per concession draft agreement are calculated in Table 5. The amount to be quoted for O and M for each annuity payment is stated in detail. For 1st annuity payment, 33.33% of quoted O and M amount should be given that is 33.33% of 1.2 crore. Calculated O and M (inflation-indexed) value is 0.47 crore.

The lowest project life cycle cost as per Eq. (1) is calculated. For case 1, static NPW is 197.95 crore and static NPW of case 2 is 195.28 crore.

5 Results and Discussion

Results were obtained in @Risk tool is to determine variable NPW. On a developed model of both cases, 1000 iterations were carried out. The calculated static NPW for case 1 is 197.95 crore. NPW of BPC of case 1 has 90% probability that value lies between 166.9 crore and 231.5 crore (Refer Fig. 2). Tornado chart gives a ranking of the risk parameters according to their effect on cash flows. BPC is the critical influencing parameter for case 1 (Refer Fig. 3). It also shows variation from the mean. NPW of BPC of case 2 has 90% probability that value lies between 165.5

Table 5 O and M calculations

Annuity	Percentage of amount quotes toward O and M (%)	Calculated O and M (in crore)
1st Annuity	33.33	0.47
2nd Annuity	33.33	0.48
3rd Annuity	33.33	0.49
4th Annuity	33.33	0.50
5th Annuity	33.33	0.51
6th Annuity	33.33	0.52
7th Annuity	50.00	0.79
8th Annuity	50.00	0.80
9th Annuity	50.00	0.82
10th Annuity	50.00	0.83
11th Annuity	50.00	0.85
12th Annuity	50.00	0.86
13th Annuity	50.00	0.88
14th Annuity	50.00	0.89
15th Annuity	66.67	1.21
16th Annuity	66.67	1.23
17th Annuity	66.67	1.25
18th Annuity	66.67	1.27
19th Annuity	66.67	1.29
20th Annuity	66.67	1.31

All values are in crore

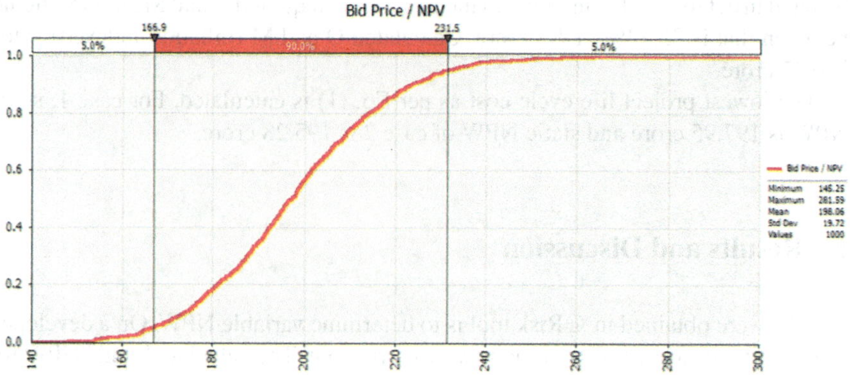

Fig. 2 NPW variation for case 1

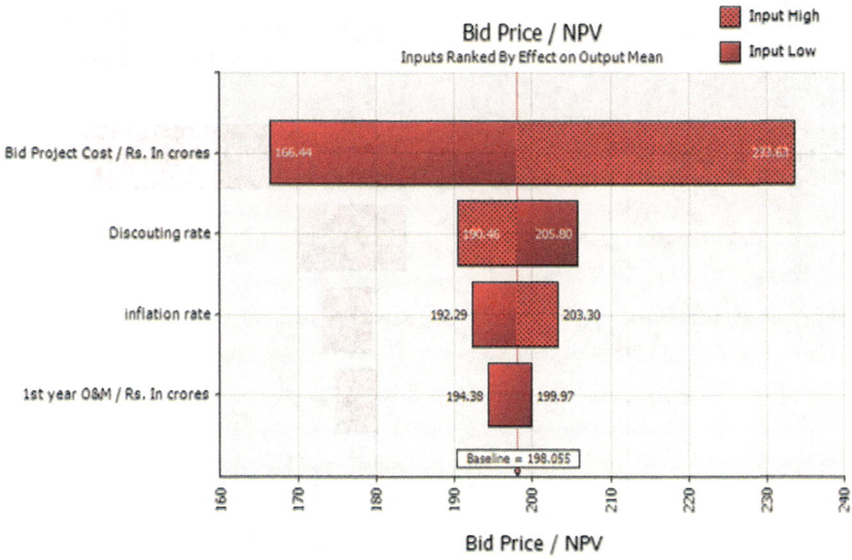

Fig. 3 Tornado chart for case 1

crore and 228.4 crore (Refer Fig. 4). Critical influencing parameter for case 2 is also coming as BPC (Refer Fig. 5).

From the results obtained through analysis, it was observed that there is not much difference between state and central policies. Table 6 summarizes comparative results between state and central policy.

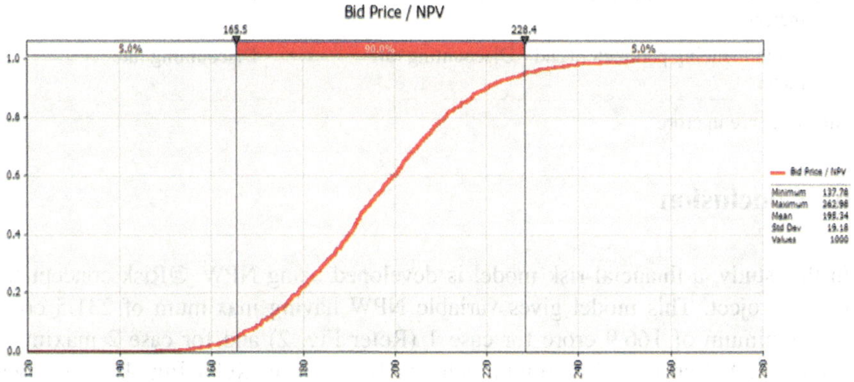

Fig. 4 NPW for case 2

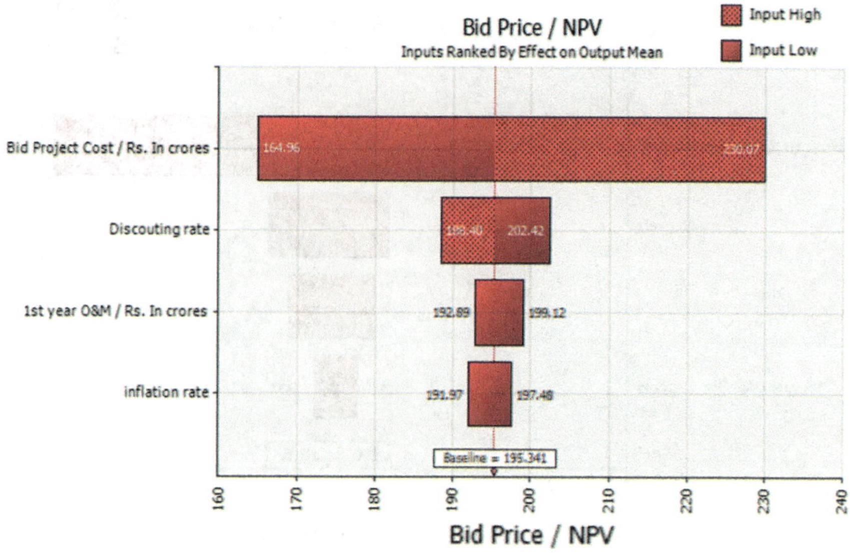

Fig. 5 Tornado chart for case 2

Table 6 Comparative results between state and central policy

	Case 1 (as per state policy)	Case 2 (as per central policy)
Static NPW	197.95	195.28
Variable NPW (Maximum value)	231.5	228.4
Variable NPW (Minimum value)	166.9	165.5
Critical influencing parameter (1st parameter)	BPC	BPC
Critical influencing parameter (2nd parameter	Discounting rate	Discounting rate

All values are in crore

6 Conclusion

In this study, a financial risk model is developed using NPW @Risk concept for HAM project. This model gives variable NPW having maximum of 231.5 crore and minimum of 166.9 crore for case 1 (Refer Fig. 2) and for case 2 maximum value is 228.4 crore and minimum value is 165.5 crore (Refer Fig. 4). This gives an idea about the uncertainty associated with risk parameters. The Tornado chart gives a critical influencing parameter. BPC is the critical influencing parameter in both cases. Mitigation techniques can be applied to reduce the effect of critical influencing parameters and maximize profit. The study also compares the NPW @Risk model between the state and central policies. The findings obtained for both cases are similar (Refer Table 6). It implies that risks associated are nearly equal irrespective of policy. It will help public sector to understand financial risk. This study can be applied to other PPP models too. Further additional risk parameters may be considered incorporating additional uncertainties which may increase the accuracy of the outcome.

References

1. GOI (2019–20) Report of Ministry of Road Transport and Highways, Government of India
2. Garg S, Mahapatra D (2019) Hybrid annuity model: hamming risk allocations in Indian highway public-private partnerships. J Public Aff 19(1):e1890
3. Singh LB, Kalidindi SN (2006) Traffic revenue risk management through annuity model of PPP road projects in India. Int J Project Manag 24(7):605–613
4. Kagne RK, Vyas GS (2020) Investigation and modeling of financial risks associated with PPP road projects in India. In: International conference on transportation and development 2020. American Society of Civil Engineers, Reston, VA
5. Kumar L, Jindal A, Velaga NR (2018) Financial risk assessment and modelling of PPP based Indian highway infrastructure projects. Transp Policy 62:2–11
6. Ye S, Tiong RLK (2000) NPV-at-risk method in infrastructure project investment evaluation. J Constr Eng Manag 126(3):227–233

Analysis of Perpetual Pavement Design Considering Subgrade CBR, Life-Cycle Cost, and CO_2 Emissions

Saurabh Kulkarni and M. S. Ranadive

Abstract The current study aims to evaluate the possibility of perpetual pavement design for the proposed ring road to the City of Pune. Perpetual Pavement concept is relatively new for Indian scenario. In India, national highways are still designed for 15–20 years of service life where as perpetual pavements can provide better service period. The Indian Road Congress guideline (IRC 37:2018) mentions various combinations of pavement design. A comparative study was carried out between proposed pavement design by Pune Metropolitan Region Development Authority and five categories of flexible pavement mentioned in IRC guidelines designed with perpetual pavement concept. The comparison is carried out on the grounds of checking the effect of raising California Bearing Ratio of subgrade soil from 10 to 15% with respect to life cycle cost assessment and carbon dioxide emissions. The study analyses whether the improvement in CBR of subgrade has any substantial change in design, cost and CO_2 emission of all five combinations of perpetual pavement. IITPAVE software was used to design perpetual pavements. KENPAVE and WESLEA software were used to validate the design. It is observed that the flexible perpetual pavement can be a better alternative considering the factors like expected service period, lifetime cost, and environmental concerns. The study also concludes that positive impact of improved subgrade CBR in terms of overall thickness, life cycle cost, and CO_2 emission depends on type of pavement combination used for perpetual pavement design.

Keywords Perpetual pavement · Life cycle cost assessment · Carbon dioxide emissions

1 Introduction

Pune is one of the largest cities and an important industrial hub in the state of Maharashtra. The traffic in and around Pune city is increasing due to increased

S. Kulkarni (✉) · M. S. Ranadive
Department of Civil Engineering, College of Engineering, Pune, India
e-mail: ssk15.civil@coep.ac.in

M. S. Ranadive et al. (eds.), *Recent Trends in Construction Technology and Management*, Lecture Notes in Civil Engineering 260,
https://doi.org/10.1007/978-981-19-2145-2_92

population and industrial growth. The bypass roads constructed for the city have now become internal part of the city limits, because of this traffic coming from outside the city and destined to another location is passing through the Pune city. This has resulted in overburden on the intra city road networks and creating traffic congestion throughout the city. The bypass roads are also experiencing heavier traffic than expected, resulting in deterioration of those roads. Therefore, to address this issue Maharashtra State Road Development Corporation (MSRDC) has decided to build circumferential 4 lane and 14 m-wide road of connecting all major highways to each other.

Government of Maharashtra has formed the Pune Metropolitan Region Development Authority (PMRDA) as the executive authority for the project [1]. The Proposed ring road Connects Pune-Mumbai, Pune-Solapur, Pune-Nagar, and Pune-Satara highway to each other bypassing the city. The service period of perpetual pavement (PP) is about 50 years. It does not require major structural repair work and to face distresses restricted to the top of the pavement, periodic maintenance with regard to surface renewal is required. India did not have any official code for the design of PP. However, with the latest publication by the IRC [2], the mechanistic method for the design of pavement to serve longer service periods can be adopted.

1.1 Objectives and Scope

- Design PP using IRC 37:2018 guidelines.
- Perform LCCA comparison between proposed pavement for ring road and five different combinations of PP.
- Determine and compare total CO_2 emissions caused by proposed pavement and different combinations of PP.
- Finding out the best suited option among five combinations of PP.
- Examine the effect of improvement in CBR of subgrade of all the five PP design combinations with respect to overall pavement thickness, total lifetime cost, and CO_2 emission.

2 Design Criteria

The compressive strain acting in the vertical direction at the top of subgrade and tensile strain acting in the horizontal direction at the bottom of asphalt layer is considered as the critical strains. The key point in PP design is to keep these strains in some particular limits. Strain level below which there is no cumulative damage over an infinite number of cycles is called as Endurance Limit (EL). The EL for horizontal tensile strain in the bituminous layer and vertical compressive strain at the top of subgrade is 70 μ and 200 μ, respectively, as per tests conducted in laboratories at Asphalt Institute at 20 °C in United States. In major part of India, the average annual pavement temperature generally is close to 35 °C. IRC 37 has proposed that for such conditions, corresponding limiting strains for the fatigue and rutting endurance limit may be considered as 80 μ and 200 μ micro strains, respectively.

2.1 *Design Traffic*

The proposed road is going to have two lanes on both side with a width of 3.75 m for each lane. For the preliminary planning purpose, the project road is divided into 19 sections. The data from the traffic survey done for the proposed ring road is used for design purpose in this paper. Almost 5976 commercial Vehicles per day are expected to be served in this project according to ring road summary report. As per IRC 37, the cumulative number of standard axles to be allowed during the design life is given by Eq. 1.

$$N = \frac{365 \times \left[(1+r)^n - 1\right] \times D \times A \times F}{r} \tag{1}$$

where

- N Cumulative number of standard axles in terms of million standard axle (msa)
- r Growth rate according to IRC (SP: 84-2009) is considered as 5%
- n Design life assumed for proposed pavement is 15 years as per PMRDA report
- D Lane distribution factor (0.75)
- A Commercial vehicles per day
- F Vehicle damage factor (6.0).

All the values are considered from IRC 37:2018 and summary report of Pune Ring Road.

According to the calculation, pavement should be designed for approximately 212 msa.

3 Information About Soil Characteristics Observed in the Region of the Proposed Project

The soils occurring in the region are mostly black soil which are classified in the four categories namely reddish-brown soil, lateritic black soil, medium light brownish black soil, and coarse shallow reddish black soil. The California Bearing Ratio (CBR) of the subgrade soil, for the design of the proposed pavement was determined as per IS:2720 Part-16. The design CBR is taken as 10%. There are many methods to improve subgrade CBR. For the design purpose, the effective resilient modulus is limited to a maximum value of 100 MPa as per the IRC guidelines. Hence, present study explores the effect of improvement in CBR from 10 to 15% in the analysis. The study also investigated whether improvement in CBR cause any beneficiary difference in pavement design, life cycle cost, and CO_2 emissions.

3.1 Resilient Modulus

For the present study, the relationship suggested in IRC 37 between resilient modulus and CBR as shown in Eq. 2 is considered.

$$M_{RS} = 17.6 * (\text{CBR})^{0.64} \quad (2)$$

where M_{RS} = Resilient modulus of subgrade soil (MPa).

Equation 3 is used here as per IRC guidelines for estimation of modulus of the granular layer.

$$M_{R\,\text{Granular}} = 0.2(h)^{0.45}\, M_{RS} \quad (3)$$

where h = Thickness of granular layer in mm.

4 Pavement Combination

The possible combinations proposed are

1. Bituminous surface course (BC) with granular sub-base (GSB) and base (Combination A).
2. BC with Cement Treated Base (CTB), Cement Treated sub-base (CTSB) and granular crack relief layer (CRL) (Combination B).
3. BC with CTSB, CTB and Stress absorbing membrane interlayer (SAMI) at the interface of base and the bituminous layer (Combination C).
4. BC with GSB, CTB and CRL (Combination D).
5. BC with CTSB and GSB (Combination E).

The SAMI is not considered as a part of structural layer for the pavement analysis. As per IS SP53:2010, SAMI may consist of elastomeric modified binder like Styrene-Butadiene Rubber (SBR) applied at the rate of minimum 1 kg/m^2, complying with Ministry of Road Transport and Highways (MoRT&H) clause number 521. As per the IRC guidelines, for protection against rutting, Viscosity Grade 40 (VG 40) bitumen shall be used for surface course and for the dense bituminous macadam (DBM). The modulus of different pavement layers as mentioned in Table 1 are considered for analysis here. Poisson's ratio for CTSB as well as for CTB is considered as 0.25 and for remaining layer types, it is considered as 0.35 as per the IRC guidelines.

Table 1 Structural layers properties as per IRC guidelines

Layer	Elastic/resilient modulus (MPa)
Subgrade	17.6 * $(\text{CBR})^{0.64}$
Unbound granular layers	$0.2(h)^{0.45}\ M_{\text{RS}}$
Granular base over CTSB sub-base	350 for crushed aggregates
CTB	5000
CRL	450
CTSB	600
Bituminous layer	3000

4.1 Fatigue Criteria

For the analysis purpose, the fatigue equation suggested in the IRC 37 guidelines was adopted.

$$N_{\text{f}} = 0.561 \times C \times 10^{-4} \times \left(\frac{1}{\varepsilon_{\text{t}}}\right)^{3.89} \times \left(\frac{1}{M_{\text{Rm}}}\right)^{0.854} \quad (4)$$

where

$C = 10^{M}$, and $M = 4.84 \times \left[\frac{V_{\text{be}}}{V_{\text{a}}+V_{\text{be}}} - 0.69\right]$

V_{a} = Percent volume of air void in the mix,

V_{be} = Percent volume of effective bitumen,

N_{f} = Fatigue life of bituminous layer,

ε_{t} = Maximum horizontal tensile strain at the bottom of the DBM,

M_{Rm} = Resilient modulus (MPa) of the bituminous mix,

C = Adjustment factor to consider the effect of effective binder volume and air void content on the fatigue life.

For the present study, the V_{a}, V_{be} values considered are 3.5% and 11.5%, respectively, as recommended in IRC guidelines. So, Eq. 4 becomes

$$N_{\text{f}} = 0.000132 \times \left(\frac{1}{\varepsilon_{\text{t}}}\right)^{3.89} \times \left(\frac{1}{M_{\text{Rm}}}\right)^{0.854} \quad (5)$$

4.2 Rutting Criteria

The equivalent number of standard axle load (80 KN) repetitions that can be served by the pavement for 90% reliability levels as per the IRC guidelines is given by Eq. 6.

$$N_R = 1.41 \times 10^{-8} \times \left(\frac{1}{\varepsilon_v}\right)^{4.5337} \quad (6)$$

where N_R = subgrade rutting life, ε_v = vertical compressive strain at the top of the subgrade.

EL provides a thickness limit for the pavement and helps to avoid extra expenditure as well as overdesign. Therefore, as shown in Tables 2 and 3, the design of pavement is carried out by trial and error with till endurance limit of the strain values closest to 80 and 200 μ, whichever first are obtained [3, 4]. For the analysis purpose, a single axle dual wheel assembly with the standard axle load of 80 KN was considered. The contact radius and tire pressure are considered as 15.5 cm and 0.56 MPa, respectively, as per the IRC guidelines.

In this study, minimum thickness permissible for CRL, sub-base and base layer is considered for all the trial sections and bituminous pavement layers are kept as variable ones for PP design. It was done so to provide sufficient stiffness in the upper pavement layers as per the concept of PP [5]. From Table 2 and putting strain values in Eqs. 5 and 6, it can be said that perpetual design provides better service period for far larger amount of traffic. The top layer acts as a protective barrier, as the distresses are restricted to the wearing course. From Table 3 and as shown in Fig. 1, it is evident that only combination A would be having greater thickness that proposed conventional pavement by PMRDA. Table 3 shows that compared to proposed conventional pavement, PP design with combination A shall have more overall thickness, which shows increase of 5.64–2.36% as subgrade CBR is improved from 10 to 15%. In case of combination B, pavement thickness is reduced by 4.83% but no change is observed even if subgrade CBR is improved from 10 to 15%. In case of combination C, pavement thickness is reduced by 22.58, 23.38, 24.1, 25, 26.61, and 27.41% as subgrade CBR is increased from 10 to 15% successively. In case of combination D, pavement thickness is reduced by 3.22%, irrespective of improvement in subgrade CBR from 10 to 15%, no change is observed in pavement thickness. In case of combination E, overall thickness is reduced by 1.61–2.41% if subgrade CBR is improved as mentioned above.

5 Life-Cycle Cost Analysis

The cost comparison for a 1 km long and 14 m wide section of Pune Ring Road for five mentioned combinations of PP was done using the standard rates in Indian rupees as per the state of Maharashtra government's schedule of rates [6]. Material

Table 2 Modulus of materials (in MPa) and corresponding strain values for pavement combinations

Trial combination	Subgrade	Sub-base	Base	Crack relief interlayer	BC + DBM	Horizontal tensile strain (μ)	Vertical compressive strain (μ)
Proposed by PMRDA	76.82	–	240.13	–	3000	151.4	245.4
Perpetual with CBR 10%							
A	76.82	–	214.45	–	3000	79.64	161.8
B	76.82	600	5000	450	3000	80	172.9
C	76.82	600	5000	–	3000	13.31	198.80
D	76.82	166.71	5000	450	3000	80	171.4
E	76.82	600	350	–	3000	78.53	171.9
Perpetual with CBR 11%							
A	81.66	–	227.96	–	3000	79.70	159.40
B	81.66	600	5000	450	3000	80	168.9
C	81.66	600	5000	–	3000	13.00	198.20
D	81.66	177.21	5000	450	3000	80	167.10
E	81.66	600	350	–	3000	78.21	167.40
Perpetual with CBR 12%							
A	86.33	–	241	–	3000	79.88	157.40
B	86.33	600	5000	450	3000	80	162
C	86.33	600	5000	–	3000	12.67	198
D	86.33	187.35	5000	450	3000	79.84	163.20
E	86.33	600	350	–	3000	79.67	166.7
Perpetual with CBR 13%							
A	90.87	–	253.67	–	3000	80.00	155.8
B	90.87	600	5000	450	3000	80	161.8
C	90.87	600	5000	–	3000	12.31	198.10
D	90.87	197.20	5000	450	3000	79.68	159.6
E	90.87	600	350	–	3000	79.49	162.9
Perpetual with CBR 14%							
A	95.29	–	266.0	–	3000	78.66	151.4
B	95.29	600	5000	450	3000	80	158.7
C	95.29	600	5000	–	3000	11.91	198.5
D	95.29	206.79	5000	450	3000	79.54	156.4
E	95.29	600	350	–	3000	79.24	159.40
Perpetual with CBR 15%							
A	99.59	–	278.0	–	3000	79.08	150.4
B	99.59	600	5000	450	3000	80	155.8

(continued)

Table 2 (continued)

Trial combination	Subgrade	Sub-base	Base	Crack relief interlayer	BC + DBM	Horizontal tensile strain (μ)	Vertical compressive strain (μ)
C	99.59	600	5000	–	3000	11.47	199.20
D	99.59	216.12	5000	450	3000	79.40	153.4
E	99.59	600	350	–	3000	79.01	156.10

Table 3 Thickness of trial combinations

Trial combination with CBR value	Sub-base (mm)	Base (mm)	CRL/SAMI (mm)	DBM (mm)	BC (mm)	Total thickness (mm)
Proposed by PMRDA (Conventional) 10%CBR	200 (GSB)	250 (WMM)	–	120	50	620
Combination A						
	GSB	WMM		DBM	BC	
10–11%	200	150		255	50	655
12%	200	150		250	50	650
13–14%	200	150		240	50	640
15%	200	150		235	50	635
Combination B						
	CTSB	CTB	CRL	DBM	BC	
10–15%	200	100	100	140	50	590
Combination C						
	CTSB	CTB	SAMI	DBM	BC	
10%	200	100		130	50	480
11%	200	100		125	50	475
12%	200	100		120	50	470
13%	200	100		115	50	465
14%	200	100		110	50	460
15%	200	100		105	50	455
Combination D						
	GSB	CTB	CRL	DBM	BC	
10–15%	200	100	100	150	50	600
Combination E						
	CTSB	Granular base course		DBM	BC	
10–11%	200	150		210	50	610
12–15%	200	150		205	50	605

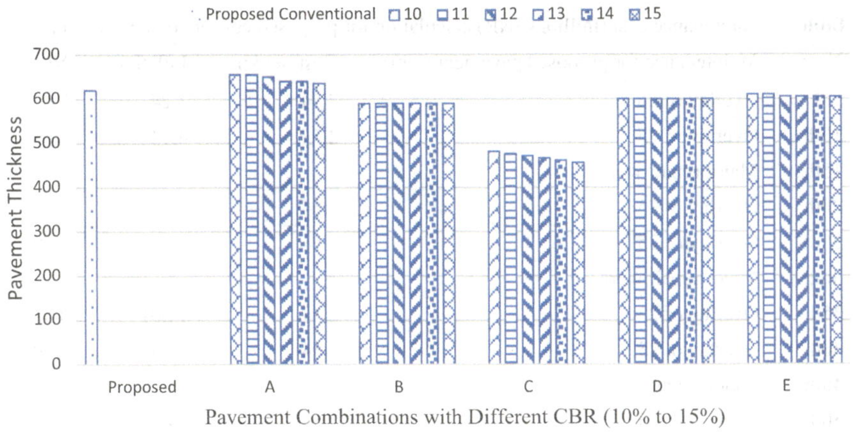

Fig. 1 Comparison of pavement thickness

Table 4 Schedule of rates

Layer	MoRT&H Clause	Rate (INR)	Unit
GSB	401	1598	m^3
WMM	406	1657	m^3
CTSB	404	1827	m^3
CTB	404	1902	m^3
CRL	401	1657	m^3
BC	509	7272	m^3
DBM	507	6742	m^3
SAMI	517	100	m^2

Specification and work requirements are considered as per MORT&H Clause [7] as shown in Table 4. In the present study, the long-term cost analysis using the Net Present Value (NPV) method has been discussed. The discount rate and an inflation rate has been adopted as 10%, and 5% respectively. MORT&H guidelines suggest that a layer of 25 mm BC should be provided once every five years. As the proposed pavement is designed for 15 years, it needs to be rebuilt after the 15th, 30th, and 45th years of service.

The present study considers limited number of parameters like cost due to initial construction, periodic maintenance, and life cycle cost for a span of 50 years. Depending on these points, cost of laying an overlay for proposed conventional pavement and five combinations of PP have been shown in Tables 5 and 6. Calculations for pavement reconstruction for proposed pavement and PP are shown in Tables 7 and 8, respectively. LCCA is carried out for 50 years to compare the lifetime cost linked with construction of one-kilometer length of PP and the proposed conventional pavement. Based on information from people related with road construction and available literature, it was found that while estimating the future cost of road

Table 5 Maintenance cost (millions INR) calculation for proposed conventional pavement

Year	Maintenance for proposed pavement (mm)	Cost per km	Inflation	NPV
5th	Overlay of 25	2.54	3.24	2.01
10th	Overlay of 25	2.54	4.14	1.60
15th	Major repair work			
20th	Overlay of 25	2.54	6.74	1.00
25th	Overlay of 25	2.54	8.60	0.79
30th	Major repair work			
35th	Overlay of 25	2.54	14.01	0.50
40th	Overlay of 25	2.54	17.88	0.40
45th	Major repair work			
50th	Overlay of 25	2.54	29.13	0.25
Total				6.55

Table 6 Maintenance cost (millions INR) for PP

Year	Maintenance for PP	Cost per km	Inflation	NPV
5th	Overlay of 25 mm	2.54	3.24	2.01
10th		2.54	4.14	1.60
15th		2.54	5.28	1.26
20th		2.54	6.74	1.00
25th		2.54	8.60	0.79
30th		2.54	10.98	0.63
35th		2.54	14.01	0.50
40th		2.54	17.88	0.40
45th		2.54	22.82	0.31
50th	Reconstruction work in 50th year			
Total				8.50

Table 7 Cost (millions INR) of pavement reconstruction for proposed conventional pavement

Year	Full pavement reconstruction	Initial construction cost per km	Inflation	NPV
15th	Proposed Conventional Pavement	26.69	38.48	18.82
30th		26.69	55.48	13.28
45th		26.69	80.00	9.37
Total				40.97

Table 8 Cost (millions INR) of pavement reconstruction (after 50 years) for proposed PP

CBR value	Initial construction cost	5.0% Inflation per annum (millions)	NPV
PP combination A			
10–11	37.11	125.67	11.60
12	36.64	124.08	11.45
13–14	35.70	120.89	11.16
15	35.23	119.30	11.01
PP combination B			
10–15	28.40	96.17	8.88
PP combination C			
10	25.16	85.20	7.86
11	24.67	83.54	7.71
12	24.20	81.95	7.56
13	23.72	80.32	7.41
14	23.25	78.73	7.27
15	22.78	77.14	7.12
PP combination D			
10–15	28.71	97.22	8.97
PP combination E			
10–11	33.38	113.04	10.43
12–15	32.91	111.44	10.29

construction, it is considered that the inflation in price happen only during 50% of the tenure of design period. The design period for the PP is 50 years. Hence, to provide an equivalent platform for comparison, LCCA is carried out for total life of 50 years as shown in Table 9. It can be observed that savings in range of 22.91–48.26% can be achieved in case of PP depending upon combination considered. Table 9 shows that any combination of PP is better option than conventional pavement even though initial construction cost is less in case of proposed pavement except compared to combination C. But if we consider life cycle cost for a span of 50 years then considerable amount of saving can be achieved by adopting any kind of PP combination as shown in Fig. 2.

6 Emission of CO_2

Embodied carbon means all the CO_2 emitted in producing and disposing materials at the end of its lifetime. The data from inventory of carbon and energy [8] as well as from the study of Auroville Earth Institute [9] was used for the calculations of embodied CO_2 as shown in Table 10. A generalized proportioning of various

Table 9 Life-cycle cost analysis for proposed conventional pavement and five combinations of PP

NPV (in millions INR)					
Pavement type and CBR	Initial construction	Overlay	Reconstruction	Total	Saving in case of PP (%)
Proposed by PMRDA	26.69	6.55	40.97	74.21	
PP combination A					
10 and 11	37.11	8.50	11.60	57.21	22.91
12	36.64	8.50	11.45	56.59	23.74
13 and 14	35.70	8.50	11.16	55.36	25.41
15	35.23	8.50	11.01	54.74	26.24
PP combination B					
10–15	28.40	8.50	8.88	45.78	38.31
PP combination C					
10	25.16	8.50	7.86	41.52	44.05
11	24.67	8.50	7.71	40.88	44.92
12	24.20	8.50	7.56	40.26	45.75
13	23.72	8.50	7.41	39.64	46.59
14	23.25	8.50	7.27	39.02	47.42
15	22.78	8.50	7.12	38.40	48.26
PP combination D					
10–15	28.71	8.50	8.97	46.18	37.77
PP combination E					
10–11	33.38	8.50	10.43	52.32	29.50
12–15	32.91	8.50	10.29	51.70	30.34

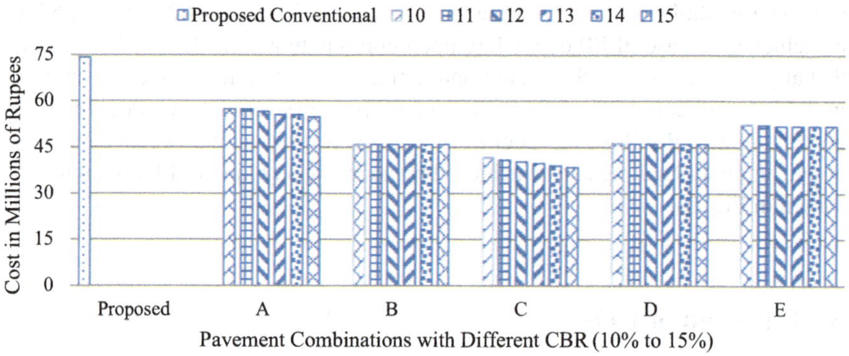

Fig. 2 Comparison of total life-cycle cost for 50 years for different pavement combination

Table 10 Embodied CO_2 (kg/kg material) of different materials

Material	Embodied CO_2
Cement	0.83
Bitumen	0.48
Coarse aggregate	0.0216
Fine aggregate	0.002

materials used in different layers of the pavement is shown in Table 11 which is based on the MoRT&H specifications, IRC SP:49 [10], IRC SP:53 [11], and IRC SP:89 [12].

Embodied CO_2 for different pavement layers is shown in Table 12. Initial total CO_2 Emission for Proposed and PP Combinations have been reported in Table 13 whereas overall CO_2 emission for span of 50 years is reported in Table 14. In case of Combination C, from the available literature, Embodied CO_2 value for SBR is considered as 2.76 kg/Kg. So, Embodied CO_2 for SAMI Layer per KM is 38.64 tons/m^3. Table 13 shows that initial total CO_2 emission of conventional pavement is less by almost 13–37%, but if we consider lifetime emissions, PP is superior option as shown in Table 14. It shows that in comparison with proposed conventional pavement for whole lifespan, combination A causes 53.16–54.42% less CO_2 emission as subgrade CBR is increased from 10 to 15% successively. Similarly, Combination B and D show a reduction of 50.30% and 55.57%, respectively. Combination E shows reduction of 49.56–49.87% as subgrade CBR is improved whereas combination C would cause reduction in CO_2 emission by 50.28–51.86% as subgrade CBR is increased from 10 to 15% successively.

Table 11 Generalized properties for various materials of flexible pavements

Properties of the mix	Pavement layers						
	CTB	CTSB	CRL	WMM	GSB	BC	DBM
Bitumen by mass (%)	–	–	–	–	–	5.5	4.5
Cement by mass (%)	2	2	–	–	–	–	–
Aggregates by mass (%)	98	98	100	100	100	94.5	95.5
Coarse fraction of total aggregates (%)	65	80	70	70	80	55	60
Fine fraction of total aggregates (%)	35	20	30	30	20	45	40
Density (kg/m^3)	2300	2300	2300	2300	2300	2400	2300

Table 12 Embodied CO_2 (in Kg) for flexible pavement layers of various combinations

Pavement layer	WMM	GSB	CTSB	CTB	BC	DBM
Embodied CO_2 for 1/m^3	36.15	40.66	78.03	71.40	92.34	79.90

Table 13 Initial total CO_2 emission for proposed conventional pavement and five combinations of PP

Proposed by PMRDA/Proposed conventional					
Total embodied CO_2 (ton)	439.24				
Total embodied CO_2 (ton)					
Combination trial	A	B	C	D	E
PP (10% CBR)	539.64	590.30	567.14	496.85	603.41
PP (11% CBR)	539.64	590.30	561.55	496.85	603.41
PP (12% CBR)	534.05	590.30	555.95	496.85	597.82
PP (13% CBR)	522.87	590.30	550.36	496.85	597.82
PP (14% CBR)	522.87	590.30	544.77	496.85	597.82
PP (15% CBR)	517.27	590.30	539.18	496.85	597.82

Table 14 Total CO_2 emission for proposed conventional pavement and five combinations of PP for 50 years span considering overlays and reconstruction

Proposed by PMRDA/Proposed conventional					
Total embodied CO_2 (ton)	1773.12				
Total embodied CO_2 (ton)					
Combination trial	A	B	C	D	E
PP (10% CBR)	830.52	881.17	858.01	787.72	894.29
PP (11% CBR)	830.52	881.17	852.42	787.72	894.29
PP (12% CBR)	824.92	881.17	846.82	787.72	888.69
PP (13% CBR)	813.74	881.17	841.23	787.72	888.69
PP (14% CBR)	813.74	881.17	835.64	787.72	888.69
PP (15% CBR)	808.14	881.17	830.05	787.72	888.69

7 Conclusion

The study led us to conclude that combination C is the most cost friendly option in addition to have least design thickness among various Combinations of PP. This also shows that strong base and sub-base layers with higher modulus can provide cost effective and strong pavement structure. Though the pavement thicknesses proposed in this study are not absolute, they give a fair idea about comparison between conventional and perpetual pavements. The concept of perpetual pavement suggests having thicker bituminous layers to keep the strain levels under limit but the study shows that providing a stable foundation, use of treated subgrade and use of high stiffness base materials are important considering the huge expenditure in road construction and maintenance. This can be related to the rising price of crude oil and therefore increased cost of bitumen in the developing country like India. Further investigation about CO_2 emissions reveals that combination D among all the five combinations is the most preferred option considering environmental aspect. The comparison among isolated perpetual pavement combination reveals that improvement in subgrade CBR combinations results in no change in overall pavement thickness, total lifetime cost, and CO_2 emission in case of combination B and D. If one considers expenditure as a major factor then the sequence of preference among the five combinations would be C, B, D, E, and A. But with respect to environmental issue the sequence of preference would be D, A, C, B, and E. India is still a developing country and signatory of Kyoto Protocol, so there can be a dilemma in choosing between least costly and most eco-friendly option. Overall, it can be said that PP should be advocated for the construction of highways in India.

References

1. Executive summary report of Pune ring road
2. Indian Road Congress IRC 37 (2012) Guidelines for design of flexible pavement
3. Ranadive M, Kulkarni S (2016) Perpetual pavement for rural roads: a concept. In: A national conference on fifteen years of PMGSY. Transportation Engineering Group, Civil Engineering Department, Indian Institute of Technology Roorkee, India
4. Kulkarni S, Ranadive M (2021) Effect of change in the resilient modulus of bituminous mix on the design of flexible perpetual pavement. In: Proceedings of the 3rd international conference on advanced technologies for societal applications, vol 1, chapter 83. Springer Nature, Switzerland AG
5. Walubita LF, Scullion T (2010) Texas perpetual pavements—new design guidelines. Report No: 0-4822-P6, Texas Department of Transportation and the Federal Highway Administration
6. Government of Maharashtra (2019–20) PWD schedule of rates
7. Ministry of Road Transport and Highways Government of India (2013) Specification for road and bridge works
8. Hammond G, Jones C (2008) Inventory of carbon & energy: ICE. Sustainable Energy Research Team, Department of Mechanical Engineering, University of Bath, UK
9. Maini S, Thautam V (2009) Embodied energy of various materials and technologies. Auroville Earth Institute (AEI), Tamil Nadu, India
10. Indian Road Congress IRC SP:49 (2010) Guidelines for the use of dry lean concrete as sub-base for rigid pavement
11. IRC SP:53 (2010) Guidelines on use of modified bitumen in road construction. Second revision, Indian Road Congress, New Delhi
12. IRC SP:89 (2010) Guidelines for soil and granular material stabilization using cement lime and fly ash. Indian Road Congress, New Delhi

Laboratory Investigation of Lateritic Soil Stabilized with Arecanut Coir Along with Cement and Its Suitability as a Modified Subgrade

B. A. Chethan, B. M. Lekha, and A. U. Ravi Shankar

Abstract If a pavement is constructed on weak soil, its lifespan drastically reduces due to the low strength induced by moisture-induced destresses. Such soils may undergo considerable changes in volume. In order to modify these properties, soil stabilization can be done. By stabilizing the soil along with the improvement in strength, its durability can be increased. Stabilization may be of chemical or mechanical type. In this investigation, lateritic soil was stabilized using 0.2–1% arecanut coir, and its compaction characteristics were evaluated. The lateritic soil is found to be nondurable. The reinforcement alone could not improve the strength and durability effectively. Therefore, 3% binding agent ordinary Portland cement (43 grade) was added to the mix. Due to cement stabilization, UCS and CBR values were improved, and the optimum values were observed at 0.6% arecanut coir dosage. The addition of cement has resulted in a change in silica, alumina, and calcium oxide contents, thereby contributing to the formation of hydration products. The samples with 1% coir and cement have completed 12 wet–dry cycles, but the weight loss observed was >14%. All the specimens showed low soil loss under freeze–thaw cycles. The performance of cured specimens under fatigue loading was satisfactory. Since the specimens could not pass wet–dry durability criteria, they can be considered for modified subgrade.

Keywords Lateritic soil · Stabilization · UCS · CBR · Durability · Fatigue

1 Introduction

In Malabar, India, laterites were used for building construction [1]. This term was cited in the eighteenth century. The lateralization process occurs due to alternate

B. A. Chethan (✉) · A. U. Ravi Shankar
Civil Engineering Department, NITK Surathkal, Mangaluru, India
e-mail: chethanba@gmail.com

B. M. Lekha
Department of Civil Engineering, KVG College of Engineering, Sullia, D.K. Dist, Karnataka, India

M. S. Ranadive et al. (eds.), *Recent Trends in Construction Technology and Management*, Lecture Notes in Civil Engineering 260,
https://doi.org/10.1007/978-981-19-2145-2_93

wet–dry seasons, where heavy rainfall, humidity, and elevated temperatures exist. India includes 60–80% red, yellowish-red laterite soils, extending from 0.1 to 0.5 m in depth. Due to their high weathering resistance, they are used as building stones. A substantial amount of aluminum (Al) and iron (Fe) is present in these soils. They are originated from igneous rocks. Leather in 1898 and Voelcker in 1893 investigated these soils [15] and stated that these soils are common in coastal areas. Weak lateritic soils can lead to pavement distresses. Therefore, the necessity of stabilization comes into the picture. Ordinary Portland cement and lime are commonly used stabilizers from the olden days. Stabilization can change the microstructure of soil due to the formation of hydration products. The lateritic soils are locally available; hence, in situ stabilization can be done at a lower cost instead of borrowing a good material. This soil is weathered naturally, where the iron, aluminum hydroxides were formed due to the chemical process [6], and may exhibit low plasticity. These hydrated oxides govern the characteristics of the deposit. If a high amount of clay is present in this soil, it may cause severe structural distress due to its low bearing capacity. The lowered plasticity may be due to sesquioxides dehydration, stronger bond formation between the particles, resisting water penetration [2].

Stabilization with a low dosage of cement is the economical and easiest way to upgrade granular soils [4, 5, 8, 22]. If the fines are <40%, cement stabilization can suit better for the granular soils. In order to improve aggregate bases, cement is found to be a suitable additive, where a linear relationship was found between cement dosage and UCS values [16]. At a cement dosage of 1 or 2%, the granular base materials performed better than recycled cement concrete materials [7]. But a higher dosage of cement may cause increased stiffness and brittle failures. An increase in compaction effort and curing period results in significant improvement of strength, and cement content reduction causes a decrease in strength [13].

India is considered to be one of the major contributors to arecanut production in the world. Karnataka and Kerala produce 72% of arecanut (482,000 tons in the cultivation area of 388,000 ha) [3]. These fibers are cheaper, naturally, abundantly available, and can be effective as reinforcing material. The hard, fibrous husk is about 30–45% of areca fruit. These fibers have rich hemicellulose content. The use of reinforcement in soil [21] may increase shear strength. Reinforced weak soil has exhibited an increase in OMC, CBR, whereas the MDD was reduced due to low fiber density [11]. The polypropylene fibers with cement showed improvement in UCS, shear strength of clayey soil. The specimens failed at higher strains due to higher ductility [20]. The fly ash and polypropylene fiber amended soil was found to be suitable for base and sub-bases [10]. The coir fiber was found to be an efficient reinforcement for improving the CBR of rural road clayey subgrade soil [14]. The fiber and sand inclusion improved the plunger penetration resistance in lithomargic clay, and resulted in higher UCS [19] values. As the arecanut coir is biodegradable, the loss of strength of reinforced soil can be expected with time. But, if a stabilizer is used along with fiber, it can compensate for strength loss, thereby not hampering the structural condition [17]. The CBR of fiber-reinforced soil with fly ash was found to be suitable for low-volume roads [9]. Similar results were observed when the soil specimens were reinforced with palm fibers [18]. Even the fiber reinforcement was found to improve CBR of black cotton soil [12].

2 Materials and Methodology

2.1 Lateritic Soil

All basic test properties of lateritic soil are determined as per relevant IS codes, a summary of which is presented in Table 1. This soil has a red color. A major fraction of the soil passes through 4.75 mm and retains over 75 micron IS sieve. Also, 32% clay portion is present in this soil; hence, this soil is classified as clayey sand (SC) as per IS code.

Laboratory chemical analysis was conducted to determine the compositions of uncemented and cemented soil. Table 2 shows that SiO_2, Al_2O_3, and CaO percentages were increased after stabilizing with cement. It indicates that silica, alumina, and CaO significantly contributed to enhance the strength of set soil.

Table 1 Basic properties of lateritic soil

Test property and values		
Specific gravity		2.45
Grain size distribution (%)	a) Gravel	9
	b) Sand	44
	c) Silt	15
	d) Clay	32
Consistency limits (%)	Liquid limit (LL)	56
	Plastic limit (PL)	29
	Plasticity index (PI)	27
IS soil classification		SC
At	IS standard Proctor compaction	IS modified Proctor compaction
MDD (g/cc)	1.68	1.91
OMC (%)	19.2	14.0
Unsoaked CBR (%)	8	13
Soaked CBR (%)	2	3
UCS (kPa)	138	206
Co-efficient of permeability (cm/s)	0.35×10^{-7}	0.20×10^{-7}

Table 2 Chemical composition of untreated and treated soil

Oxides (%)	SiO_2	Fe_2O_3	Al_2O_3	S	CaO	MgO	pH	EC (μS/cm)
Lateritic soil	55.36	6.24	4.98	0.042	0.019	0.003	5.91	1.27
Stabilized lateritic soil	58.26	5.06	7.21	0.19	2.45	2.94	8.1	4.28

Table 3 Physical properties of arecanut coir

Diameter (mm)	Length (mm)	Density (g/cc)	Young modulus (kN/mm^2)	Tensile strength (kN/m^2)
0.35	28	1.09	27	2.2

Table 4 Chemical composition of arecanut coir

Cellulose (%)	Hemicellulose (%)	Lignin (%)	Ash (%)	Pectin (%)	Wax (%)
Nil	35–64.8	13–24.8	4.4	Nil	Nil

2.2 *Arecanut Coir*

Arecanut coir was collected from Puttur Taluk, Karnataka. Dried arecanut shells have a brown color. These shells were soaked in water for a week, and coir was manually extracted. The physical and chemical properties of the arecanut coir are tabulated in Tables 3 and 4. These fibers have an aspect ratio of 80 and a specific gravity of 0.67.

The dosage of coir can have significant effects on the strength improvement of reinforced soil. A low coir dosage may be insufficient to bind the soil and may lead to low strength. A high coir dosage may be ineffective due to excess fiber concentration, which hampers strength development. Therefore, the optimum dosage was estimated based on experimental results by varying dosages from 0.2 to 1%. All strength tests, durability, and fatigue tests were performed to justify the effectiveness of fiber dosage.

2.3 *Cement*

Based on the literature, 3% of ordinary Portland cement of 43 grade was used as a stabilizer for this investigation. Cement is a substance, a binder that sets and hardens on curing. The free carbon dioxide from the air reacts with cement to bind mixed materials together. In cement manufacture, basic raw materials used are limestone or chalk, containing calcium carbonate, and clay or shale, containing silica, alumina, and iron oxide. Its basic properties are tabulated in Table 5.

Table 5 Physical properties of cement

Test conducted	Results	Requirements as per IS 8112-1989
Specific gravity	3.16	–
Normal consistency (%)	32	–
Setting time (minutes)	Initial 60	Not >30 min
	Final 225	Not >600 min
Fineness (m^2/kg)	330	Not <300 m^2/kg
Soundness (mm)—Le Chatelier test	2.50 (Expansion)	Not >10 mm

2.4 Stabilization Using Arecanut Coir with 3% Cement

The reinforced soil was not cured, as there is no chemical reaction taking place in order to improve the strength. Hence, uncemented, but fiber-reinforced specimens were tested immediately after casting. IRC SP: 72, 2015 suggests the use of a minimum of 2–3% cement for the stabilization of soils, which are to be used for modified subgrades. As the soil blended with coir can not develop any bonding at the soil-coir interface, this cement addition was found to be essential. The UCS strength was measured till 365 days of curing. At the same time, the CBR improvements were evaluated till 60 days of curing.

2.5 Sample Preparation

Sample preparations were done as per relevant Indian Standard code procedures. To the dry soil, coir was added initially and mixed using 0.5 times the OMC quantity of water. The mix is further blended with the remaining water until a homogeneous mixture is obtained. For the cemented mix preparation, same procedure was followed, but care was taken to mold the specimens within the initial setting time of cement. For conducting durability tests, ASTM codes were followed. All specimens were cured to the specified curing period, and to maintain the moisture content, samples were covered in plastic raps or kept covered in the desiccator. The samples used for soaked CBR were immersed in potable water for 4 days before testing.

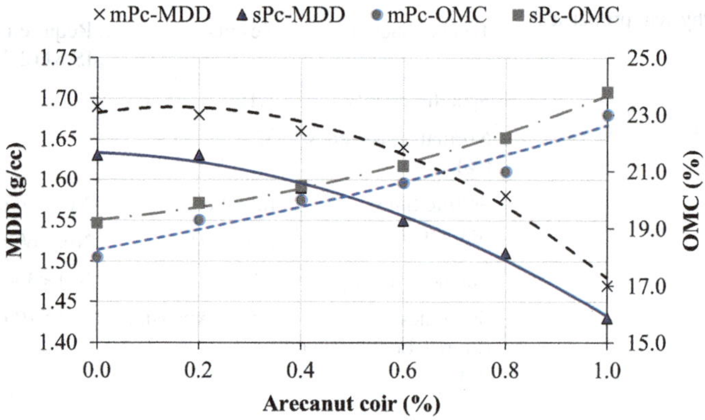

Fig. 1 Compaction properties for lateritic soil reinforced with different percentages of arecanut coir

3 Experimental Outcomes

3.1 Compaction

The arecanut coir-reinforced soil was tested to evaluate the standard and modified compaction parameters. Various dosages of fibers were tried, and the test results are depicted in Fig. 1. The increased low-density fiber dosage has resulted in a density dropdown of the mix, whereas water affinity of coir was higher, leading to an increase in the OMC of the mixes with increased dosages. Higher compaction effort has led to improved density and lowered OMC values. The same compaction characteristics were considered for 3% cement-treated lateritic soil with different arecanut coir dosages. As the dosage of cement is very low, it does not change the compaction characteristics of treated soil.

3.2 UCS

The lateritic soil, when reinforced with different percentages of arecanut coir, showed improvement in UCS to 0.6% fiber dosage (Fig. 2). A further increase in fiber dosage leads to a slight dropdown in UCS due to excess fiber concentration at certain portions of the specimen. When reinforced soils were compacted to standard Proctor density along with 3% cement, the improvement in UCS values was magnified due to the bonding effect (Fig. 3). The soil and fiber bonding was enhanced with the addition of cement due to the formation of hydration products. The higher compaction effort could improve the UCS values of mixes slightly. The observed slight improvement in UCS was due to the compact packing of mixes with lowered voids resulting in higher

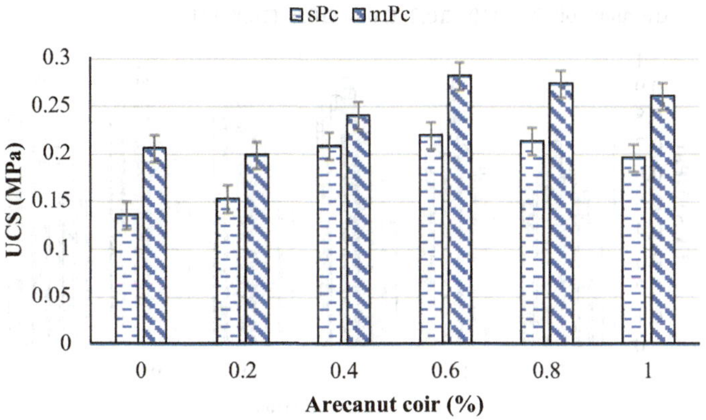

Fig. 2 UCS variation for arecanut coir-reinforced soil

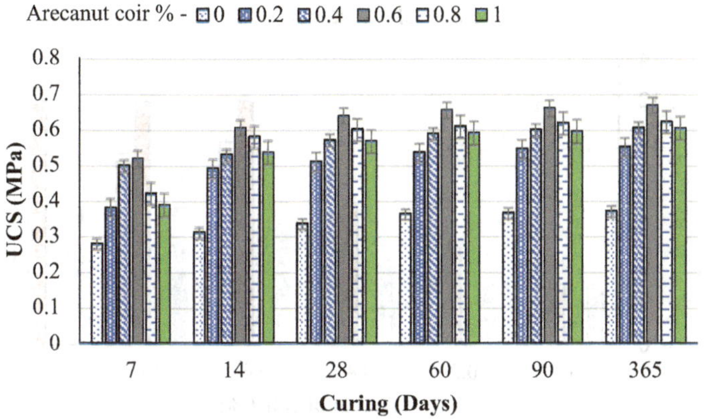

Fig. 3 UCS variation for cement and arecanut coir-treated soil under standard Proctor compaction

densities (Fig. 4). Therefore, CBR, durability, and fatigue tests were conducted on the specimens prepared at modified Proctor density.

3.3 CBR

When the soil is reinforced with coir alone, without the addition of cement, the CBR values were enhanced in comparison to unreinforced soil (Fig. 5). The addition of tensile reinforcement can improve the shear resistance of the mix, leading to a CBR hike. Further improvement in CBR strength can be expected if the soil and fibers were bonded with a cementing agent.

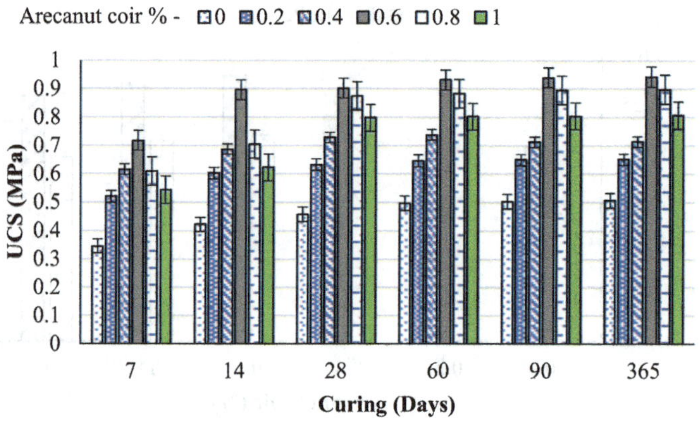

Fig. 4 UCS variation for cement and arecanut coir-treated soil under modified Proctor compaction

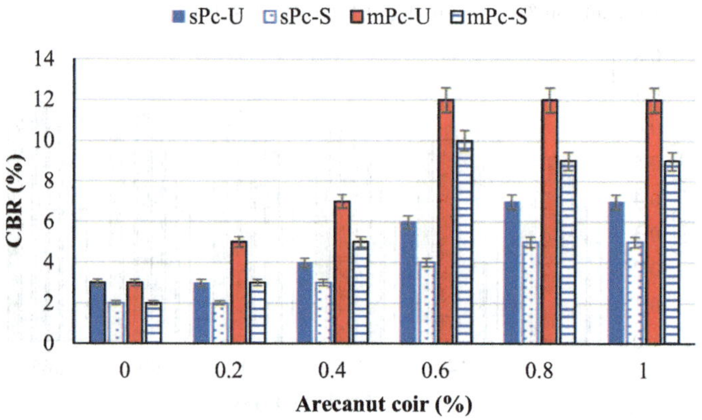

Fig. 5 CBR values for arecanut coir-treated soil

The samples cured for various days exhibited significant improvement in CBR values. Even when soaked in water, the samples could retain considerable CBR values. Unreinforced but cement-treated soil also exhibited CBR improvement. The higher CBR values were observed for mixes with 0.6% coir reinforcement (Fig. 6). This may be due to the proper distribution of reinforcement in the mix, thus improving plunger penetration resistance. Further increase in dosage could not improve CBR strength. The higher fiber dosages may cause the excess concentration of fibers, thus hampering strength improvement.

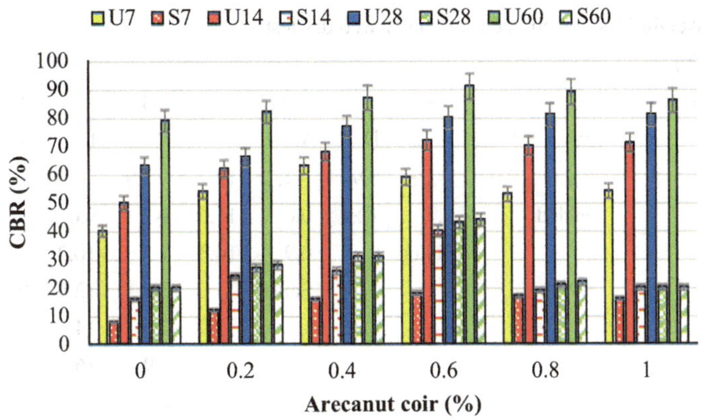

Fig. 6 Variation of CBR with curing period for 3% cement-stabilized lateritic soil at various dosages of arecanut coir reinforcement under modified Proctor compaction

3.4 *Durability*

The field performance of a pavement structure depends on the wheel loads to which it is subjected, along with variations in the climatic conditions. The changes in weather lead to changes in the temperature and moisture content of the pavement layers. These variations will affect the structural integrity and performance of the pavement. In order to evaluate the influence of weather changes, durability tests were conducted on the stabilized material. To simulate the weather changes in high rainfall areas wet–dry test was conducted as per ASTM D 559. And to resemble cold region effects, freeze–thaw tests were conducted as per ASTM D 560. The 7 days stabilized samples of UCS will be subjected to 12 cycles of durability test, which lasts for 24 days. During each cycle, the integrity of the material will be tested in terms of soil loss. A material is said to pass the durability test if it sustains 12 cycles with a weight loss not exceeding 14%.

3.4.1 Wet–Dry

The unstabilized, unreinforced lateritic soil could not pass first wet–dry cycle. The cemented samples with lower coir percentage failed with a fewer number of cycles due to insufficient reinforcement. The cement hydration resulted in the setting of the mix. As the coir dosage increased, the set soil improved the integrity due to a hike in the tensile capacity of the mix. At the highest dosage of coir, the samples could complete 12 wet–dry cycles, but with weight loss exceeding 14%. Few negative values were observed due to the weight gain of mixes during the wetting cycle. Therefore, these mixes are not suitable for base or sub-base courses of pavement. However, these mixes can be suitably used as the modified subgrade due to improvement in strength values (Table 6).

Table 6 Weight loss of stabilized soil under wet–dry test

Cement (%)	3									
Arecanut coir (%)	0.2		0.4		0.6		0.8		1	
Cycles	W	D	W	D	W	D	W	D	W	D
1	−2.0	9.0	−1.8	9.9	−1.2	10.2	−1.9	7.7	−2.4	8.9
2	Failed		0.7	12.9	0.6	11.6	−2.1	9.0	−3.3	9.2
3			Failed		0.9	14.3	1.3	10.2	−2.5	10.7
4					Failed		5.6	12.4	0.5	13.3
5							7.6	14.6	1.1	14.7
6							9.9	16.6	1.9	15.2
7							Failed		2.7	17.2
8									3.6	20.0
9									9.7	20.1
10									10.3	21.5
11									11.2	22.0
12									12.7	22.2

W Wet; *D* Dry

3.4.2 Freeze–Thaw

Improvement in the resistance to the weight loss was observed for the cement-treated soil with the increase in coir dosage under freeze–thaw test. All the specimens sustained 12 freeze–thaw cycles with weight loss within 14%. The freezing and thawing are not considered to damage the set soil. The good structural integrity of the mixes was observed after the test. Therefore, the mixes are considered durable under freeze–thaw weather conditions (Table 7).

3.5 Fatigue

Fatigue is a common phenomenon that the pavement layers undergo during their lifespan. As a result of this distress mechanism, the structural stability of the pavement will be lost. Due to repeated wheel loads, the stabilized layers may develop micro-cracks, which further continue to macrocracks. These cracks will further develop in size, where the width of the crack increases and propagates through the pavement depth. Under the repeated wheel load on cracked pavement, a sudden failure can be observed, and the integrity of the pavement is disturbed. Therefore, it becomes necessary to test the modified materials used in the pavement for their fatigue load-receiving capacity, which will be measured in terms of the number of fatigue cycles to failure under the specified loading conditions. The stabilized soil materials used

Table 7 Weight loss of stabilized soil under freeze–thaw test

Cement (%)	3									
Arecanut coir (%)	0.2		0.4		0.6		0.8		1	
Cycles	F	T	F	T	F	T	F	T	F	T
1	3.4	3.5	3.2	2.6	3.5	3.1	2.0	2.4	0.2	0.2
2	3.6	4.0	3.0	3.5	3.0	3.4	2.4	2.6	0.6	0.4
3	4.2	4.3	3.0	3.1	3.8	3.8	2.7	2.8	1.0	0.6
4	4.5	4.6	3.2	3.3	3.1	4.0	2.7	2.9	0.9	0.7
5	4.9	5.0	2.9	3.4	4.1	4.3	2.9	3.0	1.1	0.8
6	5.1	5.4	3.4	3.5	4.3	4.6	3.0	3.2	1.2	1.0
7	5.9	6.0	3.5	3.9	4.5	5.0	3.2	3.4	1.4	1.2
8	6.3	6.5	4.0	4.2	4.8	5.3	3.4	3.5	1.5	1.6
9	6.7	7.0	4.3	4.8	5.3	5.3	3.6	3.7	1.8	1.5
10	6.9	7.2	4.2	5.0	5.4	5.4	3.7	3.8	1.9	1.6
11	7.2	7.4	5.4	5.5	5.8	5.8	3.8	3.8	2.0	1.6
12	7.8	7.9	5.5	5.7	5.9	5.9	3.9	4.3	2.1	2.3

F Freeze; *T* Thaw

in this study cured for 7 days were subjected to the repeated loads of 0.33, 0.5, and 0.6 times their corresponding UCS values.

The specimen was placed in position on the loading frame. LVDTs were used to record the deformation levels. Loading is done with the help of a load cell. The data acquisition system records the load, fatigue cycles, deformation values, etc., during the experiment. Both loading and data acquisition systems work together, and testing will be stopped when the specimen fails. The failure pattern of the specimen was also noted down. At a frequency of 1 Hz, a rest period of 0.1 s selected load was applied repeatedly, which resembles a sinusoidal waveform. The fatigue life was varied depending on the effectiveness of cementation and the dosage of coir reinforcement. Untreated lateritic soil failed only after a few cycles due to lack of bonding, whereas the cemented soil, along with reinforcement, exhibited improvement in fatigue resistance with the higher fiber dosages (Fig. 7). The low stress applied was unable to cause the distress quickly; hence, the specimens could withstand more fatigue load repetitions. Therefore, reinforcement along with cementation of soil is helpful in improving the capacity to withstand repeated loads.

4 Conclusions

Based on the laboratory tests conducted, the following findings were observed.

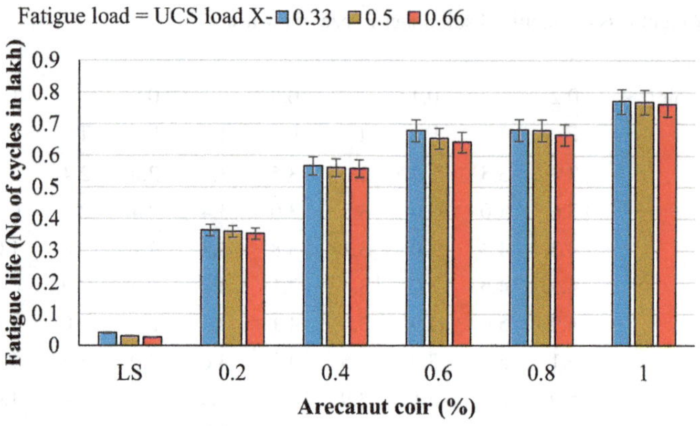

Fig. 7 Fatigue test results for lateritic soil and cement-fiber-treated lateritic soil

1. Arecanut coir tensile reinforcement alone was found to considerably improve the strength properties due to the holding of soil clods due to improved capacity. The dosage of 0.6% coir was found to be optimum for strength improvement.
2. Considerable UCS improvements resulted as the soil is treated with 3% cement and coir reinforcement due to the bonding of soil and fibers.
3. With the increase in cement, fiber dosage, and curing days, CBR values were improved. The maximum CBR improvements were observed at 3% cement along with 0.6% coir reinforcement. At further increased coir dosages, a marginal decrease in CBR was found.
4. The durability test indicates these samples can not be used as sub-base or base course layers since weight loss is more than 14%.
5. The improvement in fatigue life was observed for stabilized soil. The samples sustained >75,000 fatigue cycles at 1% coir dosage and 3% cement.

References

1. Buchanan F (1807) A journey from Madras through the countries of Mysore, Canara, and Malabar,...: for the express purpose of investigating the state of agriculture, arts, and commerce; the religion, manners, and customs; the history natural and civil, and antiquities, in the dominions of the rajah of Mysore, and the countries acquired by the honourable East India company, in the late and former wars, from Tippoo Sultan vol 2. T. Cadell and W. Davies; and Black, Parry, and Kingsbury; W. Bulmer and Company
2. Bwalya M (2006) Utilization and improvement of lateritic gravels in road bases. International Institute for Aerospace Survey and Earth Sciences (ITC), Section Engineering Geology, Delft, The Netherlands
3. Campco Arecanut‖ http://www.campco.org/index.php/arecanut. Accessed 26 Sep 2015
4. Chavva PK, Vanapalli SK, Puppala AJ, Hoyos L (2005) Evaluation of strength, resilient moduli, swell, and shrinkage characteristics of four chemically treated sulfate soils from north Texas. In: Innovations in grouting and soil improvement, pp 1–10

5. Das BM (2015) Principles of foundation engineering. Cengage Learning
6. Gidigasu MD (1971) A contribution to the study of the physico-chemical implications of tropical weathering and laterisation. Geotech Eng Bangkok 2:131–149
7. Haichert R, Kelln R, Wandzura C, Berthelot C, Guenther D (2012) Cement stabilization of conventional granular base and recycled crushed Portland cement concrete. Transp Res Rec 2310(1):121–126
8. Hicks RG (2002) Alaska soil stabilization design guide
9. Kar RK, Pradhan PK (2012) Laboratory tests of reinforced fly ash mix for use as sub-base in low volume rural roads. Indian Highways 40(1)
10. Kumar P, Singh SP (2008) Fiber-reinforced fly ash subbases in rural roads. J Transp Eng 134(4):171–180
11. Lekha KR, Sreedevi BG (2006) Coir fibre for the stabilisation of weak subgrade soils. Highway Engineering Lab, NATPAC, Thiruvananthapuram
12. Maheshwari K, Desai AK, Solanki CH (2012) Analytical modeling of flexile pavement resting on fiber reinforced clayey soil. In: Proceedings of Indian geotechnical conference, pp 1173–1176
13. Mohammad LN, Raghavandra A, Huang B (2000) Laboratory performance evaluation of cement-stabilized soil base mixtures. Transp Res Rec 1721(1):19–28
14. Mohanty B, Chauhan MS, Mittal S (2011) California bearing ratio of randomly oriented fiber reinforced clayey subgrade for rural roads. In: Proceedings of Indian geotechnical conference, Kochi, (Paper No. J-354), pp 611–614
15. Murthy RS, Pandey S (1983) Soil map of India. Natl Bur Soil
16. Peng Y, He Y (2009) Structural characteristics of cement-stabilized soil bases with 3D finite element method. Front Architect Civil Eng China 3(4):428
17. Ramaswamy SD, Aziz MA (1989) Jute geotextiles for roads. In: Proceedings of international workshop on geotextile, Banglore pp 259–270
18. Sarbaz H, Ghiassian H, Heshmati AA (2014) CBR strength of reinforced soil with natural fibres and considering environmental conditions. Int J Pavement Eng 15(7):577–583
19. Shankar AU, Chandrasekhar A, Bhat HP (2012) Experimental investigations on lithomargic clay stabilized with sand and coir. Indian Highways 40(2)
20. Tang C, Shi B, Gao W, Chen F, Cai Y (2007) Strength and mechanical behavior of short polypropylene fiber reinforced and cement stabilized clayey soil. Geotext Geomembr 25(3):194–202
21. Vidal H (1969) The principle of reinforced earth. Highw Res Rec 282
22. Zhang Z, Tao M (2008) Durability of cement stabilized low plasticity soils. J Geotech Geoenvironmental Eng 134(2):203–213

Pavement Analysis and Measurement of Distress on Concrete and Bituminous Roads Using Mobile LiDAR Technology

Prashant S. Alatgi and Sunil S. Pimplikar

Abstract Good quality of roads play a major role in depicting the development of any country. Pavement analysis and measurement of distress on roads play an important role in deciding the optimal budget allocation for the maintenance of the roads. Since the deterioration of the roads is a continuous process, the measurement of distress has to be carried out frequently on all the roads, ideally twice or thrice a year. The technology used should be fast enough to take care of the above demand and frequent inspection of the roads. Hence, automated instruments and technology, like Mobile LiDAR, have to be adopted for pavement analysis and measurement of distress (viz., cracks, potholes, International Roughness Index (IRI), etc.) on the roads. A test road of 2.8 km was selected in Pune and data was captured using "Leica Pegasus Two" mobile LiDAR instrument with pavement camera. The data processing and analysis were done using the Leica software. It was found that the measurement of distress on concrete and bituminous roads using Mobile LiDAR "Leica Pegasus Two" along with pavement cameras and required Leica softwares could be carried out with an automated process with positional accuracy better than 5 cm and quite efficiently even with the ongoing traffic conditions. Due to the in-built high precise GNSS (Global Navigation Satellite System) and IMU (Inertial Measurement Unit) system of "Leica Pegasus Two", and trajectory processing using the GPS base station data, the accurate location/position of the distress (cracks, potholes, etc.) on the field; could be precisely located and marked on the GIS maps. This will help the road maintenance engineers to locate these pavement distresses easily and efficiently using appropriate GPS enabled devices for carrying out the maintenance works efficiently and effectively.

Keywords Mobile LiDAR · 3D point cloud · International roughness index (IRI) · Inertial measurement unit (IMU) · Global navigation satellite system (GNSS)

P. S. Alatgi (✉) · S. S. Pimplikar
Dr. Vishwanath Karad MIT World Peace University, Kothrud, Pune 411038, India
e-mail: prashantalatgi@gmail.com

M. S. Ranadive et al. (eds.), *Recent Trends in Construction Technology and Management*, Lecture Notes in Civil Engineering 260,
https://doi.org/10.1007/978-981-19-2145-2_94

1 Introduction

Automated methods of pavement distress measurements have more scope than the manual methods which are very slow, have poor repeatability/reproducibility, and the traffic safety risk for the surveyor [1, 2]. Some previous authors have even used crude automated methods like video data acquired by the car's parking camera [3]. Attempt for integration of geographic information system (GIS), global positioning system (GPS), and computer vision system (CVS) was also made [4]. Whereas some have experimented with the video cameras, laser and image processing techniques [5, 6] but have not done the GPS/GNSS integration for getting the high-precise position or location of the cracks in the pavements. The paper published by Radopoulou [3] clearly states in "the further scope" to integrate GPS device to assist Road inspectors to provide the exact location of the detected defects. Some work has be previously done to determine roughness index (RI) using Bump Integrator method at the standard speed of 32 km/h [7], but have not used the Laser-based LiDAR technology for the same.

2 Objective and Improvements Planned

The objective of the present work is to perform "Pavement analysis and measurement of distress on concrete and bituminous roads using Mobile LiDAR Technology."

The improvements in the field planned w.r.t. the current status of the technology are as follows:

It is planned to give preference to the accurate location/position of the distress (cracks, potholes, etc.) on the field with better than 5 cm so that the maintenance staff can located these easily and efficiently using appropriate GPS enabled devices for carrying out the maintenance works effectively.

Emphasis has been laid to automatically capture the pavement distress on field without disturbance to the existing traffic on road and with great speed and accuracy.

Efforts have been taken to mark the position of the cracks and potholes on GIS-enabled map automatically and compare them with the actual photographs at those positions to draw appropriate inferences.

3 Method

3.1 Work Planning: Stage I

The appropriate and available LiDAR Instrument was identified for the project to capture data on site. Reconnaissance studies were done for selection of site feasible for data capture with different types of pavements; concrete, bituminous, paver

Fig. 1 Data capture on site by mobile LiDAR, "Leica Pegasus Two" for 2.8 km road in Pune

blocks, etc., with moving traffic. Estimation of resources was done (manpower, machines and material) required for the task of this project along with the time lines and the budget (cost) required to be made available.

3.2 Data Capture on Site: Stage II

Data capture was done on site using Leica Pegasus Two Mobile LiDAR system, with high-resolution pavement camera for pavement analysis and measurement of distress (cracking, potholes, etc.) on the selected road of 2.8 km near Ganga Dham Chowk in Pune city. Installation of two numbers of GPS/GNSS base stations was carried out on wooden tripod using Leica GS14 dual-frequency survey grade GNSS receivers, before start of actual data capture by the Mobile LiDAR instruments. Dense 3D point cloud data was captured by the Laser scanner Z+F 9012 and the high-definition photographs were captured in all directions. Pavement data was captured with the additional pavement camera attached to the Pegasus system. The trajectory and GNSS data was captured with the two high-precise dual-frequency GNSS receivers kept as base stations in static mode on ground along with the alignment and by the on board GNSS receiver (rover) of the "Leica Pegasus Two" system (Figs. 1 and 2).

3.3 Data Processing and Analysis: Stage III

The data captured by Mobile LiDAR: "Leica Pegasus Two" was downloaded in external hard disk of 1 Terabyte. The DGPS base station points observations in

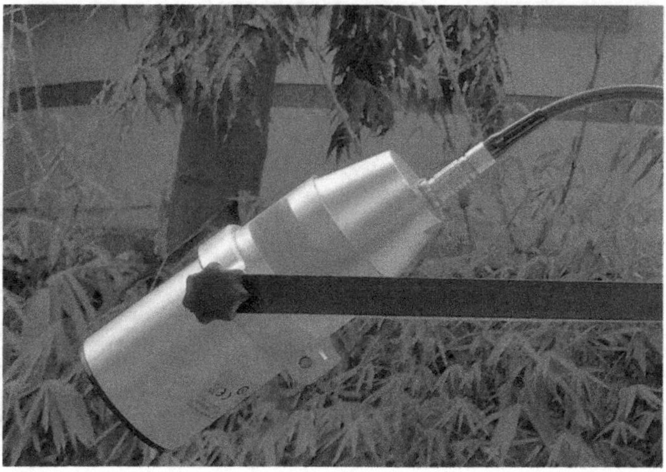

Fig. 2 Enlarged view of the pavement camera attached to the mobile LiDAR, "Leica Pegasus Two"

static mode were recorded on site and downloaded in the RAW/REINEX format. The 3D point cloud data, camera data and the DGPS raw data thus captured by the Leica Pegasus LiDAR system were then processed and analyzed on Workstations with standard workflow using licensed Leica softwares and Road Assessment tools (Figs. 3, 4, 5 and 6).

Along with the visual condition of the road, cracks and the surface distress of the roads, the IRI values in m/Km were calculated from the data captured along the 2.8 km road using the "Leica Pegasus Two" Mobile LiDAR system and compared with the maximum permissible IRI values as specified by the IRC: SP: 16–2019. It was found that all the values of IRI were found within the permissible limits under the "Good" category for cemented roads (Fig. 7) (Table 1).

4 Inferences

After the above automatic pavement distress outputs were generated, the data was compared to the actual photographs captured at that particular locations by the Leica Pegasus LiDAR instrument and suitable inferences were drawn. The accuracy of the entire data was checked by making random checks at three locations using the DGPS in RTK (Real-Time Kinematic) mode on the bases of DGPS base station coordinates. It was observed that at all the 3 locations, the positional accuracy of the data captured and pavement distress depicted (cracks, potholes), was well within 5 cm, which is sufficient for future maintenance works. The values of IRI in m/Km were found within the permissible limits under the "Good" category for cemented roads as specified by the IRC: SP: 16–2019.

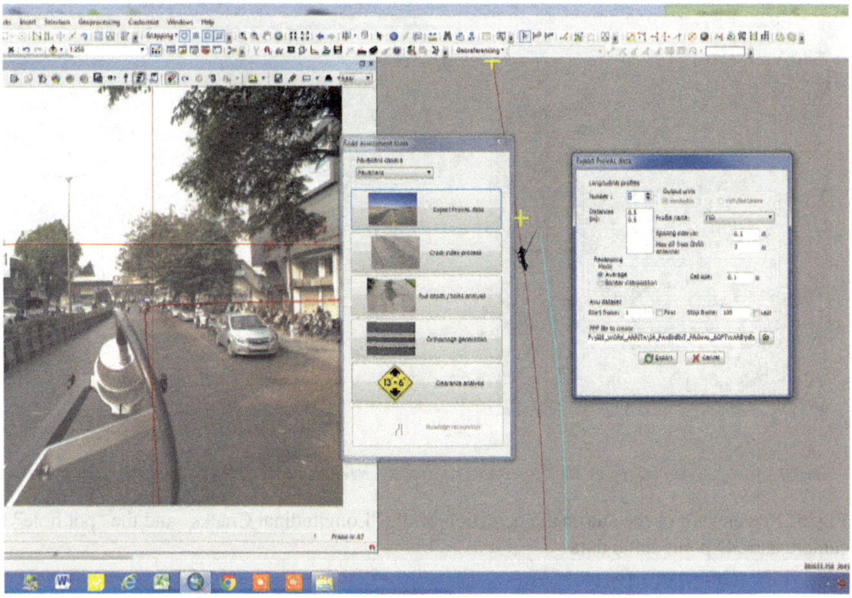

Fig. 3 Screen shot of the automatic pavement analyses using "Road assessment tools" in "Leica Map Factory"

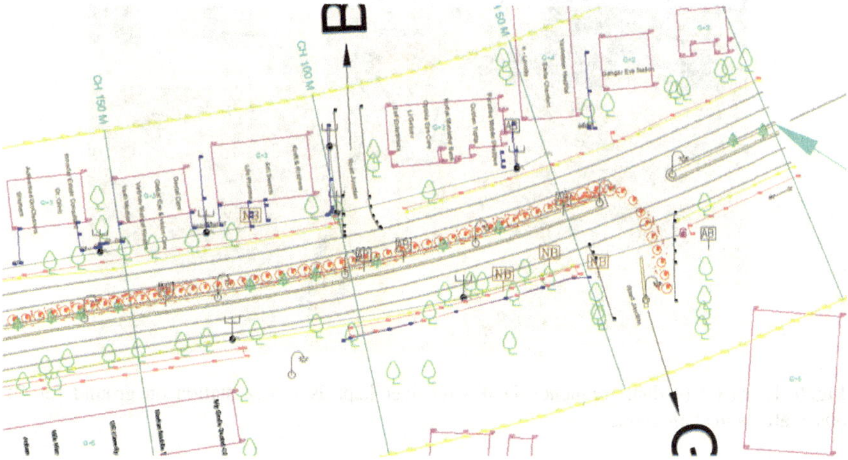

Fig. 4 Screen shot of the "Crack Index" in red circles, with the topographic features of the road

5 Conclusion

The measurement of distress on concrete and bituminous roads using Mobile LiDAR "Leica Pegasus Two" along with pavement cameras and required Leica softwares

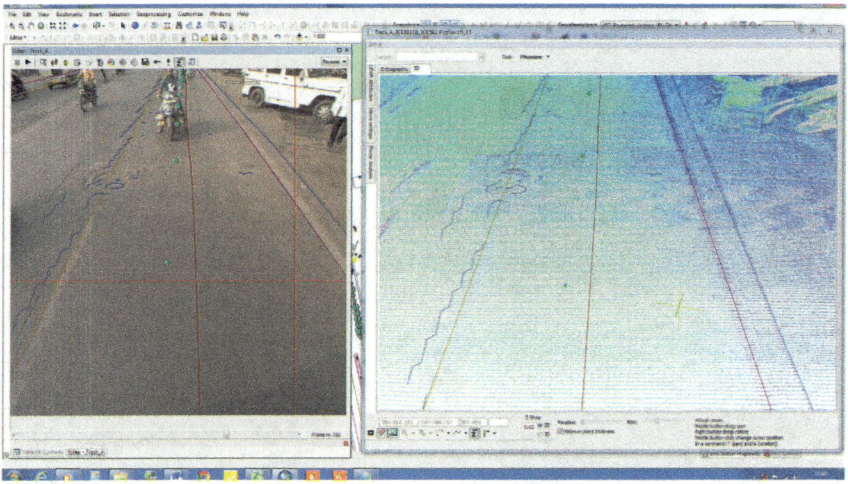

Fig. 5 Screen shot of the automatic detection of the "Longitudinal Cracks" and the "pot hole" in camera and 3D point cloud data

Fig. 6 Leica GS14 dual frequency DGPS receiver kept as a base station on ground for GPS observations in static mode

could be carried out with automated process quite efficiently and accurately even with the ongoing traffic conditions.

Due to the in-built high-precise GNSS & IMU (Inertial Measurement Unit) system of "Leica Pegasus Two", and trajectory processing using the GPS base station data, the accurate location/position of the distress (cracks, potholes, etc.) on the field; could be precisely located and marked on the GIS maps.

	C	D	E	F	G	H	I	J	K	L	M
1	Start Chainage	End Chainage	Direction	Lane Number	Lwpiri	Rwpiri	Laneiri	Speed	Survey Date	Latitude	Longitude
2	0+000.00	2+100.00	Increasing	L1	2.59	2.48	2.54	30	10/04/2021	19.456510	75.722170
3	0+100.00	2+200.00	Increasing	L1	2.43	2.54	2.49	30	10/04/2021	19.457270	75.722690
4	0+200.00	2+300.00	Increasing	L1	2.80	2.13	2.47	30	10/04/2021	19.458060	75.723130
5	0+300.00	2+400.00	Increasing	L1	2.00	2.34	2.17	30	10/04/2021	19.458900	75.723490
6	0+400.00	2+500.00	Increasing	L1	2.70	2.36	2.53	30	10/04/2021	19.459740	75.723840
7	0+500.00	2+600.00	Increasing	L1	2.11	2.48	2.30	30	10/04/2021	19.460580	75.724180
8	0+600.00	2+700.00	Increasing	L1	2.80	2.54	2.67	30	10/04/2021	19.461410	75.724560
9	0+700.00	2+800.00	Increasing	L1	2.33	2.13	2.23	30	10/04/2021	19.462210	75.725010
10	0+800.00	2+900.00	Increasing	L1	2.33	2.28	2.31	30	10/04/2021	19.463010	75.725470
11	0+900.00	3+000.00	Increasing	L1	2.09	2.34	2.22	30	10/04/2021	19.463800	75.725920
12	1+000.00	3+100.00	Increasing	L1	2.27	2.36	2.32	30	10/04/2021	19.464590	75.726380
13	1+100.00	3+200.00	Increasing	L1	2.45	2.39	2.42	30	10/04/2021	19.465380	75.726840
14	1+200.00	3+300.00	Increasing	L1	2.79	2.38	2.59	30	10/04/2021	19.466180	75.727290
15	1+300.00	3+400.00	Increasing	L1	2.41	2.36	2.39	30	10/04/2021	19.466970	75.727750
16	1+400.00	3+500.00	Increasing	L1	2.78	2.39	2.59	30	10/04/2021	19.467770	75.728200
17	1+500.00	3+600.00	Increasing	L1	2.77	2.38	2.58	30	10/04/2021	19.468560	75.728660
18	1+600.00	3+700.00	Increasing	L1	2.29	2.36	2.33	30	10/04/2021	19.469350	75.729120
19	1+700.00	3+800.00	Increasing	L1	2.79	2.39	2.59	30	10/04/2021	19.470150	75.729570
20	1+800.00	3+900.00	Increasing	L1	2.77	2.38	2.58	30	10/04/2021	19.470940	75.730020
21	1+900.00	4+000.00	Increasing	L1	2.44	2.36	2.40	30	10/04/2021	19.471740	75.730470
22	1+000.00	4+100.00	Increasing	L1	2.63	2.39	2.51	30	10/04/2021	19.472550	75.730900
23	1+100.00	4+200.00	Increasing	L1	2.33	2.38	2.36	30	10/04/2021	19.473360	75.731320
24	1+200.00	4+300.00	Increasing	L1	2.35	2.45	2.40	30	10/04/2021	19.474150	75.731780
25	1+300.00	4+400.00	Increasing	L1	2.47	2.55	2.51	30	10/04/2021	19.474940	75.732230

Fig. 7 International roughness index (IRI) values obtained in m/Km

Table 1 Maximum permissible values of roughness (RI and IRI) for SH, NH and expressways as per IRC: SP: 16–2019

S. no.	Type of surface	Condition of road surface					
		Good		Fair		Poor	
		RI	IRI	RI	IRI	RI	IRI
1	Bituminous (BC, SMA, SDBC)	<1800	<2.55	1800–2400	2.55–3.30	>2400	>3.30
2	Cemented	<2000	<2.81	2000–2400	2.81–3.30	>2400	>3.30

This will help the road maintenance engineers to locate these pavement distresses (cracks and potholes) easily and efficiently using appropriate GPS-enabled devices for carrying out the maintenance works efficiently and effectively.

Using the Leica softwares and crack detection algorithm, under "Road assessment tools" the cracks and potholes could be automatically extracted (vector form) from the captured data and marked on the GIS-enabled high-precise maps.

The values of IRI in m/Km were found within the permissible limits under the "Good" category for cemented roads as specified by the IRC: SP: 16–2019.

After analysis of the results and comparison with the actual photographs taken by the Leica Pegasus Two LiDAR system, it was found that the automatic crack detection software could detect the cracks, potholes, on concrete, bituminous roads, and even road patches with paver blocks to a great extent with high precision.

This process will save a lot of time, manpower and money, since the entire process from data capture to the output generation is automated.

6 Future Scope

Though the process is automated, human expertise is required for inspecting the end results by comparing with the actual site photographs taken by the Pegasus LiDAR system. It is required to verify and validate the actual presence of cracks detected by the automatic crack detection software. In the present study, the rumble strips/speed breakers present on ground were picked up by the software as a set of transverse cracks since the automatic crack detection algorithm may not be designed for the same. The future scope for improvement will be to fine tune and update the automatic crack detection algorithm to the area and country specific road conditions.

Acknowledgements I hereby acknowledge the leading Land Surveying and Mapping company of India "Prashant Surveys", Pune, for providing the "Leica Pegasus Two" Mobile LiDAR instrument along with the pavement camera, Leica GS14 DGPS receivers for GPS base station observations and sufficient manpower with Leica software for capturing, processing, and analysis of all the captured data for my selected 2.8 km road length in Pune.

References

1. Petra Offrell ASCE 0733-947X (2005) Repeatability in crack data collection on flexible pavements. 131(7):552
2. PENG Bo (2015) Review on automatic pavement crack image recognition algorithms. J Highw Transp Res Dev (English Ed.) 9(2):13–20
3. Radopoulou SC (2015) Patch detection for pavement assessment. Autom Constr 53:95–104
4. Obaidat MT (2006) Integration of geographic information systems and computer vision systems for pavement distress classification. Constr Build Mater 20:657–672
5. Fukuhara T (1990) Automatic pavement-distress-survey system. J Transp Eng 116(3):280–286
6. Serigos PA (ASCE) (2016) Evaluation of 3D automated systems for the measurement of pavement surface cracking. J Transp Eng 142(6):05016003
7. Pal M (2014) Pavement roughness prediction systems: a bump integrator approach. Int J Civ Environ Eng 8(12)

Laboratory Study on New Type of Self-consolidating Concrete Using Fly Ash as a Pavement Material

Bhupati Kannur and Hemant Chore

Abstract India is witnessing tremendous development in the transportation sector, especially in the construction of new expressways, upgrading the existing national and state highways to the superior standards. Road development in the rural area assumed significance in the last two decades and Pradhanmantri Gram Sadak Yojana (PMGSY), playing an important role in the development of the rural roads, is advocating the use of waste materials in the pavement construction. Such huge development in the highway sector would necessitate large amount of conventional construction materials. Therefore, there is a need to explore the possibility of utilisation of the alternative materials in the construction of the pavements. The semi-flowable self-consolidating concrete (SF-SCC) is environment friendly, efficient and economical alternative. This type of concrete is innovatively used in pavement construction replacing the conventional pavement quality concrete (PQC). This paper presents the preliminary laboratory experimental results of fresh and early strength properties of SFSCC using industrial wastes such as fly ash as the cement replacing material in the level of 20, 30 and 40%. The flow, slump, compaction factor and green strength of fresh concrete are tested. The cores extracted from the slab casted using a Mini-paver model are used to test the compressive, tensile strength and further, the beam specimen casted for flexural strength are tested. The results from the laboratory tests show that confirms the efficacy of the SFSCC so developed as the material for construction of rigid pavement as compressive strength of the SFSCC with fly ash up to 30% is more than 40 MPa and with that with 40% fly ash is more than 30 MPa. Flexural strength of all the four mixes is more than 4.5 MPa. Thus, SFSCC with fly ash up to 30% satisfies the requirement of concrete for urban road construction and the SFSCC with 40% fly ash as the material for rural rigid pavement construction.

Keywords Semiflowable self-consolidating concrete · Fly ash · Rigid pavement · Industrial waste

---

B. Kannur (✉) · H. Chore
Department of Civil Engineering, Dr B R Ambedkar National Institute of Technology Jalandhar, Jalandhar, India
e-mail: bhupatik.ce.19@nitj.ac.in

M. S. Ranadive et al. (eds.), *Recent Trends in Construction Technology and Management*, Lecture Notes in Civil Engineering 260,
https://doi.org/10.1007/978-981-19-2145-2_95

1 Introduction

There are mainly two problems due to increased population and globalisation. One is the requirement for the creation of infrastructure, which mainly depends on the concrete. Concrete is a composite material of cement, sand and aggregates, which demands the use of a huge amount of natural resources. Secondly, the production of industrial wastes in such a quantity that for the disposal of which requires acres of open lands for dumping and creates environmental pollution like land contamination, water pollution etc., posing health consequences. The solution for these two aspects lies in exploring the possibilities of using the industrial wastes in production of the concrete.

In India, 72% of electric power is being generated by the power stations using the coal/lignite and is of the opinion that this trend will be same during the near future. The Indian coal is a low grade with high ash content of about 30–45%. Managing such a huge amount of fly being generated in coal/lignite-based power plants is a big challenge. It is becoming tough to acquire lands for disposal of the ash thus generated. If not managed properly the fly ash may cause water and air pollution. The utilisation level of fly ash in different construction activities as reported during the year 1996–97 was less than 10% (6.64 metric ton). In year 2019–20, the fly ash utilisation has been tremendously increased to 83% (187.81 metric ton) due to many initiatives and policies by Government of India to increase the fly ash utilisation in construction industry like fly ash-based building materials, fly ash-based cement, roads/flyovers, low area reclamation, in construction of dams by roller compacted concrete etc. Even though a lot more fly ash is yet left unutilised and needs further attention to search for the viability to use the same. Table 1 summarises the generation and utilisation levels of fly ash in India during the Year 2019–20 [1].

Figure 1 visualises the areas of fly ash utilisation during the year 2019–20. It can be observed that the maximum fly ash is been used in manufacturing of cement (26%). Construction of roads and flyovers is also an area of substantially high amount of fly ash utilisation (9%). 17% of fly ash still unutilised.

Figure 2 explains the year wise fly ash generated and being utilised from 1996–97 to 2019–20. One can observe that the production as well as utilisation of fly ash is sharply increased. The fly ash utilisation is tremendously increased in the recent years. However, 100% fly ash generated is not being utilised. Hence, as in Fig. 2,

Table 1 Fly ash production and utilisation during 2019–20 [1]

Particulars	Year 2019–20
Number of thermal power stations considered	197
Capacity (MW)	200,691.50
Coal consumption (million tons)	678.68
Fly ash generated (million tons)	226.13
Fly ash utilised (million tons)	187.81
Utilisation (%)	83.05

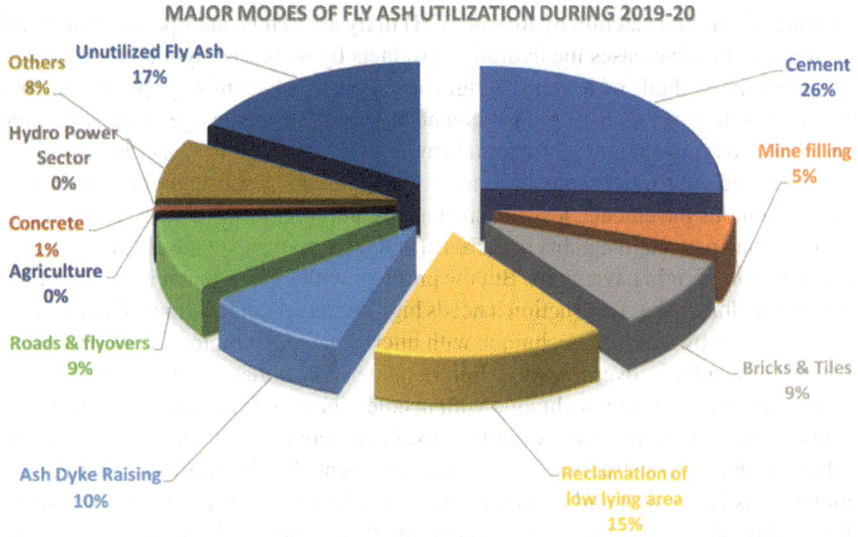

Fig. 1 Areas of fly ash utilisation during 2019–20 [1]

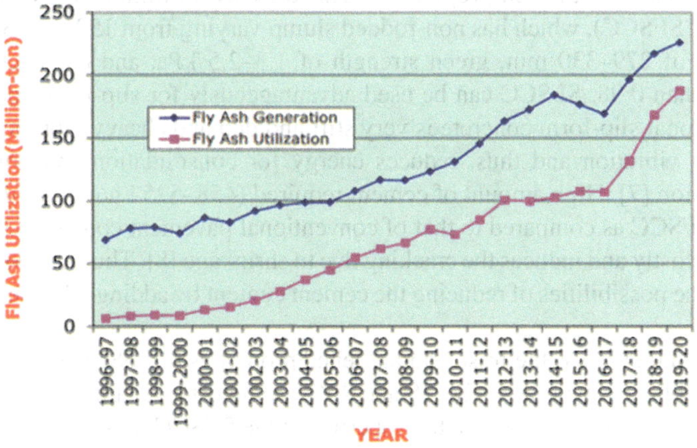

Fig. 2 Fly ash generation and utilisation from 1996–97 to 2019–20 [1]

the gap between both the generation and utilisation lines is still considerable. An expert committee formed to advice the thermal power plants to increase the fly ash utilisation recommends that plants should provide incentives through the R&D to various entities that come up with products using minimum 75% fly ash [1]. Fly ash is the best mineral admixture, which can be added easily and proves to be low-cost admixture [2]. SEM analysis of the concrete with fly ash as mineral admixture by Jindal and Ransinchung RN [3] shows the increasing amounts of calcium silicate hydrate (CSH) and calcium hydroxide (CH) in fly ash. Thus, incorporation of fly ash in concrete mix increases the hydration products by increased hydration process.

As per the Mohod and Kadam [4] rigid pavements are becoming popular compared to the flexible pavements. Rigid pavements have got the advantage of durability and resistance to failure against detrimental environment conditions. Concrete pavements are preferred over bituminous pavements in India due to substantial rise in cost of the bituminous pavements, less life span and requirement for frequent maintenance.

In India, pavement quality concrete (PQC) is being extensively used for the construction of rigid pavements. But the problem with PQC is that it is of zero or low slump and during the construction it needs high degree of compaction either by vibratory rollers or by slip form technique with internal needle vibrators. The compaction by rollers results in over consolidation or under consolidation of concrete; on the other hand, the slip form technique with needle vibrators leaves trails [5]. The solution for these problems is to develop a novel concrete mix, which needs no internal vibration during construction of concrete pavements by the slip-form paving technique. This is possible only when the concrete by slip-form paver exhibits higher flowability along with sufficient green strength to hold its shape. Sufficient green strength, the ability of the concrete to stand up and hold its shape without any edge support soon after the forward movement of slip form paver, is critical [6].

The concrete with such properties is called as semi-flowable self-consolidating concrete (SFSCC), which has non-rodded slump varying from 152 to 203 mm, flow diameter of 279–330 mm, green strength of 1.3–2.5 kPa, and compaction factor not less than 0.98. SFSCC can be used advantageously for slip-form construction. Conventional slip-form concrete is very stiff and demands heavy vibration. SFSCC needs no vibration and thus, reduces energy for consolidation and noise during construction [7]. High amount of cement required (458–525 kg/m^3) for the production of SFSCC as compared to that of conventional pavement concrete (355 kg/m^3) makes it costly and induces the cracking due to shrinkage [8]. Thus, there is a need to explore the possibilities of reducing the cement content by adding other cementitious materials.

Thus, in this study, the fresh and hardened properties of SFSCC with fly ash as a cement replacing material at 20, 30 and 40% are reported. In the present study, the fresh properties and early strength properties of the SFSCC thus produced using the supplementary cementitious material fly ash are investigated for its suitability in the construction of the rigid pavements using the slip form technology.

2 Materials

2.1 Cement

Ordinary Portland cement (OPC) of grade 43 confirming to IS 8112: 2013 [9] is used for the production of the SFSCC. Table 2 represents the physical properties of the cement used.

Table 2 Physical properties of cement

Grade	OPC 43	Requirements (IS 8112: 2013)
Sp. gravity	3.13	
Fineness modulus (retained on 90 micron sieve)	6.1%	10% maximum
Standard consistency	33%	
Initial setting time	70 min	60 min minimum
Final setting time	450 min	600 min maximum
Compressive strength		
3 days	25 MPa	23 minimum
7 days	34 MPa	33 minimum
28 days	44.8 MPa	43 minimum

Table 3 Properties of coarse aggregate

Sp. gravity	2.69	IS 2386 Part III [11]
Water absorption	0.7%	IS 2386 Part III
Impact value	12.82%	IS 2386 Part IV [12]
Crushing value	11.87%	IS 2386 Part IV

2.2 *Aggregates*

Fine aggregates used are the natural river sand confirming to Zone II and coarse aggregate of natural stone are procured from the local material supplier. The fineness modulus of the sand from the laboratory test result is found to be 2.77. Specific gravity of the sand is found to be 2.62.

Coarse aggregates of nominal size 20 and 10 mm are used and proportioned as per IRC 44: 2017 [10] keeping maximum size as 12.5 mm. Physical and mechanical properties of the aggregates used are tested in the laboratory, and the results of the same are in the Table 3.

2.3 *Mineral Admixture*

Fly ash of type c is used in the study as cement replacing material is obtained from the Thermal power plant Ropar. The specific gravity of the fly ash is found to be 1.87. Potable water is used for mixing and curing (Fig. 3).

The X-ray diffraction analysis of the fly ash is shown in Fig. 4. The sharp peak between the angles 20–30° 2θ is of Quartz (SiO_2). This implies that the fly ash contains high amount of silica and thus induces the pozzolanic activity.

(a) Cement (OPC 43) (b) Fly ash

Fig. 3 Cementitious materials used in the study

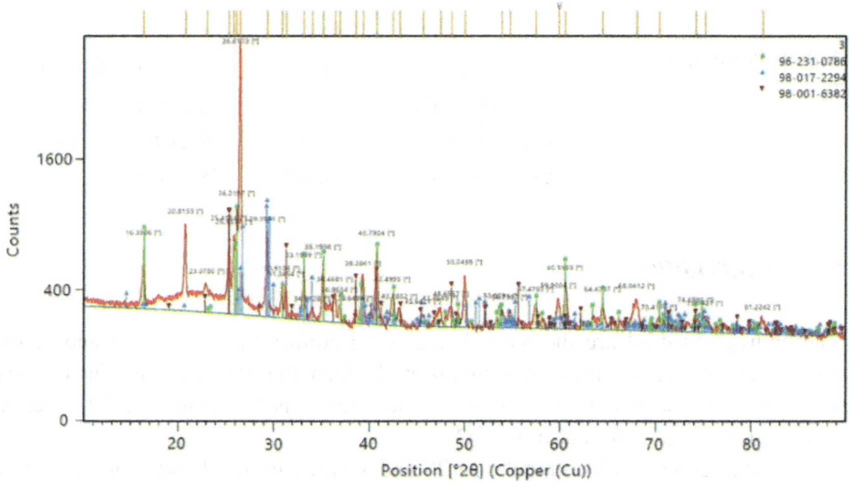

Fig. 4 XRD of fly ash used in the study

2.4 Chemical Admixtures

High range water reducer (HRWR) TamSem 53SP is used in the study. A viscosity modifying agent (VMA) TamSem VMA is used. A bonding agent TamSem SBR is also used in this work. The physical properties of the admixtures used as obtained by the data sheet from the producer are as in Table 4.

Table 4 Physical properties of admixtures used in mix

Product	TamCem 53SP(HRWR)	TamCem SBR(Bonding)
Relative density	1.09	1.02
Dry material content	–	39 ± 5%
pH	>6	–
Chloride ion content	<0.2%	–

3 Concrete Mix Design

Performance-based mix design method is used in the present study [13]. The mix proportion should aimed at achieving required slump, flow, compaction factor, green strength and other hardened properties. The mix proportion is arrived in three steps.

1. Design of Mortar:
   This step involves the determination of the *w*/*b* ratio and fine aggregate proportion. *w*/*b* ratio can be selected based on the concrete strength and the exposer condition. If no data available the *w*/*b* can be referred by ACI 211.1 and ACI 318. To determine the fine aggregate proportion, the modified flow test as per the AASHTO (ASTM C230). The acceptance criteria for the mortar design are to achieve the initial flow at 0th drop should be 10% (110 mm) and final flow at 16–18th drop to be 138% (238±5 mm).
2. Determination of Coarse Aggregate Proportion:
   After the successful design of the mortar, the next step is to determine the coarse aggregate content. The coarse aggregate amount should be selected by trial and error method using the modified slump cone test (ASTM C1611) and compaction factor test (AASHTO T126/AASHTO T23) on the unrodded concrete. The concrete thus produced should satisfy the unrodded slump of 150–200 mm, flow diameter of 275–325 mm and a regular cone shape in the fresh state. The compaction factor, ratio of unrodded to rodded density, should be minimum 0.98.
3. Mix Verification by Minipaver Test:
   A minipaver system as shown in Fig. 5 is used for the performance verification of the fresh slab extruded. For the final verification of the mix design of SFSCC, a slab extruded from the minipaver system is inspected. The performance of the fresh slab in terms of shape, surface quality, strength and durability is studied. The freshly extruded slab should be rectangular in shape with edge slump maximum of 8% of the thickness and surface quality with less than 15% surface area defects. There should be no honeycombing or segregation is found in the cross-section of the slab. The performance in terms of the strength and durability is also to be verified.

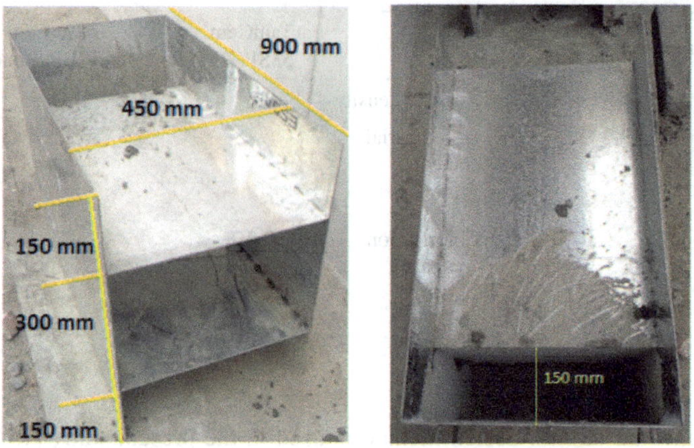

(a) Isotropic View (b) Top View

Fig. 5 Minipaver system

The mix design has been finalised for all the mixes in the study after initial trials to achieve all the stated properties of the above-mentioned mix design method stages. Figure 6 shows the steps involved in the laboratory production of SFSCC mix.

The ratio of CA: FA to satisfy the requirements of SFSCC is found to be 45:55. The finalised mix proportions are presented in the table. Figure 6 shows the different stages of the mix design process. The target compressive strength for the concrete design was kept at 40 MPa as it is aimed to use the SFSCC thus produced for the rigid pavement construction. Also to facilitate the required flexural strength of the concrete at 4.5 MPa, a bonding agent has been used in the mix.

Total 4 mixes are produced. The mix with 100% ordinary Portland cement is Control SFSCC, and the mixes with 20, 30, and 40% Fly ash as cement replacing material are indicated as SFSCCF20, SFSCC30, and SFSCC40 (Tables 5 and 6).

4 Results and Discussions

4.1 Fresh Properties of SFSCC

4.1.1 Slump and Flow

The freshly produced concrete is filled into the slump cone from a height of 300 mm from the top of the slump cone. No tamping is done by the tamping rod, the excessive concrete is struck of and levelled using the hand scoop. The slump cone is lifted vertically up without any disturbance or vibrations to the concrete filled. The slump and flow diameter of the concrete are measured. All the mixes are proportioned to

(a) Modified slump cone test

(b) Flow at 0th fall (F_0)

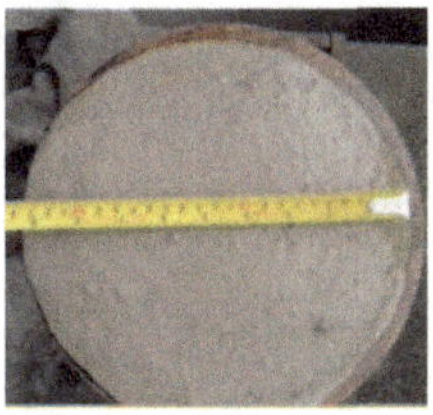

(c) Flow at 18th fall (F_{18})

(d) Non rodded slump

(e) Non rodded flow

(f) Filling for green strength

(g)Fresh Cylindre

(h) Sand loading for green strength test

(i) Fresh slab being extruded from minipaver

Fig. 6 Steps involved in laboratory production of the SFSCC

Table 5 Mix proportion (kg/m^3)

Mix	OPC	Fly ash	*w/b*	CA		FA	SBR	SP	%VMA
				20 mm	10 mm				
Control SFSCC	420	0	0.39	304	517	975	0	0.40	0.20
SFSCCF20	336	84	0.39	305	519	979	1%	0.30	0.20
SFSCCF30	308	132	0.39	296	503	950	1%	0.20	0.20
SFSCCF40	276	184	0.39	285	484	914	1%	0.20	0.20

Table 6 Fresh properties of SFSCC

Mix	Slump (mm)	Flow (mm)	Green strength (kPa)	Compaction factor
Control SFSCC	163	292	1.67	1.000
SFSCCF20	160	288	1.75	0.986
SFSCCF30	157	276	1.47	0.981
SFSCCF40	155	269	1.31	0.980
SFSCC Acceptance criteria [7, 13]	*150 to 200 mm and regular cone shape*	*250 to 300 mm*	*1.3–2.5*	*0.98 (Minimum)*

achieve the required slump and flow. It can be observed from the mix proportions that the addition of fly ash reduces the dosage of superplasticiser at the constant *w/b* ratio to achieve the required slump and flow. The reduction in the superplasticiser dosage indicates that fly ash addition increases the workability of the concrete.

4.1.2 Compaction Factor

All the mixes produced are checked for the target compaction factor of minimum 0.98. Compaction factor is determined by the ratio of non-rodded density to the rodded density. To determine the unrodded density, the container is filled with the concrete by pouring from a height of 300 mm above the top of the container.

4.1.3 Green Strength

The green strength is defined as the ratio of the load taken by the cylinder of the fresh concrete to the cross-sectional area of the concrete cylinder. Cylindrical mould of diameter 100 mm and height of 100 mm is filled with the concrete by pouring from a height of 300 mm above the top of the cylindrical mould. To maintain the height of fall, an inverted slump cone is placed over the mould.

4.1.4 Properties of Slab Extruded from the Minipaver System

Figure 7 shows the SFSCC slab extruded from the minipaver in fresh state. The shape of the slab is found to be perfect rectangular with negligible edge slump of about 0.6 cm (4%) out of the total thickness of 15 cm. The surface quality of the slab is satisfactorily good with level, no segregation and honeycombing on the top of the slab.

Fig. 7 Shape stability and surface quality of the fresh SFSCC slab extruded from the minipaver system

4.2 *Hardened Properties of SFSCC*

To test the SFSCC thus produced for its strength properties such as compressive strength and tensile strength, the cores are taken out of the slab as shown in Fig. 8 after curing for the required period using gunny bags (Table 7).

4.2.1 Compressive Strength

It is observed that the compressive strength of the concrete mixes, as in Fig. 9, is decreased with the incorporation of the fly ash compared to that of the control mix. Even though, the 28 days compressive strength of the mixes up to fly ash content of 30% was found to be more than 40 MPa. The compressive strength of the mix with 40% fly is also found to be more than 30 MPa. The decrease in the strength may due to the fact that at the early age fly ash is slow in pozzolanic action. The long-term strength is to be studied to study the fly ash effectiveness as supplementary cementitious material.

Fig. 8 Drilling cores from the slab

Table 7 Mechanical properties of SFSCC

Mix	Compressive strength (MPa)		Split tensile strength (MPa)		Flexural strength (MPa)	
	7 day	28 day	7 day	28 day	7 day	28 day
Control SFSCC	30.89	45.19	2.22	3.13	3.88	4.94
SFSCCF20	29.27	43.12	2.51	2.93	3.72	5.16
SFSCCF30	24.89	40.62	2.33	2.82	3.80	5.09
SFSCCF40	22.73	36.26	2.22	2.56	3.10	4.53

4.2.2 Split Tensile Strength

The split tensile strength results also follow the same trend as that found in case of compressive strength. With the addition of the fly ash, the tensile strength is found to decrease. The split tensile strength is very important as the resistance to cracking of the pavement is a desirable criterion (Fig. 10).

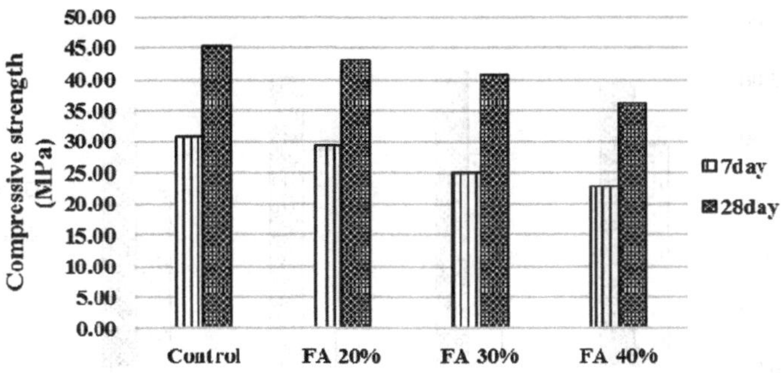

Fig. 9 Compressive strength of SFSCC mixes

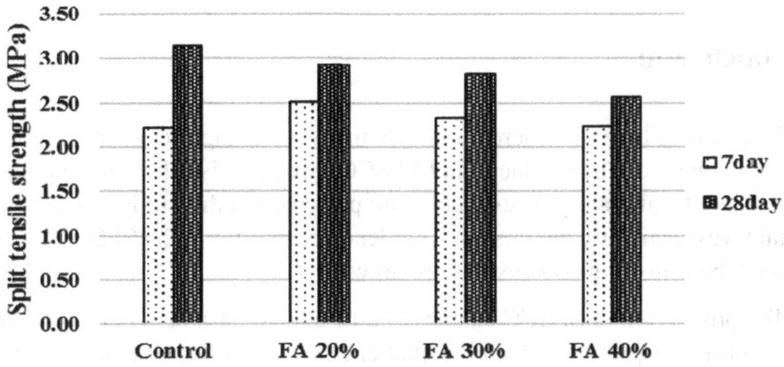

Fig. 10 Split tensile strength of SFSCC mixes

4.2.3 Flexural Strength

To test the flexural strength of the SFSCC beam, specimens of 100 mm × 100 mm × 500 mm are casted. While casting the beam specimens, the concrete is poured into the moulds from a height of 300 mm from the top of the mould to ensure same consolidation effect as that achieved in the slab extruded from the minipaver system. From Fig. 11, it can be observed that the flexural strength of all the mixes is more than 4.5 MPa satisfying the requirement for the SFSCC as material for pavement construction. But the variation in the flexural strength of all the mixes is found to be negligibly small and the trend of the flexural strength due to the incremental addition of the fly ash is not same as in the case of compressive strength and split tensile strength this phenomenon may be attributed to the effect of bonding agent used in the study and for deep understanding further study is to be done. The mix with 40% fly ash is having the lowest flexural strength of 4.53 MPa.

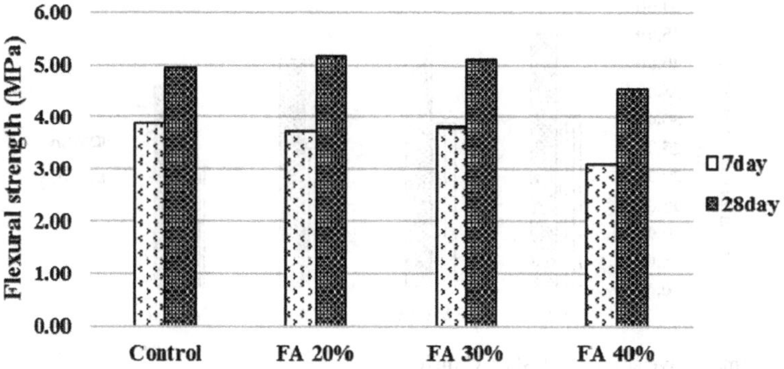

Fig. 11 Flexural strength of SFSCC mixes

5 Conclusion

SFSCC is a new type of concrete that has many advantages over the conventional pavement concrete. The production of SFSCC using the fly ash as cement replacing material and its laboratory investigation are presented in this study. From the experimental investigation on the fresh and hardened properties of the SFSCC mixes thus produced the following conclusions are drawn.

1. The production of SFSCC enables to utilise the fly ash being produced in huge quantities disposal of which is a challenging task for the thermal power plants.
2. The concrete produced using the fly ash as replacement to cement in the present study satisfies all the requirements as semi-flowable self-consolidating concrete (SFSCC).
3. Slump, flow, compaction factor and green strength of all the mixes are in line with the specifications as SFSCC for the construction of the rigid pavement using slip form technique.
4. The addition of fly ash in 20, 30 and 40% by weight of cement in the SFSCC production resulted in reduction of the compressive and split tensile strength. Even though mixes with fly ash up to 30% satisfies the compressive strength requirement as pavement quality concrete for urban roads which is M40 as per IRC 44-2008. The mix SFSCC40% can be utilised in the construction of the rigid pavement in rural roads for which the required grade is limited to M30.
5. The results of flexural strength of all the mixes fulfil the requirement for the concrete road construction as IRC 44-2008 specifies 4.5 MPa as the minimum flexural strength for concrete to be used in road construction.

SFSCC is a type of concrete that has flowability between conventionally vibrated concrete and self-consolidating concrete. As it has the semi-flowability, it also offers the shape stability in the fresh state, which enables the speeder construction using slip form technology. It also enables the use of waste materials in its production. Thus SFSCC along with industrial waste fly ash can be used in the construction of the rigid pavements by slip form paving technique.

References

1. Report on fly ash generation at coal/lignite based thermal power stations and its utilization in the country for the year 2019–20. Government of India, Ministry of Power. Central Electricity Authority, Thermal Civil Design Division, New Delhi, Nov 2020
2. Priya CC, Rao MVS, Reddy VS, Shrihari S (2020) High volume fly ash self compacting concrete with lime and silica fume as additives. In: E3S web of conferences, vol 184. ICMED 2020, p 01109. https://doi.org/10.1051/e3sconf/202018401109
3. Jindal A, Ransinchung RNGD (2018) Behavioural study of pavement quality concrete containing construction, industrial and agricultural wastes. Int J Pavement Res Technol 11(5):488–501. https://doi.org/10.1016/j.ijprt.2018.03.007
4. Mohod MV, Dr. Kadam KN (2016) A comparative study on rigid and flexible pavement: a review. IOSR J Mech Civ Eng (IOSR-JMCE) 13(3):84–88. https://doi.org/10.9790/1684-1303078488
5. Girish MG, Shetty KK, Raja AR (2018) Self-consolidating paving grade geopolymer concrete. IOP Conf. Ser Mater Sci Eng 431:092006. https://doi.org/10.1088/1757-899X/431/9/092006
6. Voigt T, Mbele J-J, Wang K, Shah SP (2010) Using fly ash, clay, and fibers for simultaneous improvement of concrete green strength and consolidatability for slip-form pavement. J Mater Civ Eng 22(2):196–206. https://doi.org/10.1061/ASCE0899-1561201022:2196
7. Lomboy GR, Wang K (2015) Semi-flowable self-consolidating concrete and its application. Int J Mater Struct Integrity 9(1/2/3):61–71. https://doi.org/10.1504/IJMSI.2015.071110
8. Wang K, Shah SP, Grove J, Taylor P, Wiegand P, Steffes R, Lomboy GR, Quanji Z, Gang L, Tregger N (2011) Self-consolidating concrete–applications for slip-form paving: phase II, Report No. DTFH61–06-H-00011 Work Plan 6, Pooled Fund Program Study TPF–5(098), National Concrete Pavement Technology Center, Iowa State University
9. IS 8112: 2013 Indian Standard, ordinary portland cement, 43 grade—specification, Second Revision
10. IRC: 44-2008 Indian Road Congress, guidelines for cement concrete mix design for pavements, Second Revision
11. IS 2386 Part III–1963, (Reaffirmed1990) Indian Standard, methods of test for aggregates for cqncrete part III specific gravity, density, voids, absorption and bulking. Eighth Reprint Mar 1997
12. IS 2386 Part IV–1963, (Reaffirmed 1990) Indian Standard, methods of test for aggregates for concrete Part IV mechanical properties. Tenth Reprint Mar 1997
13. Lomboy GR, Wang K, Taylor P, Shah SP (2012) Guidelines for design, testing, production and construction of semi-flowable SCC for slip-form paving. Int J Pavement Eng 13(3):216–225. https://doi.org/10.1080/10298436.2011.610797

FTIR Analysis for Ageing of HDPE Pyro-oil Modified Bitumen

H. P. Hadole and M. S. Ranadive

Abstract All over the world, the production of plastic is increasing over the years due to its various applications in different sectors. As the demand for plastic is increasing day by day, it consequently results in accumulation of plastic waste. Till the time various elastomer, plastomer, and rubber are used to modify binder. Here an attempt was made to modify bitumen with plastic oil also known as pyro-oil and to check ageing properties of the same. The rheological and chemical behavior of bitumen and modified bitumen is complex due to its unpredictable nature. In this investigation, bitumen was modified with pyro-oil from High-Density Polyethylene (HDPE). Pyro-oil was obtained by pyrolysis of HDPE at about 750 °C. Pyro-oil modified bitumen (PRMB) was prepared by addition of 5, 10, and 15% of HDPE pyro-oil by total weight of bitumen binder mixed at 3000 rpm for about 120 min. The oxidation process of PRMB was studied by using Fourier transform infrared spectroscopy (FTIR). The effect of short-term ageing for VG30, PRMB was studied by SuperPave test protocol and normal oven ageing. The effect of long-term ageing was studied by normal oven ageing. The FTIR results of SuperPave and normal oven aged samples were compared and change in functional group was recorded. Bitumen binders (VG30, PRMB) subjected to ageing and changes in chemical composition were analyzed using FTIR spectroscopy. FTIR results were used to validate chemical bond like C=O and S=O which leads to changes in viscosity and stiffness of binder. It was observed that ageing influences the bitumen chemistry.

Keywords Pyro-oil · FTIR · VG30 · PRMB · HDPE

H. P. Hadole (✉) · M. S. Ranadive
Department of Civil Engineering, College of Engineering, Pune, Maharashtra 411005, India
e-mail: hadole50@gmail.com

M. S. Ranadive
e-mail: msr.civil@coep.ac.in

M. S. Ranadive et al. (eds.), *Recent Trends in Construction Technology and Management*, Lecture Notes in Civil Engineering 260,
https://doi.org/10.1007/978-981-19-2145-2_96

1 Introduction

In present scenario, road infrastructure development is growing in all parts of India. In addition to this, majority of road projects has bituminous layer at top. In pavement industries, bitumen is widely used due to its viscoelastic properties. In current trend, bitumen is widely obtained from crude petroleum as byproduct. Currently, IS 15462 and IS [14] are used to check quality of supplied bitumen by different agencies of India. However, petroleum is one of the non-renewable resources, and recoverable amount of which is about to decrease in near future. As per Yao et al. [31], 90–95% by weight of bitumen contains hydrogen, carbon, heteroatoms, and metals. Heteroatoms contain nitrogen, oxygen, and sulfur which replace carbon by heat and contribute to physical and chemical behavior of bitumen. Metal atoms mainly include vanadium, nickel, and iron. Oxidation of bitumen is nothing but change in it properties due to change in temperature when it comes in contact with oxygen and oxidation of bitumen is also known as ageing of bitumen. Author continues that the important reason for such failures lies in insufficient performance-based binder specification for its selection. In performance grading, ageing analysis plays vital role. Jahromi and Khodaii [19] stated that oxidation leads to stiffer and brittle bitumen which results in rutting and fatigue cracking. This is the main reason that low- and high-temperature performance of bitumen should be improved and for the same many researchers are modifying the bitumen in different ways. Generally styrene butadiene styrene (SBS), fibers, styrene butadiene rubber crumb rubber, etc. were used for improving the properties of bitumen. Al-Hadidy and Tan [1] explained that SBS improves low-temperature performance of bitumen. On the other hand, Thodesen et al. [29] and Bischoff and Toepel [3] explained that crumb rubber could improve rheological, rutting, viscosity, and thermal cracking of bitumen. Recently for modification of bitumen lime was also used as the additive and Cheng et al. [5] studied its moisture susceptibility. Ageing of bitumen is of two types, short-term and long-term ageing. As per Hofko et al. [10], short-term ageing deals with chemical changes during mixing at elevated temperature of about 140–200 °C based upon type of asphalt. Rolling thin film oven (RTFO) is used for short-term ageing effect and pressure ageing vessel (PAV) for long-term ageing. But as explained by Behera et al. [2] normal oven can also be used to access short- and long-term effect of binder with proper temperature for particular duration concerned to specific type of binder. But as per Nivitha et al. [22], the mechanism of ageing for modified bitumen supposed to be different when compared with short- and long-term ageing as production temperature for it is above 150 °C.

As all know, world is facing major problem regarding disposal of plastic waste. Researchers from all over the world are trying to reuse plastic waste in various fields of engineering. In the same way, many researches have been done for use of plastic in flexible pavement. Researchers found out various polymers to modify the bitumen binder and checked physical, chemical, and rheological properties of the same. Here an attempt is made to use pyro-oil for modification of binder. To conclude, HDPE pyro-oil is used to modify the base bitumen to form PRMB. This modifier reacts with bitumen in chemical as well as in physical way. Such interaction leads to change in physical as well as chemical properties of base bitumen. Here some general properties such as penetration, ductility, viscosity, etc. were discussed before and after ageing. Along with this to analyze chemical properties FTIR spectroscopy was used and results are discussed in detail. Primarily, functional groups associated with oxidation are focused in the present research.

2 Literature Review

As explained by Hofko et al. [10] and Eberhardsteiner et al. [6], ageing of bitumen takes place due to oxidation of bitumen and continued that oxidative ageing is categorized into two main types, short-term and long-term ageing. Oxidation of binder during production process, in which aggregate and binder are mixed at temperature of about 150–200 °C. When bituminous mix is in service over time then corresponding oxidation of bitumen is known as long-term ageing. For materials like bitumen having very high absorption coefficient, there is problem of very low transmission which leads to unacceptable low signal-to-noise ratio. Before finding out chemical properties of modified bitumen with different modifiers it is necessary to firstly understand chemical structure of bitumen in aged and unaged condition. As per Chen et al. [4], bitumen binder consists of four fractions including saturates, asphaltenes, resins, and aromatics. Petersen [23], Harnsberger et al. [9], Nivitha et al. [22] mainly focus on the FTIR spectroscopy results to track the changes in chemical compounds after ageing for unmodified and modified bitumen.

In present scenario, ageing is the main problem pavement engineers are facing. Different researches have been coming up to decrease the ageing of bitumen and still it is always hot topic for research. Hofko et al. [10] and Eberhardsteiner et al. [6] explain about chemical sensitivity of bitumen. Shen et al. [27] concluded that ageing leads to decrease in penetration and ductility. On the other hand, viscosity, softening point, complex modulus, and creep stiffness increase. Further, author explains that this effect leads to elastic recovery, resistance to thermal cracking, and thermal fatigue becomes poor which leads to decrease in life of pavement. In the same manner, Petersen [23] stated that the ketone functional group is formed primarily from the oxidation of benzyl carbon. The ketones are formed by oxidation and asphaltenes are primarily responsible for increase in viscosity on ageing. The ageing of bitumen is caused by the processes of oxidation, steric hardening and volatilization.

Khapne et al. [20] studied HDPE pyro-oil modified bitumen (PRMB) with respect to the surface energy of the base bitumen and the modified PRMB. They prepared PRMB by adding HDPE pyro-oil in 2, 3, and 5% by weight of bitumen in the base bitumen VG30, using homogenizer to mix the pyro-oil with bitumen at a rate of 3000 rpm for 20 min. They studied the suitability of adding anti-stripping agents, hydrated lime, and fly ash, in the PRMB to enhance the resistance to moisture damage of PRMB.

Ranadive et al. [25] analyzed the effect of modifiers on the stone matrix asphalt (SMA) and bituminous concrete (BC) mixtures. Author analyzed the effects on the properties of SMA and BC modified with fibers extracted from refrigerator door panels (FERD), and of BC mixture modified with waste plastic in granular form. The optimization was done for the length of FERD in SMA and BC and for the percentage of waste plastic in AC mixtures. The lengths of fiber used were 2, 4, 6, and 8 mm, with a dosage rate of 0.3% by weight of aggregates. The granular plastic used for modification of AC mixture was 3–5 mm in size. Test such as Schellenberg drain down test, indirect tensile, and Marshall stability test were done to compare the strength and stability of the modified mixtures. The results showed an increase in both the Marshall stability and tensile strength at an optimum dose of 0.3% with a length of 6 mm of fiber, for both the SMA and BC mixtures as compared to conventional mixes.

3 Objective

The principal objective of this research is to analyze oxidative result of base binder VG30 and PRMB for different ageing conditions using FTIR spectrographs. The study of physical properties of VG30 and PRMB, chemical changes in oxidation groups due to ageing phenomenon, and compare the results of FTIR test for both short-term ageing and long-term ageing for normal oven condition. Further the analyses of results with respect to FTIR are to be studied.

4 Experimental Materials

4.1 Materials

The study was carried out with base bitumen VG30 for research work. The base bitumen was modified using HDPE pyro-oil. Pyrolysis of HDPE was done using pilot pyrolysis plant developed in TRE lab, College of Engineering, Pune, Maharashtra at about 700 °C. For the process of pyrolysis, HDPE waste was selected of about 3 kg and 68.3% by weight of pyro-oil was produced at reactor temperature of about 730–750 °C. The physical and chemical properties of HDPE pyro-oil was explained by the same author in other works [8].

Modified bitumen was prepared at temperature of about 160 °C at about 3500 rpm for 2 h. Due to this blending composition of virgin bitumen undergoes chemical changes and these chemical changes were analyzed by FTIR. 5%, 10%, and 15% HDPE pyro-oil by weight of bitumen were used for modification of base bitumen and abbreviated as PRMB5, PRMB10, and PRMB15, respectively.

The modification procedure of base bitumen with pyro-oil was explained by Hadole et al. [8], Khapne et al. [20], Suryawanshi et al. [28], and Kulkarni and Ranadive [21] in detail with other physical properties of modified bitumen.

4.2 Bitumen Ageing

The short-term ageing occurs during construction of bituminous material and long-term ageing deals with chemical changes in bitumen after 15–20 years of life and both are associated with oxidation of bitumen which leads to stiffer bituminous binder. As per SuperPave guidelines, RTFO is used for short-term ageing effect. Along with this, short-term ageing of bitumen was also conducted by normal oven method as explained by Reddy [26]. Long-term ageing of base and modified bitumen was conducted using normal oven method.

Each sample was aged for short-term ageing using Rolling Thin Film Oven Test (ASTM D2872) at 163 °C for 75 min to get short-term aged sample and denoted by AB5%, AB10%, and AB15%. The same percentage modified samples were aged by normal oven technique. For that, 10 g binder is spread in circular plate of 140 mm diameter with 650 μm of average thickness uniformly as explained by Behera et al. [2]. After preparation of sample, it is kept in preconditioned oven for 16 h at 163 °C and ageing duration is about 5 h so that result of RTFO ageing can be comparable for VG30.

For long-term ageing of base and modified bitumen, normal oven technique was used. Long-term ageing of different samples was checked after 3 days and 5 days, heated at 85 °C with allowable deviation of 2 °C. The change in chemical characterization and variation in different functional groups due to this ageing were studied using FTIR spectra.

4.3 FTIR

Fourier transform infrared spectrum plays a vital role in research work. FTIR test was used for chemical analysis. FTIR spectral data was obtained for various wavelengths. In this study, FTIR conducted using Bruker FTIR.

4.4 Test Methods

VG30 and modified bitumen samples were checked for physical properties and results were tabulated in Table 1. Habal and Singh [7] analyzed different samples which are checked for FTIR spectra using IR. Spectra were recorded from 4000 to 500 cm^{-1}. The test was conducted on base bitumen, modified bitumen, short- and long-term

Table 1 Physical properties of bitumen

Experiment	Standard	VG30	B5	B10	B15
Penetration at 25 °C (1/10 mm)	IS [13]	58	78	89	107
Softening point (°C)	IS [12]	66	52	46	38
Ductility (cm)	IS [11]	80	86	96	108
Viscosity at 60 °C (poise)	IS [16]	2850	–	–	–
Kinematic viscosity at 135 °C (cSt)	IS [17]	480	–	–	–
Viscosity at 150 °C (poise)	IS [15]	–	630	460	290
Results after rolling thin film oven test					
Loss in mass (%)	IS [18]	<1	<1	<1	<1
Softening point (°C)	IS 1205/ ASTM D36	72	63	53	44

aged bitumen, and HDPE pyro-oil. It is just to understand morphological changes when pyro-oil is used for modification of base bitumen.

5 Methodology

Complete laboratory test procedure is explained in Fig. 1.

6 Results and Discussion

6.1 Physical Properties of Bitumen and Pyro-oil

Physical properties of base and modified bitumen are explained in Table 1. It reflects penetration value increase with increase in percentage of pyro-oil which implies PRMB has less consistency than base bitumen. High-temperature performance of bitumen is observed using softening point. Softening point decreases which implies high-temperature performance of PRMB decreases. Evaluating low-temperature tensile deformation and flexibility of bitumen based on results of ductility. Higher ductility means better low-temperature performance. Increase in ductility of PRMB shows that PRMB don't harm low-temperature performance of bitumen application. But it is also important to note that, from these results, we cannot arrive at a conclusion that the PRMB has no application for high-temperature region until and unless conducting detailed study of rheological properties of PRMB. Loss of mass during short-term ageing is reported here, and it focuses on lightweight volatile compounds. Increase in loss of weight associated with increasing the average molecular weight leads to increase in bitumen viscosity. Loss of mass also has effect on indication of

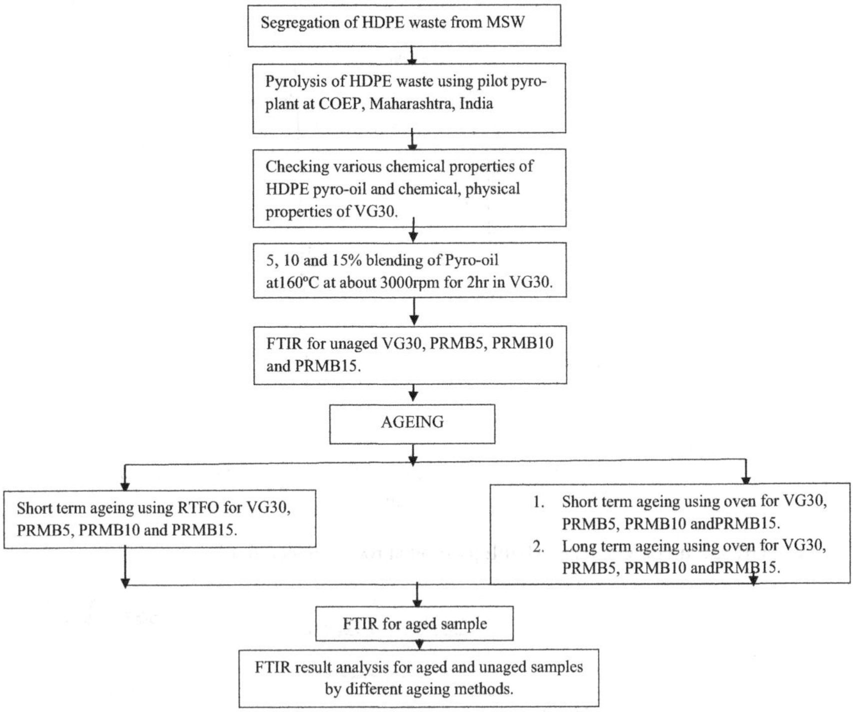

Fig. 1 Flowchart for methodology

potential bitumen binder emission during construction. The loss in mass for base bitumen and PRMB is less than 1%, which indicates that there was not much higher amount of light volatiles in pyro-oil. But lightweight fractions are more than base bitumen which might be due to pyrolysis process. It can be further concluded that there are some lightweight compounds in modified bitumen whose boiling point is lower than temperature in RTFO test.

6.2 FTIR Analysis of Base Bitumen and HDPE Pyro-oil Modified Bitumen

6.2.1 Short-Term Aged Sample by RTFO

Figures 2, 3, 4 and 5 show FTIR spectra. In the figures, horizontal axis is wave number (cm^{-1}) and vertical axis is transmission rate. From the graph, it is observed that the eminent peaks indicated for VG30 are crystal clear for bitumen. For the analysis of FTIR spectra, different groups of molecular bonds are categorized at different and specified wavelength of absorbance spectra. The four peaks in FTIR

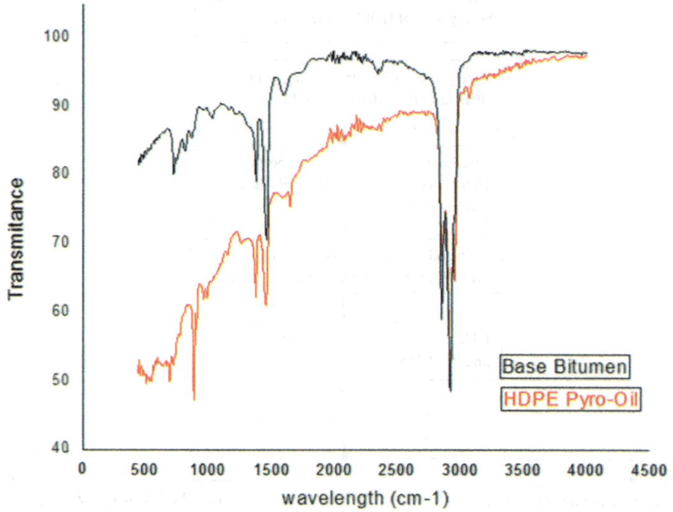

Fig. 2 FTIR spectra of VG30 and HDPE pyro-oil at room temperature

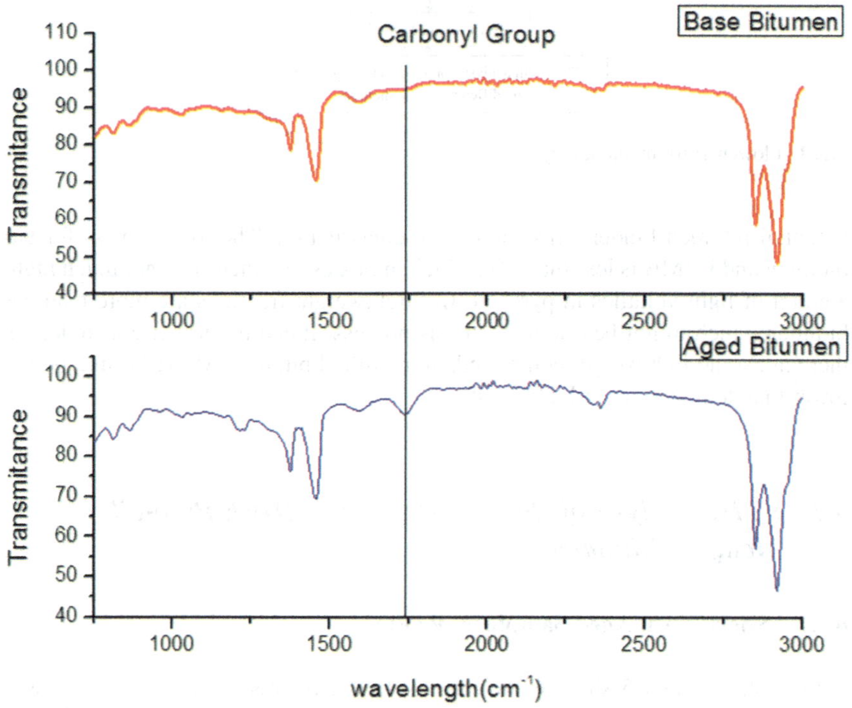

Fig. 3 FTIR spectra of VG30 (base bitumen) and short-term aged base bitumen at room temperature

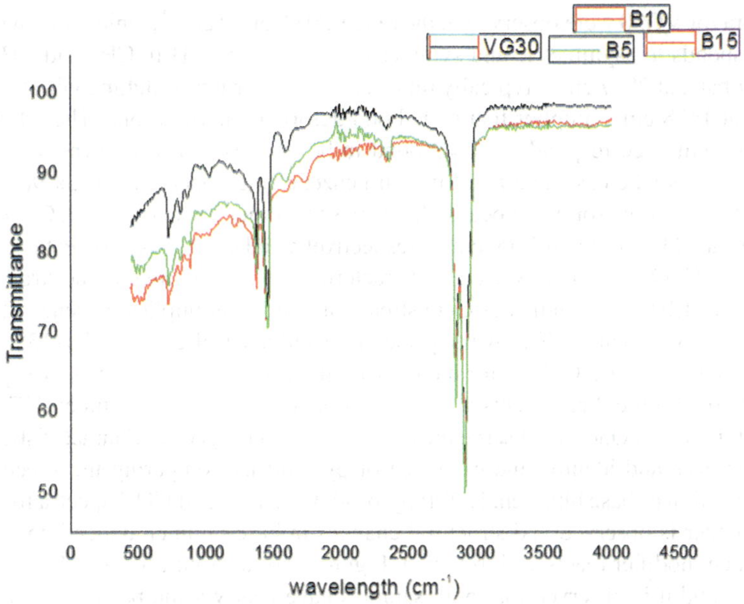

Fig. 4 FTIR spectra of VG30 (base bitumen), B5, B10 and B15 modified bitumen at room temperature

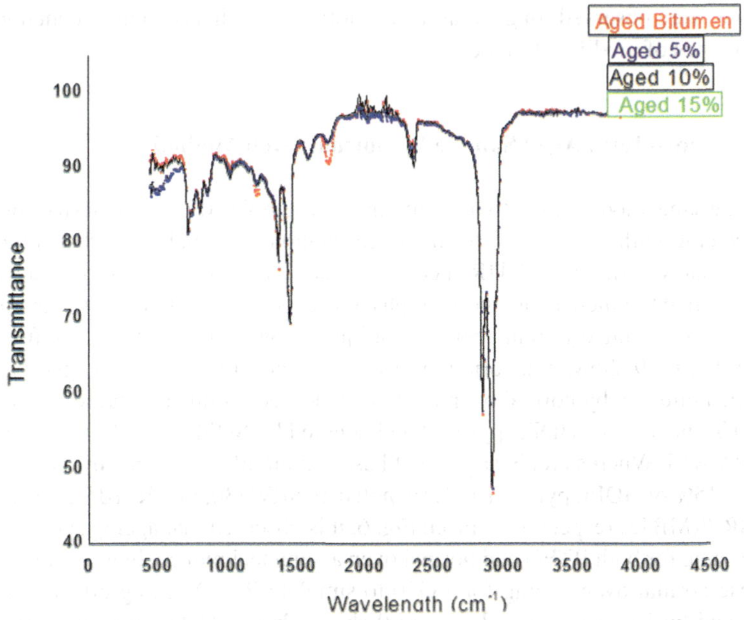

Fig. 5 FTIR spectra of short-term aged base bitumen, B5, B10 and B15 modified bitumen at room temperature

spectra for VG30 were observed in the range 2800–3000 cm^{-1}, which region mainly corresponds to asymmetric and symmetric stretches of C–H in CH_2 and CH_3. The major band at 2877 cm^{-1} typically represents hydrocarbon stretching vibrations and peak at 1458 cm^{-1} implies that C–H bond deformation vibrations. The other peak at 1600 cm^{-1} corresponds to C=C bond in benzene ring also known as vibration in aromatics whereas same bond in non-benzene ring appears for peak 960 cm^{-1}. Along with these prominent peaks, there are some other peaks like S=O, C=O which occurs at 1032 cm^{-1} and 1700 cm^{-1}, respectively, and as discussed by Petersen [24], S=O and C=O are generally used to characterize ageing in bitumen. The strong bond 1030 and 1280 cm^{-1} indicates C–O stretching vibration, implies presence of ether, alcohols, and phenols. The weak peaks at absorbance of about 1370–1380 cm^{-1} described regarding C–H deformation vibration, and this bonding is also present in PRMB. Figure 2 represents the variation in spectra for base bitumen VG30 and HDPE pyro-oil. One can observe the actual variation in spectra. Characteristics peak can observe and identify the existence of pyro-oil by comparing the spectra. For modification of base bitumen, HDPE pyro-oil was added and FTIR spectra for modified binder is observed and structural changes in base bitumen depend on the way in which modifier reacts with bitumen. Figure 3 gives spectra for base bitumen and PRMB and it is observed that both samples have nearly same peaks, which shows presence of same functional group, only intensity of some peaks vary obviously. From the same figure, this is also be verified that oxidation peaks for carbonyl and sulphoxides group change and for all other peaks remain unaltered/no prominent changes were observed. In general, it was noticed that this is common phenomenon for all types of modified bitumen.

6.2.2 Short-Term Aged Sample by Normal Oven Method

With the same modification to base bitumen, all samples were again tested for short-term ageing with the help of normal oven method as explained earlier. Figures 6, 7 and 8 show variation in FTIR spectra for both RTFO and normal oven aged base and modified bitumen. From these graphs, it is concluded that the formation of functional group during different ageing techniques shows formation of same functional group. Figure 9 shows the variation in FTIR spectra for short-term aged base and modified bitumen by normal oven method. Modified bitumen is formed by addition of 5, 10, and 15% of HDPE pyro-oil and denoted by SOPRMB5, SOPRMB10, and SOPRMB15. Whereas RTFO ageing of base and modified bitumen by addition of 5, 10, and 15% of HDPE pyro-oil and is denoted by SRVG30, SRPRMB5, SRPRMB10, and SRPRMB15, respectively. From Fig. 6, it is observed that ageing spectra is near about same for both RTFO and normal oven ageing technique which implies that we can use normal oven ageing for VG30 to simulate RTFO ageing effect as already explained by Behera et al. [2]. Figure 9 shows that as HDPE pyro-oil percentage increases sulphoxides also increases. But for relative comparison in between all these modified bitumen, one should go for calculation of different indexes using area under absorption curve. Whereas it is noted that sulphoxides group gets formed

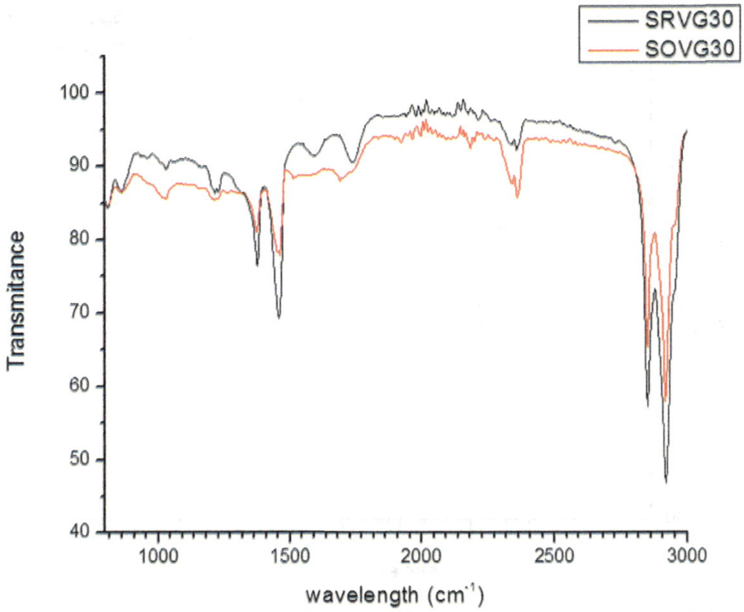

Fig. 6 FTIR spectra of short-term aged base bitumen modified bitumen by RTFO and normal oven method

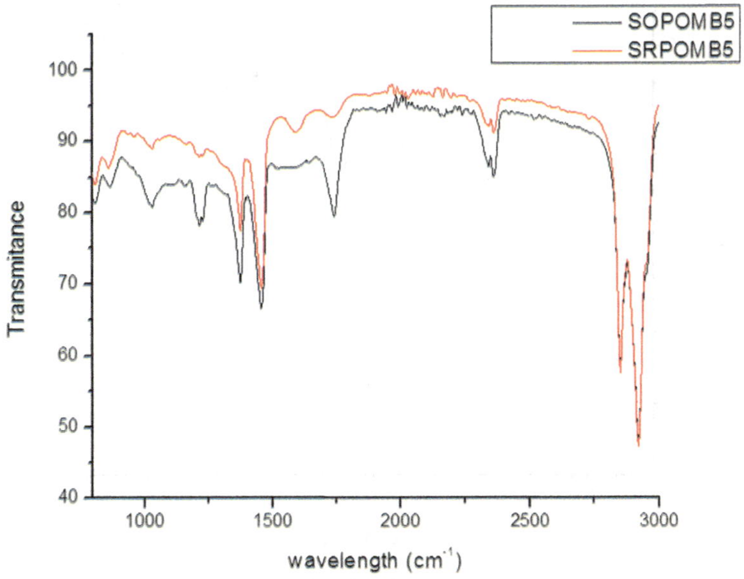

Fig. 7 FTIR spectra of short-term aged PRMB5 by RTFO and normal oven method

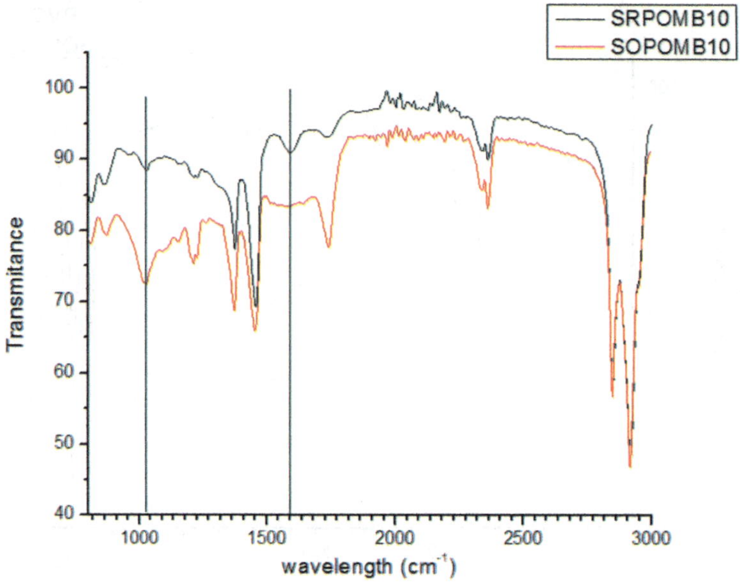

Fig. 8 FTIR spectra of short-term aged PRMB10 by RTFO and normal oven method

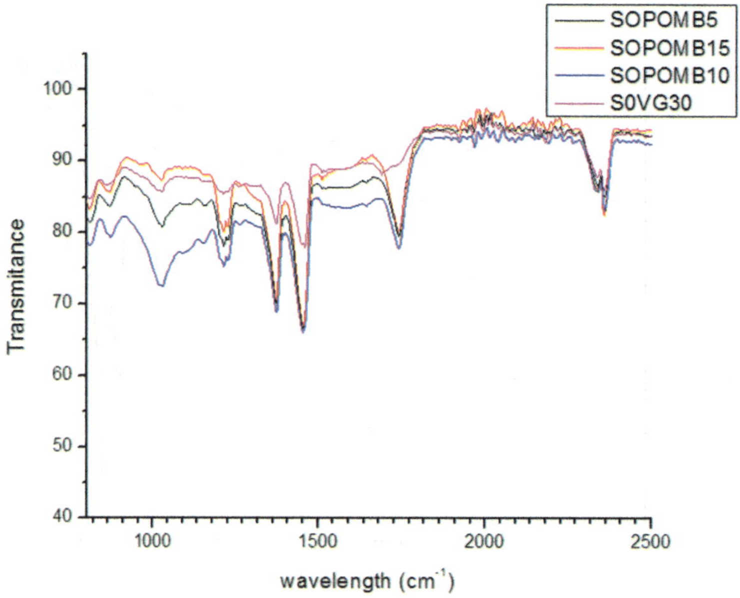

Fig. 9 FTIR spectra of short-term aged VG30, PRMB5, PRMB10, and PRMB15 normal oven method

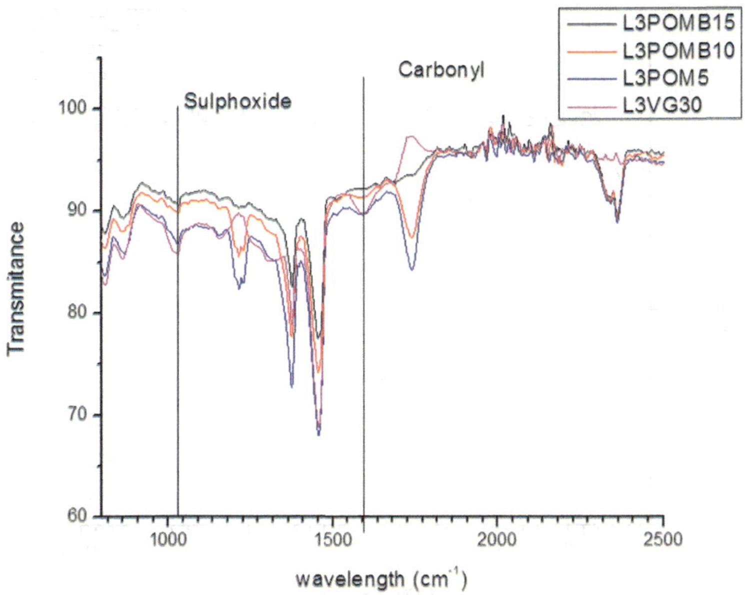

Fig. 10 FTIR spectra of 3-day long-term aged VG30, PRMB5, PRMB10, and PRMB15 by normal oven method

during short-term ageing which implies sulfur has more affinity toward oxygen than carbon.

6.2.3 Long-Term Aged Sample by Normal Oven Method

Figures 10, 11, 12 and 13 show the FTIR spectra for VG30, PRMB5, PRMB10, and PRMB15 of long-term ageing by normal oven method for 3 and 5 days at about 85 °C and denoted by L3VG30, L3PRMB5, L3PRMB10, L3PRMB15, L5VG30, L5PRMB5, L5PRMB10, and L5PRMB15, respectively. Due to the addition of HDPE pyro-oil, the variation in functional group during ageing also takes place. It is observed that with increase in ageing from 3 to 5 days carbonyl index increases. Again, to analyze further one should go for calculation of carbonyl and sulphoxides index in detail.

6.2.4 Quantification of Ageing on Unmodified and Modified Bitumen

As stated by Yang et al. [30], the ageing of bitumen during construction phase is mainly from two dimensions, the loss of volatile, and oxidation effect. The loss in volatile is calculated with the help of RTFO and already tabulated in Table 1. Along with this, oxidation of bitumen was also determined by chemical bondage

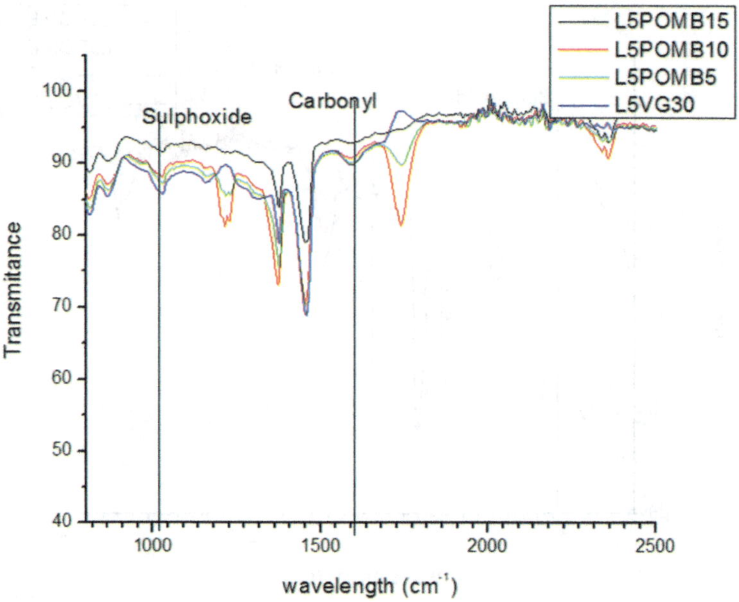

Fig. 11 FTIR spectra of 5-day long-term aged VG30, PRMB5, PRMB10, and PRMB15 by normal oven method

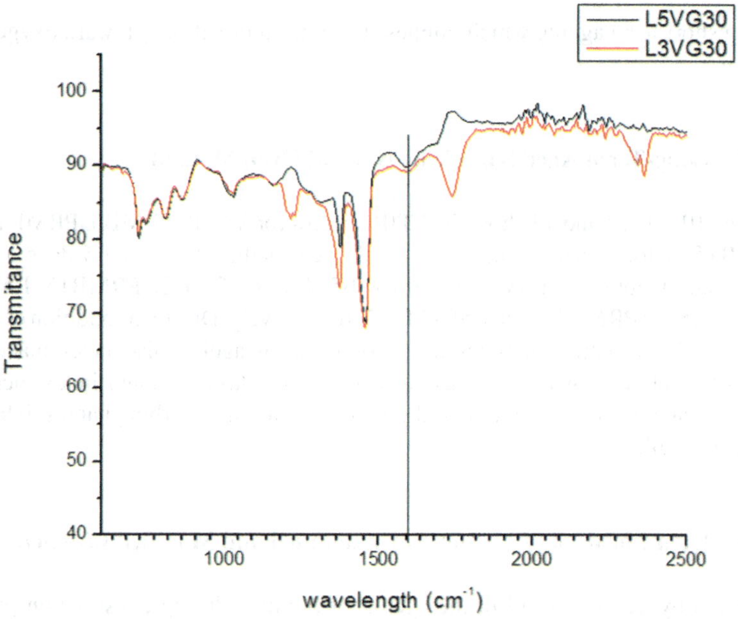

Fig. 12 FTIR spectra of 3-day long-term aged VG30, PRMB5, PRMB10, and PRMB15 by normal oven method

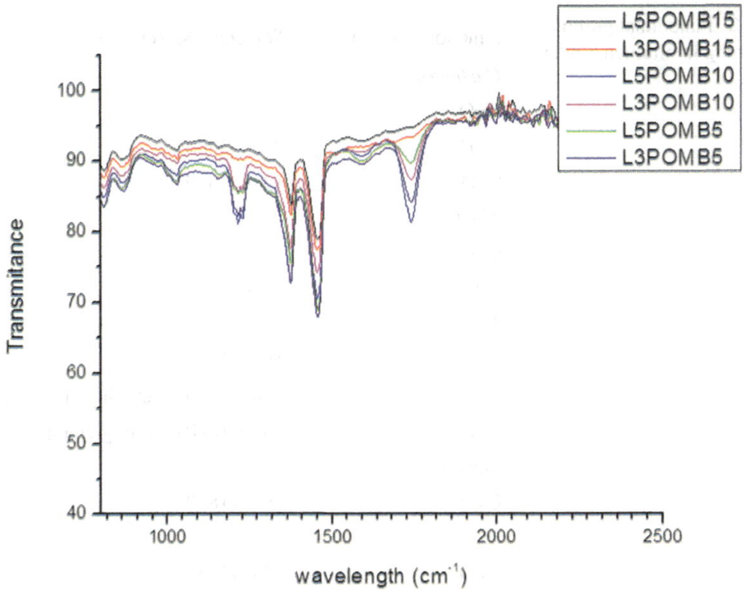

Fig. 13 FTIR spectra of 3-day long-term aged VG30, PRMB5, PRMB10, and PRMB15 by normal oven method

such as C=O and S=O in FTIR test. As explained by Nivitha et al. [22], ageing effect is monitored by peaks corresponding to carbonyl, sulphoxides, aliphaticity, and aromaticity. For the study of ageing, chemical modifications in base and modified bitumen was observed by analyzing FTIR spectra. As stated earlier, carbonyl (C=O) and sulphoxides (S=O) peaks change during oxidation of bitumen and same was observed from Fig. 3. As bitumen during modification subjected to temperature of about 160 °C for 2 h, oxidation takes place which leads to considerable amount of hydroperoxide formation. So to decrease in per hydromatic ring during the oxidation is the reason for absence of spurt during short-term ageing in modified binders. The results of FTIR graphs form Fig. 4 show that oxidation reaction lead to the formation of unsaturated bonds which may be due to addition of HDPE pyro-oil. In order to analyze the variation in functional group during ageing process, absorption peaks are recorded. These all transmission spectra converted into absorption peaks area of carbonyl, sulphoxides to calculate indices. Formation of C=O bond in carboxyl signifies the absorption of oxygen and S=O bond is formed due to the absorption of oxygen in sulfur content. But, the generation of sulphoxides mainly occurred in short-term ageing. So, it is concluded that sulfur atom has a stronger ability to absorb oxygen than carbon. C=C index decreases which implies that C=C gets affected or ruptured during ageing. The functional carbonyl groups in modified bitumen are less than those of base bitumen when ageing occurs. This indicates that this modification leads to less affinity toward oxygen. Table 2 shows important functional groups containing carbonyl in bitumen binder. It is found from the study of spectra that six

Table 2 Functional groups containing in bitumen binder

Functional group	Wavenumber (cm^{-1})
Carboxylic	
C=O	1700–1725
O–H	2500–3300
C–O	1210–1320
Aldehyde	
C=O	1720–1740
=C–H	2820–2850
Amid	
C=O	1640–1690
N–H	3100–3500 (stretching vibration)
N–H	1550–1640 (bending vibration)
Anhydride	
C=O	1800–1830
Ester	
C=O	1735–1750
C–O	1000–1300

functional group, namely, carboxylic acid, aldehyde, amide, anhydride, ester, and ketone contain carbonyl group in bitumen. This also proved and explained by different researchers. As per SuperPave report, bitumen contains around 1% of nitrogen and due to the same reason the peak of nitrogen and hydrogen bond (amide group) in bitumen not so strong but detectable. As explained by Yao et al. [31], due to the coverage of strong bands, the peaks around anhydrides are not easily observed but evidences are found to support that anhydrides are present in bitumen. From spectra, it is found that, ester formed during ageing of bitumen and carbonyl group in the ester is observed at range of 1735–1750 cm^{-1}.

7 Conclusions

In this study, HDPE pyro-oil was added to base bitumen (VG30) to investigate the effect of ageing. The base bitumen and modified bitumen were tested by FTIR to study the different functional groups under short-term and long-term ageing conditions. Based on the results of FTIR for different bitumen samples following results are summarized.

From physical properties, it was found that pyro-oil modified bitumen has better application in low-temperature region compared to hotter climate but this should be supported with rheological results. Short-term ageing by both RTFO and normal oven method gives same FTIR spectra. FTIR spectra show carbonyl area increases from unaged sample to RTFO/oven aged and thereafter maximum area under band for long-term ageing for 3 days and 5 days. It is found that pyro-oil modified bitumen has loss in mass less than 1% of base bitumen, which is as per SuperPave specifications provided for RTFO mass loss. Six functional groups amide, anhydride, aldehydes, carboxylic acid, ketone, and esters contain carbonyl group. Upper and lower wavenumber for different base and modified bitumen found to be in the range of 1600–1750 cm^{-1} for carbonyl group, 920–1060 cm^{-1} for sulphoxides. Base binder and HDPE pyro-oil exhibit physical as well as chemical interaction as peaks of some functional groups change and shift in position of peaks was recorded. Modified bitumen undergoes less ageing compared with base bitumen as carbonyl functional group is less than base bitumen.

References

1. Al-Hadidy AI, Tan YQ (2010) The effect of SBS on asphalt and SMA mixture properties. J Mater Civ Eng 1(1):156
2. Behera PK, Singh AK, Amaranatha Reddy M (2013) An alternative method for short-and long-term ageing for bitumen binders. Road Mater Pavement Des 14(2):445–457
3. Bischoff D, Toepel A (2004) Tire rubber in hot mix asphalt pavements. Wisconsin Department of Transportation, Division of Transportation Infrastructure Development, Bureau of Highway Construction, Technology Advancement Unit
4. Chen M, Leng B, Wu S, Sang Y (2014) Physical, chemical and rheological properties of waste edible vegetable oil rejuvenated asphalt binders. Constr Build Mater 66:286–298
5. Cheng J, Shen J, Xiao F (2011) Moisture susceptibility of warm-mix asphalt mixtures containing nanosized hydrated lime. J Mater Civ Eng 23(11):1552–1559
6. Eberhardsteiner L, Füssl J, Hofko B, Handle F, Hospodka M, Blab R, Grothe H (2015) Towards a microstructural model of bitumen ageing behaviour. Int J Pavement Eng 16(10):939–949
7. Habal A, Singh D (2018) Influence of recycled asphalt pavement on interfacial energy and bond strength of asphalt binder for different types of aggregates. Transp Res Rec 2672(28):154–166
8. Hadole HP, Suryawanshi SD, Khapne VA, Ranadive MS (2021) Moisture damage resistance of short-term aged pyro-oil–modified bitumen using rolling thin film oven by surface free energy approach. J Mater Civ Eng 33(10):04021268
9. Harnsberger PM et al (1993) Comparison of oxidation of SHRP asphalts by two different methods. Fuel Sci Technol Int 11(1):89–121
10. Hofko B, Handle F, Eberhardsteiner L, Hospodka M, Blab R, Füssl J, Grothe H (2015) Alternative approach toward the aging of asphalt binder. Transp Res Rec J Transp Res Board 2505(2505):24–31
11. IS 1208 Methods for testing tar and bituminous materials. Bureau of Indian Standards, New Delhi
12. IS 1205 Methods for testing tar and bituminous materials: determination of softening point. Bureau of Indian Standards, New Delhi
13. IS 1203 Methods of testing tar and-bituminous materials: determimtion of penetration
14. IS 73 (2013) Indian standard for paving bitumen-specification (fourth revision). Bureau of Indian Standards, New Delhi
15. IS 1206-Part1. Methods for testing tar and bituminous materials: determination of viscosity
16. IS 1206-Part2. Methods of testing tar and bituminous materials: determination of absolute viscosity
17. IS 1206-Part3. Methods of testing tar and bituminous materials: determination of kinematic viscosity

18. IS 9382: Methods for testing tar and bituminous materials: determination of effect of heat and air by thin film oven tests. Bureau of Indian Standards, New Delhi
19. Jahromi SG, Khodaii A (2009) Effects of nanoclay on rheological properties of bitumen binder. Constr Build Mater 23(8):2894–2904
20. Khapne V, Hadole H, Ranadive M (2020) Assessment of anti-stripping property of pyro-oil modified bituminous mixes using surface free energy approach. In: International conference on transportation and development 2020, American Society of Civil Engineers, Reston, VA, pp 127–137
21. Kulkarni SB, Ranadive MS (2020) Modified cutback as tack coat by application of pyro-oil obtained from municipal plastic waste: experimental approach. J Mater Civ Eng 32(5):04020100
22. Nivitha MR, Prasad E, Krishnan JM (2016) Ageing in modified bitumen using FTIR spectroscopy. Int J Pavement Eng 17(7):565–577
23. Petersen JC (1975) Quantitative method using differential infrared spectrometry for the determination of compound types absorbing in the carbonyl region in asphalts. Anal Chem 47(1):112–117
24. Petersen JC (2009) A review of the fundamentals of asphalt oxidation: chemical, physicochemical, physical property, and durability relationships. Transp Res Circular E-C140
25. Ranadive MS, Hadole HP, Padamwar SV (2017) Performance of stone matrix asphalt and asphaltic concrete using modifiers. J Mater Civ Eng 30(1):04017250
26. Reddy KS (2007) Investigation of rutting failure in some section of national highways-2. between KM 317 and KM 65. Transportation Engineering Section, Department of Civil Engineering, IIT Kharagpur, India
27. Shen J, Amirkhanian S, Xiao F, Tang B (2009) Influence of surface area and size of crumb rubber on high temperature properties of crumb rubber modified binders. Constr Build Mater 23(1):304–310
28. Suryawanshi SD, Hadole HP, Ranadive MS (2021) Evaluation of moisture susceptibility of pyro-oil modified bitumen by surface free energy approach. In: Ground improvement techniques: select proceedings of 7th ICRAGEE 2020, vol 118. p 363
29. Thodesen C, Shatanawi K, Amirkhanian S (2009) Effect of crumb rubber characteristics on crumb rubber modified (CRM) binder viscosity. Constr Build Mater 23(1):295–303
30. Yang X, Mills-Beale J, You Z (2017) Chemical characterization and oxidative aging of bio-asphalt and its compatibility with petroleum asphalt. J Clean Prod 142:1837–1847
31. Yao H, Dai Q, You Z (2015) Fourier transform infrared spectroscopy characterization of aging-related properties of original and nano-modified asphalt binders. Constr Build Mater 101:1078–1087

Utilization of Aluminium Refinery Residue (ARR), GGBS and Alkali Solution Mixes in Road Construction

Nityanand S. Kudachimath, Raviraj H. Mulangi, and Bhibuti Bhusan Das

Abstract Manufacturing and the construction industries are on the boom with the growing economy of the world. Aluminium and steel are produced in very large quantities compared to other metals. These industries also produce by-products that are either partially utilized or unutilized. Aluminium refinery residue (ARR) with its colour known as Redmud, produced from bauxite by Bayer process, because of its high pH demands huge storage land. In road construction, a large quantity of material is required at the lower layers. In the present work wastes from both industries are tried for reuse. Ground-granulated blast furnace slog (GGBS) from the iron and steel industry makes complex compounds with sodium silicate and sodium hydroxide to increase the strength properties of aluminium refinery residue (ARR) and is verified with basic laboratory experiments for strength and durability tests on stabilized aluminium refinery residue (ARR). Ground-granulated blast furnace slog (GGBS) of 20–30% is used to stabilize ARR, and cured for up to 28 days, and the result of the various mix has shown an increase in strength with curing periods. The highest UCS test result of 4.01 MPa was observed for 28 days cured sample. For durability tests, the treated specimens are passed through wet/dry cycle and freeze/thaw conditions for various curing periods.

Keywords Aluminium refinery residue (ARR) · Ground-granulated blast furnace slag (GGBS) · Sodium silicate (Na_2SiO_3) · Sodium hydroxide (NaOH) · Alkaline solution

N. S. Kudachimath (✉) · R. H. Mulangi · B. B. Das
Department of Civil Engineering, National Institute of Technology Karnataka, Surathkal, Mangaluru, India
e-mail: nskudachimath@gmail.com

R. H. Mulangi
e-mail: ravirajhm@nitk.edu.in

B. B. Das
e-mail: bdas@nitk.edu.in

N. S. Kudachimath
Department of Civil Engineering, Jain College of Engineering, Belagavi, Karnataka, India

M. S. Ranadive et al. (eds.), *Recent Trends in Construction Technology and Management*, Lecture Notes in Civil Engineering 260,
https://doi.org/10.1007/978-981-19-2145-2_97

Notations Used

ARR	Aluminium refinery residue
GGBS	Ground-granulated blast furnace slag
MDD	Maximum dry density
OMC	Optimum moisture content
UCS	Unconfined compressive strength

1 Introduction

1.1 About ARR

A huge quantity of waste materials is generated due to infrastructure development and industrialization. These wastes require dumping areas (land) to reduce environmental pollution. Redmud, also called bauxite residue/aluminium refinery residue (ARR), is one of the waste materials. Nearly about 1 ton of ARR is generated on extraction per ton of aluminium [1]. The extraction of ARR is followed by Bayer's process (Fig. 1). ARR is alkaline in nature and pH is about 11–13 [2]. This waste combined with alkalinity and heavy metals indirectly leads to difficulties when disposed of. Redmud consists of fine particles which consolidate slowly and contaminate the surroundings, and hence it becomes the need of the hour to reuse for other works or construction works [3].

1.2 About GGBS

GGBS is a by-product of the production of iron. The gradation of GGBS is dependent upon slag chemical composition. Hence it acts as a binding/cementitious material that leads to the formation of C–A–H and C–S–H gel [4].

Many researchers worked on soil stabilization by mixing red mud and other available industrial wastes. In 2012, Satayanarayana et al. [5] noticed that 10% of lime with red mud gave better stabilized results. Walid et al. [6] used metakaolin with sodium silicate solution, prepared new material, and named geopolymer. From experimental results, they concluded that the prepared new material behaves as an alkali activator and gives higher compressive strength results.

The present study deals with the stabilization of soil properties with the application of ARR and GGBS with activator solutions. Here, the selected activators are NaOH and Na_2SiO_3 because of their non-side effects on soil. The best proportion of ARR and GGBS is made by conducting several trials. The mechanical properties of the prepared sample are studied and explained the same.

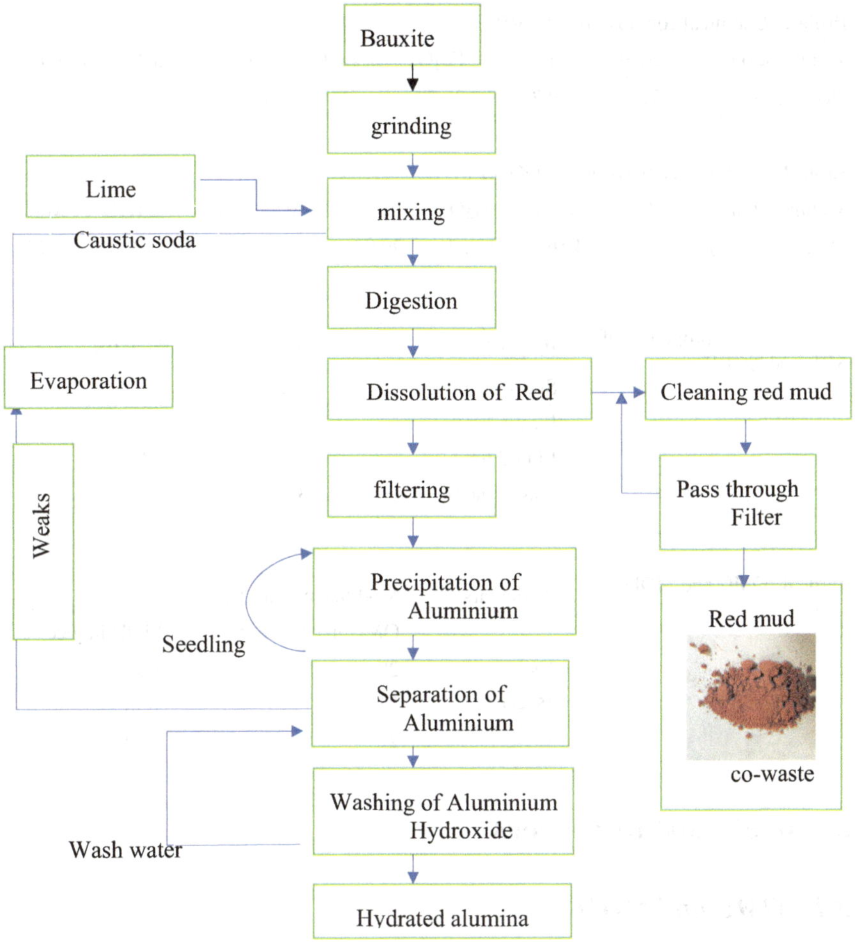

Fig.1 Bayer's process

2 Materials and Methodology

Experimentations were done by collecting AAR from HINDALCO Industries Ltd. Located in Belagavi and GGBS from JSW Ltd Bellary. Collected samples were having moisture content. To remove this, the collected sample is dried in an oven from 105 to 110 °C. The activator Na_2SiO_3 was purchased from Shanti Chemicals located in Belagavi, and dilution of NaOH is prepared 24 h prior to conducting experiments.

To examine the basic properties of the collected sample, some tests were conducted as per IS codes and the obtained results are discussed in Tables 1, 2 and 3.

Table 1 Chemical composition of ARR

Composition	Na_2O	TiO_2	Fe_2O_3	LOI	SiO_2	CaO	Al_2O_3
Percentage (%)	7.3	9.2	36.3	10.4	12.1	1.2	22.1

Table 2 Chemical composition of GGBS

Composition	TiO_2	Na_2O	Al_2O_3	FeO	CaO	SiO_2	K_2O	MgO
Percentage (%)	14.28	4.56	17.23	29.92	20.52	12.96	0.61	0.98

Table 3 Other properties of ARR and GGBS

Properties	ARR	GGBS
Specific gravity	3.2	2.9
Plastic limit	32	–
Liquid limit	37	33
Plastic index	3	–

Table 4 OMC and MDD results

Particulars	Modified proctor test	
	OMC in percentage	MDD in g/cc
20-4-1	28	1.93
25-4-1	24	1.98
30-4-1	22	1.89

3 Results and Discussions

3.1 *OMC and MDD*

The OMC and MDD were examined for the modified proctor test. Experimental results are listed in Table 4 for different mixes altering the GGBS from 20, 25 and 30% of the required total soil sample. From experimentation it is noticed that MDD obtained for 25% of GGBS is the highest, and a further increase in GGBS content caused reduction in strength and it is concluded that (25-4-1) 25% of GGBS, 4% of Na_2O content with 1 silica modulus gave a result of 1.98 g/cc.

3.2 *UCS Test*

UCS test was conducted for different proportions of GGBS and ARR by altering silica modulus from 0.5 to 1.5 and GGBS is altered from 20 to 30% by weight of the required soil sample. From experimental results, the below discussions are made (Fig. 2 and Table 5).

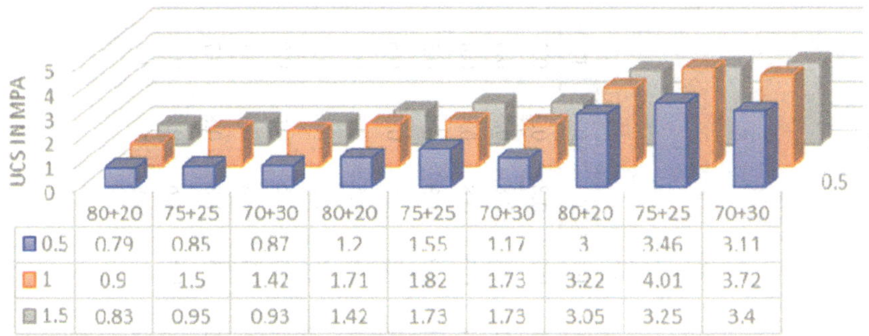

Fig. 2 UCS results for various mix proportions

Table 5 UCS test results

Proportion (ARR+GGBS) out of total required sample (in %)	Curing period (in days)	Silica modulus		
		0.5	1	1.5
80 + 20	0	0.79	0.9	0.83
	7	1.2	1.71	1.42
	28	3	3.22	3.05
75 + 25	0	0.85	1.5	0.95
	7	1.55	1.82	1.73
	28	3.46	4.01	3.25
70 + 30	0	0.87	1.42	0.93
	7	1.17	1.73	1.73
	28	3.11	3.72	3.4

From the above-obtained results, it is concluded that stabilized ARR for a combination of GGBS:Na_2O:silica modulus of 25-4-1 for curing periods of 0, 7 and 28 days gave a positive effect on the UCS test. The strength increased due to the formation of alumina silicate gel. The obtained results are compared with stabilized ARR (25-4-1, 25% GGBS, 4% Na_2O and 1% silica modulus) to find the best result among the selected mix, and it is concluded that there is an increase in strength by 66, 6 and 24.5% for 0, 7 and 28 days of curing (comparison of 20 and 25% GGBS mix). And for further increase in the amount of GGBS content, the strength is reduced by 5.3, 4.9 and 7.2% when compared with 25-4-1 (comparison of 25 and 30% GGBS mix). The best-obtained results are 1.5, 1.82 and 4.01 MPa for 0, 7 and 28 days of curing for a combination of 75% of ARR and 25% of GGBS (Table 5).

4 Durability Test

The test has been conducted over the prepared sample, during wet and dry processes. From test results shown in Table 6, it is noticed that the prepared sample cured for 0 day failed at the initial stage/cycle and the sample cured for 3 days failed at 6

Table 6 Durability test results

Number of cycles	Percentage reduced weight (%)															
	W&D test								F&T test							
	Curing period (days)															
	0		3		7		28		0		3		7		28	
	W	D	W	D	W	D	W	D	F	T	F	T	F	T	F	T
1	Failed		−10.4	8.5	−12.5	4.9	−15.4	−3.6	6	6.5	2.4	3.1	0.2	0.8	0.3	0.3
2			−9.3	9.4	−12.3	5.2	−15.3	−1.4	9.1	8.2	2.4	3.1	0.2	0.8	0.3	0.4
3			−6.5	10.2	−12.3	5.2	−14.5	2.2	9.1	9.2	2.6	3.4	0.3	0.8	0.4	0.4
4			−5.3	11	−12.1	5.4	−14.6	2.5	9.3	9.2	2.8	3.4	0.3	1.3	0.4	0.5
5			−4.7	11.5	−12.1	5.6	−14.4	2.8	9.3	9.2	3.1	3.4	0.3	1.5	0.5	0.5
6			Failed		−11.5	5.7	−13.9	3.5	9.4	9.3	3.1	4	0.4	1.5	0.5	0.6
7					−11.7	5.7	−13.5	3.9	9.5	9.3	3.6	4	0.4	1.6	0.5	0.6
8					−11.2	6.2	−12.7	4.3	9.5	9.5	3.6	4	0.5	1.6	0.6	0.6
9					−11	6.2	−11.4	4.7	9.6	9.5	3.7	4	0.7	1.8	0.7	0.7
10					−10.5	6.3	−11.4	5.5	9.6	9.7	3.8	4.2	0.8	1.8	0.8	0.7
11					−10.5	6.4	−10.	5.9	9.7	9.9	4.1	4.2	0.8	1.8	0.9	0.8
12					−10.5	6.5	−10.3	6.7	10	9.9	4.2	4.2	1.2	1.8	1	1

cycles. While the sample was cured for 7 and 28 days it passed all 12 cycles of dry and wet with a ratio weight gain of 10.5 and 10.3%, respectively. However, in the freeze–thaw test, the prepared sample passed all cycles for 0, 3, 7 and 28 days of curing with percentage weight loss of 9.9, 4.2, 1.8 and 1 correspondingly. With consideration of all the results, the stabilized ARR sample for 7 and 28 days of curing passed the dry and wet tests and hence, the considered dosage can be used for construction activities.

5 Conclusions

- Highest MDD is obtained for 25% of GGBS and 75% of ARR, and a further increase in GGBS content caused a reduction in density.
- Stabilized ARR for a combination of GGBS:Na_2O:silica modulus of 25-4-1 for curing periods of 7 and 28 days gave a positive effect on UCS.
- It is noticed that increase in strength by 66, 6 and 24.5% for 0, 7 and 28 days of curing (comparison of 20 and 25% GGBS mix). And on further increase in the amount of GGBS content, the strength is reduced by 5.3, 4.9 and 7.2% when compared with 25-4-1 (comparison of 25 and 30% of GGBS).
- The best-obtained results are 1.5, 1.82 and 4.01 MPa for 0, 7 and 28 days of curing for a combination of 75% of ARR and 25% of GGBS.
- A durability test was conducted and the prepared sample cured for the same days failed in the initial stage/cycle and the sample cured for 3 days failed at 6 cycles.
- The sample cured for 7 and 28 days passed all 12 cycles of dry and wet with a ratio weight gain of 10.5 and 10.3%, respectively.
- In the freeze and thaw test, the prepared sample passed all cycles for 0, 3, 7 and 28 days of curing with weight loss of 10, 4.2, 1.8 and 1% correspondingly.
- The stabilized ARR sample for 7 and 28 days of curing passed the dry and wet tests and hence, the considered dosage can be used for construction activities.

References

1. Samal S, Ray AK (2013) Utilization and process of redmud in India. Int J Miner Process 118:43–55
2. Newton T (2006) Effect of structure on geotechnical properties of bauxite residue. J Geotech Geoenvironmental Eng 132(2):143–151
3. Kudachimath N, Raviraj HM, Das BB (2021) Effect of Ggbs on strength of aluminium refinery residue stabilized by alkali solution. In: Recent trends in civil engineering. Springer, Singapore, pp 331–339
4. Davidovit J (1991) Inorganic polymeric new materials. J Therm Anal 37:1633–1656

5. Satayanarayana PVV, Adishesa, Ganpati Naidu (2012) Characterization of lime stabilised redmud mixfor feasibility in road construction. Int J Eng Res Dev 3(7):20–26
6. Walid H, Andreej Kovic, Alshaer M (2013) Composition and technological properties of GP based metakaolin and redmud. 52:648–654

9789811921476VOL02